Peptides,
Hormones,
and Behavior

Peptides, Hormones, and Behavior

Edited by

Charles B. Nemeroff, M.D., Ph.D.
Biological Sciences Research Center
and Department of Psychiatry
University of North Carolina

Adrian J. Dunn, Ph.D.
Department of Neuroscience
University of Florida

SP MEDICAL & SCIENTIFIC BOOKS
a division of Spectrum Publications, Inc.
New York

SPECTRUM PUBLICATIONS, INC.
175-20 Wexford Terrace
Jamaica, NY 11432

Library of Congress Cataloging in Publication Data
Main entry under title:

Peptides, hormones, and behavior.

Includes index.
1. Neuroendocrinology. 2. Neuropsychology.
I. Nemeroff, Charles B. II. Dunn, Adrian J.
[DNLM: 1. Behavior—Physiology. 2. Brain—Physiology.
3. Hormones—Physiology. 4. Peptides—Physiology.
WL 103 B419]
QP356.4.P35   152   81-8803
ISBN 0-89335-138-5

Printed in the United States of America

To Melissa, Matthew,
and Glenda

Contributors

Michael Berelowitz, M.D.
Division of Endocrinology and
Metabolism
Department of Medicine
University of Cincinnati
College of Medicine
Cincinnati, Ohio

Garth Bissette, Ph.D.
Biological Sciences Research
Center
University of North Carolina
School of Medicine
Chapel Hill, North Carolina

Floyd E. Bloom, M.D.
Arthur V. Davis Center for
Behavior Neurobiology
Salk Institute for Biologic
Studies
San Diego, California

René R. Drucker-Colín, Ph.D.
Department de Biologia
Experimental
Institute of Biology
Universidad Autonoma de
Mexico
Mexico City, Mexico

Adrian J. Dunn, Ph.D.
Department of Neuroscience
University of Florida
College of Medicine
Gainesville, Florida

Kerry S. Estes, Ph.D.
Department of Pharmacology
University of Florida
College of Medicine
Gainesville, Florida

Caleb Finch, Ph.D.
University of Southern
California
The Ethel Percy Andrus
Gerontology Center
University Park
Los Angeles, California

Lawrence A. Frohman, M.D.
Division of Endocrinology and
Metabolism
University of Cincinnati
College of Medicine
Cincinnati, Ohio

Lester D. Grant, Ph.D.
Environmental Protection
Agency
Research Triangle Park,
North Carolina

Clint A. Kilts, Ph.D.
Department of Psychiatry
Duke University Medical
Center
Durham, North Carolina

Harry E. Gray, Ph.D.
Dana Farber Cancer Institute
Harvard Medical School
Cambridge, Massachusetts

Nicholas R. Hall, Ph.D.
Department of Biochemistry
The George Washington
   University Medical Center
Washington, DC

George Koob, Ph.D.
Arthur V. Davis Center for
   Behavioral Neurobiology
Salk Institute for Biologic
   Studies
San Diego, California

Irwin J. Kopin, M.D.
Laboratory of Clinical Science
National Institute of Mental
   Health
Bethesda, Maryland

Cynthia Kuhn, Ph.D.
Department of Pharmacology
Duke University
Durham, North Carolina

Michel LeMoal, M.D.
Laboratory Neurobiology
   Comportement
Université Boredeau II
Bordeau Cedex, France

Peter T. Loosen, M.D.
Department of Psychiatry
Biological Sciences Research
   Center
University of North Carolina
Chapel Hill, North Carolina

William G. Luttge, Ph.D.
Department of Neuroscience
University of Florida
Gainesville, Florida

Daniel Luttinger, Ph.D.
Biological Sciences Research
   Center
University of North Carolina
School of Medicine
Chapel Hill, North Carolina

Tom McCown, Ph.D.
Biological Sciences Research
   Center
University of North Carolina
School of Medicine
Chapel Hill, North Carolina

Richard B. Mailman, Ph.D.
Department of Psychiatry
Biological Sciences Research
   Center
University of North Carolina
School of Medicine
Chapel Hill, North Carolina

Paul J. Manberg, Ph.D.
Burroughs-Wellcome Company
Research Triangle Park
   North Carolina

Charles B. Nemeroff, M.D., Ph.D.
Department of Psychiatry and
   the Neurobiology Program
Biological Sciences Research
   Center
University of North Carolina
School of Medicine
Chapel Hill, North Carolina

M. Ian Phillips, Ph.D.
Department of Physiology
University of Florida
College of Medicine
Gainesville, Florida

Arthur J. Prange, Jr., M.D.
Department of Psychiatry and
  the Neurobiology Program
Biological Sciences Research
  Center
University of North Carolina
School of Medicine
Chapel Hill, North Carolina

Patrick K. Randall, Ph.D.
University of Southern
  California
The Ethel Percy Andrus
  Gerontology Center
University Park
Los Angeles, California

Howard D. Rees, Ph.D.
Georgia Mental Health
Research Institute
Atlanta, Georgia

Leo P. Renaud, M.D., Ph.D.
Neurology Division
Montreal General Hospital
Montreal, Quebec, Canada

James L. Roberts, Ph.D.
Department of Biochemistry
Center for Reproductive Studies
Columbia University
College of Physicians and
  Surgeons
New York, New York

Esteban M. Rodriguez, M.D., Ph.D.
Instituto de Histologia y
  Patologia
Universidad Austral
Valdivia Casella, Chile

Saul Schanberg, M.D., Ph.D.
Department of Pharmacology
Duke University Medical
  Center
Durham, North Carolina

James W. Simpkins, Ph.D.
College of Pharmacy
University of Florida
Gainesville, Florida

Gerard A. Smith, M.D.
Department of Psychiatry
New York Hospital
Cornell Medical Center
White Plains, New York

**Tjeerd B. van Wimersma
  Greidanus, Ph.D.**
Rudolf Magnus Institute of
  Pharmacology
University of Utrecht
Utrecht, The Netherlands

Dirk H. G. Versteeg, Ph.D.
Rudolph Magnus Institute of
  Pharmacology
University of Utrecht
Utrecht, The Netherlands

Contents

Contributors   vii

Foreword   xiii
  Irwin J. Kopin

1. Design and Perspectives of Peptide Secreting Neurons   1
  Esteban M. Rodriguez

2. Distribution of Peptides in the Central Nervous System   37
  Lester D. Grant, Garth Bissette, and Charles B. Nemeroff

3. The Biosynthesis of Peptide Hormones   99
  James L. Roberts

4. The Physiological and Pharmacological Control of Anterior
  Pituitary Hormone Secretion   119
  Lawrence A. Frohman and Michael Berelowitz

5. The Neurophysiology of Hypothalamic–Pituitary Regulation and
  of Hypothalamic Hormones in Brain   173
  Leo P. Renaud

6. Effects of Hypothalamic Peptides on the Central Nervous System   217
  Charles B. Nemeroff, Garth Bissette, Paul J. Manberg,
  Daniel Luttinger, and Arthur J. Prange, Jr.

7. Effects of ACTH, β-Lipotropin, and Related Peptides on the Central
  Nervous System   273
  Adrian J. Dunn

8. The Role of Endorphins in Neurobiology, Behavior, and Psychiatric
  Disorders   349
  George Koob, Michel LeMoal, and Floyd E. Bloom

9. Neurohypophysial Hormones—Their Role in Endocrine Function and
  Behavioral Homeostasis   385
  Tjeerd B. van Wimersma Greidanus and Dirk H.G. Versteeg

10. Angiotensin and Drinking: A Model for the Study of Peptide
    Action in the Brain   423
       M. Ian Phillips

11. Gut Hormones and Feeding Behavior: Intuitions and Experiments   463
       Gerard P. Smith

12. Sleep Peptides: The Current Status   497
       René R. Drucker-Colin

13. Hormones of the Thyroid Axis and Behavior   533
       Peter T. Loosen and Arthur J. Prange, Jr.

14. Glucocorticoids and Mineralocorticoids: Actions on Brain and
    Behavior   579
       Howard D. Rees and Harry E. Gray

15. Cerebral Effects of Gonadal Steroid Hormones   645
       William G. Luttge

16. Hormones and Brain Development   775
       Cynthia Kuhn and Saul Schanberg

17. Role of Monoaminergic Neurons in the Age-Related Alterations in
    Anterior Pituitary Hormone Secretion   823
       James W. Simpkins and Kerry S. Estes

18. Neuroendocrine Mechanisms in Rodent Reproductive Aging   865
       Patrick K. Randall and Caleb Finch

19. Animal Models in Psychoneuroendocrinology   893
       Richard B. Mailman, Clint Kilts, and Tom McCown

20. Behavioral and Neuroendocrine Interactions with Immunogenesis   913
       Nicholas R. Hall

    Index   939

Foreword

Fundamental to survival of living organisms is their ability to react appropriately to their environment. Cannon (1929) recognized that "back of internal homeostatic mechanisms are powerful motivating agencies—appetites and hunger and thirst." Almost all observed behavior may be viewed as activity required to meet some physical or emotional need. "The higher in the scale of living things, the more numerous, the more perfect, and the more complicated do these regulatory agencies become." This statement by Fredricq (1885) regarding internal mechanisms is at least as valid for behavior. Adrenal medullary secretion in preparation of "fight or flight" may be considered the first described behavioral neuroendocrine response. The consequences of more prolonged stress on pituitary-adrenal cortical function and the subsequent unfolding of the means by which the brain controls the secretion of the anterior and posterior pituitary glands led to the birth of neuroendocrinology.

During the last decade, neuroendocrinology has taken a remarkable turn. Peptides which were believed at first to be involved solely in control of the pituitary by the hypothalamus were found in other areas of the brain. Other peptides were encountered in brain by their activity in competing for the high affinity binding of drugs to their receptors, and still others, first found in peripheral organs, were discovered also in brain. Perhaps even more amazing was the discovery that one or another of these peptides influence almost every aspect of behavior.

The new techniques for identifying and assaying biologically active peptides, for studying their biosynthesis, distribution and physiological roles in brain, and for examining their behavioral effects have yielded an abundance of new knowledge. This enormous body of new information has created the need for perspectives—manageable summaries of the development, state-of-the-art, and indications of the directions for future research. This volume admirably presents such perspectives and projects the enthusiasm, the sense of forward movement, and the excitement which characterize the young investigators in this rapidly accelerating field.

IRWIN J. KOPIN
NATIONAL INSTITUTE OF MENTAL HEALTH

REFERENCES

Cannon, W.B.: *Physiol. Rev. 9:* 399, 1929.
Fredricq, L.: *Arch. de Zool Exper. et Gen. iii:*/xxxv, 1885.

Design and Perspectives of Peptide Secreting Neurons

ESTEBAN M. RODRIGUEZ

EVOLUTIONARY ASPECTS

Although the emergence of the most primitive forms of nerve
cells remains a most challenging problem, it seems possible
that all neurons arose from a diffuse network of ancestor cells
displaying differentiated receptor and secretory poles (Lentz,
1968) but lacking the ability to establish synaptic contacts.

At a later phylogenetic stage, such as that of coelenterates,
a certain degree of specialization and polarization of the
nerve elements becomes evident, although an organized mass of
neurons, such as a central ganglia, is still absent. Nevertheless in this most primitive nervous system, there are already
neurons presenting clear signs of secretory activity (Scharrer,
1976). The hormonal character of this secretory material has
also been demonstrated (Lentz, 1968; Berlind, 1977).

The appearance of well-circumscribed ganglia characterizes
the nervous system of annelids. About 50 percent of the cells
in these ganglia are of the neurosecretory type (Scharrer,
1976). Since at this phylogenetically lower level endocrine
glands are still missing, the neurosecretory cells become the
only source of hormonal messengers. Processes such as reproduction and regeneration are controlled by these neurohormones.

Later in the course of evolution, arthropods were provided
with an endocrine apparatus, thus allowing a division of labor
and partially relieving neurons from performing all the endocrine functions (Scharrer, 1976). At this phylogenetic stage,
the complexity of the neuroendocrine systems increases and
storage sites for the neurosecretory products develop. In all
these lower invertebrate species the neurosecretory cells form
first-order neurosecretory systems, whereby the control over

the effector cells is exerted by a one-step process (Scharrer, 1976).

Higher invertebrate species, such as insects, possess an endocrine organ resembling the adenohypophysis of vertebrates. This organ, the corpus allatum, is under the control of neurosecretory cells. Second-order neurosecretory systems displaying a regulatory function over endocrine target tissues are thus established.

In vertebrates, the course of the evolution of the neuroendocrine system takes more than one direction. Endocrine glands with distinct hormonal properties are developed, and increasing degrees of structural (morphological and biochemical) and functional complexity are accomplished as this long-lasting evolutionary process (hundreds of million years) proceeds.

These endocrine glands, with their vast population of hormones, being themselves integrated as a system and in turn integrating vital nonendocrine processes, form one of the two integrative systems.

In all vertebrate species the first-order neurosecretory systems continue to operate, and the hypothalamo-neurohypophyseal system seems to be the most representative example.

It is the second-order neurosecretory system, represented by the hypothalamo-adenohypophyseal system, that is undergoing the most numerous adaptative changes. Thus, for example, in lower vertebrate species neurosecretory products controlling the function of the adenohypophysis are released either at synaptoid contacts between the neurosecretory fiber and the secretory cell or at connective tissue septa. Although these modalities of neuroendocrine interelationships are also found in higher vertebrates, they are confined to certain anatomical areas, such as the pars intermedia (Rodrígues and Gimenez, 1972). In these species, however, the most prominent evolutionary feature, with respect to pathways for neuroendocrine communication, is the development of a portal circulatory system linking the hypothalamus to the adenohypophysis.

Because of the existence of this vascular link, the neurosecretory mediators controlling adenohypophyseal functions can simultaneously reach several effector cells, and it is also possible for more than one neurohormonal factor to reach the same effector cell. Furthermore, because of the portal nature of the system, the hypophysiotropic factors are not diluted in the general circulation and, consequently, are delivered to the gland without delay (Scharrer, 1974b).

The mechanisms leading to the integration of the endocrine glands as a system involve the participation of the second integrator, the nervous system. The neurosecretory neuron with its dual character appears ideally equipped to bridge and integrate the two systems of integration, the nervous and the endocrine system. As a neuron, the neurosecretory cell is under the influence of neural inputs, either stimulatory or inhibitory. As a secretory cell, the neurosecretory neuron has the capacity to synthesize biologically active principles, to transport them along the axon, and to release them into the bloodstream, through which they reach the target cells.

Through the neural afferences the neurosecretory cell is connected to several areas of the brain. Because of the multiplicity of brain regions conveying information upon the neurosecretory cell, either directly or indirectly, and because of the unitary character of the neurosecretory efferent pathway, E. Scharrer (1965) regarded the second-order neurosecretory system as the "final common path" in neuroendocrine integration.

However, there is evidence clearly indicating that neurosecretory pathways originating in neurons sharing the same characteristics and location as those forming either first- or second-order neurosecretory systems, instead of producing neurovascular contacts, appear distributed in different brain areas, where they seem to contact nonneurosecretory neurons. This finding places the neurosecretory neuron in a new perspective, which might lead to new horizons in the study and understanding of the two-way relationships between brain areas specialized in secretion and those specialized in other directions.

Another new insight is that peptide molecules originally isolated from the hypothalamus and regarded as neurosecretory messengers specifically controlling certain adenohypophyseal functions have been found in perikarya located in several extrahypothalamic regions (Hökfelt et al., 1978). Even more remarkable has been the finding of one of these peptides, thyrotropin-releasing hormone (TRH), in invertebrate species in which a hypophysiotropic role can not be envisaged (Grimm-Jørgensen et al., 1975).

During the past few years numerous peptides have been isolated and characterized and their cellular distribution determined. Their cells of origin are not only distributed within the central nervous system (CNS), where they appear to originate a new modality of first-order neurosecretory systems, but also outside the brain. Remarkably, here they may be found in cells with no neuronal properties (Prange et al., 1978).

Another unexpected finding was the presence of immunoreactive adenohypophyseal hormones such as ACTH (Pelletier and Leclerc, 1979) and α-MSH (Dubé et al., 1978) in neurons located in the CNS.

This brief survey of the evolutionary aspects of the neurosecretory systems allows us to draw some conclusions that might help us understand and clarify some fundamental processes relevant to the subject of this book.

1. Cell types that do not meet all the requirements necessary to be considered primitive neurons already display distinct signs of secretory activity.
2. It appears that the development of synaptic contacts and, consequently, the polarization of nervous transmission were events that occurred at a later phylogenetic stage.
3. Before endocrine glands were developed, all the existing endocrine functions were performed by neurosecretory neurons (first-order neurosecretory system).
4. The appearance of an endocrine apparatus that took over most of the endocrine functions led to a new role for the neurosecretory neurons, that of integrating the endocrine glands as a system and, further, integrating this endocrine system with the nervous system. A second-order neurosecretory control was thus developed.
5. Biologically active neurosecretory substances not only are the phylogenetically oldest hormones but, also, they seem to represent the first signs of cellular activity of the most primitive neurons. Therefore it should not be surprising that in more evolved nervous systems, neurons located in many different regions keep this ancient secretory capacity. The evolutionary trend here would involve molecular mechanisms, resulting not only in different secretory products but also in the development of receptor molecules of multiple locations, all of which would lead to an increased diversification of the functional properties of the neurosecretory neurons.
6. The presence of the same types of biologically active principles in apparently unrelated cell types, that is, centrally located neurons and peripherally dispersed cells may be explained in ontogenetic terms, since they derive from the same neuroectodermal cells (Pearse and Takor Takor, 1976). Furthermore, it is tempting to correlate this point of departure of the ontogenetic development with that of the phylogenetic development, that is, the primitive neuron displaying secretory properties (see above).
7. Although we do not yet know the functional role of some immunoreactive adenohypophyseal hormones present in hypothalamic neurons, their dual location could be explained from the ontogenetic point of view. It should also be remembered here that in lower invertebrates all endocrine functions are performed by neurosecretory cells. It thus seems possible that after the emergence of endocrine glands, partition of labor between them and the neurosecretory cells was not fully accomplished and that, although both categor-

ies of cells may acquire distinct new properties, they also may keep some common ones.

8. The presence in invertebrates of neurohormones primarily isolated from higher vertebrates indicates that they are molecules phylogenetically older and with a wider range of functional roles than originally thought. The presence of some of them, TRH and melanocyte-stimulating, hormone (MSH), in various areas of the CNS points in the same direction. The stability of the molecular structure of a given neural messenger throughout long phylogenetic periods accompanied by shiftings in the molecule function points to the fact that, when studying evolutionary aspects of biologically active principals, the changes in location of the receptor molecules should also be considered.

9. Although it may be highly speculative, it is tempting to place, side by side, the two extremes of the phylogenetic development of the nervous and endocrine systems and to try to compare them. In the most primitive stages, where all the secretory and nervous functions are performed by the same cell type, it seems obvious that both activities are carried out in a coordinated way and that, probably, they influence each other. Experimental evidence supporting the latter possibility has been obtained, in fact, from a neurosecretory system operating in mammals (see page 27).

Despite the enormous degree of complexity resulting from the emergence of the endocrine system and the second-order neurosecretory systems on one side, and the progressive complexity of the nervous system on the other, it could still be proposed that, in vertebrates, as in the case of the pluripotential primitive neurons, the nervous and the endocrine functions are performed in a coordinated way and that probably they influence each other. The influence of the nervous system on the secretion of hormones has been proved beyond doubt, and the massive amount of work that led to this founded a solid scientific field—neuroendocrinlogy. That hormones, whatever their nature and source, may actually influence the nervous system, even in its superior functions, such as behavior and memory, is becoming progressively evident, and that possibility presents, undoubtedly, a most fascinating challenge. This book represents an effort to accept this challenge.

EVOLUTION OF THE CONCEPT OF NEUROSECRETION

The fact that the three phases of the secretory process, that is, synthesis, transport, and release in the neurosecretory neurons occur in discrete anatomical compartments of the cell

has led Pickering (1978) to state that "neurosecretory neurons
are ideal models for the study of the secretion of proteins and
peptides." Today this statement would not raise much criticism. However, when E. Scharrer presented the concept of neurosecretion more than 25 years ago, at a meeting held in London, despite the fact that a substantial amount of evidence had
accumulated by then, he faced strong and skeptical opposition.
By that time he had already spent a quarter of a century investigating the secretory properties of certain hypothalamic neurons. His first paper, which described diencephalic neurons
specialized in secretory activity to a degree comparable to
that of endocrine cells and which postulated their relationship
to hypophyseal function, was published in 1928 (Scharrer,
1928). Some years before, Speidel (1919) had reported that
what seemed to be glandular cells in the caudal spinal cord of
fish. This finding was also neglected for many years. During
the 1930s and 1940s, Ernest and Berta Scharrer continued to
obtain morphological evidence, from both vertebrates and invertebrates, which supported their view on the secretory capacity
of certain neurons. During these two decades they virtually
preached in the desert with an "almost universal rejection by
the scientific community of the validity of cytological evidence for the existence of a secretory process," as B. Scharrer
(1975) has related. This situation was to be substantially
transformed after Bargmann's inspired discovery, in 1949, that
the method originally designed by Gomori to stain the cells of
the pancreas also demonstrated the hypothalamo-neurohypophyseal
system (Fig. 1-1). Very soon after this finding, many fundamental pieces of work began to be published by different research groups.

As different methodologies became available, many different
aspects of the neurosecretory neurons were analyzed. Thus,
histochemical, electron microscopic, pharmacological, and biochemical procedures were tools used to unfold the basic design
of the neurosecretory phenomenon. In addition, several experimental approaches identified the neurosecretory material of the
hypothalamo-neurohypophyseal system with the neurohypophyseal
hormones.

Important parallel achievements indicated the existence in
vertebrates of neurosecretory systems other than the hypothal-
amic-neurohypophyseal system. Thus, evidence presented by
Benoit and Assenmacher (1953), as well as many others, indicated the existence of neurosecretory fibers ending on the portal
capillaries located in the external region of the median eminence. This and the accumulating evidence of the presence of
hypophysiotropic factors in the hypothalamus-median eminence
region progressively led to conceptual formulations, such as

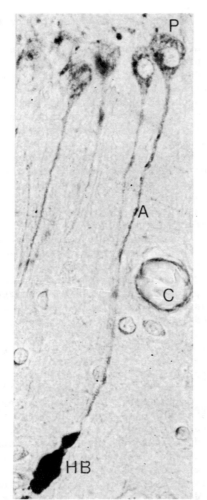

Fig. 1-1. Neurosecretory neuron stained with the Gomori
           method. P:perikaryon; A:axon; HB:Herring body;
           C:capillary.

"neuroendocrine integration" and "second-order neurosecretory
systems."
  The remarkable achievements in the field of biochemistry that
have resulted in the isolation and chemical characterization of
several hypothalamic hormones have, in turn, made possible the
development of highly specific and sensitive immunological
techniques for the quantification (radioimmunoassay) and

cellular localization (immunocytochemistry) of these hormones.
The application of these new tools is revealing so many unexpected aspects of the neurosecretory phenomenon that the
present period could be compared to that following Bargmann's
report in 1949. One has the feeling that a new, profitable
epoch has just started.

The concept of neurosecretion underwent modifications as new
aspects were being revealed by the application of newer methodologies. Originally, the neurosecretion theory was based on
the selective staining of secretory materials within certain
neurons. Later, these neurons were associated with the production of peptide hormones. Some years later the concept of
neurohaemal organ was proposed as a useful criterion to assess
whether or not a neuron was truly neurosecretory. According to
this concept the neurosecretory axons do not make synaptic
junctions with effector organs but, rather, they release their
secretory products into the bloodstream. When this proposition
was put forward, the neurosecretory products were being associated with peptidic (neurohypophyseal hormones and releasing
factors) and proteinaceous (neurophysins) molecules. The confirmation that peptide molecules are synthesized in certain
neurons and released into the bloodstream to reach distant
effector cells led to the concept of "neurohormones." The
neurohaemal concept as a criterion to identify neurosecretory
neurons was challenged by two important findings. One was the
observation of neurosecretory fibers ending on endocrine cells
or on ependymal cells. It thus became evident that the vascular release of hormones was not the sole function of neurosecretory neurons. On the other hand, catecholamine fluorescence
studies demonstrated the presence of abundant aminergic fibers
in certain neurohaemal areas. It was possible, then, that biogenic amines could be released into the bloodstream. The vascular release of chemical messengers was in all probability not
an exclusive property of peptidergic neurosecretory neurons.
The lack of a unifying criterion led Knowles and Bern (1966) to
suggest that the distinct feature of neurosecretory neurons is
that "they are engaged, directly or indirectly, in endocrine
control and may form all or part of an endocrine organ." This
functional criterion, as well as the "final common path" concept advanced by E. Scharrer in 1965 excludes neuronal systems
that, according to recent findings and applying other parameters, could be regarded as neurosecretory.

A good body of recent evidence clearly indicates that the
dividing line between conventional and neurosecretory neurons
is virtually impossible to draw. Because all neurons are capable of synthesizing and releasing chemical messengers, the
neurosecretory phenomenon, according to B. Scharrer (1974a),
should be viewed as a matter of degree. According to her

(Scharrer, 1978) the classical neurosecretory neuron differs from a conventional neuron in that it is specialized "in the manufacture of chemical mediators to a degree comparable to that of gland cells and by engaging in modes of information transfer that markedly digress from standard synaptic transmission."

However, today there seems to be a tendency to avoid the rather confusing distinction between conventional and neurosecretory neurons because they seem to represent only two grammatical structures of the language neurons use to communicate, rather than the whole vocabulary. Instead, there is a tacit agreement to designate neurons according to the chemical nature of the mediator they use, independently of the modality of the communication (neurotransmission, synaptic modulation, or hormonal). Thus, neurons are described as cholinergic, gabaminergic, aminergic, or peptidergic—or, more precisely, as dopaminergic, noradrenergic, serotoninergic—or as vasopressin-, oxytocin-, luteinizing, hormone-releasing hormone- (LH-RH), somatostatin-, endorphin-secreting neurons.

Thus the concept of neurosecretion, "once considered heretical, after reaching its golden age is now approaching anonymity as it enters the domain of modern biological thought" (Scharrer, 1975).

PEPTIDERGIC NEURONS

Nomenclature

Since the fiber systems entering the different hypophyseal regions and ending either on blood vessels or on secretory cells may transport and release different kinds of messengers, Bargmann et al. (1967) classified these systems according to the chemical nature of the messenger. Thus, they introduced the term "peptidergic neurons" to denote the classical hypothalamic neurosecretory cells and to distinguish them from the aminergic neurons, which also innervate the hypophysis. However, since the compartments where the peptidic messengers are released vary considerably (perivascular space, intercellular space at synaptoid junctions, third ventricle lumen), the only feature these hypothalamic peptidergic neurons share is that they are capable of synthesizing and releasing peptides. Therefore peptidergic neurons should be defined as nerve cells equipped to synthesize and to export peptidic messengers, which either may exert their effects locally (at synaptic or synaptoid contacts) or may be released into channels of communication (blood vessels, connective tissue septa, ventricular cavities) to reach distant effectors. According to this definition

the term peptidergic neuron will denote not only the classical
hypothalamic neurosecretory neurons but also several other neuronal systems that have, in recent years, been found to be distributed in several areas of the CNS. However, it seems rather
risky to place within the same category cells like the oxytocinergic neurons, which synthesize a peptide and a carrier protein and release them into the general circulation, and neurons
like those containing endorphins or substance P. Until more
data about the modes of synthesis, transport, and communication
operating in the nonclassical peptidergic neurons are available, it is suggested that peptidergic neurons be classified in
two large groups--those with distant effectors and those with
local effectors. The distinction between these two types of
peptidergic neurons would not only imply differences in the
extracellular pathway followed by the released peptide, but
probably could also denote differences concerning other basic
mechanisms, such as transport and release of the peptide, its
inactivation, the duration of the signal, and the properties of
the effector cells.

It seems that vasopressin and oxytocin are the only peptidic
hormones of neuronal origin that are released into the general
circulation. They also seem to be the only neuronal peptides
that are synthesized, packed, transported, and released together with larger peptide molecules, the neurophysins. The vasopressin-secreting neurons whose perikarya are localized in the
supraoptic and paraventricular nuclei and whose terminals contact the capillaries of the neural lobe would be, according to
the nomenclature suggested previously, peptidergic neurons with
distant effectors.

Recently, neurons of the paraventricular nucleus have been
reported to send their processes to the septum region, where
they end in contact with neurons of the lateral septum. The
whole neuron, including the ending, contains immunoreactive
vasopressin and neurophysin. Such neurons would represent peptidergic neurons with local effectors. The paraventricular
nucleus would then be composed of two types of vasopressinergic
neurons--those forming the paraventriculo-neurohypophyseal system (with distal effectors) and those originating the paraventriculo-septal tract (with local effectors) (see Fig. 1-2).
These two peptidergic systems, although sharing the nucleus of
origin and the same immunohistochemical properties, could actually differ with regard to several aspects, apart from that of
the relative distance of the effector cells. Do the paraventriculo-septal neurons actually transport vasopressin and neurophysin? The only evidence for a positive answer comes from
immunological methods, and it is far from being conclusive.
The presence of vasopressin or neurophysin fractions displaying

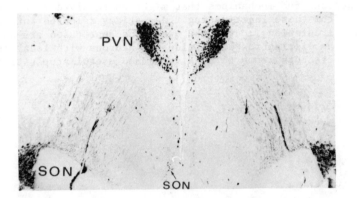

Fig. 1-2. Frontal section of the rat hypothalamus immunostained for neurophysins. PVN paraventricular nucleus, SON supraoptic nucleus, SQN suprachiasmatic
nucleus.

immunoreactivity may well be responsible for a positive immunohistochemical reaction.

Assuming that the peptides being transported in the paraventriculo-septal tract and being stored in perineuronal endings
actually are vasopressin and neurophysin, are they released
into the synaptic cleft as such or are they previously degraded
into smaller peptides? Of special interest in this respect is
the observation that the behavioral effects of vasopressin can
be attributed to fragments of the molecule, which by themselves
do not display the biological activity of the whole molecule
(vasopressin) (de Wied, 1977). If only fragments of vasopressin are released into the synaptic cleft, and neurophysin or
its fragments are not, exocytosis should be ruled out as the
mechanism of release. If fragments of both vasopressin and
neurophysin are released by exocytosis, then the presence of
peptidases in these neurosecretory granules should come under
consideration. Assuming that the paraventriculo-septal neurons
synthesize, pack, transport, and release vasopressin and neurophysin in the same way as the paraventriculo-neurohypophyseal
neurons, then the method of inactivation of these substances
after they are released must be considered.

When vasopressin and its neurophysin are released into the
general circulation, they are inactivated by degrading enzymes
localized in several organs, including the target organ. These
enzymes are intracellularly located (Johnston et al., 1975).
It is difficult to visualize how vasopressin and neurophysin
released into the interneuronal space could be inactivated.

Nevertheless, the mechanisms that might be involved most likely digress from those inactivating the peptides released into the general circulation. Until the questions just posed are answered, the distinction of peptidergic neurons with local and distant effectors seems to be a valid and useful proposition.

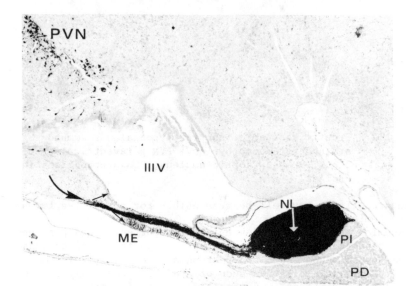

Fig. 1-3.   Saggital section of the hypothalamo-hypophysial region in the rat, immunostained for neurophysins. Immunoreactive perikarya in the paraventricular nucleus (PVN) are seen. The hypothalamo-neurohypophysial tract (bent arrow) when entering the median eminence (ME) gives a few fibers entering the third ventricle and several fibers reaching the portal vessels of the external region of the ME (small arrows). The main bundle of neurophysinergic fibers reach the neural lobe (NL). PI pars intermedia, PD pars distalis.

PEPTIDERGIC NEURONS WITH DISTANT EFFECTORS

In vertebrates, neurons of this type are found in the hypothalamus and in the caudal spinal cord (urophysis) of fish. In the hypothalamus they are grouped into well-defined nuclei or are dispersed into diffuse areas. From their site of origin they project their axons either to the neural lobe, where they release their peptides (oxytocin, vasopressin, neurophysins) to the general circulation, or to the median eminence, where the peptides (LH-RH, TRH, somatostatin) are released into the portal circulation.

The neurons of the hypothalamo-neurohypophyseal system are the peptidergic neurons most thoroughly studied (Figs. 1-2,3). The most relevant events occuring in these neurons will be described and discussed in the next section.

Hypothalamo-neurohypophyseal Peptidergic Neurons

Nature of the Secretory Products

The Hormones

The neural lobe of the hypophysis is the site of storage and release of peptidic hormones synthesized in hypothalamic nuclei. The neural lobe of each vertebrate species contains at least two hormones. These hormones are nonapeptides. In most mammalian species these two hormones are oxytocin and arginine vasopressin. Nine different neurohypophyseal nonapeptides have been found throughout the vertebrate phylum. They represent a series of natural analogues, with structural modifications occurring in positions 3, 4, and 8 of the peptide chain (for references, see Pickering, 1970).

The neurohypophyseal hormones are sequestered within membrane-bound structures observed only at the ultrastructural level and regarded as neurosecretory granules. In turn, the neurosecretory granules containing either oxytocin or vasopressin (in mammals) are sequestered in different neurons. The one neuron-one neurohypophyseal hormone hypothesis, first advanced by Heller (1966) on physiological grounds, has been further supported by ultrastructural findings (Rodríguez, 1971) and recently demonstrated through the use of immunocytochemical techniques (Dierickx et al., 1978).

The Neurophysins

In most mammalian species studied, the posterior lobe of the pituitary contains three neurophysins. Two of them are found in greater amounts than the remaining one. They are polypeptides with a molecular weight of about 10,000 and form stable complexes with the hormones at pH 5.5. The two major neurophysins have been designated as neurophysin I (NpI) and neurophysin II (NpII), according to their electrophoretic mobility. Depending on the species, NpI and NpII may be associated with either of the neurohypophyseal hormones, but there is substantial evidence indicating that within a given species each hormone is associated with one of the neurophysins. Thus in the rat, for example, vasopressin is associated with NpI and oxytocin with NpII (Pickering, 1978). The neurophysin-hormone complexes are stored within the same neurosecretory granules. Consequently, NpI and NpII should be contained in separate peptidergic neurons. This assumption is supported by immunocytochemical evidence.

Events Occurring in the Perikaryon

Synthesis

It is a well-established fact that in many cases the biosynthesis of biologically active substances involves the initial formation of biologically inactive precursors. In the case of polypeptide hormones, it has also been demonstrated that the conversion of the large inactive form into the smaller but active hormone is achieved by enzymic cleavage of specific peptide bonds. These mechanisms have been proved to operate in several endocrine cells secreting peptide hormones, especially the insulin-secreting cells of the endocrine pancreas (Tager and Steiner, 1974). Remarkably, the first piece of evidence for the formation of a precursor molecule in the biosynthesis of a peptide hormone was obtained by Sachs and Takabatake (1964) in their studies on vasopressin synthesis in the hypothalamo-neurohypophyseal system.

It is also generally accepted that peptide, protein, and glycoprotein molecules that are synthesized to be exported need to be sequestered in distinct membrane-bound compartments, the so-called secretory granules. In turn, in order to be packed these molecules have to be synthesized in organelles having the capacity to isolate them from the cytosol, to store them, and to pack them. Thus, the process leading to the synthesis of a precursor molecule of a peptide hormone (prohormone) involves the translation of the messenger RNA by ribosomes associated with the membranes of the rough endoplasmic reticulum (RER), the passage of the nascent molecule through the membrane of the RER, and its sequestration in the lumen of the RER cisternae. There is evidence that the ribosomal product is a molecule larger than the prohormone molecule (preprohormone). The additional part appears to be initially synthesized from and composed of hydrophobic residues and would promote the transport of the newly synthesized polypeptide across the membrane of the RER (Blobel and Dobberstein, 1975).

There is a good body of evidence indicating that in the peptidergic neurons of the hypothalamo-neurohypophyseal system, the synthesis of the precursor molecules does occur in ribo-

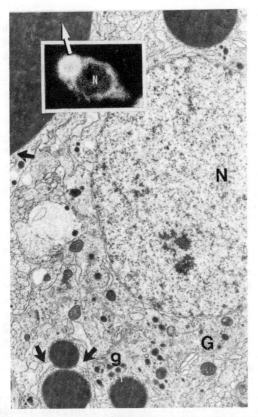

Fig. 1-4.   Neuron of the supraoptic nucleus of a lizard.   In
            addition to the elementary neurosecretory granules
            (g) there are large electron-dense droplets located
            in dilated regions of rough endoplasmic reticulum
            cisternae (black arrows).   N:nucleus; G:Golgi appa-
            ratus.   Insert: same type of neuron but processed
            for immunocytochemistry using antineurophysin serum.
            The large droplets became positive only after the
            sections had been treated with urea and trypsin.

somes associated with the RER (Sachs et al., 1969; Morris,
1978).   Evidence indicating the presence of precursor molecules
within the lumen of the RER of magnocellular peptidergic neurons has recently been obtained in our laboratory (González and
Rodríguez, 1979) (Fig. 1-4).   In 1964, Sachs and Takabatake
proposed that "the biosynthesis of vasopressin proceeds via a
bound, biologically inactive form (as part of a precursor molecule) and that the appearance of the biologically active

octapeptide occurs at a time and place removed from the initial biosynthetic events." In their review of 1969, Sachs and his colleagues discussed the possibility that the precursor is a macromolecule. They also suggested "the interesting possibility that either neurophysin and vasopressin share a common precursor or that there is a common genetic unit that controls the synthesis of the major components of the neurosecretory substance."

These inspired propositions had to wait many years to be almost fully demonstrated. Most of the evidence supporting Sachs' hypothesis of a common precursor for the neurohypophyseal hormones and their associated neurophysins comes from the elegant work by Gainer et al., (1977) and Gainer and Brownstein (1978). Using rats, they injected [^{35}S]-cysteine above the supraoptic nucleus (SON); sacrificed the animals at various times after the injection; obtained samples from the SON, median eminence, and neural lobe; and then analyzed the polypeptides that had incorporated the tracer. In the SON, radioactivity first appeared in a component with a molecular weight of 20,000. As time went on, the labeling of the 20,000 molecular weight component decreased and the radioactivity progressively increased in a different component of about 12,000 daltons, until at 24-hours postinjection virtually no radioactivity was found in the 20,000-dalton component, whereas the 12,000-dalton species had become prominent. This latter component, because it also appeared in the tract and in the neural lobe and because of several other properties analyzed by the authors, was identified as neurophysin. They concluded that the 20,000-dalton component is a precursor (or a stable intermediate state of the precursor) of neurophysin. By applying isoelectric focusing, the same authors (Gainer et al., 1977; Gainer and Brownstein, 1978) were able to identify two labeled components that in the pulse-chase paradigm behaved as precursors (20,000 daltons), two intermediate forms (17,000 daltons), and two neurophysins (12,000 daltons). The fact that Brattleboro rats had only one major protein, one intermediate species, and one neurophysin led these authors to conclude that each one of the neurohypophyseal hormones (oxytocin and vasopressin) has its own precursor-, intermediate-, and neurophysin-associated molecules.

Packaging

The presence of electron-dense material in the lumen of the Golgi cisternae, the series of images suggesting a process of pinching off at the dilated ends of these cisternae, and autoradiographic studies are all taken as evidence that the Golgi apparatus is the source of the membrane-bound structures known as elementary neurosecretory granules (Bargmann, 1966; Nishioka et al., 1970). Modifications in the development and intracellular distribution of the Golgi complexes triggered by stimuli known to promote the synthesis and release of neurohypophyseal hormones has also been related to the capacity of this organelle to pack the neurosecretory products (Morris, 1978). However, all this evidence should be considered circumstantial and other possibilities should be considered. In the magnocellular peptidergic neurons of several species, the RER cisternae display dilated portions filled with an electon-dense material that, according to histochemical and immunocytochemical methods, most likely represents storage sites of the precursor (González and Rodríguez, 1979). Whether this material is progressively translocated to the Golgi apparatus to be repacked or whether it is sequestered in membrane-bound compartments originated from the RER remains to be elucidated. It has been suggested that in the subcommissural organ the packaging process of the secretory material bypasses the Golgi apparatus (Rodríguez, 1970). Nevertheless, in endocrine and exocrine gland cells, the packaging role of the Golgi apparatus has been proved beyond doubt. Therefore the packaging of the neurosecretory material by the Golgi apparatus should be regarded as the most likely possibility. Recently, Tasso et al. (1977) have shown a peripheral ring of glycoproteins in vasopressin-containing granules. Since a well-known function of the Golgi apparatus is the glycosylation of glycoproteins, the neurosecretory granules containing glycoproteins are, almost certainly, derived from the Golgi apparatus.

The magnocellular peptidergic neurons, as do virtually all cells specialized in synthesizing export products, form secretory granules of a given size, characteristic of such cell types. It has been shown that the different neurohypophyseal hormones found throughout the vertebrate phylum are stored in granules of different sizes (Rodríguez, 1971). This poses the puzzling question of how the cell is programmed to produce granules of a given size and which are the mechanisms involved not only in the determination of the size of the granules but also in their rate of formation.

Ultrastructural, physiological, and biochemical data (Cannata and Morris, 1973; Cross et al., 1975; Gainer and Brownstein, 1978) indicate that the post-translational conversion of precursor to neurophysin (maturation) occurs within the neurosecretory granules, especially in those moving along the axon, and that most of the secretory granules stored in the perikaryon would be immature granules containing the precursor(s). The intragranular conversion of precursor into active peptides implies the presence within the granule of peptidases responsible for the enzymic cleavage(s). This in turn implies that

the enzyme(s) must be packed together with the precursor molecules. Evidence for the intragranular presence of an enzyme
has been reported by North et al (1977).

In summary, the series of events occurring in the perikaryon
of the magnocellular peptidergic neurons would include (see
Fig. 1-8):

1. Translation of messenger RNA by ribosomes associated with
   RER
2. Translocation of newly synthesized precursor (and probably
   preprohormone) to the lumen of the RER
3. Probable ribosomal synthesis and translocation of specific
   peptidases into the RER
4. Transport of precursor(s) and enzyme(s) from RER to the
   Golgi apparatus
5. Glycosylation of glycoprotein(s)
6. Packaging of all these secretory products into neurosecretory granules
7. Storage in the perikaryon of some secretory granules as an
   immature pool
8. Beginning of maturation in other secretory granules and
   their transport toward the axon.

Events Occurring in the Axon

Maturation of the Secretory Materials

As already mentioned, the newly packed secretory granules contain the precursor and probably enzyme(s). As the granules
move along the axons, their components seem to undergo important variations.

The different vasopressin-to-oxytocin ratios found for the
hypothalamus and neurohypophysis in several species, but especially in the dog, led Vogt (1953) and Lederis (1962) to suggest that synthesis of oxytocin occurs throughout the length of
the neuron. Histochemical and ultrastructural evidence also
indicated the transformation of the neurosecretory material
along the tract (Peute and van de Kamer, 1967; Cannata and
Morris, 1973).

In the same experiment performed by Gainer and Brownstein
(1978) and described earlier, they found that 2 hours after the
injection of [^{35}S]-cysteine near the supraoptic nucleus, the
median eminence (neurosecretory tract) and the neural lobe
(neurosecretory endings) had two labeled components, the
20,000-dalton component (precursor) and the 12,000-dalton component ("neurophysin peak," probably neurophysin plus hormone).
At times greater than 2 hours, these regions had only the

12,000-dalton species. The authors interpreted these results as indicative that conversion of neurophysin precursor to neurophysin occurs during axonal transport (see Fig. 1-8). There are still some relevant questions concerning the maturation process. Are vasopressin and oxytocin present in their respective neurophysin precursors? Does the conversion process involve some further modifications to the 12,000-dalton component or is it completed with the conversion of the 20,000-dalton species into the 12,000-dalton component? Which is the fate of the remaining 8000 daltons? If the enzyme responsible for the conversion is in the granules, since they are packed, what makes it active when the granules enter the domain of the axon and inactive when they enter the immature pool of the perikaryon? Does the maturation process include modifications of the glycoproteic component of the secretory granule?

Transport of the Secretory Materials

The cellulifugal axonal flow of the neurosecretory material was demonstrated by applying different experimental approaches by the pioneer neurosecretionists (see Bargmann, 1966) (Fig. 1-5). The first estimate of the transport velocity (0.2 mm/hour) was provided by radioautography (Ficq and Flament-Durand, 1963). The arrival of labeled peptides (hormones and neurophysins) at the neural lobe in rats infused with [^{35}S]-cysteine has been used as a reliable procedure to estimate the velocity at which these peptides are transported along the hypothalamo-neurohypophyseal tract. Different authors applying this procedure have estimated the transport velocity to be in the range of 2-4 mm/hour. This puts the neurohypophyseal peptides into the category of axonal substances being rapidly transported.

So far, there is no convincing evidence to explain how granules move down the axon. Compounds known to prevent the polymerization of microtubles, such as colchicine and vinblastine, also seem to disrupt the transport of the neurosecretory granules (Flament-Durand and Dustin, 1972; Rodríguez Echandía et al., 1977). Whether these two effects of colchicine are parallel phenomena or are actually interrelated in a cause-and-effect fashion is not known at present. Neurosecretory axons, as compared with neighboring nonsecretory axons, contain an even smaller population of microtubules. The uneven distribution of neurosecretory granules along the axon, which gives a beaded appearance to the latter, could be interpreted as the morphological image of a peristaltic mechanism. By observing living paraventricular neurons, Hild (1954) described a cellulifugal movement of particles. Because of the limited resolution capacity of the light microscope, these particles must have been either mitochondria or clusters of neurosecretory granules.

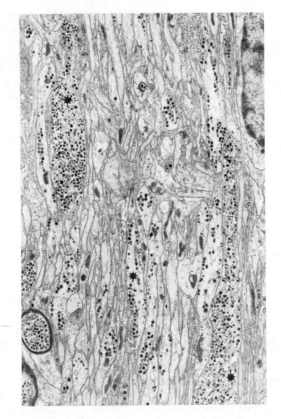

Fig. 1-5. Hypothalamo-neurohypophyseal tract after stalk
          transection. The proximal stumps of peptidergic
          fibers appear loaded with secretory granules
          (asterisks). (From Rodríguez and Dellman, 1970).

The smooth endoplasmic reticulum (SER) of neurosecretory
axons is relatively highly developed as compared with that of
nonsecretory axons. The role played by the SER in the neurosecretory axons is entirely unknown. Shortly after stalk section, the proximal stumps of the neurosecretory axons become
virtually filled with SER, and granules identical to the neurosecretory granules seem to arise from the hypertrophied SER
(Rodríguez and Dellmann, 1970).

Degradation of the Secretory Materials

The axons of the magnocellular peptidergic neurons display a
distinct structural feature, that is, the presence of large

dilatations (not to be confused with the much smaller and more numerous beads) measuring up to several microns in width, and named for Herring, who was the first author to describe them. Herring bodies contain variable amounts of secretory granules, SER, mitochondria, lysosomes, and filaments. Dellmann and Rodríguez (1970) have described a cycle of activity in Herring bodies and have suggested that they represent specialized compartments of the peptidergic neuron where neurosecretory granules are destroyed by lysosomal activity. Vilhardt (1970) has estimated that as much as 80 percent of all the hormone reaching the gland may never be released but is destroyed. There is no reasonable explanation for this uneconomic behavior of these neurons, other than that the neurohypophyseal peptides are emergency hormones. If the whole hormone store is not required, then the remaining "old" hormone would be degraded.

Events Occurring in the Preterminal and Terminal Regions of the Axon

Storage

The neural lobe represents a large store of neurohypophyseal hormones and neurophysins. In steady-state conditions the rate of disappearance of both hormones and neurophysins is about 5 percent of the total glandular content per day (Jones and Pickering, 1972). In the neural lobe, the secretory granules containing the peptides are densely packed in preterminal axonal dilatations and in the axon terminals (Fig. 1-6). There is evidence that the enormous hormonal pool of the neural lobe is not homogeneous (Sachs et al., 1969). Different experimental approaches led Sachs and his colleagues to postulate that in the dog, 10 to 20 percent of the hormonal pool is readily releasable, and that when this readily releasable pool has been discharged, the gland, if properly stimulated, continues to release the hormone(s) but at a greatly reduced rate. These authors also found that the hormone (labeled) arriving at the gland shortly after its synthesis ("new hormone") enters the readily releasable pool before equilibrating with the storage pool ("old hormone"). The same metabolic behavior was found for neurophysins (Wong and Pickering, 1976). Since each hormone and its corresponding neurophysin appear to derive from the same precursor molecule, and since the conversion takes place within the secretory granule, it seems obvious that the newly synthesized hormone and the newly synthesized neurophysin are parts of the same pool. Furthermore, since exocytosis is the accepted mechanism of release, there must be only one readily releasable pool for both peptides (hormone and neurophysin).

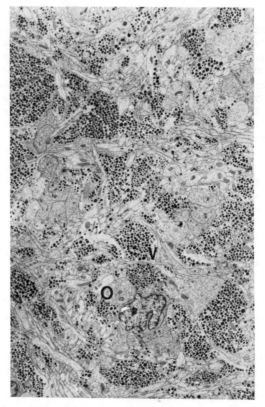

Fig. 1-6.   Lizard neural lobe. There are peptidergic endings
loaded with electron-dense granules (vasotocinergic)
and endings with electron-lucent granules (oxytocinergic) (O). (From Rodríguez, 1971).

All these considerations also lead to the conclusion that in
the neural lobe, the arriving (new) neurosecretory granules
must form a pool, which could be regarded as a "readily fusion-
able" pool of granules, since they would be the first ones fusing with the axon membrane if release were stimulated. Ultrastructural radioautography has indeed shown that radioactive
granules arriving at the gland are first found in the axon terminals and then they appear to move to the axon swellings (Heap
et al., 1975).

Release

Under both normal and experimental conditions, the rates of
disappearance of labeled hormone and labeled neurophysin are

similar (Johnston et al., 1975; Pickering et al., 1975),
strongly suggesting that the hormone and its neurophysin are
simultaneously released. Since both peptides are stored within
the same neurosecretory granules, their simultaneous release is
best explained if exocytosis is the mechanism of release. Such
a mechanism was tentatively suggested by Sachs (see discussion
in Sachs et al., 1969) and formally proposed by Douglas et al.
in 1971. Since then, the proposition that the neural lobe peptides are released by exocytosis has been largely supported by
numerous findings (for references, see Dreifuss, 1975).

It is assumed that the arrival of the action potentials at
the axon terminal induces the depolarization of the axonal membrane, thus allowing an influx of calcium ions. In turn, the
increased intracellular calcium concentration appears to initiate a series of events that eventually result in the fusion of
the granule membrane with the axon membrane and the consequent
extrusion of the granule content (Dreifuss, 1975; Thorn et al.,
1975) (see Fig. 1-8). There are still several steps in this
mechanism of hormone release that need to be elucidated. What
is the fate of the membrane of those granules undergoing exocytosis? According to Douglas and his co-workers (1971) these
membranes are recaptured as microvesicles by a compensatory
endocytosis, and this would be the origin of the small clear
vesicles found in neurosecretory endings. Much of the evidence
available does not support this hypothesis. Instead, it appears that after exocytosis, membranes are internalized as
large vacuoles of roughly the size of the neurosecretory granules (see open discussion in Walter, 1975; Pickering, 1978).
If this is true, we still have to find an explanation for the
presence, origin, and nature of the small vesicles present in
neurosecretory endings. Through which mechanisms do calcium
ions bring about exocytosis? Although the available evidence
is confusing, it would appear that microtubules and ATP are not
involved (Dreifuss, 1975). After the release mechanism is
switched off, how is the normal intracellular calcium concentration reestablished? There is evidence that calcium ions are
then sequestered into mitochondria (Thorn et al., 1975) or ex-
changed for extracellular sodium (Dreifuss, 1975) (see Fig.
1-8).

Assuming that conversion of the precursor (20,000 daltons)
occurs in the secretory granules during axonal transport, then
the granules arriving at the axon terminals should contain the
hormone (1100 daltons) and the corresponding neurophysin (about
10,000 daltons), as well as the remaining peptide chain of
about 8000 daltons or smaller fractions. Gainer and Brownstein
(1978) have detected four [^{35}S]-labeled peptides ranging in
molecular weight between 700 and 2500 daltons. If exocytosis
is the only mechanism of release, then not only the hormone and

its neurophysin are released, but also the other peptide(s)
derived from the same precursor molecule (see Fig. 1-8).

Sites of Release of Neurohypophyseal Peptides

Blood Vessels

All the considerations in the previous section with respect to
storage and release of neurohypophyseal principles apply to
those peptidergic neurons ending on blood vessels. There is no
information as to how the content of the secretory granule,
after being released by exocytosis, reaches the bloodstream.
There are several barriers that the hormones and the neurophysins have to penetrate in order to reach the capillary lumen,
namely, the perivascular connective tissue, the perivascular
basement membrane, and the endothelial cells (see Fig. 1-8).

Ependymal Cells

In several lower species, the neural lobe of the hypophysis
lacks pituicytes, and all peptidergic fibers end on ependymal
cells (Rodríguez and La Pointe, 1970; Rodríguez et al., 1978;
Weatherhead, 1978). Here the neurosecretory endings establish
synaptoid junctions and the fibers are virtually devoid of
Herring bodies. There is no information as to the mechanisms
of storage and release operating in these endings. The functional meaning of this ependymal barrier interposed between the
peptidergic endings and the blood vessels is also unknown.
However, the possibility that in these species there is transependymal transport of neurohypophyseal peptides should be considered. A similar cellular organization has been found in the
lateral regions of the mammalian median eminence, where the
peptidergic fibers (LH-RH) are separated from the portal capillaries by an ependymal barrier (Rodríguez et al., 1979).

Pars Intermedia

In several species, including mammals, the pars intermedia of
the hypophysis is innervated by peptidergic fibers (Bargmann et
al., 1967; Rodríguez and Gimenez, 1972). These fibers establish synaptoid contacts with the secretory cells of the pars
intermedia. According to recent immunocytochemical studies,
the peptidergic fibers of the pars intermedia contain the same
neurohypophyseal hormones present in the corresponding neural
lobe, and these depend on the species studied (Dierickx and
Vandesande, 1976).
The peptidergic innervation of the pars intermedia raises

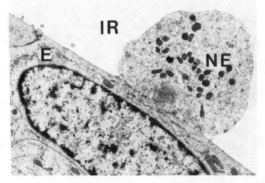

Fig. 1-7.   Rat median eminence. Neurosecretory ending (NE)
(probably oxytocinergic) in the lumen of the infundibular recess (IR). E-ependyma.

several questions essentially similar to those discussed earlier with respect to the paraventriculo-septal peptidergic system. Do the peptidergic fibers of the pars intermedia actually
contain the same peptides (neurophysins and hormones) as the
fibers ending on the capillaries of the neural lobe? A positive answer implies several new questions concerning the mechanism of release at the synaptoid junction, mode of inactivation of the different components of the secretory granules,
specializations at the receptor side (pars intermedia cell),
etc. If the answer is negative, then the questions concern the
nature of the peptide(s) stored in and released by these fibers
(MIF?), the cell of origin of these fibers, and so on.

Cerebrospinal Fluid

There is convincing evidence that, in mammals, both neurohypophyseal hormones--vasopressin and oxytocin--are present in the
cerebrospinal fluid (CSF) and that they are released into the
CSF by peptidergic fibers approaching or reaching the ventricular cavities (Figs. 1-3,7) (see Rodríguez, 1976; van Wimersma
Greidanus and de Wied, 1977). Stimuli known to cause the vascular release of neurohypophyseal hormones also induce a rise
in the CSF levels of these hormones. This observation, as well
as some morphological evidence, suggests that the fibers reaching the ventricles are collaterals of the axons of the hypothalamo-neurohypophyseal tract (Rodríguez, 1976). Several
functions have been ascribed to the neurohypophyseal hormones
present in the CSF (see Rodríguez, 1976; van Wimersma Greidanus
and de Wied, 1977; van Wimersma Greidanus and Versteeg, Chapter
9 of this volume).

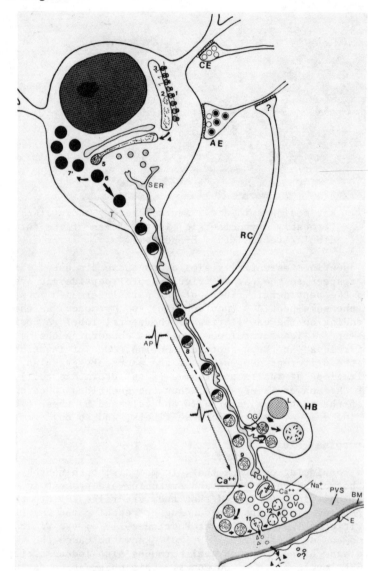

Fig. 1-8. Schematic representation of a peptidergic neuron of
the hypothalamo-neurohypophyseal system: (1) translation of messenger RNA; (2) translocation of
precursor into the lumen of the RER; (3) probable
synthesis and translocation of peptidases; (4)
transport to the Golgi cisternae; (5) probable
glycosylation in the Colgi cisternae; (6) packaging

Neuronal Properties of the Magnocellular Peptidergic Neurons

Ultrastructural studies have revealed the presence of numerous
fibers establishing synaptic contacts with these neurons.
These synapses appear to be of two types, aminergic and cholinergic. The pattern of distribution of these synapses is consistent with an aminergic inhibitory and a cholinergic excitatory input (Cross et al., 1975) (Fig. 1-8). Although there are
few differences between the spike properties of these peptidergic neurons and those of conventional neurons, it appears that
they do not have exceptional electrical properties (Cross et
al., 1975). A single antidromic stimulus inhibits the spontaneous firing of a magnocellular neuron. This is known as the
phenomenon of recurrent inhibition (Cross, 1974). It has been
pointed out that, for this phenomenon to occur, the existence
of recurrent axonal collaterals is necessary (Cross et al.,
1975). However, there is no morphological evidence of the
existence of such collaterals.
The recurrent collateral axons would inhibit the spontaneous
firing, either by a direct innervation of the peptidergic perikaryon or through internuncial cells that relay back to the
peptidergic neuron (Cross, 1974) (Fig. 1-8). The nature of the
inhibitory transmitter released by the collaterals of the peptidergic axons is unknown, although vasopressin and oxytocin
have been discarded as possibilities (Cross et al., 1975).
This implies that the magnocellular peptidergic neurons may be
capable of releasing at their terminals a transmitter substance
different from their known secretory products. This assumption
contradicts Dale's law, according to which a given neuron must
release the same chemical at any of its endings.

---

of all secretory products; (7') storage of secretory
granules as an immature pool; (7) beginning of
maturation, that is, conversion of precursor (represented black) into smaller peptides (triangles and
circles); (9) mature secretory granules containing
neurophysin-hormone complexes (triangles) and probably other peptides (circles); the arrival of action
potentials (AP) promotes the influx of Ca^{++}, thus
triggering exocytosis (10-11); (12) internalized
membranes after exocytosis. Old secretory granules
(OG) enter Herring bodies (HB) where they may be
either degraded by lysosomes (L) or transported back
to the ending. PVS-perivascular space, BM-basement
membrane, E-endothelium, RC-recurrent collateral, AE
and CE probable aminergic and cholinergic endings.

Mutual Interdependency of the Electrical and Secretory
Activities

The main function of the action potentials of a peptidergic
neuron is to provide the membrane depolarization, which in turn
promotes the entry of calcium ions into the ending, thus initiating the mechanism of release of the secretory products.

One population of magnocellular peptidergic neurons are electrically silent, others display spontaneous discharges at variable rates, and a third group shows bursts of activity alter-
nating with silent lapses (Cross, 1974). The same three
patterns have been seen in many species and under different experimental conditions. They might be related to the type of
hormone released or may correspond to a given secretory state
of the neuron (Hayward, 1977).

PEPTIDERGIC NEURONS WITH LOCAL EFFECTORS

As proposed earlier under Nomenclature, this category includes
those neurons whose peptidic messenger, instead of being released into the bloodstream or other channels of communica-
tions, exerts its effect(s) at synaptic junctions. In the CNS
of vertebrates, there are several neuronal groups that synthesize a variety of peptides and that transport them along their
processes. Many of these peptidergic fibers have been seen to
end on neurons of a different type, where they seem to establish axosomatic "peptidergic synapses." Most of these neurons
have been discovered only recently, after specific antisera and
standardized immunocytochemical methods became available.

The information about these neuronal systems is, so far,
mainly related to their anatomical distribution and the immunoreactivity of their secretory products. A striking finding is
that peptidergic neurons with local effectors seem to secrete
peptides identical to those secreted by neurons with distal
effectors and that, consequently, they release them into the
general circulation (vasopressin, oxytocin, and neurophysins)
or into the portal blood (LH-RH, TRH).

Vasopressinergic and Oxytocinergic Neurons with Local
Effectors

It has been known for many years that nerve fibers stained with
methods used to reveal neurosecretory material distribute in
many different areas of the CNS. In a large series of species,
ranging from the primitive lamprey to mammals, axons originating from the magnocellular hypothalamic nuclei have been found

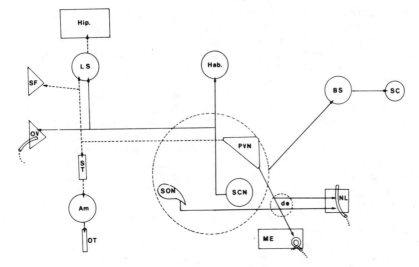

Fig. 1-9. Peptidergic pathways containing vasopressin or oxytocin and neurophysin. PVN:paraventricular nucleus;
SCN:suprachiasmatic nucleus; SON:supraoptic nucleus;
NL:neural lobe; ME:median eminence; Hab:habenula;
LS:lateral septum; Hip:hippocampus; ST:stria terminalis; Am:amygdala; OT:olfatory tract; SF:subforni-
cal organ; OV:organun vasculosum of the lamina terminalis; BS:brainstem; SC:spinal cord; de:axons of
peptidergic neurons with distant effectors.

to reach at least two different brain areas, the septal region
and the habenular region (Barry, 1961; Sterba, 1974). Recent
immunocytochemical studies have confirmed and extended these
observations. Despite the large number of publications, there
is agreement with respect to the distribution within the CNS of
peptidergic fibers originating from the magnocellular hypothalamic nuclei and a parvicellular secretory nucleus, the suprachiasmatic nucleus (SCN). The demonstration of vasopressin and
neurophysin in the SCN was a striking result of immunocytochemical studies (Vandesande et al., 1975). Oxytocin is absent
from this nucleus. The vasopressinergic neurons of the SCN
project to the lateral habenula and the lateral septum (Buijs,
1978; Sofroniew and Weindl, 1978). According to the latter
authors, the SCN also projects to areas of the brain stem. It
appears that none of the SCN neurons project to the median eminence or neural lobe (Fig. 1-9).
   Oxytocinergic and vasopressinergic neurons of the paraventricular nucleus (PVN) project to two main brain regions: (1)
the lateral septum and the hippocampus; and (2) through the

stria terminalis, to the amygdala and probably to the olfactory
tract (Buijs, 1978; Weindl et al., 1978; Zimmerman et al.,
1978). Neurophysin-containing fibers have been found in the
brain stem and spinal cord (Swanson, 1977). Fibers with similar distribution have been shown to contain oxytocin and vasopressin (Buijs, 1978; Sofroniew and Weindl, 1978). According
to Swanson (1978) and Buijs (1978), these fibers are projections of PVN neurons. However, the neurophysin-vasopressin-
containing fibers of the brain stem have been described by
Sofroniew and Weindl (1978) as descending projections of the
SCN. The peptidergic hypothalamic neurons appear to innervate,
in addition, certain circumventricular organs such as the organum vasculosum of the lamina terminalis and the subfornical
organ (Buijs, 1978; Zimmerman et al., 1978) (Fig. 1-9).

This mapping of the extrahypothalamic projections of hypothalamic peptidergic neurons obviously represents the first
step forward in the study of these neuronal systems. Certainly, other approaches similar to those used for the study of the
hypothalamo-hypophyseal system will be necessary to unfold the
design of this peptidergic circuitry.

Other neurons that probably meet the criteria to be regarded
as peptidergic neurons with local effectors will be only briefly mentioned. In several areas of the CNS and especially in
the hypothalamus, there are numerous neuronal types that, in
their perikaryon and processes, contain immunoreactive peptides. In most cases, these neurons do not end on blood capillaries but on other neurons. The list of peptides immunocytochemically demonstrated in different neurons includes: LH-RH,
TRH, somatostatin, neurotensin, substance P, enkephalins, endorphins, angiotensin II, vasoactive intestinal polypeptide,
gastrin, prolactin, ACTH, and MSH (for references, see Hökfelt
et al., 1978; Scharrer, 1978; Snyder and Innis, 1979).

REFERENCES

Bargmann, W.: Uber die neurosekretorische Verknüpfung von Hypo-
    thalamus und Neurohypophyse. Z Zellforsch 34:610-634, 1949.
Bargmann, W.: Neurosecretion. Int Rev Cytol 19:183-201, 1966.
Bargmann, W., Lindner, E., and Andres, K.H.: Uber Synapsen an
    endokrinen Epithelzellen und die Definition sekretorischer
    Neuron. Untersuchungen am Zwischenlappen der Katzenhypo-
    physe. Z Zellforsch 77:282-298, 1967.
Barry, J.: Recherches morphologiques et expérimentales sur la
    glande diéncephalique et l'appareil hypothalamo-hypophysaire.
    Ann Sci Univ Besancon, Zool Physiol Sér 2:3-133, 1961.
Benoit, J., and Assenmacher I: Rapports entre la stimulation
    sexuelle préhypophysaire et la neurosécrétion chez l'Oiseau.

Arch Anat Micr Morph Exp 42:334-386, 1953.

Berlind, A.: Cellular dynamics in invertebrate neurosecretory systems. *Int Rev Cytol 49*:171-251, 1977.

Blobel, G., and Dobberstein, B.: Transfer of proteins across membranes. I. Presence of proteolytically processed and unprocessed nascent immunoglobulin light chains on membrane-bound ribosomes of murine myeloma. *J Cell Biol 67*:835-851, 1975.

Buijs, R.M.: Intra- and extrahypothalamic vasopressin and oxytocin pathways in the rat. *Cell Tissue Res 192*:423-435, 1978.

Cannata, M.A., and Morris, J.F.: Changes in the appearance of hypothalamo-neurohypophyseal neurosecretory granules associated with their maturation. *J Endocrinology 57*:531-538, 1973.

Cross, B.A.: The neurosecretory impulse. In *Neurosecretion-- The Final Neuroendocrine Pathway*, F. Knowles and L. Vollrath, eds. Springer-Verlag, Berlin, Heidelberg, New York, 1974, pp 115-128.

Cross, B.A., Dyball, R.E.J., Dyer, R.G., Jones, C.W., Lincoln, D.W., Morris, J.F., and Pickering, B.T.: Endocrine neurons. *Recent Prog Horm Res 31*:243-294, 1975.

Dellman, H.-D., and Rodríguez, E.M.: Herring bodies: An electron microscopic study of local degeneration and regeneration of neurosecretory axons. *Z Zellforsch 111*:293-315, 1970.

Dierickx, K., and Vandesande, F.: Immuno-enzyme cytochemical demonstration of mesotocinergic nerve fibres in the pars intermedia of the amphibian hypophysis. *Cell Tissue Res 174*:25-33, 1976.

Dierickx, K., Vandesande, F., and Goossens, N.: The one neuron-one neurohypophyseal hormone hypothesis and the hypothalamic magnocellular neurosecretory system of the vertebrates. In *Biologie cellulaire des processus neurosécrétoires hypothalamiques*, J.-D Vincent and C. Kordon, eds, Edit Centre Nat Rech Sci, Paris, 1978, pp 391-398.

Douglas, W.W., Nagasawa, J., and Schulz, R.: Electron microscopic studies on the mechanism of secretion of posterior pituitary hormones and significance of microvesicles ("synaptic vesicles"): Evidence of secretion by exocytosis and formation of microvesicles as a byproduct of this process. In *Subcellular Organization and Function in Endocrine Tissues*, H. Heller and K. Lederis, eds, Cambridge University Press, London, 1971, pp 353-378.

Dreifuss, J.J.: A review on neurosecretory granules: Their contents and mechanisms of release. In *Neurophysins: Carrier of Peptide Hormones*, R. Walter, ed. New York Academy of Science, New York, 1975, pp 184-199.

Dubé, D., Lissitzky, J.C., Leclerc, R., and Pelletier, G.: Localization of α-melanocyte-stimulating hormone in rat brain and pituitary. *Endocrinology 102*:1283-1291, 1978.

Ficq, A., and Flament-Durand, J.: Autoradiography in endocrine research. In *Techniques in Endocrine Research,* P. Eckstein and F. Knowles, eds, Academic Press, New York, 1963, pp 73-85.

Flament-Durand, J., and Dustin, P.: Studies on the transport of secretory granules in the magnocellular hypothalamic neurons. *Z Zellforsch 65*:847-868, 1972.

Gainer, H., and Brownstein, M.J.: Identification of the precursors of the rat neurophysins. In *Biologie Cellulaire des Processus Neurosécrétoires Hypothalamiques,* J.-D. Vincent and C. Kordon, eds, Edit Centre Nat Rech Sci, Paris, 1978, pp 525-541.

Gainer, H., Sarne, Y., and Brownstein, M.J.: Biosynthesis and axonal transport of rat neurohypophyseal proteins and peptides. *J Cell Bio 73*:366-381, 1977.

González, C.B., and Rodríguez, E.M.: Ultrastructure and immunocytochemistry of neurons of the supraoptic and paraventricular nuclei of the lizard *Liolaemus cyanogaster:* C. Evidence for the intracisternal location of the neurophysin precursor. Submitted for publication.

Grimm-Jørgensen, Y., McKelvy, J.F., and Jackson, I.M.D.: Immunoreactive thyrotropin-releasing factor in gastropod circumoesophageal ganglia. *Nature 254*:420, 1975.

Hayward, J.N.: Functional and morphological aspects of hypothalamic neurons. *Physiol Rev 57*:574-658, 1977.

Heap, P.F., Jones, C.W., Morris, J.F., and Pickering, B.T.: Movement of neurosecretory product through the anatomical compartments of the neural lobe of the pituitary gland. *Cell Tissue Res 156*:483-497, 1975.

Heller, H.: The hormone content of the vertebrate hypothalamoneurohypophyseal system. *Br Med Bull 22*:227-231, 1966.

Hild, W.: Das morphologische, kinetische und endokrinologische Verhalten von hypothalamischem und neurohypophysärem Gewebe *in vitro. Z Zellforsch 40*:257-312, 1954.

Hökfelt, T., Johansson, O., Elde, R., Ljungdahl, A., Fuxe, K., and Goldstein, M.: Comparative topography of catecholamines and neuronal peptides in the hypothalamus. In *Biologie Cellulaire des Processus Neurosécrétoires Hypothalamiques,* J.-D. Vincent and C. Kordon, eds, Edit Centre Nat Rech Sci, Paris, 1978, pp 415-432.

Johnston, C.I., Hutchinson, J.S., Morris, B.J. and Dax, E.M.: Release and clearance of neurophysins and posterior pituitary hormones. In *Neurophysins: Carriers of Peptide Hormones,* R. Walter, ed. New York Academy Science, New York, 1975, pp 272-280.

Jones, C.W., and Pickering, B.T.: Intra-axonal transport and turnover of neurohypophyseal hormones in the rat. *J. Physiol*

*(Lond) 227:*553-564, 1972.

Knowles, F., and Bern, H.A.: The function of neurosecretion in endocrine regulation. *Nature 210:*271-272, 1966.

Lederis, K.: The distribution of vasopressin and oxytocin in hypothalamic nuclei. In *Neurosecretion,* H. Heller and R.B. Clark, eds, Academic Press, New York, 1962, pp 227-239.

Lentz, T.L.: *Primitive Nervous System,* Yale University Press, New Haven, Conn., 1968.

Morris, J.F.: Structural aspects of hormone production and storage in neurosecretion. In *Biologie Cellulaire des Processus Neurosécrétoires Hypothalamiques,* J.-D Vincent and C. Kordon, eds, Edit Centre Nat Rech Sci, Paris, 1978, pp 601-618.

Nishioka, R.S., Zambrano, D., and Bern, H.A.: Electron microscope radioautography of amino acid incorporation by supraoptic neurons of the rat. *Gen Comp Endocrinalogy 15:*477-495, 1970.

North, W.G., Valkin, H., Morris, J.F., and La Rochelle, F.T.: Evidence for metabolic conversions of rat neurophysins with neurosecretory granules of the hypothalamo-neurohypophyseal system. *Endocrinology 101:*110-118, 1977.

Pearse, A.G.E., and Takor Takor, T.: Neuroendocrine embryology and the APUD concept. *Clin Endocrinol (Oxf) 5 (Suppl.):*229-244, 1976.

Pelletier, G., and Leclerc, R.: Immunohistochemical localization of adrenocorticotropin in the rat brain. *Endocrinology 104:*1426-1433, 1979.

Peute, J., and van de Kamer, J.C.: On the histochemical differences of aldehyde-fuchsin positive material in the fibres of the hypothalamo-hypophyseal tract of *Rana temporaria. Z Zellforsch 83:*441-448, 1967.

Pickering, B.T.: Aspects of the relationships between the chemical structure and biological activity of the neurohypophyseal hormones and their synthetic structural analogues. In *Pharmacology of the Endocrine System and Related Drugs. The neurohypophysis,* H. Heller and B.T. Pickering, eds, Pergamon Press, New York, 1970, pp 81-110.

Pickering, B.T.: The neurosecretory neurone: A model system for the study of secretion. *Essays Biochem 14:*45-81, 1978.

Pickering, B.T., Jones, C.W., Burford, G.D., McPherson, M., Swann, R.W., Heap, P.F., and Morris, J.F.: The role of neurophysin proteins: Suggestions from the study of their transport and turnover. In *Neurophysins: Carriers of Peptide Hormones,* R. Walter, ed, New York Academy Science, New York, 1975, pp 15-35.

Prange, A.J., Nemeroff, Ch.B., Lipton, M.A., Breese, G.R., and Wilson, I.C.; Peptides and the central nervous system. In *Handbook of Psychopharmacology,* vol. 13, L.L. Iversen, S.D.

Iversen, and S.H. Snyder, eds, Plenum Publishing, New York, 1978, pp 1-107.

Rodríguez, E.M.: Ependymal specialization. II. Ultrastructural aspects of the apical secretion of the toad subcommissural organ. Z Zellforsch 111:15-31, 1970.

Rodríguez, E.M.: The comparative morphology of neural lobes of species with different neurohypophyseal hormones. In Subcellular Organization and Function in Endocrine Tissues, H. Heller and K. Lederis, eds, Cambridge University Press, London, 1971, pp 263-291.

Rodríguez, E.M.: The cerebrospinal fluid as a pathway in neuroendocrine integration. J Endocrinol 71:407-443, 1976.

Rodríguez, E.M., and Dellmann, H.-D: Ultrastructure and hormonal content of the proximal stump of the transected hypothalamo-hypophyseal tract of the frog (Rana pipiens). Z Zellforsch 104:449-470, 1970.

Rodríguez, E.M., and Giménez, A.R.: Comparative aspects of nervous control of pars intermedia. Gen comp Endocrinol, suppl. 3, 97-107, 1972.

Rodríguez, E.M., and La Pointe, J.: Histology and ultrastructure of the neural lobe of the lizard Klauberina riversiana. Z Zellforsch 95:37-57, 1969.

Rodríguez, E.M., Gonzalez, C.B., and Delannoy, L.: Cellular organization of the lateral and postinfundibular regions of the median eminence in the rat. Cell Tissue Res, 1979.

Rodríguez, E.M., Larsson, L, and Meurling, P.: Control of the pars intermedia of the lizard Anolis carolinensis. I. Ultrastructure of the intact neural lobe. Cell Tissue Res 186:241-258, 1978.

Rodríguez Echandía, E.L., Cavichia, J.C., and Rodríguez, E.M.: Hormonal content and ultrastructure of the hypothalamo-neurohypophyseal system of the rat after injection of vinblastine into the median eminence. J Endocrinology 73:197-205, 1977.

Sachs, H., and Takabatake, Y.: Evidence for a precursor in vasopressin biosynthesis. Endocrinology 75:943-948, 1964.

Sachs, H., Fawcett, P., Takabatake, Y., and Portanova, R.: Biosynthesis and release of vasopressin and neurophysin. Recent Prog Horm Res 25:447-484, 1969.

Scharrer, B.: The concept of neurosecretion, past and present. In Recent Studies of Hypothalamic Function. International Symposium Calgary. Karger, Basel, 1974, pp 1-7.

Scharrer, B.: The spectrum of neuroendocrine communication. In Recent Studies of Hypothalamic function. International Symposium Calgary. Karger, Basel, 1974, pp 8-16.

Scharrer, B.: Neurosecretion and its role in neuroendocrine regulation. In Pioneers in Neuroendocrinology, J. Meites, B.T. Donovan, and S.M. McCann, eds, Plenum Publishing, New York, 1975, pp 257-265.

Scharrer, B.: Neurosecretion--Comparative and evolutionary aspects. *Prog Brain Res* 45:125-136, 1976.

Scharrer, B.: Peptidergic neurons: Facts and trends. *Gen Comp Endocrinol* 34:50-62, 1978.

Scharrer, E.: Die Lichtempfindlichkeit blinder Elritzen. I. Untersuchungen über das Zwischenhirn der Fische. *Z vergl Physiol* 7:1-38, 1928.

Scharrer, E.: The final common path in neuroendocrine integration. *Arch Anat Microsc Morphol Exp* 54:359-370, 1965.

Snyder, S.H., and Innis, R.B.: Peptide neurotransmitters. *Annu Rev Biochem* 48:755-782, 1979.

Sofroniew, M.V., and Weindl, A.: Projections from the parvocellular vasopressin- and neurophysin-containing neurons of the suprachiasmatic nucleus. *Am J Anat* 153:391-430, 1978.

Speidel, C.C.: Gland-cells of internal secretion in the spinal cord of the skates. *Cargenie Inst 13 (No. 281)*:1-31, 1919.

Sterba, G. Ascending neurosecretory pathways of the peptidergic type. In *Neurosecretion--The Final Neuroendocrine Pathway,* F. Knowles and L. Vollrath, eds, Springer-Verlag, New York, 1974, pp 38-47.

Swanson, L.W.: Immunohistochemical evidence for a neurophysin-containing autonomic pathway arising in the paraventricular nucleus of the hypothalamus. *Brain Res* 128:346-353, 1977.

Tager, H.S., and Steiner, D.F.: Peptide hormones. *Annu Rev Biochem* 43:509-538, 1974.

Tasso, F., Rua, S., and Picard, D.: Cytochemical duality of neurosecretory material in the hypothalamo-posthypophyseal system of the rat, as related to hormonal content. *Cell Tissue Res* 180:11-29, 1977.

Thorn, N.A., Russell, J.T., and Vilhardt, H.: Hexosamine, calcium, and neurophysin in secretory granules and the role of calcium in hormone release. In *Neurophysins: Carriers of Peptide Hormones,* R. Walter, ed, New York Academy Science, New York, 1975, pp 202-217.

Vandesande, F., Dierickx, K., and De Ney, J.: Identification of the vasopressin-neurophysin producing neurons of the rat suprachiasmatic nuclei. *Cell Tissue Res* 156:377-380, 1975.

Vilhardt, H.: Vasopressin content and neurosecretory material in the hypothalamo-neurohypophyseal system of rats under different states of water metabolism. *Acta Endocrinology (Copenh)* 63:585-594, 1970.

Vogt, M.: Vasopressin, antidiuretic, and oxytocic activities of extracts of the dogs' hypothalamus. *Br J Pharmacol* 8:193-196, 1953.

Walter, R.: *Neurophysins: Carriers of Peptide Hormones.* New York Academy Science, New York, 1975, pp 494-512.

Weatherhead, B.: Comparative cytology of the neuro-intermediate lobe of the reptilian pituitary. *Zentralbl Veterinaermed*

7A:84-119, 1978.

de Wied, D.: Peptides and behavior. *Life Sci 20*:195-204, 1977.

Weindl, A., Sofroniew, M.V., and Schinko, I.: Distribution of vasopressin, oxytocin, neurophysin, somatostatin, and luteinizing hormone releasing hormone-producing neurons. In *Neurosecretion and Neuroendocrine Activity. Evolution, Structure and Function,* W. Bargmann, A. Oksche, A. Polenov, and B. Scharrer, eds, Springer-Verlag, Berlin, Heidelberg, 1978, pp 312-319.

Wimersma Greidanus, van Tj.B., and de Wied, D.: the physiology of the neurohypophyseal system and its relation to memory processes. In *Biochemical Correlates of Brain Structure and Function,* A.N. Davison, ed, Academic Press, New York, 1977, pp 215-248.

Wong, T.M., and Pickering, B.T.: Last in-first out in the neurohypophysis. *Gen Comp Endocrinol 29*:242-243, 1976.

Zimmerman, E.A., Stillman, M.A., Recht, L.D., Michaels, J., and Nilaver, G.: The magnocellular neurosecretory system: Pathways containing oxytocin, vasopressin and neurophysins. In *Biologie Cellulaire des Processus Neurosécrétoires Hypothalamiques,* J.-D. Vincent and C. Kordon, eds Edit Centre Nat Rech Sci, Paris, 1978, pp. 375-389.

2

Distribution of Peptides in the Central Nervous System

LESTER D. GRANT, GARTH BISSETTE,
AND CHARLES B. NEMEROFF

INTRODUCTION

Dramatic advances in understanding the neuroendocrinological
significance of neuroactive peptides and their more general
functional importance as neurotransmitters in invertebrates and
vertebrates are discussed elsewhere in this book. Of particular importance for understanding neurochemical mechanisms involved in the control of human physiological processes and
behavior, as well as related functional disorders, is the rapidly expanding literature on the roles played by neuroactive
peptides in the control of physiological and neurobehavioral
processes in mammalian species. Progress in experimental elucidation of such roles is closely linked to advances in delineation of the anatomical distribution of peptide substances in
mammalian neural and non-neural tissue.

The purpose of this paper is to review the distribution of
several neuropeptides in the central and peripheral nervous
systems of mammalian species in order to provide an introduction to morphological and histochemical characteristics of neural pathways underlying mediation of functional effects of neuropeptides. The particular peptides reviewed here are listed
in Table 2-1. This is not an exhaustive coverage of all candidate substances that might be discussed. Rather, representative peptides were selected: (1) on the basis of their having
been demonstrated to exert neurobehavioral or other effects
consistent with their likely serving as neurotransmitter substances; and (2) so as to illustrate a broad range of chemical
species and differential patterns of neural and non-neuronal
distributions (see Table 2-2). Localization patterns for several peptides not reviewed here, such as adrenocorticotropic

Table 2-1. Amino Acid Sequences of Representative Neuroactive Peptide
Substances Found in Mammalian Species

TRH   pGlu-His-Pro-NH$_2$

LH-RH   pGlu-His-Trp-Ser-Tyr-Gly-Leu-Arg-Pro-Gly-NH$_2$

Substance P   Arg-Pro-Lys-Pro-Gln-Gln-Phe-Phe-Gly-Leu-Met-NH$_2$

Neurotensin   pGlu-Leu-Tyr-Glu-Asn-Lys-Pro-Arg-Arg-Pro-Tyr-Ile-Leu-NH$_2$

Somatostatin   Ala-Gly-Cys-Lys-Asn-Phe-Phe-Trp-Lys-Thr-Phe-Thr-Ser-Cys
               |_____|

---

hormone or -endorphin and enkephalin, are discussed elsewhere
in the present volume. In discussing each substance key findings from both radioimmunoassay (RIA) and immunocytochemistry
localization studies are highlighted and, where possible,

Table 2-2. Distribution of Representative Neuroactive Peptides in
Neural and Non-neural Tissue[a]

Peptide	Pituitary Posterior	Pituitary Anterior	Hypothal amus	Cortex	Brain-stem	Spinal cord	Periph-eral nerves	Gastro-intestinal tract
TRH		+[b]	+	+	+	+		
LH-RH		+[b]	+					
Substance P			+	+	+	+	+	+
Neurotensin	+	+	+	+	+	+		+
Somatostatin	+	+[b]	+	+	+	+	+	+

---

[a] Adapted from Powell and Skrabanek (1979).

[b] target site

important relationships between the distribution of a given peptide and distribution patterns of other peptide and non-peptide neurotransmitters are noted. Accurate localization of peptides by RIA or immunocytochemical techniques is highly dependent upon the specificity of antibody reactions involved in the application of such methods. At times, immunoreactive analogues of a peptide under study or impurities in extracts of that peptide may contribute to the pattern of apparent localization for the peptide. In general, the results reviewed, then, must be recognized as possibly being affected by confounding effects due to specificity limitations associated with the localization method used.

THYROTROPIN-RELEASING HORMONE

Thyrotropin-releasing hormone (TRH) was the first hypothalamic hypophysiotropic hormone to be chemically characterized, its tripeptide structure having been determined in the laboratories of Guillemin (Burgus et al., 1969) and Schally (Boler et al., 1969). Designation of the substance as TRH was based on its identification as the neuroendocrine substance responsible for triggering the release of thyroid-stimulating hormone (TSH; thyrotropin) from anterior pituitary cells. It now appears likely that the tripeptide also affects the release of other pituitary hormones, such as prolactin, and probably serves more broadly as a neurotransmitter substance in widespread regions of the mammalian central nervous system (CNS).

The distribution of TRH in the hypothalamic-median eminence region was initially mapped by means of bioassay. Such studies demonstrated the presence of the substance in: the median eminence (ME) (Joseph et al., 1973); the dorosomedial hypothalamic nuclei and preoptic areas likely corresponding to the bed nucleus of the stria terminalis (BNST) (Krulich et al., 1974); and cerebrospinal fluid (CSF) of the third ventricle of the rat brain (Knigge et al., 1973; Knigge and Joseph, 1974). The hypothalamic distribution pattern was interpreted (Krulich et al., 1974) as suggesting the dorsomedial and preoptic hypothalamic areas as likely TRH production sites, from which the peptide is transported axonally to the ME for storage and release into the hypophyseal portal system. Alternatively, the presence of TRH in the CSF, neurons of the hypothalamic arcuate nucleus, and ependymal or glial cells lining the floor of the third ventricle led to the hypothesis (Knigge et al., 1973) that the peptide is synthesized in hypothalamic neurons, secreted into third ventricular CSF, and conducted to hypophyseal portal vessels via transependymal transport in specialized glial cells spanning the ME. Further evidence has been advanced

(Knigge, 1974; Oliver et al., 1975) in support of this hypothesis (see Rodríguez, Chapter 1).

The use of sensitive RIA procedures not only confirmed that high concentrations of TRH exist in the rat hypothalamic-median eminence region, but also indicated that notable amounts of immunoreactive TRH are detectable in many extrahypothalamic CNS areas, such as spinal cord, lower brainstem, midbrain, preoptic and septal areas, basal ganglia, and cerebral cortex (Winokur and Utiger, 1974; 1975; Brownstein et al., 1974; Jackson and Reichlin, 1974; Oliver et al., 1974a,b; Jeffcoate and White, 1975). Application of refined microdissection and RIA techniques (Brownstein et al., 1974) allowed for more specific localization of TRH in: discrete hypothalamic nuclei (arcuate, periventricular, ventromedical, and dorsomedial); the medial preoptic, dorsal, and lateral septal nuclei; and circumventricular organs of the rat brain (Kizer et al., 1976a). The latter findings are interesting in light of the ventricular transport hypothesis already discussed. Later microdissection or RIA research delineated the distribution of TRH in discrete areas of the rat spinal cord, with highest concentrations being found in the central canal and the ventral horn (Kardon et al., 1977b). Deafferentation of the hypothalamic-median eminence region of rat brain by knife cuts (Palkovits et al., 1975; Brownstein et al., 1975c; Koves and Magyar, 1975) and electrolytic lesions of the "thyrotropic" region of rat hypothalamus (Jackson and Reichlin, 1977) both markedly decreased hypothalamic-median eminence TRH, but left extrahypothalmic TRH levels intact. This suggests: (1) a relatively restricted area of TRH production in the mediobasal hypothalamus in neurons that terminate in the proximity of hypophyseal portal vessels in the ME and (2) independent in-situ production of the tripeptide in extrahypothalamic brain tissue, consistent with a more general neurotransmitter role for TRH.

Additional studies, using other types of techniques, have provided evidence confirmatory of these findings obtained with RIA methods and further substantiate the likely role of TRH as a neurotransmitter in many areas of the rat CNS. Thus, for example, immunocytochemical studies by Hökfelt et al., (1975c) revealed the presence of immunofluorescent TRH fibers in numerous spinal cord, brainstem, and forebrain regions of the rat (Table 2-3), consistent with the RIA localization results discussed. Interestingly, TRH-positive neuronal cell bodies were observed only in the dorsomedial hypothalamic nucleus by means of immunocytochemical methods used to date; this, however, does not rule out their presence in other hypothalamic or extrahypothalmic areas, such as may be revealed by improved immunocytochemical procedures, as in the case of the localization of other peptides discussed later. Ultrastructural localization

of TRH-like immunoreacitivity by means of the peroxidase-
antiperoxidase method revealed the presence of TRH in the
cytoplasm of neuronal cell bodies in the dorsomedial hypothalamic nucleus and an area dorsal to the optic chiasm, as well as
in nerve terminals in the ME, several hypothalamic nuclei, and
the spinal cord (Johansson et al., 1980). TRH immunoreactivity
in nerve terminals appeared to be confined to so-called "large
granular vesicles," approximately 1000 $\overset{o}{A}$ or more in diameter,
present in terminal boutons both in hypothalamic-median eminence and extrahypothalamic regions. In some terminals, TRH-
labeled large granular vesicles were present along with smaller
synaptic vesicles, possibly containing other "classical" neurotransmitter substances; that Johansson et al., (1980) interpret
to be evidence for the coexistence of TRH and serotonin (5-HT)
in medullary raphe neurons.

Several subcellular fractionation/RIA studies provide biochemical verification of TRH presence in terminals, demonstrating synaptosomal TRH localization and release in both rat hypothalamus-median eminence and extrahypothalamic tissue (Barnea
et al., 1975; 1976; 1978a,b; Winokur et al., 1977; Warberg et
al., 1977; Parker et al., 1977, 1978). Various neuropharmacological studies, using agents that affect catecholaminergic,
serotoninergic, or cholinergic mechanisms in the rat hypothalamus-median eminence region, however, suggest that TRH is not
contained in monoaminergic or cholinergic neurons (Nemeroff et
al., 1977; Kardon et al., 1977a; Winokur et al., 1978), but
that TRH release might be under dual monoaminergic (norepinephrine [NE] stimulation; serotoninergic inhibition) control
(Grimm and Reichlin, 1973).

Results consistent with many of these findings for TRH in the
rat CNS have been obtained for other mammalian species. Thus,
for example, Jeffcoate and White (1975) characterized TRH in
extracts not only of rat but also of rabbit and sheep hypothalamic tissues; and Bennett et al. (1975) demonstrated release of
immunoreactive TRH from sheep hypothalamic synaptosomes. Also,
Kizer et al. (1976b) mapped the distribution of TRH, biogenic
amines, and related enzymes in the bovine ME and concluded that
central neuroendocrine regulation of TRH release may involve
interactions between dopaminergic and cholinergic mechanisms.
In addition, using RIA, Oliver et al. (1974) demonstrated the
presence of TRH in human CSF; Okon and Koch (1976) mapped the
distribution of TRH in human hypothalamus, ME, and pituitary
stalk tissue; and others mapped the distribution of TRH in
extrahypothalamic human brain areas (Guansing and Murk, 1976;
Kubek et al., 1977), including cerebral cortex of the human
fetus (Aubert et al., 1977).

Still other studies (Morley et al., 1977; Leppaluoto et al.,
1978; Koivusalo and Leppaluoto, 1979) have provided evidence

Table 2-3. Summary of TRH-Containing Neural Elements in Rat Brain
and Spinal Cord Localized by Immunocytochemistry

	Immunofluorescent TRH cell bodies	Immunofluorescent TRH fibers/terminals
Spinal Cord		
Thoracic/lumbar level		
Ventral horn (anterior, medial, and lateral zones around montoneurons)	--	Moderate density
Cervical level		
Ventral horn (anterior, medial, and lateral zones around motoneurons)	--	Low density
Sympathetic lateral column	--	Low density
Medulla Oblongata		
Nucleus originalis nervi facialis	--	Moderate density
Nucleus nervi hypoglossi	--	Moderate density
Nucleus tractus solitarius	--	Moderate density
Nucleus intercalatus	--	Moderate density
Nucleus commissuralis	--	Moderate density
Nucleus ambiguus	--	Moderate density
Ventrolateral and dorsomedial reticular formation	--	Moderate density
Border zone nucleus nervi hypoglossi and nucleus dorsal motor nervi vagi	--	Moderate density
Caudal nucleus nervi hypoglossi	--	Low density
Raphe magnus and pallidus	--	Low density
Reticular formation medial and lateral to nucleus originalis nervi facialis	--	Low density
Other areas of reticular formation	--	Single fibers
Nucleus vestibularis medialis	--	Single fibers
Nucleus vestibularis spinalis	--	Single fibers
Medial lemniscus	--	Single fibers
Pons		
Nucleus originalis nervi trigemini (V)	--	Moderate density
Ventral nucleus parabrachial lateralis	--	Moderate density
Nucleus tegmental pedunculopontinus	--	Low density
Raphe region	--	Single fibers
Medial reticular formation	--	Single fibers
Periaqueductal gray	--	Single fibers
Lateral lemniscus	--	Single fibers

Table 2-3. Continued

	Immunofluorescent TRH cell bodies	Immunofluorescent TRH fibers/terminals
Nucleus trigemini nervi mesencephali	--	Single fibers
Locus ceruleus	--	Single fibers
Nucleus olivaris superior	--	Single fibers
Mesencephalon		
Nucleus originalis nervi oculomotorii	--	Moderate density
Nucleus ruber	--	Low density
Nucleus lumniscus lateralis	--	Low density
Raphe region	--	Single fibers
Paramedial reticular area	--	Single fibers
Around interpeduncular nucleus	--	Single fibers
Substantia nigra	--	Single fibers
Medial lemniscus	--	Single fibers
Diencephalon		
Hypothalamus		
Median eminence-hypophyseal stalk	--	High density
Nucleus dorsomedialis	Low density	High density
Nucleus paraventricularis parvicellular	--	High density
Perifornical region	--	High density
Nucleus ventromedial pars medialis	--	Moderate density
Periventricular area	--	Moderate density
Zona incerta	--	Moderate density
Medial forebrain bundle	--	Single fibers
Nucleus paraventricularis magnocellularis	--	Single fibers
Anterior hypothalamic nucleus	--	Single fibers
Suprachiasmatic nucleus	--	Single fibers
Basal lateral preoptic area	--	Single fibers
Telencephalon		
Neucleus accumbens	--	High density
Bed nucleus of the stria terminalis	--	Moderate density
Organum vasculosum stria terminalis	--	Moderate density
Ventral nucleus septi lateralis	--	Moderate density

for the presence of immunoreactive TRH in the rat gastrointestinal (GI) tract and other organs, analogous to dual neuronal and non-neuronal distribution of many other peptides. Similarly, the RIA demonstration of immunoreactive TRH activity in extracts of human placenta (Gibbons et al., 1975; Shambaugh et al., 1979) have stimulated interest in possible non-neuronal sources of TRH in human beings. However, Youngblood et al (1979) reported evidence for nonidentity of TRH-like immunoreactivity with synthetic pGlu-His-Pro-NH (TRH) in rat pancreas and eye, bovine and sheep pineal bodies, and human placenta. Comparison of elution profiles for TRH-like immunoreactivity with that of TRH revealed the presence of substances other than TRH in such tissues.

Youngblood et al. (1978) also reported nonidentity with synthetic TRH of TRH-like immunoreactivity in urine, serum, and extrahypothalamic brain tissue of rat. For hypothalamus and septal-preoptic samples, authentic TRH appeared to be present but constituted less than 100 percent of the immuno-reactive substances; but, for cortex, amygdala, brainstem, serum, and urine, no TRH was detectable among their respective immunoreactive constitutents. This raises the possibility that peptide substances similar to, but not identical to, TRH may have in part been localized in extrahypothalamic brain areas or nonneural tissue by means of the RIA and immunocytochemical studies already discussed. Conversely, other evidence presented by Jeffcoate and White (1975) and Krieder et al. (1979) appears to suggest that at least some of the extrahypothalamic TRH-like immunoreactivity observed in brain tissue may be attributable to TRH itself.

LUTEINIZING HORMONE-RELEASING HORMONE

The chemical structure of the decapeptide, luteinizing hormone-releasing hormone (LH-RH) or gonadotropin-releasing hormone (gonadoliberin) was first characterized in the laboratories of Guillemin (Monahan et al., 1971) and Schally (Matsuo et al., 1971). Designation of the peptide as LH-RH or GnRH reflects the fact that the hypophysiotropic substance causes the release of two anterior pituitary gonadotropins, luteinizing hormone (LH) and follicle stimulating hormone. Also, as discussed elsewhere in this volume, LH-RH exerts significant neurobehavioral effects thought to be mediated via mechanisms independent of its gonadotropin releasing effects and suggestive of possible action as a neurotransmitter substance beyond the hypothalamic-median eminence-anterior pituitary axis.

McCann et al. (1960) first demonstrated LH-RH in rat hypothalamic extracts by means of bioassay techniques. More de-

tailed regional and subcellular localization of LH-RH was not
achieved, however, until after its chemical structure was
determined and sensitive RIA (Arimura et al., 1975a) and immunocytochemical methods (Barry et al., 1973) developed. See
Sternberger and Petrali (1975) and Sternberger et al (1978) for
discussion of the specificity of immunocytochemical LH-RH localization methods.

Based on early RIA results (Palkovits et al., 1974; Deery,
1974; King et al., 1975; Araki et al., 1975; Wheaton et al.,
1975), LH-RH in rat brain appeared to be confined almost exclusively to the hypothalamic-median eminence region, except
for about 0.4-1.2 μg found in circumventricular organs
(Brownstein et al., 1976a; Kizer et al., 1976a,b). Palkovits
et al. (1974) studied the distribution of LH-RH in discrete
hypothalamic nuclei and the ME. The ME had by far the highest
LH-RH content (22.4 pg/μg protein), approximately double that
found in the arcuate nucleus. Other hypothalamic nuclei, as
well as the medial preoptic area (MPOA), were reported to
contain much lower levels of the peptide; the nucleus ventro-
medialis lateralis, for example, had about 3 percent of the
concentration found in the ME, whereas other hypothalamic
nuclei contained less than 1 percent of the ME concentration.
This pattern of differential LH-RH distribution in hypothalamus, mainly concentrated in the arcuate-median eminence region,
compared well with the hypothesized presence of LH-RH in axons
terminating in close proximity to hypophyseal portal vessels in
the ME and further suggested that cell bodies of neurons giving
rise to such axons might be confined to the arcuate nucleus.

Further evidence for LH-RH production, storage, and release
in the rat hypothalamic-median eminence region includes: (1)
visualization of immunoreactive LH-RH axons and terminals in
the arcuate-median eminence region (Barry and Dubois, 1973;
Baker et al., 1975; Kordon et al., 1974; King et al., 1974;
Hökfelt et al., 1974b; Setalo et al., 1975); (2) localization
of LH-RH in synaptosomal fractions after subcellular fractionation of hypothalamic-median eminence tissue (Barnea et al.,
1975; Taber and Karavolas, 1975; Ramirez et al., 1975); and,
(3) demonstration of production, release, and degradation of
LH-RH by hypothalamic tissue *in-vitro* or by synaptosome-
enriched extracts (Warberg et al., 1977; Sundberg and Knigge,
1978; Parker et al., 1979). The pattern of LH-RH distribution
observed corresponds to the localization of LH-RH in a tuberoinfundibular pathway projecting from the arcuate nucleus in the
mediobasal hypothalamus to the ME region. Transport of LH-RH
to the ME via such axonal projections would be consistent with
classical concepts concerning the morphological basis of hypothalamic neuroendocrine control of anterior-pituitary gonadotropin release.

Evidence suggestive of the existence of another route for
LH-RH transport to the ME, specifically via the CSF and uptake
by ependymal cells lining the base of the third ventricle, has
also been advance (Weiner et al., 1972; Knigge et al., 1973;
Ondo et al., 1973; Ben-Jonathan et al., 1974; Scott et al.,
1974; Uemura et al., 1975; Morris et al., 1975). The possibility of LH-RH ventricular transport gains further credence from
the presence of the peptide in circumventricular organs, such
as the organum vasculosum laminae terminalis (OVLT) of the
supraoptic crest, the subfornical organ, the subcommissural
organ, and the area postrema (Brownstein et al., 1975a).

In addition to the localization of LH-RH in arcuate-median
eminence region axons, the CSF, ventricular ependymal cells,
and circumventricular organs, considerable attention has been
accorded to investigation of the origin of LH-RH found in each
of these locations and to the delineation of related LH-RH-positive neural pathways. Hypothalamic deafferentation resulted
in marked (70 to 90 percent) decreases in rat hypothalamic-median eminence LH-RH content, but no change in OVLT LH-RH levels
as determined by RIA (Weiner et al., 1975; Brownstein et al.,
1975b; Brownstein et al., 1976a). This suggests that the cell
bodies of most LH-RH-positive neurons are located outside of
the mediobasal hypothalamus in the rat and that the LH-RH in
the OVLT does not originate from neurons contained in the mediobasal hypothalamus. This hypothesis is also supported by
the finding that when rats are treated with large doses of
monosodium l-glutamate (MSG) in the neonatal period, which destroys 90 percent of arcuate neurons, there is no significant
alteration in the content of LH-RH in the mediobasal hypothalamus (Nemeroff et al., 1977).

More specific delineation of cells of origin of LH-RH and
their projections, however, has been accomplished by means of
immunocytochemical studies, although not without attendant controversy. Conflicting results have been obtained by different
investigators concerning the distribution of LH-RH-positive
neuronal cell bodies, possibly due to differences in the antisera to LH-RH (Hoffman et al., 1978; Knigge et al., 1978) or
age, sex, species, stage of cycle, or other factors extant at
the time of sacrifice (Araki et al., 1975; Baker and Dermody,
1976; Wheaton and McCann, 1976; Kirsch, 1978). For example, in
the rat, some investigators report LH-RH-positive cell bodies
in the preoptic region (Setalo et al., 1976; Kirsch, 1978;
Ibata et al., 1979) or suprachiasmatic area (Setalo et al.,
1976), whereas others reported such cell bodies to be located
in the arcuate nucleus (Koslowski et al., 1975) or tanycytes of
the ME of the mouse brain (Zimmerman et al., 1974) or failed to
observe such cell bodies at all (Baker et al., 1975; Gross,
1976).

More recently, however, Hoffman et al. (1978) reported finding one population of LH-RH cell bodies in the retrochiasmatic
area, tuberal area, and arcuate nucleus using one antiserum,
but observed a second population of LH-RH perikarya scattered
throughout the medial preoptic, preoptic periventricular, and
medial septal areas using another antiserum. Immunoreactive
fibers were observed in both OVLT and the medial basal hypothalamic-median eminence region with each of the two antisera.
Of particular note are further findings reported by Burchanowski et al. (1979) demonstrating the presence in mice of LH-RH-
positive neuron cell bodies and fibers in close contact with
the surface of the third ventricle. In the medial preoptic and
suprachiasmatic areas, bipolar LH-RH neurons were seen to send
short processes to the ventricular surface and longer processes
toward the OVLT. Septal region LH-RH neurons also send their
processes toward the ventricular surface, whereas LH-RH neurons
in the nucleus of the anterior commissure and the BNST project
over the anterior commissure to form a dense plexus of fibers
in the subfornical organ. These latter results appear to support the hypothesis of LH-RH release into the ventricular system and transport via CSF to the ME or elsewhere (Knigge et
al., 1978).
The findings of LH-RH-positive neuronal cell bodies in several different fields within mouse anterior diencephalic regions,
such as the medial preoptic area and the mediobasal hypothalamus, compare well with results earlier reported for guinea pigs
and other mammals by Barry and Dubois (1974, 1975, 1976) and
Silverman and associates (Silverman, 1976; Silverman and
Denoyers, 1976; Silverman et al., 1979). In regard to the work
of Barry and Dubois, several groups of LH-RH-positive neuronal
cell bodies and associated axonal projection pathways were
delineated in the guinea pig by immunocytochemistry methods,
and some components of most of those immunoreactive LH-RH pathways have been at least partially visualized in other mammalian
species, such as rats, rabbits, cats, and dogs. These include
the following LH-RH pathways: (1) preoptico-infundibular tract;
(2) preoptic-supraoptic (or preoptic-terminal) tract; (3) hypo-
thalamo-mesencephalic tract; (4) hypothalamo-epithalamic tract;
(5) hypothalamo-amygaloid tract; and, (6) paraolfactory-
rhinencephalic tract.
As demonstrated by Barry and associates most clearly in the
guinea pig, the single largest LH-RH neural system appears to
be the preoptico-infundibular pathway, containing immunoreactive axons arising from two different groups of LH-RH-producing
perikarya: the main group being pre- and suprachiasmatic in
location and the other being in the arcuate-infundibular area.
The fibers of this tract project to the ME, where they terminate around capillaries of the hypophyseal portal plexus after

giving off numerous short and long collaterals, some of which
course caudally and terminate in the infundibular stem. Numerous, varied types of synaptic contacts appear to be made en
passage by LH-RH fibers along their trajectory from preoptic
areas to the ME (Barry and Dubois, 1973; Barry et al., 1973).
Components of extrahypophyseal pathways are less distinctly
seen but are reported by Barry and Dubois (1976) to include:
(1) a short pathway terminating at the level of the supraoptic
crest (guinea pig) or the level of the OVLT (rat and mouse);
(2) a bulbomesencephalic pathway, with axons passing over and
under the mamillary body; (3) an epithalamic pathway, with
axons passing over and under the mamillary body; (4) an epithalamic pathway, with axons running dorsally around the
habenulo-interpeduncular tract; (5) a rhinencephalic tract,
coursing rostrally into the basal rhiencephalon; and (6) an
amygdaloid tract coursing in precommissural components of the
fornix.

Results obtained by Silverman and colleagues (Silverman,
1976; Silverman and Desnoyers, 1976; Silverman et al., 1979) in
the guinea pig and other mammals, such as the mouse and hamster, confirm and extend the findings of Barry et al. As reviewed by Silverman et al. (1979), four groups of LH-RH-positive cell bodies and associated fiber tracts are observed in
the guinea pig brain (see Fig. 2-1). These include the following: (1) LH-RH-positive neurons in the medial basal hypothalamus, mainly in the arcuate nucleus, that project to the
zona externa of the ME; (2) a larger cell group in the MPOA
dorsal to the optic chiasm that project to the OVLT, the preoptic portion of the suprachiasmatic nucleus, the retrochiasmatic portion of the suprachiasmatic nucleus, the zona externa
on the lateral and ventral surfaces of the infundibular stalk,
and the zona interna throughout the ME and stalk; (the MPOA
cells also project axons through the dorsomedial hypothalamus
that converge at the level of the mamillary body and course in
the mamillary peduncle to terminate in the ventral tegmental
area (VTA) surrounding the interpeduncular nucleus [IPN]); (3)
some LH-RH neurons in the medial septal nucleus (MSN) that follow a similar course to the VTA, whereas other MSN neurons project via the stria medullaris, through the medial habenular
nucleus (MHN), and down the fasciculus retroflexus (FR) to the
medial and caudal portions of the IPN; (4) other LH-RH neurons
clustered around large blood vessels in the MSN, nucleus diagonal band, and the olfactory tubercle that project short axons
that appear to terminate primarily around those vessels. In
addition to these pathways, scattered axons of unknown origin
are reported by Silverman et al. (1979) to be seen in the
anterior colliculus and midbrain reticular formation. Ultrastructural characteristics of these axonal terminals in the ME

region include the presence of LH-RH within granules of 900 to 1200 Å size in axons of the pallisade zone and 400 to 800 Å size in terminals ending on the portal plexus (Silverman and . Desnoyers, 1976).

Turning to the localization of LH-RH in primates, LH-RH-positive cell bodies have been localized by immunocytochemistry methods within the hypothalamus at locations similar to those outlined for the guinea pig. Silverman et al. (1977, 1979), for example, demonstrated LH-RH neurons in the pre- and pericommissural region (BNST, medial preoptic nucleus), lamina terminalis and infundibular area (infundibular and premamillary nucleus, retrochiasmatic area) in the rhesus monkey. LH-RH-positive fibers, as observed by Silverman et al. (1977), tended to be more widely scattered and less well organized into discrete fiber tracts than in the guinea pig. Numerous fibers from the pericommissural area course ventrally to the OVLT, whereas others project to the septal region. Fibers from infundibular area neurons appeared to project mainly to the zona externa of the ME, and other LH-RH fibers continue down the infundibular stalk and terminate in the posterior pituitary, in a manner not observed in other species. Some LH-RH-positive fibers were observed in the MHN and FR, and a large plexus of axons occurred in the medial mamillary nucleus. The pattern of results obtained by Silverman et al. (1977, 1979) matches well that observed by Zimmerman and Antunes (1976) in rhesus monkey hypothalamic-anterior diencephalic areas.

LH-RH localization patterns similar to those observed in the rhesus monkey have also been found in other primate species, such as Macacus, Ceropithecus, *Cebus apella* and *Saimiri sciureus* (Barry, 1978, 1979; Barry et al., 1975, 1976). Barry et al. (1976) reported that LH-RH neurons form four main groups localized in the medio-basal hypothalamus, the septo-preoptic area, the anterior pericommissural area, and the perimamillary region. These neuronal groups, respectively, appear to give rise to: (1) a hypothalamo-infundibular LH-RH tract; (2) a preoptico-terminal LH-RH tract; and (3) various extrahypophyseal tracts ending in the telencephalon, the epithalamus, and mesencephalon. Detailed immunocytochemistry mapping of the trajectory of the septoepithalamo-habenular LH-RH-reactive neurons in female squirrel monkeys (*Saimiri sciureus*) and male *Cebus apella* monkeys have been reported by Barry (1978); and additional detailed study of the preoptico-terminal LH-RH tracts in the female squirrel monkey during the estrous cycle has been reported by Barry (1979). Ultrastructural characteristics of LH-RH reactive neurons in OVLT of the squirrel monkey have also been studied by Mazzuca (1977), who reported 90 to 130 nm diameter neurosecretory granules to be present in LH-RH neurons termi-

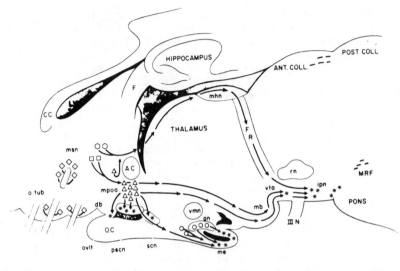

Fig. 2-1. Camera lucida drawing of parasagittal section of
guinea pig brain showing locations of LH-RH neurons
and their projections. Terminal fields are indicated
with an asterisk (*). Positive neurons (o) in the
medial basal hypothalamus, including the arcuate nucleus (an), project to the zona externa of the median eminence (me). Other neurons (Δ) in the medial
preoptic area (mpoa) project to the organum vasculosum of the lamina terminalis (ovlt), preoptic portion of the suprachiasmatic nucleus (pscn) and retrochiasmatic portion of the suprachiasmatic nucleus
(scn). LH-RH neurons in the mpoa also project to a
small extent to the zona externa on the lateral and
ventral surfaces of the infundibular stalk and to
the zona interna throughout the extent of the median
eminence and stalk. MPOA neurons also project
through the dorsomedial hypothalamus to converge at
the level of the mamillary body (mb) and especially
the fiber capsule under the mb. These fibers continue down the mamillary peduncle to the ventral
tegmental area (vta). Some LH-RH neurons (□) in the
medial septal nucleus (msn) follow the same route to
the vta as those in the mpoa. Others (O) project
via the stria medullaris, through the medial habenular nucleus (mhn) down the fasciculus retroflexus
(FR) to the medial and caudal portions of the interpeduncular nucleus (ipn). Other LH-RH neurons (◊)
are seen clustered around large blood vessels in the
medial septal nucleus, nucleus of the diagonal band
(db), and olfactory tubercle (o. tub.); their axons

nating in the OVLT and, at times, in the nearby cerebroventricular surface.

Findings similar to these results have been reported for both human fetal brain (Paulin et al., 1977; Bugnon et al., 1976, 1977a; Aksel and Tyrey, 1977) and postnatal specimens from children and adults (Barry, 1976a,b; 1977). For example, LH-RH-positive neurons are found in the mediobasal hypothalamic region (retrochiasmatic and infundibular areas, postinfundibular ME, premamillary nucleus) and, also, in the lamina terminalis, septal region, and pericommissural area. Some scattered neuronal cell bodies also appear in ventromedial and dorsomedial hypothalamic nuclei. Barry (1977) reports that cells in the mediobasal hypothalamo-infundibular LH-RH tract terminate on the primary portal plexus of the infundibulum, whereas neurons in the lamina terminalis and nearby preoptic area give rise to a preoptico-terminal tract terminating on blood vessels of the OVLT and between ependymal cells lining its ventricular surface. In addition, Okon and Koch (1977) have reported localization of LH-RH in circumventricular organs of human brain, similar to results observed in other mammalian species discussed earlier; and Bugnon et al. (1977b) described ultrastructural features of LH-RH neurons, including the presence of 60 to 130 nm diameter neurosecretory granules in terminal endings of such neurons.

Only relatively limited efforts have as yet been undertaken to delineate relationships between LH-RH neurons and other neurotransmitter systems. Results from such studies indicate the likely absence of LH-RH from catecholamine neurons of the rat brain. Intraventricular injections of 6-hydroxydopamine (6-OHDA), which destroys catecholamine neurons, failed to alter LH-RH levels (Kizer et al., 1975). On the other hand, McNeill, and Sladek (1978) used combined histochemical fluorescence-immunocytochemistry methods for simultaneous catecholamine and LH-RH localization on the same tissue sections and found that a certain band of dopamine (DA) terminal endings in the lateral ME overlaps LH-RH terminals. This suggests possible axo-axonal innervation of LH-RH terminals by DA fibers of tubero-infundibular DA systems projections originating from perikarya located

---

appear to terminate primarily around these blood vessels. Scattered axons of unknown origin are found in the anterior colliculus (Ant. Coll.) and midbrain reticular formation (MRF). Additional abbreviations are CC: corpus callosum; Post. Coll.: posterior colliculus; rn: red nucleus, vm: ventromedial nucleus. (From Silverman et al. [1979] with permission.)

in the arcuate-periventricular nuclei. At least some of these
DA neurons have been found to exhibit nuclear concentration of
estradiol (Grant and Stumpf, 1975; Heritage et al., 1977) and
testosterone (Heritage et al., 1980), as do other nearby non-
catecholamine neurons in the arcuate-periventricular nuclei,
which likely include at least some LH-RH neurons. These relationships may provide possible morphological bases for hypothesized axo-axonic regulation by DA of LH-RH release in the ME.

SUBSTANCE P

Substance P (SP) is an undecapeptide (Table 2-1) discovered by
von Euler and Gaddum (1931) as a potent hypotensive agent in
extracts of equine brain and gut. The distribution of SP was
originally studied by means of bioassay methods, as discussed
by Zetler (1970). However, since the isolation, characterization, and synthesis of SP was accomplished by Leeman and associates (Chang and Leeman, 1970; Chang et al., 1971; Traeger et
al., 1971; Leeman and Mroz, 1974), both highly sensitive RIA
measurement methods (Powell et al., 1973; Brownstein et al.,
1976b; Cuello et al., 1979) and immunocytochemical techniques
(Nilsson et al., 1974; Hökfelt et al., 1975f,g; Cuello et al.,
1979) have been developed and applied in carrying out detailed
localization of the peptide in the CNS and peripheral nervous
system (PNS) of numerous mammalian species.
It is important to note that the results obtained apply to
the localization of immunoreactive substance P (ISP), the distribution of which may vary at times from that of true endogenous SP. Evidence supporting the likely restriction of ISP
localization to endogenous SP sites includes: (1) notable similarities in SP distribution patterns as demonstrated by bioassay and immunological localization methods; and, (2) results
obtained with specificity tests in which the particular SP
antibodies used in RIA and immunocytochemistry localization
studies fail to react with structurally similar peptides.
Thus, for example, Powell et al. (1973) found that the antibody
developed by them for measuring SP in human brain did not react
with either eledoisin or physalaemin, two peptides structurally
similar to SP; and other investigators have obtained similar
results with different RIA's. However, although various RIA's
for SP do not cross react with closely related peptides, physalaemin or eledoisin (Kanazawa and Jessell, 1976; Pradelles et
al., 1979; Yanaihara et al., 1976), Brownstein et al. (1976b)
have observed a significant degree of a cross reactivity of the
SP antibody with the C-terminal octapeptide fragment (4-11).
Also, Yanaihara et al. (1976), Cannon et al. (1977), and
Pradelles et al. (1979) have shown that the SP antibody cross

reacts with SP fragments longer than the C-terminal pentapeptide; and Zetler et al. (1978) found that, while subfraction Fb of crude SP preparations contained SP-like activity, the pharmacologically active principle of crude SP subfraction Fc is actually the COOH-terminal octapeptide of cholecystokinin (CCK-8) or a closely related peptide. It is therefore conceivable that certain of the ISP localization results to be summarized may be attributable to SP-like compounds or fragments.

Substance P has been demonstrated in brain and gut of many mammalian species. In human brain, highest concentrations of the peptide, as measured both by bioassay (Zetler, 1970) and RIA (Duffy et al., 1975), are found in substantia nigra (SN) and hypothalamus; and higher concentrations occur in gray than in white matter. Subcellular fractionation studies by Duffy et al. (1975) further demonstrate that highest concentrations of SP occur in fractions containing nerve endings from bovine hypothalamus and SN extracts. Such data, together with demonstrations of high concentrations of the peptide in bovine dorsal root (Takahashi et al., 1974), in the dorsal horn of the cat spinal cord (Takahashi and Otsuka, 1975), and its differential distribution in many areas of rat brain as shown by RIA mapping methods (Brownstein et al., 1976b), provide biochemical evidence suggesting that SP is a widespread neurotransmitter in mammalian neural systems. This is further reinforced by demonstration that: (1) SP exerts potent excitatory effects when applied to spinal cord motoneurons (Konishi and Otsuka, 1974); (2) release of ISP from rat spinal cord after dorsal root stimulation is calcium dependent (Otsuka and Konishi, 1976); and (3) release of SP from rat hypothalamic brain slices occurs in response to elevated potassium levels (Jessell et al., 1976).

In pioneering immunocytochemical studies by Hökfelt and colleagues (Nilsson et al., 1974; Hökfelt et al., 1975a), ISP nerve cell bodies, fibers, and terminals were demonstrated by an indirect immunofluorescence technique in various areas of the PNS and CNS of the rat and cat, consistent with both bioassay and RIA localization findings already alluded to. The localization of ISP in the CNS by Hökfelt and colleagues included observation of high concentrations of immunofluorescent fibers and terminals in the substantia gelatinosa, the adjacent medial potion of the funiculus lateralis, and certain medial parts of the ventral horns in the cervical spinal cord. At the medulla oblongata level, high concentrations of ISP fibers and terminals were associated with the nucleus parasolitarius, extending to medial areas of the spinal trigeminal nucleus and tract. However, no such ISP neural elements were seen in the cerebellum. At the mesencephalic level, ISP fibers and terminals were seen in the SN, around the IPN, in dorsal and ventral areas of the medial geniculate body and in the periaqueductal central gray extending ventrally into the raphe region. High concentrations of ISP fibers and terminals were also seen at diencephalic levels in the dorsal, but not ventral, hypothalamus, the perifornical regions, the zona incerta, the MPOA, as well as the midline thalamic nuclei, the lateral habenular nucleus, the internal capsule, and the medial (AM) and central (AC) amygdaloid nuclei. At the striatal level, ISP fibers were seen in the septum and extending down to the nucleus tractus diagonalis and, sparsely, in the cingulate cortex. ISP neuronal cell bodies were only observed in one location in the rat CNS, that is the medial habenula. In other subsequent studies by Hökfelt and colleagues, ISP histofluorescent nerve fibers and terminals were observed in the human cerebral cortex (Hökfelt et al., 1976b), and a dense plexus of ISP fibers was seen in the ME of primates in close proximity to adenohypophyseal portal vessels (Hökfelt et al., 1978b).

The findings of Hökfelt and colleagues were confirmed and extended by the detailed studies of Cuello and Kanazawa (1978), who observed ISP neuronal cell bodies not only in spinal root ganglia and MHN but also in the IPN, caudate, putamen, and globus pallidus of rat brain. However, they did not observe ISP fibers or terminals in the ME region, unlike the results obtained in primates by Hokfelt et al. (1978b).

Considerable progress has been made in delineating specific CNS pathways comprised of ISP neuron cell bodies, fibers, and terminals and in relating such SP-containing neural systems to neurons utilizing other neurotransmitter substances. The most notable advances to date include demonstration of: (1) a SP-containing habenulo-peduncular tract; (2) SP-containing strionigral and pallidonigral pathways; and (3) ISP in certain CNS serotonin (5-HT) neurons.

Evidence for a habenulo-peduncular SP tract derives from several studies (Mroz et al., 1976; Hong et al., 1976; Emson et al., 1976), which demonstrated ipsilateral decreases in interpeduncular SP levels of 40 to 80 percent, as measured by RIA, following lesions made of the entire habenula or the MHN. Residual ISP in the IPN after complete habenular destruction appeared to be attributable to ISP cell bodies in the IPN having presently undetermined axonal projection trajectories. IPN levels of choline acetyltransferase decreased (by up to 90 percent) in parallel with decrements in interpeduncular SP following habenular lesions (Emson et al., 1976). This, and other information discussed by Mroz et al. (1976), highlights the possible existence of habenular-interpeduncular tract cholinergic neurons that may also contain and secrete SP. Alternatively, parallel SP and cholinergic projections from different habenular neurons may terminate in the IPN, as noted by Hong et al. (1976).

Neurosurgical manipulations (knife cuts, electrolytic lesions) have also been used in elucidating striatonigral and pallidonigral SP-containing tracts and to differentiate these from closely parallel descending GABAergic pathways that also innervate the SN. Initial evidence for innervation of the SN by descending SP fibers was provided by Mroz et al. (1976), who found: (1) 90 percent reduction in SP in the ipsilateral pars reticularis of rat SN following knife cuts transecting the medial forebrain bundle (MFB) and the media portion of crus cerebri at a premamillary level, and (2) a 50 percent reduction of SP levels in SN after partial globus pallidus lesions. Subsequent knife cut studies by Brownstein et al. (1977b) provided more specific delineation of striatonigral SP projections likely originating mainly from neurons in rostral portions of the striatum (caudate), in contrast to striatonigral GABAergic fibers likely originating from cell bodies more uniformally distributed in both rostral and caudal striatum and, possibly, a portion of the globus pallidus. The striatonigral SP fibers appear to be further differentiated from GABAergic nigral afferents in terms of striatal SP predominate innervation of the SN pars reticularis in contrast to GABA innervation of both SN pars compacta and reticularis (see Fig. 2-2).

Further evidence substantiating the existence of a striatonigral SP projection distinct from a closely parallel striatonigral GABA pathway was reported by Gale et al. (1977) and Kanazawa et al. (1977). The latter group also reported data pointing to the likely additional existence of pallidonigral SP projections. A subsequent study by Paxinos et al. (1978a) revealed that the striatonigral pathway likely descends in the internal capsule to reach the SN, and other studies by Paxinos et al. (1978b) and Kanazawa et al. (1980) suggest an additional striatofugal SP projection to the entopeduncular nucleus (internal pallidal segment) in the rat. Other work recently reported by Gauchy et al. (1979) confirms widespread distribution of SP in the cat SN, similar to that seen in the rat, with the zona reticularis containing approximately twice as much SP as the zona compacta. The origin of SP innervation of SN areas other than the zona reticularis, which appears to be mainly striatonigral, remains to be elucidated.

The specific patterns of innervation of the SN by both SP-containing and GABAergic neurons suggest potential interrelationships between these neurotransmitter pathways and DA neurons concentrated in various areas of the SN. Of particular interest is the likelihood that the cell bodies or dendrites of such DA neurons, which are known to give rise to an ascending nigrostriatal DA pathway, are themselves innervated by SP and GABA fibers originating in the striatum. These probable relationships, schematically depicted in highly simplified form in

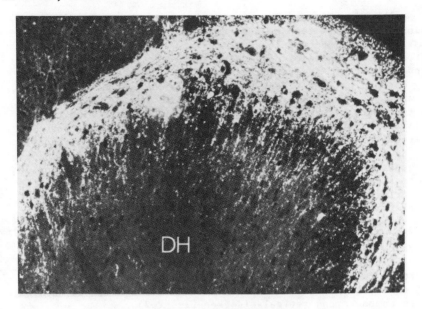

Fig. 2-2.   ISP localization in cervical spinal cord. Note
dense innervation by immunofluorescent SP-positive
fibers and terminals of the substantia gelatinosa
and adjacent areas dorsal to the dorsal horn (DH) of
the cord. (From Hökfelt et al [1975f] with
permission.)

Figure 2-2, appear to provide one basis for regulation of striatal functions, including control of extrapyramidal motor functions.

Preliminary information on the possible trajectories of other
SP-containing pathways, in diencephalic-telencephalic areas,
has also been obtained. This includes evidence obtained by
knife-cut or immunohistochemistry studies (Paxinos et al.,
1978b), which suggest that: (1) the MPOA receives a SP projection from the interstitial BNST, where ISP-positive neuron cell
bodies were visualized; and (2) the lateral septum receives a
SP projection via a region ventral to it. Paxinos et al.
(1978a) also provided knife-cut or immunocytochemical evidence
for ISP terminals innervating the dorsomedial frontal (cingulate) cortex, likely being derived from projections ascending
in the MFB from an as yet undefined but probable mesencephalic
area caudal to the IPN.

Ben-Ari et al. (1977) additionally reported RIA data that
they interpreted as suggesting the existence of an ISP-contain-

ing pathway originating from cells in the BNST and coursing via the stria terminalis to terminate in the AM nucleus. However, subsequent studies by the same group (Ben-Ari et al., 1979), using combined high performance liquid chromatography and RIA procedures, revealed that the AC and AM nuclei and the BNST not only contained authentic SP but also significant amounts of the nanopeptide C-terminal fragment (3-11); authentic SP is almost exclusively present in SN and the caudate putamen. Regardless of the precise chemical species yielding ISP-like activity in the BNST, AM and AC, the associated ISP pathway may parallel amygdalofugal neurotensin- (NT) and enkephalin-containing pathways described later in this chapter.

Additional interesting relationships between SP-containing neural elements and neurons containing other neurotransmitter substances can be discerned. For instance, evidence for ISP in rat raphe (5-HT) neurons has been obtained independently by several laboratories. Using an indirect immunofluorescence technique, Hökfelt et al. (1978a) demonstrated that numerous neuronal somata in nucleus raphe magnus, raphe obscurus, raphe pallidus, pars α of the nucleus reticularis gigantocellularis and nucleus interfascicularis hypoglossi contained both immunoreactive SP and 5-HT. Intraventricular and intracisternal injections of 5,6- or 5,7-dihydroxytryptamine (5,6-DHT; 5,7-DHT; 5-HT-depleting cytotoxic agents) caused both SP and 5-HT nerve terminals in the ventral horns of the spinal cord to disappear and allowed for visualization of enlarged SP and 5-HT fibers in the vicinity of the olivary complex. Reserpine, however, depleted only 5-HT stores in both terminals and cell bodies, suggesting possible differential mechanisms for storage, release, or reuptake of SP. Another laboratory (Chan-Palay et al., 1978) used autoradiography and fluorescence histochemistry to visualize 5-HT and immunofluorescence to demonstrate SP in identical rat brain regions on the same tissue sections. A third group (Singer et al., 1979) has recently reported RIA confirmation of reduction in SP levels in the ventral cervical spinal cord of the rat following intracisternal 5,7-DHT injection. However, after such injections no statistically significant decreases were seen in SP levels in dorsal areas of the cervical cord or in five areas of the diencephalon and telencephalon. Nor did 6-OHDA or 5,6-DHT treatments, effective in reducing spinal NE and 5-HT levels, have any effect on spinal SP. These results strongly suggest the likely existence of "5-HT/SP" neurons of supraspinal origin, that is, probably the various raphe nuclei previously listed, with descending fibers terminating in ventral cervical spinal cord.

As for ISP PNS components, Hökfelt et al. (1975a,b) showed in their initial mapping studies that about 20 percent of the

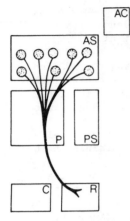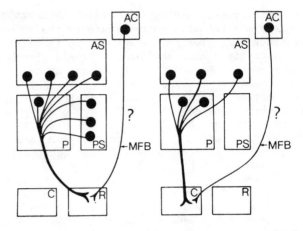

Fig. 2-3.   Diagrammatic representation of substantia nigra (SN)
            innervation by SP-containing (stippled) and GABA-
            containing (solid) neurons projecting to pars retic-
            ularis of SN.   Center: GABA neurons projecting to
            pars reticularis of SN.   Abbreviations: AC: nucleus
            accumbens; AS: anterior striatum; C: pars compacta
            of SN; MFB: medial forebrain bundle; P: pallidum;
            PS: posterior striatum; R: pars reticularis of SN.
            (From Brownstein et al. [1977b] with permission.

neuronal cell bodies, mainly of small-sized ganglion cells, in
the spinal and trigeminal ganglia exhibit SP-like immunofluorescence.   Also, fiberlike ISP structures were seen: (1)
within the ganglia; (2) forming bundles of thin, probably
unmyelinated, fibers in the sciatic nerve; and (3) forming
fibers and nerve endings in the nasal mucosa and skin, at times
surrounding sweat glands.   Other work on the cat spinal cord
(Hökfelt et al., 1975b) confirmed the localization of ISP
fibers in the dorsal horn, similar to that seen in the rat
spinal cord (Figure 2-3), and provided more detailed mapping of
ISP-positive fibers in Lissauer's fasciculus and laminae I, II,
III, VI, and VIII of the dorsal horn.   However, ISP cell bodies
of spinal ganglion neurons were observed in the cat only after
dorsal root compression or colchicine treatment.   These results
strongly suggest that SP or similar peptide is present in spinal ganglia neurons and serves as a transmitter substance in at
least some primary sensory afferents of the rat and cat (see

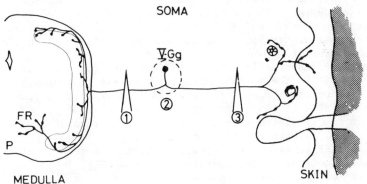

CENTRAL END(axonal)   PERIPHERAL END(dendritic)

SOMA

FR

P

MEDULLA   SKIN
OBLONGATA

Fig. 2-4. Diagrammatic depiction of SP-containing sensory neuron of trigeminal system as delineated by combined
immunohistochemical localization and surgical manipulations, including: (1) rhizotomy, (2) ganglionic
lesion, and (3) sensory denervation. Abbreviations
are FR: reticular formation; P: tractus cortico-
spinalis; V-Gg: gasserian ganglion. (From Cuello et
al. [1978] with permission.)

Figure 2-3). Also, ISP fibers, but not cell bodies, were observed by Hökfelt et al. (1977b) in sympathetic ganglia of rat,
cat, and guinea pig, suggesting a role for SP in autonomic nervous system functions as well.
    Further elaboration of evidence for SP-containing spinal and
ganglia cells has been reported. Jessell et al. (1979) demonstrated a 75 to 80 percent depletion of SP in the dorsal horn
of rat lumbar spinal cord after either unilateral section of
the sciatic nerve or dorsal rhizotomy. The SP content of dorsal root ganglia and both the peripheral and central branches
or primary sensory neurons also decreased after sciatic nerve
section. In addition, ultrastructural studies by Chan-Palay
and Palay (1977), Cuello et al. (1977), and Pickel et al.
(1977) have demonstrated, by means of immunohistochemical peroxidase staining, that ISP is associated with either large, 100
to 300 nm diameter, vesicles in rat sensory processes and dorsal horn spinal cord terminals or smaller round, 60 to 80 nm

diameter, vesicles in terminal endings within the dorsal horn, or both. Also, Gamse et al. (1979) provided RIA and immunohistochemical evidence of somatofugal axoplasmic transport of ISP in the cat vagus nerve; and Robinson et al. (1980) demonstrated ISP in the superior cervical ganglion and submaxillary gland of the rat, by means of RIA.

In other research, Cuello et al. (1978) investigated the distribution of ISP in the spinal nucleus of the rat trigeminal nerve and in the peripheral endings of ISP sensory neurons innervating the skin of the lower lip. Marked depletion of ISP occurred in the ipsilateral trigeminal spinal nucleus after rhizotomy or ganglionic lesions and was accompanied by small decreases in ISP levels in the reticular formation ventromedial to the spinal nucleus on the operated side (see Fig. 2-4). Small losses of ISP occurred in the skin of the lower lip completely disappeared after the sectioning of the mental nerve, that is, the sensory branch of the trigeminal nerve innervating the lower lip.

NEUROTENSIN

NT is a tridecapeptide (pGlu-Leu-Tyr-Glu-Asn-Lys-Pro-Arg-Arg-Pro-Tyr-Ile-Leu-OH) originally isolated from bovine hypothalamus and chemically characterized by Carraway and Leeman (1973, 1975). It is one of several peptide substances (Table 2-2) that have been localized in both the brain and gut of mammalian species. Carraway and Leeman (1976) initially identified immunoreactive neurotensin (INT) in bovine, rat, guinea pig, and rabbit hypothalami, using rabbit antisera raised against synthetic NT conjugated to either hemocyanin, thyroglobulin, or a poly Glu-Lys peptide chain. The different antisera, directed toward different parts of the NT molecule, were sensitive enough to detect 3 fmol NT, and similar quantities of INT were measured in the various species by the different antisera. Use of NT fragments and other vasoactive peptides, such as SP, xenopsin and bradykinin, helped to establish the specificity of the antisera. INT was also shown by the same laboratory to be present in calf small intestine (Kitabgi et al., 1976). The extensive electrophoretic, chromatographic, and biological assays used by these investigators provided strong evidence for their antisera being highly sensitive and specific for NT and established the presence of the substance in the brain and gut of several mammalian species.

Extensive detailed localization studies by these same and other investigators have further extended these findings. Certain groups, for example, have determined the distribution of NT in specific brain regions of several different mammalian species by means of highly sensitive RIA techniques. In one study, Kobayashi et al. (1977) used the microdissection technique of Palkovits (1975) and a specific double antibody RIA method to localize NT in discrete areas of the rat brain. The hypothalamus contained the highest concentration of INT, with the highest hypothalamic levels being found in the ME, followed by the medial and lateral preoptic nuclei, and other hypothalamic and preoptic nuclei. High INT levels were also observed in certain limbic system components, for example, nucleus accumbens, septal area, and amygdala, but not others, that is, limbic cortical areas, such as the hippocampus, cingulate cortex, and entorhinal cortex, or the olfactory bulb. The IPN and mesencephalic central gray also contained high levels of INT, in contrast to low INT levels found in the pons, medulla, central gray, spinal cord, and pineal gland.

In another set of RIA localization studies, Uhl and Snyder (1976, 1977a,b) found INT to be heterogenously distributed in calf brain in a manner essentially similar to the previously published pattern observed for rat brain. For example, the highest levels of INT were observed in medial and anterior hypothalamus and the lowest levels in much of the cerebral cortex, cerebellum, and spinal cord. Most other areas contained moderate levels of INT. The basal ganglia (caudate, globus pallidus), however, were reported to contain fairly high levels of INT, followed by the parahippocampal gyri, the anterior thalamic nucleus, and cingulate gyrus (the latter three areas being limbic system components) in a manner somewhat different from that reported for the rat. Recently, however, Dupont et al. (1979) reported RIA results indicating that extrahypothalamic INT is mainly concentrated within the accumbens, preoptic area, habenula, and amygdala of the calf brain, consistent with its distribution in rat brain.

Subcellular fractionation studies by Carraway and Leeman (1976) and Uhl and Snyder (1977b,c) demonstrated that NT is preferentially localized in synaptosomal fractions of rat and calf brain after density gradient centrifugation. Also, Uhl et al. (1977a) studied NT receptor binding in calf brain, with the highest levels of NT binding being in the dorsomedial thalamus, parahippocampal cortex, and hypothalamus, lowest levels in the cerebellum and brainstem, and intermediate values in the basal ganglia.

The regional distribution of INT in the brains of Japanese monkeys (*Macaca vuscata*) were reported by Kataoka et al. (1979) to be generally consistent with these findings for lower mammalian species, in which INT content was found to be highest in hypothalamic regions and very low in the cerebellar cortex and white matter, such as the corpus callosum. Other interspecies similarities include high INT levels in the IPN, mesencephalic

central gray, and certain thalamic nuclei. Kataoka et al.
(1979) also found moderate levels of INT in certain brainstem
and spinal cord areas, such as locus coeruleus and the anterior
and posterior columns of the spinal cord, respectively. Recently INT has been measured by RIA in human brain regions and
high levels were found in the SN, nucleus accumbens, hypothalamus, caudate, amygdala, and certain cortical regions (Manberg,
et al. in press).

Collectively, these studies provide substantial evidence for
similar distribution of NT in the CNS of several mammalian species. Also, the findings of heterogeneous NT distribution in
brain, NT synaptosomal localization, and NT receptor binding in
brain are all highly suggestive of a likely neurotransmitter
role for NT in different neural pathways within the mammalian
CNS. More conclusive evidence for localization of the peptide
in CNS neurons and delineation of specific NT-containing neural
pathways has been obtained by means of immunohistochemical
methods.

Uhl et al. (1977b, 1979) and Uhl and Snyder (1979), using immunocytochemical procedures, observed INT throughout many regions of the rat brain and spinal cord. The INT was mainly
observed in nerve fibers and terminals, but INT neuron cell
bodies were observed in scattered locations from lower brainstem to telencephalic levels. Some of the more salient findings emerging from these studies are summarized in Table 2-4.
Note that the over-all pattern of localization is highly consistent with the RIA data described previously.

The distribution of NT neuronal cell bodies, fibers, and terminals revealed by immunohistochemistry appears to correspond
well to components of previously described neuronal systems,
and efforts are underway to delineate the trajectories of NT-
containing neural pathways and their relationships to other
known neural systems. Thus, for example, Uhl and Snyder (1979)
have delineated a NT-containing neuronal pathway originating
from cell bodies in the AC nucleus, with fibers projecting via
the stria terminalis to terminate (in part) in the BNST. This
amygdalofugal NT neural pathway appears to correspond well to
the trajectory of a neural pathway, demonstrated by classic
silver staining techniques (DeOlmos, 1972), to course from
cells in the amygdala via the stria terminalis to terminate in
the BNST, as well as preoptic and hypothalamic areas.

In addition to delineating the amygdalofugal NT pathway, Uhl
et al. (1977b, 1979) have noted interesting relationships between NT-containing neural elements and components of certain
other known neurotransmitter systems. Uhl et al. (1979) point
out that striking similarities exist between the distributions
of enkephalin and NT neural elements. Both, for example, are
found in the substantia gelatinosa of the trigeminal nuclear

complex, the nucleus of the solitary tract, the parabrachial region, the periaqueductal gray, and the ventral tegmentum. Also, the amygdalofugal NT pathway appears to be strikingly similar to an enkephalin-containing pathway with cell bodies in the AC nucleus and axons projecting via the stria terminalis to terminate in the BNST (Uhl et al., 1978). Certain dissimilarities between enkephalin and NT localization, too, were pointed out by Uhl et al. (1979), who further noted some points of convergence between the localization of NT and SP.

Parallels between the distribution of NT and certain monoamines were also described by Uhl et al. (1979). The locus coeruleus, well known as a site of densely clustered perikarya of NE-containing neurons (Dahlstrom and Fuxe, 1965; Pickel et al., 1975; Swanson, 1976), also contains INT cell bodies; this suggests that either: (1) a mixture of NT and NE neurons exists within the locus coeruleus; or (2) neurons within that area may contain both NE and NT, in view of the very high density of NE neurons known to be present in the locus coeruleus. Similarly the VTA of Tsai, an area containing NT cells, is well known as the location of DA cells (group A10 of Dahlstrom and Fuxe, 1965); the periaqueductal gray not only contains NT cells but also scattered DA neurons (Swanson and Hartman, 1975); and the midbrain dorsal raphe nucleus contains both a dense cluster of 5-HT cells (Dahlstrom and Fuxe, 1965) and numerous NT neuron cell bodies. In these latter instances, the possibility exists that NT and DA or 5-HT occur in the same neurons or that interesting physiological interactions take place between proximally located NT and monoamine neurons.

Not only do parallels exist between the locations of NT neuron cell bodies and monamine cells, but also between uptake sites for steroid hormones, such as estrogens and androgens, and NT neural elements. Catecholamine neurons in several locations in rat brain have been demonstrated by a combined histochemical fluorescence and autoradiography technique, to concentrate estradiol or dihydrotestosterone in their nuclei (Grant and Stumpf, 1975; Heritage et al, 1977, 1980). Both estrogen- and androgen-concentrating NE neurons have been localized by these investigators in the region of the nucleus of the solitary tract, the nucleus commissuralis, locus coeruleus, and the parabrachial regions--all sites where INT cells have been histochemically localized as previously described. The possibility therefore exists that NT-containing neurons having cell bodies in those regions may also be estrogen or androgen target neurons, whether the NT cells are identical to the NE neurons found there or are only proximally located with the NE cells. Similarly, estradiol and dihydrotestosterone concentrating neurons have been localized (Stumpf and Sar, 1975; Sar and Stumpf, 1973) in the amygdala, among fibers of the stria terminalis,

Table 2-4. Summary of Neurotensin-Containing Elements in Rat Brain and Spinal Cord Localized by Immunocytochemistry

	Immunofluorescent NT cell bodies	Immunofluorescent NT fibers/terminals
Spinal Cord (Mid-cervical)		
Laminae I and II	--	High density
Rest of dorsal horn	--	Scattered
Ventral horn	--	Scattered
Fasciculus propius region	--	Sparse
Medulla Oblongata		
Substantia gelatinosa V	High density	High density
Nucleus commissuralis	--	High density
Nucleus of the solitary tract	Moderate density	High density
Dorsal cochlear nucleus	Occasional	Sparse
Lateral aspect nucleus spinal tract V	--	Moderate density
Pons		
Floor of fourth ventricle	--	Moderate density
Region of spinal tract V	--	Moderate density
Parabrachial nuclei	Moderate density	Moderate density
Locus ceruleus	Moderate density	Moderate density
Mesencephalon		
Dorsal raphe nucleus	Moderate density	Moderate density
Periaqueductal gray	Moderate density	Moderate density
Ventral tegmental area-Tsai	Moderate density	Moderate density
Diencephalon		
Thalamus		
Periventricular thalami nucleus	--	Moderate density
Medial thalami nucleus	--	Moderate density
Nucleus reunions	--	Moderate density
Rhomboid nuclei	--	Moderate density
Anteroventral nucleus	--	Moderate density

Table 2-4. Continued

	Immunofluorescent NT cell bodies	Immunofluorescent NT fibers/terminals
Epithalamus		
Lateral habenula	--	Moderate density
Medial habenula	--	Low density
Subthalamic area		
Zona incerta (dorsal to internal capsule)	--	Moderate density
Zona incerta (other areas)	--	Sparse
Hypothalamus		
Mamillary bodies	--	Low density
Lateral hypothalamus	--	Moderate density
Dorsomedial nucleus	--	Moderate density
Periventricular area	--	Low density
Median eminence	--	High density
Ventral anterior hypothalamus	--	High density
Preoptic areas	--	Moderate density
Telencephalon		
Septal area		
Lateral and medial nuclei	--	Moderate density
Diagonal band of Broca	--	High density
Nucleus accumbens	--	High density
Stria terminalis	--	High density
Bed nucleus of the stria terminalis	Moderate density	High density
Amygdaloid complex		
Central nucleus	Moderate density	High density
Basolateral nucleus	--	Moderate density
Basal ganglia		
Caudate putamen	--	Sparse
Globus pallidus	--	Low density
Hippocampus		
Pyramidal cell layers	--	Very sparse
Cerebral cortex		
Cingulate cortex	--	Low density
Other areas, surface layers	--	Sparse
Other areas, deepest layers	--	Low density

and in the BNST in a manner suggesting some striking overlaps with components of the newly delineated amygdalofugal NT system already described. Additional points of convergence appear to exist between steroid hormone uptake sites and locations where INT cell bodies, fibers, or terminals are located. Locations of INT neural elements listed in Table 2-4 can be compared with estrogen and androgen target sites described by Stumpf and Sar (1975) and Sar and Stumpf (1973).

Detailed review of the rapidly expanding literature on the non-neuronal localization of NT in gut of mammalian species is beyond the scope of the present chapter.

SOMATOSTATIN

Somatostatin, or somatotropin release inhibiting hormone (SRIF), is a tetradecapeptide (Table 2-1) identified as being the hypothalamic factor earlier hypothesized to be present in hypophyseal portal blood and to exert inhibitory effects on the release of growth hormone (somatotropin) from the anterior pituitary. The peptide was first isolated, chemically characterized, and demonstrated to exert somatotropin release inhibitory actions by Brazeau et al. (1973). This opened way for development of RIA (Arimura et al, 1975b) and immunocytochemical techniques for its localization.

Based on such studies, it now appears that the substance has endocrine significance beyond its effects on growth hormone release, having been reported to affect thyroid hormone secretion induced by exogenous TSH in man (Ahren et al, 1978). In addition, the peptide is found outside hypothalamic-hypophyseal tissue, having been found in both extrahypothalamic CNS regions and PNS components and, also, in non-neuronal tissue, such as the GI tract (Table 2-2). Its widespread distribution in the mammalian CNS and PNS suggests a general neurotransmitter role for SRIF, and its significance in the control of neurobehavioral functions in mammals is discussed elsewhere in the present volume. Studies of the biochemical structure of SRIF in relation to its phylogenetic and anatomical distribution in vertebrate species (Lauber et al., 1979; Zyznar et al., 1979; King and Miller, 1979), furthermore, indicate that the peptide is a phylogenetically old molecule and has been little changed structurally during the evolution of vertebrates. Also, quite interestingly, one common feature of synthesis of SRIF in various vertebrate species may be its synthesis as a metabolic product of a larger precursor molecule or prohormone (>10,000 daltons) likely synthesized via a ribosomal biosynthetic mechanism (Spiess and Vale, 1978).

In the mammalian CNS, Brownstein et al. (1975) reported the distribution of SRIF in discrete hypothalamic nuclei and extra-hypothalamic regions of rat brain, as determined by microdis-section and sensitive RIA techniques. As expected, the highest concentrations of SRIF were found in the hypothalamus; substan-tial amounts, however, were also detected in several other brain areas, including the septum and preoptic area, midbrain, brainstem, thalamus, and cortex. Within the hypothalamus, SRIF was most highly concentrated in the ME, at about seven times the concentration found in the arcuate nucleus. The latter nucleus had a markedly higher concentration (45 ng/mg protein) than any of the other hypothalamic nuclei, all of which contain significant amounts of the peptide. The next highest concen-trations were observed, in decreasing order, in the periven-tricular nucleus, the ventromedial nucleus, and the medial pre-optic nucleus (each in the 10-25 ng/mg protein range). As for extrahypothalamic brain tissue, notable levels of SRIF were observed in decreasing order, in specific areas, such as olfac-tory tubercle, amygdala, central gray, mamillary body, septum, habenula, hippocampus, dorsal raphe, IPN, and SN.

Results confirmatory of SRIF localization in the hypothalamus by RIA methods have been obtained by Wakabayashi et al. (1976) for rat brain, by Aoki et al. (1980) for dog brain, and Eckernas et al. (1978) for human brain. Deafferentation stud-ies further suggest that significant amounts of SRIF in the me-diobasal hypothalamic-median eminence region derive from axons originating from nerve cell bodies located outside that region. This is based in part on a 75 percent decrease in SRIF in the mediobasal hypothalamus of rat brain following complete knife-cut isolation of this area (Brownstein et al., 1977a); essen-tially no effect on extrahypothalamic SRIF activity was seen with such deafferentation. Interestingly, partial (frontal) deafferentation of the mediobasal hypothalamus caudal to the optic chiasm resulted in a SRIF fall (-84 percent) comparable to that seen with complete deafferentation, suggesting innerva-tion of the mediobasal hypothalamus-median eminence region by caudally projecting axons from anterior diencephalic areas. Consistent with this possibility are findings reported by Nemeroff et al. (1978), who found no decrease in hypothalamic SRIF levels following neonatal MSG treatment in the rat, re-sulting in almost complete destruction of arcuate neurons, but sparing of fibers of passage and axonal terminals in that area. Results confirming SRIF in extrahypothalamic areas include: those of Epelbaum et al. (1979), providing detailed RIA mapping of SRIH levels in amygdaloid nucei of the rat brain; and find-ings by Eckernas et al. (1978) of relatively high SRIF concen-trations in the neostriatum and the medial part of the amygda-loid complex in human brain.

That brain SRIF localized in the regions previously described by RIA methods is probably of neuronal origin is substantiated by subcellular localization studies demonstrating synaptosomal concentration of SRIF in tissue extracts from rat brain hypothalamus, striatum, cortex, and thalamus (Berelowitz et al., 1978a). Also, immunocytochemical studies of the neuronal localization of SRIF provide additional evidence for its presence as a neurosecretory or neurotransmitter substance in axonal fibers and terminals. Ultrastructural studies by Pelletier et al. (1974), for example, demonstrated the presence of immunoreactive SRIF in association with secretory granules, of 90 to 100 nm diameter, in many nerve endings located in the external layer of the rat ME proximal to primary plexus capillaries of the hypophyseal portal system.

In addition to the ultrastructural study, numerous immunocytochemical studies at the light microscrope level have contributed to delineation of the neuroanatomical distribution of SRIF fibers, terminals, and nerve cell bodies. Hökfelt et al. (1974a), for example, demonstrated, immunoreactive SRIF small dot- and fiber-like structures in the external layer of the guinea pig ME, consistent with the findings of Pelletier et al. (1974) for the rat. Hökfelt et al. (1974a) also reported SRIF nerve fibers and terminals in the ventromedial hypothalamic nucleus, the periventricular hypothalamus nucleus, and basal hypothalamic region; as well as certain amygdaloid nuclei. No immunoreactive structures were seen in the parietal cortex or hippocampus, areas identified by RIA as containing significant amounts of SRIF; nor were any SRIF-positive cell bodies visualized in the Hökfelt et al. (1974a) immunofluorescence study.

Nerve cell bodies were observed in a later study by Hökfelt et al. (1975a), specifically within the anterior portion of the rat brain hypothalamus, at the level of the suprachiasmatic nucleus and extending dorsally along the entire wall of the third ventricle but not in the paraventricular nucleus. Hökfelt et al. (1975a) also observed strongly fluorescent fibers in the internal and external layers of the ME, in the pituitary stalk and in the ventromedial, arcuate, ventral premamillary, periventricular, and suprachiasmatic nuclei of the hypothalamus, as well as in the posterior pituitary. This pattern of results suggested possible innervation of the ME, the above hypothalamic nuclei, and, possibly, the posterior pituitary by SRIF neurons with cell bodies in the anterior periventricular nucleus. Extrahypothalamic SRIF-positive fibers were also seen to abut on blood vessels in two circumventricular organs, the OVLT and subfornical organ, analogous to the pattern described for LH-RH and suggesting the possible existence of analogous extrahypothalamic SRIF systems not necessarily derived from anterior periventricular nucleus neurons. Lastly, observations by

Hökfelt et al. (1975b) of SRIF nerve fibers in the walls of the large and small intestines, SRIF-positive neuronal cell bodies in certain spinal ganglia, and SRIF-positive fibers and terminals in the substantia gelatinosa of the dorsal horn of the spinal cord suggested the likely existence of a SRIF-containing primary sensory system, analogous to SP containing afferent pathways described previously.

Many of these observations have been confirmed and extended in further studies of the rat and other mammalian species. Setalo et al. (1975) reported SRIF ME innervation patterns similar to those reported by Hökfelt et al. (1975a) and noted likely correspondence of the SRIF fibers to components of the classically defined tuberoinfundibular neural system. Also, Elde et al. (1976) reported that bilateral lesions of the periventricular nucleus resulted in markedly decreased SRIF in the ME, but no concomitant effect on LH-RH in the ventromedial or arcuate nuclei, suggesting likely independence of SRIF and LH-RH neural systems in the basal hypothalamus. Parsons et al. (1976) further delineated the distribution of SRIF systems in the rat and not only found cell bodies of SRIF neurons in the periventricular nucleus, but also a wider distribution of such perikarya in rostral hypothalamus and more caudally placed SRIF perikarya in the arcuate nucleus. The rostrally located neurons occurred in the hypothalamic suprachiasmatic and periventricular nuclei, as well as the preoptic suprachiasmatic nucleus. SRIF-positive fibers were seen in each of those nuclei and the anterior hypothalamic nucleus, the ventromedial and ventral premamillary nuclei, the ME, and the OVLT.

A subsequent immunofluorescence and autoradiography study by Alonso et al. (1978) suggested: (1) a close relationship between DA fibers and axons along the external layers of the ME; (2) the probable origin within the periventricular nucleus of some SRIF axons in the external zone of the ME; (3) exclusive innervation of the internal ME by SRIF axons from perikarya in the supraoptic nucleus; and (4) likely innervation of the external ME by SRIF axons originating from cell bodies in the medial paraventricular nucleus. In relation to the first observation Ajika et al. (1980) have recently reported evidence demonstrating distinct, although proximal, DA and SRIF cells and fibers in the hypothalamus and ME.

Dierickx and Vandesande (1979) have confirmed the presence of SRIF neurons in the rat hypothalamic arcuate nucleus and, by means of single and double immunocytochemical studies, showed that the rat supraoptic and paraventricular hypothalamic nuclei, and their accessory nuclei, do not contain magnocellular SRIF neurons. However, the parvocellular portion of the paraventricular nucleus was found to contain some SRIF-positive cells, forming a well-circumscribed periventricular cell group;

and SRIF neurons in the suprachiasmatic nucleus were found to be separate from vasopressin neurons. Numerous SRIF neurons were also observed at a distance from the third ventricle in scattered locations in the lateral hypothalamus, apparently along neurosecretory fibers that course from the paraventricular nucleus to the supraoptico-hypophyseal tract.

In another study, Krause (1979) demonstrated separation of LH-RH and SRIF neuronal elements in the OVLT, with a denser, more centrally placed complex of SRIF terminals existing in contrast to a looser, more peripheral concentration of LH-RH fibers. These findings and other evidence appear to establish the existence of SRIF-positive neuronal systems distinct from nearby dopaminergic, vasopressin, and LH-RH systems originating in anterior or basal hypothalamic regions. Lastly, studies (Shapiro et al., 1979a,b) have been reported that demonstrate the presence of SRIF in rat retina as determined by RIA, suggesting the presence of the peptide in that extension of the CNS.

Patterns of results analogous to those observed for the rat brain have been demonstrated in the fox (Bugnon et al., 1977b), the dog (Hoffman and Hayes, 1979), and human fetus (Bugnon et al., 1978). In the fox, results suggest the existence of two SRIF systems: hypothalamo-infundibular (neurophysin-negative) and hypothalamo-neurohypophyseal (neurophysin-positive). Also, Hoffman and Hayes (1979) found that SRIF fibers from dog periventricular nucleus SRIF cells project to the ME, the third ventricle, the pars nervosa of the pituitary, the OVLT, the MPOA and the BNST. SRIF cells in tuberal areas surrounding the ventromedial nucleus were found to project to the ventromedial nucleus; and SRIF cells in the dog arcuate nucleus terminated within that nucleus as well as in the ME. As noted by Hoffman and Hayes (1979), these findings suggest that SRIF may exert hormonal effects via secretion into the vasculature or CSF and via neurotransmitter or neuromodulatory effects via contacts with other neurons.

In the human fetus, Bugnon et al. (1978) observed SRIF-positive magnocellular perikarya in the paraventricular hypothalamic and supraoptic nuclei and SRIF-positive parvocellular neurons scattered in the mediobasal hypothalamus. Comparison of immunocytochemical stainings for SRIF and neurophysin on adjacent sections confirmed that some magnocellular perikarya contain both SRIF and neurophysin. Fibers derived from such neurons appear to project to the periphery of the neural lobe of the pituitary. In contrast, neurophysin negative SRIF fibers appear to course through the hypothalamus in scattered fashion, concentrate together as they enter the infundibulum, form dense bundles in the external fibrillar zone of the ME, and terminate on capillary loops of the hypophyseal portal system. Immunoreactive SRIF fibers also appeared to form extrahypophyseal pathways, a number ending in dilated fiber terminations in contact with cell bodies of neurons in the hypothalamic ventromedial nucleus and others appearing to reach posterior epithalamic regions.

In addition to these SRIF neural systems, localization of SRIF-positive neural systems in the spinal cord of guinea pig, mouse, rat, and tree shrew (*Tupaia belangeri*) was reported by Forssman (1978), using immunoenzyme techniques. Not only did Forssmann confirm earlier reports (Hökfelt et al., 1975d) of SRIF-positive nerve endings originating from primary afferent neurons and terminating in substantia gelatinosa of the spinal cord, he also demonstrated SRIF terminals around the central canal and extending into the zona intermedia, including the nucleus intermediomedialis and the columna lateralis (laminae VI and VII). Additional SRIF axons were seen in other regions of the spinal cord, for example, commissura grisea anterior and posterior. Results obtained for the tree shrew are described in more detail by Burnweit and Forssman (1979).

As for other PNS SRIF neurons, Hökfelt et al. (1975d, 1977a) demonstrated the presence of SRIF-like immunoreactivity in some sympathetic noradrenergic neurons. In the guinea pig, SRIF was observed in most of the principal ganglion cells of the coeliac-superior mesenteric ganglion, but relatively few cells of the superior cervical ganglion. The presence of SRIF and NE in the same neuron provides yet another example of the dual presence of a neuropeptide and monoamine in the same neurons as in the case of NT and certain monoamines noted earlier. Costa et al. (1977) also demonstrated SRIF in nerve cell bodies and axons of the myenteric and submucosal plexi in guinea pig gut. Extrinsic denervation resulted in decreased RIA SRIF concentrations in external muscle of the intestine but not in mucosa-submucosa containing the myenteric plexus. These and the findings of Hökfelt et al. (1977a) provide evidence that SRIF-containing neurons are interneurons within the enteric nervous system.

Of interest besides CNS and PNS localization of SRIF is delineation of its distribution in non-neural tissue. SRIF has, for example, been demonstrated by immunocytochemistry methods in certain cells (D cells) located in the periphery of pancreatic islets, in a small number of thyroid gland cells in a parafollicular position, and in endocrinelike cells of the stomach and intestine of the rat (Luft et al., 1974; Arimura et al., 1975b; Hökfelt et al., 1975a,d; and Parsons et al., 1976). SRIF localization has been reported in similar sites in: the guinea pig (Luft et al., 1974); the rabbit, cat, and pig (Alumets et al., 1977); the dog (Rufener et al., 1975a,b; Conlon et al., 1978; Kusumoto et al., 1979); the tree shrew

(Helmstaedter et al., 1977a,c); and the human (Dubois et al., 1975; Erlandsen et al., 1976; Dubois et al., 1976; Chayvialle et al., 1978). Also of interest are recent findings by Lundberg et al. (1979) of enkephalin- and SRIF-like immunoreactivity in some human adrenal medullary gland cells and pheochromocytomas, analogous to the presence of these substances in catecholamine-containing cells of the rat adrenal (Schultzberg et al., 1978). This represents yet another example of the synthesis of peptides and monoamine substances in the same cells, highlighting possible analogies between the neurochemical characteristics and the functional operation of certain types of gland cells and those of certain CNS and PNS neurons.

DISCUSSION

The neuroanatomical distributions of the peptides previously reviewed appear to provide morphological bases underlying diverse types of physiological roles played by those substances, as discussed in detail in other chapters of this book. In view of those more detailed discussions elsewhere only certain salient features of peptide localization bear comment here as contributing to unifying concepts presently emerging in regard to neuropeptide substances and their functional significance.

First, the isolation of different peptide substances from hypothalamic-median eminence tissue of many mammalian species, their chemical characterization, and their identification as having hypophysiotropic releasing factor properties confirms traditional, long-held hypotheses in the field of endocrinology regarding the existence of such releasing factors. The trajectories of certain peptide-containing neural systems, originating from neuronal perikarya located in the hypothalamus and projecting to the ME where they terminate in close proximity to hypophyseal portal vessels or extend into the infundibular stalk or pars nervosa of the pituitary, also comport well with classical views concerning the morphological distribution of such neuroendocrine pathways. However, the distribution of peptide-containing nerve cell bodies contributing to such hypophysiotropic neurosecretory pathways appears to be somewhat more widespread and complex than the simple systems earlier envisaged to arise from relatively distinctly segregated small groups of hypophysiotropic neurons mainly clustered around the third ventricle in mediobasal hypothalamic tissue.

In addition, the means by which the peptide releasing hormones are transported to the ME or infundibulum appear to be somewhat more complex than earlier conceptualized in terms of simple axonal transport from nerve cell bodies of origin to nerve terminals for storage and release in the ME. That is, it now appears that transport of at least some peptide substances involves: (1) their release into CSF from nerve endings terminating in ventricular walls; (2) their movement via the CSF; (3) their uptake by specialized ependymal cells along the base of the third ventricle; and (4) their subsequent movement via such cells or diffusion to hypophysial portal vessels in the ME.

Such ventricular transport of the physiologically active neuropeptides, possibly for purposes other than neuroendocrine regulatory functions, may also provide a route for circulation or uptake of neuropeptide stores observed in circumventricular organs sharing characteristics of the ME region in being highly vascularized and, as such, may provide a route of entry into the brain CSF for peripherally circulating peptide substances. This may include neuroactive peptides released from non-neural tissue of the GI tract or other organs and found circulating in the plasma of numerous mammalian species. Thus, the circumventricular organs may comprise an important link in the integration of both hormonal and neural activities involved in the functioning of a diffuse endocrine system utilizing selected peptides or monamines as chemical messengers, as hypothesized and discussed by numerous investigators during the past decade. Evidence indicative of nuclear concentration of steroid hormones, for example, estrogens and androgens, by cells in the vicinity of the circumventricular organs and likely by peptidergic neurons in certain brain regions, such as the hypothalamic arcuate nucleus, also adds support for peptide neuron involvement in complex endocrine feedback loops.

The demonstration of neuropeptides in circumventricular organs outside of the hypothalamus also serves to point up yet another rapidly emerging common feature in regard to the neuroanatomical localization of many peptides, that is, their widespread extrahypothalamic distribution in CNS and PNS tissue. This includes the presence of diverse neural pathways originating from cell groups widely dispersed in spinal, brainstem, and diencephalic regions of the CNS, which differentially innervate various brain and spinal cord areas. These extrahypothalamic neural pathways likely provide important morphological bases for the peptides to serve, in general, as neurotransmitters or neuromodulators in ways analogous to more traditionally accepted neurotransmitters. The functional roles played by extrahypothalamic CNS peptidergic systems appear to be highly varied and to extend beyond control of neuroendocrine events to include mediation of neurobehavioral effects and other diverse physiological functions. In the PNS, it is especially interesting to note that this includes identification of certain peptides as likely neurotransmitters in primary sensory

afferent neurons of the somatosensory system and as interneurons within certain autonomic nervous system ganglia, the neurotransmitters for which were previously entirely unknown in
the case of the former or thought to be well established as
containing only certain other well-known transmitter agents in
the case of the latter.

Also highly intriguing are findings demonstrating the presence of peptide substances within identical CNS or PNS neurons
that synthesis, store, and release monoamines, such as DA, NE,
or 5-HT, as neurotransmitters. The full functional significance of such violations of the once widely accepted, neuropharmacological principle of "one-neuron, one-transmitter"
remains to be elucidated, thus adding yet another fascinating
frontier to be addressed in the coming years as further advances in peptide research are accomplished beyond those elaborated on in the rest of the present volume.

ACKNOWLEDGEMENTS

The authors are supported by NIMH MH-32316, MH-33127, and
NICHHD HD-03110. We are grateful to Stephen Terry for preparation of this manuscript.

REFERENCES

Ahren, B., Ericksson, M., Hedner, P., Ingemansson, S., and
    Westgren, U.: Somatostatin inhibits thyroid hormone secretion
    induced by exogenous TSH in man. *J Clin Endocrinol Metab*
    47:1156-1159, 1978.
Ajika, K., Ishikawa, M., Arai, K., Okinaga, S., and
    Wakabayashi, I.: Simultaneous localization of somatostatin
    and catecholamine neuronal system in rat hypothalamus:
    Immunohistochemical identification by light and electron
    microscope. VI International Congress on Endocrinology,
    Melbourne, 1980, p. 335.
Aksel, S., and Tyrey, L.: Luteinizing hormone-releasing hormone
    in the human fetal brain. *Fertil Steril 28*:1067-1071, 1977.
Alonso, G., Balmefrezol, M., and Assenbacher, I.: Etude de
    l'innevation monoaminergic et peptidergique de l'eminence
    mediane du rat par une combinacion ser le meme hypothalamus,
    des techniques d'histofluroesesence, d'immunocyto chimile et
    de radioautographie. *C R Soc Biol (Paris) 172*:138-144, 1978.
Alumets, J., Sundler, F., and Hakanson, R.: Distribution,
    ontogeny and ultrastructure of somatostatin immunoreactive
    cells in the pancreas and gut. *Cell Tissue Res 185*:465-479,
    1977.

Alumets, J., Ekelund, M., El Munshid, H.A., Hakanson, R.,
Loren, L., and Sundler, F.: Topography of somatostatin cells
in the stomach of the rat: Possible functional significance.
Cell Tissue Rec 202:177-188, 1979.

Aoki, Y., Yamauchi, J., Takahara, J., Yunoki, S., Uneki, T.,
Hashimoto, K., Ohno, N., Ofuji, T., and Otsuka, N.: Distribution of thyrotropin releasing hormone, luteinizing hormone
releasing hormone, somatostatin and arginine vasopressin in
dog hypothalamus by radioimmunoassay. VI International Congress Endocrinology, 1980.

Araki, S., Toran-Allerand, C.D., Ferin, M., and Vande-Wiele,
R.L.: Immunoreactive gonadotropin-releasing hormone (Gn-RH)
during maturation in the rat: Ontogeny of regional hypothalamic differences. *Endocrinology 97*:693-697, 1975.

Arimura, A., Sato, H., Coy, D.H., and Schally, A.V.: Radioimmunoassay for GH release inhibiting hormone. *Proc Soc Exp
Biol Med 148*:784, 1975a.

Arimura, A., Sato, H., Dupont, A., Nishi, N., and Schally,
A.V.: Somatostatin: Abundance of immunoreactive hormone in
rat stomach and pancreas. *Science 189*:1007-1009, 1975b.

Aubert, M.L., Grumbach, M.M., and Kaplan, S.L.: The ontogenesis
of human fetal hormones. IV. Somatostatin luteinizing hormone
releasing factor, and thyrotropin releasing factor in hypothalamus and cerebral cortex of human fetuses 10-22 weeks of
age. *J Clin Endocrinol Metab 44*:1130-1141, 1977.

Baker, B.L., and Dermody, W.C.: Effect of hypophysectomy on
immunocytochemically demonstrated gonadotropin-releasing
hormone in the rat brain. *Endocrinology 98*:1116-1122, 1976.

Baker, B.L., and Yen, Y.Y.: The influence of hypophysectomy on
the store of somatostatin in the hypothalamus and pituitary
stem. *Proc Soc Exp Biol Med 151*:599-602, 1976.

Baker, B.L., Dermody, W.C., and Reel, J.R.: Distribution of
gonadotropin-releasing hormone in the rat brain as observed
with immunocytochemistry. *Endocrinology 97*:125-135, 1975.

Barnea, A., Ben-Jonathan, N., Colston, C., Johnston, J.M., and
Porter, J.C.: Differential sub-cellular compartmentalization
of thyrotropin releasing hormone (TRH) and gonadotropin releasing hormone (LRH) in hypothalamic tissue. *Proc Natl Acad
Sci USA 72*:3153-3157, 1975.

Barnea, A., Ben-Jonathan, N., and Porter, J.C.: Characterization of hypothalamic subcellular particles containing luteinizing hormone releasing hormone and thyrotropin releasing
hormone. *J Neurochem 27*:477-484, 1976.

Barnea, A., Neaves, W.B., Cho, G., and Porter, J.C.: Demonstration of a temperature-dependent association of thyrotropin
releasing hormone, α-melanocyte stimulating hormone, and luteinizing hormone releasing hormone with subneuronal particles in hypothalamic synaptosomes. *J Neurochem 31*:1125-1134,

1978a.

Barnea, A., Neaves, W.B., Cho, G., and Porter, J.C.: A subcellular pool of hypoosmotically resistant particles containing thyrotropin releasing hormone, α-melanocyte stimulating hormone, and luteinizing hormone releasing hormone in the rat hypothalamus. *J Neurochem 30*:937-948, 1978b.

Barry, J.: Characterization and topography of LH-RH neurons in the rabbit. *Neurosci Lett 2*:201-205, 1976a.

Barry, J.: Characterization of topography of LH-RH neurons in the human brain. *Neurosci Lett 3*:287-291, 1976b.

Barry, J.: Immunofluorescence study of LRF neurons in man. *Cell Tissue Res 181*:1-13, 1977.

Barry, J.: Septo-epithalamo-habenular LRF-reactive neurons in monkeys. *Brain Res 151*:183-187, 1978.

Barry, J.: Immunofluorescence study of the preoptico-terminal LRH tract in the female squirrel monkey during the estrous cycle. *Cell Tissue Res 198*:1-13, 1979.

Barry J., and Dubois, M.P.: Etude en immunofluorescence des structures hypothalamiques a competence gonadotrope. *Ann Endocrinol (Paris) 34*:735-742, 1973.

Barry, J., and Dubois, M.P.: Study of preoptico-infundibular LRF neurosecretory pathway in female guinea pig during the estrous cycle. *Neuroendocrinology 15*:200-208, 1974.

Barry, J., and Dubois, M.P.: Immunofluorescence study of LRF-producing neurons in the cat and the dog. *Neuroendocrinology 18*:290-298, 1975.

Barry, J., and Dubois, M.P.: Immunoreactive LRF neurosecretory pathways in mammals. *Acta Anat 94*:497-503, 1976.

Barry, J., Dubois, M.P., and Poulain, P.: LRF-producing cells of the mammalian hypothalamus. *Z Zellforsch 146*:351-366, 1973.

Barry, J., Poulain, P., and Carette, B.: Systematisation et efferences des neurones a LRH chez les Primates. *Ann Endocrinol (Paris) 37*:227-234, 1976.

Ben-Ari, U., Le Gal La Salle, G., and Kanazawa, I.: Regional distribution of substance P within the amygdaloid complex and bed nucleus of the stria terminalis. *Neurosci Lett 4*:299-302, 1977.

Ben-Ari, Y., Pradelles, P., Gros, C., and Dray, F.: Identification of authentic substance P in striatonigral and amygdaloid nuclei using combined high performance liquid chromatography and radioimmunoassay. *Brain Res 173*:360-363, 1979.

Ben-Jonathan, N., Mical, R.S., and Porter, J.C.: Transport of LRF from CSF to hypophysial portal and systemic blood and the release of LH. *Endocrinology 95*:18-25, 1974.

Bennett, G.W., Edwardson, J.A., Holland, D., Jeffcoate, S.L., and White, N.: Release of immunoreactive luteinizing hormone-releasing hormone and thyrotropin-releasing hormone from hypothalamic synaptosomes. *Nature* 257:323-324, 1975.

Berelowitz, M., Hudson, A., Primstone, B., Kronheim, S., and Bennett, G.W.: Subcellular localization of growth hormone release inhibiting hormone in rat hypothalamus, cerebral cortex, striatum and thalamus. *J Neurochem* 31:751-753, 1978a.

Berelowitz, M., Kronheim, S., Pimstone, B., and Shapiro, B.: Somatostatin-like immunoreactivity in rat blood. *J Clin Invest* 61:1410-1414, 1978b.

Berelowitz, M., Kronheim, S., Pimstone, B., and Sheppard, M.: Potassium stimulated calcium dependent release of immunoreactive somatostatin from incubated rat hypothalamus. *J Neurochem* 31:1537-1539, 1978c.

Blahser, S., Fellmann, D., and Bugnon, C.: Immunocytochemical demonstration of somatostatin-containing neurons in the hypothalamus of the domestic mallard. *Cell Tissue Res* 195:183-187, 1978.

Bloom, S.R., and Polak, J.M.: Motilin and neurotensin. *Clin Endocrinol Metab* 8:401-411, 1979.

Boler, J., Enzmann, F., Folkers, K., Bowers, C.Y., and Schally, A.V.: The identity of chemical and hormonal properties of the thyrotropin releasing hormone and pyroglutamyl-histidyl-proline-amide. *Biochem Biophys Res Commun* 37:705-710, 1969.

Bordi, C., and Ravazzola, M.: Endocrine cells in the intestinal metaplasia of gastric mucosa. *Am J Pathol* 96:391-398, 1979.

Brazeau, P., Vale, W., Burgus, R., Ling, N., Butcher, M., Rivier, J., and Guillemin, R.: Hypothalamic polypeptide that inhibits the secretion of immunoreactive pituitary growth hormone. *Science* 179:77-79, 1973.

Brownstein, M.J., Palkovits, M., Saavedra, J.M., Bassiri, R.M., and Utiger, R.D.: Thyrotropin-releasing hormone in specific nuclei of rat brain. *Science* 185:267-269, 1974.

Brownstein, M.J. Arimura, A., Sato, H., Schally, A.V., and Kizer, J.S.: The regional distribution of somatostatin in the rat brain. *Endocrinology* 96:1456-1461, 1975a.

Brownstein, M.J., Palkovits, M., and Kizer, J.S.: On the origin of luteinizing hormone-releasing hormone (LH-RH) in the supraoptic crest. *Life Sci* 17:679-682, 1975b.

Brownstein, M.J., Utiger, R.D., Palkovits, M., and Kizer, J.S.: Effect of hypothalamic deafferentation on thyrotropin-releasing hormone levels in rat brain. *Proc Natl Acad Sci USA* 72:4177-4179, 1975c.

Brownstein, M.M., Arimura, A., Schally, A.V., Palkovits, M., and Kizer, J.S.: The effect of surgical isolation of the hypothalamus on its luteinizing hormone-releasing hormone content. *Endocrinology* 98:662-665, 1976a.

Brownstein, M.J., Mroz, E.A., Kizer, J.S., Palkovits, M., and Leeman, S.E.: Regional distribution of substance P in the brain of the rat. *Brain Res* 116:299-305, 1976b.

Brownstein, M.J., Arimura, A., Fernandez-Durango, R., Schally, A.V., Palkovits, M., and Kizer, J.S.: The effect of hypothalamic deafferentation on somatostatin-like activity in the rat brain. *Endocrinology 100*:246, 1977a.

Brownstein, M.J., Mroz, E.A., Tappaz, M.J., and Leeman, S.E.: On the origin of substance P and glutamic acid decarboxylase (GAD) in the substantia nigra. *Brain Res 135*:315-323, 1977b.

Buchan, A.M.J., Polak, J.M., Bloom, S.R., Hobbs, S., Sullivan, S.N., and Pearse, A.G.E.: Localization and distribution of neurotensin in human intestine. *J Endocrinol 77*:41P-42P, 1978.

Bugnon, C., Bloch, B., and Fellmann, D.: Neuroendocrinologie. *C R Acad Sci (Paris) 5*:282, 1976.

Bugnon, C., Bloch, B., Lenys, D., and Fellmann, D.: Ultrastructural study of the LH-RH containing neurons in the human fetus. *Brain Res 137*:175-180, 1977a.

Bugnon, C., Bloch, B., Fellmann, D.: Etude immunocytologique des neurones hypothalamiques a LH-RH chez le foetus humain. *Brain Res 128*:249-262, 1977c.

Bugnon, C., Lenys, D., Fellmann, D., and Bloch, B.: Etude des fibres á somatostatine et des fibres á neurophysine dans l'éminence médiane du Renard, par application de la technique cytoimmunologique de double marquage. *C R Soc Biol (Paris) 171*:576-580, 1977b.

Bugnon, C., Fellmann, D., and Bloch, B.: Immunocytochemical study of the ontogenesis of the hypothalamic somatostatin-containing neurons in the human fetus. *Metabolism 27 (Suppl) 1*:1161-1165, 1978.

Burchanowski, B.J., Knigge, K.M., and Sternberger, L.A.: Rich ependymal investment of luliberin (LH-RH) fibers revealed immunocytochemically in an image like that from Golgi stain. *Proc Natl Acad Sci USA 76*:6671-6674, 1979.

Burgus, R., Dunn, T.F., Desiderio, D., and Guillemin, R.: Structure moleculaire du facteur hypothalamique hypophysiotrope TRF d'origine ovine: mise en envidence par spectrometrie de masse de la sequence PCA-His-Pro-NH$_2$. *C R Acad Sci (Paris) 269*:1870-1873, 1969.

Burnweit, C., and Forssmann, W.G.: Somatostatinergic nerves in the cervical spinal cord of the monkey. *Cell Tissue Res 200*:83-90, 1979.

Cannon, D., Skrabanek, P., Powell, D., and Harrington, M.G.: Immunological characterization of two substance P antisera with substance P fragments and analogues. *Biochem Soc Trans 5*:1736-1738, 1977.

Carraway, R., and Leeman, S.E.: The isolation of a new hypotensive peptide, neurotensin, from bovine hypothalamus. *J Biol Chem 248*:6854-6861, 1973.

Carraway, R., and Leeman, S.E.: The amino acid sequence of a hypothalamic peptide, neurotensin. *J Biol Chem 250:*1907-1911, 1975.

Carraway, R., and Leeman, S.E.: Characterization of radioimmunoassayable neurotensin in the rat. Its differential distribution in the central nervous system, small intestine and stomach. *J Biol Chem 251:*7045-7052, 1976.

Chan-Palay, V., and Palay, S.L.: Ultrastructural identification of substance P cells and their processes in rat sensory ganglia and their terminals in the spinal cord by immunocytochemistry. *Proc Natl Acad Sci USA 74:*4050-4054, 1977.

Chan-Palay, V., Jonsson, G., and Palay, S.L.: Serotonin and substance P coexist in neurons of the rat's central nervous system. *Proc Natl Acad Sci USA 75:*1582-1586, 1978.

Chang, M.M., and Leeman, S.E.: Isolation of a sialogogic peptide from bovine hypothalamie tissue and its characterization as subatance. P. *Nature J Biol Chem 245:*4784-4790, 1970.

Chang, M.M., Leeman, S.E., and Niall, H.D.: Amino-acid sequence of substance P. *New Biol 232:*86-87, 1971.

Chayvialle, J.A.P., Descos, F., Bernard, C., Martin, A., Barbe, C., and Partensky, C.: Somatostatin in mucosa of stomach and duodenum in gastroduodenal disease. *Gastroenterology 75:*13-19, 1978.

Chepurnov, S.A., Clusha, V.E., Cherpurnova, N.E., Svirskis, S.V., Jerusalimskii, B.N., and Titova, A.V.: Studies on the action of thyroliberin (TRH), melanostatin (MIF), somatostatin (SRIF) and the gastrin fragment on the limbic structure of brain. VI International Congress Endocrinology, 1980.

Conlon, J.M., Zyznar, E., Vale, W., and Unger, R.H.: Multiple forms of somatostatin-like immunoreactivity in canine pancreas. *FEBS Lett 94:*327-330, 1978.

Costa, M., Patel, Y., Furness, J.B., and Arimura, A.: Evidence that some intrinsic neurons of the intestine contain somatostatin. *Neurosci Lett 6:*215-222, 1977.

Cuello, A.C., Fiacco, M.D., and Paxinos, G.: The central and peripheral ends of the Substance P-containing sensory neurones in the rat trigeminal system. *Brain Res 152:*499-509, 1978.

Cuello, A.C., Galfre, G., and Milstein, C.: Detection of substance P in the central nervous system by a monoclonal antibody. *Proc Natl Acad Sci USA 76:*3532-3536, 1979.

Cuello, A.C., Jessell, T.M., Kanazawa, I., and Iversen, L.L.: Substance P: Localization in synaptic vesicles in rat central nervous system. *J Neurochem 29:*747-751, 1977.

Cuello, A.C., and Kanazawa, I.: The distribution of substance P immunoreactive fibers in the rat central nervous system. *J Comp Neurol 178(1):*129-156, 1978.

Dahlstrom, A., and Fuxe, K.: Evidence for the existence of monoamine-containing neurons in the central nervous system.

I. Demonstration of monoamines in the cell bodies of brain-stem neurons. *Acta Physiol Scand 62 (Suppl) 232*:1-55, 1965.

Deery, D.J.: Determination by radioimmunoassay of the luteiniz-ing hormone-releasing hormone (LH-RH) content of the hypo-thalamus of the rat and some lower vertebrates. *Gen Comp Endocrinol 24*:280-285, 1974.

DeOlmos, J.: The amygdaloid projection field in the rat as studied with the cupric-silver method. In: *The Neurobiology of the Amygdala*, B Eleftheriou, ed, Plenum Press, New York, 1972, pp 145-204.

Dierickx, K., and Vandesande, F.: Immunocytochemical localiza-tion of somatostatin-containing neurons in the rat hypothal-amus. *Cell Tissue Res 201*:349-359, 1979.

Doerr-Schott, J., and Dubois, M.P.: Immunohistochemical demon-stration of an SRIF-like system in the brain of the reptile: Lacerta muralis Laur. *Experientia 33*:947-948, 1977.

Dubois, P.M., Paulin, C., Assan, R., and Dubois, M.P.: Evidence for immunoreactive somatostatin in the endocrine cells of hu-man foetal pancreas. *Nature 256*:731-732, 1975.

Dubois, P.M., Paulin, C., and Dubois, M.P.: Gastrointestinal somatostatin cells in the human fetus. *Cell Tissue Res 166*:179-184, 1976.

Dubois, M.P., Billard, R., Breton, B., and Peter, R.E.: Compar-ative distribution of somatostatin, LH-RH, neurophysin and α-endorphin in the rainbow trout: An immunocytological study. *Gen Comp Endocrinol 37*:220-232, 1979.

Duffy, M.J., Mulhall, D., and Powell, D.: Subcellular distri-bution of substance P in bovine hypothalamus and substantia nigra. *J Neurochem 25*:305-307, 1975.

Dupont, A., Langelier, P., Merand, Y., Cote, J., and Barden, N.: Radioimmunoassay studies of the regional distribution of neurotensin, substance P, somatostatin, thyrotropin, cortico-tropin, LYS-vasopressin, and opiate peptides in bovine brain. Proceedings of the Endocrinology Society, 61st Annual Meet-ing, p 127, 1979.

Eckernas, S.A., Aquilonius, S.M., Lundqvist, G., and Roos, B.E.: Regional distribution of cholinergic and monoaminergic markers, somatostatin and substance P in the human brain. *Acta Neurol Scand 57 (Suppl 67)*:225-226, 1978.

Elde, R., Hökfelt, T., Johansson, O., Efendic, S., and Luft, R.: Somatostatin containing pathways in the nervous system. *Soc Neurosci Abst 2*:759, 1976.

Elde, R., Hökfelt, T., Johansson, O., Schultzberg, M., Efendic, S., and Luft, R.: Cellular localization of somatostatin. *Metabolism 27 (Suppl 1)*:1151-1159, 1978.

Emson, P.C., Kanazawa, I., Cuello, A.C. and Jessell, T.M.: Sub-stance P pathways in rat brain. *Biochem Soc Trans 5*:187-189, 1976.

Ensinck, J.W., Laschansky, E.C., Kanter, R.A., Fujimoto, W.Y., Koerker, D.J., and Goodner, C.H.: Somatostatin biosynthesis and release in the hypothalamus and pancreas of the rat. *Metabolism 27 (Suppl 1)*:1207-1210, 1978.

Epelbaum, J., Brazeau, P., Tsang, D., Brawer, J., and Martin, J.B.: Subcellular distribution of radioimmunoassayable somatostatin in rat brain. *Brain Res 126*:309-323, 1977.

Epelbaum, J., Arancibia, L.T., Kordon, C., Ottersen, O.P. and Ben-Ari, Y.: Regional distribution of somatostatin within the amygdaloid complex of the rat brain. *Brain Res 174*:1-3, 1979.

Erlandsen, S.L., Hegre, O.D., Parsons, J.A., McEvoy, R.C., and Elde, R.P.: Pancreatic islet cell hormones: Distribution of cell types in the islet and evidence for the presence of somatostatin and gastrin within the D-cell. *J. Histochem Cytochem 24*:883-897, 1976.

von Euler, U.S., and Gaddum, J.H.: An unidentified depressor substance in certain tissue extracts. *J Physiol (Lond) 72*:74-87, 1931.

Falkmer, S., Elde, R.P., Hellerstrom, C., Petersson, B., Efendic, S., Fohlmer, J., and Siljevall, J.B.: Some phylogenetical aspects for the occurrence of somatostatin in the gastro-entero-pancreatic endocrine system. A histological and immunocytochemical study, combined with quantitative radioimmunological assays of tissue extracts. *Arch Inst Jpn Suppl 40*:99-117, 1977.

Fernandez-Durango, R., Arimura, A., Fishback, J., and Schally, A.V.: Hypothalamic somatostatin and LH-RH after hypophysectomy, in hyper- and hypothyroidism and during anesthesia in rats. *Proc Soc Exp Biol Med 157*:235-240, 1978.

Forssmann, W.G.: A new somatostatinergic system in the mammalian spinal cord. *Neurosci Lett 10*:293-297, 1978.

Frigerio, B., Ravazola, M., Ito, S., Buffa, R., Capella, C., Solcia, E., and Orci L.: Histochemical and ultrastructural identification of neurotensin cells in the dog ileum. *Histochemistry 54*:123-131, 1977.

Gale, K., Hong, J.S., and Guidotti, A.: Presence of substance P and GABA in separate striatonigral neurons. *Brain Res 136*:371-375, 1977.

Gamse, R., Mroz, E.A., Leeman, S.E., and Lembeck, F.: The intestine as source of immunoreactive substance P in plasma of the cat. *Naunyn-Schmiedebergs Arch Pharmacol 305*:17-21, 1978.

Gamse, R., Lembeck, F., and Cuello, A.C.: Substance P in the vagus nerve: Immunochemical and immunohistochemical evidence for axoplasmic transport. *Naunyn-Schmiedebergs Arch Pharmacol 306*:37-44, 1979.

Gauchy, C., Beaujouan, J.C., Besson, M.J., Kerdelhue, B., Glowinski, J., and Michelot, R.: Topographical distribution of substance P in the cat substantia nigra. *Neurosci Lett*

12:127-131, 1979.

Gibbons, J.M., Jr., Mitnick, M., and Chieffo, V.: *In vitro* bio-synthesis of TSH- and LH-releasing factors by the human placenta. *Am J Obstet Gynecol 121*:127-131, 1975.

Gillioz, P., Giraud, P., Conte-Devolx, B., Jaquet, P., Codaccinoni, L.L., and Oliver, C.: Concentration de somatostatine dans le sang porte hypophysaire du rat. *Ann Endocrinol (Paris) 40*:435-436, 1979.

Goldsmith, P.C., Rose, J.C., Arimura, A., and Ganong, W.F.: Ultrastructural localization of somatostatin in pancreatic islets of the rat. *Endocrinology 97*:1061-1064, 1975.

Grant, L.D., and Stumpf, W.E.: Hormone uptake sites in relation to CNS catecholamine systems. In *Anatomical Neuroendocrinology,* International Conference on the Neurobiology of CNS-Hormone Interactions, W.E. Stumpf and L.D. Grant, eds, S. Karger, Basel, 1975, pp 445-463.

Grimm, Y., and Reichlin, S.: Thyrotropin-releasing hormone (TRH): Neurotransmitter regulation of secretion by mouse hypothalamic tissue *in vitro*. *Endocrinology 93*:626-631, 1973.

Gross, D.S.: Distribution of gonadotropin-releasing hormone in the mouse brain as revealed by immunohistochemistry. *Endocrinology 98*:1408-1417, 1976.

Gross, D.S., and Longer, J.D.: Developmental correlation between hypothalamic somatostatin and hypophysial growth hormone. *Cell Tissue Res 202*:251-261, 1979.

Guansing, A.R., and Murk, L.M.: Distribution of thyrotropin-releasing hormone in human brain. *Horm Metab Res 8*:493-494, 1976.

Harris, V., Michael-Conlon, J., Srikart, C.B., McKorlele, K., Schusdziarra, V., Ipp, E., and Unger, R.H.: Measurements of somatostatin-like immunoreactivity in plasma. *Clin Chim Acta 87*:275-283, 1978.

Heber, D., Marshall, J.C., and O'Dell, W.D.: GnRH membrane binding: Identification, specificity and quantification in nonpituitary tissues. *Am J Physiol 235*:E227-E230, 1978.

Heindel, J.J., Williams, E., Robison, G.A., and Strada, S.J.: Inhibition of GH rat pituitary tumor cell adenylyl cyclase activity by somatostatin. *J Cyclic Nucleotide Res 4*:453-462, 1978.

Helmstaedter, V., Feurle, G.E., and Forssmann, W.G.: Relationship of glucagon-somatostatin and gastrin-somatostatin cells in the stomach of the monkey. *Cell Tissue Res 117*:29-46, 1977a.

Helmstaedter, V., Muhlmann, G., Feurle, G.E., and Forssmann, W.G.: Immunohistochemical identification of gastrointestinal neurotensin cells in human embryos. *Cell Tissue Res 184*:315-320, 1977b.

Helmstaedter, V., Taugner, C., Feurle, G.E., and Forssmann, W.G.: Localization of neurotensin-immunoreactive cells in the small intestine of man and various mammals. *Histochemistry* 53:35-41, 1977c.

Heritage, A.S., Grant, L.D., and Stumpf, W.E.: [3]H Estradiol in catecholamine neurons of rat brainstem: Combined localization by autoradiography and formaldehyde-induced fluorescence. *J. Comp Neurol* 176:607-630, 1977.

Heritage, A.S., Stumpf, W.E., Sar, M., and Grant, L.D.: Brainstem catecholamine neurons are target sites for sex steroid hormones. *Science* 207:1377-1379, 1980.

Hoffman, G.E., and Hayes, T.A.: Somatostatin neurons and their projections in dog diencephalon. *J Comp Neurol* 186:371-392, 1979.

Hökfelt, T., Efendic, S., Johansson, O., Luft, R., and Arimura, A.: Immunohistochemical localization of somatostatin (growth hormone release-inhibiting factor) in the guinea pig brain. *Brain Res* 80:165-169, 1974a.

Hökfelt, T., Fuxe, K., Goldstein, M., Johansson, O., Park, D., Fraser, H., and Jeffcoate, S.L.: Immunofluorescence mapping of central monoamine and releasing hormone (LRH) systems. In *Anatomical Neuroendocrinology*, International Conference of the Neurobiology of CNS-Hormone Interactions, W.E. Stumpf and L.D. Grant, eds S. Karger, Basel, 1974b, pp 381-392.

Hökfelt, T., Efendic, S., Hellerstrom, C., Johansson, O., Luft, R., and Arimura, A.: Cellular localization of somatostatin in endocrine-like cells and neurons of the rat with special references to the A_1-Cells of the pancreatic islets and to the hypothalamus. *Acta Endocrinol (Copenh) (Suppl 200)*:1-41, 1975a.

Hökfelt, T., Elde, R.P., Johansson, O., Luft, R., and Arimura, A.: Immunohistochemical evidence for the presence of somatostatin, a powerful inhibitory peptide, in some primary sensory neurons. *Neurosci Lett* 1:231-235, 1975b.

Hökfelt, T., Fuxe, K., Johansson, O., Jeffcoate, S., and White, N.: Distribution of thyrotropin-releasing hormone (TRH) in the central nervous system as revealed with immunohistochemistry. *Eur J Pharmacol* 34:389-392, 1975c.

Hökfelt, T., Johansson, O., Efendic, S., Luft, R., and Arimura, A.: Are there somatostatin-containing nerves in the rat gut? Immunohistochemical evidence for a new type of peripheral nerves. *Experientia* 31:852-854, 1975d.

Hökfelt, T., Kellerth, J.O., Nilsson, G., and Pernow, B.: Experimental immunohistochemical studies on the localization and distribution of substance P in cat primary sensory neurons. *Brain Res* 100:235-252, 1975e.

Hökfelt, T., Kellerth, J.O., Nilsson, G., and Pernow, B.: Substance P: Localization in the central nervous system and in some primary sensory neurons. *Science 190*:889-890, 1975f.

Hökfelt, T., Elde, R., Johansson, O., et al.: Immunohistochemical evidence for separate populations of somatostatin-containing and substance P-containing primary afferent neurons in the rat. *Neuroscience 1*:131-136, 1976a.

Hökfelt, T., Meyerson, B., Nilsson, G., Pernow, B., and Sachs, C.: Immunohistochemical evidence for substance P-containing nerve endings in the human cortex. *Brain Res 104*:181-186, 1976b.

Hökfelt, T., Elfvin, L.G., Elde, R., Schultzberg, M., Goldstein, M., and Luft, R.: Occurrence of somatostatin-like immunoreactivity in some peripheral sympathetic noradrenergic neurons. *Proc Natl Acad Sci USA 74*:3587-3591, 1977a.

Hökfelt, T., Elfvin, L.G., Schultzberg, M., Goldstein, M., and Nilsson, G.: On the occurrence of substance P-containing fibers in sympathetic ganglia: Immunohistochemical evidence. *Brain Res 132*:29-41, 1977b.

Hökfelt, T., Ljungdahl, A., Steinbush, H., Verhofstad, A., Nilsson, G., Brodin, E., Pernow, B., and Goldstein, M.: Immunohistochemical evidence of substance P-like immunoreactivity in some 5-hydroxytrptamine-containing neurons in the rat central nervous system. *Neuroscience 3*:517-538, 1978a.

Hökfelt, T., Pernow, B., Nilsson, G., Wetterberg, L., Goldstein, M., and Jeffcoate, L.: Dense plexus of substance P immunoreactive nerve terminals in eminentia medialis of the primate hypothalamus. *Proc Natl Acad Sci USA 75*:1013-1015, 1978b.

Hong, J.S., Costa, E., and Yang, H-YT.: Effects of habenula lesions on the substance P content of various brain regions. *Brain Res 118*:523-525, 1976.

Ibata, Y., Watanbe, K., Kinoshita, H., Kubo, S., and Sano, Y.: The location of LH-RH neurons in the rat hypothalamus and their pathways to the median eminence. *Cell Tissue Res 198*:381-395, 1979.

Ito, S., Yamada, Y., Hayashi, M., Iwasaki, Y., Matsubara, Y., and Shibata, A.: Neurotensin-positive and somatostatin-positive cells in the canine gut. *Tohoku J Exp Med 127*:123-131, 1979.

Iversen, L.L., Iversen, S.D., Bloom, F.L., Douglas, C., Brown, M., and Vale, W.: Calcium-dependent release of somatostatin and neurotensin from rat brain *in vitro*. *Nature 273*:161-163, 1978.

Jackson, I.M.D., and Reichlin, S.: Thyrotropin releasing hormone (TRH): Distribution in the brain, blood and urine of the rat. *Life Sci 14*:2259-2266, 1974.

Jackson, I.M.D., and Reichlin, S.: Brain thyrotropin-releasing hormone is independent of the hypothalamus. *Nature 267*:853-854, 1977.

Jeffcoate, S.L., and White, N.: Studies on the nature of mammalian hypothalamic thyrotropin releasing hormone using immunochemical, chromatographic and enzymic techniques. *J Endocrinol 65*:83-90, 1975.

Jessell, T., Iversen, L.L., and Kanazawa, I.: Release and metabolism of substance P in rat hypothalamus. *Nature 264*:81-83, 1976.

Jessell, T., Tsunoo, A., Kanazawa, I., and Otsuka, M.: Substance P: Depletion in the dorsal horn of rat spinal cord after section of the peripheral processes of primary sensory neurons. *Brain Res 168*:247-259, 1979.

Johansson, O., Hökfelt, T., Jeffcoate, S.L., White N., and Sternberger, L.A.: Ultrastructural localization of TRH-like immunoreactivity. *Exp Brain Res 38*:1-10, 1980.

Joseph, S.A., Scott, D.E., Vaala, S.S., Knigge, K.M., and Korbisch-Dudley, G.: Localization and content of thyrotropin releasing factor (TRF) in median eminence of the hypothalamus. *Acta Endocrinol (Copenh) 74*:215-225, 1973.

Joseph, S.A., Sorrentino, S., Jr., and Sundberg, D.K.: Releasing hormones, LRF and TRF, in the cerebrospinal fluid of the third ventricle. In *Brain-Endocrine Interaction II. The Ventricular System,* Second International Symposium, Shizuoka. S. Karger, Basel, 1975, pp 306-312.

Kanazawa, I., and Jessell, T.: Post-mortem changes and regional distribution of substance P in the rat and mouse nervous system. *Brain Res 117*:362-367, 1976.

Kanazawa, I., Emson, P.C., and Cuello, A.C.: Evidence for the existence of substance P-containing fibers in striato-nigral and pallido-nigral pathways in rat brain. *Brain Res 119*:447-453, 1977.

Kanazawa, I., Mogaki, S., Muramoto, O., and Kuzuhara, S.: On the origin of substance P-containing fibers in the entopeduncular nucleus and the substantia nigra of the rat. *Brain Res 184*:481-485, 1980.

Kardon, F.C., Marcus, R.J., Winokur, A., and Utiger, R.D.: Thyrotropin-releasing hormone content of rat brain and hypothalamus: Results of endocrine and pharmacologic treatments. *Endocrinology 100*:1604-1609, 1977a.

Kardon, F.C., Winokur, A., and Utiger, R.D.: Thyrotropin-releasing hormone (TRH) in rat spinal cord. *Brain Res 122*:578-581, 1977b.

Kataoka, K., Mizuno, N., and Frohman, L.A.: Regional distribution of immunoreactive neurotensin in monkey brain. *Brain Res Bull 4*:57-60, 1979.

King, J.A., and Miller, R.P.: Phylogenetic and anatomical distribution of somatostatin in vertebrates. *Endocrinology 105*:1322-1329, 1979.

King, J.C., Parson, J.A., Erlandsen, S.L., and Williams, T.H.:

Luteinizing hormone-releasing hormone (LH-RH) pathway of the rat hypothalamus revealed by the unlabelled antibody peroxidase-antiperoxidase method. *Cell Tissue Res* 153:211-218, 1974.

King, J.C., Williams, T.H., and Arimura, A.A.: Localization of luteinizing hormone-releasing hormone in rat hypothalamus using radioimmunoassay. *J. Anat* 120:275-288, 1975.

Kirsch, B.: The distribution of LH-RH in the hypothalamus of the thirsting rat. *Cell Tissue Res* 186:135-148, 1978.

Kitabgi, P., Carraway, R., and Leeman, S.E.: Isolation of a tridecapeptide from bovine intestinal tissue and to partial characterization as neurotensin. *J Biol Chem* 251:7053-7058, 1976.

Kitabgi, P., Carraway, R., Van Rietschoten, J., Grania, C., Morgat, J., Menez, A., Leeman, S.E., and Freychet, P.: Neurotensin: Specific binding to synaptic membranes from rat brain. *Proc Natl Acad Sci USA* 74:1846-1850, 1977.

Kizer, J.S., Arimura, A., Schally, A.V., and Brownstein, M.J.: Absence of luteinizing hormone-releasing hormone (LH-RH) from catecholaminergic neurons. *Endocrinology* 96:523-525, 1975.

Kizer, J.S., Palkovits, M., and Brownstein, M.J.: Releasing factors in the circumventricular organs of the rat brain. *Endocrinology* 98:311-317, 1976a.

Kizer, J.S., Palkovits, M., Tappaz, M., Debabian, J., and Brownstein, M.J.: Distribution of releasing factors, biogenic amines, and related enzymes in the bovine median eminence. *Endocrinology* 98:685-695, 1976b.

Knigge, K.M.: Role of the ventricular system in neuroendocrine processes. Initial studies on the role of catecholamines in transport of thyrotropin releasing factor. In *Frontiers in Neurology and Neuroscience Research,* P. Seeman and G.M. Brown, eds, University of Toronto Press, Toronto, 1974, pp 40-46.

Knigge, K.M., and Joseph, S.A.: Thyrotropin releasing factor (TRF) in cerebrospinal fluid of the third ventricle of rat. *Acta Endocrinol (Copenh)* 76:209-213, 1974.

Knigge, K.M., Joseph, S.A., Silverman, A.J., and Baala, S.S.: Further observations on the structure and function of median eminence, with reference to the organization of RF-producing elements in the endocrine hypothalamus. *Prog. Brain Res* 39:7-20, 1973.

Knigge, K.M., Joseph, S.A., Schock, D., Silverman, A.J., Ching, M.C.H., Scott, D.E., Zeman, D., and Krobish-Dudley, G.: Role of the ventricular system in neuroendocrine processes: Synthesis and distribution of thyrotropin releasing factor (TRF) in the hypothalamus and third ventricle. *Can J Neurol Sci* 1:74-84, 1974.

Knigge, K.M., Joseph, S.A., and Hoffman, G.E.: Organization of

LRF- and SRIF-neurons in the endocrine hypothalamus. In *The Hypothalamus*, S. Reichlin, R.J. Baldessarini, and J.B. Martin, eds, Raven Press, New York, 1978, pp 49-67.

Kobayashi, R.M., Brown, M., and Vale, W.: Regional distribution of neurotensin and somatostatin in rat brain. *Brain Res* 126:584-588, 1977.

Koivusalo, F., and Leppaluoto, J.: High TRF immunoreactivity in purified pancreatic extacts of fetal and newborn rats. *Life Sci 24:1655-1658*, 1979.

Kordon, C., Kerdelhue, B., Pattou, E., and Jutisz, M.: Immuno-cytochemical localization of LH-RH in axons and nerve terminals of the rat median eminence. *Proc Soc Exp Biol Med 147*:122-127, 1974.

Koves, K., and Magyar, A.: On the location of TRH producing neurons: Thyroid response to PTU treatment after sterotaxic interventions in hypothalamus. *Endocrinol Exp (Bratisl)* 9:247-258, 1975.

Kozlowski, G.P., Nett, T.M., and Zimmerman, E.A.: Immunocyto-chemical localization of gonadotropin-releasing hormone (Gn-RH) and neurophysin in the brain. In *Anatomical Neuroendocrinology*, International Conference on the Neurobiology of CNS-Hormone Interactions, W.E. Stumpf and L.D. Grant, eds, S. Karger, Basel, 1975, pp 185-191.

Krause, R.: Comparative distribution of LH-RH and somatostatin in the supraoptic crest (OVLT) of the rat. *Neurosci Lett* 11:177-180, 1979.

Krieder, M.S., Winokur, A., and Utiger, R.D.: TRH immunoreactivity in rat hypothalamus and brain: Assessment by gel-filtration and thin layer chromatography. *Brain Res 171*:161-165, 1979.

Kronheim, S., Berelowitz, M., and Pimstone, B.L.: The characterization of somatostatin-like immunoreactivity in human serum. *Diabetes 27*:523-529, 1978.

Krulich, L., Quijada, M., Jefco, E., and Sundberg, D.K.: Localization of thyrotropin-releasing factor (TRF) in the hypothalamus of the rat. *Endocrinology 95*:9-17, 1974.

Kubek, M.J., Lorincz, M.A., and Wilber, J.F.: The identification of thyrotropin releasing hormone (TRH) in hypothalamic and extrahypothalamic loci of the human nervous system. *Brain Res 126*:196-200, 1977.

Kumasaka, T., Nishi, N., Yaoi, Y., Kido, Y., Saito, M., Okayasu, I., Shimizu, K., Hatakeyama, S., Sawano, S., and Kokubu, K.: Demonstration of immunoreactive somatostatin-like substance in villi and decidua in early pregnancy. *Am J Obstet Gynecol 134*:39-44, 1979.

Kusumoto, Y., Iwanaga, T., Ito, S., and Fujita, T.: Juxtaposition of somatostatin cell and parietal cell in the dog stomach. *Archivum Histol Jpn 42*:459-465, 1979.

LaBella, F.S., Havlicek, V., Pinsky, C., and Leybin, L.V.: Opiate-like naloxone-reversible effects of androsterone sulfate in rats. *Soc Neuro Sci Abstr* 3:295, 1977.

Lauber, M., Camier, M., and Cohen, P.: Immunological and biochemical characterization of distinct high molecular weight forms of neurophysin and somatostatin in mouse hypothalamus extracts. *FEBS Lett* 97:343-347, 1979.

LeClerc, R., Pelletier, G., Puviari, R., Arimura, A., and Schally, A.V.: Immunohistochemical localization of somatostatin in endocrine cells of the rat stomach. *Mol Cell Endocrinol* 4:257-261, 1976.

Leeman, S.E. and Carraway, R.: The discovery of a sialogogic peptide in bovine hypothalamic extract: Its isolation, characterization as substance P, structure and synthesis. In *Substance P*, U.S. von Euler and B. Pernow, eds, New York, Raven Press, New York, 1977, pp 5-13.

Leeman, S.E., and Mroz, E.A.: Minireview: Substance P. *Life Sci* 15:2033-2041, 1974.

Leitner, J.W., Rifkin, R.M., Maman, A., and Sussman, K.E.: Somatostatin binding to pituitary plasma membranes. *Biochem Biophys Res Commun* 87:919-927, 1979.

Leppaluoto, J., Koivusalo, F., and Kraama, R.: Thyrotropin-releasing factor: Distribution in neural and gastrointestinal tissues. *Acta Physiol Scand* 104:175-179, 1978.

LeRoith, D., Bark, H., and Glick, S.M.: Somatostatin and antidiuretic hormone secretion in dogs. *Horm Metab Res* 11:177-178, 1979.

Luft, R., Efendic, S., Hökfelt, T., Johansson, O., and Arimura, A.: Immunohistochemical evidence for the localization of somatostatin-like immunoreactivity in a cell population of the pancreatic islets. *Med Biol* 52:428-430, 1974.

Lundberg, J.M., Hamberger, B., Schultzberg, M., Hökfelt, T., Granberg, P.L., Efendic, S., Terenius, L., Goldstein, M., and Luft, R.: Enkephalin and somatostatin-like immunoreactivities in human adrenal medulla and pheochromocytoma. *Proc Natl Acad Sci USA* 76:4079-4083, 1979.

Lundquist, I., Sundler, F., Ahren, B., Alumets, J., and Hakanson, R.: Somatostatin, pancreatic polypeptide, substance P and neurotensin: Cellular distribution and effects on stimulated insulin secretion in the mouse. *Endocrinology* 104:832-838, 1979.

McCann, S.M.: A hypothalamic luteinizing-hormone-releasing factor. *Am J Physiol* 202:395, 1960.

McNeill, T.H., and Sladek, J.R. Jr.: Fluorescence-immunocytochemistry: Simultaneous localization of catecholamines and gonadotropin-releasing hormone. *Science* 200:72-74, 1978.

Manberg, P.J., Youngblood, W.W., Nemeroff, C.B., Rossor, M.N., Iverson, L.L., Prange, A.J., Jr., and Kizer, J.S.: Regional

Distribution of Neurotensin in Human Brain. *J Neurochem 38*: 1777-1180,1982.

Maruyama, T., and Ishikawa, H.V.: Somatostatin: Its inhibiting effect on the release of hormones and IgG from clonal cells strains its Ca-influx dependence. *Biochem Biophys Res Commun 74*:1083-1088, 1977.

Matso, H., Baba, Y., Nair, R.M.G., Arimura, A., and Schally, A.V.: Structure of the porcine LH and FSH-releasing hormone. *Biochem Biophys Res Commun 43*:1334-1339, 1971.

Mazzuca, M.: Immunocytochemical and ultrastructural identification of luteinizing hormone-releasing hormone (LH-RH) containing neurons in the vascular organ of the lamina terminalis (OVLT) of the squirrel monkey. *Neurosci Lett 5*:123-127, 1977.

Monohan, M., Rivier, J., Burgus, R., Amoss, M., Blackwell, R., Vale, W., and Guillemin, R.: Synthesis totale par phasa solide d'un decapeptide qui stimuli des gonadotropins hypophysaire LH et FSH. *C R Acad Sci (Paris) 273*:508-510, 1971.

Morley, J.E., Garvin, T.J., Pekary, E., and Hershman, J.M.: Thyrotropin-releasing hormone in the gastrointestinal tract. *Biochem Biophys Res Commun 79*:314-318, 1977.

Morris, M., Tandy, B., Sundberg, D.K., and Knigge, K.M.: Modification of brain and CSF LH-RH following deafferentation. *Neuroendocrinology 18*:131-135, 1975.

Mroz, E.A., Brownstein, M.J., and Leeman, S.E.: Evidence for substance P in the habenulo-interpeduncular tract. *Brain Res 113*:597-599, 1976.

Mroz, E.A., Brownstein, M.J., and Leeman, S.E.: Evidence for substance P in the striato-nigral tract. *Brain Res 125*:305-311, 1977.

Negro-Vilar, A., Ojeda, S.R., Arimura, A., and McCann, S.M.: Dopamine and norepinephrine stimulate somatostatin release by median eminence fragments *in vitro*. *Life Sci 23*:1493-1498, 1978.

Nemeroff, C.B., Konkol, R.J., Bissette, G., Youngblood, W.W., Martin, J.B., Brazeau, P., Rone, M.S., Prange, A.J., Jr., Breeze, G.R., and Kizer, J.S.: Analysis of the disruption of hypothalamic-pituitary regulation in rats treated neonatally with monosodium l-glutamate (MSG): Evidence for the involvement of tuberoinfundibular cholinergic and dopaminergic systems in neuroendocrine regulation. *Endocrinology 101*:613-622, 1977.

Nemeroff, C.B., Lipton, M.A., and Kizer, J.S.: Models of neuroendocrine regulation: Use of monosodium glutamate as an investigational tool. *Dev Neurosci 1*:102-109, 1978.

Nilsson, G., Hökfelt, T., and Pernow, B.: Distribution of substance P-like immunoreactivity in the rat central nervous system as revealed by immunohistochemistry. *Med Biol 52*:424-

427, 1974.

Nilsson, G., Pernow, B., Fischer, G.H., and Folkers, K.: Presence of substance P-like immunoreactivity in plasma from man and dog. *Acta Physiol Scand 94*:542–544, 1975.

Noe, B.D., Spiess, J., Rivier, J.E., and Vale, W.: Isolation and characterization of somatostatin from anglerfish pancreatic islet. *Endocrinology 105*:1410–1415, 1979.

Noe, B.D., Weir, G.C., and Bauer, G.E.: Biosynthesis of somatostatin in pancreatic islets. *Metabolism 27 (Suppl 1)*:1201–1205, 1978.

O'Connell, R., Skrabanek, P., Cannon, D., and Powell, D.: High sensitivity radioimmunoassay for substance P. *Ir J Med Sci 145*:392–398, 1976.

Okon, E., and Koch, Y.: Localisation of gonadotropin-releasing and thyrotropin-releasing hormones in human brain by radioimmunoassay. *Nature 263*:345–347, 1976.

Okon, E., and Koch, Y.: Localisation of gonadotropin-releasing hormone in the circumventricular organs of human brain. *Nature 268*:445–447, 1977.

Oliver, C., Charvet, J.P., Codaccioni, J.L., and Vague, J.: TRH in human CSF. *Lancet 2*:873, 1974a.

Oliver, C., Eskay, R.L., Ben-Jonathan, N., and Porter, J.C.: Distribution and concentration of TRH in the rat brain. *Endocrinology 95*:540–546, 1974b.

Oliver, C., Ben-Jonathan, N., Mical, R.S., and Porter,J.C.: Transport of thyrotropin-releasing hormone from cerebrospinal fluid to hypophyseal portal blood and the release of thyrotropin. *Endocrinology 97*:1138–1143, 1975.

Ondo, J.G., Eskay, R.L., Mical, R.S., and Porter, J.C.: Release of LH by LRF injected into the CSF: A transport role for the median eminence. *Endocrinology 93*:231–237, 1973.

Orci, L., Baetens, D., Dubois, M.P., and Rufener, C.: Evidence of the D-cell of the pancreas secreting somatostatin. *Horm Metab Res 7*:400–402, 1975.

Orci, L., Baetens, D., Rufener, C., Amherdt, M., Raviazzola, M., Studer, P., Malaisse-Lagal, F., and Unger, R.H.: Hypertrophy and hyperplasia of somatostatin-containing D-cells in diabetes. *Proc Natl Acad Sci USA 73*:1338–1342, 1976a.

Orci, L., Baetens, D., Rufener, C., Brown, M., Vale, W., and Guillemin, R.: Evidence for immunoreactive neurotensin in dog intestinal mucosa. *Life Sci 19*:559–562, 1976b.

Otsuka, M., and Konishi, S.: Release of substance P-like immunoreactivity from isolated spinal cord of newborn rat. *Nature 264*:83–84, 1976.

Palkovits, M.: Isolated removal of hypothalamic nuclei for neuroendocrinological and neurochemical studies. In *Anatomical Neuroendocrinology*, W.E. Stumpf and L.D. Grant, eds, S. Karger, Basel, 1975, pp 72–80.

Palkovits, M., Arimura, A., Brownstein, M.J., Schally, A.V., and Saavedra, J.M.: Luteinizing hormone-releasing hormone (LH-RH) content of the hypothalamic nuclei in rat. *Endocrinology 96*:554-558, 1974.

Palkovits, M., Brownstein, M.J., and Kizer, J.S.: Effect of total hypothalamic deafferentation on releasing hormone and neurotransmitter concentrations of the mediobasal hypothalamus in rat. Symposium of the International Society of Psychoneuroendocrinology, Visegard, 1975, pp 575-599.

Parker, C.R., Jr., Neaves, W.B., Barnea, A., and Porter, J.C.: Studies on the uptake of [^{3}H] thyrotropin-releasing hormone and its metabolites by synaptosome preparations of the rat brain. *Endocrinology 101*:66-75, 1977.

Parker, C.R., Jr., Neaves, W.B., Barnea, A., and Porter, J.C.: Studies on the subsynaptosomal localization of luteinizing hormone-releasing hormone and thyrotropin-releasing hormone in the rat hypothalamus. *Endocrinology 102*:1167-1175, 1978.

Parker, C.R., Jr., Foreman, M.M., and Porter, J.C.: Subcellular localization of luteinizing hormone releasing hormone degrading activity the hypothalamus. *Brain Res 174*:221-228, 1979.

Parsons, J.A., Erlandsen, S.L., Hegre, O.D., McEvoy, R.C., and Elde, R.P.: Central and peripheral localization of somatostatin immunoenzyme immunocytochemical studies. *J Histochem Cytochem 24*:872-882, 1976.

Patel, Y.C., and Zingg, H.H.: Somatostatin precursors: Evidence for existence and comparison of distribution in neural and non-neural tissues and blood of the rat. VI International Congress Endocrinology, Abstract 166, 1980, p 292.

Paulin C., Dubois, M.P., Barry, J., and Dubois, P.M.: Immunofluorescence study of LH-RH producing cells in the human fetal hypothalamus. *Cell Tissue Res 182*:341-345, 1977.

Paxinos, G., Emson, P.C., and Cuello, A.C.: The substance P projections to the frontal cortex and the substantia nigra. *Neurosci Lett 7*:127-131, 1978a.

Paxinos, G., Emson, P.C., and Cuello, A.C.: Substance P projections to the entopeduncular nucleus, the medial preoptic area and the lateral septum. *Neurosci Lett 7*:133-136, 1978b.

Pearse, A.G.E., and Polak, J.M.: Neural crest origin of the endocrine polypeptide (APUD) cells of the gastrointestinal tract and pancreas. *Gut 12*:783-788, 1971.

Pelletier, G., Labrie, F., Arimura, A., and Schally, A.V.: Electron microscopic immunohistochemical localization of growth hormone-release inhibiting hormone (somatostatin) in the rat median eminence. *Am J Anat 140*:445-450, 1974.

Pickel, V.M., Joh, T., and Reis, D.J.: Ultrastructural localization of tyrosine hydroxylase in noradrenergic neurons of brain. *Proc Natl Acad Sci USA 72*:659-663, 1975.

Pickel, V.M., Reis, D.J., and Leeman, S.E.: Ultrastructural

localization of substance P in neurons of rat spinal cord. *Brain Res 122:*534-540, 1977.

Pimstone, B.L., and Berelowitz, M.: Somatostatin-paracrine and neuromodulation peptide in gut and nervous system. *Afr Med J 53:*7-9, 1978.

Polak, J.M., Sullivan, S.N., Bloom, S.R., Buchan, A.M.J., Facer, P., Brown, M.R., and Pearse, A.G.E.: Specific localisation of neurotensin to the N cell in human intestine by radioimmunoassay and immunocytochemistry. *Nature 270:*183-184, 1977.

Powell, D., and Skrabanek, P.: Brain and gut. *Clin Endocrinol Metab 8:*299-312, 1979.

Powell, D., Leeman, S.E., Tregear, G.W., Niall, H.D., and Potts, J.D., Jr.: Radioimmunoassay for substance P. *Nature 27(New Biol 241:*252-254, 1973.

Pradayrol, L., Chayvialle, J.A., Carlquist, M., and Mutt, V.: Isolation of a porcine intestinal peptide with C-terminal somatostatin. *Biochem Biophys Res Commun 85:*701-708, 1978.

Pradelles, P., Gros, C., Humbert, J., Dray, F., and Ben-Ari, Y.: Visual deprivation decreases metenkephalin and substance P content of various forebrain structures. *Brain Res 166:*191-194, 1979.

Pugsley, T.A., and Lippmann, W.: Effect of somatostatin analogues and 17-α-dihydroequilin on rat brain opiate receptors. *Res Commun Chem Pathol Pharmacol 21:*153-156, 1978.

Ramirez, V.D., Gautron, J.P., Epelbaum, J., Pattou, E., Zamora, A., and Kordon, C.: Distribution of LH-RH in subcellular fractions of the basomedial hypothalamus. *Mol Cell Endocrinol 3:*339-350, 1975.

Remy, C., and Dubois, M.P.: Immunofluorescence of somatostatin-producing sites in the hypothalamus of the tadpole. *Cell Tissue Res 187:*315-321, 1978.

Robinson, S.E., Schwartz, J.P., and Costa, E.: Substance P in the superior cervical ganglion and the submaxillary gland of the rat. *Brain Res 182:*11-17, 1980.

Rorstad, O.P., Brownstein, M.J. and Martin, J.B.: Immunoreactive and biologically active somatostatin-like material in rat retina. *Proc Natl Acad Sci USA 76:*3019-3023, 1979.

Rouiller, D., Schusdziarra, V., Collon, J.M., Harris, V., and Unger, R.H.: Release of somatostatin-like immunoreactivity from the lower gut. *Gastroenterology 77:*700-703, 1979.

Rufener, C., Amerdt, M., Dubois, M.P., and Orci, L.: Ultrastructural immunocytochemical localization of somatostatin in rat pancreatic monolayer culture. *J Histochem Cytochem 23:*866-869, 1975a.

Rufener, C., Dubois, M.P., Malaisse-Lagae, F., and Orci, L.: Immunofluorescent reactivity to anti-somatostatin in the gastro-intestinal mucosa of the dog. *Diabetologia 11:*321-324,

1975b.

Sar, M., and Stumpf, W.E.: Autoradiographic localization of radioactivity in the rat brain after the injection of 1,2-^{3}H-testosterone. *Endocrinology 92*:251-256, 1973.

Sara, V.R., Rutherford, R., and Smythe, G.A.: The influence of maternal somatostatin administration on fetal brain cell proliferation and its relationship to serum growth hormone and brain trophin activity. *Horm Metab Res 11*:147-149, 1979.

Schonbrunn, A., and Tasjian, A.H., Jr.: Characterization of functional receptors for somatostatin in rat pituitary cells in culture. *J Biol Chem 253*:6473-6483, 1978.

Schultzberg, M., Lundberg, J.M., Hökfelt, T., Terenius, L., Brandt, J., Elde, R., and Goldstein, M.: Enkephalin-like immunoreactivity in gland cells and nerve terminals of the adrenal medulla. *Neuroscience 3*:1169-1186, 1978.

Scott, D.E., Dudley, G.K., Knigge, K.M., and Kozlowski, G.P.: In vitro analysis of the cellular localization of luteinizing hormone releasing factor (LRF) in the basal hypothalamus of the rat. *Cell Tissue Res 149*:371-378, 1974.

Setalo, G., Vigh, S., Schally, A.V., Arimura, A., and Flerko, B.: GHRH containing neural elements in the rat hypothalamus. *Brain Res 90*:352-356, 1975.

Setalo, G., Vigh, S., Schally, A.V., Arimura, A., and Flerko, B.: Immunohistological study of the origin of LH-RH containing nerve fibers of the rat hypothalamus. *Brain Res 103*:597-602, 1976.

Shambaugh, G., III, Kubek, M., and Wilber, J.F.: Thyrotropin-releasing hormone activity in the human placenta. *J Clin Endocrinol Metab 48*:483-486, 1979.

Shapiro, B., Kronheim, S., and Pimstone, B.L.: The presence of immunoreactive somatostatin in rat retina. *Horm Metab Res 11*:79-80, 1979a.

Shapiro, B., Sheppard, M.C., Kronheim, S., and Pimstone, B.L.: Tissue distribution of immunoreactive somatostatin in the South African clawed toad (Xenopus laeurs). *J Endocrinology 80*:407-408, 1979b.

Sheppard, M.C., Kronheim, S., and Pimstone, B.L.: Stimulation by growth hormone of somatostatin release from the rat hypothalamus in vitro. *Clin Endocrinol (Oxf) 9*:583-586, 1978.

Sheppard, M.C., Kronheim, S., and Pimstone, B.L.: Effect of substance P, neurotensin and the enkephalins on somatostatin release from the rat hypothalamus in vitro. *J Neurochem 32*:647-649, 1979.

Silverman, A.J.: Distribution of luteinizing hormone-releasing hormone (LH-RH) in the guinea pig brain. *Endocrinology 99*:30-41, 1976.

Silverman, A.J., and Desnoyers, P.: Ultrastructural immunocytochemical localization of luteinizing hormone-releasing hormone (LH-RH) in the median eminence of the guinea pig. *Cell Tissue Res 169:*157-166, 1976.

Silverman, A.J., Antunes, J.L., Ferin, M., and Zimmerman, E.A.: The distribution of luteinizing hormone-releasing hormone (LH-RH) in the hypothalamus of the rhesus monkey. Light microscopic studies using immunoperoxidase technique. *Endocrinology 101:*134-142, 1977.

Silverman, A.J., Krey, L.C., and Zimmerman, E.A.: A comparative study of the luteinizing hormone releasing hormone (LH-RH) neuronal networks in mammals. *Biol Reprod 20:*98-110, 1979.

Singer, E., Sperk, G., Placheta, P., and Leeman, S.E.: Reduction of substance P levels in the ventral cervical spinal cord of the rat after intracisternal 5,7-dihydroxytryptamine injection. *Brain Res 174:*363-365, 1979.

Skrabanek, P., Cannon, D., Kirrane, J., Legge, D., and Powell, D.: Circulating immunoreactive substance P in man. *Ir J Med Sci 145:*399-408, 1976.

Spiess, J., and Vale, W.: Investigation of larger forms of somatostatin in pigeon pancreas and rat brain. *Metabolism 27:*1175-1178, 1978.

Sternberger, L.A., and Petrali, J.P.: Quantitative immunocytochemistry of pituitary receptors for luteinizing hormone-releasing hormone. *Cell Tissue Res 162:*141-176, 1975.

Sternberger, L.A., Petrali, P.P., Joseph, S.A., Meyer, H.G., and Mills, K.R.: Specificity of the immunocytochemical luteinizing hormone-releasing hormone receptor reaction. *Endocrinol 102:*63-73, 1978.

Stumpf, W.E., and Sar, M.: Hormone-architectonics of the mouse brain with ^{3}H-estradiol. In *Anatomical Neuroendocrinology,* W.E. Stumpf and L.D. Grant, eds, S. Karger, Basel, 1975, pp 82-103.

Styne, D.M., Goldsmith, P.C., Burstein, S.R., Kepler, S.L., and Greenbach, M.M.: Immunoreactive somatostatin and luteinizing hormone releasing hormone in median eminence synaptosomes of the rat: detection by immunohistochemistry and quantification by radioimmunoassay. *Endocrinology 101:*1099-1103, 1977.

Sundberg, D.K., and Knigge, K.M.: Luteinizing hormone-releasing hormone (LH-RH) production and degradation by rat medial basal hypothalami *in vitro. Brain Res 139:*89-99, 1978.

Sundler, F., Alumets, J., Hakanson, R., Carraway, R.E., and Leeman, S.E.: Ultrastructure of the gut neurotensin cell. *Histochemistry 53:*25-34, 1977.

Sundler, F., Carraway, R.E., Hakanson, R., Alumets, J., and Dubois, M.P.: Immunoreactive neurotensin and somatostatin in the chicken thymus. *Cell Tissue Res 194:*367-376, 1978.

Sundler, F., Hakanson, P., Hammer, R.A., Alumets, J., Carraway, R.E., Leeman, S.E., and Zimmerman, E.A.: Immunohistochemical localization of neurotensin in endocrine cells of the gut.

Cell Tissue Res 178:313-321, 1979.

Swanson, L.W.: The locus coeruleus: A cytoarchitectonic, golgi and immunohistochemical study in the albino rat. *Brain Res 110*:39-56, 1976.

Swanson, L.W., and Hartman, B.K.: The central adrenergic system. An immunofluorescence study of the location of cell bodies and their efferent connections in the rat utilizing dopamine- -hydroxylase as a marker. *J Comp Neurol 163*:467-506, 1975.

Taber, C.A., and Karavolas, H.J.: Subcellular localization of LH releasing activity in the rat hypothalamus. *Endocrinology 96*:446-452, 1975.

Takahashi, T., Konishi, S., Powell, D., Leeman, S.E., and Otsuka, M.: Identification of the motoneuron-depolarizing peptide in the bovine dorsal root as hypothalamic substance P. *Brain Res 73*:59-69, 1974.

Takahashi, T., and Otsuka, M.: Regional distribution of substance P in the spinal cord and nerve roots of the cat and the effect of dorsal root section. *Brain Res 87*:1-11, 1975.

Tamborlane, W.V., Sherwin, R.S., Hendler, R., and Felig, P.: Metabolic effects of somatostatin in maturity-onset diabetes. *N Eng J Med 297*:181-183, 1977.

Terenius, L.: Somatostatin and ACTH are peptides with partial antagonist-like selectivity for opiate receptors. *Eur J Pharmacol 38*:211-213, 1976.

Teuwissen, B., Fauconnier, J.P., and Thomas, K.: Radioimmunoassay of LH-RH: Application to human plasma. *Gynecol Obstet Invest 9*:183-194, 1978a.

Teuwissen, B., Fauconnier, J.P., and Thomas, K.: Radioimmunoassay of LH-RH: Standard curves and specificity. *Gynecol Obstet Invest 9*:170-182, 1978b.

Traber, J., Claser, T., Brandt, M., Klebensberger, W., and Hamprecht, B.: Different receptors for somatostatin and opioids in neuroblastoma x glioma hybrid cells. *FEBS Lett 81*:351-354, 1977.

Tregear, G.W., Niall, H.D., Potts, J.T., Jr., Leeman, S.E., and Chang, M.M.: Synthesis of substance P. *Nature New Biol 232*:87-89, 1971.

Uemura, H., Asai, T., Nozaki, M., and Kobayashi, H.: Ependymal absorption of luteinizing hormone-releasing hormone injected into the third ventricle of the rat. *Cell Tissue Res 160*:443-452, 1975.

Uhl, G.R., Kuhar, M.J., and Snyder, S.H.: Neurotensin: Immunohistochemical localization in rat central nervous system. *Proc Natl Acad Sci USA 74*:4059-4063, 1977b.

Uhl, G.R., Kuhar, M.J., and Snyder, S.H.: Enkephalin-containing pathway: Amygdaloid efferents in the stria terminalis. *Brain Res 149*:223-228, 1978.

Uhl, G.R., Goodman, R.R., and Snyder, S.H.: Neurotensin-containing cell bodies, fibers and nerve terminals in the brainstem of the rat: Immunohistochemical mapping. *Brain Res* *167*:77-91, 1979.

Uhl, G.R., and Snyder, S.H.: Neurotensin receptor binding, regional and subcellular distributions favor transmitter role. *Eur J Pharmacol 41*:89-91, 1977a.

Uhl, G.R., and Snyder, S.H.: Regional and subcellular distributions of brain neurotensin. *Life Sci 19*:1827-1832, 1977b.

Uhl, G.R., and Snyder, S.H.: Neurotensin: A neuronal pathway projecting from amygdala through stria terminalis. *Brain Res 161*:522-526, 1979.

Uhl, G.R., Bennett, J.P., Jr., and Snyder, S.H.: Neurotensin, a central nervous system peptide: Apparent receptor binding in brain membranes. *Brain Res 130*:299-313, 1977a.

Wakabayashi, I., Demura, R., Kanda, M., Demura, H., and Shizume, K.: Effect of hypophysectomy on hypothalamic somatostatin content in rats. *Endocrinol Jpn 23*:439-442, 1976.

Wakabayashi, I., Miyazawa, Y., Kanda, M., Miki, N., Demura, R., Demura, H., and Shizume, K.: Stimulation of immunoreactive somatostatin release from hypothalamic synaptosomes by high (K^+) and dopamine. *Endocrinology Jpn 24*:601-604, 1977.

Warberg, J., Eskay, R.L., Barnea, A., Reynolds, H.C., and Porter, J.C.: Release of luteinizing hormone releasing hormone and thyrotropin releasing hormone from a synaptosome-enriched fraction of hypothalamic homogenates. *Endocrinology 100*:814-825, 1977.

Weiner, R.I., Pattou, E., Kerdelhue, B., and Kordon, C.: Differential effects of hypothalamic deafferentation upon luteinizing hormone-releasing hormone in the median eminence and organum vasculosum of the lamina terminalis. *Endocrinology 97*:1597-1600, 1975.

Weiner, R.L., Terkel, J., Blake, C.A., Schally, A.V., and Sawyer, C.H.: Changes in serum luteinizing hormone following intraventricular and intravenous injections of luteinizing hormone-releasing hormone in the rat. *Neuroendocrinology 10*:261-272, 1972.

Wheaton, J.E., and McCann, S.M.: Luteinizing hormone-releasing hormone in peripheral plasma and hypothalamus of normal and ovariectomized rats. *Neuroendocrinology 20*:296-310, 1976.

Wheaton, J.E., Krulich, L., and McCann, S.M.: Localization of luteinizing hormone-releasing hormone in the preoptic area and hypothalamus of the rat using radioimmunoassay. *Endocrinology 97*:30-38, 1975.

Williams, G.A., Hargis, G.K., Ensinck, J.W., Kukreja, S.C., Bowser, E.N., Chertow, B.S., and Henderson, W.J.: Role of endogenous somatostatin in the secretion of parathyroid hormone and calcitonin. *Metabolism 28*:950-954, 1979.

Winokur, A., and Utiger, R.D.: Thyrotropin-releasing hormone: Regional distribution in rat brain. *Science 185:*265-266, 1974.

Winokur, A., and Utiger, R.D.: Distribution of thyrotropin-releasing hormone (TRH) in rat brain. In *Anatomical Neuroendocrinology,* International Conference on Neurobiology of CNS-Hormone Interactions, W.E. Stumpf and L.D. Grant, eds, S. Karger, Basel, 1975, pp 328-332.

Winokur, A., Davis, R., and Utiger, R.D.,: Subcellular distribution of thyrotropin-releasing hormone (TRH) in rat brain and hypothalamus. *Brain Res 120:*423-434, 1977.

Winokur, A., Kreider, M.S., Dugan, J., and Utiger, R.D.: The effects of 6-hydroxydopamine on thyrotropin-releasing hormone in rat brain. *Brain Res 152:*203-208, 1978.

Yanaihara, C., Sato, H., Hirohashi, M., Sakagami, M., Yamamoto, K., Hashimoto, T., Yanaihara, N., Abe, K., and Kaneko, T.: Substance P radioimmunoassay using Nα-tyrosyl-substance P and demonstration of the presence of substance P-like immunoreactivities in human blood and porcine tissue extracts. *Endocrinology Jpn 23:*457-463, 1976.

Youngblood, W.W., Lipton, M.A., and Kizer, J.S.: TRH-like immunoreactivity in urine, serum and extrahypothalamic brain: non-identity with synthetic PYROGLU-HIST-PRO-NH$_2$ (TRH). *Brain Res 151:*99-116, 1978.

Youngblood, W.W., Humm, J.H., and Kizer, J.S.: TRH-like immunoreactivity in rat pancreas and eye, bovine and sheep pineals, and human placenta: nonidentity with synthetic PYROGLU-HIS-PRO-NH$_2$ (TRH). *Brain Res 163:*101-110, 1979.

Zetler, G.: Biologically active peptides (substance P). In *The Handbook of Neurochemistry,* Vol. IV, A Lajtha, ed, Plenum Press, New York, 1970, pp 135-148.

Zetler, G., Cannon, D., Powell, D., Skrabanek, P., and Vanderhaeghen, J.J.: Crude substance P from brain contains a cholecystokinin-like peptide. *Naunyn-Schmiedebergs Arch Pharmacol 305:*189-190, 1978.

Zimmerman, E.A., and Antunes, J.L.: Organization of the hypothalamic-pituitary system: Current concepts from immunohistochemical studies. *J Histochem Cytochem 24:*807-815, 1976.

Zimmerman, E.A., Hsu, K.C., Ferin, M., and Kozlowski, G.P.: Localization of gonadotropin-releasing hormone (Gn-RH) in the hypothalamus of the mouse by immunoperoxidase technique. Hypothalamic cells containing Gn-RH. *Endocrinology 95:*1-8, 1974.

Zyznar, E.S., Conlon, J.M., Schusdziarra, V., and Unger, R.H.: Properties of somatostatin-like immunoreactive polypeptides in the canine extrahypothalamic brain and stomach. *Endocrinology 105:*1426-1431, 1979.

3

The Biosynthesis of Peptide Hormones

JAMES L. ROBERTS

The peptide hormones compose a variety of different protein molecules ranging in size from 3 to 300 amino acids in length. They can exist as a single polypeptide chain or be composed of several peptide chains, such as gonadotropins. These peptides are synthesized by specific cell types in various regions of the body and are released into the surrounding medium in response to various nonhormonal and hormonal stimuli, including other peptide hormones. In order to understand how these important proteins and their physiological effects are regulated, it is important to understand how they are synthesized.

Being proteins, it is generally accepted that peptide hormones are synthesized by the ribosomal machinery using messenger RNAs (mRNAs) as templates. In general, the mRNAs for these proteins are derived from the cellular DNA via a complex series of events collectively referred to as transcription and post-transcriptional modifications. These events result in specific cytoplasmic levels of mRNA, which generally reflect the concentration of hormones. The nucleotide sequence of the peptide hormone mRNA is then translated into a specific polypeptide chain by the ribosomes, transfer RNAs, and a host of protein and nonprotein cofactors. A final series of events in the synthesis of a peptide hormone takes place when the polypeptide is modified post-translationally. These modifications may include additions to the peptide chain, such as glycosylation, phosphorylation, or acetylation, or subtractions from the peptide chain, such as occurs with proteolytic cleavages. It is these later post-translational modifications that will be discussed in more detail in this chapter, since these are the synthetic events that tend to set the peptide hormones apart from other proteins.

PEPTIDE HORMONES AS SECRETED PROTEINS

As expected from their endocrine functions, the peptide hormones can also be described as secretory proteins. These proteins are packaged into secretory granules to await the proper environmental stimuli necessary for their release into the bloodstream. The intracellular pathway of secreted proteins has been described in detail by Palade (1975) using the pancreatic zymogen-producing cell as a model. Briefly, the secretory proteins are synthesized on membrane-bound polyribosomes located on the rough endoplasmic reticulum (RER). The newly synthesized protein is sequestered into the cisternae of the RER and is subsquently transferred to a smooth membrane organelle, the Golgi apparatus. After passing through the Golgi apparatus, the secretory proteins are packaged into secretory granules to await their exit from the cell (see Fig. 3-1). Such a biosynthetic scheme would predict that a secretory protein would be present only within membrane organelles and not found free in the cellular cytosol. Indeed, subcellular fractionation studies on tumor cells producing either growth hormone (GH) or adenocorticotropin (ACTH) showed that greater than 98 percent of the peptide hormone was located in the particular fraction of the cell (Roberts and Herbert, 1977a; Bancroft, 1973). How these proteins were sequestered exclusively into the membrane organelles of the cell has been discovered only recently.

Two classes of polyribosomes can be easily identified in a cell synthesizing peptide hormones: the free polyribosomes found in the cytosol of the cell, and the membrane-bound polyribosomes found associated with the RER (Fig. 3-1). Functionally and biochemically, there is no essential difference between these two classes of ribosomes other than intracellular location. However, in general, proteins synthesized on free ribosomes are destined to become soluble cytoplasmic proteins whereas those synthesized on membrane-bound ribosomes become membrane proteins or are secreted. It must be stressed that this is a general notion to which there are exceptions.

Peptide hormones have been shown to be synthesized on membrane-bound ribosomes. For example, using established cell fractionation methods to separate free ribosomes from membrane-bound ribosomes and specific mRNA assays, it has been shown that ACTH is synthesized exclusively on membrane-bound polyribosomes (Roberts and Herbert, 1977a; Jones et al., 1978), in agreement with the previously mentioned observation that the protein itself is located in the particular fraction. Thus, there seems to be some property of secretory proteins (or their mRNAs) that causes them to be synthesized on the RER. The nature of this special property of secretory porteins had been

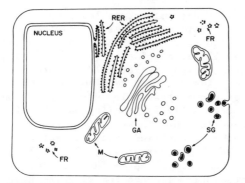

Fig. 3-1.   Model of peptide-hormone secreting cell. Peptide
            hormones are synthesized on the rough endoplasmic
            reticulum (RER) by membrane-bound polyribosomes.
            The peptide hormones are sequestered in the RER and
            are transported to the Golgi apparatus (GA) where
            post-translational events, such as glycosylation and
            proteolysis, occur. Vesicles from the GA develop
            into secretory granules (SG) in which the peptide
            hormone awaits secretion from the cell. FR: free
            ribosomes, M: mitochondria.

identified through the use of cell-free protein synthesizing
systems.

CELL-FREE SYNTHESIS OF PEPTIDE HORMONES

In the last 10 years several cell-free protein synthesizing
systems have been developed that have proved to be very useful
in studying peptide hormone biosynthesis. In these systems,
derived from both animal and plant cell lysates, little or no
post-translational modification occurs on the synthesized protein. This observation is generally attributed to the absence
of membrane material, which is removed by high-speed centrifugation during the preparation of the cell lysates. When mRNA
derived from tissues synthesizing the peptide hormone of interest is added to such systems and is translated, those
peptide hormones subsequently synthesized are larger than the
hormones found in the intact cell. One of the first observations of this nature was made during studies on the biosynthesis of parathyroid hormone (PTH) (for review, see Habener
and Potts, 1978).
    PTH, a peptide of 84 amino acids, had previously been shown
to be derived from a larger form of PTH containing a six amino

acid extension at the N-terminus, called pro-PTH. When mRNA isolated from parathyroid glands was translated in a wheat germ cell-free translation system, the only PTH protein synthesized was approximately 115 amino acids in length. Subsequent analysis of this protein, called pre-pro-PTH, showed it had an additional 25 amino acids at the N-terminus of pro-PTH. Protein sequencing studies showed that this extra peptide was highly hydrophobic. By radioactively labeling parathyroid tissue slices for short periods of time (one to three minutes), pre-pro-PTH could also be identified in the intact cell, but only in very small amounts (Habener et al., 1976). These studies suggested that in the cell pre-pro-PTH is rapidly converted to pro-PTH by a proteolytic cleavage removing the hydrophobic 25 amino acid terminal extension.

Similar hydrophobic N-terminal extensions have been found in several different secretory proteins. Translation of mRNAs coding for pancreatic secretory proteins in a wheat germ translation system and subsequent amino terminal sequence analysis of the proteins revealed highly hydrophobic N-terminal sequences (Devillers-Thiery et al., 1975). The hydrophobic N-terminal peptide has also been found in every peptide hormone in which it has been looked for. A summary of these sequences is shown in Fig. 3-2. Although these extensions vary in size between 20 and 30 amino acids, they all have in common the highly hydrophobic composition, seen most significantly in the center of the sequence. In every case, this N-terminal peptide is absent from the mature hormone, just as with PTH. Since the N-terminus would be the first part of a newly synthesized protein to emerge from the ribosome, it was suggested that this hydrophobic region could interact with the membrane of the RER and somehow "signal" that it should be sequestered within the RER (Blobel and Dobberstein, 1975a). The rapidly removed N-terminal peptide was named the signal peptide and the mechanism of its removal from the protein was soon elucidated.

Blobel and Dobberstein (1975a,b), using immunoglobulin secreting tumor cell lines, were unable to show that the removal of the signal peptide occurs while the protein is still being synthesized by the ribosome (Fig. 3-3). A specific protease responsible for cleaving the presequence from the precursor protein had been identified and named signal peptidase (Jackson and Blobel, 1977). Signal peptidase is located on the cisternal side of the RER and is believed to cleave off the signal sequence as the protein is being vectorially discharged through the RER membrane as diagrammed in Fig. 3-3. It appears the signal peptide is responsible for the initial attachment of the growing nascent peptide chain to the RER and the hydrophobic nature of the signal sequence aids in the vectorial translocation of the growing nascent protein chain through the membrane.

```
                                    -20                          -10                              -1
H₂N Met Met Ser Ala Lys | Asp Met Val Lys Val Met Ile Val Met Leu | Ala Ile Cys Phe Leu Ala Arg Ser Asp Gly | Lys    PTH

        H₂N Met Ala Leu Trp Met Arg Phe Leu Pro Leu Leu Ala Leu Leu | Val Leu Trp Glu Pro Lys Pro Ala Glu Ala | Phe   INS

H₂N Met Ala Thr Gly Ser Arg Thr Ser Leu Leu Ala Phe Gly Leu Leu | Cys Leu Pro Trp Leu Gln Glu Gly Ser Ala | Phe     GH

H₂N Met Asn Ser Glu Val Ser Ala Arg Lys Ala Gly Thr Leu Leu Leu Met Met Ser | Asn Leu Phe Cys Glu Asn Val Gln Thr | Leu   PRL

NH₂ Met Pro Arg Leu Cys Ser Ser Arg Ser Gly Ala Leu Leu Leu Ala Leu Leu Leu | Glu Ala Ser Met Glu Val Arg Gly | Trp    ACTH
                                            |_____ HYDROPHOBIC REGION _____|
```

Fig. 3-2. Summary of peptide hormone signal sequences. The N-terminal signal sequences for several peptide hormones are shown. The sequences for growth hormone (GH) (Martial et al., 1979), insulin (INS) (Villa-Komaroff et al., 1978), and adrenocorticotropin (ACTH) (Nakanishi et al., 1979) were derived from sequence analysis of their respected mRNAs. Prolactin (PRL) (McKean and Maurer, 1978) and parathyroid hormone (PTH) (Habener and Potts, 1978) signal sequences were determined from the protein itself synthesized in vitro. Note the highly hydrophobic nature of each sequence, particularly in the center. The arrow between resides 1 and -1 denotes where signal peptidase cleaves the protein.

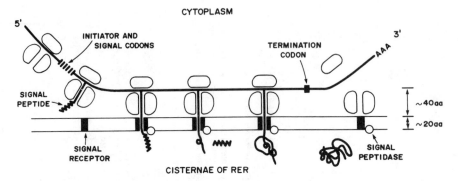

Fig. 3-3.   Signal hypothesis for secretory proteins. Many secretory proteins are found exclusively within the
membrane organelles of a cell. The signal hypothesis was formulated to explain this phenomenon
(Blobel and Dobberstein, 1975a,b). Immediately following the initiator methionine is a stretch of 20
to 30 amino acids called the signal peptide. This
signal peptide, by a mechanism as yet unknown (shown
here as the signal receptor), aids in the attachment
of the ribosome to the endoplasmic reticulum (RER)
surface. It is believed that the highly hydrophobic
nature of the signal peptide aids in its translocation across the lipid bilayer. The signal peptide
is not found attached to the secretory protein in
the cell because it is removed proteolytically by an
enzyme designated signal peptidase while it is being
translated. The ~40 aa and ~20 aa refer to the number of amino acids required to span the large subunit of a ribosome and the lipid bilayer, respectively.

POST-TRANSLATIONAL MODIFICATIONS

Minimal-to-extensive post-translational modifications of peptide hormones have been observed. On one end of the spectrum
is GH and prolactin (PRL) in which only the presequence is
removed to yield the mature hormone. On the other end of the
spectrum is the proopiomelanocortin (POMC) protein, which

undergoes extensive modifications, including proteolysis, glycosylation, and end-group blockage.

Proteolysis is a necessary event in the production of all peptide hormones inasmuch as the signal peptide must be removed. It also plays an important role in the generation of the smaller peptide hormones from their larger precursor forms. This aspect will be dealt with in detail in the discussion of insulin biosythesis.

Glycosylation is involved in the production of the gonado-tropins and thyroid-stimulating hormone. Glycosylation has also been seen in the precursor forms of a hormone even through it may be absent in the hormonal portion of the precursor. These observations have led to speculation that the carbohy-drate sidechains may play a role in determining how the pre-cursor protein is to be cleaved. A more detailed discussion of glycosylation will be made in the section dealing with the POMC protein.

The N- and C-terminal as well as the amino acid side chains can undergo several post-translational modifications, as listed in Table 3-1. It is beyond the scope of the chapter to discuss all these modifications, but they are mentioned for the sake of completeness.

PRECURSOR PEPTIDE HORMONES

In addition to having the N-terminal signal sequence already discussed, many of the peptide hormones are initially synthe-sized as larger proteins. Several examples are listed in Table 3-2. There are many possible reasons for synthesizing a small peptide initially as a larger protein. Because of the mechan-ism by which secretory proteins are sequestered within the RER (Fig. 3-3), there are minimum size requirements for the effec-tive transfer of the peptide into the RER. It takes approxi-mately 40 amino acids to span the length of the large ribosomal subunit and 20 amino acids to traverse the lipid bilayer (Blobel and Sabatini, 1970). In the signal hypothesis, the nascent chain is "pushed" into the cisternae of the RER by the continual addition of amino acids to the growing nascent chain (Blobel and Dobberstein, 1975a). If the protein terminates before at least a portion of it has entered the cisternae, then it will not be sequestered. Since the signal peptide is cleaved while it is being translated, one would expect at least 60 or more amino acids in the protein. This postulation of a minimal size for a peptide hormone appears to be well supported by experimental evidence. In every case where a peptide hor-mone precursor has been identified, the precursor contains at least 80 amino acids more than the signal peptide (Table 3-2).

Table 3-1. Post-translational Modifications Involved in Peptide
           Biosynthesis

Proteolysis
                        -endoproteolytic
                        -exoproteolytic
Glycosylation
                        -high mannose sugar side chains
                        -complex carbohydrate side chains
N- and C-terminal blocking
                        -acetylation of N-terminus
                        -amidation of C-terminus
                        -pyrrolization of N-terminal glutamic acid
Side-chain modification
                        -acetylation of ϵ-amino groups
                        -phosphorylation of serine residues
                        -sulfation of tyrosine residues

Table 3-2. Peptide Hormone Precursors

Name of hormone	No. of Amino Acids[a]	
	Hormone	Precursor[b]
Insulin	51	81
Gastrin	12	~ 80
Glucagon	29	~100
Proopiomelanocortin	39/91	239
Parathyroid hormone	84	90
Vasopressin	9	145

[a]Data for the size of the precursor are taken from the following references: insulin, glucagon, and gastrin (Patzelt et al., 1978), POMC
(Nakanishi et al., 1979), parathyroid hormone (Habener and Potts 1978), and
vasopressin (Gainer et al, 1977).

[b]This size does not include the 20 to 30 amino acids of the signal
sequence.

Such a size would permit at least 20 amino acids of the protein in the RER cisternae before termination of the peptide chain.

There can also be a functional role for the precursor form of a small polypeptide hormone. The precursor may be necessary to direct the proper three-dimensional structure of the final hormone, as in the case of insulin. Another possibility is that the nonhormonal portion of the precursor acts as a carrier protein for the hormone, as it appears to happen with vasopressin and oxytocin. A third possibility, that will be discussed in detail, is multihormone systems, such as the ACTH/LPH system in which the precursor contains several hormones.

THE PRECURSOR DIRECTS HORMONE CONFORMATION

Insulin, a pancreatic hormone, consists of two polypeptide chains (21 and 30 amino acids in the A and B chains, respectively) linked by two interchain cysteine sulfhydryl bridges and one intrachain sulfhydryl bridge, yielding a distinct three-dimensional conformation (for review, see Steiner et al., 1974). Early investigators were puzzled about how the two chains could achieve the proper sulfhydryl cross-linking necessary for biological activity. The answer was discovered when Steiner and colleagues found that insulin was synthesized as a precursor, proinsulin, with a 30 amino acid peptide connecting the A and B peptide chains. The additional peptide in proinsulin appears to aid the proper alignment of the A and B chains of insulin. If the sulfhydryl bridges of insulin and proinsulin are broken to unfold the protein and then allowed to reform, only 2 percent of the insulin molecules ever reform correctly, whereas greater than 90 percent of the proinsulin molecules regain their original conformation. A similar function for precursors to other smaller disulfide bridge-containing hormones, such as vasopressin or oxytocin, may also exist.

Of extreme interest was that at both ends of this connecting peptide were pairs of basic amino acids (lysine and arginine), which were removed during processing of the precursor to the mature hormone. Such a proteolytic event would require an initial trypsinlike endoproteolytic cleavage with a subsequent carboxypeptidase B-like exoproteolytic trimming of the remaining basic residues from the C-terminus. In studies analyzing the enzymatic activities present in insulin secretory granules, both trypsinlike and carboxypeptidaselike enzymes have been identified (Steiner et al., 1974). The use of paired basic residues to signify a specific cleavage site in the production of small peptide hormones may be a general phenomenon in that the ACTH/β-LPH precursor also follows the pattern of individual

hormonal sequences being separated by paired basic residues
(Fig. 3-6).

Precursor as a Carrier Protein

The neural nonapeptide hormones vasopressin and oxytocin are
synthesized by separate neurons in the hypothalamus and intra-
axonally transported via the median eminence to the posterior
pituitary where they are stored in granules prior to release.
Synthesized and transported along with these hormones are their
intracellular carrier proteins, the neurophysins, of which
there are two classes, one for vasopressin and one for oxytocin. The vasopressin-associated neurophysin is referred to
as Np-AVP and similarly the oxytocin-associated neurophysin is
referred to as Np-OT. Early observations that both the hormone
and its neurophysin were always found in a one to one ratio
suggested that the two peptides might be derived from a common
precursor.

Taking advantage of the fact that both vasopressin, oxytocin,
and their neurophysins are extremely rich in cysteine residues
(2 cys in the 9 amino acid hormones and 14 cys in the 93 amino
acids in bovine neurophysin), Gainer et al., (1977) were able
to study directly the biosynthesis of these proteins. They injected [^{35}S]-cysteine into the hypothalamus of the rat and then
analyzed labeled proteins from the hypothalamus, the median
eminence, and posterior pituitary. These three sections are
analogous to those in Fig. 3-4; soma, axon, and axon terminal.
For simplicity, only vasopressin synthesis will be discussed,
although oxytocin is synthesized in a similar manner. The
first vasopressin-related peptide synthesized had a molecular
weight of approximately 20,000 and was primarily seen only in
the hypothalamus. Immunological studies have confirmed that
this precursor, termed "propressophysin" contained the determinants for both Np-AVP and vasopressin (Russell et al., 1979-
a,b). With longer labeling periods, both propressophysin and
mature Np-AVP could be identified in the median eminence. Only
mature Np-AVP and vasopressin could be identified in the posterior pituitary. With these data in mind, the biosynthetic
pathway can be summarized as follows: the vasopressin neurophysin precursor is synthesized in cell bodies present in the
hypothalamus, after which the hormonal part of the precursor
along with another as yet unidentified peptide are cleaved from
the precursor during axonal transport of the peptides. During
axonal transport and granular storage in the posterior pituitary, the hormones remain tightly bound to their respective
neurophysins. This high-affinity binding can be accounted for
in that the binding of vasopressin to Np-AVP has a pH optimum

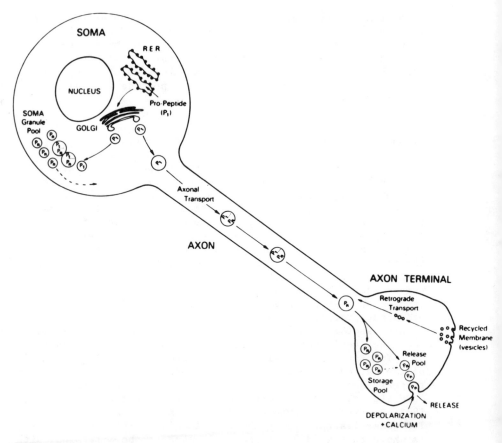

Fig. 3-4. Model of neutral peptide hormone secreting cell.
Hypothetical model of biosynthesis, translocation,
processing, and release of peptides in a peptidergic
neuron. Translocation of mRNA occurs on the rough
endoplasmic reticulum (RER) in the soma yielding a
propeptide or precursor protein molecule (P_1) in
the cisternal space of the RER. The packaging of
P_1 into secretory granules occurs in the Golgi body.
The secretory granule represents the site of post-
translational processing to smaller peptide products
($P_1 \ldots P_n$), which can occur either in the soma or
in the axon during axonal transport. The peptide
products (P_n) are released from the neuron terminals
by depolarization in the presence of extracellular
Ca^2. (Reproduced with permission from Gainer et
al., 1977).

of approximately 5.5, similar to the intragranular pH. However, after release of the granule contents into the blood (neutral pH) the hormones will no longer bind to the neurophysins. This would suggest that a possible role for the neurophysins during the synthesis and release of vasopressin or oxytocin would be to prevent the small peptide from diffusing away from the granule prior to release.

Precursor as a Mechanism of Coordinate Control

In order to guarantee the coordinate synthesis of several peptide hormones, the hormones could be synthesized together as part of a larger protein. An example of such a mechanism can be seen with β-endrophin and β-melanocyte stimulating hormone (β-MSH), two pituitary peptide hormones of 31 and 18 amino acids respectively, which are contained in the 91 amino acid β-LPH.' These two hormones are separated from each other and the rest of β-LPH by pairs of basic amino acids. This system, then, would guarantee the coordinate production of one β-endorphin molecule for every β-MSH molecule produced. Recent studies have shown that β-LPH itself is derived from a larger precursor protein (Loh and Gainer, 1977).

ACTH, another small (39 amino acid) pituitary peptide hormone, had been shown to be derived from a larger precursor protein of molecular weight approximately 30,000. Using antibodies raised against ACTH and β-endorphin, several groups have recently shown that the 30,000 molecular weight precursor to ACTH also contains the β-LPH hormone within it (Mains et al., 1977; Roberts and Herbert, 1977a,b; Nakanishi et al., 1977). The general structure of the proopiomelanocortin (POMC) protein is shown in Fig. 3-5. β-LPH is located at the C-terminal with ACTH adjacent to it in the center of the precursor.

In studying the biosynthesis of multihormone or multicomponent precursor proteins it is important to know the complete amino acid sequence of the protein. This information is helpful in determining what enzymes are responsible for cleaving the individual hormones from the precursor, in identifying the nonhormonal portions of the precursor, and in elucidating the specific sites of glycosylation. However, inherent in its function as a precursor protein, the POMC protein is rapidly broken down intracellularly into its component hormones. This makes the isolation of large quantities of the precursor protein for amino acid sequence analysis a difficult task. Another way to determine the primary sequence of a protein, which circumvents this problem, is to determine the nucleic acid sequence of the mRNA, which codes for the protein in question.

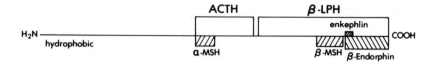

Fig. 3-5. Structure of POMC protein. β-LPH comprises the
C-terminal third of the precursor and the ACTH
sequence is adjacent to it on the N-terminal side.
There is a signal sequence at the extreme N-terminus
(hydrophobic) and the remainder of the N-terminal
end, approximately 90 amino acids, has no known
function. The sequences of ACTH and β-LPH have
also been subdivided into their component hormones,
α-MSH and β-MSH, β-endorphin, and met-enkephalin,
respectively.

Recent advances in nucleic acid sequencing techniques and
recombinant DNA technology have made such an approach quite
feasible and in certain cases the most viable alternative.
This is what has been done to determine the primary sequence of
the POMC protein. Nakanishi and colleagues (1979), using mRNA
from the intermediate lobe of bovine pituitary were able to
deduce the sequence of the POMC protein by analyzing the sequence of a DNA molecule, which was copied from the POMC mRNA.
As can be seen in Fig. 3-6, ACTH is separated from both the -
LPH sequence and amino terminal region by a pair of basic residues, similar to the C-peptide of insulin. Thus, it appears
that ACTH and β-LPH are liberated from the precursor by a
mechanism similar to that seen with proinsulin.
The pathway by which the individual hormones are cleaved from
the precursor has been studied in detail using a pituitary
tumor cell line, (Roberts et al., 1978; Mains and Eipper,
1978). A complicating feature in the processing scheme is that
in addition to proteolysis, the POMC protein is also heavily
glycosylated. A summary of the over-all processing scheme is
shown in Fig, 3-7.
The first event, as with other peptide hormones, is the
proteolytic removal of the signal peptide. Next, while the
protein is still being synthesized, glycosylation occurs at

AAGAGAACGAAGGAGGAAGGAAGAAAAGUGACCGAGGAGCGCUGAACAUCUCGCCCGGGCGCAGCGGGAGCCGCCCGAGGCAGCCUCAGCCUGCCUGGAAG
-500   -450   -400

-26
Met Pro Arg Leu Cys Ser Ser Arg Ser Gly Ala Leu Leu Leu Ala Leu Leu Gln Ala Ser Met Glu Val Arg -1 1 Trp Cys Leu Glu
AUG CCG AGA CUG UGC AGC AGU CGU UCG GGC GCC CUG CUG CUG GCC CUG CUG CAG GCC UCC AUG GAA GUG CGU UGG UGC CUG GAG
           -20   -10   -350

Ser Ser Gln Cys Gln Asp Leu Thr Thr Glu Ser Asn Leu Leu Glu Cys Ile Arg Ala Cys Lys Pro Asp Leu Ser Ala Glu Thr Pro Val
AGC AGC CAG UGU CAG GAC CUC ACC ACG GAA AGU AAC CUG CUG GAG UGC AUC CGG GCC UGC AAG CCC GAC CUC UCC GCC GAG ACG CCG GUG
   10   40   20   30
-300   -250

Phe Pro Gly Asn Gly Asp Glu Gln Pro Leu Thr Glu Asn Pro Arg Lys Tyr Val Met Gly His Phe Arg Trp Asp Arg Phe Gly Arg Arg
UUC CCC GGC AAC GGC GAU GAG CAG CCG CUG ACU GAG AAC CCC CGG AAG UAC GUC AUG GGC CAU UUC CGC UGG GAC CGU UUC GGC CGU CGG
          -200   50   60
                 -150

Asn Gly Ser Ser Ser Gly Val Gly Gly Ala Ala Gln Lys Arg Glu Glu Glu Val Ala Val Gly Glu Gly Pro Gly Pro Arg Gly Asp
AAU GGU AGC AGC AGC GGA GUU GGG GGC GCG GCC CAG AAG CGC GAG GAG GAG GUG GCG GUG GGC GAG GGC CCG GGC CCG CGC GGC GAU
          70   -100   80   -50   90

Asp Ala Glu Thr Gly Pro Arg Glu Asp Lys Arg Ser Tyr Ser Met Glu His Phe Arg Trp Gly Lys Pro Val Gly Lys Lys Arg Arg Pro
GAC GCC GAG ACG GGU CCG CGG GAG GAC AAG CGU UCU UAC UCC AUG GAA CAC UUC CGC UGG GGC AAG CCG GUG GGC AAG AAG CGG CGC CCG
   100   110   120   50
-1 1

Val Lys Val Tyr Pro Asn Gly Ala Glu Asp Glu Ser Ala Glu Ala Phe Pro Leu Glu Phe Lys Arg Glu Leu Thr Gly Glu Arg Leu Glu
GUG AAG GUG UAC CCC AAC GGC GCC GAG GAC GAG UCG GCC GAG GCC UUC CCC CUC GAA UUC AAG AGG GAG CUG ACC GGG GAG AGG CUC GAG
   130   140   150
        100

Gln Ala Arg Gly Pro Glu Ala Gln Ala Glu Ala Ala Ala Glu Lys Lys Asp Glu Gly Pro Tyr Arg Met Glu His Phe Arg Trp Gly Ser
CAG GCG CGC GGC CCC GAG GCC CAG GCC GAG GCU GCG GCC GAG AAG AAG GAC GAG GGC CCG UAC CGC AUG GAG CAC UUC CGC UGG GGC AGC
   160   170   180
        150   200

Ala Glu Ala Lys Lys Asp Ser Gly Pro Tyr Lys Met Glu His Phe Arg Trp Gly Ser Pro Pro Lys Asp Lys Arg Tyr Gly Gly Phe Met Thr
GCC GAG GCC AAG AAG GAC UCG GGC CCC UAU AAG AUG GAA CAC UUC CGC UGG GGC UCG CCC CCC AAG GAC AAG CGG UAC GGU GGC UUC AUG ACC
   190   200   210
        250   300

Ser Glu Lys Ser Gln Thr Pro Leu Val Thr Leu Phe Lys Asn Ala Ile Ile Lys Asn Ala His Lys Lys Gly Gln
UCC GAG AAG AGC CAA GAU CCC CUU UAC ACG CUU UAC AAC AAC GCC AUC AUC AAG AAC GCC CAC AAG AAG GGC CAG UGA GGGCGCAGGGGGCAG
   220   230   239
        350   400

Fig. 3-6. Sequence of POMC mRNA and protein. Primary structure of the complete amino acid sequence and part of the mRNA sequence is shown for the bovine pro-ACTH/β-LPH molecule. Signal peptidase at the arrow (↓) yielding a 26 amino acid signal sequence. The dibasic sequences shown, residues 104-105, 145-146, and 207-208, represent known cleavage sites of the molecule. ACTH is found at residue (top numbers) 106-144; β-LPH, at 147-239; and β-endorphin, at 209-239. (Data are taken from Nakanishi et al., [1979].)

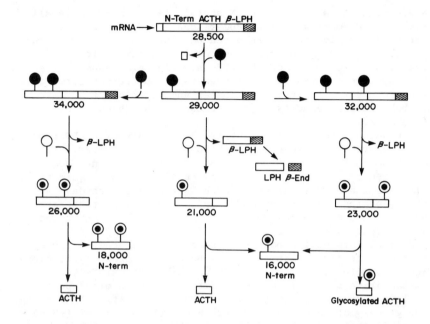

Fig. 3-7.   Pathway of processing of POMC protein. The structures of all the major precursors, intermediates,
and endproducts are shown. The numbers shown below
the structures refer to apparent molecular weights
as determined by sodium dodecyl sulfate polyacrylamide gel electrophoresis (Roberts et al., 1978).
The solid circles (♀) represent core glycosylations,
the open circles (♀) represent peripheral glycosylations, and the combined circle (♀) represents the
end-product complex carbohydrate side chain.

several points along the peptide chain. All of the carbohydrate additions to the POMC protein are classed as complex
carbohydrates and their synthesis on the protein follows a
well-defined pathway (Sharon, 1975). Briefly, the carbohydrate
is initially added as a unit to an asparagine residue, which is
two amino acids toward the N-terminal side of a serine or thre-
onine residue, for example, residues -41 to -39 in Fig. 3-6.
The sugar is covalently linked through two adjacent N-acetylglucosamine residues followed by 7 to 9 mannose residues and
several glucose residues. The glucose and most of the mannose
residues are then enzymatically removed from the carbohydrate
sidechain of the protein and the peripheral sugars are added
one by one. These sugars can be glucose, galactose,

N-acetylglucosamine, fucose, N-acetylgalactosamine, or sialic acid. Since many different sugars can be added the final carbohydrate sidechains are often heterogenous, even when the peptide backbone is homogeneous. This complex series of reactions takes place on the POMC protein before any of the hormones are cleaved from the precursor. Different forms of the precursor varying in the number and location of carbohydrate sidechains in the precursor have been identified (Fig. 3-7).

The second proteolytic event is the cleavage of β-LPH from the precursor. Subsequent to the production of β-LPH, β-endorphin can be cleaved from β-LPH. The pro-ACTH intermediates generated by the removal of β-LPH are cleaved once again to liberate the ACTH sequence from the N-terminal peptide. All of these endproducts, β-LPH, ACTH, and N-terminal peptide, are present in a one to one ratio (a portion of the β-LPH molecules are present as endorphin).

Several interesting points arise from these studies, One is that all the glycosylation occurs before any proteolytic cleavages producing the hormones. The possibility exists that the carbohydrate may be involved in directing the limited proteolytic events generating the mature hormones. After inhibiting the glycosylation of the POMC protein, Loh and Gainer (1979) found in the toad intermediate lobe that the precursor was not correctly processed. However, another group performing similar experiments with a mouse tumor cell line found that the correct proteolytic cleavages did occur (Herbert et al., 1979). Thus, it is still not clear what role, if any, glycosylation plays in the proteolytic processing of the POMC protein. Another interesting point is that not all of the POMC proteins are glycosylated; approximately half of the ACTH molecules produced are glycosylated and the rest have no carbohydrate. One explanation would be that there are several forms of the precursor, some with and some without the amino acid sequence necessary for glycosylation. Another explanation is that glycosylation occurs at random and, although all POMC proteins can be glycosylated only a portion of them actually become glycosylated. Detailed charge and sequence studies suggest that the latter possibility is the most feasible (J. Roberts, unpublished observations).

Finally, it should be noted that the pathway described applies to the tumor cell line, AtT20. The POMC precursor is expressed in several different tissues, but most notably in the anterior and intermediate lobes of the pituitary. The tumor cells appear to process the precursor identically to the anterior pituitary pro-ACTH/LPH cells. The POMC cells in the intermediate lobe of the pituitary process the precursor differently. For example, ACTH is not produced by these cells but is instead broken down into two other hormones, α-melanocyto-stimulating hormone (α-MSH) and corticotropinlike intermediate lobe peptide. The best explanation for this observation is that, although the two tissues start with a similar POMC protein, the intermediate lobe cells contain additional proteases and modifying enzymes so as to produce a different endproduct hormone.

ACKNOWLEDGMENT

The author would like to thank Mr. T.M. Mathew and Mrs. E. Freeburn for typing the manuscript and James Eberwine for critical reading of the manuscript.

REFERENCES

Bancroft, F.C.: Intracellular location of newly synthesized growth hormone. *Exp Cell Res 79*:275-278, 1973.

Blobel, G., and Dobberstein, B.: Transfer of proteins across membranes: I. Presence of proteolytically processed and un-processed nascent immunoglobulin light chains on membrane bound ribosomes of murine myeloma. *J Cell Bio 67*:835-851, 1975a.

Blobel, G., and Dobberstein, B.: Transfer of protein across membranes. II. Reconstitution of functional rough microscomes from heterologus components. *J Cell Biol 67*:852-862, 1975b.

Blobel, G., and Sabatini, D.D.: Controlled proteolysis of nascent polypeptides in rat liver cell fractions: I. Location of the polypeptides within ribosomes. *J Cell Biol 45*:130-145, 1970.

Devillers-Thiery, A., Kindt, I., Scheele, G., and Blobel, G.: Homology in amino-terminal sequence of precursors to pancreatic secretory proteins. *Proc Natl Acad Sci USA 72*:6016-5020, 1975.

Gainer, H., Sarne, Y., and Brownstein, M.J.: Biosynthesis and axonal transport of rat neurohypophysical proteins and peptides. *J Cell Biol 73*:366-381, 1977.

Habener, J.F., and Potts, J.T., Jr.: Biosynthesis of parathyroid hormone. *N Engl J Med 299*:580-585, 1978.

Habener, J.F., Potts, J.T., Jr., and Rich, A.: Pre-proparathyroid hormone: Evidence for an early biosynthetic precursor of proparathyroid hormone. *J Biol Chem 251*:3893-3899, 1976.

Herbert, E., Budarf, M.L., Phillips, M.A., Rosa, P., Policastro, P., Roberts, J., Seidah, N.G., and Chretien, M.: Presence of a pre-sequence in the common precursor to ACTH/endorphin and the role of carbohydrate in processing and secretion of ACTH and endorphin. *Ann NY Acad Sci*, 1979.

Jackson, R.C., and Blobel, G.: Post-translational cleavage of pre-secretory proteins with an extract of rough micoscomes from dog pancreas containing signal peptidase activity. *Proc Natl Acad Sci USA 74:*5598-5602, 1977.

Jones, R.E., Pulkrabek, P., and Grunberger, D.: Characterization and cell-free translation of mouse pituitary tumor mRNA which directs the synthesis of a corticotropin precursor. *Arch Biochem Biophys 188:*476-483, 1978.

Loh, Y.P., and Gainer, H.: Biosythesis, processing, and control of the release of melanotropic peptides in the neurointermediate lobe of Xenopus laevis. *J Gen Physiol 70:*37-58, 1977.

Loh, T.P., and Gainer, H.: The role of carbohydrate in the stabilization, processing, and packaging of the glycosylated adrenocorticotropin-endorpin common presursor in toad pituitaries. *Endocrinology 105:*474-487, 1979.

McKean, D.J., and Maurer, R.A.: Complete amino acid sequence of the precursor region of rat prolactin. *Biochemistry 17:*5215-5219, 1978.

Mains, R.E., and Eipper, B.A., Analysis of the common precursor to corticotropin and endorphin. *J Biol Chem 253:*5732-5744, 1978.

Mains, R.E., Eipper, B.A., and Ling, N.: Common precursor to corticotropins and endorphins. *Proc Natl Acad Sci USA 74:*3014-3018, 1977

Martial, J.A., Hallewell, R.A., Baxter, J.D., and Goodman, H.M.: Human growth hormone: Complementary DNA cloning and expression in bacteria. *Science 205:*602-607, 1979.

Nakanishi, S., Inoue, A., Taii, S., and Numa, S.: Cell-free translation product containing corticotropin and β-endrophin encoded by mRNA from anterior lobe and intermediate lobe of bovine pituitary. *FEBS Lett 84:*105-109, 1977.

Nakanishi, S., Inoue, A., Kita, T., Nakamura, M., Chang, A.C.Y., Cohen, S.N., and Numa, S.: Nucleotide sequence of cloned cDNA for bovine corticotropin-β-lipotropin precursor. *Nature 278:*423-427, 1979.

Palade, G.: Intracellular aspects of the process of protein synthesis. *Science 189:*347-358, 1975.

Roberts, J.L., and Herbert, E.: Characterization of a common precursor to corticotropin and β-lipotropin: Cell-free synthesis of the precursor and identification of corticotropin peptides in the molecule. *Proc Natl Acad Sci USA 74:*4826-4830, 1977a.

Roberts, J.L., and Herbert, E.: Characterization of a common precursor to corticotropin and β-lipotropin: Identification of β-lipotropin peptides and their arrangement relative to corticotropin in the precursor synthesized in a cell-free system. *Proc Natl Acad Sci USA 74:*5300-5304, 1977b.

Roberts, J.L., Phillips, M.J., Rosa, P.A., and Herbert, E.:
Steps involved in the processing of common precursor forms of
adrenocorticotropin and endorphin in cultures of mouse
pituitary cells. *Biochemistry 17*:3609-3621, 1978.

Russell, J.T., Brownstein, M.J., and Gainer, H.: Liberation by
trypsin of an arginine vasopressin like peptide and
neurophysin from a Mw 20,000 putative common precursor. *Proc
Natl Acad Sci USA 76*: 1977a.

Russell, J.T., Brownstein, M.J., and Gainer, H.: Biosynthesis
of the common precursors of vasopressin, oxytocin, and their
respective neurophysins. *The Role of Peptides in Neuronal
Function,* J. Baker and T. Smith, eds. Marcel Dekker, New
York, 1979b.

Sharon, N.: *Complex Carbohydrates: Their Chemistry, Biosynthesis and Functions,* Addison-Wesley Publishing Co., Reading,
Mass, 1975, pp. 65-176.

Steiner, D.F., Kemmler, W., Tager, H.S., and Peterson, J.D.:
Proteolytic processing in the biosynthesis of insulin and
other proteins. *Fed Proc 33*:2105-2115, 1974.

Villa-Komaroff, L., Efstratiadis, A., Broome, S., Lomedico, P.,
Fizard, R., Naber, S.P., Chick, W.L., and Gilbert, W.: A
bacterial clone synthesizing proinsulin. *Proc Natl Acad Sci
USA 75*:3727-3731, 1978.

Copyright © 1984 by Spectrum Publications, Inc.
Peptides, Hormones and Behavior
Edited by C.B. Nemeroff and A.J. Dunn

The Physiological and Pharmacological Control of Anterior Pituitary Hormone Secretion

LAWRENCE A. FROHMAN AND MICHAEL BERELOWITZ

INTRODUCTION

Hypothalamic control of anterior pituitary function has been recognized for more than 40 years since the pioneering work of Harris (1955) who proposed the portal vessel chemotransmitter hypothesis. He suggested that releasing and inhibiting substances of hypothalamic origin were secreted into the portal capillary plexus in the outer layer of the median eminence and transported in the long portal veins surrounding the pituitary stalk to modulate the release of anterior pituitary hormones. This concept, together with that of Scharrer and Scharrer (1940) who first suggested that nerve cells could have a secretory function, formed the basis for the establishment of neuroendocrinology as a distinct discipline, bridging the fields of neurochemistry, neuropharmacology, behavioral sciences, and endocrinology.

During the first two decades following publication of Harris' hypothesis, a large body of literature involving hypothalamic destruction and stimulation experiments began to accumulate, which provided support for the concept. The third decade saw the initiation of massive efforts that led to the isolation and structural identification of three releasing and inhibiting hormones. During the past decade, this work has continued resulting in the identification of a fourth hypophysiotropic hormone, and has been accompanied by the development of agonist and antagonist analogues of the hypothalamic hormones and an increased understanding of the physiological and pharmacological control of their secretion.

HYPOTHALAMIC-ANTERIOR PITUITARY HORMONE RELATIONSHIPS

The interaction between hypothalamic hormones and pituitary
hormones has become considerably more complex than the model
originally proposed in which each anterior pituitary hormone
was believed to be controlled by a single hypothalamic releasing or inhibiting factor. This has occurred in large part
as a result of the purification and identification of the individual factors and the recognition that more than one pituitary hormone could be affected by a single hypothalamic hormone
and, conversely, that more than one hypothalamic factor could
affect a single pituitary hormone. It has also become apparent
that at least three of the six recognized anterior pituitary
hormones are under dual (positive and negative) hypothalamic
control, a fact that has made the elucidation of neuropharmacological effects considerably more difficult. The current view
of the interrelationships between hypothalamic and anterior
pituitary hormones is shown in Fig. 4-1.

HYPOTHALAMUS

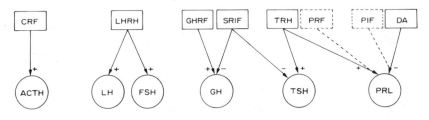

ANTERIOR PITUITARY

Fig. 4-1.   The relationship of hypothalamic hypophysiotropic
            hormones to anterior pituitary hormones. Interrupted
            lines indicate factors whose existence is at present
            controversial. Abbreviations: CRF:corticotropin-
            releasing factor, LH-RH:luteinizing hormone-releas-
            ing hormone, GHRF:growth hormone-releasing factor,
            SRIF:somatotropin release inhibiting factor (somato-
            statin), TRH:thyrotropin-releasing hormone, PRF:pro-
            lactin-releasing factor, PIF:prolactin-inhibiting
            factor, DA:dopamine.

Corticotropin-Releasing Factor

The secretion of adrenocorticotropin (ACTH) is controlled by a
single stimulatory releasing factor. Although the first to be
recognized (Saffran and Schally, 1955), structure of cortico-

tropin-releasing factor (CRF) has only recently been determined (Vale et al., 1981). It is a 41 amino acid peptide that is distinct from vasopressin which also has corticotropin-releasing activity. It is believed by some that more than one CRF may exist, although this remains to be proved. CRF is localized in the hypothalamus with highest concentrations in the median eminence. CRF activity has been detected in peripheral plasma (Yasuda and Greer, 1977) although it is uncertain as to whether this substance is identical to hypothalamic CRF. Corticotropin-releasing activity has also been identified in extracts of lung tumors associated with Cushing's syndrome (adrenocortical hyperplasia) (Upton and Amatruda, 1971).

Luteinizing Hormone-Releasing Hormone
(Gonadotropin-Releasing Hormone)

A single hypothalamic hormone, luteinizing hormone-releasing hormone (LH-RH), (or gonadotropin-releasing hormone (GnRH), is responsible for the hypothalamic control of LH and follicle-stimulating hormone (FSH) secretion. LH-RH is a decapeptide (Fig. 4-2) isolated from porcine hypothalami (Matsuo et al., 1971), which is capable of stimulating the release of both gonadotropins. Although the greater effect of LH-RH on the secretion of LH as compared to FSH led some to assume that a specific FSH-releasing factor would be found, increasing recognition of the role of gonadal steroids and peptides in the response of individual gonadotropins has provided an adequate explanation for the hypothalamic control of both gonadotropins on the basis of a single releasing hormone. LH-RH-containing neurons have been identified in the arcuate nucleus and the preoptic area. The precise role of each locus in the regulation of gonadotropin secretion exhibits considerable species variation and in human beings is still unclear. LH-RH has been identified by immunoassay or bioassay in portal-hypophyseal (Neill et al., 1977) and peripheral (Malacara et al., 1972) blood and levels are reported to be increased during the time of the midcycle ovulatory surge of LH.

Thyrotropin-Releasing Hormone

The first hypothalamic releasing hormone to be identified was thyrotropin-releasing hormone (TRH) (Boler et al., 1969, Burgess et al., 1969), which exhibits a stimulatory action on the secretion of thyroid-stimulating hormone (TSH). TRH is a tripeptide that is present in the highest concentration in the medial basal hypothalamus but is also present throughout the

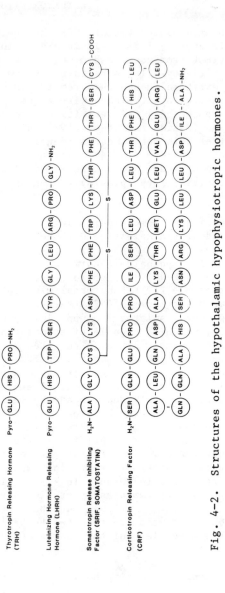

Fig. 4-2. Structures of the hypothalamic hypophysiotropic hormones.

central nervous system (CNS). Although extrahypothalamic TRH concentrations are considerably lower than those in the hypothalamus, the total weight of the brain is 100 times greater and, thus, nearly 70 percent of brain TRH is located in extrahypothalamic regions. Substances with immunological and biological similarities to TRH have been identified in retina, pineal, stomach, pancreas, and placenta. However, there is controversy as to whether these substances are truly TRH or chemically distinct substances. Using immunohistochemical stains, TRH has been identified in conjunction with TSH secretion granules in the thyrotrophs of the anterior pituitary (Childs et al., 1978). TRH-like immunoreactivity has been detected in rat plasma in concentrations of 10 to 50 pg/ml (Emerson and Utiger, 1975). TRH disappearance from plasma is extremely rapid with a half-life of three to four minutes. TRH is rapidly destroyed by heat-labile plasma enzymes, which must be inactivated before attempting its measurement in plasma. Immunoreactive TRH has also been identified in human cerebrospinal fluid (CSF) (Oliver et al., 1974) and in urine. Urinary TRH-like material, however, is chemically distinct from the hypothalamic tripeptide (Emerson et al., 1977).

Growth Hormone-Releasing Factor and Somatostatin

Growth hormone (GH) secretion is regulated by a dual control mechanism involving a stimulatory GH-releasing factor (GHRF) and an inhibitory factor (somatotropin-release inhibiting factor [SRIF], somatostatin). The preponderance of physiological studies involving hypothalamic destruction and stimulation experiments, stalk section, and injection of hypothalamic extracts indicate that the predominant neuroendocrine control of GH is stimulatory and due to GHRF. Attempts to isolate this factor have, to date, proved unsuccessful. GHRF has been identified in hypothalamic extracts from numerous species. However, peptides isolated by several laboratories and initially believed to be GHRF have been shown not to be the true releasing factor. During the isolation procedures, Brazeau et al. (1973) discovered SRIF, a GH-inhibiting factor that suggested that GH secretion is under a dual neuroendocrine control mechanism. SRIF is a tetradecapeptide that is present in highest concentrations in the median eminence and in the anterior (preoptic) region of the hypothalamus. It is also present in extrahypothalamic brain, in the spinal cord, adrenal medulla, parafollicular (C) cells of thyroid, and D cells of gastric antrum and pancreatic islets. In addition to its effect on GH secretion, SRIF inhibits the secretion of TSH, insulin, glucagon, gastrin, renin, parathyroid hormone, gastric hydrochloric

acid, and exhibits an inhibitory effect on neurotransmission.
SRIF has been detected in portal hypophyseal plasma and also in
systemic plasma, although somatostatin in peripheral circulation is believed to be due to pancreatic and gastric rather
than hypothalamic secretion. Plasma SRIF has an extremely
short half-life, one to two minutes in human beings, due to its
rapid degradation by plasma peptidases. SRIF has been identified in CSF (Patel et al., 1977) although the specific sites
within the CNS from which it originates are unknown.

Prolactin Inhibiting and Releasing Factors

The control of prolactin secretion, in contrast to that of the
other anterior pituitary hormones, is under an inhibitory CNS
mechanism. Interruption of the integrity of the hypothalamic-
pituitary axis either by stalk section or by destruction of the
medial basal hypothalamus results in an increase in prolactin
levels. The prolactin inhibitory factor (PIF) has been recognized to be under dopaminergic control. On the basis of the
presence of dopaminergic nerve terminals in the outer layer of
the median eminence (Hökfelt, 1967), the identification of
dopamine (DA) in portal hypophyseal blood in concentrations
higher than in peripheral blood (Ben-Jonathan et al., 1977),
and the demonstration of DA receptors in the anterior pituitary
(Calabro and MacLeod, 1978), it has been proposed that DA is
PIF. There is also evidence for a nondopaminergic PIF that appears to be a peptide and is located in synaptosomal fractions
of hypothalamus (Enjalbert et al, 1977). Gamma-aminobutyric
acid (GABA), also exhibits a direct inhibitory effect on prolactin secretion (Schally et al., 1977).
   The acute prolactin responses to several CNS-mediated stimuli
have not yet been explained on the basis of a change in inhibitory dopaminergic control and have been attributed to a separate prolactin-releasing factor (PRF). One candidate for PRF is
TRH, which has a stimulatory effect on prolactin comparable to
that on TSH. However, the frequent dissociation of CNS-mediated TSH and prolactin secretion, such as during nursing, stress,
or cold exposure, suggests the existence of a separate PRF.
Evidence for a PRF in hypothalamic extracts distinct from TRH
has been reported (Szabo and Frohman, 1976) although its identity remains to be determined.

Melanocyte-Stimulating Hormone Releasing and Inhibiting Factors

Hypothalamic peptides that stimulate and inhibit the release of
melanocyte-stimulating hormone (MSH) by amphibian pituitary

have been isolated and some have been structurally identified (Celis et al., 1971, Schally et al., 1973). The physiological significance of these peptides is unclear because of the recent recognition that MSH appears not to be present as a distinct hormone in the postnatal human and possibly not in several other species (Rees, 1977).

MECHANISM OF ACTION OF RELEASING FACTORS

Releasing and inhibiting factors appear to affect pituitary hormone secretion by several mechanisms. Studies performed primarily with TRH and LH-RH indicate the presence of specific, high-affinity receptors on target cells of the anterior pituitary to which they bind. There is considerable evidence that adenylate cyclase is activated and that cyclic AMP levels are increased. Whether this action occurs in parallel to that of activation of pituitary hormone secretion or is a necessary intermediate step is not yet as clear as it is in other systems, although the available data are most suggestive. The action of releasing hormones is Ca^{++}-dependent and evidence for a role of Ca^{++} as a second messenger has become convincing. Releasing hormones also appear capable of stimulating RNA and protein synthesis, leading to cellular hyperplasia and even tumor formation.

The action of the inhibitory hormones SRIF and DA are less well defined. SRIF has been reported to inhibit adenylate cyclase formation and to enhance phosphodiesterase activity, both of which mechanisms could inhibit cyclic AMP-mediated hormone release. In addition, SRIF appears to affect transmembrane Ca^{++} transport and may have other actions on exocytosis, the final process of hormone secretion involving the extrusion of secretory granules from peripheral sites in the endocrine cell. The inhibitory actions of DA are also poorly understood. DA receptors in the pituitary are recognized as being distinct from those in the brain (Rick et al., 1979) and are not coupled to an adenylate cyclase mechanism. Thus, suppression of prolactin release by DA appears to be unassociated with changes in cyclic AMP levels. Furthermore, activation of the lactotroph adenylate cyclase system results in an increase rather than in a decrease in prolactin secretion. The mechanism of DA suppression, like that of SRIF, occurs at a late stage in the secretory process, since, upon removal of the inhibitor, there is a rebound burst of hormone release (Stachura, 1976, and Leblanc et al., 1976).

The effects of most and perhaps all the releasing and inhibiting factors are modified by hormones released by target glands, such as thyroxine, cortisol, estrogens, androgens, and inhibin. The major effect of these hormones appears to involve a change in the number of receptors for releasing or inhibiting factors on the target cells, although there appear to be actions at a post-receptor level as well.

CONTROL OF NEUROSECRETION

The hypothalamic hormone-secreting cell is a specialized neuron that possesses properties of both neural and endocrine cells. It is localized within the CNS and is stimulated by the usual forms of interneural communication, that is, aminergic and peptidergic neurotransmitters. Its response, however, resembles that of an endocrine cell in that it releases preformed peptide hormones (neurosecretion) into vascular channels for transport to the pituitary. The unique function has resulted in the designation of "transducer cell" to this class of neurons (Wurtman, 1971) to indicate its hybrid nature. A discussion of the control of its secretory functions must therefore focus on the biogenic amine and peptidergic neurotransmitters. In addition, these neurons or possibly other interconnecting neurons are responsive to changing levels of steroids, thyroid hormones, and nutrients in the bloodstream and, as discussed later, peptide hormones from the pituitary and other peripheral sites.

The hypothalamic hormones are synthesized in the neural soma and transported by axoplasmic flow to specialized terminals or synaptosomes located in the outer layer of the median eminence where they are stored. The highest concentrations thus are found in the median eminence rather than in the hypothalamic loci where synthesis occurs. This differential concentration appears to distinguish them from other hypothalamic peptides that serve different functions. The biosynthetic pathways of the hypothalamic hormones remain to be clarified, although available information is most consistent with their being synthesized by ribosomal pathways involving a precursor of high molecular weight (Noe et al., 1978), than, as had been originally proposed, by nonribosomal enzymatic pathways. Hypothalamic tissue contains high concentrations of peptidases capable of destroying the hypothalamic hormones, although their role in the regulation of hormonal secretion is unclear.

The median eminence must be viewed as an extremely complex structure serving many functions. In the rat, the volume of the median eminence is about 0.4 mm^3 and in less than half of this volume there are estimated to be nearly 10^8 axon profiles and terminals with at least a dozen peptidergic and aminergic secretory granules, all in substantial concentrations (Joseph and Knigge, 1978). The fenestrated nature of the capillaries

of the median eminence allows for penetration of charged molecules normally excluded by the blood-brain barrier from the
vascular space into the intercellular space. Thus, amines and
peptides as well as many pharmacological agents can reach neurons in this brain region, and also in several other select
regions in which similar alterations of the capillary endothelial structure have been noted. Recent physiological (Oliver
et al., 1977) and anatomical (Bergland and Page, 1978) data
have provided strong evidence for the existence of retrograde
blood flow in some portal vessels from the pituitary to the
median eminence, thereby, indicating a mechanism for a short-
loop feedback of pituitary hormones to the brain. It is still
unclear as to whether such feedback occurs at the nerve terminals in the median eminence or, if not, how the peptides reach
the cell bodies that are located primarily in the regions of
the hypothalamus where the blood-brain barrier is intact.

NEUROTRANSMITTER REGULATION OF HYPOTHALAMIC HORMONE SECRETION

The hypothalamic hormone-secreting neuron can be influenced by
neurotransmitters at several sites (Fig. 4-3). These include
axo-dendritic connections (Site 1) and axo-axonic connections
involving presynaptic receptors on the hormone- containing synaptosomes (Site 3). In addition, multiple neurotransmitters
may participate in the regulation of hormone secretion through
intermediary neurons (Site 2). Neurotransmitters may also be
released directly into the portal system and modify the effect
of hypothalamic hormones on pituitary hormone secretion (Site
4), that is, DA inhibits the effect of TRH on prolactin secretion.
The availability of neuropharmacological compounds that selectively alter neurotransmitter function has led to major
advances in the understanding of the control of hypothalamic-
pituitary function (Müller et al, 1977). In particular, information from techniques that have only limited application to
human investigation has been supplemented by the use of neuropharmacological probes involving agents that can be safely
studied in man. In the following section, the synthesis, release, action, and metabolism of the biogenic amine neurotransmitters are briefly reviewed and the sites are indicated at
which neuropharmacological probes can be utilized to interfere
with neurotransmitter function. The biosynthetic pathways of
the major monoamine neurotransmitters are shown in Fig. 4-4 and
the spectrum of agents capable of modifying their effects is
summarized in Table 4-1.

Table 4-1. Pharmacological Agents Modifying Monoamine Neurotransmitters

Class	Dopamine	Norepinephrine	Serotonin	Histamine	GABA
Synthesis inhibitors	α-Methyltyrosine 3-Iodotyrosin Carbidopa Benserazide	α-Methyltyrosine 3-Iodotyrosine Carbidopa Benserazide Disulfiram Fusaric acid	p-Chlorophenyl-alanine 5-Methyl-HTP	α-Methylhistidine Hydrazines	Isoniazid Cycloserine
Precursors	DOPA	DOPA Dihydroxyphenylserine	5-HTP		
Modified precursors or false transmitters	α-Methyltyrosine α-Methyldopa	α-Methyltyrosine α-methyldopa Octopamine			
Depletors	Reserpine Guanethidine	Reserpine Guanethidine	Reserpine		
Releasers	Amphetamine	Amphetamine	Amphetamine Fenfluramine	Substance P Neurotensin	Muscimol
Uptake inhibitors	Nomifensine	Desmethylimipramine	Imipramine fluoxetine		
Degradative enzyme	Iproniazid Nialamide Pargyline	Iproniazid Nialamide Pargyline	Iproniazid Nialamide Pargyline	Imidazoles Tryptamine Iproniazid	Hydroxylamine Cycloserine Aminooxyacetic acid
Receptor agonists	Apomorphine Bromocriptine ADTN	Clonidine (α) Isoprenaline (β)	Quipazine		Muscimol Baclofen
Receptor antagonists	Pimozide Haloperidol	Phentolamine (α) Propranolol (β)	Methysergide Metergoline Cyproheptadine	Diphenhydramine (H_1,μ) Cyproheptadine (H_1) Cimetidine (H_2) Metiamide (H_2)	Picrotoxin Bicuculline

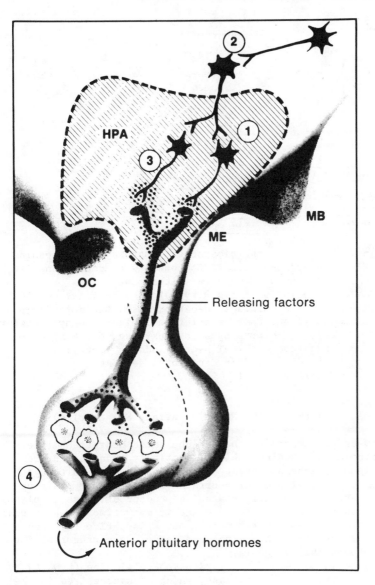

Fig. 4-3. Possible sites of neurotransmitter effects on releasing factors (see text for explanation). HPA:hy-
pophysiotropic area, OC:optic chiasm, MB:mamillary
body, and ME:median eminence. (Reprinted, by permission, from Frohman, L.A.: *N Engl J Med 286*:1391-
1397, 1972.

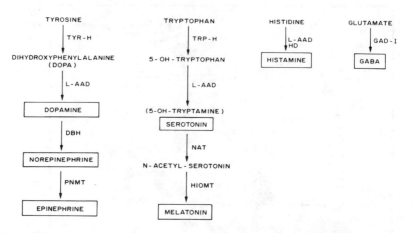

Fig. 4-4. Summary of the biosynthetic pathways of monoamine
neurotransmitters. Compounds with neurotransmitter
function are indicated within the boxes. Enzyme
abbreviations: TYR-H:tyrosine hydroxylase, L-AAD:L-
aromatic aminoacid decarboxylase, DBH:dopamine-β-hy-
droxylase, PNMT:phenylethanolamine-N-methyltransfer-
ase, TRP-H:tryptophan hydroxylase, NAT:N-acetyl-
transferase, HIOMT:hydroxyindole-O-methyl transferase, HD:histidine decarboxylase, GAD-I:glutamate
decarboxylase-I.

Catecholamines (Dopamine, Norepinephrine, Epinephrine)

The common precursor for the catecholamine neurotransmitters is
the amino acid tyrosine, which is actively transported from
blood into catecholaminergic neurons in the CNS. Although tyrosine competes with other neutral amino acids for brain uptake, tissue concentrations are independent of those in plasma.
The concentration of brain tyrosine is many times greater than
those of DA, norepinephrine (NE), and epinephrine (E), and only
a small fraction of the amino acid serves as a precursor for
catecholamine biosynthesis.
   Tyrosine is converted to dihydroxyphenylalanine (L-DOPA) by
the enzyme tyrosine hydroxylase, which is synthesized in the
neuronal cell body and transported axoplasmically to the nerve
terminal. Tyrosine hydroxylase is present in concentrations
100 to 1000 times lower than those of the other enzymes involved in catecholamine biosynthesis and represents the rate-
limiting step. It is also the enzyme most susceptible
topharmacological blockade by analogues of tyrosine, such as

α-methylparatyrosine (α-MPT). L-DOPA, which can also be transported from blood to brain, is then rapidly decarboxylated by the enzyme aromatic L-amino acid decarboxylase (L-AAD), which is not specific for L-DOPA in that it also decarboxylates other aromatic L-amino acids, such as 5-hydroxytryptophan (5-HTP), histidine, and tryptophan. This enzyme can be inhibited by L-DOPA analogues, such as α-methyldopa (Aldomet) and α-methyldopahydrazine (carbidopa). Carbidopa only partially penetrates the blood-brain barrier and, in a limited dose range, can be used selectively to block the peripheral conversion of dopa to DA while not interfering with central DA synthesis.

In dopaminergic neurons the DA thus formed is stored in secretory granules and released as a neurotransmitter, whereas in noradrenergic and adrenergic neurons it is further hydroxylated on the β-carbon by DA-β-hydroxylase (DBH) to form NE. DBH is a copper-containing enzyme localized on the membrane of the NE storage vesicles which also lacks specificity in that it hydroxylates a variety of phenylethylamines. Copper-chelating agents, such as disulfiram, are potent inhibitors of DBH and can be used to block the conversion of DA to NE. In selective pathways in which E has been shown to be involved as the neurotransmitter, the enzyme phenylethanolamine-N-methyl transferase (PNMT) methylates NE to E, using S-adenosyl-methionine as the methyl donor. The biosynthetic pathways can be entered by competitive analogues such as α-MPT and carbidopa, which are converted to analogues of the neurotransmitter, that is, α-methyl-DA and α-methyl-NE and which act as less effective or ineffective neurotransmitters (false neurotransmitters) on postsynaptic receptors. False neurotransmitters also exhibit activity on feedback inhibition of biosynthetic pathways and on inhibition of neurotransmitter reuptake and storage. The normal biosynthetic pathways can also be circumvented by the use of such agents as dihydroxyphenylserine (DOPS), which crosses the blood-brain barrier and is decarboxylated by L-AAD to form NE. Thus, DA synthesis can be blocked by inhibiting tyrosine hydroxylase and NE levels preserved by administering DOPS in order to differentiate between the two catecholamines in a particular control mechanism.

In the nerve endings, newly synthesized DA and NE are stored in secretory granules that protect these neurotransmitters from enzymatic degradation. About 20 percent of the neurotransmitter is bound to adenosine triphosphate (ATP), and there appear to be at least two distinct pools of neurotransmitters that are differentially susceptible to releasing stimuli. A classic example of a drug that interferes with catecholamine storage is reserpine, which causes a slow but constant release (leak) of the neurotransmitter from the granule and inhibits the

Mg^{++}-ATP-dependent granule uptake, resulting in a long-lasting depletion of catecholamines, and also of serotonin (5-HT).

Catecholamines are released from nerve terminals in response to nerve stimulation and membrane depolarization primarily by the vesicle membrane fusing with the cell membrane and extruding the neurotransmitter directly into the intercellular space in a Ca^{++}-dependent process. Such releasers as tyramine and amphetamine cause release by a different mechanism involving release of free cytoplasmic catecholamine. Once released into the synaptic cleft, catecholamines bind to postsynaptic receptors that appear to be similar in the hypophysiotropic neurons to those that have been demonstrated in other neural sites.

Alterations in activity of the postsynaptic receptor are readily accomplished by the use of receptor agonists and antagonists. The specificity of most of these agents is not absolute and caution must be exercised in drawing conclusions from individual experiments.

The termination of catecholamine action occurs by reuptake of the neurotransmitter into the presynaptic neuron or extraneural sites, removal into circulation, or metabolic degradation. The first mechanism is by far the most important in terminating the action of the catecholamines at the postsynaptic receptor. Once recaptured, the catecholamine is either reincorporated into the storage vesicle or is degraded. Enhancement of catecholamine effects is accomplished by drugs, such as cocaine, tricyclic antidepressants, or nomifensine, that inhibit their reuptake by presynaptic terminals.

Metabolic degradation of catecholamines is initiated by one of two enzymes: monoamine oxidase (MAO) deamination or catechol-O-methyl transferase (COMT) methylation. Intraneuronally, MAO is the principal degrading enzyme, whereas COMT is a glial and ependymal enzyme. MAO is not a single enzyme but rather a family of enzymes that have different substrate and tissue specificities. Therefore specific MAO inhibitors (MAOI's), although not currently available, will be of great interest in delineating neurotransmitter function. MAOI's commonly used include pargyline and tranylcypromine. There are presently no COMT inhibitors that are both effective and nontoxic *in vivo*.

Indoleamines (Serotonin)

The precursor of 5-HT is the amino acid tryptophan, which is actively transported into brain from blood by the same system used for tyrosine. Although tryptophan hydroxylase is the rate-limiting step in 5-HT biosynthesis, the enzyme activity is not saturated at physiological tryptophan levels. Consequent-

ly, fluctuations in plasma tryptophan concentrations and those
of other amino acids competing for the same transport system
result in corresponding changes in brain tryptophan and 5-HT.
Tryptophan is the only essential amino acid to be partially
bound by plasma albumin and it is the concentration of free
rather than total tryptophan that determines brain uptake.
Certain hormones and drugs that bind to albumin displace tryptophan from its binding sites and increase brain uptake and
5-HT levels. Brain tryptophan concentrations are also many
times greater than that of 5-HT and only a small portion of the
brain pool serves as a 5-HT precursor. 5-HTP, once formed by
tryptophan hydroxylase or transported into brain after systemic
administration, is rapidly converted to 5-hydroxytryptamine
(5-HT) by L-AAD. 5-HT functions as a neurotransmitter and in
the pineal body also serves as the precursor of melatonin by a
two-step process involving methylation and acetylation. Melatonin also appears to function as a neurotransmitter with respect to releasing factor secretion. The biosynthesis of 5-HT
can be pharmacologically inhibited by p-chlorophenylalanine
(p-CPA) which inhibits tryptophan hydroxylase (as well as exhibiting other effects on catecholamine turnover) and by administration of L-DOPA, which competes with 5-HTP for the enzyme
L-AAD.

The storage, release, and uptake of 5-HT are, in general,
similar to that of NE. Many of the same agents that release NE
also release 5-HT (amphetamine), whereas others appear relatively specific for 5-HT (fenfluramine). Tricyclic antidepressants also inhibit 5-HT uptake, although in contrast to NE,
imipramine and amitriptyline are more potent than their des-
methyl derivatives (desmethyl-imipramine and nortriptyline).
Fluoxetine appears to be a relatively specific inhibitor of
5-HT uptake. Alteration of 5-HT receptor activity can be produced by the use of 5-HT agonists (quipazine and lysergic acid
diethylamide [LSD]) and antagonists (methysergide and cyproheptadine). The latter compounds exhibit other actions on neurotransmitter function and thus have only relative specificity.

5-HT effects at the synaptic cleft are terminated by presynaptic reuptake and by metabolic degradation via MAO, which in
conjunction with aldehyde dehydrogenase results in the formation of 5-hydroxyindoleacetic acid (5-HIAA). Drugs that interfere with MAO activity also enhance 5-HT effects.

Acetylcholine

Acetylcholine (ACh) is synthesized from choline and acetyl-CoA.
Choline is taken up from CNS interstitial fluid although its
origin is unclear, since it does not penetrate the blood-brain

barrier to any extent. There is evidence suggesting that phosphatidyl choline (lecithin) serves as a precursor and, after
crossing the blood-brain barrier, is partially degraded to
choline. After uptake by a combination of a saturable carrier
and passive diffusion, choline is converted to ACh by choline
acetyltransferase. Hemicholinium, a choline analogue, inhibits
choline uptake and decreases ACh synthesis. The mechanism of
ACh release is less clear than that of the monoamines and the
question of vesicle fusion with cell membrane remains unset-
tled.

There are two types of ACh receptors, muscarinic and nicotinic, which have distinct anatomical distribution and different
physiological function. Examples of muscarinic agonists and
antagonists that cross the blood-brain barrier and exhibit CNS
actions are arecholine and atropine, respectively. Oxtremorine
exhibits nicotin-like action in the CNS; nicotinic receptor
blockers enter the brain to a very limited extent after systemic administration and have not been useful in clinical studies.

γ-Aminobutyric Acid

GABA has been found in the mammalian hypothalamus where it
functions as an inhibitory neurotransmitter. It is formed by
the decarboxylation of L-glutamate by glutamate decarboxylase
and is metabolized by transamination; both steps appear to
involve a pyridoxal-dependent enzyme. Although there are numerous agents that can inhibit both synthesis and degradation of
GABA, none are specific and none are useful clinically in the
study of the neuroendocrine effects of this compound. Bicucul-
line acts as a specific antagonist of GABA receptors, whereas
muscimol, a semirigid analogue of GABA that crosses the blood-
brain barrier, exhibits GABA-like activity.

Histamine

Histamine is synthesized within the CNS from histidine by decarboxylation via a specific histidine decarboxylase and the
nonspecific L-AAD. Both its synthesis and its enzymatic degradation can be inhibited, but not by clinically useful agents.
Alteration of histamine neurotransmission occurs primarily
through the use of histamine receptor antagonists. There are
two classes of histamine receptors: H_1 and H_2. The H_1 receptor
is stimulated by 2-methylhistamine and is inhibited by diphen-
hydramine, cyproheptadine, and promethazine, and the H_2 receptor is stimulated by 4-methylhistamine and pentagastrin and is
inhibited by cimetidine and metiamide.

Neuropeptides

The list of hypothalamic neuropeptides is currently increasing at a rapid rate. Documentation of the role of each of these peptides in the control of neuroendocrine function has just begun to occur and the evidence for their participation is currently limited to neurotensin, substance P, bombesin, vasoactive intestinal polypeptide (VIP), and enkephalin. With the exception of analogues of enkephalin, which are capable of crossing the blood-brain barrier (Von Graffenried et al., 1978), the neuroendocrine effects of the other peptides have been demonstrated only *in vitro* or after direct injection into the CNS (Frohman, 1980). Specific binding of several of the peptides to membrane fractions of brain homogenates has been reported (see Nemeroff et al., chapter 6), although the specific mechanism of altering neuroendocrine activity is for the most part unknown. Because of the availability of enkephalin antagonists, the effects of this peptide have received greater attention. There is little information concerning biosynthesis, storage, secretion, and metabolism of the hypothalamic neuropeptides. Hypothalamic peptidases capable of degrading these peptides have been demonstrated, although their specificity and thus their physiological role in the regulation of peptide function remains to be determined.

CONTROL OF SECRETION OF HYPOTHALAMIC RELEASING AND INHIBITING FACTORS

A number of increasingly sophisticated methods are available for the study of hypothalamic peptide release. *In vitro* techniques using incubation of intact medial basal hypothalamus, hypothalamic fragments, or purified hypothalamic synaptosomes have provided considerable evidence concerning the control of CRF, TRH, LH-RH, and SRIF release. *In vitro* findings have been complimented by *in vivo* studies measuring hypophyseal portal venous content of peptides and monoamines. Sensitive and specific immunoassays now exist for TRH, LH-RH, CRF, and SRIF and the release of these peptides can be accurately assessed within the limits of assay validity. GHRF and other factors have not as yet been fully characterized, although improvements in bioassay techniques now permit the measurement of these factors primarily through their effects on pituitary hormone release. This section discusses the control of releasing and inhibiting factors based on their direct measurement, whereas interpretation of those studies based only on pituitary hormone measurements is provided in the following section. A summary of the established neurotransmitter effects on hypothalamic hormone secretion is provided in Table 4-2.

Table 4-2. Effects of Neurotransmitters on Hypothalamic Hormone Secretion

	Norepinephrine[a]	Dopamine	Serotonin	Other
CRF	↓	↓,−	↑	ACh ↑ (nicotinic) GABA ↓ Melatonin ↓
LH–RH	−	↑	↓	
TRH	↑,−	↑	↓	SRIF ↓,−
SRIF	↑,−	↑	−	ACh ↑,− GABA ↑,− Neurotensin

[a] ↑ = stimulates, ↓ = inhibits, − = no effect.

Corticotropin-Releasing Factor

Neurotransmitter Regulation

ACh has been shown to stimulate CRF release *in vivo* following injection into the third ventricle (Krieger, 1975) as well as release *in vitro* from incubated hypothalami (at physiological concentrations) (Jones et al., 1976a) and hypothalamic synaptosomes (at higher doses) (Bennett and Edwardson, 1975). The release of CRF can be blocked partly by atropine *in vivo* (Hedge and Smelik, 1968) and *in vitro* (Jones et al., 1976a) and completely by hexamethonium (*in vitro*), suggesting a predominantly nicotinic receptor (Jones et al., 1976a). The neural pathways mediating stress-induced CRF release are primarily cholinergic.

5-HT stimulates ACTH release when injected centrally but not peripherally. Incubation of hypothalami *in vitro* in the presence of 5-HT results in CRF release that can be blocked by methysergide (Jones et al, 1976a,b). The response appears to be more complicated, however, since cholinergic blockade also inhibits CRF stimulation by 5-HT. The suggestion has been made that mediation is through a cholinergic interneuron with serotonergic receptors (Burden et al., 1974; Hillhouse et al., 1975).

NE inhibits the hypothalamic CRF response to ACh, an effect that is antagonized by phentolamine (Hillhouse et al., 1975;

Jones et al., 1976a), and DA has been shown in some but not all studies to be inhibitory (Edwardson and Bennett, 1974). GABA is more potent than NE in inhibiting CRF release in response to ACh or 5-HT and the effect is abolished by picrotoxin, a GABA receptor antagonist (Jones et al., 1976a,b). Melatonin inhibits CRF release (Jones et al., 1976a,b) and may be important in the circadian pattern of CRF release, which is entrained to the light-dark cycle.

Hypothalamic-Pituitary-Adrenocortical Axis

Corticosteroids inhibit ACTH secretion by two mechanisms: a concentration dependent, rapid feedback (minutes), and a delayed feedback (hours), which is concentration independent (Dallman and Yates, 1969). Studies using incubated hypothalami (Edwardson and Bennett, 1974) and anterior pituitary cells (Sayers and Portanova, 1974) have shown that the rapid feedback acts at both levels to inhibit CRF and ACTH release. Corticosteroids do not appear to inhibit CRF release via a neural mechanism but may stabilize the CRF cell membrane (Jones et al., 1976c). The delayed feedback also occurs at pituitary and hypothalamic levels (Jones et al., 1976c), but at the hypothalamus the effect is probably due to a corticosterone effect on CRF synthesis (Arimura et al., 1969).

ACTH inhibits CRF release in response to 5-HT by rat hypothalami *in vitro*, suggesting a short-loop feedback system (Jones et al., 1976b). The concentrations effective *in vitro* are similar to those occurring *in vivo* during stress, providing evidence for its physiological importance. (For a detailed review, see Rees and Gray, chapter 14.)

Luteinizing Hormone-Releasing Hormone

Neurotransmitter Regulation

The intraventricular injection of DA stimulates LH release. *In vitro* DA stimulates LH release only in the presence of co-incubated hypothalamic tissue (McCann, 1974). This effect may be explained by the stimulatory effect of DA on LH-RH release from hypothalamic synaptosomes (Bennett et al., 1975). The effect of NE on LH-RH release is controversial. Some studies have demonstrated a release of LH, whereas others have not, and NE has been reported not to stimulate LH-RH release *in vitro* from incubated synaptosomes (Bennett et al., 1975).

5-HT appears to exert an inhibitory effect on LH-RH release, since central administration results in suppression of both cyclic and tonic release of gonadotropins in rats (Kamberi, 1973).

Hypothalamic-Pituitary-Gonadal Axis

The integration of information from the ovary, pituitary, and hypothalamus are required to provide a coordinated mid-cycle LH and FSH surge. Although elevated LH-RH levels in hypophyseal-portal blood have been demonstrated at the time of the mid-cycle surge (Neill et al., 1977), the stimulus for this secretion is not clear. It may depend on progesterone positive feedback, since anti-LH-RH serum abolishes progesterone-induced gonadotropin release (Yen, 1977). Prostaglandins, neurotransmitters, or catechol-estrogens may, however, be involved in the feedback (Yen, 1977). Both a short-loop feedback inhibition of LH-RH release by LH and FSH and an ultrashort-loop feedback inhibition by LH-RH on its own release have been suggested experimentally.

Thyrotropin-Releasing Hormone

Neurotransmitter and Peptide Regulation

A stimulatory role for NE is suggested by *in vivo* observations that reserpine and noradrenergic antagonists cause TSH inhibition in the rat (Kotani et al., 1973). The stimulation of TRH release by NE from hypothalamic organ cultures *in vitro* (Hirooka et al., 1978) tends to support this role although other workers using incubated hypothalami have shown no effect of NE on TRH release (Maeda and Frohman, 1980). DA stimulates TRH release from hypothalamic synaptosomes (Bennett et al., 1975) and from incubated hypothalami (Maeda and Frohman, 1980). Similarly, bromocriptine stimulates TRH release (Maeda and Frohman, 1980).

5-HT inhibits the release of TRH *in vitro* from hypothalamic synaptosomes (Bennett et al., 1975) supporting *in vivo* data showing TSH inhibition. The release of TRH by rat hypothalami *in vitro* is also inhibited by SRIF (Hirooka et al., 1978).

Hypothalamic-Pituitary-Thyroid Axis

Acute cold exposure in rats produces a rapid burst of TSH release that results from activation of peripheral or core receptors (Martin and Reichlin, 1970). This elevation in TSH is blocked by passive immunization with anti-TRH serum, suggesting that it is secondary to TRH secretion (Szabo and Frohman, 1976).

Chronic cold exposure, in contrast, has little, if any, effect on TSH or thyroid hormone levels in the rat, although secretion and metabolic clearance rates are increased (Fortier et al., 1970). In such animals increased plasma TRH bioactivity (Redding and Schally, 1969) and decreased hypothalamic TRH content (Jackson and Reichlin, 1973) have been reported, suggesting increased TRH turnover.

In addition to the inhibitory effect of thyroid hormone on pituitary TSH secretion, feedback control may also be exerted at the hypothalamic level. Although central administration of thyroid hormone appears to inhibit TSH secretion (Belchetz et al., 1978), no consistent effect has been shown on TRH content or turnover and the TSH inhibition may be explained by increased SRIF release (Berelowitz et al., 1980).

Growth Hormone (Somatotropin)
Release-Inhibiting Factor (Somatostatin)

Neurotransmitter and Peptide Regulation

DA stimulates the release of SRIF from incubated rat hypothalami (Maeda and Frohman, 1980) and hypothalamic synaptosomes *in vitro* (Bennett et al., 1979) and into hypophyseal portal venous blood *in vivo* (Chihara et al., 1979). These observations explain the inhibitory effect of DA on GH secretion in this species, which contrasts with the stimulatory effect seen in human beings. Other neurotransmitter stimuli have resulted in contradictory findings, with some studies showing no effects and others showing stimulation of SRIF release by NE and ACh (Chihara et al., 1979) and by GABA, its precursors, and antagonists (Takahara et al., 1980).

Both neurotensin and substance P stimulate SRIF release from hypothalami *in vitro* (Sheppard et al., 1979), providing an explanation for the inhibition of GH secretion seen following central administration of either peptide. Neurotensin and bombesin, although not substance P, have been reported to stimulate somatostatin release by the hypothalamus *in vivo* (Abe et al., 1980).

Hypothalamic-Pituitary Axis

Hypothalamic-Somatotroph Axis

Passive immunization of animals with anti-SRIF serum results in an elevation in basal and stimulated GH levels suggesting that

SRIF has a physiological inhibitory effect on GH secretion
(Arimura et al., 1976). This is supported by evidence that GH
produces a dose-related stimulation of hypothalamic SRIF release in vitro, that GH-treated rats have increased hypothalamic SRIF content and release, and, conversely, that rats
treated with anti-serum to GH have decreased hypothalamic SRIF
content and release (Berelowitz et al., 1981). A possible
long-loop feedback effect of somatomedins on hypothalamic SRIF
has also recently been shown (Berelowitz et al., 1981b).

Hypothalamic—Thyrotroph Axis

Passive immunization with anti-serum to SRIF suggests a tonic
inhibitory role for SRIF in the control of TSH (Arimura and
Schally, 1976). Further support is provided by the observation
that thyroidectomized rats with elevated TSH levels exhibit
decreased hypothalamic SRIF content and release, which can be
restored to normal by triiodothyronine (T_3) replacement. T_3
although not TSH or TRH, stimulates SRIF release from normal
hypothalami in vitro, suggesting the nature of the feedback
loop (Berelowitz et al., 1980).

Growth Hormone-Releasing Factor

The control of GHRF secretion remains for the most part infer-
ential because of the lack of specific assays for this uncharacterized factor. Preliminary studies in which both SRIF and
GH measurements have been made, or performed in the presence of
SRIF anti-serum, suggest that GHRF is under a stimulatory dopaminergic control and that β-endorphin (Dupont et al., 1977),
met-enkephalin, and vasoactive intestinal polypeptide (VIP) enhance its release (Wakabayashi et al., 1980).

Prolactin-Inhibiting Factor and Prolactin-Releasing Factor

It has not yet been possible to distinguish between CNS influences on PIF and PRF. DA serves as a PIF and may stimulate
rele°ase of a separate peptide that also inhibits prolactin
secretion. 5-HT stimulation of prolactin release is believed
to occur via a PRF, although the evidence is entirely indirect,
for example, stress which is 5-HT mediated, elevates prolactin
values in the rat even in the presence of DA receptor blockade
(Valverde et al., 1973). Numerous studies have now shown a
stimulatory effect of prolactin on hypothalamic DA turnover
(Gudelsky and Moore, 1977).

CONTROL OF SECRETION OF PITUITARY HORMONES

A summary of neurotransmitter effects on the secretion of anterior pituitary hormones is provided in Table 4-3.

Table 4-3. Effects of Neurotransmitters on Anterior Pituitary
Hormone Secretion

	Norepi- nephrine[a]	Dopa- mine	Sero- tonin	Acetyl- choline	Hista- mine	Gaba	Other
ACHT	α ↑	-	↑	-	-	-	-
LH & FSH	(↑)	↓	-	-	-	-	-
TSH	↑	↓	↑	-	-	-	Neurotensin ↓
GH	α ↑ β ↓	↑	↑	↑	-	(↑)	Neurotensin ↓ Substance P ↓ Enkephalins ↑
Prolactin	-	↓	↑	H_1 H_2	↑		Neurotensin ↓ Enkephalins ↑

a ↑ = stimulates; ↓ = inhibits; - = no effect or insufficient data, () = conflicting data exists. All effects reflect CNS rather than peripheral sites of action, with the exception of dopamine. The data are derived, wherever possible, from studies in human beings.

Adrenocorticotropin

Physiological Control

Pituitary ACTH secretion is subject to a dual control system somewhat similar to that of TSH. The first level of regulation involves hypothalamic CRF through which a number of "open loop" neurogenic stimuli are regulated, such as circadian rhythm, pyrogens, pain, and anxiety. The second is a feedback control exerted by circulating corticosteroids both at hypothalamic and pituitary levels.

A diurnal pattern exists for plasma ACTH with the lowest levels occurring between 6 P.M. and 11 P.M. followed by a rise in levels in the early morning hours which peaks between 6 A.M. and 8 A.M. This pattern, in contrast to that of GH, is not sleep entrained.

Superimposed upon all other regulators of ACTH secretion is
the stimulatory influence of stress. Among the stressors known
to induce ACTH secretion are severe trauma, pyrogens, glucocy-
topenia, and anxiety.

Pharmacological Control

NE lowers ACTH levels during systemic infusion but crosses the
blood-brain barrier poorly and thus may act indirectly or at
circumventricular sites. ACTH stimulation occurs in response
to amphetamine and methylamphetamine (Rees et al., 1970) and is
completely blocked by thymoxamine, an α-adrenergic blocker
(Besser et al., 1969). ACTH elevation also occurs in response
to methoxamine, an α-adrenergic stimulant, in normal subjects
and can be inhibited by α- but not β-adrenergic blockade (Nakai
et al., 1973a). The involvement of α-adrenergic stimulation in
the ACTH response to insulin hypoglycemia is suggested by its
inhibition by phentolamine and enhancement by propranolol
(Imura et al., 1973a). In normal subjects, ACTH secretion does
not appear to be under dopaminergic control.

In human beings, support for a role of 5-HT in ACTH control
is provided by evidence that 5-HT agonists, such as 5-HTP,
stimulate ACTH (Imura et al., 1973b) and that 5-HT receptor
blockers suppress the metyrapone-induced cortisol response
(Cavagnini et al., 1975) and the inhibition of the ACTH response to insulin hypoglycemia (Plonk et al., 1974). Animal
studies suggest that 5-HT may be implicated in the circadian
ACTH rhythm (Krieger, 1975). Cyproheptadine inhibits corticosteroid rhythmicity in cats and blockade of 5-HT synthesis
with p-CPA has the same effect in rats (Van Delft et al., 1973)

There is considerable interaction between the feedback influence of circulating corticosteroids and neurotransmitter-medi-
ated impulses on CRF-secreting neurons. For example, diphenyl-
hydantoin administration, which decreases the CNS sensitivity
to steroid feedback, causes enhanced pulsatile ACTH secretion
and decreases the ACTH response to metyrapone without affecting
the responses to stress and vasopressin (Frohman, 1979).

Luteinizing Hormone and Follicle-Stimulating Hormone

Physiological Regulation

LH and FSH are present in plasma from birth. During the early
stages of puberty, plasma FSH levels increase to a greater degree than do LH levels. In the early prepubertal period (4 to
8 years) there is no evidence of pulsatile LH or FSH secretion.

LH-RH testing reveals that girls release more FSH than boys, a feature persisting throughout the prepubertal period. The hypothalamic-pituitary axis at this time is believed to be increasingly sensitive to the suppressive effect of gonadal steroids.

The late prepubertal period (7 to 9 years) is characterized by the occurrence of pulsatile LH secretion during sleep. In girls LH and FSH pulses are synchronous more frequently than in boys. At this point secretion of testosterone in boys and estradiol in girls initiates the clinical characteristics of puberty. Development of positive LH and FSH feedback to gonadal steroids in females permits the cyclic preovulatory gonadotropin surge resulting in ovulation and cyclicity becoming established by 16 to 18 years. In adult men the secretory pattern of LH is characterized by 8 to 10 secretory spikes that occur randomly through the day with no relationship to the sleep-wake cycle. Similar pulsatile secretion of LH and FSH occurs in mature women with the frequency and magnitude of pulses varying according to the phase of the menstrual cycle. With the disappearance of ovarian follicles, secretion of the major ovarian hormones stops at the menopause ($\pm$50 years). The loss of negative feedback of these hormones greatly increases secretion of both FSH and LH with a greater rise in FSH than in LH. A similar rise in FSH and LH in men is observed in the seventh and eighth decade in association with decreasing testicular function.

LH secretion during the menstrual cycle shows little variation apart from a pronounced mid-cycle peak, and a luteal nadir followed by a rise during the last days of the cycle. Blind females tend to have an earlier menarche than girls with normal vision, a finding opposite to that seen in animals, such as rats, in which blinding causes a delay in sexual maturation (Swerdloff, 1978).

Surgical stress in human males results in a transitory rise followed by fall in LH and a sustained fall in plasma testosterone. No such changes have been observed in females, although hypothalamic anovulation is frequently seen in periods of stress. Pheromones or other environmental factors may explain the synchronization of menstrual cycles seen in human females housed in dormitories (McClintock, 1971). Such airborne factors have been implicated as a cause for the synchrony of the estrous cycle, which occurs in mice housed together (Whitten, 1959).

Two major mechanisms of action of steroids regulate LH and FSH secretion. Gonadal steroids regulate the basal secretion of gonadotropins by a negative feedback mechanism; thus, decreased gonadal function leads to an increase in gonadotropin secretion. In animals LH and FSH levels after gonadectomy rise faster in the male, suggesting differences in the negative feedback effects of estrogens and androgens (McCann, 1974). A decline in sensitivity of the negative feedback mechanism occurs with time in rats, suggesting that continuous steroid presence is required for its integrity (Negro-Vilar et al., 1973). FSH release appears to be regulated by factors in addition to steroid feedback. Inhibin, a partially characterized peptide produced by the germinal epithelium, has been shown to exhibit a selective inhibition of FSH secretion (Franchimont et al., 1975).

Cyclic release of LH and FSH is triggered by the positive feedback action of steroids secreted by ovarian follicles during the final phase of rapid growth prior to ovulation. The preovulatory surge of LH is preceded by increased circulating estrogen levels in the presence of low or declining progesterone. The major site of the positive feedback appears to be hypothalamic, although both estrogen and progesterone have effects on pituitary gonadotropin secretion consisting of an initial inhibition followed by a facilitation of the response to LH-RH (Sharp and Fraser, 1978).

Gonadotropins may exert an influence on their own secretion by acting directly on the pituitary or on the hypothalamus to regulate release of LH-RH. Data to date are contradictory.

Preovulatory peaks of ACTH and cortisol secretion during the human menstrual cycle suggest a role for the adrenal in the induction of ovulation (Genazzani et al., 1975). However, normal cyclic ovulation still occurs in women with Addison's disease (adrenocortical insufficiency).

Pharmacological Control

Elegant histochemical and immunohistochemical studies in rats have demonstrated an aggregation of DA- and LH-RH-containing nerve terminals in the median eminence in proximity to NE-containing terminals, providing morphological evidence for a potential interaction (Fuxe et al., 1976). Considerable investigation into the role of catecholamines in the control of LH and FSH in this species may, however, have little relevance in man and other primates because of differences in the hypothalamic loci of control.

In man, DA infusions cause inhibition of basal LH levels by ±20 percent followed by a postinhibitory rebound on cessation of the infusion (LeBlanc et al., 1976). Similar effects have been seen after oral L-DOPA administration (in some but not all studies) and after bromocriptine (Lachelin et al. 1977). DA antagonists have also provided conflicting data: chlorpromazine has no effect on pulsatile LH secretion (Santen and Barden,

1973), whereas pimozide reduces the mid-cycle LH surge (Leppäluoto et al., 1976).

In animals NE generally stimulates LH and FSH secretion. In primates phentolamine reduces pulsatile LH secretion after ovariectomy, although neither α- nor β-adrenergic blockade affect the cyclic gonadotropin surge (Bhattacharya et al., 1972).

5-HT appears to inhibit basal and cyclic gonadotropin release in rats (McCann and Moss, 1975) in the majority of reports but there is little evidence for its role in human beings. Melatonin also inhibits LH and FSH release in rats and, by inference, may contribute to the altered gonadotropin secretion seen with pineal tumors (Motta et al. 1967).

Thyrotropin

Physiological Control

The major controlling factor in the regulation of TSH secretion is the negative feedback exerted primarily at the level of the pituitary by alterations in circulating levels of thyroid hormone. Cold exposure in lower mammals causes release of TSH (Krulich et al., 1977) but in human beings this is seen only in neonates after parturition and in young infants subjected to hypothermia (Wilbur and Baum, 1970). Cold exposure in adults is without effect on TSH secretion (Hershman et al., 1970).

Levels of TSH rise to presleep maxima and then decrease over the course of uninterrupted sleep (Parker et al., 1976).

Hypothyroidism due to thyroid disease is associated with markedly elevated TSH secretion. This increase is mediated in part by a loss of thyroid hormone inhibition at the pituitary level and in part by a decrease in hypothalamic SRIF release, permitting unopposed TRH activity. Conversely, hyperthyroidism or thyroid hormone administration causes a decrease in TSH secretion (Berelowitz et al., 1980).

Pharmacological Control

In rats the TSH response to cold is blocked by drugs inhibiting adrenergic neurotransmission, including diethyldithiocarbamate (a DBH inhibitor), reserpine, and phentolamine (Krulich et al., 1977; Tuomisto et al., 1973). In hypothyroid patients, fusaric acid decreases the elevated TSH levels (Yoshimura et al., 1977). In normal subjects phentolamine supresses the TSH response to TRH, raising the possibility of noradrenergic effects at the level of the pituitary (Nilsson et al., 1974). The adrenergic role in TSH secretion is supported by the stimulation of TSH by clonidine in rats (Annunziato et al., 1977).

L-DOPA administration has little effect on TSH secretion in normal subjects (Eddy et al., 1971) but inhibits the release of TSH after TRH (Spaulding et al., 1972) and inhibits the elevated TSH levels in primary hypothyroidism (Rapoport et al., 1973). These effects suggest a direct action of L-DOPA (or, more likely, DA) on the pituitary or could be explained by the release of a thyrotropin inhibitor, such as SRIF, which has been shown in the rat to be released by DA (Maeda and Frohman, 1980). Bromocriptine similarly decreases the elevated TSH seen in hypothyroidism (Miyai et al., 1974).

5-HT has an uncertain role in the regulation of TSH. 5-HTP (Yoshimura et al., 1973) and the 5-HT agonist metergoline (Delitala et al., 1976) have been reported to decrease plasma TSH levels in hypothyroid subjects, whereas increased plasma TSH levels have been seen after similar stimuli in experimental animals (Chen and Meites, 1975).

Neurotensin stimulates TSH secretion in the rat after systemic injection but inhibits TSH after injection into the CNS (Maeda and Frohman, 1978). Opioid peptides and their derivitives have not, to date, been shown to exert any effects on TSH secretion.

Growth Hormone

Physiological Control

A wide variety of physiological stimuli affect the secretion of GH. A diurnal pattern of GH is present under nonstressful conditions, with a stable basal plasma GH level occurring for most of the day interspersed with one or two surges three to four hours after meals and occasional sharp spikes unrelated to metabolic signals. The most consistent burst of GH secretion occurs about one hour after the onset of sleep and is correlated with the onset of sleep stages III and IV. Smaller peaks of plasma GH may occur later during the sleep period (Martin, 1973).

Plasma GH levels change with increasing age, as does the secretory pattern. High levels of GH are seen in the first days of life and decrease by 2 weeks of age. In pubertal children, although the basal plasma GH is not markedly different from adult levels, more GH peaks occur during the day. Following this period of active skeletal growth, GH levels and responsivity remain unchanged although sleep-associated GH secretion is decreased after the sixth decade (Plotnick et al., 1975).

Nutrients profoundly affect GH levels. Amino acids (of which arginine is the most potent) stimulate GH secretion (Knopf et

al., 1965), as do decreasing levels of free fatty acids
(Fineberg et al., 1972), whereas glucose causes inhibition
(Roth et al., 1963a). The stimulation of GH by insulin-induced
hypoglycemia can be blocked by the simultaneous administration
of glucose, indicating the importance of glucose metabolism in
the control of GH. These effects of amino acids and glucose
have been used clinically to study the integrity of pituitary
GH secretory control.

Exercise also stimulates GH secretion (Buckler, 1972), as do
anxiety and emotional or physical stress in man (Brown and
Reichlin, 1972). In rats, stress causes a profound reduction
in serum GH levels. The responses are greater in women than in
men, a difference attributed to estrogens, and one that underscores the modifying effects of other hormones on the control
of GH. Estrogen administration results in an increase and corticosteroid (Merimee and Rabin, 1973) and thyroid hormone
(Young, 1973) excess cause a decrease in GH responsivity.

Pharmacological Control

Although administration of E or NE alone has no effect on GH
release (Roth et al., 1963b), the importance of adrenergic
mechanisms in GH control is well established. The addition of
β-adrenergic receptor blockade to a catecholamine infusion results in increased plasma GH levels (Blackard and Hubbell,
1970) that can be inhibited by addition of phentolamine, indicating that the α-adrenergic receptor is stimulatory for GH secretion (Cavagnini and Peracchi, 1971). This is further supported by the stimulation of GH by clonidine (Lal et al.,
1975). Adrenergic receptors appear to modulate a number of
stimuli of GH secretion. α-Receptor blockade inhibits the GH
responses to insulin, arginine, vasopressin, and exercise,
whereas β-receptor blockade enhances the GH responses to insulin, glucagon, amphetamine, and E (Frohman, 1979).

DA receptors also appear to modulate a number of stimuli of
GH secretion in the human. L-DOPA stimulates GH release and
this effect is not prevented by simultaneous pharmacological
inhibition of DBH (Hidaka et al., 1973). Further support is
provided by the stimulation of GH by subemetic doses of apomorphine (Lal et al., 1973) and bromocriptine (Camanni et al.,
1975).

There is increasing evidence for a role of 5-HT in the control of GH. Although tryptophan administration has minimal
(Muller et al., 1974) or no (MacIndoe and Turkington, 1973)
stimulatory effects on GH secretion, 5-HTP stimulates GH

secretion (Imura et al., 1973b). 5-HT receptor blockers, such as cyproheptadine and methysergide, inhibit the GH responses to arginine (del Pozo and Lancranjan, 1978), insulin (Bivens et al., 1973), 5-HTP (Nakai et al., 1974), exercise (Smythe and Lazarus, 1974a), and sleep (Chihara et al., 1976). Cyproheptadine also inhibits L-DOPA-stimulated GH secretion, suggesting that it is partially 5-HT mediated (Smythe et al., 1976). Metergoline, a more recently developed specific antiserotonergic agent, appears to have no effect on basal GH or the responses to insulin or L-DOPA and has been reported to enhance rather than inhibit the response to arginine.

Melatonin has been reported to stimulate basal GH secretion (Smythe and Lazarus, 1974b) and to inhibit GH release after insulin and L-DOPA (Smythe and Lazarus, 1974a). The role of this pineal indoleamine thus awaits clarification.

Little evidence exists for a role of histamine in the control of GH in man.

Systemically administered GABA, γ-amino-β-hydroxybutyric acid (GABOB) (Fioretti et al., 1978), and muscimol, a GABA agonist (Tamminga et al., 1978), acutely increase plasma GH. However baclofen, a GABA derivative, blunts the GH response to insulin (Invitti et al., 1976), leaving uncertain the nature of GABA's role in GH secretion.

The cholinergic system appears to play a facilitatory role in sleep-related and insulin-induced GH secretion, since methscopolamine, a primarily antimuscarinic anticholinergic agent, partly inhibits both responses (Mendelson et al., 1978).

TRH has been shown to have little effect on GH secretion in normal subjects but inhibits L-DOPA- and arginine-stimulated GH release (Maeda et al., 1975, 1976). Whether this effect is a result of alterations in central neurotransmitters or a direct action on hypothalamic regulatory peptides is unclear. TRH may cause the release of GH in patients with liver disease, chronic renal failure, malnutrition, acromegaly, and psychiatric disorders.

DAMME (D-Ala^2, MePhe^4, Met(o)-ol enkephalin), an enkephalin analogue that crosses the blood-brain barrier, has been reported to stimulate GH secretion in normal persons, an effect that can be blocked by the opioid receptor antagonist, naloxone (Stubbs et al., 1978). Although this suggests that endorphins stimulate GH secretion in man, the specificity of DAMME for the enkephalin receptor remains to be established.

Neurotensin, a peptide present in high concentrations in the hypothalamus, has been shown to stimulate GH release in rats after systemic injection but to suppress GH release after direct introduction into the CSF, an effect mediated by SRIF (Maeda and Frohman, 1978).

Prolactin

Physiological Control

In normal women, but not in men, tactile stimulation of the
breast results in enhanced secretion of prolactin (Noel et al.,
1974). Stimulation of specialized receptors in the nipple and
areola, which reach the spinal cord by the intercostal nerves,
appear to be responsible.

In pregnancy, plasma prolactin rises progressively from the
first trimester (Jacobs et al., 1972), primarily as a result of
increased estrogen levels. Following parturition, estrogen and
progesterone levels fall rapidly and the unopposed action of
prolactin permits lactation from the estrogen-primed breast.
Although basal levels of prolactin return to normal during lactation, increases occur with suckling, presumably to prime the
breast for the next period of nursing (Noel et al., 1974).

Thyroid status is important in determining prolactin secretion. Subnormal levels of serum T_3 and thyroxine (T_4) in man
are associated with an increased TRH-induced prolactin response. In states of elevated thyroid hormone levels the converse occurs (Snyder et al., 1973). Surgical and other forms
of stress, exercise, amino acids, and insulin hypoglycemia also
result in prolactin secretion of uncertain significance (Adler
et al., 1975).

Pharmacological Control

Because DA has been suggested to be the PIF, numerous pharmacological studies have been performed to assess the effects of DA
on prolactin secretion in human beings and to establish its
site of action. L-DOPA has been shown to inhibit prolactin secretion, and the persistence of this effect by stimultaneous
administration of carbidopa, a peripherally acting DOPA decarboxylase inhibitor, suggested that at least part of the DA effect is centrally mediated (Fine and Frohman, 1978). It is not
clear, however, whether the DA receptors that mediate this effect are in the hypothalamus as well as in the pituitary. DA
and its agonists, including apomorphine (Martin et al., 1974)
and the ergot derivatives bromocriptine (del Pozo et al., 1972)
and lergotrile (Lemberger et al., 1974), also inhibit prolactin
secretion at the level of the pituitary, although an additional
site within the CNS has not been excluded. Conversely, drugs
interfering with normal catecholamine action, including chlorpromazine, α-methyldopa, reserpine, and other psychotropic

agents, increase prolactin levels and may result clinically in amenorrhea-galactorrhea (Frohman, 1979).

Other catecholamines do not appear to be as important in the control of prolactin secretion. Inhibition of catecholamine synthesis with α-MPT increases prolactin secretion but this effect is not overcome by DOPS. In addition, L-DOPA reduces prolactin levels even when a DBH inhibitor (diethyldithiocarbamate) is present to block conversion to NE (Donoso et al., 1971).

5-HT (Kamberi et al., 1971) and its precursors tryptophan (MacIndoe and Turkington, 1973) and 5-HTP (Kato et al., 1974) have been shown to increase prolactin secretion. Suckling-induced prolactin secretion is blocked by p-CPA, an effect that can be reversed by 5-HTP administration. Since suckling-induced prolactin secretion is thought to be mediated by a prolactin-releasing factor, it has been proposed that this system is under serotonergic control (Kordon et al., 1973).

Melatonin also stimulates prolactin secretion (Kamberi et al., 1971) and pinealectomy, which decreases melatonin levels, results in a short-term decrease in serum prolactin in rats (Relkin, 1972).

The role of cholinergic pathways in the release of prolactin in man is unclear. In animals anticholinergic drugs appear to inhibit prolactin release, for example, atropine prevents the suckling-induced prolactin elevation (Grosvenor and Turner, 1958).

Intraventricular GABA stimulates prolactin release in female rats by a central mechanism, since rats with pituitaries transplanted beneath the renal capsule fail to respond (Mioduszewski et al., 1976). In human beings GABOB and muscimol stimulate basal prolactin secretion (Tamminga et al., 1978), and baclofen increases prolactin release in response to hypoglycemia or arginine (Cavagnini et al., 1977).

Histamine has a central stimulatory effect on prolactin in animals which can be inhibited by diphenhydramne, a H_1-receptor antagonist. In man H_2-receptor antagonists, such as cimetidine (Gonzalez et al., 1980), stimulate prolactin secretion (Arakelian and Libertun, 1977). These findings suggest that H_1-receptors are involved in stimulation and H_2-receptors in inhibition of prolactin secretion.

The prolactin-releasing effect of TRH in human beings has led to the suggestion that endogenous TRH exerts a physiological role in the control of prolactin secretion. The threshold of TRH effects on prolactin is similar to that on TSH, although the secretion of the two pituitary hormones are often dissociated, indicating different mechanisms of control.

Opioid peptides (in the human, only DAMME has been studied) and morphine cause a release of prolactin, which can be blocked with naloxone (Vale et al., 1977). Naloxone also inhibits stress-induced prolactin release in rats (Van Vugt et al., 1978). These findings suggest that opiate receptors have a stimulatory role in prolactin release, possibly because they decrease DA turnover in the median eminence.

Neurotensin has been shown to stimulate prolactin after systemic administration in rats, the significance of which is uncertain (Rivier et al., 1977).

NEUROENDOCRINE DISORDERS DUE TO RELEASING FACTOR DYSFUNCTION: DISEASES OF RELEASING FACTORS

Clinical disorders of anterior pituitary function can be caused by intrinsic pituitary disease, by an interruption of the portal system that carries the releasing and inhibiting factors to the pituitary, or by alterations in releasing factor secretion associated with congenital, traumatic, inflammatory, infectious, infiltrative, or neoplastic diseases of the hypothalamus. In addition, biochemical or metabolic disorders of releasing factor-secreting neurons or of their neurotransmitter regulation can also result in disordered pituitary hormone secretion. The following section will focus on neurotransmitter and releasing factor defects that have been identified in specific disorders associated with impaired or inappropriate pituitary hormone secretion and that appear to exert a pathophysiological role in the clinical disorders.

ACTH

Hypofunction

ACTH deficiency of hypothalamic origin is seen infrequently as an isolated disorder but is often found in association with other pituitary hormone deficiencies in idiopathic hypopituitarism. The unavailability of CRF has made the diagnosis impossible to establish in the past, except by inference. The nature of the CRF or associated neurotransmitter deficiency is unknown.

Hyperfunction

Cushing's Disease (CD) is caused by bilateral adrenal hyperplasia secondary to an ACTH-secreting pituitary tumor or to corticotroph hyperplasia. The possibility that the pituitary disorder in at least some patients is secondary to excessive secretion of CRF is supported by the following lines of evidence:

1. There are reports of CD associated with CNS tumors with increased intracranial pressure (Heinbecker, 1944) and lesions have been noted at autopsy in hypothalamic paraventricular and supraoptic nuclei in patients with CD (Soffer et al., 1961).

2. Diffuse or nodular corticotroph hyperplasia has been observed in the majority of pituitaries removed from patients with CD (Ludecke et al., 1976).

3. Abnormalities in other periodic phenomena, such as sleep-associated rises in GH and prolactin and sleep electroencephalographic patterns, occur in CD and may persist after normalization of serum cortisol levels (Krieger and Glick, 1974; Krieger et al., 1976).

4. The ability of the 5-HT receptor blocker methysergide to suppress ACTH and cortisol secretion through a CNS mechanism believed to interfere with CRF production suggests that overproduction of CRF may be involved (Krieger et al., 1975). This action of methysergide has led to its use in the treatment of patients with CD as an alternative to surgery or after incomplete surgical removal of ACTH-secreting tumors. It appears to be useful in up to 50 percent of patients, but for an unexplained reason the demonstration of its effects may require one to two months of therapy.

LH and FSH

Hypofunction

Hypothalamic hypogonadism is defined as an impairment of pituitary-gonadal function attributable to deficient or disordered secretion of LH-RH. It is frequently seen as part of the idiopathic hypopituitarism syndrome and the clinical manifestations vary depending on whether the disorder occurs before or after puberty. Prepubertal females exhibit primary amenorrhea, which is clinically indistinguishable from hypopituitary hypogonadism. In some males, anosmia, nerve deafness, and visual disturbances may occur, usually on a genetic basis (Males et al., 1973). Hypoplasia of the hypothalamus, anterior commissure, or olfactory bulb has been reported in a few patients (Gauthier, 1961). In the postpubertal period, the disorder is seen primarily in females and is clinically manifested by oligomenorrhea or infertility. An alteration in the feedback sensitivity to estrogen has been suggested as the etiology and the disease tends to be self-limited. The possibility of a neurotransmitter defect appears likely, although the evidence to support this is lacking. A similar disturbance is seen in women with hyperprolactinemia and has been proposed to occur on the basis of increased hypothalamic DA turnover induced by prolactin. DA has been shown to inhibit gonadotropin secretion but whether the effect of hyperprolactinemia is mediated through this mechanism remains to be proved.

Hypothalamic hypogonadism is also seen in patients with severe disturbances in weight regulation—both starvation, including anorexia nervosa, and extreme obesity. The defects are for the most part correctable with a return to normal weight, although it may persist in patients with anorexia nervosa (Russell, 1977). Similar disturbances are seen in women participating in vigorous physical activity, i.e., ballet dancing and jogging.

Hyperfunction

Hypothalamic tumors have occasionally been associated with precocious puberty and LH-RH has recently been identified in CSF and tumor tissue from such patients (Judge et al., 1977). The tumors, histologically identified as hamartomas, consist of encapsulated nodules in the posterior hypothalamus and contain neurons with secretory granules similar to those found in the median eminence. The blood vessels in the tumor also contain the fenestrations characteristic of vessels in the median eminence, suggesting that secretory products from the tumor (LH-RH) can have access to the pituitary portal system and provide an explanation for the endocrine disturbance.

TSH

Hypothalamic (or tertiary) hypothyroidism has been seen in children in association with idiopathic GH deficiency and in adults more commonly as an isolated defect. The presumption of TRH deficiency is based on the observation of an inappropriately low plasma TSH in a hypothyroid individual who responds to TRH stimulation. It is unknown whether this disorder is due to a deficiency of TRH or to an abnormality in the control of its secretion.

GH

Hypofunction

Idiopathic GH deficiency (IGHD) is a disorder identified in childhood and manifested by impaired linear growth and absent or reduced plasma GH responses to provocative stimuli (Goodman et al., 1968). It may occur as an isolated hormone deficiency or may be associated with impaired secretion of TSH, ACTH, LH, or FSH (idiopathic hypopituitarism). The disorder is unassociated with any anatomical evidence of pituitary disease and is believed to be due to an impairment in the secretion of GHRF. Although it is currently not possible to distinguish between hypothalamic and pituitary causes of GH deficiency, the TSH and gonadotropin responses observed after injection of TRH and LH-RH in patients with idiopathic hypopituitarism argue strongly in favor of a hypothalamic defect for the GH deficiency as well. The possibility of SRIF overproduction cannot be excluded, although the absence of any impairment of the TSH response to TRH makes it unlikely.

The benign nature of this disease has precluded histopathological study of the hypothalamus, although on clinical grounds there is no suspicion of an obvious disorder. There is a single report of restoration of normal GH responses by treatment with propranolol in Japanese children with IGHD (Imura et al., 1973a), suggesting overactive β-adrenergic receptor-mediated neurotransmission. Since central β-adrenergic receptors inhibit GH secretion, a defect of this type is a distinct possibility. Unfortunately, studies in American children with IGHD have not confirmed these findings (Lee et al., 1974) and thus the disorder may be of heterogeneous origin.

Psychosocial dwarfism or emotional deprivation GH deficiency clinically resembles IGHD and occurs in children reared in home situations where there is severe emotional and social neglect (Powell et al., 1967). Such children show evidence of GH deficiency if tested immediately after being discovered. Upon removal from their environment and given the previously lacking attention and care, plasma GH responses are rapidly restored to normal (within a few days) and growth is resumed. These observations are strongly suggestive of a functional neurotransmitter disorder and in support of this hypothesis, a restoration of normal GH responsiveness to insulin hypoglycemia has been reported to occur with propranolol (Imura et al., 1971), implicating altered β-adrenergic neurotransmission.

Hyperfunction

Acromegaly is a pituitary disorder due to a GH-secreting tumor. There is at present strong evidence in support of the concept that at least some of these tumors occur secondary to an increase in the secretion of GHRF. The arguments in favor of a hypothalamic etiology (GHRF overproduction) include:
1. GH secretion in acromegaly is generally not autonomous but responds to CNS-mediated stimuli, such as glucose, insulin hypoglycemia, and arginine, often in a paradoxic manner (Lawrence et al., 1970).

2. Acromegaly has been associated with hypothalamic tumors, suggesting that overproduction of a GHRF may have been involved.

3. Ectopic production of GHRF by carcinoid or pancreatic islet tumors results in GH hypersecretion and even pituitary tumors (Zafar et al., 1979; Frohman et al., 1980). Removal of the extrapituitary tumor can result in reversal of both the GH hypersecretion and the clinical features of acromegaly.

4. GH responses to glucose and insulin, when examined after removal of GH-secreting adenomas, frequently (Decker et al., 1976) although not invariably (Hoyte and Martin, 1975) remain abnormal.

5. GH secretion in acromegaly is stimulated by α-adrenergic antagonists and β-adrenergic agonists (Cryer and Daughaday, 1974), which are believed to act within the CNS. These observations, although suggesting a hypothalamic etiology for the GH hypersecretion, do not preclude the eventual development of an autonomous pituitary tumor.

Neuropharmacological therapy has recently been utilized in the treatment of GH-secreting pituitary tumors. Although not recommended as primary therapy except under special circumstances, bromocriptine is capable of lowering GH levels in most patients (Luizzi et al., 1974). The predominant action of bromocriptine occurs at the pituitary where an inhibitory DA receptor is frequently present as a manifestation of altered gene expression by the neoplastic somatotroph. However, bromocriptine can cross the blood-brain barrier and an additional action on CNS DA receptors may also occur.

GH hypersecretion is seen in patients with poorly controlled diabetes mellitus, particularly in response to exercise (Hansen, 1970). The elevated GH levels can be restored to normal by controlling the metabolic abnormality with insulin but can also be reduced by phentolamine (Hansen, 1971), implying an overactivity of an α-adrenergic-mediated neural pathway.

GH levels in patients with Huntington's chorea are frequently elevated and responses to dopaminergic stimuli are enhanced (Caraceni et al., 1977). This observation is consistent with the postulated defect of DA metabolism in this disorder, although the precise neurotransmitter abnormality remains to be clarified.

Prolactin

Idiopathic hyperprolactinemia with or without amenorrhea and galactorrhea is believed to be a disorder of CNS etiology. It is unclear whether this disorder is separate from that of prolactin-secreting pituitary tumors or whether hyperfunctioning

lactotrophs eventually develop into foci of lactotroph hyperplasia and adenoma formation. There is considerable evidence
for altered neurotransmitter function in this spectrum of diseases and most reports do not reveal qualitative differences
between hyperprolactinemic patients with or without evidence of
a pituitary tumor. The evidence for hypothalamic defects in
this syndrome can be summarized as follows:

1. DA antagonists, which increase prolactin levels in normal
   subjects, are ineffective in patients with hyperprolactinemia and prolactin-secreting tumors, implying an absence
   of or reduction in central dopaminergic tone (Kleinberg et
   al., 1977).
2. The impairment of prolactin responses to TRH can be reproduced in normal subjects by metoclopramide (Healy and
   Burger, 1978), a DA receptor antagonist, suggesting dopaminergic tone may be required for normal prolactin responses to other stimuli.
3. L-DOPA suppresses prolactin secretion in normal subjects,
   either by peripheral conversion to DA, which acts directly
   on the pituitary, or, when accompanied by the peripheral
   L-AAD inhibitor, carbidopa, by central conversion to DA.
   In patients with hyperprolactinemia or prolactin-secreting
   tumors, concomitant administration of dopa plus carbidopa
   does not suppress prolactin secretion, suggesting an impairment in the generation of central dopaminergic-PIF
   activity (Fine and Frohman, 1978).
4. Bromocriptine decreases circulating catecholamines through
   a central mediated mechanism in normal persons but not in
   patients with prolactin-secreting tumors (Van Loon, 1978).
5. Nomifensine, a DA reuptake inhibitor, suppresses prolactin
   secretion in normal persons but not in patients with idiopathic hyperprolactinemia or prolactin-secreting tumors
   (Ferrari et al., 1980).
6. A central histaminergic defect has been suggested by the
   report that cimetidine, an H_2-receptor blocker stimulates
   prolactin release by a CNS-mediated pathway in normal persons but not in patients with tumors (Gonzalez et al.,
   1980).
7. Lactotroph hyperplasia has been found in one-third of patients with prolactin-secreting tumors (McKeel et al.,
   1978), supporting the concept that a functional disorder
   eventually leads to tumor formation.
8. There is a preliminary report demonstrating prolactin-
   releasing activity in the serum of patients with prolactin-
   secreting tumors (Garthwaite and Hagen, 1978).

Neuropharmacological treatment of hyperprolactinemia has
focused on dopaminergic agents, specifically bromocriptine
(Hökfelt and Nillius, 1978). The effect of this drug is pri-

marily on the pituitary and it is unlikely that its central dopaminergic actions are of importance in its prolactin-lowering effect. In addition, 5-HT receptor antagonists, such as metergoline, have also exhibited prolactin-lowering effects in patients with hyperprolactinemia, presumably by a CNS-mediated mechanism (Delitala et al., 1976). However, these agents also exhibit dopaminergic agonist activity and could act through the pituitary DA receptor.

REFERENCES

Abe, H., Chihara,K., Chiba, T., Matsukura, S., and Fujita, T.: Effects of various bioactive peptides on plasma immunoreactive somatostatin in rat hypophyseal portal blood. Proceedings of Sixth International Congress of Endocrinology, Melbourne, Australia. (Abstr 77), 1980.

Adler, R.A., Noel, G.L., Wartofsky, L., and Frantz, A.G.: Failure of oral water loading and intravenous hypotonic saline to suppress plasma prolactin in man. *J Clin Endocrinol Metab* 41:383-389, 1975.

Annunziato, L., DiRenzo, G., Lombardi, G., Scopacasa, F., Schettini, G., Preziosi, P., and Scapagnini, U.: The role of central noradrenergic neurons in the control of thyrotropin secretion in the rat. *Endocrinology* 100:738-744, 1977.

Arakelian, M.C., and Libertun, C.: H_1 and H_2 histamine receptor participation in the brain control of prolactin secretion in lactating rats. *Endocrinology* 100:890-895, 1977.

Arimura, A., and Schally, A.V.: Increase in basal and thyrotropin releasing hormone-stimulated secretion of thyrotropin by passive immunization with antiserum to somatostatin in rats. *Endocrinology* 98:1069-1072, 1976.

Arimura, A., Bowers, C.V., Schally, A.V., Saito, M., and Miller, M.C.: Effects of corticotropin releasing factor, dexamethasone and actinomycin D on the release of ACTH from rat pituitaries *in vivo* and *in vitro*. *Endocrinology* 85:300-311, 1969.

Arimura, A., Smith, W.D., and Schally, A.V.: Blockade of the stress induced decrease in blood GH by antisomatostatin serum in rats. *Endocrinology* 98:540-543, 1976.

Belchetz, P.E., Gredley, G., Bird, D., and Himsworth, R.L.: Regulation of thyrotrophin secretion by negative feedback of tri-iodothyronine on the hypothalamus. *J Endocrinol* 76:439-448, 1978.

Ben-Jonathan, N., Oliver, C., Weiner, H.J., Mical, R.S., and Porter, J.C.: Dopamine in hypophyseal portal plasma of the rat during the estrus cycle and throughout pregnancy. *Endocrinology* 100:452-458, 1977.

Bennett, G.W., and Edwardson, J.A.: Release of corticotropin releasing factor and other hypophysiotropic substances from isolated nerve endings (synaptosomes). *J Endocrinol 65:*33-44, 1975.

Bennett, G.W., Edwardson, J.A., Holland, D.T., Jeffcoate, S.L., and White, N.: The release of immunoreactive luteinizing hormone-releasing hormone and thyrotrophin-releasing hormone from hypothalamic synaptosomes. *Nature 257:*323-325, 1975.

Bennett, G.W., Edwardson, J.A., Marcano De Cotte, D., Berelowitz, M., Pimstone, B.L., and Kronheim, S.: Release of somatostatin from rat brain synaptosomes. *J Neurochem 32:*1127-1131, 1979.

Berelowitz, M., Maeda, K., Harris, S., and Frohman, L.A.: The effects of alterations in the pituitary-thyroid axis on hypothalamic content and *in vitro* release of somatostatin-like immunoreactivity. *Endocrinology 107:*24-29, 1980.

Berelowitz, M., Firestone S.L., and Frohman, L.A.: Effects of growth hormone excess and deficiency on hypothalamic somatostatin content and release and on tissue somatostatin distribution. *Endocrinology 109:*714-719, 1981a.

Berelowitz, M., Szabo, M., Frohman, L.A., Firestone, S., Chu, L., and Hintz, R.L.: Somatomedin-C mediates growth hormone negative feedback by effects on both the hypothalamus and the pituitary. *Science 212:*1279-1281, 1981b.

Bergland, R.M., and Page, R.B.: Can the pituitary secrete directly to the brain? (Affirmative anatomical evidence). *Endocrinology 102:*1325-1338, 1978.

Besser, G.M., Butler, P.W.P., Landon, J., and Rees, L: Influence of amphetamines on plasma corticosteroid and growth hormone levels in man. *Br Med J 4:*528-530, 1969.

Bhattacharya, A.N., Dierschke, D.J., Yamaji, J., and Knobil, E.: The pharmacologic blockade of the circhoral mode of LH secretion in the ovariectomized rhesus monkey. *Endocrinology 90:*778-786, 1972.

Bivens, C.H., Lebovitz, H.E., and Feldman, J.M.: Inhibition of hypoglycemia-induced growth hormone secretion by the serotonin antagonists cyproheptadine and methysergide. *N Engl J Med 289:*236-239, 1973.

Blackard, W.G., and Hubbell, J.H.: Stimulatory effect of exogenous catecholamines on plasma hGH concentrations in the presence of beta adrenergic blockade. *Metabolism 19:*547-552, 1970.

Boler, J., Enzmann, F., Folkers, K., Bowers, C.Y., and Schally, A.V.: The identity of chemical and hormonal properties of the thyrotropin releasing hormone and pyroglutamyl-histidyl-proline amide. *Biochem Biophys Res Commun 37:*705-710, 1969.

Brazeau, P., Vale, W., Burgus, R., Ling, N., Butcher, M., Rivier, J., and Guillemin, R.: Hypothalamic polypeptide that inhibits the secretion of immunoreactive pituitary growth hormone. *Science* *179*:77-79, 1973.

Brown, G.M., and Reichlin, S.: Psychological and neural regulation of growth hormone secretion. *Psychosom Med* *34*:45-61, 1972.

Buckler, J.M.H.: Exercise as a screening test for growth hormone release. *Acta Endocrinol (Copenh)* *69*:219-229, 1972.

Burden, J.L., Hillhouse, E.W., and Jones, M.T.: A proposed model of the transmitters involved in the control of corticotropin releasing hormone. *J Endocrinol* *63*:20-21, 1974.

Burgess, R., Dunn, T.F., Desiderio, D., and Guillemin, R.: Structure moleculaire du facteur hypothalamique hypophysiotrope TRF d'origine ovine: Mise en evidence par spectrometre de masse de la sequence PCA-His-Pro-NH$_2$. *C R Acad Sci (Paris)* *269*:1870-1873, 1969.

Calabro, M.A., and MacLeod, R.M.: Binding of dopamine to bovine anterior pituitary gland membranes. *Neuroendocrinology* *25*:32-46, 1978.

Camanni, F., Massara, F., Belforte, L., and Molinatti, G.M.: Changes in plasma growth hormone levels in normal and acromegalic subjects following administration of 2-bromo- α -ergocriptine. *J Clin Endocrinol Metab* *40*:363-366, 1975.

Caraceni, T., Panerai, A.E., Parati, E.A., Cocchi, D., and Muller, E.E.: Altered growth hormone and prolactin responses to dopaminergic stimulation in Huntington's chorea. *J Clin Endocrinol Metab* *44*:870-875, 1977.

Cavagnini, F., and Peracchi, M.: Effect of reserpine on growth hormone response to insulin hypoglycemia and to arginine infusion in normal subjects and hyperthyroid patients. *J Endocrinol* *51*:651-656, 1971.

Cavagnini, F., Panerai, A.E., Valentini, F., Bulgheroni, P., Peracchi, M., and Pinto, M.: Inhibition of ACTH response to oral and intravenous metyrapone by antiserotoninergic treatment in man. *J Clin Endocrinol Metab* *41*:143-148, 1975.

Cavagnini, F., Invitti, C., Di Landro, A., Tencani, L., Maraschini, C., and Girotti, G.: Effects of a gamma aminobutyric acid (GABA) derivative, baclofen, on growth hormone and prolactin secretion in man. *J Clin Endocrinol Metab* *45*:579-584, 1977.

Celis, M.E., Taleisnik, S., and Walter, R.: Regulation of formation and proposed structure of the factor inhibiting the release of melanocyte stimulating hormone. *Proc Natl Acad Sci USA* *68*:1428-1433, 1971.

Chen, H.J., and Meites, J.: Effects of biogenic amines and TRH on release of prolactin and TSH in the rat. *Endocrinology* *96*:10-14, 1975.

Chihara, K., Kato, Y., Maeda, K., Matsukura, S., and Imura, H.: Suppression by cyproheptadine of human growth hormone and cortisol secretion during sleep. *J Clin Invest* 57:1393-1402, 1976.

Chihara, K., Arimura, A., and Schally, A.V.: Effect of intraventricular injection of dopamine, norepinephrine, acetylcholine and 5-hydroxytryptamine on immunoreactive somatostatin release into rat hypophyseal portal blood. *Endocrinology* 104:1656-1662, 1979.

Childs, G., Kubek, M., Cole, D., Tobin, R., Lorinz, M., and Wilbur, J.: The pituitary thyrotropin secretory granule: A novel location for thyrotropin-releasing hormone in rats and man. Program of the 60th Meeting of the Endocrine Society, Miami, (Abstr 192), 1978.

Cryer, P.E., and Daughaday, W.H.: Adrenergic modulation of growth hormone secretion in acromegaly. Suppression during phentolamine and phentolamine-isoproterenol administration. *J Clin Endocrinol Metab* 39:658-663, 1974.

Dallman, M.F., and Yates, F.E.: Dynamic asymmetries in the corticosteroid feedback pathways and distribution, binding and metabolism elements of the adrenocortical system. *Ann NY Acad Sci* 156:696-721, 1969.

Decker, R.E., Epstein, J.A., Carras, R., and Rosenthal, A.D.: Transsphenoidal microsurgery for pituitary tumors. Experience with 45 cases. *Mt Sinai J Med* 43:565-577, 1976.

Delitala, G., Masala, A., Alagna, S., Devilla, L., and Lotti, G.: Growth hormone and prolactin release in acromegalic patients following metergoline administration. *J Clin Endocrinol Metab* 43:1382-1386, 1976.

Donoso, A.O., Bishop, W., Fawcett, C.P., Krulich, L., and McCann, S.M.: Effects of drugs that modify brain monoamine concentrations on plasma gonadotropin and prolactin levels in the rat. *Endocrinology* 89:774-784, 1971.

Dupont, A., Cusan, L., Caron, M., Labrie, F., and Li, C.H.: Beta-endorphin: Stimulation of growth hormone release *in vivo*. *Proc Natl Acad Sci USA* 74:358-359, 1977.

Eddy, R.L., Jones, A.L., Chakmakjian, Z.H., and Silverthorne, M.C.: Effects of levodopa on human hypophyseal trophic hormone release. *J Clin Endocrinol Metab* 33:709-712, 1971.

Edwardson, J.A., and Bennett, G.W.: Modulation of corticotrophin-releasing factor release from hypothalamic synaptosomes. *Nature* 251:425-427, 1974.

Emerson, C.H., and Utiger, R.D.: Plasma thyrotropin releasing hormone concentrations in the rat. Effect of thyroid excess and deficiency and cold exposure. *J Clin Invest* 56:1564-1570, 1975.

Emerson, C.H., Frohman, L.A., Szabo, M., and Thaakar, I.: TRH immunoreactivity in human urine: Evidence for dissociation from TRH. *J Clin Endocrinol Metab* 45:392-399, 1977.

Enjalbert, A., Moos, F., Carbonell, L., Priam, M., and Kordon,

C.: Prolactin inhibiting activity of dopamine-free subcellular fractions from rat mediobasal hypothalamus. *Neuroendocrinology* 24:147-161, 1977.

Ferrari, C., Crosignani, P.G., Caldara, R., Picciotti, M.C., Malinverni, A., Barattini, G., Rampini, P., and Telloli, P.: Failure of nomifensine administration to discriminate between tumorous and nontumorous hyperprolactinemia. *J Clin Endocrinol Metab* 50:23-27, 1980.

Fine, S.A., and Frohman, L.A.: Loss of central nervous system component of dopaminergic inhibition of prolactin secretion in patients with prolactin-secreting pituitary tumors. *J Clin Invest* 61:973-980, 1978.

Fineberg, E.S., Horand, A.A., and Merimee, T.J.: Free fatty acid concentrations and growth hormone secretion in man. *Metabolism* 21:491-498, 1972.

Fioretti, P., Melis, M.G.B., Paoletti, A.M., Pirodo, G., Caminiti, F., Capsini, G.U., and Martini, L.: γ-amino β-hydroxybutyric acid stimulates prolactin and growth hormone release in normal women. *J Clin Endocrinol Metab* 47:1336-1340, 1978.

Fortier, C., Delgado, A., Ducommon, P., Ducommon, S., Dupont, A., Jobin, M., Kraicer, J., MacIntosh-Hardt, B., Marceau, H., Miathe, P., Miathe-Voloss, C., Rerup, C., and Van Rees, G.P.: Functional interrelationships between the adenohypophysis, thyroid, adrenal cortex and gonads. *Can Med Assoc J* 103:864-874, 1970.

Franchimont, P., Chari, S., Hagelstein, M.T., and Duraiswami, S.: Existence of a follicle stimulating hormone inhibiting factor "Inhibin" in bull seminal plasma. *Nature* 257:402-404, 1975.

Frohman, L.A.: Neuroendocrine pharmacology. In *Endocrinology*, L. De Groot, et al., eds, Grune & Stratton, New York, 1979, pp 287-300.

Frohman, L.A.: Endorphins and enkephalins. *Hosp Formulary*, 1980.

Frohman, L.A., Szabo, M., Berelowitz, M., and Stachura, M.E.: Partial purification and characterization of a peptide with growth hormone-releasing activity from extrapituitary tumors in patients with acromegaly. *J Clin Invest* 65:43-54, 1980.

Fuxe, K., Hökfelt, T., Lofstrom, A., Johansson, O., Goldstein, M., Jeffcoate, S.L., White, N., and Eneroth, P.: On the role of neurotransmitters and hypothalamic hormones in the control of pituitary function and sexual behavior. In *Subcellular Mechanisms in Reproductive Endocrinology*, F. Naftolin, K.J. Ryan, and I.J. Davis, eds, Elsevier, Amsterdam, 1976, pp 193-246.

Garthwaite, T.L., and Hagen, T.C.: Plasma prolactin-releasing factor-like activity in the amenorrhea-galactorrhea syndrome.

J Clin Endocrinol Metab 47: 885-888, 1978.

Gauthier, G.: La Dysplasie olfacto-génitale (agénésie des lobes olfactifs avec absence de dévelopement gonadique à la puberté). *Acta Neuroveg (Wien) 21:* 345-394, 1961.

Genazzani, A.R., Lemarchand-Béraud, T., Aubert, N.L., and Felber, J.P.: Pattern of plasma ACTH, hGH and cortisol during the menstrual cycle. *J Clin Endocrinol Metab 41:* 431-437, 1975.

Gonzalez-Villapando, C., Szabo, M., and Frohman, L.A.: Central Nervous System--mediated stimulation of prolactin secretion by cimetidine, a histamine H_2 receptor antagonist: Impaired responsiveness in patients with prolactin-secreting tumors and idiopathic hyperprolactinemia. *J Clin Endocrinol Metab 51:* 1417-1423, 1980.

Goodman, H.G., Grumbach, M.M., and Kaplan, S.L.: Growth and growth hormone. II A Comparison of isolated growth hormone deficiency and multiple pituitary hormone deficiencies in 35 patients with idiopathic hypopituitary dwarfism. *N Engl J Med 278:* 57-68, 1968.

von Graffenried, B., del Pozo, E., Robicek, J., Krebs, E., Poldinger, W., Burmeister, P., and Kerp, L.: Effects of the synthetic enkephalin analogue FK 33-824 in man. *Nature 272:* 729-730, 1978.

Grosvenor, C.E., and Turner, C.W.: Effects of oxytocin and blocking agents upon pituitary lactogen discharge in lactating rats. *Proc Soc Exp Biol Med 97:* 463-465, 1958.

Gudelsky, G.A., and Moore, K.E.: A comparison of the effects of haloperidol on dopamine turnover in the striatum, olfactory tubercle and median eminence. *J Pharmcol Exp Therap 202:* 149-155, 1977.

Hansen, A.P.: Abnormal serum growth hormone response to exercise in juvenile diabetics. *J Clin Invest 49:* 1467-1478, 1970.

Hansen, A.P.: The effect of adrenergic receptor blockade in the exercise-induced serum growth hormone rise in normals and juvenile diabetics. *J Clin Endocrinol Metab 33:* 807-812, 1971.

Harris, G.W., *The Pituitary Gland.* Edward Arnold, London, 1955.

Healy, D.L., and Burger, H.G.: Sustained elevation of serum prolactin by metoclopramide: A clinical model of idiopathic hyperprolactinemia. *J Clin Endocrinol Metab 46:* 709-714, 1978.

Hedge, G.A., and Smelik, P.G.: Corticotrophin release: Inhibition by intrahypothalamic implants of atropine. *Science 159:* 891-892, 1968.

Heinbecker, P.: Pathogenesis of Cushing's syndrome. *Medicine 23:* 225-247, 1944.

Hershman, J.M., Read, D.G., Bailey, A.L., Norman, V.D., and Gibson, T.B.: Effect of cold exposure on serum thyrotropin. *J Clin Endocrinol Metab 30:* 430-434, 1970.

Hidaka, H., Nagasaka, A., and Takeda, A.: Fusaric (5-butylpicolinic) acid: Its effect on plasma growth hormone. *J Clin Endocrinol Metab 37*:145-147, 1973.

Hillhouse, E.W., Burden, J.L., and Jones, M.T.: The effect of various putative neurotransmitters on the release of corticotropin-releasing hormone from the hypothalamus of the rat *in vitro*. 1. The effect of acetylcholine and noradrenaline. *Neuroendocrinology 17*:1-11, 1975.

Hirooka, Y., Hollander, C.S., Suzuki, S., Ferdinand, P., and Juah, S-I.: Somatostatin inhibits release of thyrotropin releasing factor from organ cultures of rat hypothalamus. *Proc Natl Acad Sci USA 75*:4509-4513, 1978.

Hökfelt, T.: The possible ultrastructural identification of tuberoinfundibular dopamine-containing nerve endings in the median eminence of the rat. *Brain Res 5*:121-123, 1967.

Hökfelt, B., and Nillius, S.J.: The dopamine agonist bromocriptine. Theoretical and clinical aspects. *Acta Endocrinol (Copenh) 88 (Suppl 216)*:1-230, 1978.

Hoyte, K.M., and Martin, J.B.: Recovery from paradoxical growth hormone response in acromegaly after transsphenoidal selective adenomectomy. *J Clin Endocrinol Metab 41*:656-659, 1975.

Imura, H., Yoshimi, T., and Ikekubo, K.: Growth hormone secretion in a patient with deprivation dwarfism. *Endocrinol Jpn 15*:301-304, 1971.

Imura, H., Nakai, Y., Kato, Y., Yoshimoto, Y., and Moridera, K.: Effect of adrenergic agents on growth hormone and ACTH secretion. In *Endocrinology*, R.O. Scow, ed, Excerpta Medica, Amsterdam, 1973, pp 156-162.

Imura, H., Nakai, Y., and Yoshima, T.: Effect of 5-hydroxytryptophan on growth hormone and ACTH release in man. *J Clin Endocrinol Metab 36*:204-206, 1973b.

Invitti, C., Cavagnini, F., diLandro, A., and Pinto, M.: Inhibiting effect of a GABA derivative, baclofen, on growth hormone and cortisol response to insulin hypoglycemia in man. Proceedings of the Vth International Congress of Endocrinology, Hamburg, (Abstr 661), 1976.

Jackson, I.M.D., and Reichlin, S.: TRH radioimmunoassay measurements in normal and altered states of thyroid function in the rat. Program of the 49th Meeting of the American Thyroid Association, 1973, p T4.

Jacobs, L.S., Mariz, I.J., and Daughaday, W.H.: A mixed heterologous radioimmunoassay for human prolactin. *J Clin Endocrinol Metab 34*:484-490, 1972.

Jones, M.T., Hillhouse, E.W., and Burden, J.L.: Effects of various putative neurotransmitters on the secretion of corticotrophin releasing hormone from the rat hypothalamus *in vitro*. 1. A model of the neurotransmitters involved. *J Endocrinol 69*:1-10, 1976a.

Jones, M.T., Hillhouse, E.W., and Burden, J.L.: Secretion of corticotrophin-releasing factor *in vitro*. In *Frontiers of Neuroendocrinology*, W.F. Ganong and L. Martini, eds, Raven Press, New York, 1976b, pp 195-226.

Jones, M.T., Hillhouse, E.W., and Burden, J.L.: The mechanism of fast and delayed corticosteroid feedback at the hypothalamus. *J Endocrinol 69*:34p, 1976c.

Joseph, S.A., and Knigge, K.M.: The endocrine hypothalamus: Recent anatomical studies. In *The Hypothalamus*, S. Reichlin, R.J. Baldessarini, and J. Martin, eds, Raven Press, New York, 1978, pp 15-47.

Judge, D.M., Kulin, H.E., Page, R., Santen, R., and Trapukdi, S.: Hypothalamic hamartoma. *N Engl J Med 296*:7-10, 1977.

Kamberi, I.A.: The role of brain monoamines and pineal indoles in the secretion of gonadotrophins and gonadotrophin-releasing factors. *Prog Brain Res 39*:261-280, 1973.

Kamberi, I.A., Mical, R.S., and Porter, J.C.: Effects of melatonin and serotonin on the release of FSH and prolactin. *Endocrinology 88*:1288-1299, 1971.

Kato, Y., Nakai, Y., Imura, H., Chihara, K., and Ohgo, S.: Effect of 5-hydroxytryptophan (5HTP) on plasma prolactin levels in man. *J Clin Endocrinol Metab 38*:695-697, 1974.

Kleinberg, D.L., Noel, G.L., and Frantz, A.G.: Galactorrhea: A study of 235 cases, including 48 with primary tumors. *N Engl J Med 296*:589-600, 1977.

Knopf, R.F., Conn, J.W., Fajans, S.S., Floyd, J.C., Guntsche, E.M., and Rull, J.A.: Plasma growth hormone response to intravenous administration of amino acids. *J Clin Endocrinol Metab 25*:1140-1144, 1965.

Kordon, C., Blake, C.A., Terkel, J., and Sawyer, C.H.: Participation of serotonin-containing neurons in the suckling-induced rise in plasma prolactin levels in lactating rats. *Neuroendocrinology 13*:213-223, 1973.

Kotani, M., Onaya, T., and Yamada, T.: Acute increase of thyroid hormone secretion in response to cold and its inhibition by drugs which act on the autonomic or central nervous system. *Endocrinology 92*:288-294, 1973.

Krieger, D.T.: Effect of intraventricular neonatal 6-OH Dopamine or 5-6 dihydroxytryptamine administration on the circadian periodicity of plasma corticosterone levels in the rat. *Neuroendocrinology 17*:62-74, 1975.

Krieger, D.T., and Glick, S.M.: Sleep EEG stages and plasma growth hormone concentration in states of endogenous and exogenous hypercortisolemia or ACTH elevation. *J Clin Endocrinol Metab 39*:986-1000, 1974.

Krieger, D.T., Amorosa, L., and Linick, F.: Cyproheptadine-induced remission of Cushing's disease. *N Engl J Med 293*:893-896, 1975.

Krieger, D.T., Howanitz, P.J., and Frantz, A.G.: Absence of

nocturnal elevation of plasma prolactin concentrations in Cushing's disease. *J Clin Endocrinol Metab* 42:260-272, 1976.

Krulich, L., Giachetti, A., Marchlewska-Koj, A., Hefco, E., and Jameson, H.E.: On the role of the central noradrenergic and dopaminergic systems in the regulation of TSH secretion in the rat. *Endocrinology* 100:496-505, 1977.

Lachelin, G.C.L., LeBlanc, H., and Yen, S.S.C.: The inhibitory effect of dopamine agonists on LH release in women. *J Clin Endocrinol Metab* 44:728-732, 1977.

Lal, S., de la Vega, C.E., Sourkes, T.L., and Friesen, H.G.: Effect of apomorphine on growth hormone, prolactin, luteinizing hormone and follicle stimulating hormone in human serum. *J Clin Endocrinol Metab* 37:719-724, 1973.

Lal, S., Tolis, G., Martin, J.B., Brown, G.M., and Guyda, H.: Effect of clonidine on growth hormone, prolactin, luteinizing hormone, follicle-stimulating hormone and thyroid-stimulating hormone in the serum of normal men. *J Clin Endocrinol Metab* 41:827-832, 1975.

Lawrence, A.M., Goldfine, I.D., and Kirsteins, L.: Growth hormone dynamics in acromegaly. *J. Clin Endocrinol Metab* 31:239-247, 1970.

Leblanc, H., Lachelin, G.C.L., Abu-Fadil, S., and Yen, S.S.C.: Effects of dopamine infusion on pituitary hormone secretion in humans. *J Clin Endocrinol Metab* 43:668-674, 1976.

Lee, P.A., Thompson, R.G., and Blizzard, R.M.: Relationship of the adrenergic nervous system and growth hormone release in normal adults and children with various growth disorders. *Metabolism* 23:595-601, 1974.

Lemberger, L., Crabtree, R., Clemens, J., Dyke, R.W., and Woodburn, R.T.: The inhibitory effect of an ergoline derivative (lergotrile, compound 83636) on prolactin secretion in man. *J Clin Endocrinol Metab* 39:579-584, 1974.

Leppäluoto, J., Männistö, P., Ranta, T., and Linnoila, M.: Inhibition of midcycle gonadotrophin release in healthy women by pimozide and fusaric acid. *Acta Endocrinol (Copenh)* 81:455-460, 1976.

Ludecke, D., Kautzky, R., Saeger, W., and Schrader, D.: Selective removal of hypersecreting pituitary adenomas? *Acta Neurochir (Wien)* 35:27-42, 1976.

Luizzi, A., Chiodini, P.G., Bottalla, L., Cremascoli, G., Muller, E., and Silvestrini, F.: Decreased plasma growth hormone levels in acromegalics following CB 154 (2-BR-α-ergocryptine) administration. *J Clin Endocrinol Metab* 38:910-912, 1974.

McCann, S.M.: Regulation of follicle stimulating hormone and luteinizing hormone. In *Handbook of Physiology*, vol IV, part 2; *Endocrinology*. E. Knobil and W.H. Sawyer, eds, American Physiological Society, Washington, DC, 1974, pp 489-517.

McCann, S.M., and Moss, R.L.: Putative neurotransmitters involved in discharging gonadotropin-releasing neurohormones and the action of LH-releasing hormone on the CNS. *Life Sci* *16:* 833-852, 1975.

McClintock, M.K.: Menstrual synchrony and suppression. *Nature* *229:* 244-245, 1971.

MacIndoe, J.H., and Turkington, R.W.: Stimulation of human prolactin secretion by intravenous infusion of L-tryptophan. *J Clin Invest 52:* 1972-1978, 1973.

McKeel, D.W., Jr., Fowler, M., and Jacobs, L.S.: The high prevalence of prolactin cell hyperplasia in the human adenohypophysis. Program of the 60th Meeting of the Endocrine Society, Miami, (Abstr 557) 1978.

Maeda, K., Kato, Y., Chihara, K., Ohgo, S., Iwasaki, Y., and Imura, H.: Suppression by thyrotropin releasing hormone of human growth hormone release induced by L-DOPA. *J Clin Endocrinol Metab 41:* 408-411, 1975.

Maeda, K., Kato, Y., Chihara, K., Ohgo, S., Iwasaki, Y., Abe, H., and Imura, H.: Suppression by thyrotropin-releasing hormone of growth hormone release induced by arginine and insulin-induced hypoglycemia in man. *J Clin Endocrinol Metab 43:* 453-456, 1976.

Maeda, K., and Frohman, L.A.: Dissociation of systemic and central effects of neurotensin on the secretion of growth hormone, prolactin, and thyrotropin. *Endocrinology 103:* 1903-1907, 1978.

Maeda, K., and Frohman, L.A.: Release of somatostatin and thyrotropin-releasing hormone from hypothalamic fragments *in vitro*. *Endocrinology 106:* 1837-1842, 1980.

Malacara, J., Seyler, L.E., Jr., and Reichlin, S.: Luteinizing hormone-releasing factor activity in peripheral blood from women during the midcycle luteinizing hormone ovulatory surge. *J Clin Endocrinol Metab 34:* 271-278, 1972.

Males, J.L., Townsend, J.L., and Schneider, R.: Hypogonadotropic hypogonadism with anosmia--Kallmann's syndrome. A disorder of olfactory and hypothalamic function. *Arch Intern Med 131:* 501-507, 1973.

Martin, J.B.: Neural regulation of growth hormone secretion. *N Engl J Med 288:* 1384-1393, 1973.

Martin, J.B., and Reichlin, S.: Neural regulation of the pituitary thyroid axis. In *Proceedings of the 6th Midwest Conference on the Thyroid*, A.D. Kenny and R.R. Anderson, eds, University of Columbia Press, Columbia, Mo., 1970, pp 1-24.

Martin, J.B., Lal, S., Tolis, G., and Friesen, H.G.: Inhibition by apomorphine of prolactin secretion in patients with elevated serum prolactin. *J Clin Endocrinol Metab 39:* 180-182, 1974.

Matsuo, H., Baba, Y., Nair, R.M.G., Arimura, A., and Schally,

A.V.: Structure of the porcine LH- and FSH-releasing hormone. I The proposed amino acid sequence. *Biochem Biophys Res Commun 43*:1334-1339, 1971.

Mendelson, W.B., Jacobs, L.S., Sitaram, N., Wyatt, R.J., and Gillin, J.C.: Methscopolamine inhibition of sleep related growth hormone secretion. *J Clin Invest 61*:1683-1690, 1978.

Merimee, T.J., and Rabin, D.: A survey of growth hormone secretion and action. *Metabolism 22*:1235-1251, 1973.

Mioduszewski, R., Grandison, L., and Meites, J.: Stimulation of prolactin release in the rat by GABA. *Proc Soc Exp Biol Med 151*:44-46, 1976.

Miyai, K., Onishi, T., Hosokawa, M., Ishibashi, K., and Kumahara, Y.: Inhibition of thyrotropin and prolactin secretions in primary hypothyroidism by 2-Br-alpha-ergocryptine. *J Clin Endocrinol Metab 39*:391-394, 1974.

Motta, M., Fraschini, F., and Martini, L.: Endocrine effects of pineal gland and of melatonin. *Proc Soc Exp Biol Med 126*:431-435, 1967.

Müller, E.E., Brambilla, F., Cavagnini, F., Peracchi, M., and Panerai, A.: Slight effect of L-tryptophan on growth hormone release in normal subjects. *J Clin Endocrinol Metab 39*:1-5, 1974.

Müller, E.E., Nistico, G., and Scapagnini, U.: *Neurotransmitters and Anterior Pituitary Function.* Academic Press, New York, 1977, pp 1-435.

Nakai, Y., Imura, H., Yoshimi, T., and Matsukura, S.: Adrenergic control mechanism for ACTH secretion in man. *Acta Endocrinol (Copenh) 74*:263-270, 1973.

Nakai, Y., Imura, H., Sakurai, H., Kurahachi, H., and Yoshimi, T.: Effect of cyproheptadine on human growth hormone secretion. *J Clin Endocrinol Metab 38*:446-449, 1974.

Negro-Vilar, A., Ojeda, S.R., and McCann, S.M.: Evidence for changes in sensitivity to testosterone negative feedback on gonadotropin release during sexual development in the male rat. *Endocrinology 93*:729-735, 1973.

Neill, J.D., Patton, J.M., Dailey, R.A., Tsou, R.C., and Tindall, G.T.: Luteinizing hormone releasing hormone in pituitary stalk blood of rhesus monkeys: Relationship to level of LH release. *Endocrinology 101*:430-434, 1977.

Nilsson, K.O., Thorell, J.I., and Hökfelt, B.: the effect of thyrotropin releasing hormone on the release of thyrotropin and other pituitary hormones in man under basal conditions and following adrenergic blocking agents. *Acta Endocrinol (Copenh) 76*:24-34, 1974.

Noe, B.D., Fletcher, D.J., and Spiess, J.: Evidence for the existence of a biosynthetic precursor for somatostatin. *Diabetes 28*:724-730, 1978.

Noel, G.L., Dimond, R.G., Wartofsky, L, Earll, J.M., and

Frantz, A.G.: Studies of prolactin and TSH secretion by continuous infusion of small amounts of TRH. *J Clin Endocrinol Metab* 39:6-17, 1974.

Oliver, C., Charvet, J.P., Codaccioni, J.L., Vague, J., and Porter, J.C.: TRH in human CSF. *Lancet* 1:873, 1974.

Oliver, C., Mical, R.S., and Porter, J.C.: Hypothalamic-pituitary vasculature: Evidence for retrograde blood flow in the pituitary stalk. *Endocrinology* 101:598-604, 1977.

Parker, D.C., Pekary, A.E., and Hershman, J.M.: Effect of normal and reversed sleep-wake cycle upon nyctohemeral rhythmicity of plasma thyrotropin: Evidence suggestive of an inhibitory influence in sleep. *J Clin Endocrinol Metab* 43:318-329, 1976.

Patel, Y.C., Rao, K., and Reichlin, S.: Somatostatin in human cerebrospinal fluid. *N Engl J Med* 296:524-533, 1977.

Plonk, J.W., Bivens, C.H., and Feldman, J.M.: Inhibition of hypoglycemia-induced cortisol secretion by the serotonin antagonist cyproheptadine. *J Clin Endocrinol Metab* 38:836-840, 1974.

Plotnick, L.P., Thompson, R.G., Kowarski, A., DeLacerda, L., Migeon, C.J., and Blizzard, R.M.: Circadian Variation of integrated concentration of growth hormone in children and adults. *J Clin Endocrinol Metab* 40:240-247, 1975.

Porter, J.C., Eskay, R.L., Oliver, C., Ben-Jonathan, N., Warberg, J., Parker, R.C., Jr., and Barnea, A.: Release of hypothalamic hormones under *in vivo* and *in vitro* conditions. In *Advances in Experimental Medicine and Biology*. Vol. 87: *Hypothalamic Peptide Hormones and Pituitary Regulation*, J.C. Porter, ed, Plenum Press, New York, 1977, pp 181-201.

Powell, G.F., Brasel, J.A., Raiti, S., and Blizzard, R.M.: Emotional deprivation and growth retardation simulating idiopathic hypopituitarism. I. Clinical evaluation of the syndrome. *N Engl J Med* 276:1271-1284, 1967.

del Pozo, E., Brun del Re, R., Varga, L., and Friesen, H.: The inhibition of prolactin secretion in man by CB-154 (2-Br-α-ergocryptine). *J Clin Endocrinol Metab* 35:768-771, 1972.

del Pozo, E., and Lancranjan, I.: Clinical use of drugs modifying the release of anterior pituitary hormones. In *Frontiers in Neuroendocrinology*, Vol 5. W.F. Ganong and L. Martini, eds, Raven Press, New York, 1978, pp 207-247.

Rapoport, B., Refetoff, S., Fang, V.S., and Friesen, H.G.: Suppression of serum thyrotropin by L-DOPA in chronic hypothyroidism: Interrelationships in the regulation of TSH and prolactin secretion. *J Clin Endocrinol Metab* 36:256-262, 1973.

Redding, T.W., and Schally, A.V.: Studies on the thyrotropin releasing hormone activity in peripheral blood. *Proc Soc Exp Biol Med* 131:421-425, 1969.

Rees, L., Butler, P.W.P., Gosling, C., and Besser, G.M.: Adrenergic blockade and the corticosteroid and growth hormone responses to methylamphetamine. *Nature* 228:565-566, 1970.

Relkin, R.: Rat pituitary and plasma prolactin levels after pinealectomy. *J Endocrinol* 53:179-180, 1972.

Rick, J., Szabo, M., Payne, P., Kovathana, N., Cannon, J.G., and Frohman, L.A.: Prolactin-suppressive effects of two aminotetralin analogs of dopamine: Their use in the characterization of the pituitary dopamine receptor. *Endocrinology* 104:1234-1242, 1979.

Rivier, C., Brown, M., and Vale, W.: Effect of neurotensin, substance P and morphine sulphate on the secretion of prolactin and growth hormone in the rat. *Endocrinology* 100:751-754, 1977.

Roth, J., Glick, S.M., Yalow, R.S., and Berson, S.A.: Hypoglycemia: A potent stimulus to the secretion of growth hormone. *Science* 140:987-988, 1963a.

Roth, J., Glick, S.M., Yalow, R.S., and Berson, S.A.: Secretion of human growth hormone: Physiologic and experimental modification. *Metabolism* 12:577-579, 1963b.

Russell, G.F.M.: General management of anorexia nervosa and difficulties in assessing the efficacy of treatment. In *Anorexia Nervosa*, R. Vigersky, ed, Raven Press, New York, 1977, pp 277-289.

Saffran, M., and Schally, A.V.: The release of corticotrophin by anterior pituitary tissue *in vitro*. *Can J Biochem* 33:408-415, 1955.

Santen, R.J., and Barden, C.W.: Episodic luteinizing hormone secretion in man. Pulse analysis, clinical interpretation, physiologic mechanisms. *J Clin Invest* 52:2617-2627, 1973.

Sayers, G., and Portanova, R.: Secretion of ACTH by isolated anterior pituitary cells: Kinetics of stimulation by corticotropin-releasing factor and of inhibition by corticosterone. *Endocrinology* 94:1723-1730, 1974.

Schally, A.V., Arimura, A., and Kastin, A.J.: Hypothalamic regulatory hormones. *Science* 179:341-350, 1973.

Schally, A.V., Redding, T.W., Arimura, A., duPont, A., and Linthicum, G.L.: Isolation of gamma-amino butyric acid from pig hypothalami and demonstration of its prolactin release-inhibiting (PIF) activity *in vivo* and *in vitro*. *Endocrinology* 100:681-691, 1977.

Scharrer, E., and Scharrer, B.: Secretory cells within the hypothalamus. In *The Hypothalamus*, vol 20, J.H. Fulton, ed, Hafner, New York, 1940, pp 170-174.

Sharp, P.J., and Fraser, H.M.: Control of reproduction. In *The Endocrine Hypothalamus*. S.L. Jeffcoate and J.S.M. Hutchinson, eds, Academic Press, London, 1978, pp 271-332.

Sheppard, M.C., Kronheim, S., and Pimstone, B.L.: Effect of substance P, neurotensin and the enkephalins on somatostatin release from the rat hypothalamus *in vitro*. *J Neurochem* *32*:182-184, 1979.

Smythe, G.A., and Lazarus, L.: Suppression of human growth hormone secretion by melatonin and cyptoheptadine. *J Clin Invest* *54*:116-121, 1974a.

Smythe, G.A., and Lazarus, L.: Growth hormone responses to melatonin in man. *Science 184*:1373-1374, 1974b.

Smythe, G.A., Compton, P.J., and Lazarus, L.: Serotoninergic control of human growth hormone secretion: The actions of L-DOPA and 2-Bromo-α-ergocryptine. In *Growth Hormone and Related Peptides*. A. Pecile, and E.E. Müller, eds, Exerpta Medica, Amsterdam, 1976, pp 222-235.

Snyder, P.J., Jacobs, L.S., Utiger, R.D., and Daughaday, W.H.: Thyroid hormone inhibition of the prolactin response to thyrotropin releasing hormone. *J Clin Invest 52*:2324-2329, 1973.

Soffer, L.J., Iannaccone, A., Gabrilove, J.L.: Cushing's syndrome. *Am J Med 30*:129-146, 1961.

Spaulding, S.W., Burrow, G.N., Donabedian, R., and Van Woert, M.: L-DOPA suppression of thyrotropin releasing hormone response in man. *J Clin Endocrinol Metab 35*:182-185, 1972.

Stachura, M.E.: Influence of synthetic somatostatin upon growth hormone release from perifused pituitaries. *Endocrinology 99*:678-683, 1976.

Stubbs, W.A., Delitala, G., Jones, A., Jeffcoate, W.J., Edwards, C.R.W., Ratter, S.J., Besser, G.M., Bloom, S.R., and Alberti, K.G.M.M.: Hormonal and metabolic responses to an enkephalin analogue in normal man. *Lancet 2*:1225-1227, 1978.

Swerdloff, R.S.: Physiological control of puberty. *Med Clin North Am 62*:351-367, 1978.

Szabo, M., and Frohman, L.A.: Dissociation of prolactin-releasing activity from thyrotropin-releasing hormone in porcine stalk median eminence. *Endocrinology 98*:1451-1459, 1976.

Takahara, J.S., Yunoki, S., Hosogi, H., Yakushiji, W., Kageyama, J., and Ofugi, T.: Concomitant increases in serum growth hormone and hypothalamic somatostatin in rats after injection of γ-aminobutyric acid, aminooxyacetic acid or γ-hydoxybutyric acid. *Endocrinology 106*:343-348, 1980.

Tamminga, C.A., Neophytides, A., Chase, T.N., and Frohman, L.A.: Stimulation of prolactin and growth hormone secretion by muscimol, a gamma-aminobutyric acid agonist. *J Clin Endocrinol Metab 47*:1348-1350, 1978.

Tuomisto, J., Ranta, T., Saarinen, A., Männistö P., and Leppäluoto, T.: Neurotransmission and secretion of thyroid stimulating hormone. *Lancet 2*:510-511, 1973.

Upton, G.V., and Amatruda, Jr., T.T.: Evidence for the presence of tumor peptides with corticotropin-releasing-factor-like activity in the ectopic ACTH syndrome. *N Engl J Med 285*:419-424, 1971.

Vale, W., Rivier, C., and Brown, M.: Regulatory peptides of the hypothalamus. *Annu Rev Physiol 39:* 473–527, 1977.

Vale, W., Spiess, J., Rivier, C., and Rivier, J.: Characterization of a 41–Residue ovine hypothalamic peptide that stimulates secretion of corticotropin and β-endorphin. *Science 213:* 1394–1397, 1981.

Valverde C., Chieffo V., and Reichlin, S.: Failure of reserpine to block ether–induced release of prolactin: Physiological evidence that stress induced prolactin release is not caused by acute inhibition of PIF secretion. *Life Sci 12:* 327–335, 1973.

Van Delft, A.M.L., Kaplanski, J., and Smelik, P.G.: Circadian periodicity of pituitary-adrenal function after p-chlorophenylalanine administration in the rat. *J Endocrinol 59:* 465–474, 1973.

Van Loon, G.R.: A defect in catecholamine neurones in patients with prolactin-secreting pituitary adenomas. *Lancet 2:* 868–871, 1978.

Van Vugt, D.A., Bruni, J.F., and Meites, J.: Naloxone inhibition of stress-induced increase in prolactin secretion. *Life Sci 22:* 85–89, 1978.

Wakabayashi, I., Miki, N., Ohmura, E., Kanda, M., Miyoshi, H., Demura, R., and Shuzume, K.: Effects of chlorpromazine, naloxone and hypothalamic ventromedial lesions on vasoactive intestinal polypeptide-induced growth hormone and prolactin release in rats. Proceedings of 6th International Congress of Endocrinology. Melbourne, Australia. (Abstr 337), 1980.

Whitten, W.K.: Occurrence of anestrous in mice caged in groups. *J Endocrinol 18:* 102–107, 1959.

Wilbur, J.F., and Baum, D.: Elevation of plasma TSH during surgical hypothermia. *J Clin Endocrinol Metab 31:* 372–375, 1970.

Wurtman, R.J.: Brain monoamines and endocrine function. *Neurosci Res Progr Bull 9:* 172–297, 1971.

Yasuda, N., and Greer, M.: Demonstration of corticotropin-releasing activity in rat and human peripheral blood. *Acta Endocrinol (Copenh) 84:* 1–10, 1977.

Yen, S.S.C.: Neuroendocrine aspects of the regulation of cyclic gonadotropin release in women. In *Clinical Neuroendocrinology,* L. Martini and G.M. Besser, eds, Academic Press, New York, 1977, pp 175–196.

Yoshimura, M., Ochi, Y., Miyazaki, Y., Shiomi, K., and Hachiya, T.: Effect of L-5-HTP on release of growth hormone, TSH and insulin. *Endocrinol Jpn 20:* 135–141, 1973.

Yoshimura, M., Hachiya, T., Ochi, Y., Nagasaka, A., Takeda, A., Hidaka, H., Refetoff, S., and Fang, V.S.: Suppression of elevated serum TSH levels in hypothyroidism by fusaric acid. *J Clin Endocrinol Metab 45:* 95–98, 1977.

Young, R.T.T.: Effect of propranolol on plasma growth hormone

response in insulin-induced hypoglycemia in thyrotoxic pa-
tients. *J Clin Endocrinol Metab 37:* 968-971, 1973.
Zafar, M.S., Mellinger, R.C., Fine, G., Szabo, M., and Frohman,
L.A.: Acromegaly associated with a bronchial carcinoid tumor.
Evidence for ectopic production of growth hormone-releasing
activity. *J Clin Endocrinol Metab 48:* 66-71, 1979.

5

The Neurophysiology of Hypothalamic–Pituitary Regulation and of Hypothalamic Hormones in Brain

LEO P. RENAUD

INTRODUCTION

From a historical viewpoint, the pioneering efforts of Ernst and Berta Scharrar, Geoffrey Harris, and John Green are largely responsible for the development of our present concepts on the role of the hypothalamus in the regulation of neurohypophyseal and adenohypophyseal secretion. Several chapters in this volume detail our current understanding of the physiology and pharmacology in this area. From a neurophysiological viewpoint, the studies of these early investigators also prompted the development of electrophysiological techniques for the identification of hypothalamic "neuroendocrine" neurons and subsequent examination of the electrical aspects of hypothalamic-pituitary regulation. With the discovery that hypothalamic peptide "release" or "release-inhibiting" factors was distributed within neural tissue in extrahypothalamic brain (and in certain non-neural tissues), there have been numerous investigations centered on the possible role of endogenous brain peptides, especially in interneuronal communication.

This chapter is intended to provide a brief review of developments in two areas of current interest: (1) the neurophysiology of hypothalamic regulation of pituitary secretion, obtained through analysis of the activity of hypothalamic "peptidergic" neurons related to anterior and posterior pituitary secretion, and (2) the role of peptides in brain, obtained from an electrophysiological analysis of the actions of three hypothalamic peptides, that is, thyrotrophin-releasing hormone(TRH), luteinizing hormone-releasing hormone (LH–RH), and somatostatin, on neural and adenohypophyseal tissue. Since this chapter is not intended as an exhaustive review, the reader is referred to

several current reviews for further details (for peptidergic
neurons: Cross et al., 1975; Moss, 1976; Hayward, 1977; Renaud,
1978; Richard et al., 1978; for peptides actions: Barker, 1977;
Iversen et al., 1978; Renaud and Padjen, 1978; Barker and
Smith, 1980; Liebeskind et al., 1979).

NEUROPHYSIOLOGY OF HYPOTHALAMIC PEPTIDERGIC NEURONS

For the sake of simplicity of description, two general categories of hypothalamic peptidergic neurons can be recognized with
electrophysiological techniques (Fig. 5-1): *neurohypophyseal*
neurons project to the posterior pituitary and are engaged in
the secretion of oxytocin and vasopressin (and possibly other
peptides) into the systemic circulation; *tuberoinfundibular*
neurons project to the median eminence, where they liberate releasing or release-inhibiting factors that govern the secretion
of anterior pituitary hormones. In each instance, the electrophysiological identification of these cells depends upon evidence for antidromic activation following stimulation of their
nerve terminals within the posterior pituitary or median eminence, respectively (Renaud, 1978). However, it is important
to mention that their identification as "peptidergic" neurons
is tentative and based on the assumption that neurons with
these projections are *most probably* peptide-producing neurons.
Unfortunately, the required correlation between single cell recordings and immunocytochemical identification has proved to be
extremely difficult in mammalian tissue with current *in vivo*
methods. Therefore the following description will pertain to
neurons most appropriately described as *putative* peptidergic
neurons.

Neurohypophyseal Neurons

Neurons in the supraoptic and paraventricular nuclei are responsible for the synthesis and release of oxytocin and vasopressin (Bisset, 1976). Since these nuclei are considered to
have differentiated from the preoptic nucleus of lower vertebrates (Pickford and Atz, 1957), electrophysiological studies
performed on these peptide neurons in the goldfish provided the
first detailed information on the neuronal nature of neurosecretory neurons (Kandel, 1964; Hayward, 1974). Extracellular
and intracellular recordings from supraoptic and paraventricular neurosecretory neurons in higher species have also confirmed that neurosecretory neurons do indeed display electrical
properties similar to those recorded from other central neurons
(see Cross et al., 1975; Hayward, 1977; Renaud, 1978; for re-

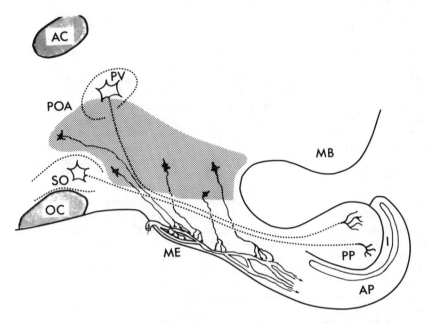

Fig. 5-1. A simplified sketch of the rat hypothalamus in saggital section to illustrate the two classic neurosecretory pathways related to the pituitary: the
neurohypophyseal tract is indicated as interrupted
lines originating from magnocellular paraventricular
(PV) and supraoptic (SO) neurons, terminating in the
posterior pituitary (PP); the tuberoinfundibular
pathway is seen to originate from parvicellular neurons in the medial hypothalamus, preoptic area
(POA), and mamillary body region (MB), with axon
terminations on portal capillaries in the median
eminence (ME) for the release of factors that regulate anterior pituitary (AP) secretion. Abbreviations: AC:anterior commissure; OC:optic chiasm;
I:inter-mediate lobe. (From Renaud, 1978.)

view). The ability to record these neurons under conditions
appropriate to the secretion of oxytocin or vasopressin has
provided evidence in favor of a direct correlation between neuronal activation (action potential invasion of axon terminals
in the neurohypophysis) and hormone release. In agreement with
immunohistochemical observations (Zimmerman et al., 1978),
electrophysiological evidence has been obtained for a distinction between oxytocin-secreting and vasopressin-secreting neurons. These data are reviewed in the following sections.

Oxytocinergic Neurons

The milk ejection reflex is thought to result from the episodic
release of oxytocin from the neurohypophysis. Recordings obtained from identified supraoptic and paraventricular neurosecretory neurons in the anesthetized lactating female rat with
suckling pups has indicated that approximately 50 percent of
the neurosecretory neurons display an irregular or random pattern of action-potential activity that is interrupted every few
minutes by an abupt synchronized increase in spike discharge
frequency (up to 30 to 40 Hz) lasting two to four seconds and
followed within 12 to 15 seconds by a rise in intramammary
pressure and milk injection (Fig. 5-2) (Lincoln and Wakerley,
1974, 1975; Poulain et al., 1977). This response of the mammary gland also occurs 10 to 15 seconds after the intravenous
injection of synthetic oxytocin or electrical stimulation of
the pituitary stalk at 50 Hz (Wakerley and Lincoln, 1973;
Lincoln and Wakerley, 1975; Cross et al., 1975), indicating
that the episodic neuronal patterns already described most
probably represent the activity of "oxytocinergic" neurons.

In the lactating animal, vaginal distension also provokes an
increase in the activity of supraoptic neurosecretory neurons
and a rise in intramammary pressure (Dreifuss et al., 1976).
The magnitude of both the unit activity and the intramammary
pressure increase is less than that associated with sucking.
The specificity of this particular stimulus as a means of
identification of oxytocinergic neurons has been questioned,
since vaginal distension will also produce a rise in the activity of both randomly firing neurons and phasically firing
neurons; the later are usually related to the release of vasopressin (Dreifuss et al., 1976).

Oxytocin is a potent stimulator of myometrial activity at
term (Cross, 1958) and is considered to be of importance for
both the onset and the course of labor. Recordings from paraventricular neurosecretory neurons during labor have revealed
an increase in the mean spontaneous firing rate of all neurosecretory neurons, suggesting that a tonic release of both neuro-
hyphyseal hormones occurs during labor (Boer and Nolten, 1978).
However, to date, there has been no consistent relationship
demonstrated between uterine contraction and specific neurosecretory cellular activity.

Vasopressinergic Neurons

It has long been known that hypothalamic neurosecretory neurons
are sensitive to the intracarotid injection of hyperosmolar solutions, leading to the postulate that such activation prompts

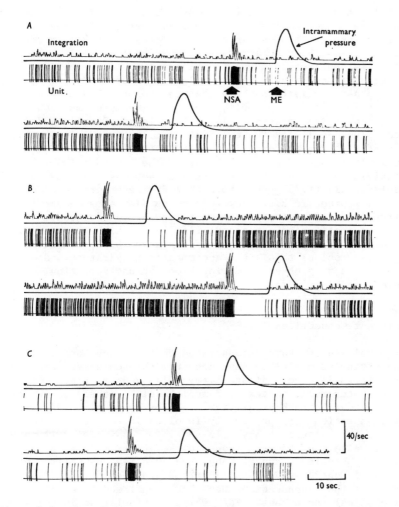

Fig. 5-2. The three sets of polygraph records illustrate activity patterns for three different supraoptic oxytocinergic neurons. For each row, the top and bottom records illustrate unit activity, and the intermittent appearance of high-frequency bursts of neurosecretory activity (NSA), followed within a few
seconds by a rise in intramammary pressure (illustrated in the central trace) that precipitates milk
ejection (ME). Note the stereotyped nature of the
secretory response and the relatively uniform latency to milk ejection. (Lincoln and Wakerley, 1975).

the release of vasopressin (Cross and Harris, 1952; Cross and
Green, 1959; Harris, 1955). In fact, vasopressin release is
probably the result of an interaction between several factors,
including osmotic, volumetric, and behavioral stimuli (Hayward,
1975). In contrast to oxytocin-releasing neurons, supraoptic
and paraventricular vasopressin-releasing neurons display a
unique phasic activity pattern that can be provoked by several
vasopressin-releasing stimuli, that is, osmotic stimuli and
dehydration (Arnauld et al., 1975; Brimble and Dyball, 1977),
blood volume depletion (Wakerley et al., 1975; Poulain et al.,
1977) (Fig. 5-3), and carotid occlusion (Dreifuss et al.,
1976). This phasic activity appears to be one functional state
of activity of these "vasopressinergic" neurons. For example,
Arnauld et al. (1975) have shown that during progressive dehydration, supraoptic neurosecretory cells display a change in
their pattern of activity from an irregular low frequency discharge to phasic discharges and, finally, to high frequency
discharges, in association with an increase in plasma osmolar-
ity; a reversal of this trend accompanies rehydration. Similar
changes have been observed during blood volume depletion
(Poulain et al., 1977).

Afferent Connections

Anatomical studies have demonstrated that the supraoptic and
paraventricular nuclei receive connections from several structures, including the brainstem, amygdala, septum, hippocampus,
olfactory tubercle, cortex, and mediobasal hypothalamus (Woods
et al., 1969; Léranth et al., 1975; Zaborsky et al., 1975).
These connections are presumably involved in the reflex mechanisms for the release of oxytocin and vasopressin, as well as
for the release of these hormones evoked by electrical stimulation in hypothalamic and extrahypothalamic regions (Bisset et
al., 1971; Negoro and Holland, 1972; Hayward et al., 1977).
Certain of these connections have been confirmed by electrophysiological techniques. For example, stimulation in the septum, amygdala, midbrain reticular formation, and hippocampus
can evoke short latency alterations in excitability of supraoptic and paraventricular neurosecretory neurons (Negoro and
Holland, 1972; Negoro et al., 1973; Koizumi and Yamashita,
1972; Pittman et al., 1980; Renaud and Arnauld, 1979). Peripheral afferents are also important in the release of oxytocin
and vasopressin (Dyball, 1968); for example, vagal stimulation
will produce vasopressin release, but presumably this effect is
mediated through multisynaptic pathways, since supraoptic neurosecretory neurons display only long latency excitations to
such stimuli (Barker et al., 1971). However, it is clear from

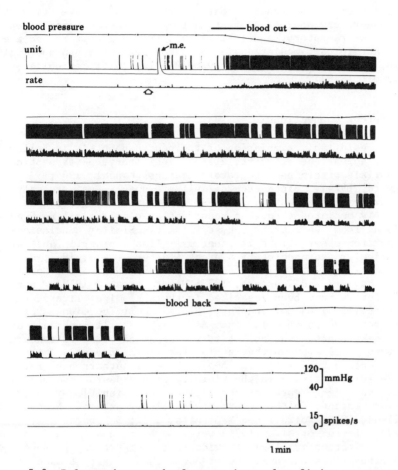

Fig. 5-3. Polygraph records from an irregular firing neurosecretory neuron in the supraoptic nucleus of a lactating female rat to illustrate an absence of response
to milk ejection (me), but a typical response to
hemorrhage with the appearance of high bursts of
phasic activity, characteristic of vasopressinergic
neurons. Blood pressure is illustrated in the upper
trace, unit activity in the second and fourth
traces, and intramammary pressure in the third
trace. (From Poulain et al., 1977.)

a series of recent studies that supraoptic and paraventricular neurosecretory neurons do respond to activation of chemoreceptors, baroreceptors, and left atrial stretch receptors and are at least in part mediated via the vagus nerve (Kannan and Yagi, 1978; Koizumi and Yamashita, 1978; Harris, 1979; Yamashita and Koizumi, 1979).

Efferent Connections

In the goldfish, stimulation of the neurohypophysis is followed by the appearance of short latency, inhibitory, postsynaptic potentials within neurosecretory neurons (Kandel, 1964; Hayward, 1974). Since such stimuli presumably activate only the axons of the neurohypophyseal tract, this inhibition is probably mediated through a recurrent axon collateral pathway arising from the neurohypophyseal tract. Similar conclusions arise from observations in other mammalian supraoptic and paraventricular neurosecretory neurons (Barker et al., 1971; Nicoll and Barker, 1971; Dreifuss and Kelly, 1972; Negoro and Holland, 1972; Hayward and Jennings, 1973; Negoro et al., 1973). Such collaterals have been described in the anatomical literature (Christ, 1966; Lu Qui and Fox, 1976). Their functional role is a source of much speculation. Also, it is still uncertain as to whether these axon collaterals synapse directly onto neurosecretory cells or whether an interneuron is present in the recurrent pathway (see Fig. 5-5). Several authors have recorded cellular activity in the vicinity of the neurosecretory neurons that might represent the activation of inhibitory interneurons engaged in the recurrent inhibitory pathway (Negoro and Holland, 1972; Koizumi and Yamashita, 1972; Negoro et al., 1973). Barker er al., (1971) were, however, unable to demonstrate electrophysiological evidence to suggest the presence of inhibitory interneurons, and proposed that recurrent axon collaterals in the neurohypophyseal tract might directly mediate inhibition of neurosecretory neurons. Their failure to block recurrent inhibition with a variety of antagonists, and the observation that vasopressin itself had a depressant effect on the excitability of neruosecretory neurons supported this postulate. However, the finding that recurrent inhibition persists in the neurohypophyseal system in the homozygous vasopressin-deficient Brattleboro rat (Dyball, 1974) and the observation that oxytocin tends to enhance the excitability of neruosecretory neurons (Moss et al., 1972) has argued against the consideration of these peptides as direct mediators of recurrent inhibition in the neurohypophyseal system.

In vitro studies in the bullfrog hypothalamo-hypophyseal system have indicated that stimulation of the neurohypophysis activated both recurrent inhibitory and recurrent facilitatory connections (Koizumi et al., 1973). A precise neural network that would account for these observations remains uncertain.

Recent anatomical observations have demonstrated pathways containing vasopressin, oxytocin, and their associated neurophysins in areas extending beyond the neurohypophyseal tract and include the amygdala, midbrain, spinal cord, and median eminence (Swanson, 1977; Elde and Hökfelt, 1978; Zimmerman et al., 1978; Swanson and McKellar, 1979). This has raised the question as to whether the same neuron might project simultaneously to the neurohypophysis and to one or more of these regions. Studies in our laboratory have noted that some neurons in the paraventricular nucleus and periventricular regions demonstrate simultaneous antidromic activation from both the neurohypophysis and the median eminence, indicating that some cells send axons to both sites (Pittman et al., 1978). Such observations may indicate functional interactions between the neurohypophyseal system and the tuberoinfundibular neuronal system. Additional studies have noted that some neurons in the periventricular region, but not within the paraventricular nucleus, project to both the neurohypophysis and the midbrain region (Pittman et al., 1980). Failure to detect evidence of simultaneous projections to the neurohypophysis and to other extrahypothalamic regions from paraventricular neruosecretory neurons suggests that the extrahypothalamic peptidergic pathways from this nucleus are, in general, anatomically separate entities from the neurohypophyseal tract.

Tuberoinfundibular Neurons

For purposes of description, the term "tuberoinfundibular" neuron is defined as any neuron that demonstrates antidromic activation following electrical stimulation of the surface of the median eminence near the junction with the pituitary stalk.

Location And Activity Patterns

In the rat and guinea pig, tuberoinfundibular neurons have been located throughout the medial hypothalamus and medial preoptic area by electrophysiological techniques (Makara et al., 1972; Sawaki and Yagi, 1973; Poulain, 1977; Renaud et al., 1978) (Fig. 5-4). In general, the activity of spontaneously active tuberoinfundibular neurons lacks specific patterns, with one possible exception—some tuberoinfundibular neurons in the paraventricular nucleus display phasic activity patterns similar to those described for vasopressinergic neurons (Blume er al.,

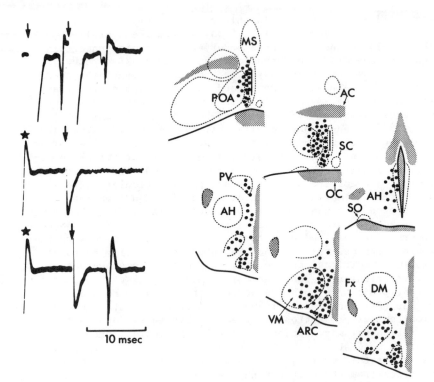

Fig. 5-4. Identification and location of tuberoinfundibular
neurons. The oscilloscope sweeps on the left illustrate antidromic invasion of a neuron in the ventromedial nucleus following stimulation of the surface
of the median eminence (at arrows). This neuron
demonstrates constant latency responses to paired
median eminence stimulation at frequencies greater
than 150 Hz (upper trace), and collisions at appropriate intervals between spontaneous (star) action
potentials and antidromic evoked action potentials
(lower traces). On the right, the black dots located throughout the medial hypothalamus and the
medial preoptic area represent the distribution of
tuberoinfundibular neurons identified using antidromic invasion criteria. (From Renaud, 1978.)

1978). If the latter tuberoinfundibular neurons can indeed be
associated with vasopressin release in the median eminence,

this would provide the first indication in the tuberoinfundibular system that activity can differentiate a particular category of neuron.

Afferent Connections

A wealth of evidence based on stimulation and lesion experiments indicates that several extrahypothalamic sites, including
the amygdala, preoptic area, hippocampus, and brainstem, can
influence adenohypophyseal secretion, presumably through their
ability to alter the activity patterns of tuberoinfundibular
neurons (Döcke, 1974; Ellendorf et al., 1973; Martin, 1972;
Martin et al., 1973; Zolovick, 1972). Recent electrophysiological studies in the rat hypothalamus have supported this proposal. Thus, electrical stimulation in the amygdala, lateral
septum, preoptic-anterior hypothalamic area, and dorsal hippocampus have been shown to evoke short latency changes in the
excitability of tuberoinfundibular neurons in the mediobasal
hypothalamus (Renaud, 1976a,b, 1977a,b; Renaud et al., 1978).
On the other hand, stimulation in the midbrain periaqueductal
gray and medial reticular formation does not appear to modify
the activity of mediobasal hypothalamic tuberoinfundibular
neurons (Pittman et al., 1979) (Fig 5-5).

Efferent Connections

Electrophysiological studies on tuberoinfundibular neurons suggest that some of these cells have central axon collaterals in
addition to their median eminence projection. Some of these
collaterals directly or indirectly activate a local postsynaptic inhibitory mechanism (Renaud, 1976a,b); others project to
adjacent regions of the hypothalamus, that is, anterior hypothalamic area and paraventricular nucleus (Harris and Sanghera,
1974; Renaud and Martin, 1975a; Renaud, 1976a). Finally, some
collaterals appear to extend to extrahypothalamic regions, such
as the medial preoptic area, midline thalamic nuclei, amygdala,
and lateral septum (Renaud, 1976b; Renaud et al., 1978) (Fig.
5-5). Indications for these connections arise from the electrophysiological demonstration of simultaneous antidromic activation from both the median eminence and from one of the areas
mentioned previously. At present, little is known of the functional significance of such extrahypothalamic or certain intrahypothalamic axon collaterals, but one might postulate that
such pathways serve to inform other brain areas of the state of
activity in the tuberoinfundibular system in the form of an
"ultra-short" feedback loop. It is also noteworthy that some

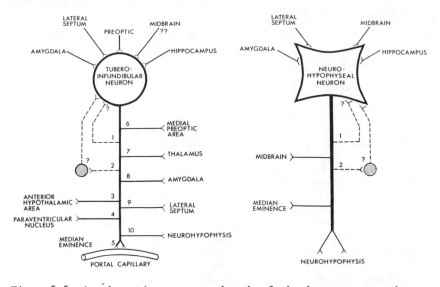

Fig. 5-5. A schematic summary sketch of the known connections
of tuberoinfundibular neurons (on the left) and neurohypophyseal neurons (on the right). The upper
portion of each figure indicates sites known to send
projections to the tuberoinfundibular or neurohypophyseal neurons respectively. The heavy vertical
lines refer to the main peptidergic axons projecting
to the median eminence portal capillary and neurohypophysis, respectively, thus identifying each cell
type. Additional efferent projections are indicated
as axon collaterals of this main connection. Intrahypothalamic axon collaterals in recurrent pathways
that either directly or indirectly (through local
interneurons) engage the parent neuron, and whose
transmitter agent is unknown (?), are depicted by
numbers 1 and 2 in each system. Additional axon
collaterals extend both within the hypothalamus and
to extrahypothalamic areas, as indicated. (Adapted
from Renaud et al., 1979b.)

of the extrahypothalamic regions that receive such axon collaterals in turn provide afferent fibers to tuberoinfundibular
neurons. The evidence for such collaterals in putative

peptidergic pathways supports the notion (Dale, 1935) that hypothalamic releasing factors may be released from terminals of their central connections in association with activity in the tuberoinfundibular system. The following sections present some evidence that such peptides may indeed have a functional role at synaptic junctions in both the central and peripheral nervous system.

NEUROPHYSIOLOGY OF HYPOTHALAMIC HORMONES

Thyrotropin-Releasing Hormone

TRH was the first of the hypothalamic releasing factors to be isolated and structurally characterized (Nair et al., 1970; Burgus et al., 1970). Radioimmunoassayable TRH is reported to be present both within the hypothalamus and in widely distributed sites of the central nervous system (CNS) in several species (Brownstein et al., 1974; Winokur and Utiger, 1974; Jackson and Reichlin, 1974; Oliver et al., 1974). Recent reports have provided evidence that extrahypothalamic TRH-like immunoreactivity had been found within neurons and nerve fibers in specific CNS regions (Elde and Hökfelt, 1978). With subcellular fractionation studies, TRH has been found within and released from synaptosomal preparations (Bennett et al., 1975; Edwardson and Bennett, 1977; Winokur et al., 1977; Barnea et al., 1978; Jeffcoate et al., 1979). High-affinity TRH binding has been noted in rat brain subcellular fractions (Burt and Snyder, 1975). TRH biosynthesis occurs within neural tissue (McKelvy et al., 1975; McKelvy and Epelbaum, 1978; Hersh and McKelvy, 1979). These observations have heightened speculation as to the possible neurotransmitter role of TRH in extrahypothalamic brain. Reports listed elsewhere in this volume provide evidence that TRH can influence neural tissue. Systemic, intracerebral, or intracerebroventricular administrations of TRH induce actions or effects that appear to be unrelated to its role as a hypothalamic releasing factor. Thus, TRH may be involved in central temperature regulating processes. TRH-induced hypothermia has been reported (Metcalf, 1974), although the majority of reports suggest that TRH has a stimulatory action to increase body temperature and to reverse drug-induced narcosis and hypothermia (Breese et al., 1975; Brown and Vale, 1975; Carino et al., 1976; Cott et al., 1976; Brown et al., 1977; Holaday et al., 1978).

The following sections will review the electrophysiological aspects related to TRH action in adenohypophysis and neural tissue.

Adenohypophysis

The release of TRH into the pituitary portal circulation and
its subsequent contact with appropriate cells in the adenohypophysis, that is, thyrotrophs and mammotrophs, is presumed to
produce a calcium-dependent stimulus-secretion coupling with
subsequent hormone release (Douglas, 1968, 1974). The events
associated with this process have been examined with electrophysiological techniques in both *in vivo* and *in vitro* preparations. Intracarotid injections of synthetic TRH were found to
be associated with a predominantly depolarizing response during
intracellular recording from some cells in the intact adenohypophysis, although these cells did not demonstrate action potentials (York et al., 1975). On the other hand, a marked
increase in the frequency of calicum-dependent action potential
activity has been described during extracellular and intracellular recordings from dissociated adenohypophyseal cells maintained in culture, and from a clonal anterior pituitary cell
line (Kidokoro, 1975; Taraskevich and Douglas, 1977; Dufy et
al., 1979). The short latency noted between the TRH applications and cellular responses (Fig. 5-6) are likely to be more
than casually related to the prompt TRH-induced secretion of
thyroid-stimulating hormone (TSH) and prolactin from dis-persed
and cultured rat anterior pituitary cells (Nakano et al.,
1976).

Spinal Cord

Systemic injections of TRH induce or increase tremor, shaking,
and rotational behavior (Prange er al., 1974; Schenkel-Hulliger
et al., 1974; Cohn et al., 1975; Wei et al., 1975; Kruse, 1976;
Cooper and Boyer, 1978) and stimulate colonic activity (Smith
et al., 1977). This increase in motoricity suggests that at
least a part of the action of TRH is to facilitate the activity
of lower motoneurons. TRH has been found in the spinal cord of
the rat, especially concentrated in the anterior horn (Hökfelt
et al., 1975c; Kardon et al., 1977). Evidence for a direct and
excitatory action of TRH on motoneurons in the rat has not yet
been reported. There is, however, indirect evidence that TRH
does stimulate motoneuron activity in the cat spinal cord;
intravenous TRH administration (2 mg/kg) is followed by an
increase in electromyographic activity and muscle tonus, an increase in the spontaneous electrical activity recorded from
lumbar (L6) ventral roots, and a marked increase in monosynap-
tic and polysynaptic evoked reflex potentials recorded from the
L6 ventral root following stimulation of L6 dorsal root (Cooper
and Boyer, 1978).

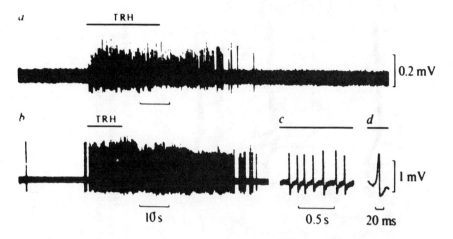

Fig. 5-6. The upper and lower traces illustrate action potentials induced by TRH in two adenohypophyseal cells. TRH is applied at the time illustrated by the horizontal bar. The upper record illustrates the response of an electrically silent cell to 50 nM TRH, while the lower spontaneously active cell responds to 5 nM TRH. (From Taraskevich and Douglas, 1977.)

TRH has been detected in frog brain (Jackson and Reichlin, 1974), but no histochemical studies have yet been reported on its specific location within amphibian spinal cord. Nevertheless, the hemisected frog spinal cord has been found ideally suited for quantitative neuropharmacological studies of peptide actions using sucrose gap and intracellular recording techniques. In this preparation, TRH (10^{-4} –10^{-3}M) can be shown to produce rather weak depolarizations, presumably through both direct and indirect effects on motoneurons (Nicoll, 1977, 1978) (Fig. 5-7). The 3-methyl-histidine TRH analoque is equal to, or more potent than, TRH, whereas the 1-methyl-histidine TRH analoque is inactive; intracellular records suggest that the depolarization is associated with a decrease in membrane resistance, probably due to an increase in sodium ion conductance.

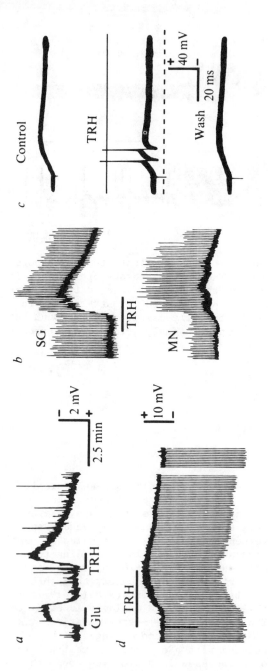

Fig. 5-7. Exitatory actions of TRH in the frog spinal cord. Upper deflections in (A) represent depolarization of motoneurons to glutamate and TRH recorded with sucrose gap from a ventral root. In (B), the top trace records ventral root, and the bottom trace records intracellular responses in a motoneuron during TRH application. In (C), intracellular recordings demonstrate the weak depolarizing effect during TRH application (central record) with the occurrence of action potentials. In (D), records indicate a slight decrease in membrane resistance during TRH application, in association with depolarization during intracellular recording. (From Nicoll, 1977.)

It is worth noting that the excitatory action of TRH is weaker that that of glutamate, and considerably weaker than that of another peptide substance P, suggesting that TRH may function in a background manner to facilitate transmission through pathways that subserve basic reflex activity.

Central Neurons

The electrophysiological studies of Steiner (1972, 1975) yielded some of the earlier evidence that TRH had central effects independent of its actions on adenohypophyseal tissue. Steiner noted that microiontophoretic application of TRH was associated with a decrease in the excitability of a certain percentage of neurons, an effect that was usually opposite to the effects of the thyroid hormones triiodothyronine (T_3) and L-thyroxine (T_4) on these neurons (Davidoff and Ruskin, 1972). This depressant effect of TRH on both spontaneous and, especially, L-glutamate-evoked activity has been confirmed in the cerebrum, cerebellum, hypothalamus, and brainstem (Dyer and Dyball, 1974; Renaud and Martin, 1975b,c; Renaud et al., 1975, 1976; Winokur and Beckman, 1978). In most instances, the decrease in excitability is abrupt in onset and readily reversible. At present there are no data on TRH actions during intracellular recording in mammalian CNS, and one can only surmise that some of the observed effects may represent a *direct* action of TRH. There is evidence that suggests TRH may have another additional mode of action, that is, to *modify* the response of neurons to other neurotransmitters and, thus, act as a potential neuromodulator. In the cerebral cortex, for example, it has been reported that TRH and certain TRH analogues can enhance the excitatory actions of acetylcholine, but not of glutamate, on specific cholinoceptive neurons (Yarbrough, 1976, 1978; Yarbrough et al., 1978; Yarbrough and Singh, 1978). This observation has *not* been confirmed by others (Winokur and Beckman, 1978; Renaud et al., 1979a; but cf. Braitman et al., 1979). Moreover, our recent observations (Renaud et al., 1979a) indicate that TRH and TRH-like analogues may have a selective action to attenuate L-glutamate- (but not L-aspartate- or acetylcholine-) evoked activity of cortical neurons (Figs. 5-8, 5-9). Thus, endogenous TRH or TRH-like peptides may have a specific influence modifying synaptic activity at glutamate-activated receptor sites.

Luteinizing Hormone-Releasing Hormone

The decapeptide (LH-RH) was the second hypothalamic factor to be isolated and to receive structural characterization (Amoss

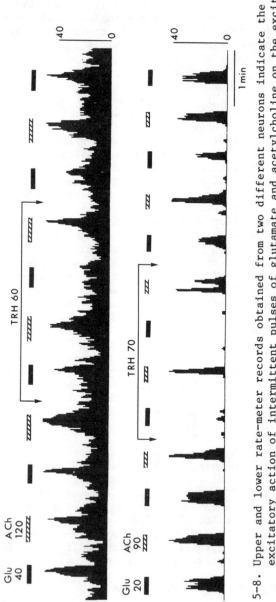

Fig. 5-8. Upper and lower rate-meter records obtained from two different neurons indicate the excitatory action of intermittent pulses of glutamate and acetylcholine on the excitability of neurons in the rat cerebral cortex. Numbers refer to microiontophoretic application currents. Note that during simultaneous application of TRH, the glutamate- but not the acetylcholine-evoked responses are diminished reversibly. (From Renaud et al., 1979. Copyright 1979 by the American Association for the Advancement of Science.)

et al., 1971; Schally et al., 1971). Shortly thereafter, it was reported that minute amounts of LH–RH administered subcutaneously could potentiate sexual behavior in hypophysectomized, ovariectomized female rats (Moss and McCann, 1973; Pfaff, 1973). Suspicion that LH–RH, or a related peptide, might have effects on neural tissue, possibly through a neurotransmitter or neurohormonal type of action, is strengthened by the detection of its presence within neural tissue by radioimmunoassay (see chapter 2 by Grant) and its location within nerve cells and fibers by immunohistochemistry (Elde and Höfelt, 1978). Its presence within and release from synaptosome fractions of brain (Edwardson and Bennett, 1977) supports this notion.

Microiontophoretic methods have been utilized to apply LH–RH and related peptides to central neurons in several areas. LH–RH appears to exert a predominantly depressant action on the excitability of neurons in the organum vasculosum lamina terminalis (Felix and Phillips, 1979), the cerebellar cortex, and ventromedial hypothalamus (Dyer and Dyball, 1974; Steiner, 1975; Renaud et al., 1975, 1976; Renaud, 1976b). Others have noted that certain neurons in the hypothalamus and elsewhere may be excited by LH–RH (Kawakami and Sakuma, 1974, 1976; Kawakami and Kimura, 1975; Moss et al., 1978); LH–RH may have a specific excitatory action on neurons in the arcuate–ventromedial hypothalamic region, which do not send fibers to the median eminence, that is, nontuberoinfundibular neurons (Moss et al., 1978). It is perhaps worthy of note that some LH–RH analogues with little or no ability to release LH from anterior pituitary may still have powerful actions on central neurons (Renaud, 1977b) (Fig. 5–10), in support of the possibility that one may eventually be able to dissociate CNS effects of LH–RH–like substances from their neuroendocrine actions.

Recently studies by Kuffler and his colleagues (Jan et al., 1979) have indicated that certain presympathetic ganglionic fibers of the frog spinal cord may contain and release an LH–RH–like peptide that appears responsible for a very prolonged excitatory action within sympathetic ganglia of that species.

Somatostatin

Since the tetradecapeptide somatostatin was first isolated and structurally characterized (Brazeau et al., 1973), the development of specific sensitive radioimmunoassays and immunohistochemical procedures have led to the detection of somatostatin in both neural and non–neural tissues (Hökfelt et al., 1975a; Elde and Hökfelt, 1978; see also chapter 2 by Grant et al.). With the availability of synthetic somatostatin and somatostatin analogues, a wide variety of studies indicate that somato-

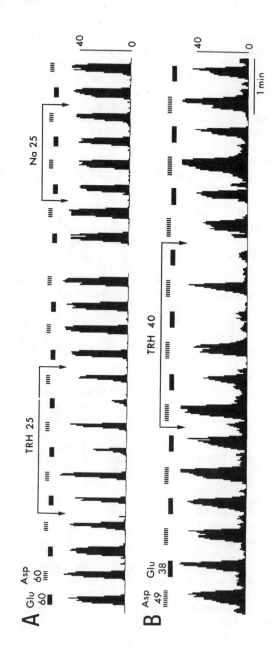

Fig. 5–9. The upper and lower rate–meter records from two different rat cerebral cortical neurons indicate the selective depressant effect of TRH on glutamate– but not aspartate– evoked excitations. The figure includes a current control sequence (sodium ions) in the upper right. (From Renaud et al., 1979. Copyright 1979 by the American Association for the Advancement of Science.)

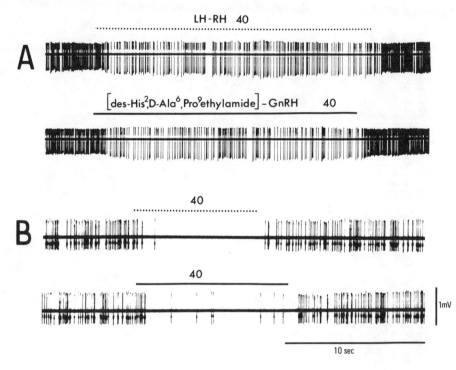

Fig. 5-10. Oscilloscope records from two different ventromedial
hypothalamic neurons illustrate the similarity in
the depressant effect of LH-RH and the LH-RH analogue applied by microiontophoresis with similar
currents. Note the reversibility in the action of
the peptides. (Renaud, unpublished observations.)

statin has functions unrelated to its action as an anterior
pituitary growth hormone release inhibiting factor. Of particular interest is the possibility that somatostatin may have a
neurotransmitter or neuromodulatory role. The following section examines specific examples in which electrophysiology has
contributed to our understanding (or confusion) of the possible
role of somatostatin.

Peripheral Nervous System

Somatostatin is present in the peripheral nervous system (Elde
and Hökfelt, 1978). Some of the nerve cell bodies and axons of

the myenteric and submucous plexuses display somatostatin-like immunoreactivity, suggesting that these may be interneurons in the enteric nervous system (Costa et al., 1977). Ganglia of the myenteric plexus of the guinea pig ileum maintained *in vitro* (North and Williams, 1977) can be recorded with extracellular methods. Advantages of this type of preparation over *in vivo* CNS studies are simplicity and ability to apply substances in known concentration to the perfusing solution and to eliminate transsynaptic influences by the use of calcium-free solutions. Somatostatin (300 pM–1 µM) inhibits the spontaneous discharge of the majority of myenteric neurons, even after removal of extracellular calcium (Williams and North, 1978) (Fig. 5-11), presumably through a direct action on the cell. The activity of many of these same neurons can also be depressed by enkephalin and β-endorphin (North and Williams, 1976); these actions can be reversibly antagonized by naloxone (up to 1 µM), although naloxone has been shown to have no effect on the action of somatostatin.

An apparently different effect of somatostatin occurs at the myenteric plexus nerve terminals on longitudinal smooth muscle, where somatostatin (0.7–2.1 µM) is reported to inhibit the electrically induced release of acetylcholine by the isolated myenteric plexus, without modifying the contractile response of smooth muscle to exogenous acetylcholine (Guillemin, 1976). The response demonstrates tachyphylaxis and is not modified by opiate antagonists. This inhibition of acetylcholine release may provide at least partial explanation for the effects of somatostatin in decreasing gastrointestinal contractility (Guillemin and Gerich, 1976). Intracellular recordings from guinea pig taenia coli have revealed only a weak hyperpolarization by somatostatin in a proportion of cells (quoted in Williams and North, 1978). Therefore it would seem that somatostatin in the gut has at least two separate roles: (1) a direct (postsynaptic) effect on myenteric neurons to decrease their activity, compatible with its role as a potential neurotransmitter; (2) a decrease in the release of acetylcholine from presynaptic nerves, an action suggesting that somatostatin may have a neuromodulatory function at this site.

Spinal Cord

Somatostatin-containing neurons and nerve fibers are present in several specific regions of the CNS. In the spinal cord, smaller dorsal root ganglia neurons demonstrate intense somatostatin-like immunoreactivity that can be traced into the dorsal horn of the spinal cord (Hökfelt et al., 1975b; Hökfelt et al., 1976). Somatostatin can be released in a calcium-dependent

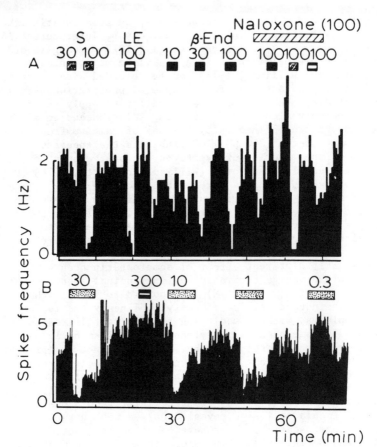

Fig. 5-11. Frequency and time histograms of the activity of two myenteric neurons during perfusion with somatostatin (S), leucine enkephalin (LE), and β-endorphin (β-end). During exposure to naloxone, the effect of β-endorphin and of enkephalin are markedly reduced, whereas the inhibition caused by somatostatin is un-affected. The records in the lower trace were obtained from a neuron recorded in a calcium-free solution that contained magnesium. Concentrations indicated are in nM. (From Williams and North, 1978.)

manner from cell cultures of these primary sensory neurons (Mudge et al., 1977), suggesting that this peptide may have some role in primary afferent transmission. Applications of somatostatin by microiontophoresis (20 to 80 nA) to nociceptive

responsive neurons in the dorsal horn of the spinal cat is usually associated with a decrease in their excitability (Randič
and Miletic, 1978). We have noted similar observations from a
small number of lamina IV-V, low-threshold, unidentified dorsal
horn neurons in the Dial anesthetized cat (unpublished observations) (Fig. 5-12). These incomplete observations would lead
one to believe that somatostatin may participate in synaptic
transmission at terminals of small (high-threshold?) primary
afferent fibers, and possibly in other primary sensory fibers.

Further studies with somatostatin have been conducted in the
isolated hemisected and perfused frog spinal cord, where somatostatin levels are especially high in the dorsal quadrant
(Padjen, 1977; Renaud and Padjen, 1978). In this preparation,
somatostatin ($2 \times 10^{-6}-10^{-5}M$) evokes two types of responses:
(1) an immediate but rather weak hyperpolarization of motoneurons, accompanied by a decrease in the response of dorsal and
ventral roots to glutamate; (2) a delayed (10 to 20 minute)
dose-dependent increase in the size of the ventral root potentials evoked by dorsal root stimulation, outlasting the application period by 30 to 40 minutes, with no evidence of desensitization or tachyphylaxis. The initial response appears to
be a direct (postsynaptic) action of somatostatin, whereas the
later responses probably represent an indirect or presynaptic
action at a site yet to be determined. Admittedly, it is difficult and perhaps impossible to reconcile the events recorded
in vivo in the mammalian spinal cord with those seen *in vitro*
in the amphibian spinal cord. In fact, differences in synaptic
circuitry may preclude any meaningful association between the
two sets of data, at least until further details are known.

Central Neurons

The results of investigations with somatostatin at higher levels of the neural axis suggest that this peptide may have both
depressant and excitatory roles. The site of testing, experimental conditions, and data interpretation appear to be determinants in the mode of action of somatostatin. Behavioral
experiments employing systemic, intracerebral, or intracerebroventricular somatostatin infusions are reported to increase
pentobarbital anesthesia time (Prange et al., 1974), decrease
spontaneous motor activity (Segal and Mandell, 1974; Cohn and
Cohn, 1975; Koranyi et al., 1977), lower the LD_{50} of barbiturates, inhibit strychnine-induced seizure activity, and increase strychnine LD_{50} (Brown and Vale, 1975). These results
suggest that somatostatin may be a CNS depressant. This impression is in agreement with some electrophysiological observations. Intraperitoneal injection of somatostatin is followed

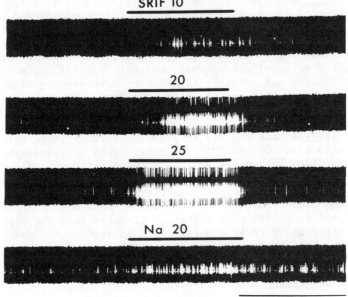

Fig. 5-12. Oscilloscope records of the extracellular spike
         discharges from a spontaneously active neuron in
         lamina IV of the spinal cord of the DIAL anesthe-
         tized cat. Somatostatin applied as a cation during
         the period indicated by the horizontal lines and at
         the currents illustrated is associated with a dose-
         dependent decrease in the activity of this neuron.
         The lowest trace illustrates no response to sodium
         ions applied as a current control. (From Y. Lamour,
         H.W. Blume, Q.J. Pittman, and L. Renaud, unpublished
         observations.)

by a decrease in the firing frequency of neurons in the brainstem reticular formation (Koranyi et al., 1977). The microiontophoretic application of somatostatin in the cerebral and
cerebellar cortices, hypothalamus, and spinal cord is associated with a decrease in the activity of a certain population of
neurons (Renaud et al., 1975, 1976; Randić and Miletic, 1978).
   In contrast, others have reported excitatory responses to somatostatin. Thus, intraperitoneal, intraventricular, and intracerebral somatostatin infusions in lower doses (0.1 to 0.1 µg)

produce a reduction in slow-wave and rapid-eye-movement sleep, whereas higher doses (1.10 µg) may produce hyperkinesia, tremor and rigidity, catatonia, and tonic-clonic seizures (Rezak et al., 1976a,b; 1977a). It is therefore interesting to note that when somatostatin was applied by microiontophoresis to sensori-motor cortical neurons in the awake habituated rabbit, almost 60 percent of neurons displayed an associated rise in excitability (Ioffe et al., 1978) Somewhat similar results have been reported in the cortex of the urethane or chloral hydrate anesthetized rat (H.Olpe, personal communication). Although some of the differences reported in the behavioral responses to somatostatin infusions could be due to interpretation of a response pattern (Rezak et al., 1977b), the virtual opposite responses noted to the microiontophoretic applications of somatostatin remain puzzling. Several factors could account for these differences, including technical problems associated with the delivery of the peptide by a microiontophoretic method, breakdown of peptide before or during the microiontophoretic ejection, effects of (or the lack of) different anesthetics.

Partly because of the disenchantment with *in vivo* microiontophoretic experiments and the associated difficulties in ensuring adequate delivery of drugs to the test site, loss of stability in recording conditions, and so on, others have turned to *in vitro* recording methods. These studies are particulary well advanced in the brain slice preparation, specifically within the hippocampus (Dingledine et al., 1977). In the dorsal hippocampus, CA1 and CA2 pyramidal neurons are surrounded by a high density of nerve terminals containing somatostatin-like immunoreactivity (Petrusz et al., 1977). This site should therefore represent an ideal location for examining possible postsynaptic effects of somatostatin. Accordingly, intracellular recordings have been obtained from CA1 and CA2 cells during application of somatostatin near the recording site by microiontophoresis (negative current), by direct pressure injection, or by local application of some droplets. During applications of somatostatin and the synthetic analogues D-Trp^8-somatostatin and Arg^9-somatostatin, rapid dose-dependent and reversible depolarizing actions have been noted consistently (Dodd et al., 1978b; Dodd and Kelly, 1978). In terms of rapidity of onset, the excitatory actions of somatostatin resemble those evoked by the putative excitatory neurotransmitter glutamate; in contrast, the onset of excitation in these same neurons by acetycholine is considerably delayed, whether the site of application is somatic or dendritic, and prolonged beyond the period of application by several tens of seconds (Dingledine et al., 1977; Dodd et al., 1978a). However, unlike the effects of glutamate or acetycholine, somatostatin applications were not accompanied by significant changes in membrane resistance or

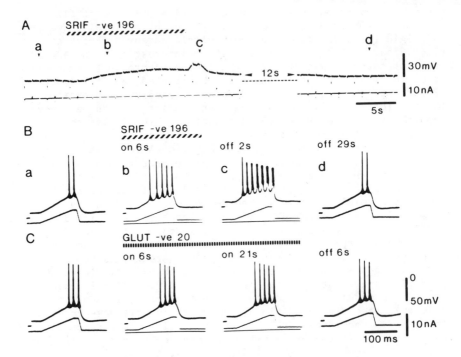

Fig. 5-13. The effect of somatostatin on a CA1 pyramidal neuron
recorded with an intracellular microelectrode in a
rat hippocampal slice preparation. Somatostatin
(SRIF) applied with microiontophoresis by negative
current produced a depolarization, with no change in
membrane resistance, as indicated by the upper trace
in (A). In (B), detailed records selected from the
tracing in (A), at the periods indicated, illustrate
the effect of somatostatin on the excitability of
this cell, with the progressive increase in the number of action potentials on each intracellularly applied current ramp. In (C), records illustrate the
excitation of the same cell by L-glutamate. (From
Dodd and Kelly, 1978.)

conductance (Dodd et al., 1978b), indicating that the actions
of somatostatin, at least at this site, may be voltage independent, or at least different from the voltage-dependent conductance changes usually associated with the actions of the
more conventional neurotransmitter agents (Krnjević, 1974; see
also Fig. 5-13).

A second type of *in vitro* approach to peptide actions has centered on the use of tissue cultures of dissociated cortical (Dichter and Delfs, 1980) and spinal neurons (MacDonald, 1980). Intracellular recordings from cultured cortical neurons suggest that somatostatin has no direct effects on membrane potential, membrane conductance, or action potential characteristics, but may activate presynaptic terminals. On the other hand, in cultured spinal neurons, two effects are observed. First, somatostatin has a direct postsynaptic hyperpolarizing action, comparable with the notion that somatostatin can act as an inhibitory *neurotransmitter*. Second, somatostatin can evoke both excitatory and inhibitory postsynaptic potentials, again suggesting that it may activate presynaptic terminals. This latter action would suggest that this peptide also subserves some form of *neuromodulator* function.

In summary, although one can conclude from the available physiological data that somatostatin does influence the excitability of peripheral and central neurons, its mode or mechanism of action(s) remain uncertain. Until further information becomes available, the discrepancies noted in different studies with somatostatin may be partially resolved by suggesting that somatostatin's actions are different, according to the location of testing and experimental conditions. If one is to regard somatostatin as a neurotransmitter, then the observations of a somatostatin-evoked depolarization without a detectable change in membrane resistance is sufficient to suggest that at its location within the hippocampus, somatostatin does not act as a conventional neurotransmitter substance. Similar comments apply for some of the indirect effects of the peptide on cultured neurons.

Other Peptides And Hormones

The central and peripheral nervous systems contain additional peptides whose actions have been examined with electrophysiological methods: vasopressin, angiotensin, substance P, leucine and methionine enkephalin, β-endorphin, neurotensin, vasoactive intestinal peptide, and cholecystokinin. Anterior pituitary and other hormones are also present in brain tissue and influence neuronal excitability: LH (Sanghera et al., 1978), ACTH, and adrenal steroids (Mandelbrod et al., 1974; Steiner, 1975; Kelly et al., 1977a,b), thyroid hormones (Davidoff and Ruskin, 1972; Steiner, 1975), estrogen (Kelly et al., 1976; 1977a,b), and testosterone (Yamada, 1979). For further details, the reader is requested to consult the reviews cited earlier and original journal reports.

COMMENT

The past decade has been witness to a dramatic increase in our understanding of hypothalamic pituitary relationships. The modest contributions provided by neurophysiology have been reviewed in the first part of this chapter. Although no one approach provides all answers to a particular problem under investigation, the neurophysiological approach offers the investigator a knowledge of cellular location in association with the ability to observe spontaneous or induced activity patterns, a correlation with particular stimuli or behavioral responses, evidence of specific afferent and efferent connections, and neuropharmacology. However, the approach is tedious, and improvement is required in several areas, such as development of methods to correlate identified peptidergic neuronal activity with specific peptide content and with behaviorally meaningful events. It is evident that peptide-containing neurons are present throughout the peripheral and CNS and that each area in which these endogenous peptides are present requires careful evaluation as to the functional role of these specific neurons in each instance. The evidence already available leaves little doubt that peptides are important in neural function. The reader is well advised to maintain an open mind as to whether they function as neurotransmitters, neuromodulators, or neurohormones, since different mechanisms of peptide action may prevail at a specific location. Furthermore, what is applicable to one species may not necessarily relate to another, and cautious interpretation is advised.

ACKNOWLEDGMENT

The author is grateful to the Conseil de la recherche en santé du Québec and the Canadian Medical Research Council for financial support, and to Drs. Q.J. Pittman, H.W. Blume, R.E. Kearney, S. LaFontaine, Y. Lamour, E. Arnauld, B. Layton, C. Bouillé and H. Geller for their collaboration in his laboratory. Mr. William Ellis provided expert technical assistance and Mrs. M. Walker typed this manuscript.

REFERENCES

Amoss, M., Burgus, R., Blackwell, R., Vale, W., Fellows, R., and Guillemin, R.: Purification, amino- acid composition and N-terminus of hypothalamic luteinizing-hormone releasing factor (LRF) of ovine origin. *Biochem Biophys Res Commun* 44:205-210, 1971.

Arnauld, E., Dufy, B., and Vincent, J.D.: Hypothalamic supra-optic neurones: rates and patterns of action potential firing during water deprivation in the unanesthetized monkey. *Brain Res 100*:315-325, 1975.

Barker, J.F., Physiological roles of peptides in the nervous system. In *Peptides in Neurobiology*, H. Gainer, ed Plenum Press, New York, 1977, pp 295-343.

Barker, J.F., and Smith, T.G., Jr.: The Role of Peptides in Neuronal Function, Marcel Dekker, New York, 1980.

Barker, J.L., Crayton, J.W., and Nicoll, R.A.: Antidromic and orthodromic responses of paraventricular and supraoptic neurosecretory cells. *Brain Res 33*:353-366, 1971.

Barnea, A., Parker, C.R., Neaves, W.B., Cho, G., Oliver, C., and Porter, J.C.: Subcellular distribution of neurohormones in the central nervous system. In *Biologie cellulaire des processus neurosécrétoires hypothalamiques (Cell biology of hypothalamic neurosecretion)*, J.-D. Vincent and C. Kordon, eds Editions du Centre Nationale de la recherche Scientifique, Paris, 1978, pp 357-372.

Bennett, G.W., Edwardson, J.A., Holland, D.T., Jeffcoate, S.L., and White, N.: The release of luteinizing hormone-releasing hormone and thyrotropin-releasing hormone from hypothalamic synaptosomes. *Nature 275*:323-325, 1975.

Bissett, G.W. Neurohypophyseal hormones. In *Peptide Hormones,* J.A. Parsons, ed The MacMillan Press Ltd., London, 1976, pp 145-177.

Bissett, G.W., Clark, B.J., and Errington, M.L.: The hypothalamic neurosecretory pathways for the release of oxytocin and vasopressin in the cat. *J Physiol (Lond) 217:111-131, 1971.*

Blume, H.W., Pittman, Q.J., and Renaud, L.P.: Electrophysiological indications of a 'vasopressinergic' innervation of the median eminence. *Brain Res 155*:153-158, 1978.

Boer, K., and Nolten, J.W.L.: Hypothalamic paraventricular unit activity during labour in the rat. *J Endocrinol 76:155-163. 1978.*

Braitman, D.J., Auker, C.R., and Carpenter, D.O.: Thyrotropin-releasing hormone modulates the response to several neurotransmitters in somatosensory cortex. *Soc Neurosci Abstr 5*:584, 1979.

Brazeau, P., Vale, W., Burgus, R., Ling, N., Butcher, M., Rivier, J., and Guillemin, R.: Hypothalamic polypeptide that inhibits the secretion of immunoreactive pituitary growth hormone. *Science 197:77-79, 1973.*

Breese, G.R., Cott, J.M., Cooper, B.R., Prange, A.J., Lipton, M.A., and Plotnikoff, N.P.: Effects of thyrotropin-releasing hormone (TRH) on the actions of pentobarbital and other centrally acting drugs. *J. Pharmacol Exp Ther 193*:11-22, 1975.

Brimble, M.J., and Dyball, R.E.J.: Characterizations of the

responses of oxytocin and vasopressin-secreting neurones in the supraoptic nucleus to osmotic stimulation. *J Physiol (Lond)* 271:253-271, 1977.

Brown, M., and Vale, W.: Central nervous system effects of hypothalamic peptides. *Endocrinology* 96:1333-1336, 1975.

Brown, M., Rivier, J., and Vale, W.: Actions of bombesin, thyrotropin-releasing factor, prostaglandin E_2, and naloxone on thermoregulation in the rat. *Life Sci* 20:1681-1688, 1977.

Brownstein, M.J., Palkovits, M., Saavedra, J.M., Bassiri, R.M., and Utiger, R.D.: Thyrotropin-releasing hormone in specific nuclei of rat brain. *Science 185:267-269, 1974.*

Burgus, R., Dunn, T.F., Desiderio, D., Ward, D.N., Vale, W., and Guillemin, R.: Characterization of ovine hypothalamic hypophysiotropic TSH-releasing factor. *Nature* 226:321-325, 1970.

Burt, D.R., and Snyder, S.H.: Thyrotropin-releasing hormone (TRH) apparent receptor binding in rat brain membranes. *Brain Res* 93:309-328, 1975.

Carino, M.A., Smith, J.R., Weick, B.G., and Horita, A.: Effects of thyrotropin-releasing hormone (TRH) microinjected into various brain areas of conscious and pentobarbital-pretreated rabbits. *Life Sci* 19:1687-1692, 1976.

Christ, J.F. Nerve supply, blood supply and cytology of the neurohypophysis. In *The Pituitary Gland,* G.W. Harris and B.J. Donovan, eds Butterworths, London, 1966, pp 62-130.

Cohn, M.L., and Cohn, M.: 'Barrel rotation' induced by somatostatin in the non-lesioned rat. *Brain Res* 96:138-141, 1975.

Cohn, M.L., Cohn, M., and Taylor, F.H.: Thyrotropin-releasing factor (TRH) regulation in the non-lesioned rat. *Brain Res* 96:134-137, 1975.

Cooper, B.R., and Boyer, C.E.: Stimulant action of thyrotropin-releasing hormone on cat spinal cord. *Neuropharmacology* 17:153-156, 1978.

Costa, M., Patel, Y., Furness, J.B., and Arimura, A.: Evidence that some intrinsic neurons of the intestine contain somatostatin. *Neurosci Lett* 6:215-222, 1977.

Cott, J.M., Breese, G.R., Cooper, B.R., Barlow, T.S., and Prange, A.J.: Investigations into the mechanism of reduction of ethanol sleep by thyrotropin-releasing hormone (TRH). *J Pharmacol Exp Ther* 196:594-604, 1976.

Cross, B.A.: The motility and reactivity of the oestrogenised rabbit uterus in vivo, with comparative observations on milk ejection. *J Endocrinol* 16:261-276, 1958.

Cross, B.A., and Green, J.D.: Activity of single neurons in the hypothalamus: Effect of osmotic and other stimuli. *J Physiol (Lond)* 148:554-569, 1959.

Cross, B.A., and Harris, G.W.: The role of the neurohypophysis in the milk-ejection reflex. *J Endocrinol* 8:148-161, 1952.

Cross, B.A., Dyball, R.E.J., Dyer, R.G., Jones, C.W., Lincoln, D.W., Morris, J.F., and Pickering, B.T.: Endocrine neurons. *Recent Prog Horm Res* 31:243-294, 1975.

Dale, H.A.: Pharmacology and nerve endings. *Proc R Soc Med* 28:319-332, 1935.

Davidoff, R.A., and Ruskin, H.M.: The effects of microelectrophoretically applied thyroid hormone on single cat nervous system neurons. *Neurology* 22:467-472, 1972.

Dichter, M.A., and Delfs, J.R.: Somatostatin and cortical neurons in cell culture. In *Neurosecretion and Brain Peptides: Implications for Brain Function and Neurological Disease,* J.B. Martin, S. Reichlin, and K.L. Bick, eds Raven Press, New York, 1981. pp 145-157.

Dingledine, R., Dodd, J., and Kelly, J.S.: Ach-evoked excitation of cortical neurons. *J Physiol* 273:79-80P, 1977.

Döcke, F.: Differential effects of amygdaloid and hippocampal lesions on female puberty. *Neuroendocrinology* 14:345-350, 1974.

Dodd, J., and Kelly, J.S.: Is somatostatin an excitatory transmitter in the hippocampus? *Nature* 273:674-675, 1978.

Dodd, J., Dingledine, R., and Kelly, J.S.: Intracellular recordings from CA-3 pyramidal neurones of hippocampal slices and the action of iontophoretic acetylcholine. In *Iontophoresis and Transmitter Mechanisms in the Mammalian Central Nervous System,* R.W.Ryall and J.S. Kelly, eds Elsevier/North Holland Biomedical Press, New York, 1978a, pp 182-184.

Dodd, J., Kelly, J.S., and Rivier, J.E.: Excitation of rat hippocampal pyramidal neurons by somatostatin. *J Physiol (Lond)* 282:16-17P, 1978b.

Douglas, W.W.: Stimulus-secretion coupling: The concept and clues from chromaffin and other cells. *Br J Pharmacol* 34:451-474, 1968.

Douglas, W.W.: Involvement of calcium in exocytosis and the exocytosis-vesiculation sequence. *Biochem Soc Symp* 39:1-28, 1974.

Dreifuss, J.J., and Kelly, J.S.: Recurrent inhibition of antidromically identified rat supraoptic neurons. *J Physiol (Lond)* 200:87-103, 1972.

Dreifuss, J.J., Harris, M.C., and Tribollet, E.: Excitation of phasically firing hypothalamic supraoptic neurones by carotid occulsion in rats. *J Physiol (Lond)* 257:337-354, 1976.

Dreifuss, J.J., Tribollet, E., and Baertschi, A.J.: Excitation of supraoptic neurones by vaginal distention in lactating rats: Correlation with neurohypophyseal hormone release. *Brain Res* 113:600-605, 1976.

Dufy, B., Vincent, J-D., Fleury, H., Du Pasquier, P., Gourdiji, D., and Tixier-Vidal, A.: Membrane effects of thyrotropin-releasing hormone and estrogen shown by intracellular recording from pituitary cells. *Science 204:*509-511, 1979.

Dyball, R.E.J.: Stimuli for the release of neurohypophyseal hormones. *Br J Pharmacol 33:*319-328, 1968.

Dyball, R.E.J.: Single unit activity in the hypothalamo-neurohypophyseal system of Brattleboro rats. *J Endocrinol 60:*135-143, 1974.

Dyer, R.G., and Dyball, R.E.J.: Evidence for a direct effect of LRF and TRF on single unit activity in the rostral hypothalamus. *Nature 252:*486-488, 1974.

Edwardson, J.A., and Bennett, G.W.: Hypothalamic hormones and mechanisms of neuroendocrine integration. In *Biologically Active Substances: Exploration and Exploitation,* D.A. Hems, ed John Wiley & Sons, New York, 1977. pp 281-299.

Elde, R., and Hökfelt, T.: Distribution of hypothalamic hormones and other peptides in the brain. In *Frontiers in Neuroendocrinology,* Vol 5, W.F. Ganong and L. Martini, eds Raven Press, New York, 1978, pp 1-33.

Ellendorf, F., Colombo, J.A., Blake, C.A., Whitmoyer, D.I., and Sawyer, C.H.: Effects of electrical stimulation of the amygdala on gonadotropin release and ovulation in the rat. *Proc Soc Exp Biol Med 142:*417-420, 1973.

Felix, D., and Phillips, M.I.: Inhibitory effects of luteinizing hormone releasing hormone (LH-RH) on neurons in the organum vasculosum lamina terminalis (OVLT). *Brain Res 169:*204-208, 1979.

Guillemin, R.: Somatostatin inhibits the release of acetycholine induced electrically in the myenteric plexus. *Endocrinology 99:*1653, 1976.

Guillemin, R., and Gerich, J.E.: Somatostatin: Physiological and clinical significance. *Annu Rev Med 27:*379-388, 1976.

Harris, G.W.: *Neural Control of Pituitary Gland.* London, Edward Arnold, 1955.

Harris, M.C.: Effects of chemoreceptor and baroreceptor stimulation on the discharge of hypothalamic supraoptic neurones in rats. *J Endocrinol 82:*115-125, 1979.

Harris, M.C., and Sanghera, M.: Projection of medial basal hypothalamic neurons to the preoptic anterior hypothalamic areas and the paraventricular nucleus in the rat. *Brain Res 81:*402-411, 1974.

Hayward, J.N.: Physiological and morphological identification of hypothalamic magnocellular neuroendocrine cells in goldfish preoptic nucleus. *J Physiol (Lond) 239:*103-124, 1974.

Hayward, J.N.: Neural control of the posterior pituitary. *Ann Rev Physiol 37:*191-210, 1975.

Hayward, J.N.: Functional and morphological aspects of hypothalamic neurons. *Physiol Rev 57:*574-658, 1977.

Hayward, J.N., and Jennings, D.P.: Activity of magnocellular neuroendocrine cells in the hypothalamus of unanesthetized monkeys. II Osmosensitivity of functional cell types in the supraoptic nucleus and the internuclear zone. *J Physiol (Lond)* 232:545-572, 1973.

Hayward, J.N., Murgas, K., Pavasuthipaisit, K., Perez-Lopez, F.R., and Sofroniew, M.V.: Temporal patterns of vasopressin release following electrical stimulation of the amygdala and the neuroendocrine pathway in the monkey. *Neuroendocrinology* 23:61-75, 1977.

Hersh, L.B., and McKelvy, J.F.: Enzymes involved in the degradation of thyrotropin-releasing hormone (TRH) and luteinizing hormone-releasing hormone (LHRH) in bovine brain. *Brain Res* 168:553-564, 1979.

Hökfelt, T., Efendic, S., Hellerstrom, C., Johansson, O., Luft, R., and Arimura, A.: Cellular localization of somatostatin in endocrine-like cells and neurons of the rat with special reference to A_1-cells of the pancreatic islet and to the hypothalamus. *Acta Endocrinol (Suppl 200)* 80:1-40, 1975a.

Hökfelt, T., Elde, R., Johansson, O., Luft., and Arimura, A.: Immunohistochemical evidence for the presence of somatostatin, a powerful inhibitory peptide, in some primary sensory neurons. *Neurosci Lett* 1:231-235, 1975b.

Hökfelt, T., Fuxe, K., Johansson, O., Jeffcoate, S., and White, N.: Thyrotropin-releasing hormone (TRH)-containing nerve terminals in certain brain stem nuclei and in the spinal cord. *Neurosci Lett* 1:133-139, 1975c.

Hökfelt, T., Elde, R., Johnsson, O., Luft, R., Nilsson, G., and Arimura, A.: Immunohistochemical evidence for separate populations of somatostatin-containing and substance P-containing primary afferent neurons in the rat. *Neuroscience* 1:131-136, 1976.

Holaday, J.W., Tseug, L.-F., Loh, H.H., and Li, C.H.: Thyrotropin-releasing hormone antagonizes β-endorphin hypothermia and catalepsy. *Life Sci* 22:1537-1544, 1978.

Ioffe, S., Havlicek, V., Friesen, H., and Chernick, V.: Effect of somatostatin (SRIF) and L-glutamate on neurons of the sensorimotor cortex in awake habituated rabbits. *Brain Res* 153:414-148, 1978.

Iversen, L.L., Nicoll, R.A., and Vale, W.W.: Neurobiology of peptides. *Neurosci Res Program Bull* 16:211-370, 1978.

Jackson, I.M.D., and Reichlin, S.: Thyrotropin-releasing hormone (TRH)--distribution in hypothalamic and extrahypothalamic brain tissues of mammalian and submammalian chordates. *Endocrinology* 95:854-862, 1974.

Jan, Y.N., Jan, L.Y., and Kuffler, S.W.: A peptide as a possible transmitter in sympathetic ganglia of the frog. *Proc Natl Acad Sci USA* 76:1501-1506, 1979.

Jeffcoate, S.L., White, N., Bennett, G.W., Edwardson, J.A., Griffiths, E.C., Forbes, R., and Kelly, J.A.: Studies on the release and degradation of hypothalamic releasing hormones by the hypothalamus and other CNS areas *in vitro*. In *Central Regulation of the Endocrine System*, K. Fuxe, T. Hökfelt and R. Luft, eds Plenum Press, New York, 1979, pp 61-70.

Kandel, E.R.: Electrical properties of hypothalamic neuroendocrine cells. *J Gen Physiol* 47:691-717, 1964.

Kannan, H., and Yagi, K.: Supraoptic neurosecretory neurons: Evidence for the existence of converging inputs from carotid baroreceptors and osmoreceptors. *Brain Res* 145:385-390, 1978.

Kardon, F.C., Winokur, A., and Utiger, R.D.: Thyrotropin-releasing hormone (TRH) in rat spinal cord. *Brain Res* 122:578-581, 1977.

Kawakami, M., and Kimura, F.: Possible roles of CNS estrogen-neuron systems in the control of gonadotropin release. In *Anatomical Endocrinology*, W.E. Stumpf and L.D. Grant, eds Karger, Basel, 1975, pp 216-231.

Kawakami, M., and Sakuma, Y.: Responses of hypothalamic neurons to the microiontophoresis of LHRH, LH, and FSH under various levels of circulating ovarian hormones. *Neuroendocrinology* 15: 290-307, 1974.

Kawakami, M., and Sakuma, Y.: Electrophysiological evidence for possible participation of periventricular neurons in anterior pituitary regulation. *Brain Res* 101:79-94, 1976.

Kelly, M.J., Moss, R.L., and Dudley, C.A.: Differential sensitivity of preoptic-septal neurons to microelectrophoresed estrogen during the estrous cucle. *Brain Res* 114:152-157, 1976.

Kelly, M.J., Moss, R.L., Dudley, C.A., and Fawcett, C.P.: The specificity of the response of preoptic-septal area neurons to estrogen: 17 α-estradiol versus 17 β-estradiol and the response of extrahypothalamic neurons. *Exp Brain Res* 30:43-52, 1977a.

Kelly, M.J., Moss, R.L., and Dudley, C.A.: The effects of microelectrophoretically applied estrogen, cortisol, and acetylcholine on medial preoptic-septal unit activity throughout the estrous cycle of the female rat. *Exp Brain Res* 30:53-64, 1977b.

Kidokoro, Y.: Spontaneous calcium action potentials in a clonal pituitary cell line and their relationship to prolactin secretion. *Nature* 258:741-742, 1975.

Koizumi, K., and Yamashita, H.: Studies of antidromically identified neurosecretory cells of the hypothalamus by intracellular and extracellular recordings. *J Physiol (Lond)* 221:683-705, 1972.

Koizumi, K., and Yamashita, H.: Influence of atrial stretch receptors on hypothalamic neurosecretory neurones. *J Physiol (Lond)* 285:341-358, 1978.

Koizumi, K., Ishikawa, T., and McC.Brooks, G.: The existence of facilitatory axon collaterals in neurosecretory cells of the hypothalamus. *Brain Res 63*:408–413, 1973.

Koranyi, L., Whimoyer, D.I., and Sawyer, C.H.: Effect of thyrotropin-releasing hormone, luteinizing hormone-releasing hormone, and somatostatin on neuronal activity of brain stem reticular formation and hippocampus in the female rat. *Exp Neurol 57*:807–816, 1977.

Krnjević, K.: Chemical nature of synaptic transmission in vertebrates. *Physiol Rev 54*:418–540, 1974.

Kruse, H.: Thyrotropin-releasing hormone (TRH): Restoration of oxotremorine tremor in mice. *Naunyn-Schmiedebergs Arch Pharmacol 294*:39–45, 1976.

Léranth, Cs., Zaborszky, L., Marton, J., and Palkovits, M.: Quantitative studies on the supraoptic nucleus in the rat. I Synaptic organization. *Exp Brain Res 22*:509–523, 1975.

Liebeskind, J.C., Dismukes, R.K., Barker, J.L., Berger, P.A., Creese, I., Dunn, A.J., Segal, D.S., Stein, L., and Vale, W.W.: Peptides and behavior: A critical analysis of research strategies. *Neurosci Res Program Bull 16*:489–635, 1979.

Lincoln, D.W., and Wakerley, J.B.: Electrophysiological evidence for the activation of supraoptic neurons during the release of oxytocin. *J Physiol (Lond) 242*:533–554, 1974.

Lincoln, D.W., and Wakerley, J.B.: Factors governing the periodic activation of supraoptic and paraventricular neurosecretory cells during suckling in the rat. *J Physiol (Lond) 250*:443–461, 1975.

Lu Qui, I.J., and Fox, C.A.: The supraoptic nucleus and the supraoptic-hypophyseal tract in the monkey (*Macacca mulatta).* *J. Comp Neurol 168*:7–40, 1976.

MacDonald, R.L. and Nowak, L.: Somatostatin and substance P actions on spinal cord neurons in primary dissociated cell culture. In *Neurosecretion and Brain Peptides: Implications for Brain Function and Neurological Disease,* J.B. Martin, S Reichlin, and K.L. Bick, eds Raven Press, New York, 1981, pp 159–173.

McKelvy, J., and Epelbaum, J.: Biosynthesis, packaging, transport and release of brain peptides. In *The Hypothalamus,* S. Reichlin, R.J. Baldessarini, and J.B. Martin, eds Raven Press, New York, 1978, pp. 195–211.

McKelvy, J.F., Sheridan, M., Joseph, S., Phelps, C.H., and Perrie, S.: Biosynthesis of thyrotropin-releasing hormone in organ cultures of the guinea pig median eminence. *Endocrinology 97*:908–918, 1975.

Makara, G.B., Harris, M.C., and Spyer, K.M.: Identification and distribution of tuberoinfundibular neurons. *Brain Res 40*:283–290, 1972.

Mandelbrod, I., Feldman, S., and Werman, R.: Inhibition of firing is the primary effect of microelectrophoresis of cortisol to units in the rat tuberal hypothalamus. *Brain Res 80*:303–

315, 1974.

Martin, J.B.: Plasma growth hormone (GH) response to hypothalamic electrical stimulation. *Endocrinology 91:*107-115, 1972.

Martin, J.B., Kontor, J., and Mead, P.: Plasma GH responses to hypothalamic, hippocampal, and amygdaloid electrical stimulation: Effects of variation in stimulus parameters and treatment with α-methyl-p-tyrosine (αMT). *Endocrinology 92:*1354-1361, 1973.

Metcalf, G.: TRH: A possible mediation of thermoregulation. *Nature 252:*310-311, 1974.

Moss, R.L.: Unit responses in preoptic and arcuate neurons related to anterior pituitary function. In *Frontiers in Neuroendocrinology,* Vol 4, L. Martini and W.F. Ganong, eds Raven Press, New York, 1976, pp 95-128.

Moss, R.L., and McCann, S.M.: Induction of mating behavior in rats by luteinizing hormone-releasing factor. *Science 181:*177-179, 1973.

Moss, R.L., Dyball, R.E.J., and Cross, B.A.: Excitation of antidromically identified neurosecretory cells in the paraventricular nucleus by oxytocin applied iontophoretically. *Exp Neurol 34:*95-102, 1972.

Moss, R.L., Dudley, C.A., and Kelly, M.J.: Hypothalamic polypeptide releasing hormones: Modifiers of neuronal activity. *Neuropharmacology 17:*87-93, 1978.

Mudge, A.W., Fishback, G.D., and Leeman, S.E.: The release of immunoreactive substance P and somatostatin from sensory neurons in dissociated cell cultures. *Soc Neurosci Abstr 3:*410, 1977.

Nair, R.M.G., Barrett, J.F., Bowers, C.Y., and Schally, A.V.: Structure of porcine thyrotropin-releasing hormone. *Biochemistry 9:*1103-1106, 1970.

Nakano, H., Fawcett, C.P., and McCann, S.M.: Enzymatic dissociation and short-term culture of isolated rat pituitary cells for studies on the control of hormone secretion. *Endocrinology 48:*278-288, 1976.

Negoro, H. and Holland, R.C.: Inhibition of unit activity in the hypothalamic paraventricular nucleus following antidromic activation. *Brain Res 42:*385-402, 1972.

Negoro, H., Vieseeuwan, S., and Holland, R.C.: Inhibition and excitation of units in paraventricular nucleus after stimulation of the septum, amygdala, and neurohypophysis. *Brain Res 57:*479-483, 1973.

Nicoll, R.A.: Excitatory actions of TRH on spinal motoneurons. *Nature 265:*242-243, 1977.

Nicoll, R.: The action of thyrotropin-releasing hormone, substance P and related peptides on frog spinal motoneurons. *J Pharmacol Exp Ther 207:*817-824, 1978.

Nicoll, R.A., and Barker, J.L.: The pharmacology of recurrent inhibition in the supraoptic neurosecretory system. *Brain Res* 35:501-511, 1971.

North, R.A., and Williams, J.T.: Enkephalin inhibits firing of myenteric neurones. *Nature* 264:460-461, 1976.

North, R.A., and Williams, J.T.: Extracellular recording from the guinea pig myenteric plexus and the action of morphine. *Eur J Pharmacol* 45:23-33, 1977.

Oliver, C., Eskay, R.L., Ben-Jonathan, N., and Porter, J.C.: Distribution and concentration of TRH in rat brain. *Endocrinology* 95:540-546, 1974.

Padjen, A.L.: Effects of somatostatin in frog spinal cord. *Soc Neurosci Abstr* 3:411, 1977.

Petrusz, P., Sar, M., Grossman, G.H., and Kizer, J.S.: Synaptic terminals with somatostatin-like immunoreactivity in the rat brain. *Brain Res* 137:181-187, 1977.

Pfaff, D.W.: Luteinizing hormone-releasing factor (LRF) potentiates lordorsis behavior in hypophysectomized ovariectomized female rats. *Science* 182:1148-1149, 1973.

Pickford, G.E., and Atz, J.W.: *The physiology of the pituitary gland of fishes*. New York Zoological Society, New York, 1957.

Pittman, Q.J., Blume, H.W., and Renaud, L.P.: Electrophysiological indications that individual hypothalamic neurons innervate both median eminence and neurohypophysis. *Brain Res* 157:364-468, 1978.

Pittman, Q.J., Blume, H.W., Kearney, R.E., and Renaud, L.P.: Influence of midbrain stimulation on the excitability of neurons in the medial hypothalamus of the rat. *Brain Res* 174:39-53, 1979.

Pittman, Q.J., Blume, H.W., and Renaud, L.P.: Connections of the hypothalamic paraventricular nucleus with the neurohypophysis, median eminence, amygdala, lateral septum and midbrain periaqueductal gray: an electrophysiological study in the rat. *Brain Res* 215:15-28, 1981.

Poulain, P.: Septal afferents to the arcuate-median eminence region in the guinea pig: Correlative electrophysiological and horseradish peroxidase studies. *Brain Res* 137:150-158, 1977.

Poulain, D.A., Wakerley, J.B., and Dyball, R.E.J.: Electrophysiological differentiation of oxytocin- and vasopressin-secreting neurones. *Proc Roy Soc Lond (Biol)* 196:367-384, 1977.

Prange, A.J. Jr., Breese, G.R., Cott, J.M., Martin, B.R., Cooper, B.R., Wilson, I.C., and Plotnikoff, N.P.: Thyrotropin releasing hormone: Antagonism of pentobarbital in rodents. *Life Sci* 14:447-455, 1974.

Randić, M., and Miletic, V.: Actions of peptides on cat dorsal horn neurones activated by noxious stimuli. In *Iontophoresis and Transmitter Mechanisms in the Mammalian Central Nervous System*, R.W. Ryall and J.S. Kelly, eds Elsevier/North Holland

Biomedical Press, New York, 1978, pp 124-126.

Renaud, L.P.: Tuberoinfundibular neurons in the basomedial hypothalamus of the rat: Electrophysiological evidence for axon collaterals to hypothalamic and extrahypothalamic areas. *Brain Res* 105:59-72, 1976a.

Renaud, L.P.: Influence of amygdala stimulation on the activity of identified tuberoinfundibular neurones in the rat hypothalamus. *J Physiol (Lond)* 260:237-252, 1976b.

Renaud, L.P.: Influence of medial preoptic-anterior hypothalamic area stimulation on the excitability of mediobasal hypothalamic neurones in the rat. *J Physiol (Lond)* 264:541-564, 1977a.

Renaud, L.P.: TRH, LHRH and somatostatin: Distribution and physiological action in neural tissues. In *Approaches to the Cell Biology of Neurons*, W.M. Cowan and J.A. Ferrendelli, eds Society for Neuroscience, Bethesda, Md., 1977b, pp 269-290.

Renaud, L.P.: Neurophysiological organization of the endocrine hypothalamus. In *The Hypothalamus*, S. Reichlin, R.J. Baldessarini, and J.B. Martin, eds Raven Press, New York, 1978, pp 269-301.

Renaud, L.P., and Arnauld, E.: Supraoptic phasic neurosecretory neurons: Response to stimulation of amygdala and dorsal hippocampus, and sensitivity to microiontophoresis of amino acids and noradrenaline. *Soc Neurosci Abst* 5:233, 1979.

Renaud, L.P., and Martin, J.B.: Electrophysiological studies of connections of hypothalamic ventromedial nucleus neurons in rat--evidence for a role in neuroendocrine regulation. *Brain Res* 93:145-151, 1975a.

Renaud, L.P., and Martin, J.B.: Microiontophoresis of thyrotropin-releasing hormone (TRH). Effects on the activity of central neurons. In *Anatomical Endocrinology*, W.E. Stumpf and L.D. Grant, eds Basel, Karger, 1975b.

Renaud, L.P., and Padjen, A.: Electrophysiological analysis of peptide actions in neural tissue. In *Centrally Acting Peptides,* J. Hughes, ed MacMillan Press, London, 1978, pp 59-84.

Renaud, L.P., and Martin, J.B.: Thyrotropin-releasing hormone (TRH)--depressant action on central neuronal activity. *Brain Res* 86:150-154, 1975c.

Renaud, L.P., Martin, J.B., and Brazeau, P.: Depressant action of TRH, LHRH, and somatostatin on activity of central neurones. *Nature* 255:233-235, 1975.

Renaud, L.P., Martin, J.B., and Brazeau, P.: Hypothalamic releasing factors: Physiological evidence for a regulatory action on central neurons and pathways for their distribution in brain. *Pharmacol Biochem Behav (Suppl 1)* 5:171-178, 1976.

Renaud, L.P., Blume, H.W., and Pittman, Q.J.: Neurophysiology and neuropharmacology of the hypothalamic tuberoinfundibular system. In *Frontiers in Neuroendocrinology, Vol 5*, W.F.

Ganong and L. Martini, eds Raven Press, New York, 1978, pp 135-162.

Renaud, L.P., Blume, H.W., Pittman, Q.T., Lamour, Y., and Tan, A.T.: Thyrotropin-releasing hormone (TRH) selectively depresses glutamate excitation of cerebral cortical neurons. *Science* 205:1275-1277, 1979a.

Renaud, L.P., Pittman, Q.J., and Blume, H.W.: Neurophysiology of hypothalamic peptidergic neurons. In *Central Regulation of the Endocrine System*, K. Füxe, T. Hökfelt, and R. Luft, eds New York, Plenum Press, 1979b.

Rezak, M., Havlicek, V., Hughes, K.R., and Friesen, H.: Central site of action of somatostatin (SRIF): Role of hippocampus. *Neuropharmacology* 15:499-504, 1976a.

Rezak, M., Havlicek, V., Hughes, K.R., and Friesen, H.: Cortical administration of somatostatin (SRIF): Effect on sleep and motor function. *Pharmacol Biochem Behav* 5:73-77, 1976b.

Rezak, M., Havlicek, V., Hughes, K.R., and Friesen, H.: Behavioural and motor excitation and inhibition induced by the administration of small and large doses of somatostain into the amygdala. *Neuropharmacology* 16:157-162, 1977a.

Rezak, M., Havlicek, V., Leybin, L., Pinsky, C., Kroeger, E.A., Hughes, K.R., and Friesen, H.: Neostriatal administration of somatostatin: Differential effect of small and large doses on behavior and motor control. *Can J Physiol Pharmacol* 55:234-242, 1977b.

Richard, P., Freund-Mercier, M.J., and Moos, F.: Les neurones hypothalamiques ayant une fonction endocrine. Identification, localisation, caractéristiques électrophysiologiques et contrôle hormonal. *J Physiol (Paris)* 74:61-112, 1978.

Sanghera, M., Harris, M.C., and Morgan, R.A.: Effects of microiontophoretic and intravenous application of gonadotrophic hormones in the discharge of medial-basal hypothalamic neurones in rats. *Brain Res* 140:63-74, 1978.

Sawaki, Y., and Yagi, K.: Electrophysiological identification of cell bodies of the tuberoinfundibular neurones in the rat. *J Physiol (Lond)* 230:75-85, 1973.

Schally, A.V., Arimura, A., Baba, Y., Nair, R.M.G., Matsua, J., Redding, T.W., Debeljur, L., and White, W.F.: Isolation and properites of FSH and LH-releasing hormone. *Biochem Biophys Res Commun* 43:393-399, 1971.

Schenkel-Hulliger, L., Koella, W.P., Hartmann, A., and Maître, L.: Tremorgenic effect of thyrotropin-releasing hormone in rats. *Experientia* 30:1168-1170, 1974.

Segal, D.S., and Mandell, A.J.: Differential behavioral effects of hypothalamic polypeptides. In *The Thyroid Axis, Drugs and Behavior*, A.J. Prange, Jr., ed New York: Raven Press, 1974, pp 129-133.

Smith, J.R., La Hann, T.R., Chesnut, R.M., Carino, M.H., and

Horita, A.: Thyrotropin-releasing hormone: Stimulation of colonic activity following intracerebroventricular administration. *Science 196*:660-662, 1977.

Steiner, F.A.: Effects of locally applied hormones and neurotransmitters on hypothalamic neurons. Proceedings 4th International Congress of Endocrinology, Washington, 1972, Series 273. Amsterdam: *Exerpta Medica*, 1973, pp 202-204.

Steiner, F.A.: Electrophysiological mapping of brain sites sensitive to corticosteroids, ACTH, and hypothalamic releasing hormones. In *Anatomical Neuroendocrinology*, W.E. Stumpf and L.D. Grant, eds Basel, Karger, 1975, pp 270-275.

Swanson, L.W.: Immunohistochemical evidence for a neurophysin-containing autonomic pathway arising in the paraventricular nucleus of the hypothalamus. *Brain Res 128*:346-353, 1977.

Swanson, L.W., and McKellar, S.: The distribution of oxytocin- and neurophysin-stained fibers in the spinal cord of the rat and monkey. *J Comp Neurol 188*:87-106, 1979.

Taraskevich, P.S., and Douglas, W.W.: Action potentials occur in cells of the normal anterior pituitary gland and are stimulated by the hypophysiotropic peptide thyrotropin-releasing hormone. *Proc Nat Acad Sci USA 74*:4064-4067, 1977.

Wakerley, J.B. and Lincoln, D.W.: The milk ejection reflex of the rat: A 20-40-fold accceleration in the firing of paraventricular neurones during oxytocin release. *J Endocrinol 57*:477-493, 1973.

Wakerley, J.B., Poulain, D.A., Dyball, R.E.J., and Cross, B.A.: Activity of phasic neurosecretory cells during haemorrhage. *Nature 258*:82-84, 1975.

Wei, E., Sigel, S., and Way, E.L.: Thyrotropin-releasing hormone and shaking behavior in rat. *Nature 253*:739-740, 1975.

Williams, J.T., and North, R.A.: Inhibition of firing of myenteric neurones by somatostatin. *Brain Res 155*:165-168, 1978.

Winokur, A., and Beckman, A.L.: Effects of thyrotropin-releasing hormone, norepinephrine, and acetylcholine on the activity of neurons in the hypothalamus, septum, and cerebral cortex of the rat. *Brain Res 150*:205-209, 1978.

Winokur, A., and Utiger, R.D.: Thyrotropin-releasing hormone: Regional distribution in rat brain. *Science 185*:265-267, 1974.

Winokur, A., Davis, R., and Utiger, R.D.: Subcellular distribution of thyrotropin-releasing hormone (TRH) in rat brain and hypothalamus. *Brain Res 120*:423-434, 1977.

Woods, W.H., Holland, R.C., and Powell, E.W.: Connections of cerebral structures functioning in neurohypophyseal hormone release. *Brain Res 12*:26-46, 1969.

Yamada, Y.: Effects of testosterone on unit activity in rat hypothalamus and septum. *Brain Res 172*:165-168, 1979.

Yamashita, H., and Koizumi, K.: Influence of carotid and aortic baroreceptors on neurosecretory neurons in supraoptic nuclei. *Brain Res* *170*:259–277, 1979.

Yarbrough, G.G.: TRH potentiates excitatory actions of acetylcholine on cerebral cortical neurones. *Nature 263*:523–524, 1976.

Yarbrough, G.G.: Studies on the neuropharmacology of thyrotropin-releasing hormone (TRH) and a new TRH analog. *Eur J Pharmacol 48*:19–27, 1978.

Yarbrough, G.G., Haubrich, D.R., and Schmidt, D.E.: Thyrotropin-releasing hormone (TRH) and MK-771 interactions with CNS cholinergic mechanisms. In *Iontophoresis and Transmitter Mechanisms in the Mammalian Central Nervous System,* R.W. Ryall and J.S. Kelly, eds Elsvier/North Holland, amsterdam, 1978, pp 136–138.

York, D.H., Baker, F.L., and Kraicer, J.: The effect of synthetic TRH on transmembrane potential and membrane resistance of adenohypophyseal cells. *Can J Physiol Pharmacol 53*:777–786, 1975.

Zaborszky, L., Léranth, Cs., Makara, G.B., and Palkovits, M.: Quantitiative studies in the supraoptic nucleus in the rat. II Afferent fiber connections. *Exp Brain Res 22*:525–540, 1975.

Zimmerman, E.A., Stillman, M.A., Recht, L.D., Michaels, J., and Nilaver, G.: The magnocellular neurosecretory system: Pathways containing oxytocin, vasopressin, and neurophysins. In *Biologie cellulaire des precessus neurosecretoires hypothalamiques,* J.-D. Vincent and C. Kordon, eds Editions du Centre Nationale de la recherche Scientifique, Paris, 1978, pp 375–389.

Zolovick, A.J.: Effects of lesions and electrical stimulation of the amygdala on hypothalamic hypophyseal regulation. In *The Neurobiology of Amygdala,* B.E. Eleftheriou, ed New York: Plenum Press, 1972, pp 643–684.

6

Effects of Hypothalamic Peptides on the Central Nervous System

CHARLES B. NEMEROFF, GARTH BISSETTE,
PAUL J. MANBERG, DANIEL LUTTINGER,
AND ARTHUR J. PRANGE, JR.

INTRODUCTION

The dramatic advances in neuroendocrinology of the past 25 years have clearly established that the brain influences endocrine homeostasis by secretion of hypothalamic release and release-inhibiting factors (see chapter 4 by Frohman and Berelowitz). The chemical identities of these substances, which long eluded researchers, have been shown, with perhaps one exception, prolactin-inhibiting factor (PIF), to be peptides (Table 6-1). Problems in the assay of these peptides have been described in detail in a recent volume devoted to radioimmunoassay (RIA) (Jaffe and Behrman, 1978). The elaboration of accurate assay systems has led to the knowledge that these "hypothalamic" peptides exist in other regions of the brain as well as in peripheral organs. Using direct RIA, several groups have reported, for example, substantial quantities of thyrotropin-releasing hormone (TRH) in the septum, preoptic area, and several circumventricular organs (Brownstein et al., 1974; Winokur and Utiger, 1974). This initial work served as an impetus to study the effects of exogenous TRH administration, systemically and directly into the central nervous system (CNS), on behavioral and pharmacological responses of laboratory animals. Over the past decade there has been an almost exponential increase in research concerning direct brain effects of endogenous peptides (see Prange et al., 1978a,b; Moss, 1979; Klee, 1979, for reviews). It is now clear that the brain contains many peptides. Some are concerned with neuroendocrine function, although surely not exclusively; others have demonstrable CNS actions and clearly do not serve as releasing hormones.

217

Table 6-1. Amino Acid Sequence of Hypothalamic Peptides

TRH	pGlu-His-Pro-NH$_2$
MIF-I	Pro-Leu-Gly-NH$_2$
LH-RH	pGlu-His-Trp-Ser-Tyr-Gly-Leu-Arg-Pro-Gly-NH$_2$
SRIF	H-Ala-Gly-Cys-Lys-Asn-Phe-Phe-Trp-Lys-Thr-Phe-Thr-Ser-Cys-OH
	I_____I
SP	H-Arg-Pro-Lys-Pro-Gln-Gln-Phe-Phe-Gly-Leu-Met-NH$_2$
NT	pGlu-Leu-Tyr-Glu-Asn-Lys-Pro-Arg-Arg-Pro-Tyr-Ile-Leu-OH

The purpose of the present chapter is to review briefly certain of the literature concerning brain effects of endogenous
neuropeptides. In other places, we (Prange et al., 1978a,b)
have extensively reviewed the literature; here, we have focused
on specific works and have attempted to review them critically.
For the sake of brevity we have concentrated on the following
peptides: TRH, luteinizing-hormone-releasing hormone (LH-RH),
neurotensin (NT), melanocyte-stimulating hormone release-inhibiting factor (MIF), substance P (SP), and somatastatin (SRIF).

THYROTROPIN-RELEASING HORMONE

TRH, a tripeptide (Table 6-1), was the first hypothalamic peptide isolated and characterized (Burgus et al., 1969; Bøler et
al., 1969). As already noted, TRH has since been found to be
present in other brain regions as well (Brownstein et al.,
1974; Winokur and Utiger, 1974; Youngblood et al., 1978). A
detailed summary of the heterogeneous distribution of TRH in
the CNS can be found in chapter 2 by Grant et al. in this volume. Brain TRH has been found to be concentrated in a synap-
tosome-rich fraction after density gradient centrifugation
(Barnea et al., 1975; Bennett et al., 1975; Winokur et al.,
1977) from which it is released by depolarization in a Ca^{2+}-de-
pendent fashion (Bennett et al., 1975; Schaeffer et al., 1977).
Further, high-affinity, saturable binding sites for TRH have
been found in rat brain (Burt and Snyder, 1975). In sheep
brain, the pharmacological characteristics of the high-affinity
binding sites for TRH agree remarkably well with those obtained
from pituitary (Burt and Taylor, 1980). However, the popula-

tion of TRH receptors in the brain of both sheep and rats differs somewhat from the TRH receptors in pituitary. Aside from high-affinity binding sites, there also exist low-affinity, saturable binding sites for TRH in brain, but not pituitary. Like the distribution of TRH in brain, the binding sites for TRH also appear to be heterogeneously distributed.

Many researchers have reported that TRH, like most neuropeptides, is rapidly degraded in tissue extracts and plasma. For example, Saperstein et al. (1975) demonstrated that TRH can be degraded by hypothalamic extracts from pigs or rats. It was further shown that extrahypothalamic brain regions are also capable of degradng TRH. The substrate specificity in these areas is similar, but not identical, to those observed in hypothalamic extracts.

Iontophoretic administration of TRH affects the firing rate of neurons in a variety of brain regions. Steiner (1975), Dyer and Dyball (1974), and Renaud (chapter 5) have all presented evidence that TRH can directly decrease (and rarely increase) the frequency of both spontaneous and stimulated, for example, glutamate-induced, discharge of CNS neurons.

The fact that TRH is heterogeneously distributed in brain and preferentially found in synaptosomes from which it can be released, taken together with the presence of TRH binding sites in the brain and the effects of iontophoretic administration of the tripeptide, provide presumptive evidence that TRH may act as an endogenous modulator of CNS function. Further support for the concept that TRH is involved in CNS function has been derived from behavioral studies that have examined the effects of exogenous TRH administration. The discussion that follows, rather than being comprehensive, is selective of the data available. The interested reader may wish to refer to recent reviews of this subject (Prange et al., 1978a; Yarbrough, 1979; Morley, 1979).

Hine et al. (1973) observed that IV administration of TRH in dogs produced several effects indicative of sympathetic activation, including mydriasis, vasoconstriction, increased respiratory rate, and general restlessness. Other authors have also noted a generalized behavioral excitation in rats, mice (Kruse, 1975), rabbits (Carino et al., 1976), cats (Myers et al., 1977), and fowl (Nistico et al., 1978). In most species, cats being a notable exception, this excitation results in a measurable increase in locomotor activity (Segal and Mandell, 1974; Havlicek et al., 1976; Vogel et al., 1979). The increase in locomotor activity produced by peripherally administered TRH is quantitatively and qualitatively different from that produced by d-amphetamine, a CNS stimulant. Amphetamine produces significantly greater increases in locomotor behavior than does TRH. Furthermore, the most notable effect observed after administration of TRH to rats is an increase in grooming of the face and forepaws, with only a small but significant increase in locomotor activity and sniffing behavior (Costall et al., 1979; Ervin, Schmitz, Nemeroff, and Prange, 1981). This is in contrast to the effects of d-amphetamine, which produces marked increases in locomotor activity, rearing, and sniffing behavior in rats. Consistent with the observed differences between TRH and d-amphetamine is the observation that the discriminative properties of TRH and d-amphetamine are different (Jones et al., 1978). The behavioral profile induced by TRH is not observed after thyroid-stimulating hormone (TSH) or thyroid hormone administration, suggesting that activation of the hypothalamic-pituitary-thyroid axis is not involved in mediating the behavioral effects of the tripeptide.

Local injections of TRH into the nucleus accumbens have been reported to produce increased locomotor activity in rats (Miyamoto and Nagawa, 1977; Heal and Green, 1979). It had previously been shown that dopamine (DA) administration into the nucleus accumbens is capable of producing increases in locomotor behavior. Data obtained in experiments analyzing several pharmacological approaches indicated that TRH-induced increases in locomotor activity may be mediated by CNS DA systems. Haloperidol and pimozide, two DA antagonists, blocked TRH-induced increases in locomotor behavior. DA antagonists block this TRH effect whether injected peripherally or directly into the nucleus accumbens. Administration of a monoamine oxidase inhibitor (MAOI), tranylcypromine, potentiates TRH-induced increases in locomotor activity, and similar results have been obtained when tranylcypromine is administered prior to intra-accumbens injections of DA. TRH appears to be acting presynaptically to induce DA release, because destruction of DA terminals in the nucleus accumbens with 6-hydroxydopamine (6-OHDA) abolishes TRH-induced increases in locomotor activity. Intracaudate injections of TRH were without effect on DA-mediated behaviors, suggesting a certain degree of regional specificity for a TRH-DA interaction. TRH-stimulated DA release from the nucleus accumbens, but not the caudate nucleus, has recently been observed in vitro (Kerwin and Pycock, 1979). These data, taken together, indicate that TRH in the nucleus accumbens may modulate dopaminergic neuronal activity.

TRH has been reported to cause an increase in body temperature in several laboratory animals (Carino et al., 1976; Nisticŏ et al., 1978; Horita et al, 1976). In rabbits, TRH, administered centrally or peripherally, produces hyperthermia in a dose-dependent manner. When injected into the hypothalamus, small quantities of TRH induce a significant hyperthermia in rabbits (Horita et al., 1976). Conversely, in cats, TRH produces hypothermia (Metcalf, 1974). Further study localized the site of action for TRH-induced hypothermia in the cat to mesencephalic regions; TRH injected directly into the hypothalamus was without effect on body temperature.

After peripheral or central administration, TRH reduces food and water intake (Barlow et al., 1975; Vijayan and McCann, 1977; Vogel et al., 1979). This effect of TRH is also dose-dependent. In rats receiving TRH peripherally, the inhibition of feeding is short-lived. Intraventricular injections of TRH were found to be much more potent than IP injections in inhibiting food intake (Vogel et al., 1979). This difference in potency suggests a CNS locus of action. The fact that TSH does not inhibit feeding suggests that TRH-induced release of TSH is not responsible for the action of TRH on appetitive behavior. TRH has recently been reported to attenuate the suppression of licking behavior maintained by punishment, and this effect was antagonized by naloxone and amphetamine (Vogel et al., 1980).

A well-studied effect of TRH is its ability to antagonize pentobarbital-induced sedation and hypothermia (Prange et al., 1974; 1975). This analeptic effect of TRH is not limited to pentobarbital, since the sedative and hypothermic effects of other barbiturates, ethanol, chloral hydrate, reserpine, chlorpromazine and diazepam are also partially antagonized by TRH (Breese et al., 1974; 1975; Prange et al., 1975). TRH antagonism of pentobarbital-induced sedation and hypothermia has been demonstrated in several mammalian species including rats, mice, gerbils, guinea pigs, rabbits, and monkeys (Breese et al., 1975; Carino et al., 1976; Kraemer et al., 1976). Several lines of evidence indicate that this analeptic effect of TRH is mediated in the CNS. Like the behavioral effects already described, a much lower dose of TRH is required to antagonize pentobarbital-induced sedation and hypothermia when the peptide is administered directly into the CNS. The septal area is one anatomical site that is extremely sensitive to these effects of TRH (Kalivas and Horita, 1979).

Use of structural analogues of TRH suggest that their ability to antagonize ethanol-induced sedation and hypothermia is unrelated to their ability to release TSH from the pituitary (Cott et al., 1976). This is inferred from a lack of correlation between the ability of the structural analogues of TRH to decrease the magnitude of sedation and hypothermia and their ability to release TSH from the anterior pituitary. Furthermore, neither TSH nor thyroid hormones share with TRH its analeptic properties. Finally, TRH has been shown to reverse pentobarbital-induced sedation in hypophysectomized rats, clearly demonstrating that the pituitary is not required for mediation of the effect of the tripeptide.

The antagonism of the effects of CNS depressants by TRH is consistent with the data described concerning TRH-induced

behavioral excitation in non-drug-treated animals. In fact,
TRH has been reported to reverse a natural CNS depression, that
is, hibernation (Stanton et al., 1980). A small quantity of
TRH injected into the hippocampus, but not the cortex, aroused
ground squirrels from hibernation.

The neurochemical basis underlying the behavioral effects of
TRH has not yet been elucidated. However, several neurotransmitters, especially the catecholamines, have been implicated in
mediating the effects of TRH. Early data supporting this hypothesis, albeit indirect, include the ability of TRH to potentiate the effects of dihydroxyphenylalanine (L-DOPA), a precursor
of catecholamines (Plotnikoff et al., 1972a). The L-DOPA test,
a putative screening procedure for antidepressant agents, consists of treating animals with pargyline (an MAOI) and L-DOPA.
The animals are then administered the test substance (TRH) or
vehicle, and the effect on activity is scored by an observer
ignorant of the treatment regimen. TRH significantly increases
the amount of behavioral excitation in this test. This effect
of TRH is neither thyroid nor pituitary dependent, since removal of either gland does not alter this TRH effect. When TRH is
injected into the brain, considerably smaller doses are needed
to elicit DOPA potentiation (Huidobro-Toro et al., 1975).

Certain biochemical evidence indicates that TRH is capable of
increasing brain catecholamine turnover. TRH does not alter
endogenous concentrations of brain norepinephrine (NE) or DA
(Horst and Spirt, 1974; Keller et al., 1974; Breese et al.,
1975). However, several studies indicate that TRH does increase catecholamine turnover. Peripherally administered TRH
has been reported to increase DA turnover, as measured with
α-methyl-p-tyrosine, a catecholamine synthesis inhibitor (Marek
and Haubrich, 1977). This effect was observed in all brain regions examined and was not dependent on the presence of an intact thyroid. However, Keller et al. (1974), using radioactive
precursor incorporation into DA, reported that DA turnover was
unaffected by peripherally administered TRH. The difference in
methodology may account for this discrepancy. The evidence
that TRH increases NE turnover is somewhat more consistent. NE
turnover, estimated by several different methods of measurement, has consistently been found to be increased after TRH
treatment. These methods include: radioactive precursor incorporation into NE; increased 3-methoxy-4-hydroxyphenylethyleneg-
lycol (MHPG), an NE catabolite; and increased rate of decline
of NE after synthesis inhibition (Keller et al., 1974;
Constantinides et al., 1974; Horst and Spirt, 1974; Marek and
Haubrich, 1977). However, no effect of either acute or chronic
TRH treatment on the activity of regional brain tyrosine hydroxylase activity was found in adult rats (Nemeroff et al.,
1977a).

An interaction between TRH and cholinergic systems has also been postulated, largely from data obtained in pharmacological experiments. For example, cholinergic antagonists, such as atropine, block the reversal of ethanol and pentobarbital sedation by TRH (Cott et al., 1976; Breese et al., 1975). Consistent with this are studies that indicate that TRH enhances the excitatory activity of acetylcholine (ACh) and carbachol on the spontaneous firing of cortical neurons (Yarbrough, 1976; 1978; 1979; Yarbrough and Singh, 1978); however, other investigators have been unable to confirm these findings (Winokur and Beckman, 1978; Pittman et al., 1978; Renaud, chapter 5). TRH has been reported to decrease the content and increase the turnover rate of ACh in the parietal cortex (Malthe-Sorenssen et al., 1978). At present, it appears likely that the interactions with cholinergic systems play a role in the analeptic effects of TRH (Yarbrough, 1979), although they do not appear to mediate several of its other actions.

Finally, agents that presumably act as γ-aminobutyric acid (GABA) agonists (γ-hydroxybutyric acid, baclofen, amino-oxyacetic acid) have also been shown to be effective in blocking the analeptic and tremorogenic effects of TRH in rats and mice (Cott and engel, 1977; Yarbrough, 1979), although the specificity of these agents in affecting GABA systems has been questioned.

In summary, TRH is capable of eliciting a variety of behavioral responses. When the matter has been studied, these responses have been found to be independent of the hypothalamic-pituitary-thyroid axis. TRH appears to interact with several neurotransmitter systems. The effects of TRH in human beings are discussed in chapter 13 by Loosen and Prange.

PRO-LEU-GLY-NH$_2$ (PLG, MIF-I)

This tripeptide, originally isolated from bovine hypothalamus (Nair et al., 1971; Celis et al., 1971), has been shown in certain species to inhibit the release of melanocyte-stimulating hormone (MSH) from the pituitary gland. Whether this substance is the major physiological regulator of MSH release remains a matter of controversy. Thus, Kastin et al. (1971) reported that MIF-I inhibits MSH release from frog pituitary; Bowe et al. (1971) found no such effect. However, recent evidence seems to support the contention of a role for MIF-I in the regulation of MSH secretion (Vivas and Celis, 1975). The amino acid sequence of MIF-I is identical to the C-terminal sequence of oxytocin, and evidence has been presented that suggests that the tripeptide is formed by clevage from a precursor oxytocin molecule (Celis et al., 1971). However, despite its early

isolation and characterization, lack of a suitable RIA procedure has limited the analysis of the CNS distribution of this
tripeptide. Peptidases that degrade MIF-I have been found in
the brain and plasma of animals and man (Bennett and McMartin,
1979; Neidle et al., 1980). The paragraphs that follow focus
on the behavioral effects of MIF-I in animals and man, and on
the few neurochemical studies performed with this tripeptide.

Behavioral and Neurochemical Effects in Animals

Behavioral studies on the effects of MIF-I in animals provided
a rational basis for the subsequent clinical trials of this
compound. Three different research groups (Plotnikoff et al.,
1971, 1974a; Huidobro-Toro et al., 1974, 1975; Voith, 1977)
have demonstrated that MIF-I, like TRH, potentiates the effects
of pargyline+L-DOPA treatment in normal and hypophysectomized
(HPX) mice and rats. MIF-I was considerably more potent than
TRH in this regard. This interaction naturally led to an investigation of the effects of MIF-I in a number of screening
tests for antiparkinsonian drugs. However, Carolei et al.
(1977), utilizing an invertebrate system (Planaria) purported
to be a sensitive paradigm for assessing dopaminergic and anticholinergic activity of drugs, found no significant effects of
MIF-I after either acute or chronic treatment. MIF-I has also
been found to be effective in antagonizing oxotremorine-induced
tremors (Plotnikoff et al., 1972b; Plotnikoff and Kastin, 1974;
Castensson et al., 1974) and fluphenazine-induced catalepsy in
rats (Voith, 1977). MIF-I potentiates the actions of L-DOPA in
antagonizing harmine-induced tremors (Huidobro-Toro et al.,
1975) and deserpidine-induced sedation (Plotnikoff et al.,
1973). In contrast, Bjorkman et al., (1980), in a "blind"
study, found no significant effect of MIF-I on the effects of
oxotremorine, fluphenazine-induced catalepsy, or amphetamine-
induced hyperactivity in rats. MIF-I has also been reported to
potentiate apomorphine-induced hyperactivity (Barbeau et al.,
1975) and rotational behavior in rats. When administered to
undrugged rats, MIF-I has been found to exert no significant
effect on locomotor activity (Kastin et al., 1973). Additional
reported behavioral effects of MIF-I include electroencephalographic (EEG) activation, increased mounting behavior, and weak
analgesic activity in rats (Plotnikoff and Kastin, 1974), and
induction of sterotypies in cats (North et al., 1973).
    Several of these effects are suggestive of prodopaminergic
activity, a concept further supported by the reported increase
in DA synthesis in the striatum (Freidman et al., 1973) and DA
turnover in the caudate nucleus (Versteeg et al., 1978) follow-

ing MIF-I administration. Increases in NE turnover were also noted. No effect on steady-state levels of DA, NE, or serotonin (5-HT) have been noted in rat brain after MIF-I treatment (Plotnikoff et al., 1974a; Spirtes et al., 1975). Antithetically, the increase in DA synthesis was prevented by HPX, whereas the L-DOPA potentiating action was unaffected after this surgical procedure, indicating a dissociation of these effects. However, others have observed no effect of the peptide on measures of brain DA activity (Kostrzewa et al., 1975). Pugsley and Lippmann (1977) reported that MIF-I and a potent analogue of MIF-I significantly increase brain DA levels and turnover. Recently, Snider et al. (1980) have reported that a synthetic analogue of MIF (Pro-D-Leu-Gly-NH$_2$) that is resistant to degradation *in vivo* significantly potentiates L-DOPA-induced behaviors in two experimental models of Parkinson's disease. The reported clinical efficacy of MIF-I in patients with Parkinson's disease (see the following section) might be construed as providing supportive evidence of a prodopaminergic effect in man.

MIF-I (10 to 20 mg/kg) has been reported to antagonize morphine-induced catalepsy, whereas higher doses are ineffective (Chiu and Mishra, 1979). Similarly, Kastin et al. (1979) recently reported that MIF-I blocks the effects of enkephalin and morphine in the tail-flick test for analgesia, but not in the vas deferens assay. Two conflicting reports have appeared concerning the effects of MIF-I on the development of physical dependence to morphine. Van Ree and de Wied (1976) showed that MIF-I would facilitate the development of physical dependence on morphine in rats, whereas Walter et al. (1979) found that the peptide has the opposite effect in mice. Although species differences may be involved, these conflicting results indicate that a cautious approach should always be maintained when interpreting the significance of research based on pharmacological interactions.

Effects of MIF-I in Human Beings

The activity of this tripeptide has been examined in several neurological disorders and in mental depressions. Ehrensing and Kastin (1974) treated 18 depressed women in a double-blind study. Six received MIF-I, 60 mg/day orally; six received 150 mg; and six received placebo. Marked improvement from the lower dose of the tripeptide was obseved. In a second study these authors treated 24 bipolar or unipolar depressed patients. Patients received MIF-I orally (75 or 750 mg/day) or placebo, for six days. Again, only the smaller dose of the tripeptide produced a significant antidepressant effect (Ehrensing and Kastin, 1978).

Studies are now sufficient to allow a comparison between a
set of findings from the animal laboratory and a set of findings from the neurological clinic. In Parkinson's disease, the
role of DA in pathophysiology and the role of L-DOPA, its precursor, in treatment are acknowledged. As recounted in chapter
13 of this volume by Loosen and Prange, TRH, although it may
favorably affect the mental state of patients with Parkinson's
disease, fails to benefit their neurological condition. In
contrast, MIF-I appears at least partially effective in this
clinical condition, just as it is effective in a relevant animal model (Barbeau et al., 1975).

In an early study of 16 parkinsonian patients, it appeared
that the infusion of 10 mg of MIF-I exerted antiparkinsonian
activity and might reduce the severity of dyskinesia caused by
L-DOPA therapy (Kastin and Barbeau, 1972). Chase and his colleagues (1974) gave 20 mg of the peptide to six otherwise untreated patients and found it beneficial in three. Therapeutic
effects were less clear in five patients undergoing L-DOPA
treatment, and dyskinesia was increased in three. Fischer et
al. (1974) gave infusions of up to 30 mg daily to 10 unselected, otherwise untreated patients and found slight, but statistically significant, improvement after two weeks. Barbeau
(1975) found that an intravenous bolus of 200 mg of MIF-I
greatly potentiated the effects of oral L-DOPA therapy.
Gerstenbrand et al. (1975) found that infusions of 400 mg/day
benefited a series of 10 patients. The only negative study of
MIF-I in Parkinson's disease is the most recent one. In a placebo-controlled design, Carceni et al. (1979) found that 200 mg
intravenously had no effect, but their patients were already
receiving what they regarded as optimum L-DOPA treatment.

Woods and Chase (1973) addressed directly the question of
possible effects of MIF-I on L-DOPA-induced dyskinesia and concluded that the peptide, at least after brief infusion, tends
to exacerbate the condition. Guided by the earlier observation
that MIF-I might reduce L-DOPA-induced dyskinesia, which resembles tardive dyskinesia, and by the observation that MIF-I reduced tardive dyskinesia in one patient (Ehrensing et al.,
1974), Ehrensing and Kasting (1977) studied oral MIF-I in 13
patients with tardive dyskinesia. Temporary improvement was
observed after small, but not larger, doses. However, the
authors concluded that a placebo effect was the most likely
explanation.

LUTEINIZING HORMONE-RELEASING HORMONE

Behavioral Effects in Animals

The induction of lordosis by the systemic or central injection of LH-RH in extremely small quantities is perhaps the best example of a CNS effect of a hypothalamic-releasing hormone that is clearly unrelated to its adenohypophyseal actions. This phenomenon is discussed in detail; other CNS effects of this peptide also will be reviewed briefly.

In 1973, two groups of investigators--Moss and McCann in Dallas and Pfaff in New York--using slightly different paradigms, reported that the SC injection of extremely small quantities of LH-RH-induced lordosis in female rats that were either estrogen-primed, hypophysectomized and ovariectomized, or estrogen-primed and ovariectomized. Since this response was not observed in identically treated animals after injection of LH, follicle-stimulating hormone (FSH), or TRH, the suggestion was made that LH-RH may play a physiological role in the induction of mating behavior (Pfaff, 1973; Moss and McCann, 1973). In later studies, Moss et al (1975) found that as little as 150 ng of LH-RH SC produced a significant increase in lordotic behavior. Although LH-RH, like most small neuropeptides, is rapidly degraded *in vivo*, Moss and his co-workers found that LH-RH-induced lordosis persisted for relatively long periods of time. After 500 ng of LH-RH, the increased lordosis persisted for more than eight hours (Fig. 6-1). Direct CNS injection of the decapeptide into the medial preoptic area or the arcuate-ventromedial nucleus region (medial-basal hypothalamus) induced lordosis, whereas injections into the lateral hypothalamus or cerebral cortex produced no such effect. In addition, LH-RH infusions into the midbrain central gray of estrogen-primed, ovariectomized female rats produced a significant increase in lordosis when compared to vehicle-injected controls (Riskind and Moss, 1979; Sakuma and Pfaff, 1980). Moss recently reported that the decapeptide may be endogenous to this area. Corroborative evidence for a physiological role for LH-RH in inducing lordosis is found in the report of Kozlowski and Hofstetter (1978) that the infusion of antiserum to LH-RH into the third ventricle of estrogen- and progesterone-primed female rats produces a *decrease* in lordosis. Recently, Sakuma and Pfaff (1980) confirmed and extended these observations; LH-RH injections into the mesencephalic central gray facilitated the incidence of lordosis, whereas LH-RH antiserum completely blocked the behavior.

Cheng (1977) demonstrated that this effect of LH-RH is not species specific. Ovariectomized ring doves that were estrogen-primed and then treated with LH-RH exhibited typical female mating behavior; LH or progesterone failed to induce this courtship behavior.

Moss' group has also examined the effects of LH-RH on male sexual behavior. In intact male rats, LH-RH injection significantly reduced the latency to first intromission and

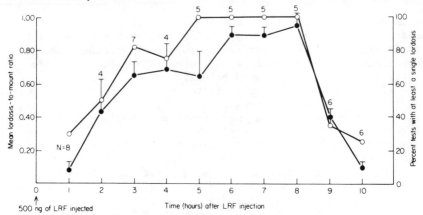

Fig. 6-1. Time course for the initiation and maintenance of
the LRF(LH-RH)-facilitated lordosis response. Solid
dots give lordosis-to-mount ratio; circles give percentage of animals showing at least one lordosis;
vertical bars give standard error. (Reproduced from
Moss et al., 1975, with permission.)

ejaculation as compared to vehicle-treated controls; no effect
on the number of mounts or lordosis to mount ratio was observed. In castrated males that were subsequently testosterone-
primed at a dose too low to induce mating behavior, 500 ng of
LH-RH significantly decreased the time required to achieve
ejaculation when compared with vehicle-treated controls (Moss
et al., 1975; Moss and Foreman, 1976).

Although TRH by itself exerts no significant effect on lordosis in female rats, the tripeptide, when injected into animals
that have received LH-RH, significantly reduces subsequent
mating behavior (Moss, 1978; 1979). Foreman and Moss (1979)
found that arcuate-ventromedial hypothalamus or medial preoptic
area infusions of TRH or LH significantly reduced LH-RH-induced
lordosis; FSH injections exerted no effect in either brain region. The observed inhibitory action of LH in this paradigm
may reflect physiological coordination of the endocrinological
and behavioral processes mediated by the hypothalamic-pitui-
tary-gonadal axis, whereas the TRH effect is more difficult to
interpret.

We have recently found (Bissette, Nemeroff, Kizer, and
Prange, unpublished observations) that lesions of the organum
vasculosum laminae terminalis (OVLT), an area known to contain
high concentration of the decapeptide, do not abolish the induction of lordosis by systemically injected LH-RH in the female rat. These data indicate that the OVLT is probably not

the neuroanatomical locus mediating these effects of the peptide. However, the peptide has recently been shown to inhibit
the firing rate of neurons in this circumventricular organ
after iontophoretic application (Felix and Phillips, 1979).

Lordosis in ovariectomized, estrogen-primed female rats has
been reported to be increased after stimulation of hypothalamic
DA receptors (with DA or apomorphine) and diminished after
haloperidol blockade of DA receptors (Foreman and Moss, 1979).
Haloperidol also totally abolished LH-RH-induced lordosis. The
electrophysiological effects of LH-RH are discussed in chapter
5 by Renaud and in a recent review article (Moss, 1979).

Few studies have been concerned with evaluating other behavioral effects of LH-RH. These studies generally have used
relatively high doses of the peptide when compared to the quantities required to produce lordosis. We (Bissette et al.,
1978) have noted that, although peripherally (IP) administered
LH-RH exerts no significant effect on the duration of narcosis
induced by peripherally administered pentobarbital, intracisternal (IC) injection of the decapeptide in microgram quantities does significantly reduce barbiturate-induced sedation.
Nemeroff et al. (1979) reported that approximately 1 μg of LH-
RH (IC) in the mouse produced significant hyperthermia at an
ambient temperature of 23°C, whereas, at 4°C, it produced a
small but non-significant decrease in body temperature
(Bissette et al., 1976). Lomax et al. (1980) reported that LH-
RH (5 μg) produced hypothermia in rats after preoptic area injections. Plotnikoff et al. (1976) have reported that the
injection of high doses of LH-RH (1-8 mg/kg IP), like MIF-I and
TRH, potentiated the behavioral effects of L-DOPA and 5-hy-
droxytryptophan (5-HTP) in pargyline-pretreated animals.

Behavioral Effects in Human Beings

This decapeptide has been used as an independent variable in
several disease states. Both behavioral changes and endocrine
changes have been studied as dependent variables. In the paragraphs that follow, we will be concerned only with the former.

In a small group of depressed patients, Benkert (1975) found
that LH-RH, like TRH, exerted a slight antidepressant effect.
However, this effect was not statistically significant relative
to the effect of saline.

Consistent with findings in animals, as noted previously, LH-
RH probably exerts effects on sexual behavior in human beings.
Such behavior, of course, is complexly determined, and in the
human, in contrast to animals, it has not been possible to establish conclusively that behavioral effects following LH-RH
administration are independent of its endocrine effects.

Mortimer and his colleagues (1974) suggested that the peptide may indeed produce a behavioral effect independent of endocrine effects. Six of seven hypogonadal men noted increased sexual potency shortly after starting daily treatment, and this occurred before their low levels of androgens had changed. McAdoo et al. (1978) gave LH-RH to normal males and found an "increase in alertness, decrease in anxiety and fatigue, and an increased speed of performance on automatized motor tasks. These patients also reported an increase in sexual arousal. La Ferla and his colleagues (1978) showed that the viewing of an erotic film resulted in enhanced LH secretion. After LH-RH injection, the magnitude of the LH response was correlated with self-evaluation of sexual arousal. Schwarzstein et al. (1975) treated four men with daily injections of LH-RH for an extended time. Before treatment, all four men were sterile and showed diminished sperm counts; three had normal libido, and one had diminished libido. During treatment, all experienced increased libido, although the sperm count increased substantially only in two. After completion of treatment, the two who had shown the greatest increases in sperm count showed a new, abrupt increase. This latter observation may be of particular interest in light of Benkert's (1975) experience. He gave LH-RH by nasal spray to six impotent men. Although the over-all results were equivocal, the clearest positive results associated with LH-RH treatment occurred immediately after its cessation.

SOMATOSTATIN

SRIF, a tetradecapeptide (Table 6-1) that inhibits the secretion of growth hormone from the anterior pituitary (Vale et al., 1975, for review), was isolated from bovine hypothalamus by Brazeau et al. (1973, 1974). SRIF exerts an inhibitory effect on the secretion of a variety of hormones, including TRH, growth hormone, thyrotropin, thyroid hormone, insulin, glucagon, gastrin, vasoactine intestinal polypeptide, calcitonin, parathyroid hormone, prolactin, and gastric inhibitory peptide (Lucke et al., 1976; Vale et al., 1975; Hirooka et al., 1978; Pimstone et al., 1976; Gerich et al., 1976). SRIF also inhibits the secretion of vasopressin. The effect of SRIF on pancreatic hormone secretion has led to attempts to determine whether this peptide may in some way be involved in the etiology of diabetes mellitus; at present, this subject is controversial (Unger et al., 1977; Felig et al., 1976; Gerich et al., 1976). The role of SRIF in endocrine homeostasis will not be reviewed here; several excellent reviews have already been cited.

Bioassay (Vale et al., 1975), RIA (Arimura et al., 1975, 1978; Diel et al., 1977; Yanaihara et al., 1978; Nakagawa et al., 1978), and radioreceptor assay (Ogawa et al., 1976) techniques have been developed to measure SRIF in biological fluids. Using these techniques, SRIF has been assayed in the brain, spinal cord, blood, and a variety of other tissues, including chorionic villi and decidua (Kumasaka et al., 1979), thyroid, gastrointestinal tissues, pancreas, lymph nodes, thymus, and salivary gland (Hökfelt et al., 1975). Large, higher-molecular weight forms of SRIF immunoreactivity have also been detected in brain and pancreas, but not yet fully characterized (Dupont and Alvarado-Urkina, 1976; Spiess and Vale, 1978; Millar, 1978; Conlon et al., 1978). This material may represent large prohormone precursor forms of the peptide.

The release of SRIF has also been the subject of a considerable number of studies. Wakabayashi et al. (1977) found that, like many other putative neurotransmitters, SRIF is released from rat hypothalamic synaptosomes by high K^+ concentration in a CA^{2+}-dependent manner. These results have now been confirmed using rat hypothalamic fragments (Iversen et al., 1978; Berelowitz et al., 1978) and neurohypophyseal tissue (Patel et al., 1977). Neurotransmitter candidates, including DA, NE, and NT, have been reported to release SRIF from brain tissue *in vitro*; the physiological significance of these findings is unclear.

A SRIF-binding protein has been found in the cytosol of a variety of organs, including brain (Ogawa et al., 1977; 1978). The binding of radiolabeled SRIF to bovine pituitary plasma membranes has been described (Leitner et al., 1979). It is saturable, specific, and temperature dependent. We are not aware of any studies that have utilized radiolabeled SRIF to assess binding sites in the brain. Controversy in this field concerns whether SRIF has affinity for opiate receptors. Terenius (1976), using synaptic membranes derived from rat brain, reported that SRIF, and adrenocorticotropin (ACTH), exhibit partial antagonist-like selectivity for opiate receptors. These data have been confirmed (Pugsley and Lippmann, 1978). In contrast, Traber et al. (1977) reported that neuroblastoma X glioma hybrid cells possess different receptors for SRIF and opioids.

The presence of enzymes that degrade or inactivate SRIF in brain and plasma have been described (Benuck and Marks, 1976; Griffiths et al., 1976, 1977; McMartin and Purdon, 1978; Sheppard et al., 1979). The biological potency of SRIF and chemically modified analogues is clearly related to their relative rates of biodegradation (Marks et al., 1976). McMartin and Purdon (1978) reported that one minute after IV administration in the rat, 80 to 90 percent of the injected SRIF was

converted to [des-Ala¹]-SRIF, a biologically active form of the
peptide. These workers suggested that the majority of SRIF is
eventually converted to biologically inactive forms in tissue
and not in the circulation. Sheppard et al (1979) obtained
evidence in man that the kidney is one major site of metabolic
clearance of SRIF.

Several studies have examined CNS effects of SRIF, and this
subject has been reviewed (Kastin et al., 1978). Electrophysiological effects of this peptide have been found and will not
be discussed, since they are amply reviewed by Renaud in chapter 5 and elsewhere (Renaud, 1977). Unlike the effects of LH-
RH and angiotensin, which are linked to specific spheres of
behavioral effects (that is, sexual behavior and drinking,
respectively), there seems to be no specific sphere of behavioral effect to which SRIF has been postulated to play a major
role. In an early study, Segal and Mandell (1974) found that
the infusion of SRIF into the lateral ventricle of freely moving rats significantly reduced their spontaneous locomotor activity. In another study, Plotnikoff et al. (1974b) reported
that high does of SRIF (1 to 4 mg/kp IP), like TRH, MIF, and
LH-RH, significantly potentiated the effects of L-DOPA in pargyline-pretreated mice. The tetradecapeptide did not alter the
effects of oxotremorine, 5-HTP, footshock-induced fighting, or
audiogenic-seizure threshold. In several studies (see Prange
et al., 1979, for review) we have noted a slight but significant potentiation of pentobarbial-induced narcosis and lethality (Prange et al., 1975; Nemeroff et al., 1977b). However, in
other studies (Bissette et al., 1978) we have been unable to
confirm these findings. Brown and Vale (1975) reported that IV
SRIF significantly decreased the LD_{50} of pentobarbital and increased the LD_{50} of strychnine in the rat. Cohn (1975) found
that intracerebroventricular (ICV) SRIF (5 to 45 μg) significantly potentiated amobarbital-induced sedation in the rat.
Chihara et al. (1978a) found that ICV injection of antiserum to
SRIF, a putative tool to inactivate endogenous SRIF, exerted
effects opposite to those observed after administration of
SRIF. Anti-SRIF serum increased the LD_{50} of a CNS depressant--
pentobarbital--and a CNS convulsant--strychine.

SRIF has also been reported to exert behavioral effects in
the untreated intact laboratory animal. Cohn and Cohn (1975),
using freehand lateral ventricle injection into unanesthetized
rats, reported that SRIF (25-50 μg) induced "barrel rotation,"
that is rolling in a left-handed direction. This effect was
found to be blocked by atropine but not by haloperidol or apomorphine pretreatment.

In intact and hypophysectomized rats with lateral ventricular
cannulae, Havlicek et al. (1976) reported that SRIF administration (10 μg) produced marked behavioral excitation associated

with significant reduction in both slow-wave and rapid-eye-movement (REM) sleep. SRIF also produced intense motor excitation and stereotypy, characterized by prolonged compulsive scratching and circular movements, and distrubances of gait, balance, and motor coordination. At higher doses SRIF-induced paraplegia-in-extension and actual tonic-clonic seizures. Rezek et al. (1976a) reported that the infusion of SRIF (>10 ng) into the dorsal hippocampus of the rat produced a "freezing response" characterized by vibrissae trembling, lower jaw quivering, and a tremor of the masseter muscles. Doses greater than 100 ng SRIF into this same site also produced steroltyped oral movements (chewing and licking) and a significant reduction in REM sleep, whereas infusion of 1.0 μg of SRIF markedly increased the intensity, frequency, and duration of freezing, facial tremor, and oral stereotypies, and three of five animals exhibited ravenous eating. Both slow-wave and REM sleep were markedly reduced. A dissociation between the electroencephalogram (EEG) patterns and the behavioral state of the animal was noted. After 5.0 μg SRIF, the animals exhibited profound loss of balance and coordination; and after 10 μg, these workers observed changes similar to those of Cohn and Cohn (1975) after ICV SRIF, such as barrel rolling, paraplegia, and tonic-clonic seizures. Of interest was their finding that SRIF analogues with little endocrine activity did not exert such effects after intrahippocamal infusion.

Rezek et al. (1977a) found that SRIF induces biphasic effects on behavior and motor control after intra-amygdaloid injection, the effect depending upon the dose. Low doses (10 to 100 ng) induced behavioral excitation associated with a variety of tremors and stereotypies, whereas higher doses (1 to 10 μg) resulted in motor incoordination. Once again, dissociation between EEG pattern and behavior was noted, and SRIF analogues without endocrinological activity exerted no effects comparable to those described after administration of SRIF. Rezek et al. (1976b, 1977b) found effects similar to those observed after hippocampal and amygdaloid injections: behavioral excitation with reduction of REM and slow-wave sleep at low doses and loss of motor coordination at high doses. Thus, with minor exceptions, similar effects have been noted after SRIF administration into several CNS sites, including the hippocampus, amygdala, neostriatum, cortex, and lateral ventricle. Such data would appear to indicate that SRIF either is diffusing to the neuroanatomical locus mediating these effects or, alternatively, SRIF may be able to induce these effects by acting at all of these sites.

Recently, Havlicek et al. (1976) reported that the ICV injection of 1 to 10 μg SRIF in the rat produces significant analgesia in the hot water tail-immersion test. We (Nemeroff et al., 1979) have not been able to confirm this finding using a similar test for evaluating analgesia in mice.

However, in contrast to the findings cited, it is of interest to note the report of Gordon et al. (1978), who found no significant behavioral changes after 10 to 600 µg SRIF (ICV) in the conscious goat.

A role for endogenous SRIF, at least in endocrine responses to stress, has been clearly demonstrated. Two research groups (Terry et al., 1976; Arimura et al., 1976) have demonstrated that IV administration of antiserum to neutralize endogenous SRIF results in abolition of the stress-induced inhibition of growth hormone secretion in the rat.

Neurochemical sequelae of SRIF action have been sought. In the peripheral nervous system, SRIF has been shown to produce a myriad of inhibitory effects. Guillemin (1976) reported an inhibition of ACh release in the electrically stimulated myenteric plexus-longitudinal muscle of the guinea pig ileum by SRIF *in vitro*. This effect was not blocked by naloxone, the opiate antagonist. Tachyphylaxis to this effect was observed after repeated SRIF treatment. This work was extended by Magnan et al. (1979), who showed that SRIF also inhibits the electrically evoked contractions of the rat vas deferens, an adrenergically mediated response. Similar findings were obtained by Cohn et al. (1978) in a variety of smooth muscle preparations. This body of data suggests that SRIF inhibits the release of both cholinergic and adrenergic neurotransmitters in these peripheral systems.

Enock and Cohn (1975) reported that SRIF increases brain cyclic AMP levels both *in vivo* and *in vitro*. This finding has now been confirmed after ICV SRIF in the rat; Herschl et al. (1977) reported that 10 µg SRIF produces a significant increase in cyclic AMP levels in the hippocampus, neostriatum, and cerebral cortex.

Tan et al. (1977) reported that *in vitro* SRIF, but not TRH, enhances glutamate-induced-Ca^{2+} accumulation in guniea pig cortical synaptosomes. SRIF also inhibited the release of radiolabeled Ca^{2+} in this preparation. Nemeth and Cooper (1979) found no effect of SRIF *in vitro* on ACh release from rat hippocampal synaptosomes under physiological conditions, but they did note increased [^{3}H]-ACh efflux in the presence of high concentrations of the peptide. Since this effect was also associated with efflux of the enzyme lactic dehydrogenase, these investigators suggested that such effects of high-dose SRIF are due to nonspecific membrane damage.

Few studies have examined the effects of SRIF on brain neurotransmitter function. Garcia-Sevilla et al. (1979) reported that 20 µg SRIF administered ICV significantly increased the turnover of regional brain 5-HT, DA, and NE; evidence of both increased biosynthesis and utilization of these monoamines was obtained. Wood et al. (1979) reported that ICV SRIF (25 µg), as well as α-MSH and ACTH, increased the turnover rate of ACh in the rat hippocampus. Since transsection of the cholinergic fibers to the hippocampus did not abolish this neuropeptide-induced alteration in cholinergic neurotransmission, it was suggested that these neuropeptides interact directly with hippocampal receptors. The studies described in this section all indicate that pharmacological doses of SRIF can produce alterations in behavior, neurochemical measures, and responses to pharmacological agents. Whether such effects signify an important neuromodulatory role for SRIF is uncertain at this time.

NEUROTENSIN

NT is a tridecapeptide (Table 6-1), which was originally isolated from bovine hypothalamus by Carraway and Leeman (1973).

NT localization has been investigated utilizing two approaches: immunohistochemistry and RIA. Specific problems related to immunoassay of NT have recently been discussed by Carraway (1978). NT localization is discussed in detail in chapter 2 by Grant et al. in this volume. In summary, immunoreactive NT has been detected in substantial quantities in the hypothalamus, preoptic area, nucleus accumbens, septum, amygdala, habenula, interpeduncular nucleus, and in the substantia gelantinosa of the spinal cord and trigeminal nerve nuclei.

In addition to its heterogeneous CNS distribution, other data support a neurotransmitter or neuromodulatory role for NT. High-affinity, reversible, and saturable binding of radiolabeled NT to CNS membranes has been reported by several groups (Uhl et al., 1977; Lazarus et al., 1977; Kitabgi et al., 1977). Recently, Iversen et al. (1978) have described K^+-induced release of immunoreactive NT from hypothalamic slices *in vitro*; this NT release was found to be Ca^{2+}-dependent. Finally, the degradation of NT has been studied in tissue extracts of the rat; kidney, liver, and brain peptidases degrade NT in a time- and temperature-dependent manner (Dupont and Merand, 1978).

Although NT administered peripherally or centrally can modify the secretion of many anterior pituitary hormones, insulin and glucagon (Makino et al., 1973; Rivier et al., 1977a; Maeda and Frohman, 1978; Nagai and Frohman, 1976; Carraway et al., 1976), at present, it appears that the peptide does not act as a classical hypothalamic releasing or inhibiting hormone. Nevertheless, NT does produce a variety of endocrine, hemodynamic, and behavioral actions that may provide clues to its physiological function. Detailed reviews of the effects of exogenously administered NT have appeared elsewhere (Bissette et al., 1978; Prange et al., 1979; Nemeroff, 1980). In this section, we shall briefly review the major effects observed after CNS administration of this peptide.

The profound CNS effects of NT were first discovered while evaluating the effect of the peptide on pentobarbital-induced sedation in mice (Nemeroff et al., 1976, 1977b; Bissette et al., 1978). Unlike TRH, which significantly shortens the length of sleeping time induced by pentobarbital, IC NT potentiated the effects of this barbiturate. Subsequent studies utilizing [^{3}H]-pentobarbital revealed that IC NT apparently inhibits the rate of metabolic clearance of pentobarbital. In addition, Luttinger et al. (1980a) reported that IC NT enhances ethanol-induced sedation and hypothermia in mice. This effect is not associated with any significant alteration in blood or brain ethanol levels. The effect of NT on barbiturate-induced sedation led to studies that demonstrated that centrally administered NT produces a marked dose-dependent hypothermia in mice, an effect that was enhanced when the animals were placed in a cold environment (4°C) (Bissette et al., 1976). This hypothermic effect observed after central, but not peripheral, NT administration has been confirmed by other investigators (Clineschmidt and McGuffin, 1977, Brown and Vale, 1980). In addition to its hypothermic effect in adult rodents, NT also has been shown to induce hypothermia after IC administration in cold-exposed young (18 days) and aged (>2 years) rats (Figure 6-2). Structure-activity studies utilizing various NT analogues have demonstrated that the C-terminal regions (especially the two arginine moieties at position 8 and 9) of the tridecapeptide are essential for hypothermic activity (Rivier et al., 1977b; Loosen et al., 1978).

A number of studies have been undertaken in an attempt to understand the mechanisms underlying the hypothermic effect of NT. We evaluated the effect of centrally administered NT on the body temperature of various representatives of the vertebrate kingdom. As expected, IC NT had no effect on the colonic temperature of several poikilotherms (fish, frogs, lizards). However, NT did produce hypothermia in the mouse, rat, gerbil, guinea pig, hamster, and monkey. Of the homeotherms tested, only rabbits, pigeons, and two obligate hibernators (woodchuck and ground squirrel) were unresponsive to IC NT (Prange et al., 1979; Metcalf et al, 1980). Data from this study provided provocative evidence concerning one possible mechanism by which NT produces its characteristic hypothermic effect. It appears that NT produces a significant hypothermic response only in those species that require activation of their basal metabolic rate (BMR) in order to maintain body temperature at the ambient temperatures at which they were tested (4°C or 23°C). Thus,

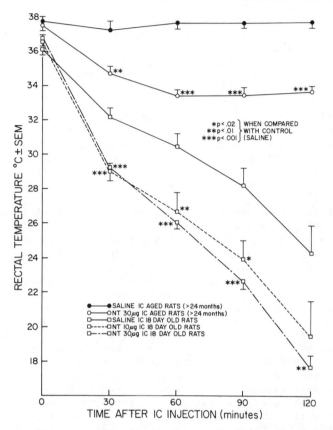

Fig. 6-2. The effect of intracisternal administration of neurotensin on colonic temperature of young (18 days
old) and old (24 months old) rats exposed to a 4°C
ambient temperature. Neurotensin produced a significant hypothermic effect in both young and old rats.
(Dunnett's test for multiple comparisons.)

the hamster and guinea pig exhibited hypothermic responses to
NT only at 4°C, whereas the rabbit and the two hibernators,
which do not increase BMR until ambient temperatures are well
below 4°C (Precht et al., 1973), remained unresponsive to NT.
Therefore NT may be acting as an inhibitor of BMR activation.
  Attempts have also been made to block the hypothermic effects
of NT using pharmacological techniques (Nemeroff et al., 1980).
Pretreatment of rats with α-adrenergic (phenoxybenzamine), β-
adrenergic (propranalol), cholinergic (atropine), or opiate

(naloxone and naltrexone) antagonists exerted no significant effect on NT-induced hypothermia. However, reduction of the functional activity of CNS DA systems, either by selective 6-OHDA-induced DA depletions or by pretreatment with a DA antagonist (haloperidol), potentiated the hypothermic effets of NT. The observed synergistic effect of NT and haloperidol with regard to body temperatures indicates that NT might possess neuroleptic-like activity. Consistent with this concept, NT stimulates prolactin release (Rivier et al., 1977a), decreases locomotor activity (Nemeroff et al., 1977b), potentiates the actions of a barbiturate, and increases brain monoamine turnover (Garcia-Sevilla et al., 1978). All of these effects are shared by neuroleptic drugs, such as chlorpromazine and haloperidol.

In light of these considerations, we studied the effects of NT in a number of screening tests for neuroleptic agents. For example, we found that NT, like chlorpromazine, produces muscle relaxation in the Joulou-Courvoisier traction test in mice (Osbahr et al., 1979), a screening test for neuroleptic agents. In a more sensitive and specific test system, NT, either ICV or directly injected into the nucleus accumbens, blocked the forward locomotion and rearing produced by systemic d-amphetamine administration (Lipton et al., 1979; Nemeroff, 1980). Since amphetamine is believed to act via a release of DA from nerve terminals, this study provides evidence that NT may act as an antidopaminergic agent. Current studies in this laboratory are investigating the mechanisms underlying this peptide-catecholamine interaction. Recently, we have noted that centrally administered NT possesses significant activity in two screening tests for neuroleptics. NT produced a diminished rate of self-stimulation after intra-accumbens injections and also exhibited neuroleptic-like activity in an active avoidance paradigm (Luttinger et al., 1980b).

An interaction between NT and another endogenous neuropeptide, TRH, has been noted during the course of our studies. TRH partially but significantly reverses the hypothermia produced by NT (Nemeroff et al., 1980; Brown and Vale, 1980). Furthermore, we have found that TRH will also block both the muscle-relaxing effects of NT and NT-induced enhancement of ethanol-induced sedation (Osbahr et al., 1979; Luttinger et al., 1980a). Both Nemeroff et al. (1980) and Brown and Vale (1980) have found that ICV NT significantly reduces serum TSH levels in the rat. Our study was performed in thyroidectomized animals, the latter study in cold-exposed rats. Although the mechanism of the antagonism between these two neuropeptides remains unclear, it is interesting to note that TRH is generally thought to exhibit prodopaminergic characteristics, and NT, antidopaminergic properties. Whether these effects truly represent physiologically important neuromodulatory activity of these peptides remains to be determined.

Finally, NT has also been shown to exhibit antinociceptive activity in a number of analgesic screening tests (Clineschmidt and McGuffin, 1977; Clineschmidt et al., 1979; Nemeroff et al., 1979). We have found that on an equimolar basis, NT is more potent than morphine or the enkephalins, but less potent than β-endorphin, in producing antinociceptive activity after central administration (Nemeroff et al., 1979). This analgesic-like activity is one of several actions shared by NT and the opiate peptides. Figure 6-3 illustrates the antinocisponsive and hypothermic effects of 1 μg of NT administered IC to mice, at an ambient temperature of 23°C. Both β-endorphin and NT induce hypothermia and elevations in prolactin secretion, and appear to possess somewhat similar CNS distributions based on immunohistochemical studies. However, distinct differences have been noted (for instance NT-induced hypothermia and antinociception are not blocked by the opiate antagonist naloxone), indicating that these peptides possess distinct pharmacobehavioral profiles.

In Figure 6-4, the known effects of centrally administered NT are summarized. These data concerning the CNS effects of NT suggest that this peptide may play a physiological role in the regulation of a variety of important functions, that is, thermoregulation, nociception, locomotor activity, and endocrine regulation.

SUBSTANCE P

SP was discovered by von Euler and Gaddum (1931) in extracts of horse brain and intestine and was characterized by its hypotensive (*in vivo*) and gut-contracting (*in vitro*) properties in rabbits. These effects were not abolished by atropine. For comprehensive reviews of the work on SP prior to its chemical characterization, see Zetler (1970) and Lembeck and Zetler (1971). Chang and Leeman (1970) purified SP from extracts of bovine hypothalamus and later sequenced (Tregar et al., 1971) and synthesized the peptide (Chang et al., 1971). SP was found to be an undecapeptide, whose sequence is shown in Table 6-1. The localization of SP is discussed in detail in chapter 2 by Grant et al. in this volume. In brief, substance P is distributed heterogeneously in the brain and spinal cord of a variety of mammalian species and is concentrated in the synaptosomal fraction after density-gradient centrifugation. Incubated rat brain synaptosomes have been shown to release SP in response to electrical stimulation and K^+. The latter release mechanism was found to be Ca^{2+}-dependent (Lembeck et al., 1977).

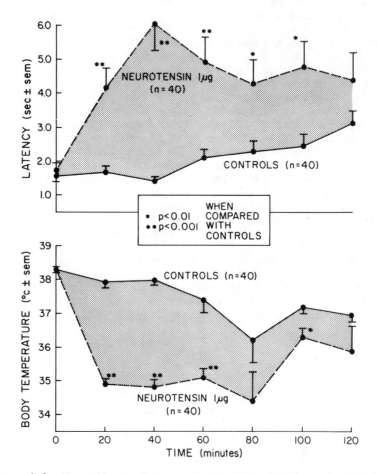

Fig. 6-3. The effect of intracisternal injection of vehicle
(0.9 percent sodium chloride) or neurotensin (1µg)
in the mouse on: (1) Top - response to noxious stimulus (tail immersion into a 48°C water bath), and
(2) Bottom - colonic temperature. Neurotensin produced a significant antiociceptive and hypothermic
response. (Student's T test, two tailed.)

Specific binding of [³H]-SP to synaptosomes from rabbit brain
was highest in midbrain and dorsal spinal cord (Nakata et al.,
1977). Using the [¹²⁵I-Tyr⁸]-SP analogue, the characteristics

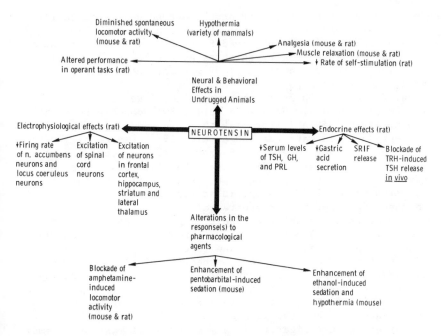

Fig. 6-4. Schematic diagram summarizing the effects of neurotensin when administered directly into the central
nervous system.

of this binding have been investigated (Mayer et al., 1979).
SP binding was reversible, saturable, highly specific, and pH
dependent. Phosphatidyl serine has been suggested to be the
binding substrate for endogenous SP and the binding of SP to
this lipid and to phosphatidyl ethanolamine is sensitive to
physiological Ca^{2+} and Na^+ concentrations, but not to K^+
(Lembeck et al., 1979). Using both RIA and immunohistofluorescence for localization of SP in vagus nerves of cats and
rabbits, Gamse et al. (1979) saw proximal accumulation of SP
after ligation of the nerve, which was imitated by colchicine
treatment, suggesting microtubule-dependent axoplasmic transport of immunoreactive SP.

SP has been shown to be inactivated in brain by neutral endo-
peptidases (Benuck and Marks, 1975). There is controversy over
the degradation of SP in plasma. Cannon et al. (1977) observed
a heat-labile component of plasma that degraded SP. This process was inhibited by SQ 20881, a nonapeptide inhibitor of an-
giotensin-converting enzyme. More recently, this group has
shown marked differences in the plasma inactivation of endoge-

endogenous versus synthetic SP and have hypothesized a carrier
protein that protects endogenous SP (Cannon et al., 1979).
Bury and Mashford (1977) observed SP degradation only in whole
blood and postulated the erythrocyte as the locus of inactivation. The plasma degradation of SP was not inhibited by baci-
tracin, whereas this peptidase inhibitor was effective in reducing SP degradation by brain homogenates (Cuello et al.,
1978; Lembeck et al., 1977). Recently, SQ 20881 has been shown
to be an effective inhibitor of both brain and blood peptidases
that degrade SP in rats (Cannon et al., 1977; Lee et al.,1979).

In an early study, Stern and Hadzovic (1973) reported that
peripherally administered SP crosses the blood-brain barrier.
Segawa et al. (1976) reported that uptake of [^{3}H]-SP into the
crude mitochondrial (P$_2$) subcellular fraction differs from that
of acknowledged putative neurotransmitter agents by: (1) being
nontemperature-dependent, (2) exhibiting a low pellet to medium
ration ($\approx$1), and (3) being taken up primarily by mitochondria
and not synaptosomes. Such uptake is generally considered to
be due to passive diffusion or nonspecific binding.

Electrophysiological data have provided the most compelling
evidence for consideration of SP as a neurotransmitter.
Lembeck proposed that SP might be a transmitter of sensory afferents, and this hypothesis has been supported by the findings
of a number of investigators in a variety of elegant paradigms.
As early as 1963, Krivoy et al. (1963a) had demonstrated SP-
induced potentiation of dorsal root ganglion discharge in
decerebrate cats. However, this work was performed with crude
SP extracts of horse intestine, and LSD was required to raise
the discharge threshold of this preparation (Krivoy et al.,
1963b).

The pioneering work of Otsuka and co-workers has done much to
establish the role of SP as a neurotransmitter in the dorsal
root system. For excellent reviews of this subject, see Otsuka
et al. (1975) and Otsuka and Konishi (1976). Using isolated
spinal cords from frogs or newborn rats (Otsuka et al., 1972a),
this group demonstrated the ability of extracts of bovine dorsal root ganglia to depolarize spinal motoneurons. They also
demonstrated release of immunoreactive SP from neonatal rat
dorsal root after electrical stimulation (Otsuka and Konishi,
1976). Nicoll (1978) reported that isolated, hemisected spinal
cord motoneurons in the frog are depolarized by SP. NT and
bombesin also depolarized this preparation, but their actions
could be blocked by tetrodotoxin, indicating a possible action
through interneurons, as opposed to the seemingly direct actions of SP. Henry et al. (1975) reported that 45 percent of
the cat dorsal horn neurons tested were excited by SP applied
iontophoretically. The activity of Renshaw cells was depressed
by SP; ACh had mixed actions (some cells excited, some

depressed, and some unchanged) on SP-excitable cells. Henry
(1976) also found that SP-excitable cells in cat dorsal horn
were excited by a noxious (heat) stimulus. Phyllis and
Limacher (1974) reported the presence of SP- and ACh-excitable
Betz cells in rat cerebral cortex. Although the action of ACh
could be blocked by atropine, SP was still able to discharge
the atropinized neurons.

Randič and Miletič (1977) found that noxious stimuli (tactile
or thermal) would excite neurons in the cat dorsal horn that
were also responsive to SP. These neurons were stimulated by
A, C, or δ fibers, while neurons responding to light tactile
stimuli were not depolarized by SP iontophoresis. Sastry
(1979) and Zieglgansberger and Tulloch (1979) confirmed these
findings. A current controversy in this area is whether SP
produces changes in membrane resistance. Certain investigators
have found no change (Zieglgansberger and Tulloch, 1979); others have noted increases in membrane resistance (Katayama and
North, 1978), and Dun and Karczmar (1979) reported a decrease
in membrane resistance. These seeming contradictions are probably due to differences in the experimental preparation and
tissues utilized.

Several studies have sought to establish CNS effects of SP by
microiontophoretic injection of the peptide into discrete brain
regions. Davies and Dray (1976) reported that 26 of 34 tested
cells in the rat substantia nigra were excited by microiontophoretic application of SP. These cells were also excited by
L-glutamate and ACh, and inhibited by GABA. Guyenet and
Aghajanian (1977) applied SP to individual neurons of rat locus
coeruleus and found excitation in more than 80 percent of
tested cells. A similar proportion of these cells was also excited by ACh. Further work in the locus coeruleus (Guyenet and
Aghajanian, 1979) differentiated the SP and ACh responses by
blockade of the latter with the muscarinic antagonist scopolamine. Scopolamine blocked neither excitation by SP and L-
glutamate nor inhibition by GABA and met-enkephalin. Naloxone
was found to block met-enkephalin-induced inhibition without
blocking SP-induced excitation, thus intimating separate receptors for ACh, SP, and met-enkephalin in rat locus coeruleus
neurons. La Salle and Ben-Ari (1977) found SP-excitable cells
in the medial, but not lateral, amygdaloid nucleus and in the
putamen of the rat. SP excitation was characteristically slow
in onset and of prolonged duration. L-glutamate would excite
these same neurons, but the effect was without the long latency
characteristically observed with SP. Sastry (1978) found cells
in the rat interpeduncular nucleus that increased their firing
rate for long periods (80 to 400 msec) after electrical stimulation of the habenula (12 to 26 msec before). ACh or SP alone
was able to increase the rate at which the interpeduncular

neurons were discharged during habenula stimulation; applied
together, they synergistically increased this rate in several
cells. Atropine antagonized the effect of both ACh and habenula stimulation but did not significantly alter the response to
SP. Mayer and MacLeod (1979) reported that electrical stimulation of the bed nucleus of the stria terminalis of rats would
excite neurons of the medial preoptic nucleus; 72 percent of
these preoptic neurons were also excited by SP. TRH, ACh and
L-glutamate also excited certain of these cells, but only TRH
produced changes in the firing rate and amplitude similar in
onset and duration to those of SP.

Thus, it would appear that SP fulfills the criteria for consideration as a neurotransmitter. It is synthesized in cells,
mechanisms exist for its degradation, it is found in discrete
brain regions, and its application to these regions in near
physiological concentrations triggers their activation. In
particular, the available data strongly support the hypothesis
that SP is indeed a neurotransmitter agent of primary afferent
neurons. Thus, as outlined earlier, the undecapeptide exerts a
powerful depolarizing action on neurons in the spinal cord, the
concentration of the peptide in the dorsal horn is substantially reduced after dorsal root section (Fig. 6-5), and SP accumulates on the ganglionic side of the ligature after dorsal root
ligation.

Many studies have focused on the effects of SP on neurotransmitter function in the CNS. Magnusson et al. (1976) were the
first to examine the effects of SP on monoaminergic systems in
brain. Intracerebroventricularly administered SP (25 to 60 μg)
was found to induce rotational behavior, and this treatment
significantly increased the formation of DOPA (after decarboxylase inhibition) in all brain regions examined. Furthermore,
after monoamine synthesis inhibition, SP-treated rats exhibited
accelerated disappearance of DA, NE, and 5-HT. These data suggest that SP increases the activity of monoaminergic systems in
the rat brain. Cheramy et al. (1977) infused SP into the substantia nigra for one hour and measured the release of $[^3H]$-DA
synthesized from $[^3H]$-tyrosine. SP increased $[^3H]$-DA release
in the caudate nucleus, a result consistent with an increase in
DA turnover. Waldmeier et al. (1978) measured levels of homovanillic acid and dihydroxyphenylacetic acid, the two major DA
metabolites in the corpus striatum, after intranigral SP administration. SP increased the formation of both metabolites,
providing further evidence of increased activity in the nigroneostriatal DA system after SP treatment. Starr (1978) found
no effect of SP (10^{-9}-10^{-5}M) on the uptake of radiolabeled DA,
5-HT, or GABA in slices of striatum or substantia nigra *in
vitro*. SP potentiated the K^+-stimulated release of $[^3H]$-DA and
$[^3H]$-5HT, but not $[^3H]$-GABA; however, the observed changes were

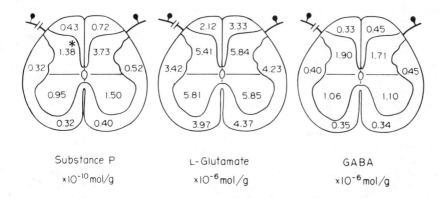

Fig. 6-5. Effect of dorsal root section on substance P, GABA, and L-glutamate in cat spinal cords. Dorsal roots (below L5) were unilaterally sectioned, and after the survival time of 11-12 days, the distributions of these substances in the spinal cords (L5-S1) were examined. In each map, the lesioned side is represented on the left and the intact side on the right. Each value represents the mean of three to six determinations performed in five operated cats. *P <0.01 compared with the intact side. (Reproduced from Otsuka et al., 1975, with permission.)

very small. Gale et al. (1978) have recently reported that both nigral GABA and SP play a role in the activation of striatal tyrosine hydroxylase observed after haloperidol treatment, and they concluded that striatonigral SP fibers comprise a major input to nigral DA neurons.

Other studies have evaluated the effect of psychoactive agents on SP content. Pettibone et al. (1978) reported that injection of rats with a high dose of d-amphetamine (10 mg/kg IP) produced a small but significant elevation in striatal immunoreactive SP levels. Haloperidol pretreatment abolished the effect of d-amphetamine, indicating that this effect of d-amphetamine is probably mediated by DA systems. Hong et al. (1978) found no effect of chronic haloperidol treatment on striatal or hypothalamic levels of SP but did observe a significant decrease in the substantia nigra content of the peptide. Pimozide, another neuroleptic agent, produced similar effects.

Results from the two sets of studies just cited suggest that SP produces activation of brain amine systems, particularly the nigroneostriatal DA system. Furthermore, DA activators, like

amphetamine, increase brain SP levels, whereas neuroleptics reduce SP levels in the CNS. In this light, it is of interest to
note the brief report of Elliot and Glen (1978), who found that
the spasmogenic effects of SP on the guinea pig ileum, although
unaffected by a wide variety of pharmaceutical agents, were inhibited by neuroleptics, including haloperidol, droperidol, and
pimozide.

Antagonism of SP has been investigated with a putative SP
blocker, Lioresal (baclofen, p-chorophenyl—GABA) and a variety
of depressant agents. Lioresal, a GABA and phenylethylamine
derivative, was first reported to inhibit the SP depolarization
of isolated rat spinal cord by Saito et al. (1975). Fotherby
et al. (1976) reported that SP-induced contraction of guinea
pig ileum is blocked by Lioresal or NE. the action of both
Lioresal and NE was inhibited by phentolamine, an α-adrenergic
blocker. The effect of Lioresal was not found to be very potent or specific. In a related paper, Phyllis (1976) evaluated
the specificity of Lioresal as an antagonist of the electrophysiological effects induced by SP. Drugs were microiontopho-
retically applied onto cerebral cortical cells of rat, and subsequent observation of changes in spontaneous firing rates was
used as the criterion of activity. Lioresal inhibited the increases in neuronal firing rates induced by SP and ACh. The
drug was therefore shown to be rather nonspecific in its depressant effect. Thus, a specific antagonist of SP that can be
used as a tool to unravel its role in the nervous system has
yet to be discovered.

Tiru-Chelvam (1973), using two bioassay methods (guinea pig
ileum and rat blood pressure), studied the effects of snake
venom administration on SP levels in brain, gut, and blood in
mice *in vivo* and *in vitro*. Significant decreases in brain and
gut SP levels were observed under both experimental conditions.
This, coupled with purported similarities in the effects of
parenterally administered SP and snake venom, provide evidence
that SP may mediate certain of the effects of administered
snake venom.

Jessell and Iversen (1977) reported that opiate analgesics,
including morphine, a stable enkephalin analogue (D-Ala^2-met-
enkephalin amide), and β-endorphin, inhibit substance P release
from rat trigeminal nucleus after perfusion of tissue slices
with K$^+$(47 mM).

Several reports have appeared concerning behavioral effects
after SP administration to laboratory animals. In a very early
study, Stern and Hadzovic (1973) reported that SP (0.5 mg/kg
IM) abolishes the abstinence syndrome in morphine-treated mice,
tranquilized aggressive mice, but exerted no effect on strychnine-induced seizures. In a later study, this same group

(Stern et al., 1976) reported that SP abolished the proconvulsive action of morphine on pentylenetetrazol-induced seizures in mice.

Goldstein and Malick (1977) reported that SP in high doses (60 to 120 µg/rat) injected into the medial forebrain bundle caused a significant depression in the rate of self-stimulation from that site. Naloxone, an opiate antagonist, abolished the SP effect, suggesting an interaction of SP with endogenous opiate systems.

James and Starr (1977) have reported that as little as 1 ng of SP, when injected into the substantia nigra but not the posterior thalamus of rats, produces circling behavior. Similar results have been presented by Olpe and Koella (1977), who observed a dose-dependent increase in contralateral rotational behavior after SP injection.

Kelley and Iversen (1978) reported that bilateral substantia nigra SP injections produce increased locomotor activity and, after repeated injection, grooming. Rondeau et al. (1978) observed increased locomotor activity (in photocell cages) after ICV SP (0.31 to 1.25 µg). At higher doses of the peptide (40 to 80 µg), immobility, rigidity, and barrel rotation were observed. A low dose of the peptide exerted no effect on behavioral responses to d-amphetamine or apomorphine. Kelley and Iversen (1978) reported that infusion of SP into the ventral tegmental area significantly increased locomotor activity; this effect was blocked by haloperidol infusion into the nucleus accumbens or by lesioning the A-10 pathway with 6-OHDa.

Garcia-Sevilla et al. (1979) observed increased locomotor activity after ICV SP, which was associated with increased turnover of CNS DA and NE. The alterations in catecholamine levels were blocked by naloxone, while locomotor activity was potentiated by this opiate receptor blocker.

Further evidence of a locomotor role for SP comes from work of Rondeau et al. (1979), in which rats were rendered hypokinetic by 6-OHDA injections into the hypothalamus. Intracerebroventricularly administered SP increased locomotor activity, while TRH and SRIF were without effect in this experimental preparation. Recently, Katz (1979) reported that SP (12.5-100 µg) administered ICV in the mouse suppressed spontaneous locomotor activity and also elicited grooming. Both effects were dose dependent and were not altered by pretreatment with the opiate antagonist naloxone.

Evered et al. (1977) reported that substance P and eledoisin induced drinking in pigeons after intracranial injection, whereas in the rat, SP inhibits drinking induced by angiotensin II, carbachol, water deprivation, or sodium chloride load (De Caro et al., 1978).

Both analgesia and hyperalgesia have been reported after CNS and peripheral SP administration (Malick and Goldstein, 1978; Fredrickson et al., 1978; Nemeroff et al., 1979; Stewart et al., 1976). Growcott and Shaw (1978) were unable to induce analgesia in mice with as much as 200 μg SP injected IP. Goldstein and Malick (1979) have recently been unable to observe analgesia using the rat tail-flick and mouse hot-plate paradigms at doses as high as 10 mg/kg of SP administered IP. Mohrland and Gebhart (1979) found that low doses of SP (<0.25 mg/kg) are necessary to observe significant analgesia after IP injection in rats, and this was only observed in the tail-flick paradigm. Analgesia was not evident in rats using the hot-plate test at doses of SP as high as 1 mg/kg IP. Oehme et al. (1980) have recently isolated two important variables in determining the effects of intravenous SP on responses to noxious stimuli. One is dose; low doses produce hyperalgesia, higher doses produce analgesia. Second is the animals' baseline responses to the noxious stimulus; SP produces analgesia in animals that are relatively sensitive to the noxious stimulus, and hyperalgesia in animals that are relatively insensitive to the noxious stimulus.

DISCUSSION

Neuropeptides exert a variety of endocrine, behavioral, and pharmacological effects. At present, research concerning the CNS effects of peptides is primarily at the level of descriptive phenomenology, for instance, peptide X administered by route Y induces responses Z. In many cases we know how the peptide does not act. However, for most peptides, we are still ignorant of their neuroanatomical as well as their molecular loci of action. Nevertheless, even at this early stage of investigation, certain generalizations can be advanced.

The hypothalamic release and release-inhibiting peptides have behavioral and pharmacological actions, at least some of which are not mediated by their endocrine actions. These effects are not abolished by either exirpation of the anterior pituitary (their direct endocrine target) or by removal of their final target endocrine gland, such as thyroid, adrenal, or gonads. In addition, administration of the pituitary trophic hormone, whose secretion is stimulated by the releasing factor under study, does not mimic the behavioral effects of the releasing factor. The mechanisms by which the neuropeptides exert their CNS effects remain obscure.

The hypothalamic peptides discussed in this chapter fulfill many of the criteria for being considered as neurotransmitters

or neuromodulators. Thus far, it can be stated that the peptides are generally heterogeneously distributed in brain;
released from synaptosomes or brain slices by depolarizing
stimuli; have high-affinity, saturable binding sites to brain
membranes; alter rates of neuronal firing after iontophoretic
application; and are rapidly inactivated by brain.

Although hypothalamic peptides fulfill many criteria for neurotransmitters, there are several interesting findings that indicate they may not act by the same mechanisms as "classical"
neurotransmitters, such as ACh and NE. Exogenous administration of classical neurotransmitters generally produces behavioral effects of short duration. Many of the behavioral effects elicited by administration of hypothalamic peptides have
a much longer duration of action, and this is probably not due
to a slower rate of metabolism. For example, LH-RH is rapidly
metabolized, yet the LH-RH-induced increase in lordosis persists for more than eight hours (Moss et al., 1975). The mechanism by which LH-RH increases lordotic behavior for such a
long time after injection is not understood. Perhaps a small
quantity of LH-RH is protected from degradation by being tightly bound to receptors. It is also conceivable that LH-RH triggers biochemical events that can last for eight hours. Iontophoretic administration of neuropeptides can produce alterations in neuronal firing that last for longer time periods than
with classical neurotransmitters (Barker, 1978). SP, when ion-
tophoresed onto responsive neurons, causes a slower onset excitation that is of longer duration than that seen with classical
neurotransmitters. This is consistent with the previously mentioned duration of action of behavioral effects. It is important to note that not all effects of neuropeptides are long-
lasting. Some are of relatively short duration, which is typical of classical neurotransmitters.

Peptidergic neurons appear to be different from classically
described neurons. It is currently assumed that the peptides
are synthesized by ribosomes in the cell body and then transported to the axonal terminals; classical transmitters are believed to be synthesized enzymatically. To date, no one has
been able to demonstrate an active peptide uptake mechanism in
neurons. Most, but not all classical neurotransmitters are
capable of being taken up by the neuron.

Many peptides produce similar effects under certain circumstances. One technique used to study peptides is the investigation of drug interaction, that is, the effects of peptides
on drug-induced behaviors. Several hypothalamic peptides,
including TRH, MIF-I, LH-RH, and SRIF, lower the threshold for
L-DOPA-induced locomotor activity in pargyline-pretreated mice.
Likewise, the duration of pentobarbital-induced sedation is

shortened by the central administration of several peptides, including LH-RH, oxytocin, TRH, $ACTH_{1-24}$, angiotensin I, SP, bradykinin, and angiotensin II (Bissette et al., 1978). The duration of sedation is increased by NT. Similar overlapping actions of neuropeptides have also been observed in several other behavioral tests, for instance LH-RH and $ACTH_{4-10}$ have comparable effects and are equipotent in inhibiting the extinction of a pole-jump avoidance response (De Wied et al., 1975). This does not necessarily argue for a lack of specificity of the effects of neuropeptides, but rather may emphasize the complexity of peptide integration of behavior. In addition, most of the behavioral screening paradigms that have been used provide only gross measures of over-all behavioral activity. For example, TRH was initially thought to stimulate locomotor activity in a manner comparable to that of amphetamine. More detailed analysis has shown that the stimulated activity by TRH is different from that which occurs after amphetamine administration (Costall et al., 1979). These findings highlight the importance of detailed observations of the experimental animals.

Although many effects of neuropeptides are shared by more than one peptide, the relative potency of the peptides on either a weight or molar basis is often different. Peptides can have profoundly different effects in different species, for example, centrally administered TRH produces hyperthermia in rats and rabbits and hypothermia in cats. A consideration of variables, such as dose, route of administration, treatment interval, species, age, sex, and even season, may be important in characterizing CNS effects of hypothalamic peptides.

Certain of the hypothalamic peptides have behavioral actions that seemingly complement their endocrine effects. Thus, LH-RH induces the release of pituitary gonadotropins, as well as directly inducing lordosis. Both actions of LH-RH can be considered complementary or harmonious in that they both contribute in the regulation within a particular behavioral arena, that is, reproductive behavior.

At present, several problems exist when conducting research with neuropeptides. The question of whether the CNS actions of the peptides are best regarded as an aspect of pharmacology or physiology is fundamental at this point. the vast majority of studies on the CNS effects of neuropeptides examine the effects of exogenous administration of the peptide. Additional approaches are needed to determine if endogenous neuropeptide actions are similar to those elicited by exogenous neuropeptide administration. To answer this question, specific peptide antagonists are needed; at present, for the neuropeptides discussed in this chapter, no antagonists exist. A promising approach has been the use of antisera as antagonists for peptides. Antisera to LH-RH injected into the mesencephalic central gray reduces the amount of lordotic behavior in rats (Sakuma and Pfaff, 1980). Other researchers have also successfully used antisera to peptides as antagonists (Dunn et al., 1979; van Wimersma Greidanus et al., 1975). However, problems of interpretation exist with the use of antisera as peptide antagonists. Diffusion of the large antibody molecules within the CNS is also probably quite limited; thus antisera must ideally be placed directly into the active site within the brain. Finally, the immunoreactive binding site of a peptide is not necessarily the receptor binding area, and thus an antibody:peptide complex could conceivably still be biologically active. Ideally, one would prefer also to be able to measure turnover of the peptide during different behaviors and in response to different environmental events. At present, the available techniques for assessing neuropeptide turnover seem to be inadequate for the task. This is, undoubtedly, partly because the route of biosynthesis of almost all of the neuropeptides remains obscure.

A consideration of the dose utilized might be expected to be useful in determining whether a behavior elicited by neuropeptide administration is pharmacological or physiological. At the present state of knowledge, such a conideration, at least in absolute terms, has not been very helpful in the appraisal of peptide effects. For example, 1 mg/kg TRH given peripherally may seem to be too large for its central effects to be considered physiological. But what portion of this dose is quickly destroyed? What part of the remainder is taken up by brain? And what part of that is delivered to critical sites of action? All must be considered when determining the physiological relevance of the dose used.

In agreement with the data obtained concerning the behavioral actions of neuropeptides in animals, comparable results have been observed in human beings. By this we refer to the growing awareness in both the basic science and clinical research laboratories that neuropeptides possess potent biological activity, which is partly expressed by behavioral changes in animals and man. Whether such activity might be usefully employed in the development of therapeutic agents for the treatment of a specific behavioral or neurological disorder is not yet clear. We might expect that the previously mentioned clinical trials using MIF-I and LH-RH represent the beginnings of tests on a new wave of peptidergic drugs. Refinements in methodology will undoubtedly increase our understanding of the behavioral effects of these peptides and hopefully will lead to an elucidation of their physiological functions.

ACKNOWLEDGMENTS

This work was supported in part by National Institute of Mental Health grants MHCRC MH-33127, MH-32316, MH-34121; a Career Scientist Award to AJP (MH-22536); a Neurobiology Program grant MH-14277; by National Institute of Child Health and Human Development grant HD-03110; and by the Alcohol Research Authority of North Carolina. We are grateful to Dori Yarborough, Jeanine Wheless, and JoAnne Robb for help in preparation of this review.

REFERENCES

Arimura, A., Sato, H., Coy, D.H., and Schally, A.V.: Radioimmunoassay for GH-release inhibiting hormone. *Proc Soc Exp Biol Med 148*:784-789, 1975.

Arimura, A., Smith, W.D., and Schally, A.V.: Blockade of the stress-induced decrease in blood GH antisomatostatin serum in rats. *Endocrinology 98*:540-543, 1976.

Arimura, A., Lundquist, G., Rothman, J., Chang, R., Fernandez-Durango, R., Elde, R., Coy, D.H., Meyers, C., and Schally, A.V.: Radioimmunoassay of somatostatin. *Metabolism 27(Suppl 1)*:1139-1144, 1978.

Barbeau, A.: Potentiation of levodopa effect by intravenous L-prolyl-L-leucyl-glycine amide in men. *Lancet 2*:683-689, 1975.

Barbeau, A., Burnett, C., Strother, E., and Butterworth, R.F.: MIF potentiation of apomorphine-induced hyperactivity. *Clin Res 23*:641A, 1975.

Barker, J.L.: Evidence for diverse cellular roles of peptides in neuronal function. In *Peptide and Behavior: A Critical Analysis of Research Strategies*. *Neurosci Res Program Bull 16*:535-553, 1978.

Barlow, T.S., Cooper, B.R., Breese, G.R., Prange, A.J., Jr., and Lipton, M.A.: Effects of thyrotropin-releasing hormone on behavior: Evidence for an anorexic-like action. *Soc Neurosci Abstr 1*:334, 1975.

Barnea, A., Ben-Johnathan, N., Colston, C., Johnston, J.M., and Porter, J.C.: Differential subcellular compartmentalization of thyrotropin-releasing hormone (TRH) and gonadotropin releasing hormone (LRH) in hypothalamic tissue. *Proc Natl Acad Sci USA 72*:3153-3159, 1975.

Benkert, O.: Studies on pituitary hormones and releasing hormones in depression and sexual impotence. In *Hormones, Homeostasis, and the Brain. Progress in Brain Research 42*, W.H. Gispen, Tj. B. van Wimersma Greidanus, B. Bohus, and D. deWied, eds. Elsevier, Amsterdam, 1975, pp 25-36.

Bennett, G.W., Edwardson, J.A., Holland, D., Jeffcoate, S.L., and White, N.: Release of immunoreactive luteinizing hormone-releasing hormone and thyrotropin-releasing hormone from hypothalamic synaptosomes. *Pharmacologist 18*:184, 1975.

Bennett, H.P.J., and McMartin, C.: Peptide hormones and their analoques: Distribution, clearance from the circulation and inactivation *in vivo*. *Pharmacol Rev 30*:247-292, 1979.

Benuck, M., and Marks, N.: Enzymatic inactivation of Substance P by a partially purified enzyme from rat brain. *Biochem Biophy Res Comm 65*:153-160, 1975.

Benuck, M., and Marks, N.: Differences in the degradation of hypothalamic releasing factors by rat and human serum. *Life Sci 19*:1271, 1976.

Berelowitz, M., Kronheim, S., Pimstone, B., and Sheppard, M.: Potassium-stimulated calcium dependent release of immunoreactive somatostatin from incubated rat hypothalmus. *J Neurochem 31*:1537-1539, 1981.

Bissette, G., Nemeroff, C.B., Loosen, P.T., Prange, A.J., Jr., and Lipton, M.A.: Hypothermia and intolerance to cold induced by intracisternal administration of the hypothalamic peptide neurotensin. *Nature 262*:607-609, 1976.

Bissette, G., Nemeroff, C.B., Loosen, P.T., Breese, G.R., Burnett, G., Lipton, M.A., and Prange, A.J., Jr.: Modification of pentobarbital-induced sedation by natural and synthetic peptides. *Neuropharmacology 17*:229-237, 1978.

Bjorkman, S., Lewander, T., and Zetterstrom, T.: MIF (Pro-Leu-Gly-NH$_2$): Failure to affect oxotremorine effects in mice and rats as well as fluphenazine catalepsy or amphetamine hyperactivity in rats. *J Pharm Pharmacol 32*:296-297, 1980.

Bøler, J., Enzmann, F., Folkers, K., Bowers, C.Y., and Schally, A.V.: The identity of chemical and hormonal properties of the thyrotropin releasing hormone and pyroglutamyl-histidyl-proline-amide. *Biochem Biophy Res Comm 37*:705, 1969.

Bower, S.A., Hadley, M.E., and Hruby, V.J.: Comparative MSH release-inhibiting activities of tocinoic acid (the ring of oxytocin), and L-Pro-L-Leu-Gly-NH$_2$ (the side chain of oxytocin). *Biochem Biophys Res Comm 45*:1185-1191, 1971.

Brazeau, P., Vale, W., Burgus, R., Ling, N., Butcher, M., Rivier, J., and Guillemin, R.: Hypothalamic polypeptide that inhibits the secretion of immunoreactive pituitary growth hormone. *Science 179*:77-79, 1973.

Brazeau, P., Vale, W., Burgus, R., and Guillemin, R.: Isolation of somatostatin (a somatotropin release inhibitory factor) of ovine hypothalamic origin. *Can J Biochem 52*:1067-1072, 1974.

Breese, G.R., Cott, J.M., Cooper, B.R., Prange, A.J., Jr., and Lipton, M.A.: Antagonism of ethanol narcosis by thyrotropin-releasing hormone. *Life Sci 14*:1053-1063, 1974.

Breese, G.R., Cott, J.M., Cooper, B.R., Prange, A.J., Jr., Lipton, M.A., and Plotnikoff, N.P.: Effects of thyrotropin-releasing hormone (TRH) on the actions of pentobarbital and other centrally acting drugs. *J Pharmacol Exp Ther 193*:11-22, 1975.

Brown, M.R., and Vale, W.: Central nervous system effects of hypothalamic peptides. *Endocrinology 96*:1333-1336, 1975.

Brown, M.R., and Vale, W., 1980. Peptides and thermoregulation. In *Thermoregulatory Mechanisms and Their Therapeutic Implications,* B. Cox, P. Lomax, A.S. Milton, and E. Schonbaum, eds. S. Karger, Basel, 1980, pp 186-194.

Brownstein, M.J., Palkovits, M., Saavedra, J.M., Bassiri, R.M., and Utiger, R.D.: Thyrotropin-releasing hormone in specific nuclei of rat brain. *Science 185*:267-269, 1974.

Burgus, R., Dunn, T.F., Desiderio, D.M., and Guilleman, R.: Structure moleculaire du facteur hypothalamique hypophysio-trope TRF d'origine ovine: Mise en evidence par spectrometrie de masse de la sequence PCA-His-Pro-NH$_2$. *CR Acad Sci (Paris) [D] 269*:1870-1873, 1969.

Burt, D.R., and Snyder, S.H.: Thyrotropin-releasing hormone (TRH): Apparent receptor binding in rat brain membranes. *Brain Res 93*:309-328, 1975.

Burt, D.R., and Taylor, R.L.: Binding sites for thyrotropin-releasing hormone in sheep nucleus accumbens resemble pituitary receptors. *Endocrinology 106*:1416-1423, 1980.

Bury, R.W., and Mashford, M.L.: The stability of synthetic Substance P in blood. *Eur J Pharmacol 45*:257-260, 1977.

Cannon, D.E.B., Skrabanek, P., and Powell, D.: Substance P degradation in plasma and its partial prevention by heat inactivation and enzyme inhibitors. *Ir J Med Sci 146*:314, 1977.

Cannon, D.E.B., Skrabanek, P., and Powell, D.: Difference in behavior between synthetic and endogenous Substance P in human plasma. *Naunyn Schmiedebergs Arch Pharmacol 307*:251-255, 1979.

Carceni, T., Parati, E.A., Girotti, F., Celano, I., Frigerio, C., Cocchi, D., and Muller, E.E.: Failure of MIF-I to affect behavioral responses in patients with Parkinson's disease under L-DOPA therapy. *Psychopharmacology 63*:217-222, 1979.

Carino, M.A., Smith, J.R., Weick, B.B., and Horita, A.: Effects of thyrotropin-releasing hormone (TRH) microinjected into various brain areas of conscious and pentobarbital-pretreated rabbits. *Life Sci 19*:1687-1692, 1976.

Carolei, A., Margotta, V., and Palladini, G.: Melanocyte-stimulating hormone release-inhibiting factor (MIF): Lack of dopaminergic and anticholinergic activity. *Neuroendocrinology 23*:129-132, 1977.

Carraway, R.E.: Neurotensin and related substances. In *Methods of Hormone Radioimmunoassay,* 2nd. ed., B.M. Jaffe and H.R. Behrmar, eds. Academic Press, New York, 1978, pp 139-169.

Carraway, R.E., and Leeman, S.E.: The isolation of a new

hypotensive peptide, neurotensin, from bovine hypothalami. *J Biol Chem 248:*6854-6861, 1973.

Carraway, R.E., Demers, L.M., and Leeman, S.E.: Hyperglycemic effect of neurotensin, a hypothalamic peptide. *Endocrinology 99:*1452-1462, 1976.

Castensson, S., Sievertsson, H., Linoeke, B., and Sum, C.Y.: Studies on the inhibition of oxytremorine induced tremor by a melanocyte-stimulating hormone release-inhibiting factor, thyrotropin releasing hormone, and related peptides. *Febs Lett 44:*101, 1974.

Celis, M.F., Taleisnik, S., and Walter, R.: Regulation of formation and proposed structure of the factor inhibiting the release of melanocyte-stimulating hormone. *Proc Natl Acad Sci USA 68:*1428-1433, 1971.

Chang, M.M., and Leeman, S.E.: Isolation of a sialogogic peptide from bovine hypothalamic tissue and its characterization as Substance P. *J Biol Chem 245:* 4784-4790, 1970.

Chang, M.M., Leeman, S.E., and Niall, H.D.: Synthesis of Substance P. *Nature 232:*87-89, 1971.

Chase, T.N., Woods, M.B., Lipton, M.A., and Morris, C.E.: Hypothalamic releasing factors and Parkinson's disease. *Arch Neurol 31:*55-56, 1974.

Cheng, M-F.: Role of gonadotropin releasing hormone in the reproductive behavior of female ring doves (*Strepelia risoria*). *J Endocrinol 74:*37-45, 1977.

Cheramy, A., Nieoullon, A., Michelot, R., and Glowinski, J.: Effects of intranigral application of dopamine and Substance P on the *in vitro* release of newly synthesized [^{3}H]-dopamine in the ipsilateral caudate nucleus of the cat. *Neurosci Lett 4:*105-109, 1977.

Chihara, K., Arimura, A., Chihara, M., and Schally, A.V.: Effect of intraventricular administration of antisomatostatin γ-globulin on the lethal dose-50 of strychnine and pentobarbital in rats. *Endocrinology 103:*912-916, 1978a.

Chihara, K., Arimura, A., Coy, D.H., and Schally, A.V.: Studies on the interaction of endorphins, Substance P, and endogenous somatostatin in growth hormone and prolactin release in rats. *Endocrinology 102:*281-290, 1978b.

Chiu, S., and Mishra, R.K.: Antagonism of morphine-induced catalepsy by L-prolyl-L-leucyl-glycinamide. *Eur J Pharmacol 53:*119-125, 1979.

Clineschmidt, B.V., and McGuffin, J.C.: Neurotensin administered intracisternally inhibits responsiveness of mice to noxious stimuli. *Eur J Pharmacol 46:*395-396, 1977.

Clineschmidt, B.V., McGuffin, J.C., and Bunting, P.B.: Neurotensin: Antinocisponsive action in rodents. *Eur J Pharmacol 54:*129-139, 1979.

Cohn, M.L.: Cyclic AMP, thyrotropin releasing factor and somatostatin: Key factors in the regulation of the duration of
narcosis. In *Molecular Mechanism of Anesthesia*, B.R. Find,
ed. Raven Press, New York, 1975, pp 485–500.

Cohn, M.L., and Cohn, M.: Barrel rotation' induced by somatostatin in the nonlesioned rat. *Brain Res 96*:138–141, 1975.

Cohn, M.L., Rosing, E., Wiley, K.S., and Slater, I.H.: Somatostatin inhibits adrenergic and cholinergic neurotransmission in smooth muscle. *Life Sci 32*:1659–1664, 1978.

Conlon, J.M., Zyznar, E., Vale, W., and Unger, R.H.: Multiple
forms of somatostatin-like immunoreactivity in canine pancreas. *FEBS Letters 94*:327–330, 1978.

Constantinidis, J., Geissbuhler, F., Gaillard, J.M.,
Hovaguimian, Th., and Tissot, R.: Enhancement of cerebral
noradrenaline turnover by thyrotropin-releasing hormone:
Evidence by fluorescence histochemistry. *Experientia 30*:1182,
1974.

Costall, B., Siu-Chun, G.H., Metcalf, G., and Naylor, R.J.: A
study of the changes in motor behavior caused by TRH on
intracerebral injection. *Eur J Pharmacol 53*:143–150, 1979.

Cott, J., and Engel, J.: Antagonism of the analeptic activity
of thyrotropin-releasing hormone (TRH) by agents which enhance GABA transmission. *Psychopharmacology 52*:145–149, 1977.

Cott, J.M., Breese, G.R., Cooper, B.R., Barlow, T.S., and
Prange, A.J., Jr.: Investigations into the mechanism of re-
duction of ethanol sleep by thyrotropin-releasing hormone
(TRH). *J Pharmacol Exp Therap 196*:594–604, 1976.

Cuello, A.C., Emson, P., Del Fiacco, M., Gale, J., Iversen,
L.L., Jessell, T.M., Kanazawa, I., Pakinos, G., and Quik, M.:
Distribution and release of Substance P. In *Centrally Acting
Peptides*, J. Hughes, ed. The MacMillan Press, Basingstoke,
England, 1978, pp 135–155.

Davies, J., and Dray, A.: Substance P in the substantia nigra.
Brain Res 107:623–627, 1976.

De Caro, G., Massi, M., and Micassi, L.G.: Antidipsogenic effect of intracranial injections of substance P in rats. *J
Physiol (Lond) 279*:133–140, 1978.

Diel, F., Schneider, E., and Quabbe, H.J.: Development of a
radioimmunoassay for cyclic somatostatin: Antibody production, comparative radioiodination and dose-response curve. *J
Clin Chem Clin Biochem 15*:669–677, 1977.

Dun, N.J., and Karczmar, A.G.: Actions of substance P on sympathetic neurons. *Neuropharmacology 18*:215–218, 1979.

Dunn, A.J., Green, E.J., and Isaacson, R.L.: Intracerebral
adrenocorticotropic hormone mediates novelty-induced grooming
in the rat. *Science 203*:281–283, 1979.

Dupont, A., and Alvarado-Urkina, G.: Conversion of pancreatic
somatostatin without peptide bond cleavage into somatostatin

tetradecapeptide. *Life Sci 19:*1431-1434, 1976.

Dupont, A., and Merand, Y.: Enzyme inactivation of neurotensin by hypothalamic and brain extracts of the rat. *Life Sci 22:*1623-1630, 1978.

Dyer, R.G., and Dyball, R.E.: Evidence for a direct effect of LRF and TRF on single unit activity in the rostral hypothalamus. *Nature 252:*486-488, 1974.

Ehrensing, R.H., and Kastin, A.J.: Melanocyte-stimulating hormone-release inhibiting hormone as an antidepressant. *Arch Gen Psychiatry 30:*63-65, 1974.

Ehrensing, R.H., and Kastin, A.J.: Dose-related biphasic effect of prolyl-leucyl-glycinamide (MIF-I) in depression. *Am J Psychiatry 135:*566, 1978.

Ehrensing, R.H., Kastin, A.J., Larsores, P.A., and Bishop, G.A.: Melanocyte-stimulating hormone release-inhibiting factor-I and tardive dyskinesia. *Dis Nerv System 38(4):*303-307, 1977.

Elliot, J.M., and Glenn, J.B.: The effects of some analgesic and neuroleptic drugs on the spasmogenic actions of substance P on guinea pig ileum. *J Pharm Pharmacol 30:*578-579, 1978.

Enock, D., and Cohn, M.L.: Somatostatin (SRIF) effects in vivo and in vitro on cyclic AMP concentration in rat brain. *Soc Neurosci Abst 1:*451, 1975.

Ervin, G.N., Schmitz, S.A., Nemeroff, C.B., and Prange, A.J., Jr.: Thyrotropin-releasing hormone and amphetamine produce different patterns of behavioral excitation in rats. *Eur J Pharmacol 72:*35-43, 1981.

von Euler, U.S., and Gaddum, J.H.: An unidentified depressor substance in certain tissue extracts. *J Physiol (Lond) 72:*74-87, 1931.

Evered, M.D., Fitzsimons, J.T., and De Caro, G.: Drinking behavior induced by intracranial injections of eledoisin and substance P in the pigeon. *Nature 268:*332-333, 1977.

Felig, P., Wahren, J., Sherwin, R., and Hendler, B.: Insulin, glucagon, and somatostatin in normal physiology and diabetes mellitus. *Diabetes 25:*1091-1099, 1976.

Felix, D., and Phillips, M.I.: Inhibitory effects of luteinizing hormone-releasing hormone (LH-RH) on neurons in the organum vasculosum lamina terminalis (OVLT). *Brain Res 169:*204-208, 1979.

Fischer, P.A., Schneider, E., Jacobi, P., and Maxion, H.: Effect of melanocyte-stimulating hormone-release inhibiting factor (MIF) in Parkinson's syndrome. *Eur Neurol 12:*360-368, 1974.

Foreman, M.M., and Moss, R.L.: Role of gonadotropins and releasing hormones in hypothalamic control of lordosis behavior in ovariectomized, estrogen-primed rats. *J Comp Physiol Psychol 93:*556-565, 1979.

258 Nemeroff, et al.

Fotherby, K.S., Morrish, N.J., and Ryall, R.W.: Is Lioresal (baclofen) an antagonist of substance P? *Brain Res 113:*210-213, 1976.

Fredrickson, R.C.A., Burgis, V., Harrel, C.E., and Edwards, T.D.: Dual actions of substance P on nociception: Possible role of endogenous opioids. *Science 199:*1359-1362, 1978.

Friedman, E., Friedman, J., and Gershon, S.: Dopamine synthesis: Stimulation by a hypothalamic factor. *Science 182:*831, 1973.

Gale, K., Costa, E., Toffano, G., Hong, J.S., and Guidotti, A.: Evidence for a role of nigral γ-aminobutyric acid and substance P in the haloperidol-induced activation of striatal tyrosine hydroxylase. *J Pharm Exp Ther 206:*29-37, 1978.

Gamse, R., Lembeck, F., and Cuello, A.C.: Substance P in the vagus nerve. *Naunyn Schmiedebergs Arch Pharmacol 306:*37-44, 1979.

Garcia-Sevilla, J.A., Magnusson, T., Carlsson, A., Leban, J., and Folkers, K.: Neurotensin and its amide analogue [Gln⁴]-neurotensin: Effects of brain monamine turnover. *Naunyn Schmiedebergs Arch Pharmacol 305:*213-218, 1978.

Garcia-Sevilla, J.A., Magnusson, T., and Carlsson, A.: Opposite effects of naloxone on substance P-induced changes in brain Dopa synthesis and in locomotor activity in rats. *J Neural Transm 45:*185-193, 1979.

Gerich, J.E., Charles, M.A., and Grodsy, G.M.: Regulation of pancreatic insulin and glucagon secretion. *Ann Rev Physiol 38:*353-388, 1976.

Gerstenbrand, F., Binder, H., Kozma, C., Pusch, St., and Reisner, T.: Infusions Therapie mit MIF (Melanocyte Inhibiting Facotr) beim Parkinson-Syndrom. *Wien Klin Wochenschr 87(24):*822-823, 1975.

Goldstein, J.M., and Malick, J.B.: Effect of substance P on medial forebrain bundle self-stimulation in rats following intracerebral administration. *Pharmacol Biochem Behav 7:*475-478, 1977.

Goldstein, J.M., and Malick, J.M.: Lack of analgesic activity of substance P following intraperitoneal administration. *Life Sci 25:*431-436, 1979.

Gordon, A., Eriksson, L., Blom, A.K., Taskinen, M.R., and Gyhrquist, F.: Lack of behavioral effects following intraventricular infusion of somatostatin in the conscious goat. *Pharmacol Biochem Behav 9:*255-257, 1978.

Griffiths, E.C., Holland, D.T., and Jeffcoate, S.L.: The presence of somatostatin-inactivating peptidases in the hypothalamus and other brain areas of the rat. *J Physiol (Lond) 259:*50-51P, 1976.

Griffiths, E.C., Jeffcoate, S.L., and Holland, D.T.: Inactivation of somatostatin by peptidases in different areas of the rat brain. *Acta Endocrinol 85:*1, 1977.

Growcott, J.W., and Shaw, J.S.: Failure of substance P to produce analgesia in the mouse. *Br J Pharmacol 66:*129P, 1979.

Guillemin, R.: Somatostatin inhibits the release of acetycholine induced electrically in the myenteric plexus. *Endocrinology 99:*1653-1654, 1976.

Guyenet, P.G., and Aghajanian, G.K.: Excitation of neurons in the locus coeruleus by substance P and related peptides. *Brain Res 136:*178-184, 1977.

Guyenet, P.G., and Aghajanian, G.K.: ACh, substance P and met-enkephalin in the locus coeruleus: Pharmacologic evidence for independent sites of action. *Eur J Pharmacol 53:*319-328, 1979.

Havlicek, V., Rezek, M., and Friesen, H.: Somatostatin and thyrotropin-releasing hormone: Central effect on sleep and motor system. *Pharmacol Biochem Behav 4:*455-459, 1976.

Havlicek, V., Rezek, M., Leybin, L., and Friesen, H.: Analgesic effect of cerebroventricular administration of somatostatin (SRIF). *Fed Proc 36:*363, 1977.

Heal, D.J., and Green, A.R.: Administration of thyrotropin-releasing hormone (TRH) to rats releases dopamine in nucleus accumbens but not nucleus caudatus. *Neuropharmacology 18:*23-31, 1979.

Henry, J.L.: Effects of substance P on functionally identified units in cat spinal cord. *Brain Res 114:*439-451, 1976.

Henry, J.L., Krnjević, K., and Morris, M.E.: Substance P and spinal neurones. *Can J Physiol Pharmacol 53:*423-432, 1975.

Herschl, R., Havlicek, V., Rezek, M., and Kroeger, E.: Cerebroventricular administration of somatostatin (SRIF): Effect on central levels of cyclic AMP. *Life Sci 20:*821-826, 1977.

Hine, B., Sanghvi, L., and Gershon, S.: Evaluation of thyrotropin-releasing hormone as a potential antidepressant agent in the conscious dog. *Life Sci 13:*1789-1797, 1973.

Hirooka, Y., Hollander, C.S., Suzuki, S., Ferdinand, P., and Juan, S-I: Somatostatin inhibits release of thyrotropin-releasing factor from organ cultures of rat hypothalamus. *Proc Natl Acad Sci USA 75:*4509-4513, 1978.

Hökfelt, T., Efendics, S., Hellerstrom, C., Johansson, O., Luft, R., and Arumira, A.: Cellular localization of somatostatin in endocrine-like cells and neurons of the rat with special references to the A_1-cells of the pancreatic islets and to the hypothalamus. *Acta Endocrinol 80 (Suppl 200):*1-41, 1975.

Hong, J.S., Yang, H-Y.T., and Costa, E.: Substance P content of substantia nigra after chronic treatment with antischizophrenic drugs. *Neuropharmacology 17:*83-85, 1978.

Horita, A., Carino, M.A., and Smith, J.R.: Effects of thyrotropin-releasing hormone on the central nervous system of the rabbit. *Pharmacol Biochem Behav 5:*111-116, 1976.

Horst, W.D., and Spirt, N.: A possible mechanism for the anti-depressant activity of thyrotropin-releasing hormone. *Life Sci 15:*1073-1082, 1974.

Huidobro-Toro, J.P., Scotti de Carolis, A., and Longo, V.G.: Action of two hypothalamic factors (TRH, MIF) and of angiotensin II on the behavioral effects of L-DOPA and 5-hydroxytryptophan in mice. *Pharmacol Biochem Behav 2:*105-109, 1974.

Huidobro-Toro, J.P., Scotti de Carolis, A., and Longo, V.G.: Intensification of central catecholaminergic and serotonergic processes by the hypothalamic factors MIF and TRF and by angiotensin II. *Pharmacol Biochem Behav 3:*235-242, 1975.

Iversen, L.L., Iversen, S.D., Bloom, F.E., Douglas, C., Brown, M., and Vale, W.: Calcium-dependent release of somatostatin and neurotensin from rat brain *in vitro. Nature 273:*161-163, 1978.

Jaffe, B.M., and Behrman, H.R., eds. *Methods of Hormone Radio-immunoassay* 2nd ed. Academic Press, New York, 1978.

James, T.A., and Starr, M.S.: Behavioral and biochemical effects of substance P injected into the substantia nigra of the rat. *J Pharm Pharmacol 29:*181-182, 1977.

Jessell, T.M., and Iversen, L.L.: Opiate analgesics inhibit substance P release from rat trigeminal nucleus. *Nature 268:*549-551, 1977.

Jones, C.N., Grant, L.D., and Prange, A.J., Jr.: Stimulus properties of thyrotropin-releasing hormone. *Psychopharmacology 59:*217-224, 1978.

Kalivas, P.W., and Horita, A.: Thyrotropin-releasing hormone: Central site of action in antagonism of pentobarbital narcosis. *Nature 278:*461-463, 1979.

Kastin, A.J., and Barbeau, A.: Preliminary clinical studies with L-prolyl-L-leucyl-glycinamide in Parkinson's disease. *Can Med Assoc J 107:*1079-1081, 1972.

Kastin, A.J., Schally, A.V., and Viosca, S.: Inhibition of MSH release in frogs by direct application of L-Prolyl-L-leucylglycinamide to the pituitary. *Pro Soc Exp Biol Med 137:*1437-1439, 1971.

Kastin, A.J., Miller, M.C., Ferrell, L., and Schally, A.V.: General activity in intact and hypophysectomized rats after administration of melanocyte-stimulating hormone (MSH), melantonin, and Pro-Leu-Gly-NH$_2$. *Physiol Behav 10:*399-401, 1973.

Kastin, A.J., Coy, D.H., Jacquet, Y., Schally, A.V., and Plotnikoff, N.P.: CNS effects of somatostatin. *Metabolism 27 (Suppl 1):*1247-1252, 1978.

Kastin, A.J., Olson, R.D., Ehrensing, R.H., Berzas, M.D., Schally, A.V., and Coy, D.H.: MIF-I's differential actions as an opiate antagonist. *Pharmacol Biochem Behav 11:*721-723,

1979.

Katayama, Y., and North, R.A.: Does substance P mediate slow synaptic excitation within the myenteric plexus. *Nature* 274:387-388, 1978.

Keller, H.H., Bartholini, C., and Pletscher, A.: Enhancement of cerebral noradrenaline turnover by thyrotropin-releasing hormone. *Nature* 248:528-529, 1974.

Kelley, A.E., and Iversen, S.D.: Behavioral response to bilateral injections of substance P into the substantia nigra of the rat. *Brain Res* 158:474-478, 1978.

Kerwin, R.W., and Pycock, C.J.: Thyrotropin-releasing hormone stimulates release of ³H-dopamine from slices of rat nucleus accumbens *in vitro. Br J Pharmac* 67:323-325, 1979.

Kitabgi, P., Carraway, R., Van Rietschoten, J., Granier, C., Morgat, J.L., Menez, A., Leeman, S., and Freychet, P.: Neurotensin: Specific binding to synaptic membranes from rat brain. *Proc Natl Acad Sci USA* 74:1846-1850, 1977.

Klee, W.A.: Peptides of the central nervous system. *Adv Protein Chem* 33:243-286, 1979.

Kozlowski, G., and Hofstetter, G.: Cellular and subcellular localization and behavioral effects of gonadotropin-releasing hormone (GrRH) in the rat. In *Brain-Endocrine Interaction. III Neural Hormones and Reproduction.* D.E. Scott, ed. S. Karger, Basel, 1978, pp 138-153.

Kostrzewa, R.M., Kastin, A.J., and Spirtes, M.A.: α-MSH and MIF-I effects on catecholamine levels and synthesis in various rat brain areas. *Pharmacol Biochem Behav* 3:1017-1023, 1975.

Kraemer, G.W., Mueller, R., Breese, G.R., Prange, A.J., Jr., Lewis, J.K., Morrison, H., and McKinney, W.T., Jr.: Thyrotropin-releasing hormone: Antagonism of pentobarbital narcosis in the monkey. *Pharmacol Biochem Behav* 4:709-712, 1976.

Krivoy, W.A., and Kroeger, D.C.: The neurogenic activity of high potency substance P. *Experientia* 19:366-367, 1963b.

Krivoy, W.A., Lane, M., and Kroeger, D.C.: The actions of certain polypeptides on synaptic transmission. *Ann Acad Sci* 104:312-329, 1963b.

Kruse, H.: Thyrotropin-releasing hormone: Interaction with chlorpromazine in mice, rats, and rabbits. *J Pharmacol (Paris)* 6:249-268, 1975.

Kumasaka, T., Nishi, N., Yaoi, Y., Rido, Y., Saito, M., Okayasu, I., Shimizof, K., Hatakeyama, S., Sawano, S., and Kokubu, K.: Demonstration of immunoreactive somatostatin-like substance in villi and desidua in early pregnancy. *Am J Obstet Gynecol* 97:343-347, 1979.

LaFerla, T.T., Anderson, D.L., and Schalch, D.S.: Psychoendocrine response to sexual arousal in human males. *Psychosom Med* 40:166-172, 1978.

LaSalle, G.L.G., and Ben-Ari, Y.: Microiontophoretic effects of substance P on neurons of the medial amygdala and putamen of the rat. *Brain Res 135:*174-179, 1977.

Lazarus, L.H., Brown, M.R., and Perrin, M.H.: Distribution, localization, and characterization of neurotensin binding sites in the rat brain. *Neuropharmacology 36:*625-629, 1977.

Lee, C.M., Arregui, A., and Iversen, L.L.: Substance P degradation by rat brain peptidases: Inhibition by SQ 20881. *Biochem Pharmacol 28:*553-556, 1979.

Leitner, J.W., Rifkin, R.M., Maman, A., and Sussmer, K.E.: Somatostatin binding to pituitary plasma membranes. *Biochem Biophys Res Comm 87:*919-927, 1979.

Lembeck, F., and Zetler, G.: Substance P. In *International Encyclopedia of Pharmacology and Therapeutics*, Section 72, Volume I, J.M. Walker, ed. Pergamon Press, Oxford, 1971, pp 29-71.

Lembeck, F., Mayer, N., and Schindler, G.: Substance P in rat brain synaptosomes. *Naunyn Schmiedebergs Arch Pharmacol 301:*17-22, 1977.

Lembeck, F., Saria, A., and Mayer, N.: Substance P: Model studies of its binding to phospholipids. *Naunyn Schmiedebergs Arch Pharmacol 306:*189-194, 1979.

Lipton, M.A., Ervin, G.N., Birkemo, L.S., Nemeroff, C.B., and Prange, A.J., Jr.: Neurotensin-neuroleptic similarites: An example of peptide-catecholamine interactions. In *Catecholamines: Basic and Clinical Frontiers*, E. Usdin, I.J. Kopin, and J.D. Barchas, eds. Pergamon Press, New York, 1979, pp 657-662.

Lomax, P., Bajorek, J.G., Chesanek, W., and Tataryn, I.V.: Thermoregulatory effects of luteinizing hormone-releasing hormone in the rat. In *Thermoregulatory Mechanisms and Their Therapeutic Implications*, B. Cox, P. Lomax, A.S. Milton, and E. Schonbaum, eds. S. Karger, Basel, 1980, pp 208-211.

Loosen, P.T., Nemeroff, C.B., Bissette, G., Burnett, G., Prange, A.J., Jr., and Lipton, M.A.: Neurotensin-induced hypothermia in the rat: Structure-activity studies. *Neuropharmacology 17:*109-113, 1978.

Lucke, C., Mitzkat, H.J., and von zur Muhlen, A.: Somatostatin. *Klin Wochenschr 54:*293-301, 1976.

Luttinger, D., Mason, G.A., Frye, G.D., Osbahr, A.J., III, Nemeroff, C.B., and Prange, A.J., Jr.: Potentiation of ethanol-induced narcosis and hypothermia by neurotensin. *Trans Amer Soc Neurochem 11:*355, 1980a.

Luttinger, D., Wiggins, R., King, R.A., Nemeroff, C.B., and Prange, A.J., Jr.: Inhibition of conditioned avoidance responding and rewarding electrical self-stimulation of the brain on centrally administered neurotensin. *Soc Neurosci Abst 6:*283-287, 1980b.

McAdoo, B.C., Doering, C.H., Kraemer, H.C., Dessert, N., Brodie, H.K.H., and Hamburg, D.A.: A study of the effect of gonadotropin releasing hormone on human mood and behavior. *Phychosom Med 40:* 199–209, 1978.

McMartin, C., and Purdon, G.E.: Early fate of somatostatin in the circulation of the rat after intravenous injection. *J Endocrinol 77:* 67–74, 1978.

Maeda, K., and Frohman, L.A.: Dissociation of systemic and central effects of neurotensin on the secretion of growth hormone, prolactin, and thyrotropin. *Endocrinology 103:* 1903–1909, 1978.

Magnan, J., Regoli, D., Quirion, R., Lemaire, S., St-Pierre, S., and Rioux, F.: Studies on the inhibitory action of somatostatin in the electrically stimulated rat vas deferens. *Eur J Pharmacol 55:* 347–354, 1979.

Magnusson, T., Carlsson, A., Fisher, G.H., Chang, D., and Folkers, K.: Effect of synthetic substance P on monoaminergic mechanisms in brain. *J Neural Transm 38:* 87–93, 1976.

Makino, T., Carraway, R., Leeman, S., and Greep, R.O.: *In vitro* and *in vivo* effects of newly purified hypothalamic tridecapeptide on rat LH and FSH releases. *Soc Study Repro Abst 26:* 1973.

Malick, J.B., and Goldstein, J.M.: Analgesic activity of substance P following intracerebral administration in rats. *Life Sci 23:* 835–844, 1978.

Malthe-Sorenssen, D., Wood, P.L., Cheney, D.L., and Costa, E.: Modulation of the turnover rate of acetylcholine in rat brain by intraventricular injections of thyrotropin-releasing hormone, somatostatin, neurotensin, and angiotensin II. *J Neurochem 31:* 685–691, 1978.

Merek, K., and Haubrich, D.R.: Thyrotropin-releasing hormone-increased catabolism of catecholamines in brains of thyroidectomized rats. *Biochemical Pharmacol 26:* 1817–1818, 1977.

Marks, N., Stern, S., and Benuck, M.: Correlation between biological potency and biodegradation of a somatostatin analogue. *Nature 261:* 511–512, 1976.

Mayer, M.L., and MacLeod, N.K.: The excitatory action of substance P and stimulation of the stria terminalis bed nucleus on preoptic neurones. *Brain Res 166:* 206–210, 1979.

Mayer, N., Lembeck, F., Saria, A., and Gamse, R.: Substance P: characteristics of binding to synaptic vesicles of rat brain. *Naunyn Schmiedebergs Arch Pharmacol 306:* 45–51, 1979.

Metcalf, G.: TRH: A possible mediator of thermoregulation. *Nature 252:* 310–311, 1974.

Metcalf, G., Dettmar, P., and Watson, T.: The role of neuropeptides in thermoregulation. In *Thermoregulatory Mechanisms and Their Therapeutic Implications in Thermoregulation*, B. Cox, P. Lomax, A.S. Milton, and E. Schonbaum, eds. S. Karger,

Basel, 1980, pp 175-179.

Millar, R.P.: Somatostatin immunoreactive peptides of higher molecular weight in ovine hypothalamic extracts. *J Endocrinol* 77:429-430, 1978.

Miyamato, M., and Nagawa, Y.: Mesolimbic involvement in the locomotor stimulant action of thyrotropin-releasing hormone (TRH) in rats. *Eur J Pharmacol* 44:143-152, 1977.

Mohrland, J.S., and Gebhart, G.F.: Substance P-induced analgesia in the rat. *Brain Res* 171:556-559, 1979.

Morley, J.E.: Minireview: Extrahypothalamic thyrotropin-releasing hormone (TRH)--its distribution and its functions. *Life Sci* 25:1539-1550, 1979.

Mortimer, C.H., McNeilly, A.S., Fisher, R.A., Murray, M.A.F., and Besser, G.M.: Gonadotropin-releasing hormone therapy in hypogonadal males with hypothalamic or pituitary dysfunction. *Br Med J* 4:617-621, 1974.

Moss, R.L.: Relationship between the central regulation of gonadotropin and mating behavior in female rats. In *Reproductive Behavior*, W. Montagna and W.A. Sadler, eds. Plenum Press, New York, 1975, pp 55-76.

Moss, R.L.: Effects of hypothalamic peptides on sex behavior in animals and man. In *Psychopharmacology: A Generation of Progress*, M.A. Lipton, A. Dimascio, and K.F. Killam, eds. Raven Press, New York, 1978, pp 431-440.

Moss, R.L.: Actions of hypothalamic hypophysiotrophic hormones on the brain. *Ann Rev Physiol* 41:617-631, 1979.

Moss, R.L., and Foreman, M.M.: Potentiation of lordosis behavior by intrahypothalamic infusion of synthetic luteinizing hormone-releasing hormone. *Neuroendocrinology* 20:176-181, 1976.

Moss, R.L., and McCann, S.M.: Induction of mating behavior in rats by luteinizing hormone-releasing factor. *Science* 181:177-179, 1973.

Moss, R.L., McCann, S.M., and Dudley, C.A.: Releasing hormones in sexual behavior. *Prog Brain Res* 42:37-46, 1975.

Myers, R.D., Metcalf, G., and Rice, J.C.: Identification by microinjection of TRH-sensitive sites in the cat's brain stem that mediate respiratory, temperature, and other autonomic changes. *Brain Res* 126:105-115, 1977.

Nagai, K., and Frohman, L.A.: Hyperglycemia and hyperglucagonemia following neurotensin administration. *Life Sci* 19:273-280, 1976.

Nair, R.M.G., Kastin, A.J., and Schally, A.V.: Isolation and structure of another hypothalamic peptide possessing MSH-release-inhibiting activity. *Biochem Biophys Res Commun* 43:1376, 1971.

Nakagawa, K., Obara, T., Matsubara, M., and Horikana, H.: Enzyme-linked immunoassay of somatostatin. *Endocrinol Japon*

25:197-199, 1978.

Nakata, Y., Kusaka, Y., Segawa, T., Yajima, H., and Kitagawa, K.: Substance P: Regional distribution and specific binding to synaptic membranes in rabbit central nervous system. *Life Sci* 22:259-268, 1977.

Neidle, A., Yessaian, N., and Lajtha, A.: The degradation of Pro-Leu-Gly-NH$_2$ (MIF) by mouse brain. *Trans Am Soc Neurochem* 11:92, 1980.

Nemeroff, C.B.: Neurotensin: Perchance an endogenous neuroleptic? *Biol Psychiatry* 15:283-302, 1980a.

Nemeroff, C.B., Bissette, G., Prange, A.J., Jr., Loosen, P.T., and Lipton, M.A.: Centrally administered neurotensin potentiates the depressant actions of pentobarbital. *Proc Endocrine Soc Ann Meeting* 312: 1976.

Nemeroff, C.B., Diez, J.A., Bissette, G., Prange, A.J., Jr., Harrell, L.E., and Lipton, M.A.: Lack of effect of chronically administered thyrotropin-releasing hormone (TRH) on regional rat brain tyrosine hydroxylase activity. *Pharmacol Biochem Behav* 6:467-469, 1977a.

Nemeroff, C.B., Bissette, G., Prange, A.J., Jr., Loosen, P.T., Barlow, T.S., and Lipton, M.A.: Neurotensin: Central nervous system effects of a hypothalamic peptide. *Brain Res* 128:485-496, 1977b.

Nemeroff, C.B., Osbahr, A.J., III, Manberg, P.J., Ervin, G.N., and Prange, A.J., Jr.: Alterations in nociception and body temperature after intracisternally administered neurotensin, β-endorphin, other endogenous peptides, and morphine. *Proc Natl Acad Sci USA* 76:5368-5371, 1979.

Nemeroff, C.B., Bissette, G., Manberg, P.J., Osbahr, A.J., III, Breese, G.R., and Prange, A.J., Jr.: Neurotensin-induced hypothermia: Evidence for an interaction with dopaminergic systems and the hypothalamic-pituitary-thyroid axis. *Brain Res* 195:69-84, 1980b.

Nemeth, E.F., and Cooper, J.R.: Effect of somatostatin on acetylcholine release from rat hippocampal synaptosomes. *Brain Res* 165:166-170, 1979.

Nicoll, R.A.: The action of thyrotropin-releasing hormone, substance P, and related peptides on frog spinal motoneurons. *J Pharmocol Exp The* 207:817-824, 1978.

Nistico, G., Rotiroti, D., de Sarro, A., and Stephenson, J.D.: Behavioral, electrocortical and body temperature effects after intracerebral infusion of TRH in fowls. *Eur J Pharmacol* 50:253-260, 1978.

North, R.B., Harik, S.L., and Snyder, S.H.: L-leucyl-glycinamide (PLG): Influences on locomotor and stereotyped behavior in cats. *Brain Res* 63:435-439, 1973.

Oehme, P., Hilse, H., Morgenstern, E., and Gores, E.: Substance P: Does it produce analgesia or hyperalgesia? *Science*

208:305-307, 1980.

Ogawa, N., Friesen, H.G., Martin, J.B., and Brazeau, P.: Radio-receptor assay for somatostatin. *Proc Endocrine Soc*, 58th Annual Meeting, 154: 1976.

Ogawa, N., Thompson, T., Friesen, H.G., Martin, J.B., and Brazeau, P.: Properties of soluble somatostatin-binding protein. *Biochem J 165*:269-277, 1977.

Ogawa, N., Thompson, T., and Friesen, H.G.: Characteristics of a somatostatin-binding protein. *Can J Physiol Pharmacol 56*:48-53, 1978.

Olpe, A.R., and Koella, W.P.: Rotary behavior in rats by intra-nigral application of substance P and an eledoisin fragment. *Brain Res 126*:576-579, 1977.

Osbahr, A.J., III, Nemeroff, C.B., Manberg, P.J., and Prange, A.J., Jr.: Centrally administered neurotensin: Activity in the Julou-Courvoisier muscle relaxation test in mice. *Eur J Pharmacol 54*:299-302, 1979.

Otsuka, M., and Konishi, S.: Substance P and excitatory trans-mitter of primary sensory neurons. *Cold Spring Harbor Symp Quant Biol 40*:135-143, 1976.

Otsuka, M., Konishi, S., and Takahashi, T.: The presence of a motoneurone-depolarizing peptide in bovine dorsal roots of spinal nerve. *Proc Japan Acad 48*:342-346, 1972a.

Otsuka, M., Konishi, S., and Takanhashi, T.: A further study of the motoneurone-depolarizing peptide extracted from dorsal roots of bovine spinal nerves. *Proc Japan Acad 48*:747-752, 1972b.

Otsuka, M., Konishi, S., and Takanhashi, T.: Hypothalamic sub-stance P as a candidate for transmitter of primary afferent neurons. *Fed Proc 34*:1922-1928, 1975.

Patel, Y.C., Zingy, H.H., and Dreifuss, J.J.: Calcium-dependent somatostatin secretion from rat neurohypophysis *in vitro*. *Nature 267*:852-853, 1977.

Pettibone, D.J., Wurtman, R.J., and Leeman, S.E.: d-Amphetamine administration reduces substance P concentration in the rat striatum. *Biochem Pharmacol 27*:839-842, 1978.

Pfaff, D.W.: Luteinizing hormone-releasing factor potentiates lordosis behavior in hypophysectomized ovariectomized female rats. *Science 182*:1148-1149, 1973.

Phyllis, J.W.: Is β-(4-chlorophenyl)-GABA a specific antagonist of substance P on cerebral cortical neurons? *Experientia 32*:593-594, 1976.

Phyllis, J.W., and Limacher, J.J.: Substance P excitation of cerebral cortical Betz cells. *Brain Res 69*:158-163, 1974.

Pimstone, B.L., Berelowitz, M., and Kronheim, S.: Somatostatin. *S Afr Med J 50*:1471-1474, 1976.

Pittman, Q.J., Blume, H.W., and Renaud, L.P: Depressant effect of thyrotropin-releasing hormone (TRH) and TRH analogs on central neuronal excitability. *Can Fed Biol Sci* 21:2, 1978.

Plotnikoff, N.P., and Kastin, A.J.: Pharmacological studies with a tripeptide L-prolyl-L-leucyl-glycinamide. *Arch Int Pharmacodyn Ther* 211:211-224, 1974.

Plotnikoff, N.P., Kastin, A.J., Anderson, M.S., and Schally, A.V.: Dopa potentiation by a hypothalamic factor, MSH release-inhibiting hormone (MIF). *Life Sci* 10:1279-1283, 1971.

Plotnikoff, N.P., Prange, A.J., Jr., Breese, G.R., Anderson, M.S., and Wilson, I.C.: Thyrotropin-releasing hormone: Enhancement of DOPA activity by a hypothalamic hormone. *Science* 178:417-418, 1972a.

Plotnikoff, N.P., Kastin, A.J., Anderson, M.S., and Schally, A.V.: Oxotremorine antagonism by a hypothalamic hormone, melanocyte stimulating-hormone release-inhibiting hormone (MIF). *Proc Soc Exp Biol Med* 140:811-814, 1972b.

Plotnikoff, N.P., Kastin, A.J., Anderson, M.S., and Schally, A.V.: Deserpidine antagonism by a tripeptide, L-prolyl-L-leucyl-glycinamide. *Neuroendocrinology* 11:67-71, 1973.

Plotnikoff, N.P., Minard, F.N., and Kastin, A.J.: Dopa potentiation in ablated animals and brain levels of biogenic amines in intact animals after L-prolyl-L-leucyl-glycinamide. *Neuroendocrinology* 14:271-279, 1974a.

Plotnikoff, N.P., Kastin, A.J., and Schally, A.V.: Growth hormone release-inhibiting hormone: Neuropharmacological studies. *Pharmacol Biochem Behav* 2:693-696, 1974b.

Plotnikoff, N.P., white, W.F., Kastin, A.J., and Schally, A.V.: Gonadotropin-releasing hormone (GNRH). Neuropharmacological studies. *Life Sci* 17:1685-1692, 1976.

Prange, A.J., Jr., Breese, G.R., Cott, J.M., Martin, B.R., Cooper, B.R., Wilson, I.C., and Plotnikoff, N.P.: Thyrotropin-releasing hormone: Antagonism of pentobarbital in rodents. *Life Sci* 14:447-455, 1974.

Prange, A.J., Jr., Breese, G.R., Jahnke, G.D., Martin, B.R., Cooper, B.R., Cott, J.M., Wilson, I.C., Alltop, L.B., Lipton, M.A., Bissette, G., Nemeroff, C.B., and Loosen, P.T.: Modification of pentobarbital effects by natural and synthetic polypeptides: Discusion of brain and pituitary effects. *Life Sci* 16:1907-1914, 1975.

Prange, A.J., Jr., Nemeroff, C.B., Lipton, M.A., Breese, G.R., and Wilson, I.C.: Peptides and the central nervous system. In *Handbook of Psychopharmacology*, L.L. Iversen, S.D. Iversen, and S.H. Snyder, eds. Plenum Press, New York, 1978a, pp 1-107.

Prange, A.J., Jr., Nemeroff, C.B., and Lipton, M.A.: Behavioral effects of peptides: Basic and clinical studies. In *Psychopharmacology: A Generation of Progress*, M.A. Lipton, A. DiMascio, and K.F. Killam, eds. Raven Press, New York, 1978b, pp 441-458.

Prange, A.J., Jr., Nemeroff, C.B., Bissette, G., Manberg, P.J., Osbahr, A.J., III, Burnett, G.B., Loosen, P.T., and Kraemer, G.W.: Neurotensin: Distribution of hypothermic response in mammalian and submammalian vertebrates. *Pharmacol Biochem Behav* 11:473-477, 1979.

Precht, H., Christopherson, J., Hensel, H., and Larcher, W.: *Temperature and Life*. Springer-Verlag, New York, 1973.

Pugsley, T.A., and Lippmann, W.: Synthetic melanocyte-stimulating hormone release-inhibiting factor (MIF). III Effect of L-prolyl-L-leucyl-glycinamide on biogenic amine turnover. *Arzneim Forsch* 27:2293-2296, 1977.

Pugsley, T.A., and Lippmann, W.: Effects of somatostatin analogues and 17-α-dihydroequilin on rat brain opiate receptors. *Res Commun Chem Pathol Pharmacol* 21:153-156, 1978.

Randić, M., and Miletić, V.: Effect of Substance P in cat dorsal horn neurons activated by noxious stimuli. *Brain Res* 128:164-169, 1977.

van Ree, J.M., and de Wied, D.: L-prolyl-L-leucyl-glycinamide (PLG) facilitates morphine dependence. *Life Sci* 19:1331, 1976.

Renaud, L.P.: TRH, LH-RH, and somatostatin: Distribution and physiological action in neural tissue. In *Society for Neuroscience Symposia II: Approaches to the Cell Biology of Neurons,* W.M. Cowan and J.A. Ferendelli, eds. Society of Neurosience, Bethesda, 1977, pp 265-290.

Rezek, M., Havelicek, V., Hughes, K.R., and Friesen, H.: Central site of action of somatostatin (SRIF): Role of hippocampus. *Neuropharmacology* 15:499-504, 1976a.

Rezek, M., Havlicek, V., Hughes, K.R., and Friesen, H.: Cortical administration of somatostatin (SRIF): Effect on sleep and motor behavior. *Pharmacol Biochem Behav* 5:73-77, 1976b.

Rezek, M., Havlicek, V., Hughes, K.R., and Friesen, H.: Behavioral and motor excitation and inhibition induced the administration of small and large doses of somatostatin into the amygdala. *Neuropharmacology* 16:157-162, 1977a.

Rezek, M., Havlicek, V., Leybin, L., Pinsky, C., Kroeger, E.A., Hughes, K.R., and Friesen, H.: Neostriatal administration of somatostatin: Differential effect of small and large doses on behavior and motor control. *Can J Physiol Pharmacol* 55:234-242, 1977b.

Riskind, P., and Moss, R.L.: Midbrain central grey: LH-RH infusion enhances lordotic behavior in estrogen-primed ovariectomized rats. *Brain Res Bull* 4:203-205, 1979.

Rivier, C., Brown, M., and Vale, W.: Effect of neurotensin, substance P, and morphine sulfate on the secretion of prolactin and growth hormone in the rat. *Endocrinology* 100:751-754, 1977a.

Rivier, J.E., Lazarus, L.H., Perrin, M.H., and Brown, M.R.:

Neurotensin analogues. Structure-activity relationships. *J Med Chem 20*:1409-1411, 1977b.

Rondeau, D.B., Jolicoeur, F.B., Belanger, F., and Barbeau, A.: Motor activity induced by substance P in rats. *Pharmacol Biochem Behav 9*:769-775, 1978.

Rondeau, D.B., Jolicoeur, F.B., Belanger, F., and Barbeau, A.: Effect of brain peptides on hypokinesia produced by anterolateral hypothalamic 6-OHDA lesions in rats. *Pharmacol Biochem Behav 10*:943-946, 1979.

Saito, K., Konishi, S., and Otsuka, M.: Antagonism between Lioresal and substance P in rat spinal cord. *Brain Res 96*:177-180, 1975.

Sakuma, Y., and Pfaff, D.W.: LH-RH in the mesencephalic central grey can potentiate lordosis reflex on female rats. *Nature 283*:566-567, 1980.

Saperstein, R., Mothon, S., and Reichlin, S.: Enzymatic degradation of TRH and LRH by hypothalamic extracts. *Fed Proc 34*:239, 1975.

Sastry, B.R.: Effects of substance P, acetylcholine, and stimulation of habenula on rat interpeduncular neuronal activity. *Brain Res 144*:404-410, 1978.

Sastry, B.R.: Substance P effects on spinal nociceptive neurones. *Life Sci 24*:2169-2178, 1979.

Schaeffer, J.M., Axelrod, J., and Brownstein, M.J.: Regional differences in dopamine-mediated release of TRH-like material from synaptosomes. *Brain Res 138*:571-574, 1977.

Schwarzstein, L., Aparicio, N.J., Turner, D., Calamera, J.C., Mancini, R., and Schally, A.V.: Use of synthetic luteinizing hormone-releasing hormone in treatment fo oligospermic men: A preliminary report. *Fertil Steril 26(4)*:331-336, 1975.

Segal, D.S., and Mandell, A.J.: Differential behavioral effects of hypothalamic polypeptides. In *The Thyroid Axis, Drugs, and Behavior*, A.J. Prange, Jr., ed. Raven Press, New York, 1974, pp 129-133.

Segawa, T., Nakata, Y., Nakamura, K., Yajima, H., and Kitagawa, K.: Substance P in the central nervous system of rabbits: Uptake system differs from putative transmitters. *Jap J Pharmacol 26*:757, 1976.

Sheppard, M., Shapiro, B., Pimstone, B., Kronheim, S., Berelowitz, M., and Gregory, M.: Metabolic clearance and plasma half-disappearance time of exogenous somatostatin in man. *J Clin Endocrinol Metab*:50-53, 1979.

Snider, S.R., Matthew, E., Roberts, P., Zimmerman, E.A., Cort, J.H., and Hollander, S.: Potentiation of levodopa-induced behavior by a synthetic MIF (Pro-Leu-Gly-NH$_2$) analog in two experimental models of Parkinsonism, *Neurology 30*:366, 1980.

Spiess, J., and Vale, W.: Investigation of larger forms of somatostatin in pigeon pancreas and rat brain. *Metabolism 271*

(Suppl 1):1175-1178, 1978.

Spirtes, M.A., Kostrezewa, R.M., and Kastin, A.J.: α-MSH and MIF-I effects on serotonin levels and accumulation in various rat brain areas. Pharmacol Biochem Behav 3:1011-1015, 1975.

Stanton, T.L., Winokur, A., and Beckman, A.L.: Reversal of natural CNS depression by TRH action in the hippocampus. Brain Res 181:470-475, 1980.

Starr, M.J.: Investigation of possible interactions between substance P and transmitter mechanisms in the substantia nigra and corpus striatum of the rat. J Pharm Pharmacol 30:359-363, 1978.

Steiner, F.A.: Electrophysiological mapping of brain sites sensitive to corticosteroids, ACTH, and hypothalamic releasing hormones. In Anatomical Neuroendocrinology, W.E. Stumpf and L.D. Grant, eds. S. Karger, Basel, 1975, pp 270-275.

Stern, P., and Hadzovic, J.: Pharmacological analysis of central actions of synthetic substance P. Arch Int Pharmacodyn Ther 202:259-262, 1973.

Stern, P., Hadzovic, J., and Radivojevic, M.: The inhibition of the effects of morphine by synthetic substance P. Experientia 32:1326-1327, 1976.

Stewart, J.M., Getto, C.J., Neldner, K., Reeve, E.B., Krivoy, W.A., and Zimmerman, E.: Substance P and analgesia. Nature 262:784-785, 1976.

Tan, A.T., Tsang, D., Renaud, L.P., and Martin, J.B.: Effect of somatostatin on calcium transport in guinea pig cortex synaptosomes. Brain Res 123:193-196, 1977.

Terenius, L.: Somatostatin and ACTH are peptides with partial antagonist-like selectivity for opiate receptors. Eur J Pharmacol 38:211-213, 1976.

Terry, L.C., Willoughby, J.O., Brazeau, P., Martin, J.B., and Patel, Y.C.: Antiserum to somatostatin prevents stress-induced inhibition of growth hormone secretion in the rat. Science 192:565-566, 1976.

Tiru-Chelvam, R.: Tissue levels of substance P associated with experimental snake venom poisoning. Br J Exp Path 54:524-533, 1973.

Traber, J., Glaser, T., Brandt, M., Klebensberger, W., and Hamprecht, B.: Different receptors for somatostatin and opioids in neuroblastoma X glioma hybrid cells. FEBS Lett 81:351-354, 1977.

Tregear, G.W., Niall, H.D., Potts, J.T., Jr., Leeman, S.E., and Chang, M.M.: Amino acid sequence of substance P. Nature 232:86-89, 1971.

Uhl, G.R., Bennett, J.P., Jr., and Snyder, S.H.: Neurotensin, a central nervous system peptide: Apparent receptor binding in brain membranes. Brain Res 130:299-313, 1977.

Unger, R.H., Ipp, E., Schusdziarra, V., and Orci, L.: Hypothesis: Physiological role of pancreatic somatostatin and the contribution of D-cell disorders to diabetes mellitus. *Life Sci 20:*2081-2086, 1977.

Vale, W., Brazeau, P., Rivier, C., Brown, M., Boss, B., Rivier, J., Burgus, R., Ling, N., and Guillemin, R.: Somatostatin. *Recent Prog Horm Res 31:*365-397, 1975.

Versteeg, D.H.G., Tanaka, M., Dekloet, E.R., Van Ree, J.M., and de Wied, D.: L-prolyl-L-leucyl-glycinamide (PLG) regional effects on alpha-MPT-induced catecholamine disappearance in rat brain. *Brain Res 143:*561-566, 1978.

Vijayan, E., and McCann, S.M.: Suppression of feeding and drinking activity in rats following intraventricular injection of thyrotropin-releasing hormone (TRH). *Endocrinology 100:*1727-1730, 1977.

Vivas, A., and Celis, M.E.: Effects of Pro-Leu-Gly-NH$_2$ and tocinoic acid on the secretion of melanocyte-stimulating hormone from rat pituitaries incubated *in vitro*. In *Peptides: Chemistry, Structure and Biology*, R. Walter and J. Meienhofer, eds. Ann Arbor Science Publishers, Ann Arbor, Mich., 1975, pp 777-785.

Vogel, R.A., Cooper, B.R., Barlow, T.S., Prange, A.J., Jr., Mueller, R.A., and Breese, G.R.: Effects of thyrotropin-releasing hormone on locomotor activity, operant performance, and ingestive behavior. *J Pharmacol Exp Ther 208:*161-168, 1979.

Vogel, R.A., Frye, G.D., Wilson, J.H., Kuhn, C.M., Mailman, R.B., Mueller, R.A., and Breese, G.R.: Attenuation of the effect of punishment by thyrotropin-releasing hormone: Comparisons with chlordiazepoxide. *J Pharmacol Exp Ther 212:*153-161, 1980.

Voith, K.: Synthetic MIF analogues. II Dopa potentiation and fluphenazine antagonism. *Arzneim Forsch 27:*2290-2293, 1977.

Wakabayashi, I., Miyazawa, Y., Kanda, M., Miki, N., Demura, R., Demura, H., and Shizume, K.: Stimulation of immunoreactive somatostatin release from hypothalamic synaptosomes by high (K$^+$) and dopamine. *Endocrinol Jpn 24:*601-604, 1977.

Waldmeier, P.C., Kam, R., and Stocklin, K.: Increased dopamine metabolism in rat striatum after infusions of substance P into the substantia nigra. *Brain Res 159:*223-227, 1978.

Walter, R., Ritzmann, R.F., Bhargava, H.N., and Flexner, L.B.: Prolyl-leucyl-glycine and derivatives block development of physical dependence on morphine in mice. *Pro Natl Acad Sci USA 76(1):*518-520, 1979.

de Wied, D., Witter, A., and Grevery, H.D.: Behaviorally active ACTH analogues. *Biochem Pharmacol 24:*1463-1468, 1975.

van Wimersma Greidanus, T.B., Dogterom, J., and de Wied, D.: Intraventricular administration of anti-vasopressin serum inhibits memory consolidation in rats. *Life Sci 16:* 637–644, 1975.

Winokur, A., and Beckman, A.L.: Effects of thyrotropin-releasing hormone, norepinephrine and acetylocholine on the activity of neurons in the hypothalamus, septum, and cerebral cortex of the rat. *Brain Res 150:* 205–209, 1978.

Winokur, A., and Utiger, R.D.: Thyrotropin-releasing hormone: Regional distribution in rat brain. *Science 185:* 265–266, 1974.

Winokur, A., Davis, R., and Utiger, R.D.: Subcellular distribution of thyrotropin-releasing hormone (TRH) in rat brain and hypothalamus. *Brain Res 120:* 423–434, 1977.

Wood, P.L., Cherey, D.L., and Costa, E.: Modulation of the turnover of hippocampal acetylcholine by neuropeptides: Possible site of action of α-melanocyte-stimulating hormone, adrenocorticotrophic hormone, and somatostatin. *J Pharmacol Exp Ther 209:* 97–103, 1979.

Woods, A.C., and Chase, T.N.: MIF: Effect of levodopa dyskinesias in man. *Lancet 2:* 513, 1973.

Yanaihara, N., Sato, H., Sakura, N., and Yanaihara, C.: Somatostatin radioimmunoassay with ^{125}I-N^a-tyrosyl-somatostatin. *Endocrinol Jpn 25:* 95–103, 1978.

Yarbrough, G.G.: TRH potentiates excitatory actions of acetylcholine on cerebral cortical neurons. *Nature 263:* 523–524, 1976.

Yarbrough, G.G.: Studies on the neuropharmacology of thyrotropin-releasing hormone (TRH) and a new TRH analog. *Eur J Pharmacol 48:* 19–27, 1978.

Yarbrough, G.G.: On the neuropharmacology of thyrotropin-releasing hormone (TRH). *Prog Neurobiol 12:* 291–312, 1979.

Yarbrough, G.G., and Singh, D.K.: Intravenous thyrotropin-releasing hormone (TRH) enhances the excitatory actions of acetylcholine (ACh) on rat cortical neurons. *Experientia 34:* 390, 1978.

Youngblood, W.W., Lipton, M.A., and Kizer, J.S.: TRH-like immunoreactivity in urine, serum, and extrahypothalamic brain: Non-identity with synthetic Pyr-Glu-Hist-Pro-NH$_2$ (TRH). *Brain Res 151:* 99–115, 1978.

Zetler, G.: Biologically active peptides (substance P). In *Handbook of Neurochemistry*, Volume IV, A. Lajtha, ed. Plenum Press, New York, 1970.

Zieglgansberger, W., and Tulloch, I.F.: Effects of substance P on neurones in the dorsal horn of the spinal cord of the cat. *Brain Res 166:* 273–282, 1979.

Copyright © 1984 by Spectrum Publications, Inc.
Peptides, Hormones and Behavior
Edited by C.B. Nemeroff and A.J. Dunn

7

Effects of ACTH, β-Lipotropin, and Related Peptides on the Central Nervous System

ADRIAN J. DUNN

INTRODUCTION

The hormones to be discussed in this chapter can all be considered to be derived from the high molecular weight precursor, pro-opiomelanocortin (POMC). As described by Roberts
in chapter 3 of this volume, POMC is first split into adrenocorticotropin (ACTH) and β-lipotropin (β-LPH). ACTH can
then further be cleaved to form alpha-melanocyte-stimulating
hormone (α-MSH) ([N-AcetylSer$_1$]ACTH$_{1-13}$) and corticotropinlike intermediate lobe peptide (CLIP, ACTH$_{18-39}$). On the
other hand, β-LPH can be cleaved to form β-MSH (β-LPH$_{41-58}$),
β-endorphin (β-LPH$_{61-91}$), α-endorphin (β-LPH$_{61-76}$), γ-endorphin (β-LPH$_{61-77}$), and methionine-enkephalin (met-enkephalin,
β-LPH$_{61-65}$). This hierarchy is depicted diagrammatically
in Fig. 7-1 (see Fig. 3-6 in chapter 3 by Roberts for the
respective sequences). There is now good evidence that
little if any met-enkephalin is derived physiologically
from β-LPH; instead it is derived from other, as yet poorly
characterized, high molecular weight precursor(s) (Lewis
et al., 1980b). Although the existence of a β-LPH or   :
a β-endorphin containing leucine at position 65, which would
therefore correspond to leucine enkephalin (leu-enkephalin),
has been speculated upon, efforts to detect it have not been
fruitful (Guillemin et al., 1977b; Rubinstein et al., 1978).
The authenticity of the report that a hemodialysate from a
schizophrenic patient contained [Leu^5]β-endorphin has not
been established. Recently, two new endorphins have been
isolated from brain and pituitary: dynorphin and α-neo-
endorphin. Both peptides contain leu-enkephalin at their N-
terminus, but they are not thought to be precursors to

273

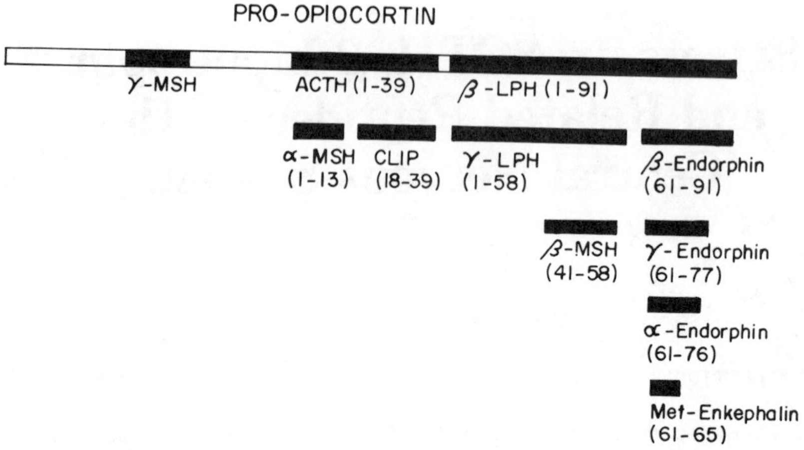

PRO-OPIOCORTIN

Fig. 7-1 Pro-opiomelanocortin (POMC)-derived polypeptide.
There is evidence that the anterior lobe produces
primarily ACTH and β-LPH with some β-endorphin;
whereas the intermediate lobe produces primarily α-
MSH and β-endorphin. There is no good evidence for
the release from the adenohypophysis of α- or γ-
endorphin or met-enkephalin.

leu-enkephalin because they are much more potent as opiate
agonists. Dynorphin is Tyr-Gly-Gly-Phe-Leu-Arg-Ile-Arg-Pro-
Lys-Leu-Lys-Trp-Asp-Asn-Gln (Goldstein et al., 1981), and α-
neo-endorphin is Tyr-Gly-Gly-Phe-Leu-Arg-Lys-Tyr-Pro-Lys
(Minamino et al., 1981).

In this review, the term "endorphin" will be used in the
sense defined by Simon as any endogenous opiate (Adler, 1980),
thereby including the enkephalins as endorphins. Detailed
aspects of the effects of endorphins on animals and humans as
they relate to psychiatric therapy are covered in chapter 8 by
Koob et al.

LOCALIZATION

As would be expected, the POMC-derived peptides are all located
in the same cells. However, the enkephalins are located in a
separate population of cells (Bloom et al., 1978; Watson et
al., 1978; Gramsch et al., 1979) (see Fig. 7-2).

Cells containing ACTH, β-LPH, β-endorphin, and α-MSH are
found in two principal locations: the pituitary and the
hypothalamus. In the pituitary, ACTH and β-LPH are found in

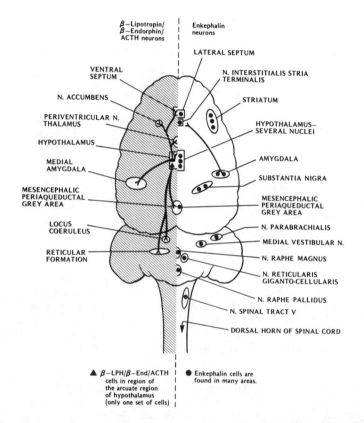

β−Lipotropin/
β−Endorphin/
ACTH neurons

Enkephalin
neurons

LATERAL SEPTUM

VENTRAL
SEPTUM

N. INTERSTITIALIS STRIA
TERMINALIS

N. ACCUMBENS

STRIATUM

PERIVENTRICULAR N.
THALAMUS

HYPOTHALAMUS−
SEVERAL NUCLEI

HYPOTHALAMUS

MEDIAL
AMYGDALA

AMYGDALA

SUBSTANTIA NIGRA

MESENCEPHALIC
PERIAQUEDUCTAL
GREY AREA

MESENCEPHALIC
PERIAQUEDUCTAL
GREY AREA

LOCUS
COERULEUS

N. PARABRACHIALIS

MEDIAL VESTIBULAR N.

RETICULAR
FORMATION

N. RAPHE MAGNUS

N. RETICULARIS
GIGANTO-CELLULARIS

N. RAPHE PALLIDUS

N. SPINAL TRACT V

DORSAL HORN OF SPINAL CORD

▲ β−LPH/β−End/ACTH
cells in region of
the arcuate region
of hypothalamus
(only one set of cells)

● Enkephalin cells are
found in many areas.

Fig. 7-2.   Immunocytochemistry of opiate peptides and related
substances in rat brain. Schematic horizontal
section of the distribution of the ACTH/β-LPH-
(left), and the enkephalin-containing (right) fibers
in rat brain. (Reproduced from Barchas et al.
(1978) with the permissions of the author and publisher. Copyright 1978 by the American Association
for the Advancement of Science.)

specific cells of the anterior lobe, and in most cells of
the intermediate lobe in animals in which this is present.
There is some evidence that the processing POMC is different in the two lobes: in the anterior lobe ACTH,β-LPH,
and some β-endorphin are produced, whereas the intermediate
lobe seems to be geared to the production of α-MSH (and
presumably CLIP) and β-endorphin (Gianoulakis et al., 1979).
The presence of β-endorphin in the anterior lobe is not

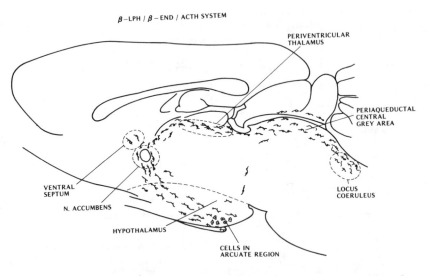

β−LPH / β − END / ACTH SYSTEM

PERIVENTRICULAR THALAMUS

PERIAQUEDUCTAL CENTRAL GREY AREA

VENTRAL SEPTUM

N. ACCUMBENS

LOCUS COERULEUS

HYPOTHALAMUS

CELLS IN ARCUATE REGION

Fig. 7-3.   The ACTH/β-LPH system of rat brain.   Parasagittal
section showing the cells and pathways of the ACTH/
β-LPH-containing neuronal system from rat brain.
(Reproduced from Watson and Barchas (1979) with the
permissions of the author and publisher.

artifactual, even though it can be produced from β-LPH during
extraction (Rossier et al., 1979a).   The β-endorphin found in
the intermediate lobe is acetylated on the N-terminal tyrosine
(Smyth et al., 1979).

     In the hypothalamus, ACTH-, β-LPH-, β-endorphin- and α-MSH-
containing cells are located in one group of cell bodies in the
ventromedial and arcuate nuclei (Figs. 7-2, 7-3).   Any cell
that contains one of the peptides, also contains all the others
(Watson et al., 1978; Bloch et al., 1979; but see later).
These cells are distinct from those containing luteinizing
hormone-releasing hormone (LH-RH), somatostatin (SRIF),
neurophysin, oxytocin, vasopressin, dopamine (DA) and the
enkephalins (Bloch et al., 1979).   The varicose axons project
caudally, running ventrally to the cerebral aqueduct into the
pons.   Terminals are extensive in midline nuclear areas of the
thalamus, periaqueductal gray, locus coeruleus, and the nucleus
of the solitary tract, with some in the dorsal raphe.   Fibers
were not seen caudal to the locus coeruleus (Bloom et al.,
1978).   There are, however, extensive projections within the
hypothalamus, particularly to the medial preoptic, anterior

hypothalamic, dorsomedial, and periventricular nuclei, and the median eminence, and less extensive ones to the paraventricular and arcuate nuclei, and the mammillary bodies. Fibers also course from the hypothalamus into the stria terminalis, the lateral septum, nucleus accumbens and the amygdala (Watson et al., 1978; Bloom et al., 1978; Jacobowitz and O'Donohue, 1978). Jacobowitz and O'Donohue (1978) noted a remarkable coincidence of α-MSH fibers and noradrenergic axonal projections. Potential interactions between endorphins and norepinephrine (NE) have recently been studied by Watson et al. (1980) using antisera to β-endorphin, leu-enkephalin and dopamine- β - hydroxylase.

Watson and Akil (1980), using colchicine to enhance the fluorescence of cell bodies, have recently described a group of cells containing immunoreactive α-MSH, but not β-endorphin or β-LPH, in the hypothalamus lateral to the dorsal limit of the third ventricle. This cell group extended laterally into the zona incerta, between the fornix and the mammillothalamic tract and ventrally to the hypothalamic sulcus. This group of cells was not observed by others, including Jacobowitz and O'Donohue (1978), O'Donohue et al. (1979) and Bloch et al. (1979).

β-MSH, at least in humans, is now considered to be an extraction artifact (Bloomfield et al., 1974), except in one reported pituitary abnormality.

The localization of the enkephalins is very different and much more widespread. Several reports indicate the presence of enkephalins in all lobes of the pituitary, but the highest concentration is in the pars nervosa (Gramsch et al., 1979; Rossier et al., 1980). Enkephalins are present throughout the brain, but the highest concentrations are found in the globus pallidus, caudate nucleus, anterior hypothalamus, amygdala, and periaqueductal and periventricular gray areas (Simantov et al., 1976; Miller et al., 1978). Immunohistochemical studies confirm this distribution with cell bodies and axons found throughout the brain and spinal cord (Hökfelt et al., 1977a; Sar et al., 1978). Although they are concentrated in structures possibly related to pain, such as the periaqueductal gray, nucleus solitarius, the spinal trigeminal nucleus, and the substantia gelatinosa, they are also found extensively outside these areas (Hökfelt et al., 1977a,b). The distribution of enkephalins largely parallels that of opiate receptors (Simantov et al., 1976), although the correlation is not perfect (Sar et al., 1978). Enkephalins are apparently contained mostly in short-axoned neurons, although there are a few exceptions (Cuello and Paxinos, 1978). Both enkephalins are also found in the adrenal medulla (Viveros et al., 1979) and the sympathetic nervous system and gut (Lundberg et al., 1979). Enkephalins, like opiate receptors, are largely localized to the synaptosomal fraction (Osborne et al., 1978).

Whereas initially the ratio of leu- to met-enkephalin was thought to show species and regional variations, most recent studies have found very similar distributions for both peptides, although the concentration of met-enkephalin is always higher (three- to five-fold) than that of leu-enkephalin (Yang et al., 1977; Sar et al., 1978). This does not exclude the possibility of differential physiological potencies for the two enkephalins (Vaught and Takemori, 1979). A report that leu- and met-enkephalin are located in separate neurons (Larsson et al., 1979) contradicts the evidence that they are contained in a common precursor (Lewis et al., 1980b).

Recently, met-enkephalin was detected in human plasma and in cerebrospinal fluid (CSF) (Clement-Jones et al., 1980), but the concentrations were much lower than those of the endorphins.

Dynorphin and α-neo-endorphin are most concentrated in the posterior pituitary, but are also found in the anterior pituitary, hypothalamus, and throughout the brain (Höllt et al., 1980; Minamino et al., 1981). Recent evidence indicates that both peptides may be contained in the same cells with cell bodies in the hypothalamus and widespread fibers and terminals especially dense in the substantia nigra, medial forebrain bundle, internal capsule, nucleus accumbens and medulla (Weber et al., 1982). Whereas the system is clearly distinct from the β-endorphin system, there is considerable overlap with the enkephalinergic systems.

PHYSIOLOGY

The classical endocrine function of ACTH is to stimulate the release of glucocorticosteroids from the adrenal cortex (see chapter 14 by Rees and Gray). It also has trophic activity on both the adrenal cortex and the medulla. ACTH is released episodically throughout the day, but with more frequent episodes in the period preceding and during the active part of the day, that is night for rats, and day for most humans. Its release is dramatically enhanced during stress. Stress also increases the plasma concentration of β-LPH, and β-endorphin (Guillemin et al., 1977a). Plasma ACTH and β-endorphin immunoreactivity rise concomitantly during stress in approximately equimolar amounts (Guillemin et al., 1977a). It has been reported that there is a differential release of ACTH and MSH during stress (Sandman et al., 1973b); both hormones are released during physical stress, whereas only MSH is released during "psychological" stress. However, a recent study by de Rotte et al. (1982) failed to find an increase of plasma α-MSH following footshock, although there was an elevation following "psychologic" stress. As is the case with ACTH, the increases in plasma β-endorphin and β-LPH can be prevented by the synthetic glucocorticoid, dexamethasone (Guillemin et al., 1977a; Krieger et al., 1979b). After adrenalectomy, the plasma concentration of β-endorphin increases in parallel with that of ACTH (Guillemin et al., 1977a). The latter has long been considered to reflect the lack of feedback control by circulating glucocorticoids. Increases of plasma ACTH, β-LPH, and β-endorphin also occur in parallel in Nelson's syndrome, Addison's disease, and Cushing's disease (Höllt et al., 1979; Krieger et al., 1979b).

The origin of plasma ACTH and β-endorphin in the rat appears to be the anterior rather than the intermediate lobe. Adrenalectomy increased the β-endorphin content of the anterior lobe, whereas chronic dexamethasone or stress decreased it (Rossier et al., 1979a). None of these treatments affected the β-endorphin content of the rat intermediate lobe. Similarly, adrenalectomy increased the ACTH content of the anterior but not the neurointermediate lobe (van Dijk, 1979).

Both β-endorphin and β-LPH are present in the plasma of rats and humans, but normally β-endorphin predominates (Csontos et al., 1979; Akil et al., 1979). Surprisingly, most plasma β-endorphin is probably released as such, because exogenous purified β-LPH is not converted significantly to β-endorphin after injection into humans (Suda et al., 1978), although β-LPH is converted to β-endorphin in plasma incubated *in vitro* (Höllt et al., 1979). In humans, circulating β-endorphin is normally very low, but according to one report it rises during pregancy (Akil et al., 1979), whereas another report found an increase only during labor (Csontos et al., 1979). This may correlate with the presence of an intermediate lobe in pregnant human females (see Akil et al., 1979), especially since the human fetus, which also has an intermediate lobe, also exhibits a high plasma β-endorphin content (Csontos et al., 1979).

The control of cerebral ACTH and β-LPH release is much less clear. A most recent study indicates that human CSF β-endorphin-like immunoreactivity is about an order of magnitude higher than plasma, and that only 20 percent is β-endorphin, the remainder being β-LPH (Nakao et al., 1980).

A complicating factor is the potential for transport of pituitary peptides into the brain in the capillaries of the hypothalamo-pituitary complex (Oliver et al., 1977). Once present in the median eminence region, tanycytes (modified ependymal cells, see chapter 1 by Rodriguez) may be able to transport such peptides as ACTH and β-endorphin into third ventricle CSF. Pituitary-hypothalamus transport of a radiolabeled ACTH analogue (ORG 2766), [Met^4SO_2,D-Lys^8,Phe^9]-$ACTH_{4-9}$ and of β-endorphin has been demonstrated (Mezey and

Palkovits, 1982). The transport did not occur if the pituitary stalk was transected, or if the analogue was injected into the subarachnoidal CSF of the sella turcica.

Hypothalamic ACTH content is not altered by hypophysectomy, adrenalectomy, stress, or chronic corticosterone or dexamethasone treatments (Krieger et al., 1979a; van Dijk, 1979). However, van Dijk (1979) found that adrenalectomy caused a temporary decrease (one to three days) in hippocampal ACTH, which was potentiated by hypophysectomy, although hypophysectomy alone had no effect. Hypophysectomy did not alter hypothalamic β-endorphin (Watson et al., 1978), but prolonged footshock stress decreased it slightly (Rossier et al., 1977). One human patient with no detectable β-endorphin in plasma during glucocorticoid therapy had a normal CSF β-endorphin content (Nakao et al., 1980).

The physiology of the release of enkephalins is even less clear. Repeated footshock (1 mA, 1 sec duration, 12 shocks/minute) to rats for 30 minutes decreased leu-enkephalin specifically in the hypothalamus (Rossier et al., 1978). It is presumed that this decrease reflected increased release followed by extracellular degradation. Chronic treatment with neuroleptics (Hong et al., 1978), lithium (Gillin et al., 1978), or electroconvulsive shock (ECS) (Hong et al., 1979) increased the enkephalin content of rat striatum and nucleus accumbens. Presumably, these increases reflect increased synthesis compensating for increased release. Curiously, peripherally injected morphine rapidly doubles the leu-enkephalin content of CSF, an effect that is prevented by naloxone (Bergmann et al., 1978). Morphine similarly elevates plasma leu-enkephalin in dogs (Laasberg et al., 1980). An early report that morphine-tolerant rats exhibit increased cerebral enkephalin has not been reproduced (Childers et al., 1977).

The endocrine effects of endorphins will not be dwelt on here; they have been reviewed recently by Van Vugt and Meites (1980) and Morley (1981). In general, endorphins have actions similar to morphine; the most pronounced effects being stimulation of prolactin (PRL), and inhibitions of growth hormone (GH) (except in primates), LH, follicle-stimulating hormone (FSH), thyroid-stimulating hormone (TSH), and oxytocin secretion. The situation with regard to ACTH and vasopressin (AVP) is complicated. Whereas classically, morphine stimulates release of AVP, accounting for the antidiuretic properties of the drug (see Firemark and Weitzman, 1979; Van Vugt and Meites, 1980; Morley, 1981), more recent data suggest that endorphins inhibit AVP release, especially in response to stressful stimuli (van Wimersma Greidanus et al., 1979b; Knepel et al., 1982). Morphine and the endorphins stimulate ACTH release and corticosteroid production, however, naloxone stimulates ACTH (and β-endorphin) release (see Morley, 1981). An enkephalinergic pathway from the hypothalamus terminates in the neurohypophysis and may regulate vasopressin and/or oxytocin secretion (Rossier et al., 1980) but AVP neurons have also been reported to contain dynorphin (Watson et al., 1981). In humans, endorphins decrease plasma ACTH, cortisol, LH and FSH, but increase PRL and GH. Naloxone (10 mg IV) increased serum cortisol, LH, FSH, and insulin, but did not alter GH, PRL, TSH, arginine vasopressin (AVP), glucagon, triiodothyronine or tetraiodothyronine (Morley, 1981). A surprising finding was that in the rat, β-endorphin (4 µg/kg IV) induced ornithine decarboxylase (ODC) in the kidney, but not in several other tissues, including the adrenal, liver, gut, and the striatum (Haddox and Russell, 1979).

ACTH has a slight depressant effect on blood pressure (Nakamura et al., 1976). The effect is present in adrenal-ectomized rats and can be elicited with intact ACTH, $[AIB^1]$-$ACTH_{1-18}-NH_2$, but not $[AIB^1]ACTH_{1-10}$, $ACTH_{11-18}-NH_2$, α-MSH or β-MSH.

The electrophysiology of ACTH and the endorphins is discussed in chapter 5 by Renaud, but a summary is in order here. There is also an excellent review of endorphins and single cell responses by North (1978).

Like morphine, the enkephalins and β-endorphin injected intraventricularly induce seizure activity (Frenk et al., 1978, Henriksen et al., 1978). These effects are attenuated or prevented by prior systemic administration of naloxone. The epileptiform activity seems to be distinct from the analgesia, because either can be observed separately, and because, in general, lower doses of peptide (10 µg enkephalin, 1 to 2 µg β-endorphin) are necessary for the production of seizures.

ACTH is generally excitatory. Systemic administration of ACTH or $ACTH_{1-24}$ generally increased the firing rate of individual diencephalic neurons (Van Delft and Kitay, 1972), and local administration of $ACTH_{1-24}$ did the same in the hypothalamus (Steiner et al., 1969). Corticosteroids and dexamethasone generally had effects opposite to those of ACTH (see chapter 14 by Rees and Gray). Complex changes and frequency shifts were also observed with α-MSH and $ACTH_{4-10}$ on the cortical electroencephalogram (EEG) (Miller et al., 1974, 1977a), on hippocampal theta activity (Urban and de Wied, 1976), and on visual evoked potentials (Miller et al., 1976). Most recently, $ACTH_{1-24}$ was found to cause a long-lasting potentiation of transmittor release from motor neurons in the frog (Johnston et al., 1983).

Methodological Considerations

Before reviewing the data on the behavior and biochemistry of these hormones, it is appropriate to discuss the methodology.

Experimental manipulations include the classical endocrinological ablation and substitution experiments. Hypophysectomy
removes the pituitary source of ACTH, MSH, β-LPH and β-endorphin, but unless performed very carefully, ACTH-producing
cells may persist and multiply within the sella turcica (Moldow
and Yalow, 1978). The principal problem with hypophysectomy is
that all the pituitary hormones are lost. The release of
pituitary ACTH and β-endorphin can be suppressed with exogenous
glucocorticoids or dexamethasone, but these may have their own
independent effects (see chapter 14 by Rees and Gray).

The hypothalamic POMC system can be lesioned by classical techniques. However, a novel and perhaps more convenient procedure
is to use monosodium glutamate (MSG). High doses of MSG administered peripherally to neonatal rats cause lesions of
perikarya but not fibers in the arcuate nucleus and severely
deplete cerebral ACTH, β-endorphin and α-MSH (Krieger et al.,
1979c). Unfortunately, this procedure also depletes hypothalamic DA and choline acetyltransferase and presumably has
other effects, but these may be no more serious than after
electrical lesions.

When peptides are administered peripherally, degradation is a
serious problem and plasma half-lives are relatively short
(minutes). Slow release preparations, such as zinc phosphate
(ZP) gels (de Wied, 1966) or minipumps can be used, but an increasingly important method is the use of analogues with amino
acid substitutions designed to retard or prevent proteolytic
action. Thus the ACTH analogue, ORG 2766 ($[Met^4SO_2,D-Lys^8,$
$Phe^9]ACTH_{4-9}$) has 1000 times the half-life of ACTH, and
$[D-Ala^2,D-Leu^5]$enkephalin (DADLE) has 100 times the half-life
of leu-enkephalin.

Secondary effects of the administered peptides must also be
controlled. The adrenocortical activity of ACTH can be prevented by prior adrenalectomy, but this not only removes
circulating glucocorticoids, which may play a "permissive" role, but also depletes mineralocorticoids and circulating catecholamines and elevates plasma ACTH. For these
reasons, ACTH analogues that allegedly lack adrenocortical
activity, such as $ACTH_{1-10}$, $ACTH_{1-16}$, and α-MSH, have been
used, but these may not exert all the extra-adrenal activities
of ACTH. However, it is not absolutely clear that these analogues do lack adrenocortical activity. The active site for
adrenal steroidogenesis is contained in the $ACTH_{4-10}$ sequence.
Brain and Evans (1977) found that $ACTH_{1-10}$ elevated plasma
corticosterone in mice, and Dunn et al. (1978a,b) found a
similar effect of $ACTH_{4-10}$. Furthermore, in rats, α-MSH
elevated plasma corticosterone when the animals were not also
exposed to a novel environment (Datta and King, 1980).

A further problem is that any opiate agonist that comes into contact with an opiate receptor will elicit physiological, behavioral, and chemical effects. Such an effect does not imply that the agonist, peptide or otherwise, is normally active on this receptor. A physiological function can only be demonstrated by carefully documenting the endogenous presence of the ligand in the vicinity of the receptor in sufficient concentrations and under appropriate physiological conditions.

An issue that has been overdebated is the permeability of the brain to peripherally administered peptides (see also Kastin et al., 1979a). It has been argued that the blood-brain barrier prevents access of peptides to the brain (Cornford et al., 1978). This argument is not only wrong, it is also irrelevant. There is now ample evidence that peptides administered peripherally do enter the brain, albeit in small amounts (Greenberg et al., 1976; Verhoef and Witter, 1976; Kastin et al., 1979a; Rapoport et al., 1980). Moreover, many important areas of the brain, including the median eminence region of the hypothalamus and several other periventricular regions, do not exhibit a blood-brain barrier, nor does the entire pituitary, the infundibulum, and the pineal. Furthermore, as discussed previously most of the peptides have an independent existence within the brain. The study by Cornford et al. (1978) showed that the cerebral uptake of a number of labeled peptides, including leu- and met-enkephalin, was only a small proportion of that of water. However, the 2 to 3 percent extraction fraction shown is more than adequate to deliver active amounts of peptide to the brain. A recent study by Rapoport et al., (1980) found significant brain uptake of several opioid peptides and suggests that the brain uptake index used by Cornford et al. is an inappropriate method for studying the uptake of compounds with relatively low permeability.

For the foregoing reasons, intracerebral administration of peptides has been used for many experiments. Sometimes the peptides are administered into the ventricular CSF and allowed to diffuse into the tissue (there is no effective CSF-brain barrier), but in other cases cannulae have been inserted in discrete cerebral locations.

A final manipulation of the POMC system is stress. As already indicated, stress stimulates the release of ACTH, β-LPH, and β-endorphin from the pituitary, but as yet the effect on the central nervous system (CNS) neurons is unknown.

An important technique in endorphin research is the use of specific opiate antagonists, naloxone, and to a lesser extent, naltrexone. Naloxone "reversibility" is an important criterion for an opiate effect. However, as Sawynok et al. (1979) have pointed out, naloxone antagonism is a necessary but not a sufficient criterion for opiate activity. Naloxone sensitivity of a phenomenon does not prove the involvement of an endogenous opiate, a deduction unfortunately all too common in the current plethora of literature on endorphins. In addition, it is worth pointing out that there are activities of morphine and other opiates that are not naloxone reversible. It is even possible to produce paradoxical effects with naloxone and naltrexone. Naltrexone has some agonistic properties, as does naloxone at high doses. Furthermore, the drugs may stimulate the release of endorphins (by blocking negative feedback control) so that endorphin receptors not blocked by naloxone may be activated.

Most research with peptides has been performed with synthesized material. However, some of the earlier studies with ACTH and MSH used the purified natural material, usually porcine. In these cases, the injected peptide will be designated ACTH (as opposed to $ACTH_{1-39}$) and doses are specified in units rather than moles or grams. There is no evidence that $ACTH_{1-24}$ lacks any of the biological activity of natural ACTH ($ACTH_{1-39}$).

BEHAVIOR

Locomotor Activity

Like morphine, opioid peptides can produce changes in activity levels. In one report, peripheral administration (100 μg IP) of α-, β- or γ-endorphin or their $D-Ala^2$ derivatives did not alter locomotor activity of rats in an open field, $[D-Ala^2]\gamma$-endorphin increased sexual arousal, γ-endorphin decreased the time to travel to the wall, and $[D-Ala^2]\gamma$-endorphin increased defecation, the last two indicative of "heightened emotionality" (Veith et al., 1978b). However, $[D-Ala^2]$met- or $[D-Ala^2]$leu-enkephalin (6-50 μg ICV) increased activity in mice, an effect that was reversed by naloxone (Katz et al., 1978). And, β-endorphin, $[D-Ala^2]$met-enkephalin or $[D-Met^2,Pro^5]$enkephalin-amide had a biphasic effect in rats, decrease then increase (Browne and Segal, 1980). $ACTH_{4-10}$ (7.5-15 μg ICV) antagonized morphine-induced behavioral activation (Katz, 1979b). Recently, Stinus et al. (1980) showed that direct infusion of β-endorphin (2 μg) into the ventral tegmental area increased locomotor activity in rats. The effect was prevented by naloxone treatment (1 mg/kg) or 6-hydroxydopamine lesions of the mesocortical DA projection. α- and γ-endorphin and $[Des-Tyr^1]\gamma$-endorphin (DTγE) had similar effects but were less potent than β-endorphin.

Naloxone (0.5 - 25 mg/kg SC) reduced locomotion and altered interactions with environmental stimuli (Arnsten and Segal, 1979). High doses of naloxone (4 to 8 mg/kg) reduced "ex-

ploratory" activity (head-dipping) in a hole-board apparatus
(Katz and Gelbart, 1978) but low doses (0.5 mg/kg) increased
stimulus investigation (Arnsten and Segal, 1979). Reductions
of head-dipping are also observed with $ACTH_{1-24}$, $ACTH_{4-10}$, [D-
$Phe^7]ACTH_{4-10}$, p-chlorophenylalanine, chlordiazepoxide, and
ethanol (File, 1978). In cats, β-endorphin (50 g/kg IV) increased licking but decreased locomotion (Catlin et al., 1978).

ACTH decreased locomotor activity of rats in the morning when
endogenous levels were low, but not in the afternoon (Ley and
Corson, 1973). The effect was also present following
adrenalectomy.

Ingestion

Strains of genetically obese mice (ob/ob) and rats (fa/fa) were
found to have increased concentration of β-endorphin in their
pituitaries and plasma (rats only) (Margules et al., 1978).
Pituitary α-MSH and ACTH are also elevated (Rossier et al.,
1979b). The obesity that is caused by overeating can be prevented with naloxone. Naloxone also decreased food intake in
rats made hyperphagic with ventromedial hypothalamic lesions,
and to a lesser extent in normal rats (King et al., 1979).
Obese hirsute women also exhibit elevated plasma β-endorphin
(Givens et al., 1980). Although Margules et al., (1978) argued
that these data suggest a causative role for β-endorphin in
obesity, Rossier et al. (1979b) pointed out that the obesity
precedes the increased hypophyseal β-endorphin content.

ICV enkephalins have been reported to exhibit antidipsogenic
effects (de Caro et al., 1979), an effect that may be related
to effects on AVP secreation (see above).

Stretching and Yawning

Ferrari et al., (1963) first reported that intraventricular
injection of ACTH induced a behavioral syndrome, including
increased stretchings and yawnings (SYS), and spontaneous
penile erections. The effect was observed in rats, rabbits,
mice, cats, and dogs. There is some dispute over the sequence
of ACTH active in stretching and yawning. Ferrari et al.
(1963) found ACTH, α-MSH, and β-MSH to be active in dogs,
whereas Baldwin et al. (1974) found $ACTH_{4-10}$ to be less potent
than $ACTH_{1-24}$ in rabbits. Gispen et al. (1975) performed a
detailed structure-function analysis in rats; whereas, $ACTH_{1-24}$,
α-MSH, β-MSH, $ACTH_{1-16}$, $ACTH_{5-16}$, $ACTH_{5-14}$, and $ACTH_{1-13}-NH_2$
were active, $ACTH_{11-24}$, $ACTH_{1-10}$, $ACTH_{4-10}$, and ORG 2766 were
not. Curiously, $[D-Phe^7]ACTH_{1-10}$ and $[D-Phe^7]ACTH_{4-10}$ were

active. The results with $ACTH_{4-10}$ and $[D-Phe^7]ACTH_{4-10}$ were confirmed by Rees et al. (1976) in mice. Gispen et al. (1975) did not find any evidence for penile erections following ACTH or endorphins. Stretching and yawning may also be initiated by other agents, including zinc and the glutamate receptor antagonist, glutamic acid diethyl ester (GDEE, 25 to 500 mg IP) (Lanthorn and Isaacson, 1979), naloxone (1-5 mg/kg SC) (Bertolini et al., 1978). Ferrari et al. (1963) found that SYS induced by ACTH was blocked by chlorpromazine, atropine, and scopolamine, but not by reserpine, amphetamine, or serotonin.

Grooming

Although Ferrari et al. (1963) noted increased grooming in rats and rabbits before the stretching and yawning initiated by ICV ACTH, Gispen et al. (1975) first studied this phenomenon in detail. The excessive grooming showed the same structure-function relationships as SYS, namely that $ACTH_{1-13}-NH_2$ and $ACTH_{5-16}-NH_2$ elicited it, but $ACTH_{1-10}$, $ACTH_{4-10}$, and ORG 2766 did not. However, much lower doses of ACTH were necessary, the ED_{50} for $ACTH_{1-24}$ being between 10 and 100 ng per rat. $ACTH_{4-10}$ did not antagonize the grooming produced by $[D-Phe^7]ACTH_{4-10}$ (Wiegant et al., 1978), indicating that the "grooming" receptor is probably distinct from that involved in avoidance behavior. Curiously, although $ACTH_{4-10}$, $ACTH_{4-9}$, $ACTH_{4-8}$, and $ACTH_{4-6}$ are inactive, $ACTH_{4-7}$ does promote grooming, but $ACTH_{5-7}$ does not (Wiegant and Gispen, 1977). Rees et al. (1976) confirmed the grooming activity of $ACTH_{1-24}$ and $[D-Phe^7]ACTH_{4-10}$ in mice, along with the lack of activity of $ACTH_{4-10}$.

The grooming response is considered to be elicited by direct action on the brain, because of its short latency, and because it is not elicited by peripherally administrated $ACTH_{1-24}$ up to 5 mg/kg (Gispen et al., 1975). However, one report indicates that 100 µg of β-endorphin (IP) increased grooming activity in rats (Veith et al., 1978b). ACTH elicits grooming in male and female rats, and also in hypophysectomized, adrenalectomized, and castrated male rats (Gispen et al., 1975). The behavioral response is apparently specific to grooming, which competes with other ongoing behaviors, such as locomotion or exploration (Isaacson and Green, 1978). The ACTH-elicited grooming is not abnormal in type, nor is it stereotyped, the proportion of time spent on various parts of the body and the sequence of parts groomed being indistinguishable from natural grooming (Gispen and Isaacson, 1981).

The finding that the specific opiate antagonists naloxone and naltrexone block ACTH-induced grooming (Gispen and Wiegant,

1976) suggested the involvement of an opiate receptor. Subsequently, it was found that low doses of morphine (50 to 500 μg ICV) (Gispen and Wiegant, 1976) elicited grooming, as did β-endorphin, but not α-endorphin, β-LPH, or the enkephalins (Gispen et al., 1976). It is therefore likely that the "grooming" receptor is an opiate receptor, because the ability of ACTH derivatives to elicit grooming closely parallels their ability to counteract morphine-induced analgesia and to compete for opiate receptors with dihydromorphine (Wiegant et al., 1977a). Also, ketocyclazocine produced excessive grooming along with wet-dog shakes (Lanthorn et al., 1979). Most recently, Dunn et al. (1981b) have shown that naloxazone, an irreversible opiate antagonist, prevents ACTH-induced grooming with a time course that parallels its effects on morphine-induced antinociception.

ACTH-induced grooming may involve brain DA because it is inhibited by haloperidol and other DA-receptor antagonists (Wiegant et al., 1977b; Guild and Dunn, 1982). Also $ACTH_{1-24}$ elicits grooming when injected into the substantia nigra but not the striatum, whereas haloperidol injected into the striatum but not the substantia nigra inhibits grooming elicited by ICV $ACTH_{1-24}$ (Wiegant et al., 1977b).

Some important observations suggest that grooming behavior may occur under certain circumstances during stress, and that this may be a consequence of the endogenous release of ACTH, and perhaps β-endorphin. Grooming is elicited in rats by mild stress, such as transfer to a novel environment (Colbern et al., 1978; Jolles et al., 1979) or "non-traumatic" noise-light stress (Katz and Roth, 1979). This increased grooming is impaired by low doses of naloxone (0.2 mg/kg: Green et al., 1979), which do not alter baseline (home cage) grooming activity (Gispen and Wiegant, 1976; Green et al., 1979) or locomotor or exploratory activities (hole-board) (Green et al., 1979). Novelty-induced grooming is also prevented by haloperodol at doses (0.2 mg/kg SC) that did not alter locomotor or exploratory activities (Green et al., 1979). In one report, hypophysectomy severely inhibited novel-cage-induced grooming (Dunn et al., 1979), whereas in another it did not (Jolles et al., 1979). However, in the latter study, hypophysectomized rats showed higher home-cage grooming. Most important, ICV (but not SC) administration of antiserum to ACTH also inhibited novel-cage-induced grooming (Dunn et al., 1979). The antiserum used was specific to ACTH and did not interact with $ACTH_{1-10}$, α-MSH, β-LPH, β-endorphin, or the enkephalins. Novelty-induced grooming is also diminished by benzodiazepines, which do not affect ACTH- or β-endorphin-induced grooming in rats or mice (Dunn et al., 1981a). This would be consistent with a decreased stress response.

ACTH-induced grooming exhibits other opiate like characteristics. Although there is no loss of grooming potency with daily injections of ACTH (Colbern et al., 1978), the potency is diminished for up to 18 hours after a single injection (Jolles et al., 1978). There is complete cross reactivity between $ACTH_{1-24}$, $[D-Phe^7]ACTH_{4-10}$, β-endorphin, and morphine; however, prevention of the grooming with naloxone also blocks the "tolerance" (Jolles et al., 1978). Furthermore, the peptide injections may elicit a stress response, since sequences of ACTH injected ICV that elicit grooming also elevate plasma corticosterone, even though those same sequences, such as $ACTH_{1-16}$, and $[D-Phe^7]ACTH_{4-10}$, have no adrenocortical activity injected systemically (Wiegant et al., 1979b). Notably, morphine elicits a similar response (Van Vugt and Meites, 1980; Morley, 1981).

The site of action of ACTH in eliciting grooming is not known, but appears to be distinct from the site involved in eliciting stretching and yawning. Electrolytic lesions of the septum, preoptic area, mammillary bodies, amygdala, posterior thalamus, and dorsal or ventral hippocampus did not affect $ACTH_{1-24}$-induced grooming (ICV), but aspiration of the hippocampus did (Colbern et al., 1977). Amygdaloid or hippocampal lesions enhanced SYS. Small doses of $ACTH_{1-24}$ injected into the substantia nigra elicited grooming, whereas large doses in the striatum did not (Wiegant et al., 1977b). Grooming is also elicited shortly after third ventricle application of ACTH, whereas SYS is considerably delayed, and Gessa et al. (1967) considered it to be a fourth ventricle phenomenon. Recent data show that $ACTH_{1-24}$ injected into the lateral or third ventricle effectively induces grooming in rats in which the cerebral aqueduct has been blocked, indicating a third ventricle site. Using cold cream injections to block potential active sites, the crucial areas were found to be in the anteroventral part of the third ventricle (Dunn and Hurd, unpublished observation).

The significance of ACTH/endorphin-induced grooming is obscure. Nonessential grooming is considered by many to be a displacement behavior, which means that it is not functionally directed but has neutral value as perceived by other animals. One could speculate that displacement behaviors have biological value during stress. Thus stress elicits grooming, unless some more important behavior, such as fight or flight, overrides it. This would have the advantage of not signaling weakness or indecision to real or potential adversaries. If this speculation is correct, it is not obvious why the behavior should be ACTH- or β-endorphin mediated. It is significant, however, that enkephalins do not elicit grooming.

Catalepsy, Catatonia, and Explosive Motor Behavior

Two papers published simultaneously in 1976 indicated that β-

endorphin administered intraventricularly produced a state of
muscular rigidity often known as catalepsy. Bloom et al.
(1976) reported a catatonia was specific to β-endorphin (25 to
50 μg ICV); α-endorphin, β-endorphin, and met-enkephalin did
not show this response even though each of these compounds
caused loss of the corneal reflex. All these opiate peptides,
but not morphine, also induced acute "wet-dog" shakes, normally
a symptom of opiate withdrawal. Both the catatonia and the
wet-dog shakes were prevented by prior treatment with naloxone.
Jacquet and Marks (1976) reported that β-endorphin (2 or 4 μg
ICV) induced catalepsy, which was also observed after α-endorphin (50 μg), met-enkephalin (80 or 150 μg) and leu-enkephalin
(54 or 100 μg). They emphasized the similarities between
opiate- and neuroleptic-induced catalepsy. But the two forms
of catalepsy can be distinguished behaviorally (Segal et al.,
1977; de Ryck et al., 1980) or by striatal lesions, so that
Browne and Segal (1980) made a distinction between opiate-
induced catatonia and neuroleptic-induced catalepsy. Tseng et
al. (1979) reported catalepsy after DADLE, but the ED_{50} (25-50
μg ICV) was ten times higher than β-endorphin, and the
enkephalin-induced catalepsy lacked the muscular rigidity
caused by β-endorphin.

Subsequently, Jacquet (1978) reported that injection of
morphine (10 μg) or $ACTH_{1-24}$ (25-50 μg) into the periaqueductal
gray area produced explosive motor behavior (EMB), whereas
β-endorphin (4 μg) did not. Although the analgesia produced by
all three compounds was prevented by naloxone, the EMB produced
by morphine or ACTH was not. Since the EMB was accompanied by
other symptoms of opiate withdrawal, the authors suggested that
opiate withdrawal was caused by action of opiates on an
endogenous ACTH receptor insensitive to naloxone. α-MSH or
$ACTH_{4-10}$ also produced these withdrawal symptoms, but not EMB.
Since the doses of ACTH necessary to produce the EMB were very
high, it is doubtful if the receptor was a specific ACTH receptor. Nevertheless, there is ample evidence for multiple
opiate receptors, some of which may not be affected by naloxone
(Childers, 1980).

The possibility that abnormalities of endorphins might be
responsible for certain psychiatric disorders, and the use of
endorphins to treat psychiatric disorders is discussed in
chapter 8 by Koob et al. However, I will discuss the evidence
that certain endorphins have neuroleptic-like properties.

De Wied et al. (1978b) first reported that $β-LPH_{62-77}$
DTγE had certain neuroleptic-like properties. Principally,
these involved a positive "grip test," related to various
catalepsy tests and also shown by haloperidol. In addition,
γ-endorphin, DTγE (30 ng SC or 0.3 ng ICV) and low doses of
haloperidol facilitated extinction of pole-jump avoidance and

attenuated passive avoidance behavior. The peptides did not
affect open-field behavior even at much higher doses. The
minimum sequence necessary for these effects of DTγE is
β-LPH$_{66-77}$, clearly indicating a distinction from opiate-like
activity (de Wied et al., 1980). These observations led de
Wied et al. (1978b) to speculate that DTγE or some closely
related compound might be an endogenous neuroleptic. Support
for this concept arises from the reported antipsychotic
activity of DTγE in schizophrenic patients (see Koob et al.,
chapter 8).

Weinberger et al. (1979) severely criticized the proposal
that DTγE had neuroleptic properties. They reported that
DTγE had no significant effect on several behavioral tests used
to characterize neuroleptics, including amphetamine-induced
crossover activity, the time spent on the vertical grid or
horizontal bar, or on compartment entries or stimulus contact
or duration in the multicompartment chamber test. However,
DTγE did produce a few of the effects of haloperidol; notably
it suppressed rearings in the multicompartment chamber test,
and prolonged the latencies (not significantly) in the horizontal bar and vertical grid tests. Further support for the
DA-antagonist properties of DTγE are the self-stimulation
studies of Dorsa et al. (1979) and the receptor binding studies
of Pedigo et al. (1979b) (see later).

Thermoregulation

Bloom et al. (1976) first reported temperature changes in rats
injected intraventricularly with β-endorphin. In general, low
doses (1 to 10 μg) increase body temperature, whereas higher
doses (20 to 50 μg) markedly decrease it (Tseng et al., 1979;
Martin and Bacino, 1979). The enkephalins themselves do not
significantly alter body temperature, but DADLE produces a mild
hyperthermia in rats and hamsters (Tseng et al., 1979), and the
highly potent [D-Ala^2,MePhe^4,Met(O)^5-ol]enkephalin (FK 33-824)
produced both the low-dose hyperthermia and the high-dose hypothermia, even when administered subcutaneously (Bläsig et al.,
1979). Naloxone blocks the hypothermic response (Holaday et
al., 1978a) and according to many studies also the hyperthermic
response (Bläsig et al., 1979; Martin and Bacino, 1979). Small
doses of β-endorphin injected into the preoptic area cause
hyperthermia, which is naloxone sensitive in the early phases
and indomethacin-sensitive later (Martin and Bacino, 1979).
This later phase may thus be due to prostaglandin release
caused by the cannulation procedure and is probably unrelated
to opiates. β-Endorphin, like morphine, also produced a
naloxone-sensitive hypothermia when injected into the

subarachnoid space surrounding the spinal cord (Martin and Bacino, 1979).

Holaday et al. (1978a) suggested a role for the endorphins in thermoregulation, and that behaviors elicited by ICV endorphins were thermoregulatory behaviors, such as salivation and grooming. However, naloxone did not reduce pyrogen-induced hyperthermia or alter the adaptation to acute cold stress (Bläsig et al., 1979). Naloxone did, however, interfere with adaptation to heat (Holaday et al., 1978b). The stress of handling produces a hyperthermia, which is naloxone reversible (Bläsig et al., 1978). Since the handling produced an elevation of plasma β-endorphin, it was suggested that endorphins may cause hyperthermia when released during stress. However, they do not appear to perform an obligatory role in thermoregulation.

Doses of α-MSH or $ACTH_{1-24}$ greater than 1 to 25 μg ICV produced a slight (1°C) hypothermia in rabbits (Lipton and Glyn, 1980). $ACTH_{1-10}$ was ineffective.

Antinociception ("Analgesia")

The most important medicinal property of opiates is their ability to decrease pain. In the laboratory this is measured by a decreased responsiveness of animals to pain, normally applied heat, as in the "tail-flick" or "hot-plate" tests. Technically these tests measure nociception, or responsivity to pain; however, it is common to refer to the apparent decreased sensitivity to pain reflected by increased response latencies as "analgesia," even though pain perception is not necessarily lost, or even measured in this assay. Thus, in the following discussion, as in the literature it reviews, analgesia generally refers to increased tail-flick or hot-plate latencies.

The endorphins share the "analgesic" properties of opiates. Most investigators have detected elevated pain thresholds on the tail-flick or hot-plate tests after intraventricular injection of α-, β-, or γ-endorphin, or met- or leu-enkephalin in rats, mice, and cats (for reviews, see Terenius, 1978; Olson et al., 1979a), but the effect requires high doses of peptide and is often transitory. β-Endorphin is much more potent (100-fold) than the enkephalins, and met-enkephalin is two to three times more potent than leu-enkephalin. The potency of both β-endorphin and the enkephalins can, however, be increased dramatically by amino acid substitutions that prolong the biological stability of the molecules. Thus DADLE is as potent as β-endorphin (Wei et al., 1977). On a molar basis β-endorphin and the "stabilized" enkephalins are more potent than morphine. Inhibitors of peptide degradation, such as

captopril, also potentiate the analgesia (Stine et al., 1980).
Intrathecal β-endorphin (3 mg) produced profound and long-
lasting analgesia in human beings (Oyama et al., 1980).
Peripheral administration of the endorphins has generally been
less successful, although analgesia has been reported following
IV β-endorphin (Tseng et al., 1976), or stabilized enkephalins
(Kastin et al., 1979b), some of which are active orally, FK
33-824 (Roemer et al., 1977) and [D-Thr^2,Thz^5]enkephalinamide
(Tseng et al., 1978). The analgesic effects are in all cases
naloxone reversible.

ACTH appears to have an opposite effect on pain perception.
Gispen et al. (1976) showed that ACTH$_{1-24}$, ACTH$_{1-16}$, ACTH$_{5-16}$,
ACTH$_{5-14}$, and [D-Phe^7]ACTH$_{4-10}$ all significantly inhibited
morphine-induced analgesia (hot plate) in rats, as did ACTH$_{4-10}$
at high doses. ACTH$_{11-16}$ and ACTH$_{11-24}$ were inactive. This
potency correlated well with the ability of the ACTH sequences
to displace [^{3}H]dihydromorphine from cerebral opiate receptors.
Consistent with this, Bertolini et al. (1979) reported that ICV
ACTH$_{1-24}$ (20-50 µg) produced a hyperalgesia in two tests. This
hyperalgesia was antagonized by morphine and potentiated by
naloxone.

These results have led to speculation that the endorphins may
be endogenous analgesics. Basically two kinds of experiments
have been cited as evidence for this. Firstly, treatments
known to cause antinociception have been shown to result in
endorphin release. Secondly, the antinociception caused by
such treatments has been shown to be prevented by naloxone
(so-called "naloxone-reversible"). The experimental evidence
in both cases is equivocal, and it appears that the release of
endorphins can only partly explain endogenous or stress-induced
analgesia.

Following the initial demonstration by Madden et al. (1977)
that footshock increased tail-flick latencies in rats, a number
of experimenters have demonstrated that increased pain thresholds accompany stress in rats and mice (see reviews by Amir et
al., 1980; Bodnar et al., 1980a; Lewis et al., 1980a; Riley et
al., 1980). Stressors used include electric footshock, immobilization, cold-water swimming, glucoprivic stress (2-deoxy-
glucose or insulin administration), or hypertension. In one of
these studies, the cerebral content of endorphins (uncharacterized) increased in parallel with the analgesia (Madden et al.,
1977). Cold-water swimming, like morphine, induced Straub tail
elevation, an effect that was naloxone reversible (Katz,
1979a). Whether or not the "stress-induced analgesia" is prevented by naloxone depends upon the stressor, the test of
analgesia, and perhaps the species and investigator (see Bodnar
et al., 1980a; Lewis et al., 1980a). Naloxone blocked the effects of footshock stress in some studies (Madden et al., 1977;

Chesher and Chan, 1977); however, Lewis et al. (1980a) found that it blocked the analgesia (tail flick) following intermittent footshock (for 30 min), but not continuous footshock for 3 min. Naloxone also blocked antinociception following immobilizaton stress (Amir and Amit, 1978), food deprivation (McGivern et al., 1979), and hypertensive stress (Zamir and Segal, 1979), but was partially (Bodnar et al., 1980a) or totally (Lal et al., 1978) ineffective against cold-water swim analgesia and was totally ineffective for 2-deoxyglucose-induced analgesia (Bodnar et al., 1980a). Habituation of the analgesia to repeated stress occurred (Madden et al., 1977; Bodnar et al., 1978).

Cross tolerance with morphine varied; morphine tolerant mice did not exhibit footshock-induced analgesia (Chesher and Chan, 1977), but morphine-tolerant rats showed full cold-water swim analgesia and partial 2-deoxyglucose-induced analgesia (Spiaggia et al., 1979).

In a related paradigm, conditioned fear (of prior footshock) was shown to increase tail-flick latencies in rats (Chance et al., 1978). This change occurred with a concomitant decrease in available enkephalin-binding sites, thought to reflect increased occupancy by endogenous enkephalins. Interestingly, this antinociception was not prevented by hypophysectomy, naloxone (20 mg/kg IP), naltrexone (1 mg/kg IP), or diazepam (2.5 mg/kg IP), nor was there any cross tolerance with morphine (see review by Chance, 1980).

Electrical stimulation of the periaqueductal gray area induced analgesia in rats that was partly blocked by naloxone in some but not all studies (see Chance, 1980; Bodnar et al., 1980a). However, decreased pain sensitivity following electrical stimulation of the central gray in humans was prevented by naloxone (Hosubuchi et al., 1977). Moreover, similar stimulation resulted in the release of enkephalins (Akil et al., 1978a) and β-endorphin (Akil et al., 1978b; Hosubuchi et al., 1979) into CSF in parallel with the decreased pain sensitivity. The increased release of β-endorphin (more than 10-fold) far exceeded the increased release of enkephalin (two-fold). In other human studies, patients with chronic organic pain showed decreased CSF contents of uncharacterized endorphins (Almay et al., 1978) or enkephalins (Akil et al., 1978a). Electrical stimulation of the locus coeruleus and substantia nigra also induced an analgesia that was prevented by naloxone, by contrast with the self-stimulation produced at these same sites (Sandberg and Segal, 1978).

Endorphins have also been implicated in acupuncture analgesia; most studies finding electroacupuncture-induced analgesia to be prevented by naloxone and other opiate antagonists (see Terenius, 1978; Cheng and Pomeranz, 1980). The

analgesia was accompanied by an increased endorphin content of
mouse brain (Ho et al., 1978) and human CSF (Sjölund et al.,
1977) and plasma (Malizia et al., 1979). A recent parametric
study in mice found that 200 Hz stimulation (of the forepaw)
was more analgesic than 4 Hz stimulation, but naloxone
prevented only the analgesia produced by 4 Hz stimulation
(Cheng and Pomeranz, 1979). Analgesia induced by hypnosis is
controversial; no effect of naloxone was observed by Goldstein
and Hilgard (1975) and Barber and Mayer (1977), whereas Frid
and Singer (1979) observed marked reversal of analgesia.

Naloxone has been used to determine whether endorphins are
involved in "baseline" detection of pain. Following the observations of Jacob et al. (1974), most studies in "unstressed"
rodents have found that naloxone increased the sensitivity to
pain, but with a few exceptions (see Goldstein, 1979). However in humans, naloxone appears ineffective (see Grevert and
Goldstein, 1978), although one study found that pain-insensi-
tive subjects showed analgesia, whereas pain-sensitive subjects
showed hyperalgesia (Buchsbaum et al., 1977). The most likely
explanation is that endorphins are not released tonically, but
under conditions of arousal or stress, released endorphins decrease pain sensitivity. The handling and placing of rodents
in novel environments may be stressful enough to release
endorphins (see Dunn et al., 1979). In a series of patients
undergoing dental treatment, naloxone sensitivity has been
shown, but only in subjects showing significant "placebo-
induced" analgesia (Levine et al., 1978). Even within this
group there was a biphasic effect; low doses (< 2 mg IV) were
analgesic, whereas high doses (8 to 10 mg) were hyperalgesic
(Levine et al., 1979).

A further complication is the involvement of the pituitary
and adrenal in stress-induced analgesia. Various authors have
shown that hypophysectomy or adrenalectomy potentiates the
opiate activity of morphine (see Bodnar et al., 1980a). Hypophysectomy also potentiates the hyperthermic, cataleptic, and
analgesic effects of β-endorphin (ICV or IV), whereas adrenalectomy potentiates the effects of β-endorphin when administered IV; it does not affect them when β-endorphin is given
ICV (Holaday et al., 1977, 1979a). Dexamethasone reversed the
effect of adrenalectomy, which in the case of morphine is
probably due to decreased metabolism by liver enzymes induced
by dexamethasone (Holaday et al., 1979a). The increased
sensitivity to peripheral β-endorphin is unexplained. Chronic
ACTH treatment partially reverses the effect of hypophysectomy,
part of which is due to the lack of adrenal glucocorticoids
(Holaday et al., 1979b).

Hypophysectomy blocked analgesia induced by immobilization
(Amir and Amit, 1979). It also decreased analgesia caused by

cold-water or insulin stress, but potentiated that caused by
2-deoxyglucose stress or morphine (Bodnar et al., 1980a). The
latter might implicate release of brain endorphins, but 2-
deoxyglucose analgesia was not naloxone reversible. Acu-
puncture-induced analgesia was also prevented by hypophysectomy
(Pomeranz at al., 1977). Dexamethasone also blocked naloxone-
sensitive analgesia induced by footshock (Lewis et al., (1980).
In a recent study, Bodnar et al. (1980b) showed that neonatal treatment with MSG, which severely depleted hypothalamic POMC neurons, completely blocked the analgesia resulting from cold-water swim stress but failed to alter that
due to food deprivation or 2-deoxyglucose. Brattleboro rats,
which lack AVP (see chapter 9 by van Wimersma Greidanus and
Versteeg), are hypersensitive to footshock but do not show
cold-water swim-induced analgesia, although morphine-induced
analgesia is normal. Chronic vasopressin treatments reversed
these abnormalities (Bodnar et al., 1980a).
    Clearly, all analgesic phenomena cannot be explained by the
release of endorphins. Nevertheless, some undoubtedly are.
One factor adding to the complexity may be the release in
parallel with β-endorphin of ACTH, which is hyperalgesic.
Currently, studies rely almost exclusively on the use of systemic naloxone or naltrexone, usually in single arbitrary
doses. Local microinjections of opiate antagonists, including
antisera to enkephalins and β-endorphin will produce more
interesting and interpretable data. It is also clear that
endocrine secretions can influence algesia. However, although
in some cases the data are consistent with pituitary secretions
mediating the analgesia, this is not the only mechanism of
stress-induced analgesia and there are complex interactions
between the pituitary-adrenal hormones and the effects of
endorphins and stress.

Self-Stimulation

Belluzzi and Stein suggested that endogenous opiates may play a
role in reward mechanisms (Stein and Belluzzi, 1979). As evidence they showed that rats would bar press to receive intraventricular injections of morphine, or leu- or met-enkephalin
(1 to 10 µg per injection). Sites in the brain rich in
enkephalins are good sites for self-stimulation, including
sites not previously shown to be active in this respect, such
as nucleus paratenialis thalami and globus pallidus. And,
self-stimulation in the central gray is blocked by naloxone.
Such a role for endorphins would be consistent with the eu-
phoric properties of morphine and the apparently immutable
tolerance. FK 33-824 also supported self-stimulation (Mello

and Mendelson, 1978). In partial support of Stein and
Belluzzi, van Ree et al. (1979) showed self-stimulation for
β-endorphin but were unable to obtain any effect with met-
enkephalin. Dorsa et al. (1979) reported that DTγE, like
haloperidol, decreased self-stimulation with ventral tegmental
area electrode placements, whereas α-endorphin, like amphetamine, enhanced self-stimulation. Self-administration of
ethanol was also decreased by naltrexone (Altshuler et al.,
1980). Self-administration of ACTH and $ACTH_{4-10}$ (IV) has also
been reported (Jouhaneau-Bowers and Le Magnen, 1979).

Learning and Extinction of Learning

Active Avoidance

Hypophysectomized rats are deficient in the acquisition of
active avoidance; Appelzweig and Baudry (1955) first demonstrated this by using a shuttle box and showed that the deficit
could be reversed by chronic treatment with ACTH. De Wied
(1964) replicated this effect and subsequently showed that the
action of ACTH was independent of adrenal activation with the
use of analogues of ACTH (see de Wied et al., 1975; de Wied,
1977). The sequence $ACTH_{4-10}$ (1 μg SC) was fully active,
although it lacked significant adrenocortical activity.
Whereas these effects of ACTH were all chronic, Gold et al.
(1977) showed that the deficiency of hypophysectomized rats in
learning an aversive Y-maze discrimination could be reversed by
post-training ACTH injections (0.3 IU).

Under appropriate conditions, ACTH can stimulate acquisition
of active avoidance in intact animals. Pretraining ACTH stimulated acquisition of avoidance behavior in both intact and
adrenalectomized rats (Bohus et al., 1968). Similar effects
were obtained by Beatty et al. (1970) with ACTH and Stratton
and Kastin (1974) with α-MSH on shuttle-box behavior. Post-
training $ACTH_{4-10}$ (0.3 mg/kg SC) stimulated acquisition of
T-maze learning in mice, whereas $[D-Phe^{7}]-ACTH_{4-10}$ (1-3 mg/kg)
impaired it (Flood et al., 1976). Post-training ACTH in rats
altered Y-maze acquisition in a dose-dependent manner; low
doses (1 IU/kg) stimulated, whereas high doses (1000 IU/kg)
inhibited retention (Sands and Wright, 1979). α-MSH stimulated
acquisition of reversal learning in rats (Sandman et al., 1972,
1973a). The acquisition was only facilitated on an intradimensional shift; α-MSH impaired performance of an extradimensional shift (Sandman et al., 1974). Intradimensional shifts
involve changing the correct stimulus, but in the same dimen-
sion (modality), such as color. An extradimensional shift

might involve changing a color discrimination to one for shape, for example.

More recently, peripheral administration of enkephalins has been reported to impair acquisition of active avoidance in rats; an effect that was not prevented by reasonable doses of naloxone (Rigter et al., 1980). Post-training administration of leu-enkephalin or β-endorphin (10 μg/kg IP), like morphine (1 mg/kg IP), impaired retention for shuttle box in rats (Izquierdo et al., 1980); naloxone (0.4 mg/kg IP) facilitated performance in the same tasks (Izquierdo, 1979); see also Messing et al., 1979).

Much work has concerned extinction of active avoidance, which is facilitated by removal of the entire pituitary or the posterior lobe (de Wied, 1965). ACTH can reverse this deficit (de Wied, 1965) and retard extinction in intact rats (Murphy and Miller, 1955; de Wied, 1966; Bohus et al., 1968). The effects were clearly extra-adrenal, since ACTH delayed extinction in adrenalectomized rats (Miller and Ogawa, 1962), and analogues of ACTH weakly active on the adrenal cortex, such as ACTH$_{4-10}$, were fully active (Greven and de Wied, 1973). Furthermore, the effect of glucocorticoids on extinction was opposite; corticosterone and dexamethasone facilitated extinction in normal and hypophysectomized rats (de Wied, 1967). [D-Phe^7]ACTH$_{4-10}$ also facilitated extinction (Greven and de Wied, 1973). The minimum sequence for retarding extinction is ACTH$_{4-7}$, and analogues of greatly increased potency (100 times) have been synthesized, such as [Met^{4}SO$_2$,D-Lys^8,Phe^9]-ACTH$_{4-9}$ (ORG 2766) (de Wied et al., 1975). Surprisingly, it has recently been reported that ACTH$_{11-24}$ and ACTH$_{25-39}$ also delay extinction, although 10-fold higher doses are necessary (Greven and de Wied, 1977). α-MSH also retards extinction of active avoidance behavior (Sandman et al., 1971). Kendler et al. (1976) have also shown that ACTH prolongs extinction in a lithium chloride-induced taste-aversion paradigm. Consistent with a physiological role for ACTH and/or α-MSH in extinction, antisera to either of these molecules injected ICV facilitated extinction, especially when both antisera were used (van Wimersma Greidanus et al., 1978).

ACTH and MSH peptides are not the only ones that can affect active avoidance. Vasopressin can also restore deficient acquisition in hypophysectomized rats and retard extinction in intact ones, and it is generally more potent than ACTH in these respects (see de Wied, 1977); van Wimersma Greidanus and Versteeg, chapter 9). However, ACTH$_{4-10}$ stimulates passive avoidance behavior in Brattleboro rats, so the effect is not mediated by vasopressin release (Bailey and Weiss, 1978). Moreover, POMC-derived peptides, other than ACTH$_{4-10}$(β-LPH$_{47-53}$) and α-MSH are active. Met-enkephalin, α- and β-endorphins, and

β-LPH$_{61-69}$(SC) were as active as or better than ACTH$_{4-10}$ in facilitating extinction of pole-jump avoidance (de Wied et al., 1978a), α-endorphin being the most active (30 times as active as ACTH$_{4-10}$, 0.1 g being effective). Strikingly, γ-endorphin and DTγE (β-LPH$_{62-77}$) facilitated extinction, the latter being about three times as active as the former (de Wied et al., 1978b). Other active peptides include LH-RH and thyrotropin-releasing hormone (TRH) at doses similar to effective doses of ACTH (de Wied et al., 1975) leading one to doubt the specificity of this effect. Moreover, naltrexone also facilitated extinction and this effect was antagonized by ACTH$_{4-10}$, ORG 2766, and α- and β-endorphin (de Wied et al., 1978b).

These results have been controversial, ironically because of the "minute" peripheral doses of peptides involved. However, the opposite effects of α- and β-endorphin on the extinction of pole-jump avoidance have recently been confirmed by Le Moal et al. (1979) (see chapter 8 by Koob et al.).

Studies on the localization of the ACTH action on the extinction of active avoidance behavior have implicated the caudal thalamus. Bilateral lesions of the parafascicular nucleus abolished the ability of ACTH peptides to retard extinction (Bohus and de Wied, 1967). Van Wimersma Greidanus and de Wied (1971) found that local administration of ACTH$_{1-10}$ into the rostral mesencephalon or caudal thalamus retarded extinction of pole-jump avoidance, whereas [D-Phe[7]]ACTH$_{1-10}$ facilitated extinction in these same loci. However, the peptides were also active anywhere in the ventricular system. More recently, lesions of the rostral septum, dorsal hippocampus, amygdala, or fornix (see van Wimersma Greidanus et al., 1979a) were also found to block the effect of ACTH$_{4-10}$ on extinction.

Passive Avoidance

ACTH$_{1-24}$ given prior to acquisition improved retention of passive avoidance in rats, but only at low footshock intensities (Lissák and Bohus, 1972). Flood et al. (1976) found similar effects in mice; post-training injections of ACTH$_{4-10}$ (0.1-1.0 mg/kg SC) at low footshock levels facilitated retention in a time- and dose-dependent manner, whereas [D-Phe[7]]ACTH$_{4-10}$ (0.3-3.0 mg/kg) was ineffective. At high footshock levels, [D-Phe[7]]ACTH$_{4-10}$ (0.3-3.0 mg/kg) inhibited retention, time and dose dependently, whereas ACTH$_{4-10}$ was ineffective. Gold and Van Buskirk (1976) showed that post-training injections of ACTH could facilitate subsequent performance in rats, but that there were interactions between the footshock intensity and the dose of ACTH. An inverted U-shaped dose-response curve was obtained in which moderate doses of ACTH facilitated performance, and high doses inhibited it. At higher footshock intensities, when performance was improved, ACTH was only inhibitory. Gold and McGaugh (1977) interpreted these results in terms of hormonal modulation of memory storage; the endogenous release of hormones added to the effects of exogenous ACTH. ORG 2766 (5 mg/kg SC) facilitated acquisition only when administered prior to training; immediately post-training or one hour prior to retention testing was ineffective (Martinez et al., 1979). These results contrast with those of the Utrecht group who found that ACTH$_{4-10}$ facilitated passive avoidance performance only when administered shortly before retention testing (van Wimersma Greidanus, 1977; see also Bailey and Weiss, 1978). In support of this, antisera to α-MSH or ACTH$_{1-24}$, injected ICV shortly before retention testing, impaired performance, but were ineffective after training (van Wimersma Greidanus et al., 1978). This is consistent with a release of ACTH during retention testing in proportion to the step-through latency, that is, the extent of retention (van Wimersma Greidanus et al., 1977).

Gray (1975) argued that the effect of ACTH on passive avoidance was a case of state-dependency, but his data did not substantiate a simple state-dependency model.

α-Endorphin also facilitated passive avoidance when given (SC) to rats 1 hour before retention testing, being 10 times more potent than ACTH$_{4-10}$ (de Wied et al., 1978a). Injected immediately after training, α-endorphin also facilitated performance, whereas γ-endorphin had no effect, and DTγE inhibited it (de Wied et al., 1978b). Subsequently, Kovács et al. showed that DTαE (β-LPH$_{62-76}$) had an action similar to α-endorphin, either immediately post-training or prior to testing, DTαE facilitated performance, whereas DTγE inhibited it (see Bohus, 1980). It was suggested that this might explain the varied results with β-endorphin, which normally facilitated performance but could also inhibit it. *In vivo*, β-endorphin is degraded first to γ-endorphin and then to α-endorphin, giving mixed results depending on the dose and the extent of the degradation. Naltrexone did not alter the facilitation by β-endorphin given prior to retention. In a similar paradigm, met- but not leu-enkephalin facilitated performance given immediately after training, and this effect was not prevented by naloxone (Stein and Belluzzi, 1979). Surprisingly, Messing et al., (1979) reported that naloxone (0.5 mg/kg IP) given post-training facilitated performance.

ACTH peptides also affect extinction of passive avoidance. Levine and Jones (1965) found that ACTH prolonged avoidance of a water-reinforced bar press task after punishment. Anderson et al. (1968) confirmed this result in hypophysectomized rats.

α-MSH retarded extinction of step-down passive avoidance in rats (Datta and King, 1977). Surprisingly, $ACTH_{4-10}$ did not retard extinction of passive avoidance when administered one hour before the first retention test, but $[D-Phe^7]ACTH_{4-10}$ did (Greven and de Wied, 1973). Moreover, $ACTH_{4-10}$, $[D-Phe^7]-ACTH_{4-10}$ and α-MSH all delayed extinction of a conditioned taste aversion (Rigter and Popping, 1976).

Peptides also reverse amnesia for passive avoidance caused by a variety of agents. Flexner and Flexner showed that puromycin-induced amnesia in mice could be reversed by corticotropin gel, but the effect was later attributed to des-glycinamide lysine vasopressin (DG-LVP) (Lande et al., 1972). Rigter et al. (1975) then showed that CO_2-induced amnesia for passive avoidance in mice could be reversed by $ACTH_{4-10}$ or $[D-Phe^7]-ACTH_{4-10}$ but not $ACTH_{11-24}$ given one hour before retention, but not before testing. Similar results were found for ORG 2766. This contrasted with analogues of vasopressin which were active, administered before training or testing (Rigter et al., 1975). Subsequently, an effect similar to ACTH-analogues was shown for β-LPH or leu-enkephalin, but met-enkephalin was active given either before training or testing (Rigter, 1978). The effect of enkephalins given prior to testing was not naloxone reversible. Other investigators have found ACTH given immediately after training to be effective in reversing ECS-induced (Keyes, 1974) or anisomycin-induced amnesia ($ACTH_{4-10}$) (Flood et al., 1976). In the latter study, $[D-Phe^7]ACTH_{4-10}$ had a hyperamnesic effect.

Appetitive Learning

POMC peptides also influence appetitive behavior. Facilitation of acquisition of complex tasks has been reported for α-MSH (Stratton and Kastin, 1975), $ACTH_{4-10}$ (Isaacson et al., 1976), met-enkephalin and $[D-Ala^2]$enkephalin (Kastin et al., 1976), and $[D-Ala^2,F5Phe^4]$enkephalinamide (Olson et al., 1979a). However, like active avoidance training, extinction seems to be more sensitive. Extinction of food-deprivation tasks was retarded by α-MSH (Sandman et al., 1969; Kastin et al., 1974), ACTH (Guth et al., 1971), ACTH, $ACTH_{4-10}$ and $ACTH_{1-24}$, but enhanced by corticosterone or $[D-Phe^7]ACTH_{4-10}$ (Garrud et al., 1974) (see also de Wied, 1977). Extinction of sexually motivated behavior was also retarded by $ACTH_{4-10}$ (Bohus et al., 1975). There is a conflict of data regarding extinction of a water-running task. Le Moal et al. (1979) reported that DTαE and DTγE both retarded extinction; however, Kovács et al. (see Bohus, 1980) found a facilitation.

Operant Learning

ACTH peptides also affect operant behavior. ACTH enhanced conditioned suppression but only in the morning (Schneider et al., 1974), acquisition of a lever-pressing response (Guth et al., 1971), and free operant avoidance (Wertheim et al., 1967). ACTH also retarded extinction of frustrative nonreward (Gray and Garrud, 1977). In this case, the effect closely resembled that on the extinction of active avoidance; $ACTH_{1-24}$ and $ACTH_{4-10}$ retarded extinction, whereas $[D-Phe^7]ACTH_{4-10}$ and corticosterone had opposite effects.

Imprinting

Landsberg and Weiss (1976) reported that stress or ACTH administration inhibited imprinting in ducklings, an effect they attributed to corticosterone. In a more careful analysis, Martin (1979) showed that $ACTH_{1-10}$ facilitated the imprinting approach response, whereas $ACTH_{1-24}$ did not. This would be explained by corticosterone antagonism of the primary ACTH effect.

Effects of ACTH have been reported in a variety of species other than mammals and birds. Enhancement of learning or habituation responses has been reported in lizards (Stratton and Kastin, 1975), toads (Horn et al., 1979), and beetles (Shelman et al., 1978).

General Discussion

In the foregoing a variety of evidence has been summarized that ACTH, MSH, and the endorphins can influence learning. It is clear that hypophysectomized animals have learning deficits that can be corrected by chronic ACTH treatment, suggesting that ACTH or MSH has a trophic or maintenance effect on learning mechanisms. However, the role of ACTH is not essential, because hypophysectomized animals can still learn at least some tasks, and because AVP can also reverse the effects of hypophysectomy. The effects of ACTH/MSH on acquisition were less dramatic and appear only when the learning would otherwise have been weak. It appears that under these circumstances ACTH can substitute for behavioral arousal, although it is not clear whether this is a physiological role of ACTH. Catecholaminergic drugs, like amphetamines, or arousal by tail-pinch would have similar effects.

Facilitatory effects of ACTH or MSH were commonly observed when the hormones were administered prior to testing. Since this is unlikely to be specific to the learning, it seems that once again the role of ACTH/MSH is that of arousal. The work of Sandman and Kastin and their colleagues is particularly relevant. In animals and man in reversal learning, ACTH and MSH facilitate intradimensional shifts, but inhibit extradimensional shifts. This suggests a focusing on the task at hand, paying careful attention to certain dimensions of the stimulus but not others. This analysis led to the concept that ACTH/MSH increases selective attention (Kastin et al., 1975; LaHoste et al., 1980). Currently, this hypothesis appears to explain very well most of the experimental data (see discussion by LaHoste et al., 1980).

The extinction data are by far the most difficult to interpret. Retardation of extinction can be regarded as perseverance or even the inability to "unlearn" a task. Although extinction of active avoidance is extremely sensitive to minute doses of peptides and is therefore excellent for screening purposes, it is not readily adapted to behavioral analysis. There is also some serious question about its specificity. Although initially Greven and de Wied (1973) indicated that the effect was specific for the ACTH$_{4-7}$ sequence, subsequently, not only have other sequences in the ACTH molecule been shown to be active, but other hormones, AVP, LH-RH and TRH, and several endorphins too (de Wied et al., 1975; de Wied and Gispen, 1977). The effect of AVP is particularly unusual, because one injection is sufficient, whereas repeated injections of ACTH and the other peptides are necessary (de Wied, 1977). The effect of endorphins, since it is not prevented by naloxone, is probably not mediated by an opiate receptor.

In particular instances, the effects of the peptide administration can be considered as aversive or reinforcing stimuli (see LaHostè et al., 1980; and Riley et al., 1980 for discussion). Nevertheless, there is sufficient evidence to indicate that the effects of peptides are not task specific. In fact, the consistency of the peptide effects across tasks is rather remarkable. In general, these effects are consistent with an action of ACTH/MSH to increase arousal or selective attention. There is no good evidence for an effect on consolidation or retrieval distinct from this, nor for a specific effect on learning.

Anxiety

Chronic treatment with benzodiazepines (anxiolytics) stabilizes social interaction in rats in a situation in which unfamiliar stimuli decrease the interaction. Since ACTH$_{1-24}$ (50-75 µg/kg IP) decreased the social interaction, File and Vellucci (1978) suggested that ACTH may be anxiogenic and that benzodiazepines may act by blocking the ACTH receptor. But benzodiazepines do not affect ACTH-induced grooming (Dunn et al., 1981a). The effect of $ACTH_{1-24}$ is apparently independent of the adrenals, since $ACTH_{4-10}$ (40 µg/kg, IP) is effective, and corticosterone is not. Also, $ACTH_{1-24}$ (1.25 µg), $ACTH_{4-10}$, and $[D-Phe^7]$ $ACTH_{4-10}$ (1.6 µg) were effective ICV (File and Clarke, 1980). An anxiogenic role for ACTH would be consistent with its induction of explosive motor behavior (Jacquet, 1978). However, in humans, $ACTH_{4-10}$ reduced anxiety (Miller et al., 1974, 1976; Sandman et al., 1975), and Sahgal et al. (1979) found no effect of this peptide on a conflict schedule in pigeons. The apparent complexity of the results almost certainly reflects the differences between the behavioral paradigms.

Aggression

There is ample evidence for a role of pituitary-adrenal hormones in aggressive behavior (see Leshner, 1978, pp 85-87 for a review). In general, circulating glucocorticoids correlate positively with aggression (see Rees and Gray, chapter 14). However, ACTH administered acutely stimulates fighting between male mice, and since $ACTH_{1-10}$ is active there may be an extra-adrenal component to this activity (Brain and Evans, 1977). By contrast, chronic ACTH decreases aggressiveness, and this may account for the effects of adrenalectomy (Leshner, 1978). Both ACTH and glucocorticoids are active on submissive behavior, corticosterone increases avoidance of attack, whereas ACTH enhanced retention of the response (Moyer and Leshner, 1976). In defeat, decreases in ACTH apparently mediate the loss of aggressiveness, whereas decreases of corticosterone increase submission (Nock and Leshner, 1976).

Human Behavior

Effects of ACTH-like peptides on human behavior generally resemble those in animals (Bohus, 1979). Kastin et al. first postulated that α-MSH might increase attention in man. Later they showed that performance in intradimensional reversals was improved, whereas that in extradimensional shifts was impaired (Kastin et al., 1975). Systematic studies of a number of parameters following $ACTH_{4-10}$ were initiated by Miller et al. (1974). Male medical students showed improved performance in visual memory tests and a decreased anxiety level following 10 mg $ACTH_{4-10}$ (IV) but not 0.5 mg $ACTH_{1-24}$. These results with $ACTH_{4-10}$ were confirmed by Sandman et al. (1975) and Miller et al. (1976, 1980) using male subjects. When female subjects

were used, visual memory was not increased, but verbal memory
was, and anxiety was not decreased (Veith et al., 1978a). In a
study designed to resolve this conflict, male and female subjects both showed improvements with $ACTH_{4-10}$ in an item recognition test (Ward et al., 1979). There were no sex differences
and no effect on anxiety. These data were taken to indicate
that the peptide induced increased attention (Sandman et al.,
1977; Ward et al., 1979; see also van Riezen et al., 1977,
LaHoste et al., 1980). This was consistent with the ability of
$ACTH_{4-10}$ or ORG2766 to counteract the usual decay in performance observed in reaction tasks (Gaillard and Varey, 1979).
$ACTH_{4-10}$ did not affect either the acquisition or the extinction of a conditioned avoidance response (Miller et al.,
1977b). $ACTH_{4-10}$ has also been found to have small beneficial
effects on the elderly (Miller et al., 1980), the mentally
retarded, and those with organic brain syndrome (Branconnier et
al., 1979), but it did not prevent the amnestic effects of ECS
(for other references, see Bohus, 1979). Chronic treatment
with ORG2766 (5-10 mg SC) improved attention and performance in
retarded adults (Sandman et al., 1980) Unexpectedly, the drug
also increased communication and sociability in these subjects.
   Few studies have yet been performed with endorphins in
humans. β-Endorphin (440 µg/kg over 30 minutes IV) produced
some analgesia and mild improvement of mood in two out of three
terminal cancer patients (Catlin et al., 1977). It also
reversed the symptoms in two patients undergoing withdrawal
from methadone. Neither catalepsy nor euphoria was observed.
When the potent enkephalin analogue, FK 33-824, was given to 40
healthy males (0.1-1.2 mg IM), the major symptoms were "a
feeling of 'heaviness' in all the muscles, often combined with
a feeling of oppression on the chest or tightness of the
throat" (van Graffenried et al., 1978). No cardiovascular
changes were noted, nor were any of the clinical symptoms of
morphine observed. Dose-dependent increases in PRL and GH
secretions were observed with some significant changes in the
EEG spectrum.
   For effects on schizophrenia and other clinical states, see
Koob et al. (chapter 8) and the review by Olson et al. (1980).

NEUROCHEMICAL EFFECTS

A comprehensive review of the neurochemical effects of ACTH and
its analogues was reported by Dunn and Gispen (1977), including
a speculative model for ACTH action on brain cells.
   Cerebral receptors for β-endorphin (Law et al., 1979) and the
enkephalins (Childers et al., 1979) have been identified in
brain. Their distribution and properties largely parallel

those of receptors for other opiate ligands, although there may be some important differences (Childers et al., 1979; Chang et al., 1979). The current view is that β-endorphin is the endogenous ligand for the μ-receptor, the enkephalins for the δ-receptor, and dynorphin for the κ-receptor (Childers, 1980; Chavkin et al., 1982). It is pertinent that not all β-endorphin receptors bind naloxone (Hazum et al., 1979). Binding of enkephalin to synaptosomes has been reported to be inhibited by neuroleptics (Somozo, 1978), although enkephalins or β-endorphin did not inhibit [^{3}H]haloperidol binding (Meltzer and So, 1979).

To date, there has been no real success in identifying ACTH or MSH receptors. This may be because the concentration of such receptors is so low as to be close to or below the limit of detection by current techniques (Witter, 1980). Studies with iodinated ligands are probably invalid, since iodination destroys the biological activity of MSH (Heward et al., 1979). ACTH has a weak affinity for opiate receptors, determined by [^{3}H]dihydromorphine or [^{3}H]naltrexone binding (Terenius et al., 1975), and for DA receptors, determined by [^{3}H]haloperidol binding (Czlonkowski et al., 1978). Akil et al. (1980) determined that $ACTH_{1-24}$ was a far better competitor for β-endorphin binding than it was for naloxone binding. Uptake of a labeled ACTH analogue ([H]ORG 2766) showed some small preference for the septal area (Verhoef et al., 1977; Rees et al., 1980).

Cerebral Blood Flow and Glucose Uptake

Cerebral blood flow is altered by α-MSH and ORG 2766 (40 μg/kg IV) (Goldman et al., 1975, 1979); in both cases there was a general decrease in cerebral blood flow with a selective sparing of the occipital cortex. Glucose uptake measured by the 2-deoxyglucose procedure was not altered following peripheral injections (SC) of $ACTH_{4-10}$, [DPhe^7]$ACTH_{4-10}$, α-MSH, $ACTH_{1-24}$ or ORG 2766 in mice (Delanoy and Dunn, 1978). However, ICV $ACTH_{1-24}$ (1 μg) consistently caused a relative decrease in deoxyglucose uptake in pyriform cortex (Dunn et al., 1980; Dunn and Hurd, 1982). Less consistent decreases were observed in the olfactory bulb, amygdala, and frontal cortex, and increases in thalamus and cerebellum. Many of these changes were correlated with the grooming behavior elicited by this peptide, and may well be associated with it because no regional changes were observed following administration of $ACTH_{4-10}$ (1 μg ICV) or by $ACTH_{1-24}$ in naloxone-pretreated mice (Dunn and Hurd, 1982). McCulloch et al. (1982) found that chronic ORG 2766 produced significant

increases in deoxyglucose accumulation in parts of the
hippocampus, anterior nucleus of the thalamus, and cingulate
cortex of the rat, although Dunn and Hurd (unpublished observations) found a significant regional decrease only in the
septal area of the mouse. β-Endorphin (4-8 μg ICV) decreased
glucose uptake in the superior colliculus, and both the medial
and lateral geniculate bodies (Sakurada et al., 1978). However,
Dunn and Hurd (1982) did not observe any regional changes
following naloxone (1 mg/kg SC) in mice.

RNA and Protein Synthesis
[For a detailed review, see Dunn and Schotman, 1981.]

Hypophysectomy decreased the RNA content of the rat brain stem
(Gispen et al., 1972), and decreased the incorporation of
[^{3}H]uridine into RNA (Gispen et al., 1970). These decreases
were not reversed by ACTH (Schotman et al., 1972, 1976).
    In intact animals, high doses of $ACTH_{1-24}$ (50 IU/kg SC)
slightly decreased the brainstem content of RNA and the incorporation of [^{3}H]uridine into brainstem RNA in mice (Jakoubek
et al., 1972) and rats (Gispen and Schotman, 1976), but $ACTH_{1-10}$
(0.2 or 0.5 μg/kg SC) was ineffective (Gispen and Schotman,
1976). Curiously, in adrenalectomized rats, $ACTH_{1-24}$ stimulated the [^{3}H]uridine incorporation into RNA, whereas $ACTH_{1-10}$
did not (Gispen and Schotman, 1976). Lower doses of ACTH (10
or 100 mU SC) did not alter the [^{3}H]uridine incorporation into
mouse brain RNA (Dunn and Kinnier, see Dunn and Gispen, 1977).
Unfortunately the incorporation of precursors into RNA is a
very insensitive technique and it is entirely possible that
ACTH and its analogues stimulated the production of particular
types of RNA, such as specific messenger RNAs. These could
only be detected, if at all, by extensive hybridization studies
that have yet to be performed. The interpretation of RNA
labeling data is also fraught with difficulties, since, when
short pulses of tracers are used, the principal type of RNA
labeled is heterogeneous nuclear (hn)RNA, which is a small
proportion of the total. Moreover, this hnRNA has a very short
half-life (10 to 15 minutes), so that its true synthesis rate
cannot be measured, and, although some of it is undoubtedly a
precursor to messenger RNA, much of it never enters the cytoplasm but is rapidly degraded (Dunn, 1976).
    Related to these RNA effects may be the metabolism of poly-
amines. In a recent study, SC $ACTH_{1-24}$ increased the putrescine and spermidine but not the spermine content of mouse
telencelphalon 6 to 24 hours after injection (Tintner et al.,
1979). A lesser effect occurred with $ACTH_{4-10}$ and the effect
was dependent on intact adrenals, although it was not mimicked

by corticosterone. ODC activity was also increased 6 hours following ICV $ACTH_{1-24}$ (Tintner et al., unpublished observations).

Hypophysectomy decreases the incorporation of amino acids precursors into protein *in vivo* (Versteeg et al., 1972). This can most likely be interpreted as a deficit in protein synthesis, since diencephalic slices *in vitro* show a similar defect (Reith et al., 1974), as do cell-free systems prepared from hypophysectomized rats (see Dunn and Schotman, 1981). Chronic treatment with $ACTH_{1-10}$ or $ACTH_{4-10}$ (10 μg ZP SC, every other day for 12 days) can reverse the *in vivo* protein synthesis deficit (Schotman et al., 1972). Moreover, $ACTH_{1-10}$ *in vitro* could reverse the deficit in brain slices (Reith et al., 1974). [D-Phe⁷]$ACTH_{4-10}$ only exacerbated the *in vivo* deficit (Schotman et al., 1972), whereas *in vitro* this analogue had no effect on the incorporation by brain slices (Reith et al., 1975b).

The proportion of ribosomes found as polyribosomes was decreased in the brains of hypophysectomized rats (Gispen et al., 1970). Neither chronic $ACTH_{1-10}$ treatment (0.2 mg/kg ZP SC, every other day for 10 days), nor shuttle-box training alone altered the decreased polyribosomal aggregation. However, chronic ACTH treatment in conjunction with conditioned avoidance training restored the normal state of ribosomal aggregation in hypophysectomized rats (Gispen et al., 1971). Thus, the restoration of the acquisition of avoidance behavior occurred in parallel with the restored polyribosomal aggregation. Taken together these results suggest that at least a part of the deficit in hypophysectomized rats is due to a lack of ACTH.

Effects of ACTH on cerebral protein synthesis have also been observed in intact animals. In rats and mice, ACTH (50 IU/kg SC) increased the incorporation of [¹⁴C]leucine into brain proteins (Semiginovský and Jakoubek, 1971). In mice, amino acid incorporation into brain proteins was increased by ACTH or α-MSH (1-3 mg/kg IP) (Rudman et al., 1974). $ACTH_{1-24}$ (0.6 mg/kg SC) increased the incorporation of [³H]lysine into brain and liver proteins, whereas $ACTH_{4-10}$ (0.3 mg/kg SC) was only effective on brain proteins (Dunn et al., 1978a). Administered ICV, $ACTH_{1-24}$ (15 μg) also increased the incorporation of [³H]lysine into brain protein (Rees et al., 1976); [D-Phe⁷] $ACTH_{4-10}$ (5 μg) had a similar effect, but $ACTH_{4-10}$ did not, paralleling the effects of the peptides on the induced stretching and yawning and excessing grooming behaviors.

Some of the studies have attempted to relate the effects of ACTH to those of stress. The responses of amino acid uptake and incorporation into protein in the brains and livers of mice to $ACTH_{1-24}$ injection closely resemble those to footshock or other stressors (Dunn et al., 1978a). That ACTH may be the

physiological mediator of the stress-induced changes is
suggested, since they were blocked by hypophysectomy (see Dunn
and Schotman, 1981) but not by adrenalectomy and were not
mimicked by corticosterone (Rees and Dunn, 1977). However,
ACTH may not completely account for the responses. In rats
similar parallels between the effects of ACTH and stress have
been observed. ACTH treatment *in vivo*, like stress, reduced
the incorporation of amino acids into protein by brain slices
(Jakoubek et al., 1970). However, the major factor involved in
this effect (an increased uptake of free amino acids) was
apparently due to corticosterone, since adrenalectomy prevented
this stress-induced effect and enhanced the labeling of protein
(Semiginovský et al., 1974). This suggests that ACTH may
enhance protein synthesis, whereas corticosterone not only
increases precursor uptake, but also interferes with the
response to ACTH.

Zomzely-Neurath and Keller (1977) reported that acute treatment of rats with $[Met^4SO_2,D-Lys^8,Phe^9]ACTH_{4-10}$ increased the
cerebral content of neuron-specific protein (NSP, identical
with antigen-α, 14.3.2, and enolase) and S-100. This effect of
the ACTH analogue mimicked the effect of avoidance training on
NSP.

Analysis of the proteins labeled *in vivo* by polyacrylamide
gel electrophoresis suggests that a rather broad spectrum of
soluble and membrane proteins is synthesized. Hypophysectomy
produced a rather general decrease in the labeling of most
protein species separated in this way, and chronic treatment of
hypophysectomized rats with $ACTH_{1-10}$ produced a nonselective
increase in labeling (Reith et al., 1975a). A variety of
different treatments of mice with ACTH ($ACTH_{1-24}$ or ORG 2766,
SC or ICV) did not alter the polyacrylamide gel pattern of
brain proteins synthesized (Dunn and Gildersleeve, 1980).

A consensus suggests then that the rate of protein synthesis
in brain can be altered by ACTH or $ACTH_{4-10}$. Since only ACTH
altered RNA metabolism, it could be speculated that, whereas
the $ACTH_{4-10}$ sequence is sufficient to produce translational
effects, the full $ACTH_{1-24}$ sequence is necessary for transcriptional ones. Although biochemistry texts emphasize control
of gene expression at the transcriptional level (so that the
synthesis of proteins is regulated by the availability of
messenger RNA), in eukaryotic systems, extensive evidence
exists for control at the translational level. It is possible
that feedback control exerted at the ribosome could secondarily
alter messenger RNA production. Thus translational control
could affect both RNA and protein metabolism.

No effects of endorphins on nervous system RNA or protein
metabolism have been reported.

Cyclic Nucleotides and Protein Phosphorylation.
[See recent reviews by Gispen et al. (1979) and Wiegant et al. (1981).]
Currently it is believed that peptide hormones are most likely to interact with membrane receptors on the outer surface of cells and not to penetrate them. Following such reception, an intracellular adenylate cyclase is presumed to be activated and the cyclic AMP (cAMP) so produced acts as an intracellular second messenger, promoting various metabolic activities, principally by altering the phosphorylation state of crucial proteins. This is the presumed sequence of events in cells of the adrenal cortex, where the cAMP apparently activates the synthesis of certain proteins, permitting the synthesis of glucocorticoids.

Massive doses of ACTH or β-MSH (5-500 µg) administered intrathecally to rabbits increased the CSF concentration of cAMP (Rudman and Isaacs, 1975) but not that of cGMP (Rudman, 1976). A subsequent report indicated an effect of intrathecal ACTH on the choroid plexus (Rudman et al., 1977), but $ACTH_{1-24}$, α-MSH and β-MSH all stimulated cAMP in slices of circumventricular, but not other brain structures (Rudman, 1978). Moderate doses of $ACTH_{1-16}-NH_2$ (1 µg ICV) increased cAMP only in the septum. In a broken cell preparation of rat subcortical tissue, $ACTH_{1-24}$ stimulated cAMP formation at low doses (10^{-5} M) but inhibited it at higher doses (Wiegant et al., 1979a). The same concentration of $ACTH_{1-24}$ stimulated cAMP but not cGMP in striatal slices. *In vivo* $ACTH_{1-24}$ (ICV) stimulated cAMP principally in the septum (Wiegant et al., 1979a). Chronic α-MSH treatment elevated cortical cAMP but not cGMP (see Gispen et al., 1979; Wiegant et al., 1981).

In general, the endorphins inhibit the accumulation of cAMP: in striatal slices (Minneman and Iversen, 1976), in particulate fractions from cortex (Wollemann et al., 1979), and in neuroblastoma-glioma hybrids (see Klee, 1979). However, in brainstem particulate fractions, met-enkephalin stimulated, whereas β-endorphin inhibited, cAMP production (Wollemann et al., 1979). These effects were generally reversed by naloxone. One study (Minneman and Iversen, 1976) found that met-enkephalin greatly increased cGMP production by striatal slices, but this report has not been replicated.

In all cases the effects of the peptides on cyclic nucleotides could have been mediated via the action of any one of a number of neurotransmitters. The plethora of neurotransmitter-sensitive adenylate cyclases present in the brain will not render the interpretation of the data easy.

ACTH peptides alter the phosphorylation of synaptic plasma membrane proteins. $ACTH_{1-24}$ *in vitro* decreased the phosphorylation of certain low molecular weight proteins, in contrast to cAMP, which increased the phosphorylation of three protein bands of higher molecular weight (Zwiers et al., 1976). Thus the *in vitro* effect of ACTH is apparently not mediated by cAMP. The effect of ACTH is apparently on the protein kinase; dephosphorylation was not affected and the isolated kinase was inhibited by ACTH$_{1-24}$ (Gispen et al., 1979). The ability of peptides to inhibit the protein phosphorylation paralleled their ability to elicit grooming; ACTH$_{1-16}$ was active but ACTH was not. *In vivo* treatment with ACTH$_{1-24}$ (ICV) also affected the phosphorylation of some of the same bands, but generally in the reverse direction, so that *in vitro* phosphorylation was increased. This could reflect a dephosphorylation *in vivo*, so that more sites were available for *in vitro* phosphorylation (Gispen et al., 1979). These effects of ACTH on synaptic plasma membrane protein phosphorylation may relate to earlier reports of changes during avoidance training.

With respect to the endorphins, Davis and Ehrlich (1979) reported that both met- and leu-enkephalin inhibited the *in vitro* phosphorylation of two specific membrane proteins. These are the same proteins that were affected by chronic morphine treatment *in vivo* but not by cAMP treatment *in vitro*. Subsequently, β-endorphin (10^{-8} to 10^{-10} M) was shown to produce a general increase in phosphorylation except for the two specific bands, which were again inhibited. Naloxone prevented the stimulation but not the inhibition (Ehrlich et al., 1980).

Neurotransmitter Metabolism
[For a recent comprehensive review, see Versteeg (1980).]

Acetylcholine

The ubiquity of opiate receptors and cholinergic neurons would suggest complicated effects of endorphins on acetylcholine (ACh), and such is the case. Met- or leu-enkephalin (ICV 2 µg) decreased ACh release from cortex and this effect was naloxone reversible (Jhamandas et al., 1977). Met-enkephalin decreased release of ACh from hippocampal slices (naloxone-reversible) (Subramanian et al., 1977), whereas it and β-endorphin increased ACh release from striatal slices (not naloxone reversible) (Harsing et al., 1978). Moroni et al. (1978) showed that β-endorphin (1–10 µg ICV) induced analgesia and decreased the turnover rate of ACh in hippocampus, nucleus accumbens, globus pallidus, and cortex, but not the striatum. These effects were prevented by naltrexone and not observed after α-endorphin. Neither choline nor ACh contents were altered, which conflicts with the study of Botticelli and Wurtman (1979), who found β-endorphin or [D-Ala^2]met-enkephalinamide (10 μg ICV) to increase hippocampal ACh content, an effect that was not prevented by naloxone. Intraseptal injection of β-endorphin (2.5 μg) did not produce analgesia, and decreased ACh turnover in the hippocampus, but not the cortex or striatum (Moroni et al., 1978). These effects were prevented by naltrexone.

By contrast, α-MSH or ACTH$_{1-24}$ (10 μg ICV) increased ACh turnover in the hippocampus, but not cortex, striatum, diencephalon, or brainstem (Wood et al., 1978). The changes in ACh turnover corresponded roughly in time with increased SYS and were not observed after ORG 2766, which does not produce SYS. The specificity of the correlation is doubtful, however, because in a subsequent study, SRIF (10 μg ICV) was also shown to increase hippocampal ACh turnover, but it produced barrel rotation not SYS (Wood et al., 1979). Intraseptal α-MSH or ACTH$_{1-24}$ did not elicit the behavior nor the increase of ACh turnover, neither of which were affected by cingulate or entorhinal lesions after ICV injections. This suggests that the effects of α-MSH and ACTH$_{1-24}$ were directly on the hippocampus. A cholinergic activation of SYS has been noted by others (Yamada and Furukawa, 1980).

Catecholamines

Hypophysectomy did not alter the cerebral content of either DA or NE, but the "turnover" rate as measured by the decline of the amines following inhibition of synthesis by α-methyl-ρ-tyrosine (α-MPT) was decreased (Versteeg et al., 1972). According to these authors, chronic ACTH$_{1-10}$ treatment (20 μg ZP SC, every other day for 13 days) of hypophysectomized rats did not alter the decreased rates of turnover, but Kostrzewa et al. (1975) found that α-MSH (100 μg/kg IP, daily for 3 days) decreased NE turnover in all brain regions and DA turnover in the striatum. Adrenalectomy has been reported to increase NE turnover, measured by the decline of histofluorescence following α-MPT (Fuxe et al., 1970); and it is possible that this was due to the elevation of ACTH that follows adrenalectomy. Corticosterone reversed the effect of adrenalectomy (Fuxe et al., 1970). Adrenalectomy also accelerated the rate of loss of ^{3}H from the brain following ICV injection of [^{3}H]NE (Endröczi et al., 1976).

Treatment of intact rats with ACTH$_{1-24}$ (30 μg/rat SC) increased NE in the locus coeruleus, and decreased it in the ventromedial and arcuate hypothalamic nuclei 3 hours later (Fekete et al., 1978). DA was not changed in any structure assayed, and none of the changes in NE occurred following corticosterone treatment. Kostrzewa et al. (1975) also found a decrease in hypothalamic NE following acute or chronic α-MSH

(100 µg/kg IP). However, Datta and King (1980) found reductions in whole brain DA and NE following α-MSH (10 µg IP, daily for 3 or 5 days).

Versteeg (1973) reported that treatment of rats with ACTH$_{4-10}$ (80 µg ZP, SC 1 day previously) increased NE but not DA turnover (measured by the α-MPT procedure) in whole brain; [D-Phe^7] ACTH$_{4-10}$ was ineffective. These results agree with earlier histofluorescence data (see Versteeg, 1980). In partial agreement with this, Leonard (1974), in very similar experiments, found that chronic ACTH$_{4-10}$ or [D-Phe^7]ACTH$_{4-10}$ (10 µg SC, daily for 13 days) increased the turnover of NE in rat forebrain and hindbrain, whereas ACTH$_{4-10}$ increased DA turnover in midbrain. Kostrzewa et al. (1975) found that α-MSH (100 µg/kg IP, daily for 3 days) increased the turnover of NE but not DA, and that only in midbrain.

Measuring catecholamine synthesis from [^{3}H]tyrosine, Versteeg found that repeated ACTH$_{4-10}$ (25 or 50 µg ZP SC, 72 and 24 hours previously) increased total catecholamine synthesis in whole brain (Versteeg and Wurtman, 1975). Using a similar technique, Leonard et al. (1976) did not find changes of DA or NE synthesis in any of seven brain regions following chronic α-MSH (30 µg SC, daily for 10 days). However, in mice, Iuvone et al. (1978) found that chronic treatment with ACTH$_{1-24}$, ACTH$_{4-10}$, [D-Phe^7]ACTH$_{4-10}$, but not α- or β-MSH (0.3 µmoles/kg ZP SC, daily for 3 days) increased DA synthesis in whole brain. NE synthesis was not significantly affected, but its content was significantly decreased by ACTH$_{1-24}$. Under the same conditions, neither ACTH$_{1-24}$ nor ACTH$_{4-10}$ affected tyrosine hydroxylase activity from any of eight mouse brain regions assayed *in vitro* (Dunn et al., 1978b). A single acute injection of ACTH$_{1-24}$ (0.5 mg/kg SC) increased [^{3}H]tyrosine incorporation into both DA and NE of whole mouse brain (Dunn, Delanoy and Kramarcy, unpublished observations). The effect of ACTH$_{4-10}$ did not occur in adrenalectomized animals (Versteeg and Wurtman, 1975; Iuvone et al., 1978), perhaps because elevated ACTH secretion masked the effect.

Administered intraventricularly, ACTH$_{4-10}$ or ACTH$_{1-24}$ (0.1 - 5 µg) accelerated the loss of [^{3}H]NE (previously injected ICV) from neocortex, hypothalamus, and hippocampus (Endröczi et al., 1976). ACTH$_{4-10}$ had a similar effect when implanted into the locus coeruleus (Endröczi, 1977). Intraventricular ACTH$_{1-24}$ (1 µg), accelerated the synthesis of DA but not NE from [^{3}H]tyrosine measured in frontal cortex slices *in vitro* (Delanoy et al., 1982). Striatal DA synthesis was decreased but not significantly, whereas hippocampal NE synthesis was unaffected.

Lichtensteiger et al. have examined changes in DA-cells in

the substantia nigra and arcuate nucleus by quantitative histofluorescence. Acute injection of α-MSH (10-100 µg/kg IP) altered the fluorescence of both the substantia nigra and the arcuate nucleus in a manner consistent with increased firing, the latter being the more sensitive. $ACTH_{1-24}$ had a lesser effect on nigrostriatal neurons, but did not affect tuberoinfundibular ones (Lichtensteiger and Lienhart, 1977). Lesions of the area postrema or potential ascending tracts emanating from it blocked the response to α-MSH in both systems. These results are consistent with a diffuse activation of DA-neurons by ACTH and MSH, mediated by brainstem ascending projection systems. Subsequently, Lichtensteiger and Monnet (1979) showed that ORG 2766, like $ACTH_{4-10}$, affected only the nigral neurons. $α-MSH_{11-13}$ affected both arcuate nucleus and substantia nigra neurons. There were some interactions between these changes and the sex of the animals.

$ACTH_{4-10}$ *in vitro* potentiated the dopaminergic inhibitory postsynaptic potential (IPSP) observed in frog sympathetic ganglia (Wouters and Van den Bercken, 1979).

Following intraventricular administration of met-enkephalin or [D-Ala^2]-met-enkephalinamide, the cerebral or striatal content of dihydroxyphenylacetic acid (DOPAC) and homovanillic acid (HVA) was increased (Algeri et al., 1978a; Biggio et al., 1978). Similar results were observed following β-endorphin (Van Loon and Kim, 1978; Berney and Hornykiewicz, 1977). These effects were prevented by naloxone treatment. When DA turnover is measured by α-MPT technique, β-endorphin (Van Loon and Kim, 1978; Deyo et al., 1979) and the enkephalins (Algeri et al., 1978a; Deyo et al., 1979) increased the apparent turnover. Similar results were obtained by studying the accumulation of DOPA or the decline of DA and NE following aromatic amino acid decarboxylase inhibition (Garcia-Sevillá et al., 1980), by the decline of DOPAC and HVA following monoamine oxidase (MAO) inhibition (Van Loon and Kim, 1978), or the synthesis or DA from [^{3}H]tyrosine (Algeri et al., 1978a). β-Endorphin *in vitro* did not affect the conversion of [^{3}H]tyrosine to DA in synaptosomes (Segal et al., 1977).

Although these effects resemble those seen after DA-receptor blockers, such as haloperidol, neither morphine, leu-enkephalin, or β-endorphin affected [^{3}H]spiroperidol binding to membranes of bovine anterior pituitary (Meltzer and So, 1979). DTγE did not alter the binding of [^{3}H]-spiroperidol *in vitro* to membranes of striatum, nucleus accumbens, or frontal cortex (van Ree et al., 1978; Pedigo et al., 1979a). However, DTγE administered SC, like haloperidol, inhibited the *in vivo* binding of [^{3}H]spiroperidol to hypothalamus, striatum, and nucleus accumbens, but not frontal cortex or cerebellum (Pedigo et al., 1979b). Walczak et al. (1979) reported naloxoneblockable inhibition by met-enkephalin, [D-Ala^2]met-enkephalin, and [D-Met^2,Pro^5]enkephalinamide of DA-sensitive adenylate cyclase from monkey amygdala, although Algeri et al. (1978a) found no effect of morphine or [D-Ala^2]met-enkephalinamide on this enzyme from rat striatum.

Some brain regions show a different response. Hypothalamic DA turnover (α-MPT or MAOI) was decreased by β-endorphin without a change in basal levels (Van Loon et al., 1980). Median eminency DA turnover measured by the α-MPT technique was decreased by morphine, β-endorphin, and DADLE (Deyo et al., 1979; Van Vugt et al., 1979). A similar effect was observed in frontal cortex (Deyo et al., 1979) and midbrain (Izumi et al., 1977).

Versteeg et al. (1978, 1979) have presented turnover data using the α-MPT technique following ICV injection of lower doses of endorphins. Met-enkephalin (120 ng) increased the apparent turnover of DA in the arcuate nucleus and decreased it in the dorsal central gray. Apparent NE turnover was increased in the medial preoptic and central amygdaloid nuclei, but decreased in the ventral central gray and A$_2$-region (Versteeg et al., 1978). Following α-endorphin, β-endorphin or DTγE (100 ng ICV) different and rather complex changes were observed. Generally, α-endorphin decreased DA and NE turnover, whereas β-endorphin and DTγE had relatively few effects. In particular, β-endorphin did not effect DA turnover in the striatum (Versteeg et al., 1979). The number of statistically significant effects was small relative to the number of comparisons, but the major effects were replicated in a subsequent dose-response analysis with DTγE and DTαE (Versteeg et al., 1982).

The data from slice experiments are complex. Generally, morphine and the endorphins, if anything, inhibited DA and NE release from brain slices (Arbilla and Langer, 1978).

The apparent increase in striatal DA release caused by opiates and endorphins is difficult to interpret behaviorally. Dopaminergic-receptor blockers are known to produce similar biochemical responses, so it should not be assumed that endorphins are dopaminergic agonists. The apparent paradox occurs because feedback controls activate DA mechanisms to overcome the receptor blockade. However, [D-Ala^2]enkephalin was active in the "DOPA-potentiation" text, suggesting that opiates can be dopaminergic agonists (Plotnikoff et al., 1976). A further complication is the opposite effects found in striatum and median eminenc In the latter structure, the evidence is entirely consistent with the opiate receptors being presynaptic on DA neurons and inhibiting DA release. Since the DA-neurons in turn inhibit PRL release, probably because the DA released is prolactin-inhibiting factor (PIF), (see chapter 4 by Frohman and

Berelowitz), opiates increase PRL release and naloxone inhibits it. Since in the few studies performed, frontal cortex DA turnover is also increased by opiates, a similar mechanism may operate there. This mechanism would be consistent with most electrophysiological studies that find opiates to exert primarily inhibitory effects on neurons. In the striatum a similar mechanism may operate, but opiate receptors also occur on nondopaminergic neurons. Feedback controls may thus activate DA release. Because the striatum contains much of the brain DA, and because it has a high concentration of opiate receptors readily accessible from the ventricles, it may obscure other actions in whole brain studies.

Serotonin

The cerebral content of serotonin (5-HT) was not altered by hypophysectomy, but the turnover rate, estimated by the decline of 5-hydroxyindoleacetic acid (5-HIAA) or the accumulation of 5-HT following MAO inhibition, was reduced (Versteeg et al., 1972; Spirtes et al., 1975). Chronic ACTH$_{1-10}$ did not alter this diminished turnover rate in hypophysectomized rats (Versteeg et al., 1972), but α-MSH (500 μg/kg IP, daily for 3 days) increased it (Spirtes et al., 1975).

In intact rats, Leonard (1974) found small decreases in the content and turnover of 5-HT following chronic ACTH$_{4-10}$ or [D-Phe^7]ACTH$_{4-10}$ (10 μg daily SC for 13 days); whereas Spirtes et al. (1975) found no effect with α-MSH. Kovács et al. (1976) found that acute ACTH (2 and 4 U, 30 minutes previously) increased the hypothalamic and mesencephalic 5-HT content in intact but not in adrenalectomized rats, suggesting an effect of corticosterone. Acute ACTH$_{1-24}$ (75 μg/kg IP) increased 5-HIAA in hippocampus, hypothalamus, and midbrain but not cortex 10 minutes after injection; 5-HT was not changed in any region (File and Vellucci, 1978). Leonard et al. (1976) found that α-MSH accelerated the conversion of [^{3}H]tryptophan to 5-HT in rat cortex, but not in other brain regions. Ramaekers et al. (1978) found an increase in hippocampal 5-HT one hour following ACTH$_{4-10}$ or [D-Phe^7]ACTH$_{4-10}$ but not ACTH$_{11-24}$. However, these changes were only associated with reversal of CO_2-induced amnesia; peptide treatment alone was ineffective. There is a complex relationship between cerebral 5-HT and adrenal corticosteroids (see Rees and Gray, chapter 14).

[D-Ala^2]met-enkephalin (25 μg ICV) increased whole brain 5-HIAA (Algeri et al., 1978b). The effect was most pronounced in "limbic forebrain" and was prevented by naloxone. Consistent with this, high doses (ICV) of β-endorphin, met-enkephalin, leu-enkephalin, [D-Ala^2]met-enkephalinmide and FK 33-824 increased 5-HT synthesis (measured by the accumulation of 5-hydroxytryptophan (5-HTP) following decarboxylase inhibition) in most brain regions (Garcia-Sevillá et al., 1980). Van Loon and de Souza (1978) found that intracisternal β-endorphin (15 μg) increased hypothalamic and brainstem 5-HT (see also Izumi et al., 1977) and 5-HIAA, but decreased hippocampal 5-HIAA. After MAO inhibition, β-endorphin decreased 5-HT accumulation in all three regions. The results were partially compatible with an increase of the release of 5-HT in hypothalamus and brainstem and a decrease in hippocampus. Repeated injections of β-endorphin had similar but larger effects (Van Loon et al., 1978).

5-HT is also involved in the analgesic properties of opioids. Raphe lesions, reserpine, or dihydroxytryptamine lesions all reduce opiate-induced analgesia (Fields and Basbaum, 1978), which can be partially reversed by intracerebral 5-HT (Lee et al., 1978).

Gamma-Aminobutyric Acid

Acute or chronic treatment with $ACTH_{4-10}$ or $[D-Phe^7]ACTH_{4-10}$ decreased gamma-aminobutyric acid (GABA) content in fore-, mid-, and hindbrain (Leonard, 1974). However, repeated treatments of mice with $ACTH_{1-24}$, $ACTH_{4-10}$, lysine vasopressin (LVP) (ZP SC) or corticosterone did not alter glutamic acid decarboxylase (GAD) activity in several brain regions (Dunn et al., 1978b).

β-Endorphin (10 μg ICV) decreased the GABA content of diencephalon and brainstem, but not the cerebral hemispheres (Nisticó et al., 1979). GAD activity was not altered, but the degradative enzyme GABA-transaminase (GABA-T) was increased in diencephalon and brainstem. These effects were prevented by naloxone.

Moroni et al., (1979) estimated GABA turnover by measuring the relative enrichment of $[^{13}C]$glutamate and $[^{13}C]$GABA following administration of $[U-^{13}C]$glucose. Using this assay, β-endorphin (10 μg ICV) was found to increase GABA turnover in the globus pallidus and substantia nigra, but to decrease it in the striatum. These results were interpreted as a a primary effect on the caudate to decrease GABA release, resulting in compensatory increases in GABA projections to the substantia nigra and globus pallidus.

DISCUSSION—CLASSIFICATION OF RECEPTORS FOR DIFFERENT BEHAVIORS AND NEUROCHEMICAL RESPONSES CORRELATED WITH SPECIFIC BEHAVIORS

Clearly, as might be predicted, the neurochemical effects of the peptides are complex. An important question is to what

extent the neurochemical responses can be correlated with
specific behaviors. A particularly useful way of doing this is
to compare the structure-function relationships. In Table 7-1
are summarized the reported structure-function relationships
between the peptides and the various behaviors. Table 7-2
summarizes the reported neurochemical responses.

It is convenient to discuss the relationships on a behavioral
basis, the most reasonable analysis defining receptors by their
function. Greven and de Wied (1977) attempted to classify the
behavioral effects of ACTH in a Venn diagram showing overlapping relationships for the effects on grooming, avoidance
behavior, and displacement of dihydromorphine from opiate
receptors.

Undoubtedly in experiments in which peptides are administered, ACTH and its analogues are producing some effects on
opiate receptors, and the endorphins and their analogues may be
producing some effects on ACTH receptors. It is not clear to
what extent these interactions may be physiological, rather
than "pharmacological" artifacts. Using the criterion of
naloxone antagonism, we may distinguish opiate from nonopiate
receptors. Thus the effects on learning in general appear to
be exerted via ACTH/MSH receptors, whereas opiate receptors
appear to be involved in grooming, thermoregulation, and
analgesia. Theoretically, six possible combinations of
receptor activity can exist. Table 7-3 lists these combinations and it can be seen that at least one of the behaviors
fits each combination. Even this classification is an over-
simplification, because, for example, we know that more than
one opiate receptor is involved in thermoregulation, and the
effects of ACTH/MSH on learning are too complex to be explained
by one receptor. Table 7-3 also includes the data on the activity of $ACTH_{4-10}$ in the various behaviors. Given that $ACTH_{4-10}$
has no affinity for the opiate receptor, one would predict that
its positive effects would parallel those of $ACTH_{1-24}$ when
these are not naloxone sensitive, and this is in fact the case.

The data associating neurochemical changes with particular
behaviors are at present fragmentary. Many of the effects of
ACTH on RNA and protein synthesis appear to be of a long-term
nature, probably related to the trophic activity of the
hormones. However, some data do suggest a link with avoidance
learning, (polyribosomal aggregation: Gispen et al., 1971) or
grooming (protein synthesis: Rees et al., 1976, Schotman et
al., see Dunn and Schotman, 1981). Protein phosphorylation may
also be linked to grooming (Gispen et al., 1979) and possibly
to other opiate functions. Turning to the neurotransmitters,
there are some data linking increases of ACh turnover to SYS
(Wood et al., 1979). The catecholamine changes have not yet
been investigated thoroughly enough, but there may be

Table 7-1. Behavioral Effects of Various Analogues of ACTH and Endorphins

Behavior	ACTH or ACTH$_{1-24}$	α-MSH ACTH$_{4-10}$	ORG 2766[a] ACTH$_{4-10}$	ACTH$_{4-10}$D[b]	ACTH$_{11-24}$	Morphine	Enkephalins	β-LPH	β-End[c]	α-End[c]	γ-End[c]	DTγE[d]	NX[e] Reversible	NX[f]
Locomotor activity	-	-		-		±[g]	±[g]	±[g]					+	-
Food Intake						-	-							-
Water Intake														
Stretching and yawning syndrome	+	0/+[h]	0	+	0	±[j]	0	+						+
Grooming	+	0	0	+	0	+	0	+	0				+	
Catalepsy-Catatonia						+	+/0	+	+/0	0			+	+
Explosive motor behavior	+	0							0	0				
Temperature	+	0				±[i]	±[i]	0[i]	0	0			-	
Analgesia	-	-			0	+	+	±[l]	+[l]	+	0		+	+
Self-stimulation	+	+				+	+	+	+	+	-		+	+
Active avoidance Acquisition	+/±[j]	+	+	-	-	-	-	-	-	+		-	-?	+
Active avoidance Extinction	-	-	-	+	+								+?	+
Passive avoidance Acquisition	+	+	+	-		+	+		+	+		+		+
Passive avoidance Extinction	-	0/-	-	-					0	0				
Appetitive learning Acquisition	+	+	+	+				+						
Appetitive learning Extinction	-	-	-	-										
Amnesia reversal	+	+	+/-	+	0	+	+	+			+/-		-	
Imprinting	-													
Anxiety	+	+/-/0		+		-	+				+/-			
Aggression	+	+		+										

a [Met$^+$SO$_2$,D-Lys^8,Phe^9]ACTH$_{4-10}$;

b [D-Phe^7]ACTH$_{4-10}$;

c β-, α-, γ-endorphin;

d Des-tyr-γ-endorphin (β-LPH$_{62-77}$);

e Effects of peptides are naloxone reversible;

f Naloxone (or naltrexone);

g Hypoactivity followed by hyperactivity;

h A stroke separates different results from different reports;

i Low-dose hyperthermia, high-dose hypothermia;

j Low-dose facilitation, high-dose Inhibition;

k + facilitates, - retards;

l Due to corticosterone.

Table 7-2. Neurochemical Effects of Various Analogues of ACTH and Endorphins

Neurochemical	Peptide									
	ACTH or ACTH$_{1-24}$	α-MSH ACTH$_{4-10}$	ORG 2766[a]	ACTH$_{4-10}$ D[b]	ACTH$_{11-24}$	Morphine	Enkephalins	β-End[c]	α-End[c]	NX[d] reversible
Blood flow	-/+[e] /0	-	-							
RNA synthesis	+	0								
Polyamines	+	+								
Protein synthesis	+[f]	+		+/0/-			-			
cAMP	+/±[f]	+		0	0	-	+	-		
cGMP	0	0				+	+			
Phosphoproteins	-	0					-	0/+		-
ACh content	0	0	0							
ACh turnover	+	+	0					-	0	+
DA content	0	0/-					0	0		
DA turnover	+/0	0/+	+	0			±[g]	±[g]	-	
NE content	0/-	0/-		0			0	0		
NE turnover	+/0	+/0		0				±	-	
5-HT content	0	0		-				±		+
5-HT turnover	+	+/0		-			+	± -		+
GABA content				-						
GABA turnover	+/0			-				±		

a [Met⁴SO₂,D-Lys⁸,Phe⁹]ACTH$_{4-10}$
b [D-Phe⁷]ACTH$_{4-10}$
c β-, α-endorphin
d Effects of peptides are prevented by naloxone
e Increased in adrenalectomized rats;
f low dose facilitation, high dose inhibition
g + in striatum, - in median eminence and frontal cortex.

Table 7-3. Classification of Behaviors in Terms of Receptor Activities

ACTH	Endorphins	Naloxone-sensitive	Behaviors	ACTH
+	0	0	Explosive motor behavior (Aggression?)	0
+	+	0	Extinction	+
+	+	+	Grooming (Self-stimulation?) (Stretching and yawning syndrome?) (Hypothermia?)	0
+	–	0	Acquisition	+
0	+	+	Catalepsy Hyperthermia (Hypothermia?)	0
–	+	+	Analgesia Locomotor activity Anxiety	–

associations between the effects of ACTH on avoidance behavior and those on DA and/or NE metabolism. Effects on DA may also be related to the catatonic (van Loon and Kim, 1978) and motor (Stinus et al., 1980) effects of endorphins. 5-HT may be linked to their analgesic effects (Fields and Basbaum, 1978). All of these associations are, however, rather tenuous. In general, the neurochemical studies, like many of the behavioral ones, have been performed with high doses of the peptides, and there have been relatively few attempts to correlate neuro- chemistry and behavior in the same animals. The future in this area lies in carefully performed studies involving such correlations, ideally with localized measurements and applications of peptide, antagonists, and specific antisera.

ACKNOWLEDGMENTS

I am grateful to many scientific colleagues for their assistance in compiling this manuscript. The patience and forebearance of the typists, Vicki Durrance and Debe Martin is deeply appreciated.

REFERENCES

Adler, M.W.: Opioid peptides. *Life Sci* 26:497-510, 1980.
Akil, H., Richardson, D.E., Hughes, J., and Barchas, J.D.:
  Enkephalin-like material elevated in ventricular cerebro

spinal fluid of pain patients after analgetic focal stimulation. *Science 201*:463-465, 1978a.

Akil, H., Richardson, D.E., Barchas, J.D., and Li, C.H.: Apnd: Appearance of β-endorphin-like immunoreactivity in human ventricular cerebrospinal fluid upon analgesic electrical stimulation. *Proc Natl Acad Sci USA 75*:5170-5172, 1978b.

Akil, H., Watson, S.J., Barchas, J.D., and Li, C.H.: β-endorphin immunoreactivity in rat and human blood: Radioimmunoassay, comparative levels and physiological alterations. *Life Sci 24*:1659-1666, 1979.

Akil, H., Hewlett, W.A., Barchas, J.D., and Li, C.H.: Binding of ^{3}H-β-endorphin to rat brain membranes: characterization of opiate properties and interaction with ACTH. *Europ J Pharmacol 64*:1-8, 1980.

Algeri, S., Brunello, N., Calderini, G., and Consolazione, A.: Effect of enkephalins on catecholamine metabolism in rat CNS. *Adv Biochem Psychopharmacol 18*:199-210, 1978a.

Algeri, A., Consolazione, A., Calderini, G., Achilli, G., Puche Canas, E., and Garattini, S.: Effect of the administration of (d-ala)$_2$methionine-enkephalin on the serotonin metabolism in rat brain. *Experientia 34*:1488-1489, 1978b.

Almay, B.G.L., Johansson, F., von Knorring, L., Terenius, L., and Wahlstrom, A.: Endorphins in chronic pain. I Differences in CSF endorphin levels between organic and psychogenic pain syndromes. *Pain 5*:153-162, 1978.

Altshuler, H.L., Phillips, P.E., and Feinhandler, D.A.: Alteration of ethanol self-administration by naltrexone. *Life Sci 26*:679-688, 1980.

Amir, S., and Amit, Z.: Endogenous opioid ligands may mediate stress-induced changes in the affective properties of pain related behavior in rats. *Life Sci 23*:1143-1152, 1978.

Amir, S., and Amit, Z.: The pituitary gland mediates acute and chronic pain responsiveness in stressed and nonstressed rats. *Life Sci 24*:439-448, 1979.

Amir, S., Brown, Z.W., and Amit, Z.: The role of endorphins in stress: Evidence and speculations. *Neurosci Biobehav Rev 4*:77-86, 1980.

Anderson, D.C., Winn, W., and Tam, T.: Adrenocorticotrophic hormone and acquisition of a passive avoidance response. *J Comp Physiol Psychol 66*:497-499, 1968.

Applezweig, M.H., and Baudry, F.D.: The pituitary-adrenocortical system in avoidance learning. *Psychol Rep 1*:417-420, 1955.

Arbilla, S., and Langer, S.Z.: Morphine and β-endorphin inhibit release of noradrenaline from cerebral cortex but not of dopamine from rat striatum. *Nature 271*:559-561, 1978.

Arnsten, A.T., and Segal, D.S.: Naloxone alters locomotion and interaction with environmental stimuli. *Life Sci*

25:1035–1042, 1979.

Bailey, W.H., and Weiss, J.M.: Effect of ACTH$_{4-10}$ on passive avoidance of rats lacking vasopressin (Brattleboro strain). Horm Behav 10:22–29, 1978.

Baldwin, D.M., Hawn, C.K., and Sawyer, C.H.: Effects of intraventricular infusions of ACTH$_{1-24}$ and ACTH$_{4-10}$ on LH release, ovulation, and behavior in the rabbit. Brain Res 80:291–301, 1974.

Barber, J., and Mayer, D.: Evaluation of efficacy and neural mechanism of a hypnotic analgesia procedure in experimental and clinical dental pain. Pain 1:41–48, 1977.

Barchas, J.D., Akil H., Elliott, G.R., Holman, R.B., and Watson, S.J.: Behavioral neurochemistry: Neuroregulators and behavioral states. Science 200:964–973, 1978.

Beatty, P.A., Beatty, W.W., Bowman, R.E., and Gilchrist, J.C.: The effect of ACTH, adrenalectomy and dexamethasone on the acquisition of an avoidance response in rats. Physiol Behav 5:939–944, 1970.

Bergmann, F., Altstetter, R., and Weissman, B.A.: In vivo interaction of morphine and endogenous opiate-like peptides. Life Sci 23:2601–2608, 1978.

Berney, S., and Hornykiewicz, O.: The effect of β-endorphin and met-enkephalin on striatal dopamine metabolism and catalepsy: Comparison with morphine. Commun Psychopharmacol 1:597–604, 1977.

Bertolini, A., Genedane, S., and Castelli, M.: Behavioral effects of naloxone in rats. Experientia 34:771–772, 1978.

Bertolini, A., Poggioli, R., and Ferrari, W.: ACTH-induced hyperalgesia in rats. Experientia 35:1216–1217, 1979.

Biggio, G., Casu, M., Corda, M.G., di Bello, C., and Gessa, G.L.: Stimulation of dopamine synthesis in caudate nucleus by intrastriatal enkephalins and antagonism by naloxone. Science 200:552–554, 1978.

Bläsig, J., Höllt, V., Bauerle, U., and Herz, A.: Involvement of endorphins in emotional hyperthermia of rats. Life Sci 23:2525–2532, 1978.

Bläsig, J., Bauerle, U., and Herz, A.: Endorphin-induced hypothermia: Characterization of the exogenously and endogenously induced effects. Naunyn-Schmiedebergs Archs Pharmac 309:137–143, 1979.

Bloch, B., Bugnon, C., Fellman, P., Lenys, P., and Gouget, A.: Neurons of the rat hypothalamus reactive with antisera against endorphins, ACTH, MSH and β-LPH. Cell Tissue Res 204:1–15, 1979.

Bloom, F., Segal, D., Guillemin, R., and Ling, N.: Endorphins: Profound behavioral effects in rat suggest new etiological factors in mental illness. Science 194:630–632, 1976.

Bloom, F., Battenberg, E., Rossier, J., Ling, N., and

Guillemin, R.: Neurons containing β-endorphin in rat brain exist separately from those containing enkephalin: Immunocytochemical studies. *Proc Natl Acad Sci USA 75*:1591-1595, 1978.

Bloomfield, G.A., Scott, A.P., Lowry. P.J., Gilkes, J.J.H., and Rees, L.H.: A reappraisal of human β-MSH. *Nature 252*:492-493, 1974.

Bodnar, R.J., Kelly, D.D., Spiaggia, A., and Glusman, M.: Stress-induced analgesia: Adaptation following chronic cold water swims. *Bull Psychonom Soc 11*:337-340, 1978.

Bodnar, R.J., Kelly, D.P., Brutus, M., and Glusman, M.: Stress-induced analgesia: Neural and hormonal determinants. *Neurosci Biobehav Rev 4*:87-100, 1980a.

Bodnar, R.J., Abrams, G.M., Zimmerman, E.A., Krieger, D.T., Nicholson, G., and Kizer, J.S.: Neonatal monosodium glutamate: Effects upon analgesic responsivity and immunocytochemical ACTH/β-lipotropin. *Neuroendocrinology 30*:280-284, 1980b.

Bohus, B.: Effects of ACTH-like neuropeptides on animal behavior and man. *Pharmacology 18*:113-122, 1979.

Bohus, B.: Endorphins and behavioral adaptation. In *Psychoneuroendocrinology and Abnormal Behavior*, J. Mendelwicz and H.M. van Praag, eds. *Adv Biol Psychiatry 5*:7-19, 1980.

Bohus, B., and de Wied, D.: Failure of α-MSH to delay extinction of conditioned avoidance in rats with lesions in the parafascicular nuclei of the thalamus. *Physiol Behav 2*:221-223, 1967.

Bohus, B., Nyakas, C., and Endröczi, E.: Effects of adrenocorticotropic hormone on avoidance behavior of intact and adrenalectomized rats. *Int J Pharmacol 7*:307-314, 1968.

Bohus, B., van Wimersma Greidanus, Tj.B., and de Wied, D.: Behavioral and endocrine responses of rats with hereditary hypothalamic diabetes insipidus (Brattleboro strain). *Physiol Behav 14*:609-615, 1975.

Botticelli, L.J., and Wurtman, R.J.: β-Endorphin administration increases hippocampal acetylcholine levels. *Life Sci 24*:1799-1804, 1979.

Brain, P.F., and Evans, A.E.: Acute influences of some ACTH-related peptides on fighting and adrenocortical activity in male laboratory mice. *Pharmacol Biochem Behav 7*:425-434, 1977.

Branconnier, R.J., Cole, J.O., and Gardos, G.: ACTH$_{4-10}$ in the amelioration of neuropsychological symptomatology associated with senile organic brain syndrome. *Psychopharmacology 61*:161-165, 1979.

Browne, R.G., and Segal, D.S.: Behavioral activating effects of opiates and opioid peptides. *Biol Psychiatry 15*:77-86, 1980.

Buchsbaum, M.S., Davis, G.C., and Bunney, W.E., Jr.: Naloxone alters pain perception and somatosensory evoked potentials in normal subjects. *Nature* 270:620–622, 1977.

de Caro, G., Micossi, L.G., and Venturi, F.: Drinking behaviour induced by intracerebroventricular administration of enkephalins to rats. *Nature* 277:51–53, 1979.

Catlin, D.H., Hui, K.K., Loh, H.H., and Li, C.H.: Pharmacologic activity of β-endorphin in man. *Commun Psychopharmacol* 1:493–500, 1977.

Catlin, D.H., George, R., and Li, C.H.: β-Endorphin: Pharmacologic and behavioral activity in cats after low intravenous doses. *Life Sci* 23:2147–2154, 1978.

Chance, W.T.: Autoanalgesia: Opiate and non-opiate mechanisms. *Neurosci Biobehav Rev* 4:55–67, 1980.

Chance, W.T., White, A.C., Krynock, G.M., and Rosencrans, J.A.: Conditional fear-induced antinociception and decreased binding of [^{3}H]N-leu-enkephalin to rat brain. *Brain Res* 141:371–374, 1978.

Chang, K.-J., Cooper, B.R., Hazum, E., and Cuatrecasas, P.: Multiple opiate receptors: Different regional distribution in the brain and differential binding of opiates and opioid peptides. *Mol Pharmacol* 16:91–104, 1979.

Chavkin, C., James, I.F., and Goldstein, A.: Dynorphin is a specific endogenous ligand of the κ opioid receptor. <u>Science</u> 215:413–415, 1982.

Cheng, R.S.S., and Pomeranz, B.: Electroacupuncture analgesia could be mediated by at least two pain-relieving mechanisms: Endorphin and non-endorphin systems. *Life Sci* 25:1957–1962, 1979.

Cheng, R.S.S., and Pomeranz, B.H.: Electroacupuncture analgesia is mediated by stereospecific opiate receptors and is reversed by antagonists of type I receptor. *Life Sci* 26:631–638, 1980.

Chesher, G.B., and Chan, B.: Footshock induced analgesia in mice: Its reversal by naloxone and cross tolerance with morphine. *Life Sci* 21:1569–1574, 1977.

Childers, S.R.: Endorphin receptors. In *Neurotransmitter Receptors. Part I: Receptors and Recognition,* S.J. Enna and H. Yamamura, eds. Chapman-Hall, London, 1980, pp. 107–147.

Childers, S.R., Simantov, R., and Snyder, S.H.: Enkephalin: Radioimmunoassay and radioreceptor assay in morphine dependent rats. *Eur J Pharmacol* 46:289–293, 1977.

Childers, S.R., Creese, I., Snowman, A.M., and Snyder, S.H.: Opiate receptor binding affected differentially by opiates and opioid peptides. *Eur J Pharmacol* 55:11–18, 1979.

Clement-Jones, V., Lowry, P.J., Rees, L.H., and Besser, G.M.: Met-enkephalin circulates in human plasma. *Nature* 283:295–297, 1980.

Colbern, D., Isaacson, R.L., Bohus, B., and Gispen, W.H.:

Limbic-midbrain lesions and ACTH-induced excessive grooming. *Life Sci* 21:393-402, 1977.

Colbern, D.L., Isaacson, R.L., Green, E.J., and Gispen, W.H.: Repeated intraventricular injections of $ACTH_{1-24}$: The effects of home or novel environments on excessive grooming. *Behav Biol* 23:381-387, 1978.

Cornford, E.M., Braun, L.D., Crane, P.D., and Oldendorf, W.H.: Blood-brain barrier restriction of peptides and the low uptake of enkephalins. *Endocrinology* 103:1297-1303, 1978.

Csontos, K., Rust, M., Höllt, V., Mahr, W., Kromer, W., and Teschemacher, H.J.: Elevated phasma β-endorphin levels in pregnant women and their neonates. *Life Sci* 25:835-844, 1979.

Cuello, A.C., and Paxinos, G.: Evidence for a long leu-enkephalin striopallidal pathway in rat brain. *Nature* 271:178-180, 1978.

Czlonkowski, A., Höllt, V., and Herz, A.: Binding of opiates and endogenous opioid peptides to neuroleptic receptor sites in the corpus striatum. *Life Sci* 22:953-962, 1978.

Datta, P.C., and King, M.G.: Effects of melanocyte-stimulating hormone (MSH) and melatonin on passive avoidance and on an emotional response. *Pharmacol Biochem Behav* 6:449-452, 1977.

Datta, P.C., and King, M.G.: α-MSH and novelty-induced emotional responses in rats: Associated changes in plasma corticosterone and brain catecholamines. *J Comp Physiol Psychol* 94:324-336, 1980.

Davis, L.G., and Ehrlich, Y.G.: Opioid peptides and protein phosphorylation. *Adv Biochem Psychopharmacol* 116:233-244, 1979.

Delanoy, R.L. and Dunn, A.J.: Mouse brain deoxyglucose uptake after footshock, ACTH-analogs, α-MSH, cortocosterone or lysine vasopressin. *Pharmacol Biochem Behav* 9:21-26, 1978.

Delanoy, R.L., Kramarcy, N.R., and Dunn, A.J.: $ACTH_{1-24}$ and lysine vasopressin selectively activate dopamine synthesis in frontal cortex. *Brain Res* 231:117-129, 1982.

Deyo, S.N., Swift, R.M., and Miller, R.J.: Morphine and endorphins modulate dopamine turnover in rat median eminence. *Proc Natl Acad Sci USA* 76:3006-3009, 1979.

Dorsa, D.M., Van Ree, J.M., and de Wied, D.: Effects of [Des-Tyr[1]]-α-endorphin and α-endorphin on substantia nigra self-stimulation. *Pharmacol Biochem Behav* 10:899-905, 1979.

Dunn, A.J.: Biochemical correlates of training: A discussion of the evidence. In *Neural Mechanisms of Learning and Memory*, M. Rosenzweig and E.L. Bennett, eds MIT Press, Boston, 1976, pp 311-320.

Dunn, A.J., and Gildersleeve, N.B.: Corticotrophin-induced changes in protein labelling: lack of molecular specificity. *Pharmacol Biochem Behav* 13:823-827, 1980.

Dunn, A.J., and Gispen, W.H.: How ACTH acts on the brain.

Biobehav Rev 1:15-23, 1977.

Dunn, A.J., and Hurd, R.W.: Regional deoxyglucose uptake in mouse brain following ACTH peptides and naloxone. *Pharmacol Biochem Behav 17:*0000, 1982.

Dunn, A.J., and Schotman, P.: Effects of ACTH and related peptides on cerebral RNA and protein synthesis. *Pharmacol Ther 12*:353-372, 1981.

Dunn, A.J., Rees, H.D., and Iuvone, P.M.: ACTH and the stress-induced changes of lysine incorporation into brain and liver protein. *Pharmacol Biochem Behav 8*:455-465, 1978a.

Dunn, A.J., Gildersleeve, N.B., and Gray, H.E.: Mouse brain tyrosine hydroxylase and glutamic acid decarboxylase following treatment with adrenocorticotrophic hormone, vasopressin and corticosterone. *J Neurochem 31*:977-982, 1978b.

Dunn, A.J., Green, E.J., and Isaacson, R.L.: Intracerebral adrenocorticotropic hormone mediates novelty-induced grooming in the rat. *Science 203*:281-283, 1979.

Dunn, A.J., Steelman, S., and Delanoy, R.L.: Intraventricular ACTH and vasopressin cause regionally specific changes in cerebral deoyglucose uptake. *J Neurosci Res 5*:485-495, 1980.

Dunn, A.J., Guild, A.L., Kramarcy, N.R., and Ware, M.D.: Benzodiazepines decrease grooming in response to novelty but not ACTH or β-endorphin. *Pharmacol Biochem Behav 15*:605-608, 1981a.

Dunn, A.J., Childers, S.R., Kramarcy, N.R., and Villiger, J.W.: ACTH-induced grooming involves high-affinity opiate receptors. *Behav Neural Biol 31*:105-109, 1981b.

Ehrlich, Y.H., Davis, L.G., Keen, P., and Brunngraber, E.G.: Endorphin-regulated protein phosphorylation in brain membranes. *Life Sci 26*:1765-1772, 1980.

Endröczi, E.: Brain mechanisms involved in ACTH-induced changes of exploratory activity and conditioned avoidance behavior. In *Neuropeptide Influences on the Brain and Behavior*, L.H. Miller, C.A. Sandman, and A.J. Kastin, eds Raven Press, New York, 1977, pp 179-187.

Endröczi, E., Hraschek, A., Nyakas, C., and Szabo, G.: Effect of ACTH and its fragment on brain catecholamines in intact and adrenalectomized rats. *Endokrinologie 68*:51-59, 1976.

Fekete, M.I., Stark, E., Herman, J.P., Palkovits, M., and Kanyicska, B.: Catecholamine concentration of various brain nuclei of the rat as affected by ACTH and corticosterone. *Neurosci Lett 10*:153-158, 1978.

Ferrari, W., Gessa, G.L., and Vargiu, L.: Behavioral effects induced by intracisternally injected ACTH and MSH. *Ann NY Acad Sci 104*:330-345, 1963.

Fields, H.L., and Basbaum, A.I.: Brain control of spinal pain transmission neurons. *Annu Rev Physiol 40*:217-248, 1978.

File, S.E.: ACTH, but not corticosterone impairs habituation and reduces exploration. *Pharmacol Biochem Behav* 9:161-166, 1978.

File, S.E., and Clarke, A.: Intraventricular ACTH reduces social interaction in male rats. *Pharmacol Biochem Behav* 12:711-715, 1980.

File, S.E. and Vellucci, S.V.: Studies on the role of ACTH and 5-HT in anxiety using an animal model. *J Pharm Pharmacol* 30:105-110, 1978.

Firemark, H.M., and Weitzman, R.E.: Effects of beta-endorphin, morphine and naloxone on arginine vasopressin secretion and the electroencephalogram. *Neuroscience* 4:1895-1902, 1979.

Flood, J.F., Jarvik, M.E., Bennett, E.L., and Orme, A.E.: Effects of ACTH peptide fragments on memory formation. *Pharmacol Biochem Behav* 5 (Suppl 1): pp 41-51, 1976.

Frenk, H., Urca, G., and Liebeskind, J.C.: Epileptic properties of leucine- and methionine-enkephalin: Comparison with morphine and reversibility by naloxone. *Brain Res* 147:327-337, 1978.

Frid, M., and Singer, G.: Hypnotic analgesia in conditions of stress is partially reversed by naloxone. *Psychopharmacology* 63:211-215, 1979.

Fuxe, K., Corrodi, H., Hökfelt, T., and Jonsson, G.: Central monoamine neurons and pituitary-adrenal activity. *Prog Brain Res* 32:42-56, 1970.

Gaillard, A.W.K., and Varey, C.A.: Some effects of an ACTH$_{4-9}$ analogue (ORG 2766) on human performance. *Physiol Behav* 23:79-84, 1979.

Garcia-Sevillá, J.A., Magnusson, T., and Carlsson, A.: Effects of enkephalins and two enzyme resistant analogues on monoamine synthesis and metabolism in rat brain. *Naunyn-Schmiedebergs Arch Pharm* 310:211-218, 1980.

Garrud, P., Gray, J.A., and de Wied, D.: Pituitary-adrenal hormones and extinction of rewarded behavior in the rat. *Physiol Behav* 12:109-119, 1974.

Gessa, G.L., Pisano, M., Vargiu, L., Crabai, F., and Ferrari, W.: Stretching and yawning movements after intracerebral injection of ACTH. *Rev Can Biol* 26:229-236, 1967.

Gianoulakis, C., Seidh, N.G., Routhier, R., and Chretien, M.: Biosynthesis and characterization of adrenocorticotropic hormone, alpha-melanocyte-stimulating hormone, and an NH$_2$-terminal fragment of the adrenocorticotropic hormone/ beta-lipotropin precursor from rat pars intermedia. *J Biol Chem* 254:11903-11906, 1979.

Gillin, J.C., Mendelson, W.B., Sitaram, N., and Wyatt, R.J.: The neuropharmacology of sleep and wakefulness. *Annu Rev Pharmacol Toxicol* 18:563-579, 1978.

Gispen, W.H., and Isaacson, R.L.: ACTH-induced excessive grooming in the rat. *Pharmacol Ther* 12:209-246, 1981.

Gispen, W.H., and Schotman, P.: ACTH and brain RNA: Changes in content and labelling of RNA in the brain stem. *Neuroendocrinology* 21:97-110, 1976.

Gispen, W.H., and Wiegant, V.M.: Opiate antagonists suppress $ACTH_{1-24}$-induced excessive grooming in the rat. *Neurosci Lett* 2:159-164, 1976.

Gispen, W.H., de Wied, D., Schotman, P., and Jansz, H.S.: Effects of hypophysectomy on RNA metabolism in rat brain stem. *J Neurochem* 17:751-761, 1970.

Gispen, W.H., de Wied, D., Schotman, P., and Jansz, H.S.: Brain stem polysomes and avoidance performance of hypophysectomized rats subjected to peptide treatment. *Brain Res* 31:341-351, 1971.

Gispen, W.H., Schotman, P., and de Kloet, E.R.: Brain RNA and hypophysectomy: A topographical study. *Neuroendocrinology* 9:285-296, 1972.

Gispen, W.H., Wiegant, V.M., Greven, H.M., and de Wied, D: The induction of excessive grooming in the rat by intraventricular application of peptides derived from ACTH: Structure-activity studies. *Life Sci* 17:645-652, 1975.

Gispen, W.H., Wiegant, V.M., Bradbury, A.F., Hulme, E.C., Smyth, D.G., Snell, C.R., and de Wied, D.: Induction of excessive grooming in the rat by fragments of lipotropin. *Nature* 264:794-795, 1976.

Gispen, W.H., Zwiers, H., Wiegant, V.M., Schotman, P., and Wilson, J.E.: The behaviorally active neuropeptide ACTH as neurohormone and neuromodulator: The role of cyclic nucleotides and membrane phosphoproteins. *Adv Biochem Psychopharmacol* 116:199-224, 1979.

Givens, J.R., Wiedemann, E., Andersen, R.N., and Kitabchi, A.E.: β-Endorphin and β-lipotropin plasma levels in hirsute women: Correlation with body weight. *J Clin Endocr Metab* 50:975-976, 1980.

Gold, P.E., and McGaugh, J.L.: Hormones and memory. In *Neuropeptide Influences on the Brain and Behavior*, L.H. Miller, C.A. Sandman, and A.J. Kastin, eds Raven Press, New York, 1977, pp 127-143.

Gold, P.E., and Van Buskirk, R.: Effects of posttrial hormone injection on memory processes. *Horm Behav* 7:509-517, 1976.

Gold, P.E., Rose, R.P., Spanis, C.W., and Hankins, L.L.: Retention deficit for avoidance training in hypophysectomized rats: Time-dependent enhancement of retention performance with post-training ACTH injections. *Horm Behav* 8:363-371, 1977.

Goldman, H., Sandman, C.A., Kastin, A.J., and Murphy, S.: MSH affects regional perfusion of the brain. *Pharmacol Biochem Behav* 3:661-664, 1975.

Goldman, H., Murphy, S., Schneider, D.R., and Felt, B.T.:

Cerebral blood flow after treatment with ORG-2766, a potent analogue of ACTH$_{4-9}$. *Pharmacol Biochem Behav 10*:883-887, 1979.

Goldstein, A.: Endorphins and pain: A critical review. In *Mechanisms of Pain and Analgesic Compounds*, R.F. Beers and E.G. Bassett, eds. Raven Press, New York, 1979, pp 249-262.

Goldstein, A., and Hilgard, E.R.: Failure of the opiate antagonist naloxone to modify hypnotic analgesia. *Proc Natl Acad Sci USA 72*:2041-2043, 1975.

Goldstein, A., Fischli, W., Lowney, L.I., Hunkapiller, M., and Hood, L.: Porcine pituitary dynorphin: complete amino acid sequence of the biologically active heptadecapeptide. *Proc Nat Acad Sci USA 78*:7219-7233, 1981.

von Graffenried, B., del Pozo, E., Roubicek, J., Krebs, E., Pöldinger, W., Burmeister, P., and Kerp, L.: Effects of the synthetic enkephalin analogue FK 33-824 in man. *Nature 272*:729-730, 1978.

Gramsch, C., Höllt, V., Mehraein, P., Pasi, A., and Herz, A.: Regional distribution of methionine-enkephalin-and beta-endorphin-like immunoreactivity in human brain and pituitary. *Brain Res 171*:261-270, 1979.

Gray, J.A., and Garrud, P.: Adrenopituitary hormones and frustrative nonreward. In *Neuropeptide Influences on the Brain and Behavior*, L.H. Miller, C.A. Sandman, and A.J. Kastin, eds. Raven Press, New York, 1977, pp 201-212.

Gray, P.: Effect of adrenocorticotropic hormone on conditioned avoidance in rats interpreted as state-dependent learning. *J Comp Physiol Psychol 88*:281-284, 1975.

Green, E.J., Isaacson, R.L., Dunn, A.J., and Lanthorn, T.H.: Naloxone and haloperidol reduce grooming occurring as an aftereffect of novelty. *Behav Neural Biol 27*:546-551, 1979.

Greenberg, R., Whalley, C.E., Jourdikian, F., Mendelson, I.S., Walter, R., Nikolics, K., Coy, D.H., Schally, A.V., and Kastin, A.J.: Peptides readily penetrate the blood-brain barrier: Uptake of peptides by synaptosomes is passive. *Pharmacol Biochem Behav 5* (Suppl 1):151-158, 1976.

Greven, H.M., and de Wied, D.: The influence of peptides derived from corticotrophin (ACTH) on performance. Structure activity studies. *Progr Brain Res 39*:429-442, 1973.

Greven, H.M., and de Wied, D.: Influence of peptides structurally related to ACTH and MSH on active avoidance behavior in rats. *Front Horm Res 4*:140-152, 1977.

Grevert, P., and Goldstein, A.: Endorphins--naloxone fails to alter experimental pain or mood in humans. *Science 199*:1093-1095, 1978.

Guild, A.L., and Dunn, A.J.: Dopamine involvement in ACTH-induced grooming behavior. *Pharmac Biochem Behav 17*:31-36, 1982.

Guillemin, R., Vargo, T., Rossier, J., Minick, S., Ling, N., Rivier, C., Vole, W., and Bloom, F.: β-Endorphin and adrenocorticotropin are secreted concomitantly by the pituitary gland. *Science 197*:1367-1369, 1977a.

Guillemin, R., Ling, N., Lazarus, L., Burgus, R., Minick, S., Bloom, F., Nicoll, R., Siggins, G., and Segal, D.: Endorphins, novel peptides of brain and hypophyseal origin and opiate-like activity--biochemical and biologic studies. *Ann NY Acad Sci 297*:131-157, 1977b.

Guth, S., Levine, S., and Seward, J.P.: Appetitive acquisition and extinction effects with exogenous ACTH. *Physiol Behav 7*:195-200, 1971.

Haddox, M.K., and Russell, D.H.: β-Endorphin is a kidney trophic hormone. *Life Sci 25*:615-629, 1979.

Harsing, L.G., Jr., Vizi, E.S., and Knoll, J.: Increase by enkephalin of acetycholine release from striatal slices of the rat. *Pol J Pharmacol Pharm 30*:387-395, 1978.

Hazum, E., Chang, K.-J., and Cuatrecasas, P.: Specific non-opiate receptors for β-endorphin. *Science 205*:1033-1035, 1979.

Henriksen, S.J., Bloom, F.E., McCoy, F., Ling, N., and Guillemin, R.: β-endorphin induces nonconvulsive limbic seizures. *Proc Natl Acad Sci USA 75*:5221-5225, 1978.

Heward, C.B., Yang, Y.C.S., Sawyer, T.K., Bregman, M.D., Fuller, B.B., Hurky, V.J., and Hadley, M.E.: Iodination associated inactivation of β-melanocyte stimulating hormone. *Biochem Biophys Res Commun 88*:266-273, 1979.

Ho, W.K.K., Wen, H.L., Lam, S., and Ma, L.: The influence of electro-acupuncture on naloxone-induced morphine withdrawal in mice: Elevation of brain opiate-like activity. *Eur J Pharmacol 49*:197-199, 1978.

Hökfelt, T., Elde, R., Johansson, O., Terenius, L., and Steen, L.: The distribution of enkephalin-immunoreactive cell bodies in the rat central nervous system. *Neurosci Lett 5*:25-31, 1977a.

Hökfelt, T., Ljungdahl, A., Terenius, L., Elde, R., and Nilsson, G.: Immunohistochemical analysis of peptide pathways possibly related to pain and analgesia: Enkephalin and substance P. *Proc Natl Acad Sci USA 74*:3081-3085, 1977b.

Holaday, J.W., Law, P.-Y, Tseng, L.-F, Loh, H.H., and Li, C.H.: β-Endorphin: pituitary and adrenal glands modulate its action. *Proc Natl Acad Sci USA 74*:4628-4632, 1977.

Holaday, J.W., Loh, H.H., and Li, C.H.: Unique behavioral effects of β-endorphin and their relationship to thermoregulation and hypothalamic function. *Life Sci 22*:1525-1536, 1978a.

Holaday, J.W., Wei, E., Loh, H.H., and Li, C.H.: Endorphins may function in heat adaptation. *Proc Natl Acad Sci USA*

75:2923-2927, 1978b.

Holaday, J.W., Law, P.-Y., Loh, H.H., and Li, C.H.: Adrenal steroids indirectly modulate morphine and β-endorphin effects. J Pharmaco Exp Ther 208:176-183, 1979a.

Holaday, J.W., Dallman, M.F., and Loh, H.H.: Effects of hypophysectomy and ACTH on opiate tolerance and physical dependence. Life Sci 24:771-782, 1979b.

Höllt, V., Miller, O.A., and Fahlbusch, R.: β-Endorphin in human plasma: Basal and pathologically elevated levels. Life Sci 25:37-44, 1979.

Höllt, V., Haarmann, I., Bovermann, K., Jerlicz, M., and Herz, A.: Dynorphin-related immunoreactive peptides in rat brain and pituitary. Neurosci Lett 18:149-153, 1980.

Hong, J.S., Yang, H.-Y.T., Fratta, W., and Costa, E.: Rat striatal methionine-enkephalin content after chronic treatment with cataleptogenic and noncataleptogenic antischizophrenic drugs. J Pharmacol Exp Ther 205:141-147, 1978.

Hong, J.S., Gillin, J.C., Yang, H.-Y.T., and Costa, E.: Repeated electroconvulsive shocks and the brain content of endorphins. Brain Res 177:273-278, 1979.

Horn, E., Greiner, B., and Horn, I.: The effect of ACTH on habituation of the turning reaction in the Toad Bufo bufo L. J Comp Physiol 131:129-135, 1979.

Hosobuchi, Y., and Li, C.H.: The analgesic activity of human β-endorphin in man. Commun Psychopharmacol 2:33-37, 1978.

Hosobuchi, Y., Adams, J.E., and Linchitz, R.: Pain relief by electrical stimulation of the central gray matter in humans and its reversal by naloxone. Science 197:183-186, 1977.

Isaacson, R.L., and Green, E.L.: The effect of ACTH$_{1-24}$ on locomotion, exploration, rearing and grooming. Behav Biol 24:118-122, 1978.

Isaacson, R.L., Dunn, A.J., Rees, H.E., and Waldock, B.: ACTH$_{4-10}$ and improved use of information in rats. Physiol Psychol 4:159-162, 1976.

Iuvone, P.M., Morasco, J., Delanoy, R.L., and Dunn, A.J.: Peptides and the conversion of [^{3}H]tyrosine to catecholamines: Effects of ACTH-analogues, melanocyte-stimulating hormones and lysine-vasopressin. Brain Res 139:131-139, 1978.

Izquierdo, I.: Effect of naloxone and morphine on various forms of memory in the rat: Possible role of endogenous opiate mechanisms in memory consolidation. Psychopharmacology 66:199-203, 1979.

Izquierdo, I., Paiva, A.C.M., and Elisabetsky, E.: Posttraining intraperitoneal administration of leu-enkephalin and β-endorphin causes retrograde amnesia for two different tasks in rats. Behav Neural Biol 28:246-250, 1980.

Izumi, K., Motomatsu, T., Chrétien, M., Butterworth, R.F., Lis, M., Seidah, N., and Barbeau, A.: β-Endorphin induced akinesia in rats: Effect of apomorphine and α-methyl-p-tyrosine and related modifications of dopamine turnover in the basal ganglion. *Life Sci 20*:1149-1156, 1977.

Jacob, J.J., Tremblay, E.D., and Colombel, M.C.: Facilitation de reactions nociceptives pour la naloxone chez la souris et chez le rat. *Psychopharmacology 37*:217-223, 1974.

Jacobowitz, D.M., and O'Donohue, T.L.: α-Melanocyte stimulating hormone: Immunohistochemical identification and mapping in neurons of rat brain. *Proc Natl Acad Sci USA 75*:6300-6304, 1978.

Jacquet, Y.F.: Opiate effects after adrenocorticotropin or β-endorphin injection in the periaqueductal gray matter of rats. *Science 220:*1032-1034, 1978.

Jacquet, Y.F., and Marks, N.: C-fragment of β-lipotropin: An endogenous neuroleptic or antipsychotogen. *Science 194*:632-635, 1976.

Jakoubek, B., Semiginovský, B., Kraus, M., and Erdossová, R.: The alterations of protein metabolism of the brain cortex induced by anticipation stress and ACTH. *Life Sci 9* Part 1:1169-1179, 1970.

Jakoubek, B., Buresová, M., Hajék, I., Etrychova, J., Pavlík, A., and Dedicova, A.: Effect of ACTH on the synthesis of rapidly labelled RNA in the nervous system of mice. *Brain Res 43*:417-428, 1972.

Jhamandas, K., Sawynok, J., and Sutak, M.: Enkephalin effects on release of brain acetycholine. *Nature 269*:433-434, 1977.

Johnston, M.F., Kravitz, E.A., Meiri, H., and Rahamimoff, R.: Adrenocorticotropic hormone cause long-lasting potentiation of transmittor release from frog motor neuron terminals. *Science* 220:1071-1072, 1983.

Jolles, J., Wiegant, V.M., and Gispen, W.H.: Reduced behavioral effectiveness of ACTH$_{1-24}$ after a second administration: Interaction with opiates. *Neurosci Lett 9*:261-266, 1978.

Jolles, J., Rompa-Barendregt, J., and Gispen, W.H.: Novelty and grooming behavior in the rat. *Behav Neural Biol 25*:563-572, 1979.

Jouhaneau-Bowers, M., and Le Magnen, J.: ACTH self-administration in rats. *Pharmacol Biochem Behav 10*:325-328, 1979.

Kastin, A.J., Dempsey, G.L., LeBlanc, B., Dyster-Aas, K., and Schally, A.V.: Extinction of an appetitive operant response after administration of MSH. *Horm Behav 5*:135-139, 1974.

Kastin, A., Sandman, C., Stratton, L., Schally, A., and Miller, L.: Behavioral and electrographic changes in rat and man after MSH. *Progr Brain Res 42*:143-150, 1975.

Kastin, A.J., Scollan, E.L., King, M.G., Schally, A.V., and Coy, D.H.: Enkephalin and a potent analog facilitate maze performance after intraperitoneal administration in rats. *Pharmacol Biochem Behav 5*:691-695, 1976.

Kastin, A.J., Olson, R.D., Schally, A.V., and Coy, D.H.: CNS effects of peripherally administered brain peptides. *Life Sci 25*:401-414, 1979a.

Kastin, A.J., Jemison, M.T., and Coy, D.H.: Analgesia after

peripheral administration of enkephalin and endorphin analogues. *Pharmacol Biochem Behav 11*:713-716, 1979b.

Katz, R.J.: Stress induced Straub tail elevation: Further
behavioral evidence in rats for the involvement of endorphins
in stress. *Neurosci Lett 13*:249-252, 1979a.

Katz, R.J.: ACTH$_{4-10}$ antagonism of morphine-induced behavioral
activation in the mouse. *Eur J Pharmacol 53*:383-385, 1979b.

Katz, R.J., and Gelbart, J.: Endogenous opiates and behavioral
responses to environmental novelty. *Behav Biol 24*:338-348,
1978.

Katz, R.J., and Roth, K.A.: Stress-induced grooming in the
rat--an endorphin mediated syndrome. *Neurosci Lett 13*:209-
212, 1979.

Katz, R.J., Carroll, B.J., and Baldrighi, G.: Behavioral
activation by enkephalins in mice. *Pharmacol Biochem Behav
8*:493-496, 1978.

Kendler, K., Hennessy, J.W., Smotherman, W.P., and Levine, S.:
An ACTH effect on recovery from conditioned taste aversion.
Behav Biol 17:225-229, 1976.

Keyes, J.B.: Effect of ACTH on ECS-produced amnesia of a
passive avoidance task. *Physiol Psychol 2*:307-309, 1974.

King, B.M., Castellanos, F.X., Kastin, A.J., Berzas, M.C.,
Mauk, M.D., Olson, G.A., and Olson, R.D.: Naloxone-induced
suppression of food intake in normal and hypothalamic obese
rats. *Pharmacol Biochem Behav 11*:729-732, 1979.

Klee, W.A.: Opioid peptides as modulators of cyclic AMP levels.
Adv Biochem Psychopharmacol 116:225-231, 1979.

Knepel, W., Nutto, D., and Hertting, G.: Evidence for
inhibition by β-endorphin of vasopressin release during foot
shock-induced stress in the rat. *Neuroendocrinology
34*:353-356, 1982.

Koffer, K.B., Berney, S., and Hornykiewicz, O.: The role of the
corpus striatum in neuroleptic- and narcotic-induced
catalepsy. *Eur J Pharmacol 47*:81-86, 1978.

Kostrzewa, R.M., Kastin, A.J., and Spirtes, M.: α-MSH and MIF-I
effects on catecholamine levels and synthesis in various rat
brain areas. *Pharmacol Biochem Behav 3*:1017-1023, 1975.

Kovács, G.L., Telegdy, G., and Lissák, K.: 5-hydroxytryptamine
and the mediation of pituitary-adrenocortical hormones in the
extinction of active avoidance behaviour. *Psychoneuroendocrinology 1*:219-230, 1976.

Krieger, D.T., Liotta, A.S., Hauser, H., and Brownstein, M.J.:
Effect of stress adrenocorticotropin or corticosteroid
treatment, adrenalectomy, or hypophysectomy on hypothalamic
immunoreactive adrenocorticotropin concentrations. *Endocrinology 105*:737-742, 1979a.

Krieger, D.T., Liotta, A.S., Suda, T., Goodgold, A., and
Condon, E.: Human plasma immunoreactive lipotropin and ad

renocorticotropin in normal subjects and in patients with pituitary-adrenal disease. *J Clin Endocrinol Metab* *48*:566-571, 1979b.

Krieger, D.T., Liotta, A.S., Nicholsen, G., and Kizer, J.S.: Brain ACTH and endorphin reduced in rats with monosodium glutamate induced arcuate nuclear lesions. *Nature 278*:562-563, 1979c.

Laasberg, L.H., Johnson, E.E., and Hedley–Whyte, J.: Effect of morphine and naloxone on leu-enkephalin–like immunoreactivity in dogs. *J Pharmacol Exp Ther 212*:496-502, 1980.

LaHoste, G.J., Olson, G.A., Kastin, A.J., and Olson, R.D.: Behavioral effects of melanocyte stimulating hormone. *Neurosci Biobehav Rev 4*:9-16, 1980.

Lal, H., Spaulding, T., and Fielding, S.: Swim-stress induced analgesia and lack of its naloxone antagonism. *Commun Psychopharmacol 2*:263-266, 1978.

Lande, S., Flexner, J.B., and Flexner, L.B.: Effect of corticotropin and desglycinamide[9]-lysine vasopressin on suppression of memory by puromycin. *Proc Natl Acad Sci USA 67*:558-560, 1972.

Landsberg, J.-W., and Weiss, J.: Stress and increase of the corticosterone level prevent imprinting in ducklings. *Behaviour L 57*:3-4, 1976.

Lanthorn, T.H., and Isaacson, R.L.: Stretching and yawning: A role of glutamate. *Psychopharmacology 65*:317-318, 1979.

Lanthorn, T.H., Smith, M.A., and Isaacson, R.L.: Wet-dog shaking in the rat: Possible involvement of a kappa opiate receptor. *Neuropharmacolology 18*:743-745, 1979.

Larsson, L.I., Childers, S., and Snyder, S.H.: Met- and leu-enkephalin immunoreactivity in separate neurons. *Nature 282*:407-410, 1979.

Law, P.-Y., Loh, H.H., and Li, C.H.: Properties and localization of β-endorphin receptor in rat brain. *Proc Natl Acad Sci USA 76*:5455-5459, 1979.

Lee, R.L., Sewell, R.D.E., and Spencer, P.S.J.: Importance of 5-hydroxytryptamine in the antinociceptive activity of the leucine-enkephalin derivative, D-ala$_1$-leu^5-enkephalin (BW.180C), in the rat. *Eur J Pharmacol 47*:251-253, 1978.

Le Moal, M.L., Koob, G.F., and Bloom, F.E.: Endorphins and extinction: Differential actions on appetitive and adverse tasks. *Life Sci 24*:1631-1636, 1979.

Leonard, B.E.: The effect of two synthetic ACTH analogues on the metabolism of biogenic amines in the rat brain. *Arch Int Pharmacodyn Ther 207*:242-253, 1974.

Leonard, B.E., Kafoe, W.F., Thody, A.F., and Shuster, S.: The effect of α-melanocyte stimulating hormone (α-MSH) on the metabolism of biogenic amines in the rat brain. *J Neurosci Res 2*:39-45, 1976.

Leshner, A.I.: An Introduction to Behavioral Endocrinology. Oxford University Press, New York, 1978.

Levine, J.D., Gordon, N.C., and Fields, N.L.: The mechanism of placebo analgesia. Lancet 2:654-657, 1978.

Levine, J.D., Gordon, N.C., and Fields, H.L.: Naloxone dose dependently produces analgesia and hyperalgesia in post-operative pain. Nature 278:740-741, 1979.

Levine, S., and Jones, L.E.: Adrenocorticotropic hormone (ACTH) and passive avoidance learning. J Comp Physiol Psychol 59:357-360, 1965.

Lewis, J.W., Canon, J.T., and Liebeskind, J.C.: Opioid and nonopioid mechanisms of stress analgesia. Science 208:623-625, 1980a.

Lewis, R.V., Stern, A.S., Kimura, S., Rossier, J., Stein, S., and Udenfriend, S.: An about 50,000-dalton protein in adrenal medulla: a common precursor of [met]- and [leu]enkephalin. Science 208:1459-1461, 1980b.

Ley, K.F., and Corson, J.A.: Effects of ACTH adrenalectomy and time of day on emotional activity of the rat. Behav Biol 9:111-115, 1973.

Lichtensteiger, W., and Lienhart, R.: Response of mesencephalic and hypothalamic dopamine neurones to α-MSH: Mediated by area postrema. Nature 266:635-637, 1977,

Lichtensteiger, W., and Monnet, F.: Differential response of dopamine neurons to α-melanotropin and analogues in relation to their endocrine and behavioral potency. Life Sci 25:2079-2087, 1979.

Lipton, J.M., and Glyn, J.R.: Central administration of peptides alters thermoregulation in the rabbit. Peptides 1:15-18, 1980.

Lissák, K., and Bohus, B: Pituitary hormones and avoidance behavior of the rat. Int J Psychobiol 2:103-115, 1972.

Lundberg, J.M., Hamberger, S., Schültzberg, M., Hökfelt, T., Gramberg, P.-O., Efendic, S., Terenius, L., Goldstein, M., and Luft, R.: Enkephalin- and somatostatin-like immunore-activities in human adrenal medulla and pheochromocytoma. Proc Natl Acad Sci USA 76:4079-4083, 1979.

Madden, J., Akil, A., Patrick, R.L., and Barchas, J.D.: Stress-induced parallel changes in central opioid levels and pain responsiveness in the rat. Nature 265:358-360, 1977.

Malizia, E., Andreucci, G., Paolucci, D., Crescenzi, F., Fabbri, A., and Fraioli, F.: Electroacupuncture and peripheral β-endorphin and ACTH levels. Lancet 2:535-536, 1979.

Margules, D.L., Moisset, B., Lewis, M.J., Shiuya, H., and Pert, C.B.: β-Endorphin is associated with over-eating in genetically obese mice (ob/ob) and rats (fa/fa). Science 202:989, 1978.

Martin, G.E., and Bacino. C.B.: Action of intracerebrally

injected β-endorphin on the rat's core temperature. *Eur J Pharmacol* 59:227-236, 1979.

Martin, J.T.: Imprinting behaviour: Influence of vasopressin and ACTH analogues. *Psychoneuroendocrinology* 3: 261-269, 1979.

Martinez, J.L., Vasquez, B.J., Jensen, R.A., Soumireu-Mourat, B., and McGaugh, J.L.: ACTH₄₋₉ analogue (ORG 2766) facilitates aquisition of inhibitory avoidance response in rats. *Pharmacol Biochem Behav* 10:145-147, 1979.

McCulloch, J., Kelly, P.A.T., and van Delft, A.M.L.: Alterations in local cerebral glucose utilization during chronic treatment with an ACTH₄₋₉ analog. *Eur J Pharmacol* 78:151-158, 1982.

McGivern, R., Berka, C., Berntson, G.G., Walker, J.M., and Sandman, C.: Effect of naloxone on analgesia induced by food deprivation. *Life Sci* 25:885-888, 1979.

Mello, N.K., and Mendelson, J.H.: Self administration of an enkephalin analogue by rhesus monkey. *Pharmacol Biochem Behav* 9:579-586, 1978.

Meltzer, H.Y., and So, R.: Effect of morphine, β-endorphin and leu-enkephalin on ³H-spiroperidol binding to bovine pituitary membranes. *Life Sci* 25:531-536, 1979.

Messing, R.B., Jensen, R.A., Martinez, J.L., Jr., Spiehler, V.R., Vasquez, B.J., Soumireu-Mourat, B., Liang, K.C., and McGaugh, J.L.: Naloxone enhancement of memory. *Behav Neural Biol* 27:266-275, 1979.

Mezey, E., and Palkovits, M.: Two-way transport in the hypo-thalamolypophyseal system. In *Frontiers in Neuroendocrinology*. Vol. 7, W.F. Gronong and L. Martini, eds, Raven Press, New York, pp 1-29, 1982.

Miller, L.H., Kastin, A.J., Sandman, C.A., Fink, M., Van Veen, W.J.: Polypeptide influences on attention, memory and anxiety in man. *Pharmacol Biochem Behav* 2:663-668, 1974.

Miller, L.H., Harris, L.C., Van Riezen, H., and Kastin, A.J.: Neuroheptapeptide influence on attention and memory in man. *Pharmac Biochem Behav* 5 (Suppl 1):17-22, 1976.

Miller. L.H., Kastin, A.J., and Sandman, C.A.: Psychobiological actions of MSH in man. *Front Horm Res* 4:153-161, 1977a.

Miller, L.H., Fischer, C., Groves, G.A., Rudrauff, M.E., and Kastin, A.J.: MSH/ACTH₄₋₁₀ influences on the CAR in human subjects: A negative finding. *Pharmacol Biochem Behav* 7:417-420, 1977b.

Miller, L.H., Groves, G.A., Bopp, M.J., and Kastin, A.J.: A neuroheptapeptide influence on cognitive functioning in the elderly. *Peptides* 1:55-57, 1980.

Miller, R.E., and Ogawa, N.: The effect of adrenocorticotrophic hormone (ACTH) on avoidance conditioning in the adrenal-ectomized rat. *J Comp Physiol Psychol* 55:211-213, 1962.

Miller, R.J., Chang, K,-J., Cooper, B., Cuatrecasas, P.: Radioimmunoassay and characterization of enkephalins in rat tissue. *J Biol Chem 253*:531-538, 1978.

Minamino, N., Kitamura, K., Hayashi, Kangawa, K., and Matsuo, H.: Regional distribution of α-neo-endorphin in rat brain and pituitary. *Biochem Biophys Res Commun 102*:226-234, 1981.

Minneman, K.P., and Iversen, L.L.: Enkephalin and opiate narcotics increase cyclic GMP accumulation in slices of the neostriatum. *Nature 268*:313-314, 1976.

Moldow, R., and Yalow, R.S.: Extrahypophysial distribution of corticotropin as a function of brain size. *Proc Natl Acad Sci USA 75*:994-998, 1978.

Morley, J.E.: The endocrinology of the opiates and opioid peptides. *Metabolism 30*:195-209, 1981.

Moroni, F., Cheney, D.L., and Costa, E.: The turnover rate of acetycholine in brain nuclei of rats injected intraventricularly and intraseptally with alpha and beta endorphin. *Neuropharmacology 17*:191-196, 1978.

Moroni, F., Peralta, E., Cheney, D.L., and Costa, E.: On the regulation of γ-aminobutyric acid neurons in caudatus, pallidus and nigra: Effects of opioids and dopamine agonists. *J Pharmacol Exp Ther 208*:190-194, 1979.

Moyer, J.A., and Leshner, A.I.: Pituitary-adrenal effects on avoidance-of-attack in mice: Separation of the effects of ACTH and corticosterone. *Physiol Behav 17*:297-301, 1976.

Murphy, J.V., and Miller, R.E.: The effect of adrenocorticotropic hormone (ACTH) on avoidance conditioning in the rat. *J Comp Physiol Psychol 48*:47-49, 1955.

Nakamura, M., Matsuda, S., Ueda, M., and Tanaka, A: Depressor effect of synthetic peptides related to ACTH on blood pressure in rats. *Experientia 32*:368-369, 1976.

Nakao, N., Nakai, Y., Oki, S., Matsubara, S., Konishi, T., Nishitani, H., and Imura, H,: Immunoreactive β-endorphin in human cerebrospinal fluid. *J Clin Endocrinol Metab 50*:230-233, 1980.

Nisticò, G., DiGiorgio, R.M., Naccari, F., and Calatroni, A.: Effects of intraventricular beta-endorphin on GABA system in some areas of chick brain. *Res Commun Chem Pathol Pharmacol 26*:469-478, 1979.

Nock, B.L., and Leshner, A.I.: Hormonal mediation of the effects of defeat on agonistic responding in mice. *Physiol Behav 17*:111-119, 1976.

North, R.A.: Opiates opioid peptides and single neurons. *Life Sci 24*:1527-1546, 1978.

O'Donohue, T.L., Miller, R.L., and Jacobowitz, D.M.: Identification, characterization and stereotaxic mapping of intraneuronal α-melanocyte-stimulating hormone—like immunoreactive peptides in discrete regions of the rat brain. *Brain Res 176*:101-123, 1979.

Oliver, C., Mical, R.S., and Porter, J.C.: Hypothalamic-pituitary vasculature: Evidence for retrograde blood flow in the pituitary stalk. *Endocrinology 101*:598-604, 1977.

Olson, G.A., Olson, R.D., Kastin, A.J., and Coy, D.H.: Endogenous opiates: Through 1978. *Neurosci Biobehav Rev 3*:285-299, 1979b.

Olson, G.A., Olson, R.D., Kastin, A.J., Green, M.T., Roig-Smith, R., Hill, C.W., and Coy, D.H.: Effects of an enkephalin analog on complex learning in the rhesus monkey. *Pharmacol Biochem Behav 11*:341-345, 1979a.

Olson, R.D., Kastin, A.J., Olson, G.A., and Coy, D.H.: Behavioral effects after systemic injection of opiate peptides. *Psychoneuroendocrinology 5*:47-52, 1980.

Osborne, H., Höllt, V., and Herz, A.: Subcellular distribution of enkephalins and endogenous opioid activity in rat brain. *Life Sci 22*:611-618, 1978.

Oyama, T., Jin, T., Yamaya, R., Ling, N., and Guillemin, R.: Profound analgesic effects of β-endorphin in man. *Lancet 122*, 1980.

Pedigo, N.W., Ling, N.C., Reisine, T.D., and Yamamura, H.I.: Examination of des-tyrosine 1-γ-endorphin activity at ^{3}H-spiroperidol binding sites in rat brain. *Life Sci 24*:1645-1650, 1979a.

Pedigo, N.W., Schallert, T., Overstreet, D.H., Ling, N.C., Ragan, P., Reisine, T.D., and Yamamura, H.I.: Inhibition of in vivo ^{3}H-spiperone binding by the proposed antipsychotic des-tyr-γ-endorphin. *Eur J Pharmacol 60*:359-364, 1979b.

Plotnikoff, N.P., Kastin, A.J., Cox, D.H., Christensen, C.W., Schally, A.V., and Spirtes, M.A.: Neuropharmacological actions of enkephalins after systemic administration. *Life Sci 19*:1283-1288, 1976.

Pomeranz, B., Cheng, R., and Law, P.: Acupuncture reduces electrophysiological and behavioral responses to noxious stimuli, pituitary is implicated. *Exp Neurol 54*:172-178, 1977.

Ramaekers, F., Rigter, H., and Leonard, B.E.: Parallel changes in behaviour and hippocampal monoamine metabolism after administration of ACTH-analogues. *Pharmacol Biochem Behav 8*:547-551, 1978.

Rapoport, S.I., Klee, W.A., Pettigrew, K.D., and Ohno, K.: Entry of opioid peptides into the central nervous system. *Science 207*:84-86, 1980.

van Ree, J.M., Witter, A., and Leysen, J.E.: Interaction of des-tyrosine-γ-endorphin (DTγE, β-LPH62-77) with neuroleptic binding sites in various areas of rat brain. *Eur J Pharmacol 52*:411-413, 1978.

van Ree, J.M., Smyth, D.G., and Colpaut, F.C.: Dependence creating properties of lipotropin C-fragment (β-endorphin). Evidence for its internal control of behavior. *Life Sci*

24:495-502, 1979.

Rees, H.D., and Dunn, A.J.: The role of pituitary-adrenal system in the footshock-induced increase of [³H]lysine incorporation into mouse brain and liver proteins. *Brain Res 120*:317-325, 1977.

Rees, H.D., Dunn, A.J., and Iuvone, P.M.: Behavioral and biochemical responses of mice to the intraventricular administration of ACTH analogs and lysine vasopressin. *Life Sci 18*:1333-1340, 1976.

Rees, H.D., Verhoef, J., Witter, A., Gispen, W.H., and de Wied, D.: Autoradiographic studies with a behaviorally potent ³H-ACTH₄₋₉ analog in the brain after intraventricular injection in rats. *Brain Res Bull 5*:509-514, 1980.

Reith, M.E.A., Schotman, P., and Gispen, W.H.: Hypophysectomy, ACTH₁₋₁₀ and *in vitro* protein synthesis in rat brain stem slices. *Brain Res 81*:571-575, 1974.

Reith, M.E.A., Schotman, P., and Gispen, W.H.: Incorporation of [³H]leucine into brain stem protein fractions: The effect of a behaviorally active, N-terminal of ACTH in hypophysectomized rats. *Neurobiology 5*:355-368, 1975a.

Reith, M.E.A., Schotman, P., and Gispen, W.H.: The neurotropic action of ACTH: Effects of ACTH-like peptides on the incorporation of leucine into protein of brain stem slices from hypophysectomized rats. *Neurosci Lett 1*:55-59, 1975b.

van Riezen, H., Rigter, H., and de Wied, D.: Possible significance of ACTH fragments for human mental performance. *Behav Biol 20*:311-324, 1977.

Rigter, H.: Attenuation of amnesia in rats by systemically administered enkephalins. *Science 200*:83-85, 1978.

Rigter, H., and Popping, A.: Hormonal influences on the extinction of conditioned taste aversion. *Psychopharmacology 46*:255-261, 1976.

Rigter, H., Elbertse, R., and van Riezen, H.: Time-dependent anti-amnesic effect of ACTH₄₋₁₀ and desglycinamide-lysine vasopressin. *Progr Brain Res 42*:163-171, 1975.

Rigter, H., Hannan, T.J., Messing, R.B., Martinez, J.L., Vasquez, B.J., Jensen, R.A., Veliquette, J., and McGaugh, J.L.: Enkephalins interfere with acquisition of an active avoidance response. *Life Sci 26*:337-345, 1980.

Riley, A.J., Zellner, D.A., and Duncan, H.J.: The role of endorphins in animal learning and behavior. *Neurosci Biobehav Rev 4*:69-76, 1980.

Roemer, D., Buescher, H.H., Hill, R.C., Pless, J., Bauer, W., Cardinaux, F., Closse, A., Hauser, D., and Huguenin, R.: A synthetic enkephalin analogue with prolonged parenteral and oral analgestic activity. *Nature 268*:547-549, 1977.

Rossier, J., French, E.D., Rivier, C., Ling, N., Guillemin, R., and Bloom, F.E.: Foot-shock induced stress increases β-endorphin levels in blood but not brain. *Nature 270*:618-620, 1977.

Rossier, J., Guillemin, R., and Bloom, F.: Foot shock induced stress decreases leu-enkephalin immunoreactivity in rat hypothalamus. *Eur J Pharmacol 48*:465-466, 1978.

Rossier, J., French, E., Cros, C., Minick, S., Guillemin, R., and Bloom, F.E.: Adrenalectomy, dexamethasone or stress alters opioid peptides levels in rat anterior pituitary but not intermediate lobe or brain. *Life Sci 25*:2105-2112, 1979a.

Rossier, J., Rogers, J., Shibaski, T., Guillemin, R., and Bloom, F.E.: Opioid peptides and α-melanocyte-stimulating hormone in genetically obese (ob/ob) mice during development. *Proc Natl Acad Sci USA 76*:2077-2080, 1979b.

Rossier, J., Pittman, Q., Bloom, F., and Guillemin, R.: Distribution of opioid peptides in the pituitary: A new hypothalamic-pars nervosa enkephalinergic pathway. *Fed Proc 39*:2555-2560, 1980.

de Rotte, A.A., van Egmond, M.A.H., and van Wimersma Greidanus, Tj.B.: α-MSH levels in cerebrospinal fluid and blood of rats during behavioral manipulations. *Physiol Behav 28*:765-768, 1982.

Rubinstein, M., Stein, S., and Udenfriend, S.: Characterization of pro-opiocortin, a precursor to opioid peptides and corticotropin. *Proc Natl Acad Sci USA 75*:669-671, 1978.

Rudman, D.: Injection of melatonin into cisterna magna increases concentration of 3',5'-cyclic guanosine monophosphate in cerebrospinal fluid. *Neuroendocrinology 20*:235-242, 1976.

Rudman, D.: Effect of melanotropic peptides on adenosine 3', 5'-monophosphate accumulation by regions of rabbit brain. *Endocrinology 103*:1556-1561, 1978.

Rudman, D., and Isaacs, J.W.: Effect of intrathecal injection of melanotropic-lipolytic peptides on the concentration of 3',5'-cyclic adenosine monophosphate in cerebrospinal fluid. *Endocrinology 97*:1476-1480, 1975.

Rudman, D., Scott, J.W., del Rio, A.E., Houser, D.H., and Sheen, S.: Effect of melanotropic peptides on protein synthesis in mouse brain. *Am J Physiol 226*:687-692, 1974.

Rudman, D., Hollins, B.M., Lewis, N.C., and Scott, J.W.: Effects of hormones on 3',5'-cyclic adenosine monophosphate in choroid plexus. *Am J Physiol 232*, E353-E357, (1977).

de Ryck, M., Schallert, T., and Teitelbaum, P.: Morphine versus haloperidol catalepsy in the rat: A behavioral analysis of postural support mechanisms. *Brain Res 201*:143-172, 1980.

Sahgal, A., Iversen, S.D., Lewis, M.E., and Trimnell, L.E.: Effects of ACTH$_{4-10}$ on a conflict schedule in pigeons. *Commun Psychopharmacol 3*:211-216, 1979.

Sakurada, O., Sokoloff, L., and Jacquet, Y.F.: Local cerebral glucose utilization following injection of β-endorphin into periaqueductal gray matter in the rat. *Brain Res* *153*:403-407, 1978.

Sandberg, D.E., and Segal, M.: Pharmacological analysis of analgesia and self-stimulation elicited by electrical stimulation of catecholamine nuclei in the rat brain. *Brain Res* *152*:529-542, 1978.

Sandman, C.A., Kastin, A.J., and Schally, A.V.: Melanocyte-stimulating hormone and appetitive behavior, *Experientia* *25*:1001-1002, 1969.

Sandman, C.A., Kastin, A.J., and Schally, A.V.: Behavioral inhibition as modified by melanocyte-stimulating hormone (MSH) and light-dark conditions. *Physiol behav* *6*:45-48, 1971.

Sandman, C.A., Miller, L.H., Kastin, A.J., and Schally, A.V.: A neuroendocrine influence on attention and memory. *J Comp Physiol Psychol* *80*:54-58. 1972.

Sandman, C.A., Alexander, V.D., and Kastin, A.J.: Neuroendocrine influences on visual discrimination and reversal learning in the albino and hooded rat. *Physiol Behav* *11*:613-617, 1973a.

Sandman, C.A., Kastin, A.J., Schally, A.V., Kendall, J.W., and Miller, L.J.: Neuroendocrine responses to physical and psychological stress. *J Comp Physiol Psychol* *84*:386-390, 1973b.

Sandman, C.A., Beckwith, B.E., and Gittis, M.M.: Melanocyte-stimulating hormone (MSH) and overtraining effects on extradimensional shift (EDS) learning. *Physiol Behav* *13*:163166, 1974.

Sandman, C.A., George, J., Nolan, J., van Riezen, H., and Kastin, A.: Enhancement of attention in man with ACTH/MSH$_{4-10}$. *Physiol Behav* *15*:427-431, 1975.

Sandman, C.A., George, J., McCanne, T.R., Nolan, J.D., Kaswan, J., and Kastin, A.J.: MSH/ACTH$_{4-10}$ influences behavioral and physiological measures of attention. *J Clin Endocrinol Metab* *44*:884-891, 1977.

Sandman, C.A., Walker, B.B., and Lawton, C.A.: An analog of MSH/ACTH$_{4-9}$ enhances personal and environmental awareness in mentally retarded adults. *Peptides* *1*:109-114, 1980.

Sands, S.F., and Wright, A.A.: Enhancement and disruption of retention performance by ACTH in a choice task. *Behav Neural Biol* *27*:413-422, 1979.

Sar, M., Stumpf, W.E., Miller, R.J., Chang, K.-J., and Cuatrecasas, P.: Immunohistochemical localization of enkephalin in rat brain and spinal cord. *J Comp Neurol* *182*:17-38, 1978.

Sawynok, J., Pinksy, C., and LaBella, F.S.: On the specificity of naloxone as an opiate antagonist. *Life Sci* *25*:1621-1632, 1979.

Schneider, A.M., Weinberg, J., and Weissberg, R.: Effects of

ACTH on conditioned suppression: A time and strength of conditioning analysis. *Physiol Behav* 13:633-636, 1974.

Schotman, P., Gispen, W.H., Janz, W.S., and de Wied, D.: Effects of ACTH analogues on macromolecule metabolism in the brain stem of hypophysectomized rats. *Brain Res* 46:347-362, 1972.

Schotman, P., Reith, M.E.A., van Wimersma Griedanus, Tj.B., Gispen, W.H., and de Wied, D.: Hypothalamic and pituitary peptide hormones and the central nervous system. With special reference to the neurochemical effects of ACTH. In *Molecular and Functional Neurobiology*, W.H. Gispen, ed. Elsevier, Amsterdam, 1976, pp 310-366.

Segal, D.S., Browne, R.G., Bloom, F., Ling, N., and Guillemin, R.: β-Endorphin: Endogenous opiate or neuroleptic? *Science* 198:411-414, 1977.

Semiginovský, B., and Jakoubek, B.: Effects of ACTH on the incorporation of L-[U-^{14}C]leucine into the brain and spinal cord of mice. *Brain Res* 35:319-323, 1971.

Semiginovský, B., Jakoubek, B., Kraus, M., and Erdösova, R.: Stress-induced changes of the amino acid uptake and protein synthesis in the brain cortex slices of rats: Effects of adrenalectomy. *Physiol Bohemoslov* 23:503-510, 1974.

Shelman, I.M., Ponomareva-Stepnaya, M.A., Maksimova, L.A., Nezovibati'ko, V.N., and Ashmarin, I.P.: The effect of adrenocorticotropic hormone and tripeptides glu-his-L-phe and glu-his-D-phe on the imprinting in the beetle *Tenerio molitor*. *J Biol Biochem Physiol* 14:398-400, 1978.

Simantov, R., Kuhar, M.J., Pasternak, G.W., and Snyder, S.H.: The regional distribution of a morphine-like factor enkephalin in monkey brain. *Brain Res* 106:189-197, 1976.

Sjölund, B., Terenius, L., Eriksson, M.: Increased cerebrospinal fluid levels of endorphins after electro-acupuncture. *Acta Physiol Scand* 100:382-384, 1977.

Smyth, D.G., Massey, D.E., Zakarian, S., and Finnie, M.D.A.: Endorphins are stored in biologically active and inactive forms: Isolation of α-N-acetyl peptides. *Nature* 279:252-254, 1979.

Somoza, E.: Influence of neuroleptides on the binding of met-enkephalin, morphine and dihydromorphine to synaptosome-enriched fractions of rat brain. *Neuropharmacology* 17:577-581, 1978.

Spiaggia, A., Bodnar, R.J., Kelly, D.D., and Glusman, M.: Opiate and non-opiate mechanisms of stress-induced analgesia: Cross-tolerance between stressors. *Pharmacol Biochem Behav* 10:761-765, 1979.

Spirtes, M.A., Kostrzewa, R.M., and Kastin, A.J.: α-MSH and MIF-I effects on serotonin levels and accumulation in various rat brain areas. *Pharmacol Biochem Behav* 3:1011-1015, 1975.

Stein, L., and Belluzzi, J.D.: Brain endorphins: Possible role in reward and memory formation. *Fed Proc 38*:2468-2474, 1979.

Steiner, F.A., Ruf, K., and Akert, K.: Steroid-sensitive neurons in rat brain: Anatomical localization and responses to neurohumors and ACTH. *Brain Res 12*:74-85, 1969.

Stine, S.M., Yang, H.-Y.T., and Costa, E.: Inhibition of in situ metabolism of [^{3}H](met^5)-enkephalin and potentiation of (met^5)-enkephalin analgesia by captopril. *Brain Res 188*:295-299, 1980.

Stinus, L., Koob, G.F., Ling, N., Bloom, F.E., and Le Moal, M.: Locomotor activation induced by infusion of endorphins into the ventral tegmental area: Evidence for opiate-dopamine interactions. *Proc Natl Acas Sci USA 77*:2323-2327, 1980.

Stratton, L.O., and Kastin, A.J.: Avoidance learning at two levels in receiving MSH. *Horm Behav 5*:149-155, 1974.

Stratton, L.O., and Kastin, A.J.: Increased acquisition of a complex appetitive task after MSH and MIF. *Pharmacol Biochem Behav 3*:901-904, 1975.

Subramanian, N., Mitznegg, P., Sprugel, W., Domschke, W., Domschke, S., Wunsch, E., and Demling, L.: Influence of enkephalin on K$^+$-evoked efflux of putative neurotransmitters in rat brain. Selective inhibition of acetylcholine and dopamine release. *Naunyn Schmiedebergs Arch Pharmacol 299*:163-165, 1977.

Suda, T., Liotta, A.S., and Krieger, D.T.,: β-Endorphin is not detectable in plasma from normal human subjects. *Science 202*:221-223, 1978.

Terenius, L.: Endogenous peptides and analgesia. *Annu Rev Pharmacol Toxicol 18*:189-204, 1978.

Terenius, L., Gispen, W.H., and de Wied, D.: ACTH-like peptides and opiate receptors in the rat brain: Structure activity studies. *Eur J. Pharmacol 33*:395-399, 1975.

Tintner, R., Dunn, A.J., Iuvone, P.M., Shukla, J.B., and Rennert, O.M.: Corticotrophin increases cerebral polyamine content. *J Neurochem 33*:1067-1073, 1979.

Tseng, L.F., Loh, H.H., and Li, C.H.: β-endorphin as a potent analgesic by intravenous injection. *Nature 263*:239-240, 1976.

Tseng, L.-F., Loh, H.H., and Li, C.H.: [D-Thr$_2$,Thz$_5$]-enkephalinamide: A potent analgesic by subcutaneous and oral administration. *Life Sci 23*:2053-2056, 1978.

Tseng, L.-F., Ostwald, T.J., Loh, H.H., and Li, C.H.: Behavioral activities of opioid peptides and morphine sulfate in golden hamsters and rats. *Psychopharmacology 64*:215-218, 1979.

Urban, I., and de Wied, D.: Changes in excitability of the theta activity generating substrate by ACTH$_{4-10}$ in the rat. *Exp Brain Res 24*:325-334, 1976.

Van Delft, A.M.L., and Kitay, J.I.: Effects of ACTH on simple unit activity in the diencephalon of intact and hypophysectomized rats. *Neuroendocrinology* 9:188-196, 1972.

Van Dijk, A.M.A.: Determination of radioimmunoassayable ACTH in plasma and/or brain tissue of rats as influenced by endocrine and behavioral manipulations. Ph.D. Thesis, University of Utrecht, 1979.

Van Loon, G.R., and De Souza, E.B.: Effects of β-endorphin on brain serotonin metabolism. *Life Sci* 12:971-978, 1978.

Van Loon, G.R., and Kim, G.: Dopaminergic mediation of β-endorphin-induced catalepsy. *Res Commun Chem Pathol Pharmacol* 21:37-44, 1978.

Van Loon, G.R., De Souza, E.B., and Kim, C.: Alterations in brain dopamine and serotonin metabolism during the development of tolerance to human β-endorphin in rats. *Can J Physiol Pharmacol* 56:1067-1071, 1978.

Van Loon, G.R., Ho, D., and Kim, C.: Beta-endorphin-induced decrease in hypothalamic dopamine turnover. *Endocrinology* 106:76-80, 1980.

Van Vugt, D.A., and Meites, J.: Influence of endogenous opiates on anterior pituitary function. *Fed Proc* 39:2533-2538, 1980.

Van Vugt, D.A., Bruni, J.F., Sylvester, P.W., Chem, H.T., Ieiri, T., and Meites, J.: Interaction between opiates and hypothalamic dopamine on prolaction release. *Life Sci* 24:2361-2368, 1979.

Vaught, J.L., and Takemori, A.E.: A further characterization of the differential effects of leucine enkephalin, methionine enkephalin and their analogs on morphine-induced analgesia. *J Pharmacol Exp Ther* 211:280-283, 1979.

Veith, J.L., Sandman, C.A., George, J.M., and Stevens, V.C.: Effects of MSH/ACTH$_{4-10}$ on memory attention and endogenous hormone levels in women. *Physiol Behav* 20:43-50, 1978a.

Veith, J.L., Sandman, C.A., and Walker, J.M.: Systemic administration of endorphins selectively alters open field behavior in rats. *Physiol Behav* 20:539-542, 1978b.

Verhoef, J., and Witter, A.: *In vivo* fate of a behaviorally active ACTH$_{4-9}$ analog in rats after systemic administration. *Pharmacol Biochem Behav* 4:583-590, 1976.

Verhoef, J., Palkovits, M., and Witter, A.: Specific uptake of a behaviorally potent [^{3}H]ACTH$_{4-9}$ analog in septal area after intraventricular injection in rats. *Brain Res* 131:117-128, 1977.

Versteeg, D.: Effects of two ACTH-analogs on noradrenaline metabolism in rat brain. *Brain Res* 49:483-485, 1973.

Versteeg, D.H.G.: Interactions of peptides related to ACTH, MSH, and β-LPH with neurotransmitters in the brain. *Pharmacol Ther* 11:535-588, 1980.

Versteeg, D.H.G., and Wurtman, R.J.: Effects of ACTH$_{4-10}$ on the rate of synthesis of [^{3}H]catecholamines in the brains of intact, hypophysectomized and adrenalectomized rats. *Brain Res 93*:552-557, 1975.

Versteeg, D.H.G., Gispen, W.H., Schotman, P., Witter, A., and de Wied, D.: Hypophysectomy and rat brain metabolism: Effects of synthetic ACTH analogs. *Adv Biochem Psychophamacol 6*:219-239, 1972.

Versteeg, D.H.G., de Kloet, E.R., and de Wied, D.: Interaction of endorphins with brain catecholamine systems. In *Characteristics and Function of Opioids*. J.M. van Ree and L. Terenius, eds. Elsevier, Amsterdam, 1978, pp 323-331.

Versteeg, D.H.G., de Kloet, E.R., and de Wied, D.: Effects of α-endorphin, β-endorphin and (des-tyr$_1$)-γ-endorphin on α-MPT-induced catecholamine disappearance in discrete regions of the rat brain. *Brain Res 179*:85-92, 1979.

Versteeg, D.H.G., Kovåcs, G.L., Bohus, B., de Kloet, E.R., and de Wied, D.: Effect of des-tyr^1-γ-endorphin and des-tyr^1-α-endorphin on a α-MPT-induced catecholamine disappearance in rat brain nuclei: a dose-response study. *Brain Res 231*:343-351, 1982.

Viveros, O.H., Diliberto, E.J., Hazum, E., and Chong, K.-J.: Opiate-like materials in the adrenal-medulla: Evidence for storage and secretion with catecholamines. *Mol Pharmacol 16*:1101-1108, 1979.

Walczak, S.A., Wilkening, D., and Makman, M.H.: Interaction of morphine, etorphine and enkephalins with dopamine-stimulated adenylate cyclase of monkey amygdala. *Brain Res 160*:105-116, 1979.

Ward, M.M., Sandman, C.A., George, J.M., and Shulman, H.,: MSH/ACTH$_{4-10}$ in men and women: Effects upon performance of an attention and memory task. *Physiol Behav 22*:669-673, 1979.

Watson, S.J., and Akil, H.: α-MSH in rat brain: Occurrence within and outside of β-endorphin neurons. *Brain Res 182*:217-223, 1980.

Watson, S.J., and Barchas, J.D.: Anatomy of the endogenous opioid peptides and related substances: The enkephalins, β-endorphin, β-lipotropin, and ACTH. In *Mechanisms of Pain and Analgesic Compounds*, R.F. Beers, and E.G. Bassett, eds. Raven Press, New York, 1979, pp 227-237.

Watson, S.J., Richard, C.W., and Barchas, J.D.: Adrenocorticotrophin in rat brain. Immunocytochemical localization in cells and axons. *Science 200*:1180-1182, 1978.

Watson. S.J., Richard, C.W., Ciaranello, R.D., and Barchas, J.D.: Interaction of opiate peptide and noradrenalin systems: Light microscopic studies. *Peptides 1*:23-30, 1980.

Weber, E., Roth, K.A., and Barchas, J.D.: Immunohistochemical distribution of α-neo-endorphin/dynorphin neuronal systems in rat brain: evidence for colocalization. *Proc Nat Acad Sci USA 79*:3062-3066, 1982.

346 Dunn

Wei, E.T., Tseng, L.F., Loh, H.H., and Li, C.H.: Comparison of
the behavioral effects of β-endorphin and enkephalin analogs.
Life Sci 21:321-328, 1977.
Weinberger, S.B., Arnsten, A., and Segal, D.S.: Des-tyrosine[1]-
γ-endorphin and haloperidol: Behavioral and biochemical
differentiation. *Life Sci 24*:1637-1644, 1979.
Wertheim, G.A., Conner, R.L., and Levine, S.: Adrenocortical
influence on free-operant behavior. *J Exp Anal Behav*
10:555-563, 1967.
de Wied, D.: Influence of anterior pituitary on avoidance
learning and escape behavior. *Am J Physiol 207*:255-259, 1964.
de Wied, D.: The influence of the posterior and intermediate
lobe of the pituitary and pituitary peptides in the maintenance of a conditioned avoidance response in rats. *Int J*
Neuropharmacol 4:157-167, 1965.
de Wied, D.: Inhibitory effects of ACTH and related peptides on
extinction of conditioned avoidance behavior in rats. *Proc*
Soc Exp Biol Med 122:28-32, 1966.
de Wied, D.: Opposite effects of ACTH and glucocorticoids on
extinction of conditioned avoidance in rats. *Exc Med Int*
Congr Ser 132:945-951, 1967.
de Wied, D.: Peptides and behavior. *Life Sci 20*:195-204, 1977.
de Wied, D., and Gispen, W.H.: Behavioral effects of peptides.
In *Peptides in Neurobiology*, H. Gainer, ed. Plenum Press, New
York, 1977, pp 397-448.
de Wied, D., Witter, A., and Greven, H.M.: Behaviorally active
ACTH analogues. *Biochem Pharmacol 24*:1463-1468, 1975.
de Wied, D., Bohus, B., van Ree, J.M., and Urban, I.: Behavioral and electrophysiological effects of peptides related to
lipotropin. *J Pharmacol Exp Ther 204*:570-580, 1978a.
de Wied, D., Kovács, G.L., Bohus, B., van Ree, J.M., and
Greven, H.M.: Neuroleptic activity of the neuropeptide β-LPH
(Des-Tyr[1]]γ-endorphin; DTγE). *Eur J Pharmacol 49*:427-436,
1978b.
de Wied, D., van Ree, J.M., and Greven, H.M.: Neuroleptic-like
activity of peptide related to [Des-Tyr[1]]γ-endorphin:
Structure activity studies. *Life Sci 26*:1575-1579, 1980.
Wiegant, V.M., and Gispen, W.H.: ACTH-induced excessive
grooming in the rat: Latent activity of $ACTH_{4-10}$. *Behav Biol*
19:554-558, 1977.
Wiegant, V.M., Gispen, W.H., Terenius, L., and de Wied, D.:
ACTH-like peptides and morphine: Interaction at the level of
the CNS. *Psychoneuroendocrinology 2*:63-69, 1977a.
Wiegant, V.M., Cools, A.R., and Gispen, W.H.: ACTH-induced
excessive grooming involves brain dopamine. *Eur J Pharmacol*
41:343-345, 1977b.
Wiegant, V.M., Colbern, D., van Wimersma Greidanus, T.B., and
Gispen, W.H.: Differential behavioral effects of

ACTH $_{4-10}$ and [D-Phe^7]ACTH $_{4-10}$. *Brain Res Bull* *3*:167-170, 1978.

Wiegant, V.M., Dunn, A.J., Schotman, P., and Gispen, W.H.: ACTH-like neurotropic peptides: Possible regulators of rat brain cyclic AMP. *Brain Res* *168*:565-584, 1979a.

Wiegant, V.M., Jolles, J., Colbern, D.L., Zimmermann, E., and Gispen, W.H.: Intracerebroventricular ACTH activates the pituitary-adrenal system: Dissociation from a behavioral response. *Life Sci* *25*:1791-1796, 1979b.

Wiegant, V.W., Zwiers, H., and Gispen, W.H.: Neuropeptides and brain cAMP and phosphoproteins. *Pharmacol Ther* *12*:463-490, 1981.

van Wimersma Graidanus, Tj.B.: Effects of MSH and related peptides on avoidance behavior in rats. *Front Horm Res* *4*:129-139, 1977.

van Wimersma Greidanus, Tj.B., and de Wied, D.: Effects of systemic and intracerebral administration of two opposite acting ACTH-related peptides on extinction of conditioned avoidance behavior. *Neuroendocrinology* *7*:291-301, 1971.

van Wimersma Greidanus, Tj.B., Rees, L.H., Scott, A.P., Lowry, P.J., and de Wied, D.: ACTH release during passive avoidance behavior. *Brain Res Bull* *2*:101-104, 1977.

van Wimersma Greidanus, Tj.B., van Dijk, A.M.A., de Rotte, A.A., Goedermans, J.H.J., Croiset, G., and Thody, A.J.: Involvement of ACTH and MSH in active and passive avoidance behavior. *Brain Res Bull* *3*:227-230, 1978.

van Wimersma Greidanus, Tj.B., Croiset, G., and Schilling, G.A.: Fornix transection: Discrimination between neuropeptide effects on attention and memory. *Brain Res Bull* *4*:625-629, 1979a.

van Wimersma Greidanus, Tj.B., Thody, T.J., Verspaget, H., de Rotte, G.A., Goedemans, H.J.B., Croiset, G., and van Ree, J.M.: Effects of morphine and β-endorphin on basal and elevated plasma levels of α-MSH and vasopressin. *Life Sci* *24*:579-586, 1976b.

Witter, A: On the presence of receptors for ACTH neuropeptides in the brain. In *Receptors for Neurotransmitters and Peptide Hormones*, G. Pepen, M.J. Kuhar, and S.J. Enna, eds. Raven Press, New York, 1980; pp 407-414.

Wollemann, M., Szebeni, A., Bajusz, S., and Graf, L.: Effects of met-enkephalin and (D-Met^2,Pro^5)-enkephalinamide on the adenylate cyclase activity of rat brain. *Neurochem Res* *4*:627-631, 1979.

Wood, P.L., Malthe-Sorenssen, D., Cheney, D.L., and Costa, E.: Increase of hippocampal acetylcholine turnover rate and the stretching-yawning syndrome elicited by alpha-MSH and ACTH. *Life Sci* *20*:673-678, 1978.

Wood, P.L., Cheney, D.L., and Costa, E.: Modulation of the

turnover rate of hippocampal acetylcholine by neuropeptides: Possible site of action of α-melanocyte-stimulating hormone, adrenocorticotrophic hormone and somatostatin. *J Pharmacol Exp Ther 209*:97-103, 1979.

Wouters, W., and van den Bercken, J.: Effects of ACTH$_{4-10}$ on synaptic transmission in frog sympathetic ganglion. *Eur J Pharmacol 57*:353-363, 1979.

Yamada, K., and Furukawa, T.: Direct evidence of involvement of dopaminergic inhibition and cholinergic activation in yawning. *Psychopharmacology 67*:39-43, 1980.

Yang, H.-Y., Hong, J.S., and Costa, E.: Regional distribution of leu and met enkephalins in rat brain. *Neuropharmacology 16*:303-307, 1977.

Zamir, N., and Segal, M.: Hypertension-induced analgesia: Changes in pain sensitivity in experimental hypertensive rats. *Brain Res 160*:170-173, 1979.

Zomzely-Neurath, C., and Keller, A.: The different forms of brain enolase: Isolation, characterization, cell specificity and physiological significances. In *Mechanisms, Regulation and Special Functions of Protein Synthesis in the Brain*, S. Roberts, A. Lajtha, and W.H. Gispen, eds. Elsevier, Amsterdam, 1977, pp 279-298.

Zwiers, H., Veldhuis, H.J., Schotman, P., and Gispen, W.H.: ACTH, cyclic nucleotides and brain protein phosphorylation in vitro. *Neurochem Res 1*:669-677, 1976.

The Role of Endorphins in Neurobiology, Behavior, and Psychiatric Disorders

GEORGE KOOB, MICHEL LEMOAL, AND FLOYD E. BLOOM

THE ROLE OF PEPTIDES IN NEUROBIOLOGY

An impressive number of naturally occurring peptides has
recently been found in the brain and many of them have been
localized to brain regions presumptively significant for behavior, such as neocortex (Emson and Lindvall, 1979). Prominent among these peptides and likely to play important roles
in the cerebral cortex, especially of primates, are such
unlikely candidates as vasoactive intestinal polypeptide (VIP)
(see Emson and Lindvall, 1979), the terminal octapeptide of
cholecystokinin (CCK) (Innis et al, 1979; Pinget et al., 1978;
Straus and Yalow, 1978), and substance P (Gale et al., 1978;
Hökfelt et al., 1976; Paxinos et al., 1978). Other peptides,
such as pancreatic polypeptide, enkephalin, neurotensin
(Kobayashi et al., 1977; Uhl et al., 1977), somatostatin
(Kobayashi et al., 1977; Finley et al., 1978), and bradykinin
(Correa et al., 1979) may participate in cortical mechanisms to
a lesser extent, judging from the density of immunocytochemically detectable fibers and of the amount of radioimmunoas-
sayable peptide.

The impetus behind all these discoveries arises in part from
the improvement in methods for isolating biologically active
substances, purifying them through bioassays and general
chemical methodologies, and then determining their amino acid
sequence (see Guillemin, 1978; Iversen et al., 1978). The
determination of the primary structure of a neuropeptide
permits its synthesis so that its physiological properties and
cellular distribution can then be determined through the use of
radioimmunoassays and immunocytochemistry. Thus, this imposing
catalogue of new neuropeptides,

350 Koob et al.

which has recently been accumulated, also includes reassignment
to the central nervous system (CNS) of some peptides previously
thought to be restricted to gut or endocrine glands. Studies
of their function in the brain may perhaps provide at least
partial answers to the larger question as to their functions in
all the locations in which they are found. The extent to which
the most studied neuropeptides satisfy the general criteria
used to identify neurotransmitters are indicated in Table 8-1.

Table 8-1. A Summary of the Information Currently Available on
Neuropeptides[a]

Peptide	Localized to neurons	Actions on neurons	Released from neurons	Synaptic mimicry
Thyrotropin-releasing hormone	Yes	Inhibitory	Yes	Not shown
Enkephalins and β-endorphin	Yes / Yes	Inhibitory and excitatory	Yes / Yes	Not shown / Not shown
Angiotensin II	Yes	Excitatory	Not known	Not shown
Oxytocin and vasopressin	Yes	Inhibitory	Yes	Not shown
Luteinizing hormone-releasing hormone	Yes	Inhibitory and excitatory	Not known	Not shown
Substance P	Yes	Excitatory	Yes	Possibly shown
Neurotensin	Yes	Excitatory	Yes	Not shown
Somatostatin	Yes	Inhibitory	Yes	Not shown
Vasoactive intestinal peptide	Yes	Excitatory	Yes	Not shown

[a] Based on standard properties of neurotransmitters: (1) presence within
neurons (regionally specific content in chemical assays, or by cyto-
chemistry); (2) actions on neurons (generally based upon iontophoretic
tests); (3) release from neurons through voltage and Ca^{2+}-dependent
mechanisms (based on studies **in vitro** or **in vivo**); and (4) synaptic mimicry
(in which the effects of the peptide are shown to be identical with the
effects produced on a target neuron by selective stimulation of afferents).

Peptides as Neuromodulators?

Although much more is yet to be established concerning the
cytological distribution of these peptides, work on the
mechanisms by which they produce their effects has ever more
room for progress. Because of the effects of these peptides on
spinal neurons *in vivo* or *in vitro*, which are not rigorously
established as synaptic targets, some workers have suggested
that a new mode of neurotransmitter action, termed "neuromodulation," is necessary to explain the observed effects of
these peptides (see Barker et al., 1978; Zieglgänsberger and
Bayerl, 1976; Zieglgänsberger and Tulloch, 1979). In this
sense, neuromodulation is defined as "the action of a substance, that is, a peptide, that by itself produces neither a
conductance nor a membrane potential change and yet alters the
conductance change produced by a "true neurotransmitter," such
as gamma aminobutyric acid (GABA). The location of the various
peptide and amino acid receptors involved in these interactive
responses, that is which cells and where on their surfaces, is
as yet unknown (see Zieglgänsberger and Champagnat, 1979).
Presumptive peptide neuromodulation has been reported in the
CNS and in peripheral tissues for interactions between substance P and nicotinic cholinergic receptor responses in cortex
(Phillis and Limacher, 1974; Krnjevic and Lekic, 1977) and in
adrenal medulla (Livett et al., 1979), and for CNS interactions
between enkephalin receptors and excitations due to glutamate
(Barker et al., 1979; Zieglgänsberger and Champagnat, 1979;
Zieglgänsberger and Tulloch, 1978). In one report on spinal
neurons, the excitation produced by substance P was also found
to modulate responses induced by iontophoresis of opiates and
opiate peptides (see Belcher and Ryall, 1977; Davies and Dray,
1977), indicating a peptide-peptide interaction. The peptide
in these cases appears to act by preventing the "real transmitter," that is glutamate or acetylcholine, from activating
their normal ion channels, but through a noncompetitive
mechanism (Barker et al., 1978; Stallcup and Patrick, 1980).
In more abstract terms, these peptide receptor actions may be
specified as "dis-enabling" modulations, preventing the effect
of the other amino acid, amine, or peptide. This "dis-enable"
action is the opposite functional abstraction to the "enabling"
modulation that we have attributed to central synaptic actions
of norepinephrine, and possibly other monoamines (see Bloom,
1979a,b, 1980).
However, on intrinsic neurons of the autonomic nervous system
where the functional site is more easily associated with the
afferent peptidergic fibers, these same peptides produce
"classical" membrane potential and conductance changes (North,
1979), making creation of a new class of mechanistic responses

less pressing. No intracellular recording testing for receptors has yet been done in behaviorally pertinent regions,
such as cortex, with peptides known to be present in these
regions. In no case has a cortical peptide been tested on a
cortical pathway in which the neurotransmitter is known. It is
therefore difficult to know whether the proposed neuromodulatory action is typical for all peptides at every site where
they act or only at selective sites for some peptides. Whether
these interactions between some peptides and other neurotransmitters will be paired in specific "dis-enabling" combinations
in specific target areas remains an intriguing question for
future research.

The excitatory action described in cortex for VIP (Phillis et
al., 1978) is potent, provided that the tests are done at well-
separated intervals. This specific response property suggests
that the intracortical neurons that react immunocytochemically
for VIP content (see Emson and Lindvall, 1979) fire infrequently. The same possibility may also hold for VIP-cyclase
interactions (Borghi et al., 1979; also see Nathanson, 1977).
Finally, yet another caveat: although long response latencies
to peptides are frequently reported in iontophoretic tests,
this lag may well be an artefact of the technique, rather than
an indication of slow onset or another aspect of a unique
neurotransmitter action (see Guyenet et al., 1979).

Neurobiology of Endorphins

Although each peptide undoubtedly has a separate story, current
research is epitomized by the experiments on the endorphins and
enkephalins, a family of previously unrecognized neuropeptides
that was detected, isolated, and eventually identified on the
basis of their ability to interact with receptors mediating responses to opiates. The explosion of work on endorphins began
with attempts to isolate and characterize the opiate receptor,
which ultimately demonstrated the existence of a high-affinity
binding site in synaptic membranes with selective opioid recognition properties (Kosterlitz et al., 1977). Hughes,
Kosterlitz, and their colleagues in Aberdeen next demonstrated
that brain extracts contained two substances that competed in
the opiate receptor assays and showed opiate-like activity *in
vitro* (Hughes et al., 1975). These substances were shown to be
two pentapeptides, named enkephalins, which shared common N-
terminal tetrapeptide sequences, (Tyr-Gly-Gly-Phe-X) and varied
only in the C-terminal position (Met or Leu): hence called
met[5]-enkephalin and leu[5]-enkephalin (Hughes et al., 1975). The
entire structure of met[5]-enkephalin is contained within a 91-
amino acid pituitary hormone, β-lipotropin (β-LPH), isolated

and sequenced by Li et al. (1965), but which shows no opioid activity per se. Almost immediately after this, other groups reported the isolation, purification, chemical structures, and synthetic confirmation of three additional endorphin peptides: α-endorphin (β-LPH$_{61-76}$), γ-endorphin (β-LPH$_{61-77}$), and β-endorphin (β-LPH$_{61-91}$; also called "C-fragment").

Specific immune sera for the various endorphins and enkephalins evolved into immunocytochemical and radioimmunoassay tests. These technical developments showed that the brain, pituitary, and gastrointestinal tract each contain enkephalin and β-endorphin, but not within the same cells (Bloom et al., 1978, see also chapter 7 by Dunn). Iontophoretic tests with enkephalins and β-endorphin suggest that neurons throughout the CNS can be depressed by these peptides; these effects were shown to be qualitatively identical to the effects of stereo-specific opiate agonists and could be selectively antagonized with the pure antagonist naloxone (Zieglgänsberger et al., 1978). An important apparent exception is the excitation of hippocampal pyramidal neurons by opiates, which is also a stereo-specific, naloxone-reversible action (see Nicoll et al., 1977). However, more recent studies (Zieglgänsberger et al., 1979) indicate that this limbic excitation--also observable in electroencephalographic (EEG) records *in vivo* (Henriksen et al., 1978) and in isolated hippocampal fragments transplanted to differentiate in oculo (Taylor et al., 1979)--derives from a disinhibition: the opiates directly depress tonically active inhibitory interneurons releasing the pyramidal cells.

Although the role of these opioid peptides in central and peripheral functions regulating behavior is far from clear, the possible involvement of these potent natural substances in abnormal behavioral states seems likely. We have therefore focused on this set of endorphin peptides in the following sections. More recent developments in this field may be found in our other reviews (see Koob, LeMoal and Bloom, 1981; Koob and Bloom, 1982, 1983).

THE ROLE OF ENDORPHINS IN ANIMAL BEHAVIOR

Two major strategies have been used to examine the behavioral significance of the opioid peptides. First, direct receptor stimulation using peripheral or central injections of endorphins and endorphin analogues has led to a suggestion of a role in such diverse effects as analgesia, reward mechanisms, and memory. A second, more indirect, approach infers opioid peptide function on the basis of antagonism by naloxone. Recent results using this latter approach have implicated the opioids in feeding and consummatory behavior and general behavioral arousal. Both approaches make major assumptions. In the first case it is assumed that the endorphins act directly on specific opioid receptors to produce their behavioral effects. The specificity of such receptors is still unknown, even at sites that contain endogenous endorphins or enkephalins. Recent behavioral results with endorphins (vide infra) have pushed these concepts of specificity to their limit, even questioning the criterion of naloxone antagonism by which opioid activity is often defined. In the latter case, assumptions must be made regarding the pharmacological specificity of the antagonist itself, until other means can be developed by which to validate the antagonism of an endogenous neurotransmitter or neuro-regulator. For example, in high concentrations naloxone can apparently also antagonize GABA (Dingledine et al., 1978).

Behavioral Effects of Centrally Injected Endorphins

Opiates, such as morphine and heroin, produce a variety of behavioral effects, such as analgesia, behavioral activation at low doses, sedation at high doses, and euphoria. All of these effects can be seen after the administration of endorphins. The closer the peptide is injected to the active site for such actions, the lower the dose that is required. For example, for analgesia, less endorphin is required by the intraventricular route than by systemic injection, and even less is needed for direct intracerebral injections than for intraventricular in-jections (Belluzzi et al., 1976; Takagi et al., 1978; Henriksen et al., 1978)

This may also be true for the stimulant actions of the en-dorphins. β-Endorphin and met^5-enkephalin produce no effect on locomotor activity when injected peripherally (Bloom et al., 1976; Segal et al., 1979), but produce opiate-like activation with intraventricular injection (Segal et al., 1979; Katz et al., 1978). More recent work has localized at least one site of this activation to stimulation of the mesolimbic dopamine (DA) system. Injections of D-ala-enkephalin analogues and β-endorphin and endorphin analogues directly into the vicinity of DA cell bodies of the ventral tegmental area of the midbrain produce a behavioral activation virtually identical to the activation seen with systemic opioids (Broekkamp et al., 1979; Joyce and Iversen, 1979; Stinus et al., 1980; Kelley et al., 1980), (see Fig. 8-1). Here, the rats show increased sniffing and grooming interrupted by bursts of locomotor activity in which the animals literally leap and hop about the cage. This activation is naloxone reversible, for both the enkephalin and endorphin analogues (Kelley et al., 1980; Stinus et al., 1980) (see Fig. 8-2), appears to be dependent on transmission in the mesolimbic DA system in that this activation can be blocked by

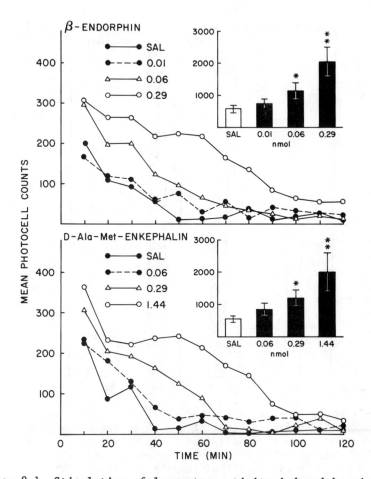

Fig. 8-1. Stimulation of locomotor activity induced by simultaneous bilateral infusion into the ventral tegmental
area (VTA) of β-endorphins (top panel or D-ala²-met
enkephalin (bottom panel). Bar graphs indicate total
scores. One microliter containing 0.04, 0.20, and
1.00 μg of peptide (total infusion) was injected over
2.5 minutes (for details, see Stinus et al., 1980).
Analysis of variance showed a significant group
effect for both β-endorphin (F = 12.086, df = 3,18, p
<0.01) and D-ala-met-enkephalin (F = 5.889, df =
3,18, p < 0.01). *Significantly different from
saline, p < 0.05, Student's t test; (n = 7). (Joyce,
E.M., Strecker, R.L., Le Moal, M., and Koob, G.F.,
unpublished results.)

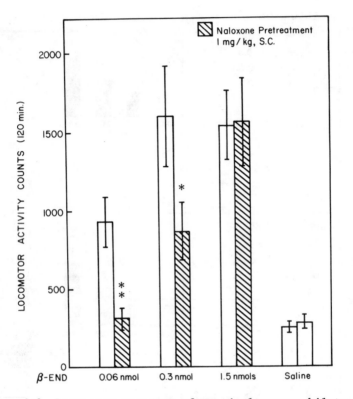

Fig. 8-2. Locomotor response after simultaneous bilateral VTA
infusions of a total of 0.2, 1.0, and 5.0 μg β-endorphin corresponding to 0.06, 0.30, and 1.5 nmol,
respectively; following naloxone (1 mg/kg, SC). n ≥6
in each group. *Significantly different from endorphin without naloxone, p < 0.05, group effect, two-way
analysis of variance with repeated measures on one
factor, **p < 0.01. (Taken from Stinus et al., 1980.)

low doses of neuroleptics and by 6-hydroxydopamine (6-OHDA)
lesions of the mesolimbic DA system. (Kelley et al., 1980;
Stinus et al., 1980). Of some considerable interest is the
fact that β-endorphin on a molar basis is equally or more
potent than D-ala-met-enkephalin in producing this activation;
doses as low as 0.06 nmoles of β-endorphin produce a significant increase in locomotor activity, (Fig. 8-2). Yet no
β-endorphin fibers or terminals have been localized to this
region of the brain (Bloom et al., 1978). This could suggest
that β-endorphin stimulates met-enkephalin receptors. Further

work using endorphin analogues without the N-terminal enkephalin moiety (β-LPH$_{61-65}$) may be necessary to resolve this
issue. This opioid-DA interaction provides an intriguing in-
terface by which endorphin systems could produce abnormal
behavior, particularly in view of the already proposed role of
the mesolimbic DA system in the pathophysiology of schizophrenia (Stevens and Livermore, 1978).

Behavioral Effects of Peripherally Injected Endorphins

A completely different line of research has explored the
effects of peripheral injections of endorphins and endorphin
analogues on the extinction of active and passive avoidance
behavior (de Wied et al., 1978a,b). Based on the pioneering
work of de Wied and associates using ACTH and ACTH analogues
(Bohus and de Wied, 1966), peripheral injections of met-
enkephalin in doses as low as 30 μg per rat have been shown to
delay the extinction of a pole-jump avoidance task, facilitate
the retention of passive avoidance (de Wied et al., 1978a,b),
and reverse carbon dioxide induced amnesia (Rigter, 1978).
Similar effects have been observed using β-endorphin and
α-endorphin (β-LPH$_{61-76}$) a presumptive breakdown product of
β-endorphin. However, γ-endorphin (β-endorphin (β-LPH$_{61-77}$)
another presumptive endorphin by-product, produced the opposite
results: γ-endorphin facilitated extinction of the pole-jump
avoidance (de Wied et al., 1978b). These effects appear to be
unrelated to changes in gross activity in that open-field
behavior is largely unaffected by these small doses (de Wied et
al., 1978a) or even doses 100 times greater (Veith et al.,
1978). An analogue of γ-endorphin, Des-Try-γ endorphin (DTγE)
(β-LPH$_{62-77}$) appeared to be even more potent (de Wied et
al., 1978b) in facilitating extinction of active avoidance.
This property, and the fact that DTγE appeared to have "cataleptogenic" properties similar to haloperidol (de Wied et al.,
1978b) prompted de Wied to speculate that DTγE could be an
endogenous neuroleptic. However, no other laboratories have
yet been able to produce catalepsy using DTγE, (Weinberger et
al., 1979; Pert, A., personal communication), nor has it been
localized to the brain in significant quantities. Whether this
inability to produce catalepsy with DTγE is due to a difference
in the drug synthesis or experimental procedure remains to be
determined. In addition, these behavioral effects appear to be
independent of classical opiate receptors in that naltrexone
pretreatment did not block the effects of β-endorphin on extinction of pole-jumping avoidance behavior (de Wied et al.,
1978a). Further evidence for a dissociation between the
classical opiate effects and the effects of peripheral

injection of endorphins can be seen in a study where met-
enkephalin and its D-ala-analogue, D-ala^2-met-enkephalin,
injected intraperitoneally facilitated maze performance (Kastin
et al., 1976). Morphine actually produced the opposite effect,
slowing running times and increasing errors in the maze.

Many studies have attempted to examine the effects of peripherally
administered endorphins on avoidance behavior and, in general, the
results appear to be consistent with the findings using carbon
dioxide-induced amnesia and the extinction of active avoidance.
Treatment with alpha endorphin, one hour post-training or pre-
retention, significantly increased the latency to re-enter an
aversive test chamber, whereas under the same conditons, DTγE
decreased the latency to re-enter (de Wied, et al., 1978b). Similar
results were observed in our laboratory where we have observed that
injection of 10 μg of α endorphin per rat significantly inhibited
extinction in rats trained for 3 days but γ endorphin significantly
facilitated extinction in rats that were trained for 4 days (see Fig
8-3). Consistent with these results, in a passive avoidance test,
pre-retention injection (1 hour) of α endorphin increased the latenc
for re-entry whereas γ endorphin had the opposite effect decreasing
re-entry latency (see Fig. 8-4).

In contrast to the above results, opioid peptides, centrally or
peripherally injected immediately post-training, inhibited retentior
(Izquierdo, et al., 1980), suggesting to the authors that the
endorphins mediate an endogenous amnesic mechanism (Izquierdo, et
al., 1980, 1981). These effects were observed after peripheral
post-training injections of β endorphin, methionine-enkephalin,
leucine enkephalin and des-tyrosine methionine enkephalin; and
were reversed by naloxone (Izquierdo and Dias, 1981). Similar
effects were observed for post-training intraventricular injections
of nanogram amounts of β endorphin and met-enkephalin (Lucion, et
al., 1982), and the authors have speculated that these effects
were mediated by CNS opiate receptors. However, there is little
evidence to support the contention that peripherally derived opioid
peptides cross the blood brain barrier (Houghten et al., 1980).

Others have shown that methionine and leucine enkephalin (40–400
μg/kg) and [D-leu^5] enkephalin (4.0 μg/kg) actually impaired acquis
of an active avoidance task when injected intraperitoneally (i.p.)
immediately before acquisition, but produced no effect when injecte
15 min before training (Izquierdo et al., 1980; Rigter, et al., 198
b). However, these peptides facilitated acquisition of an inhibito
avoidance task suggesting to the authors that the enkephalins produ
a behavioral change associated more with an increase in fear or
arousal than amnesia (Rigter, et al., 1980b). Some of these enkeph
effects can be reversed by adrenal demedullation (Martinez, et al.,
1981) and appear to be correlated with an action on delta (δ) opioi
receptors (Rigter, et al., 1981). These results appear, therefore,
to be more opiate-like, appear to be peripherally mediated and thus
contrast with other non-opiate like peptide actions cited above.

Table 8-2.   Training and Testing Procedure for Pole-Jump Avoidance

Day 1	10 trials acquisition--shaping employed
Day 2	10 trials acquisition
Day 3	10 trials acquisition
Day 4	0 hours--10 trials Extinction I--immediately followed by injection of saline or peptide[a]
	2 hours 10 trials Extinction II
	4 hours 10 trials Extinction III
	6 hours 10 trials Extinction IV

[a]Only those animals that attained the criteria were injected.

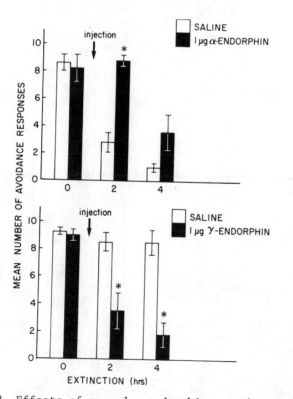

Fig. 8-3. Effects of α- and γ-endorphine on the extinction of
pole-jump avoidance behavior. Peptides (1 μg/rat)
were injected immediately after the first extinction
session. *Significantly different from saline,
p < 0.05. (Student's t test); n = 5 for top panel,
n = 4 for bottom panel. (Taken from LeMoal, et al.,
1979.)

Table 8-3. Training and testing procedure for passive avoidance

Day 1	Two minutes familiarization with the box
Day 2	Three trials followed by shock on the third trial
	Intertrial interval approximately 5 minutes
Day 3	Injections were made 1 hour prior to the test

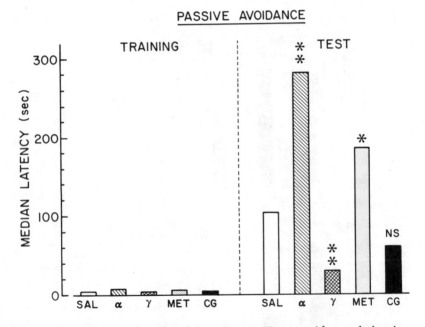

Fig. 8-4. Effects of endorphins on passive avoidance behavior.
Values represent median latency to enter a dark box
during the last training trial (left Panel) and during
the subsequent test trial 24-hours later (right panel).
Peptides were injected one hour before the test trial.
*Significantly different from saline, p <0.05
(Mann-Whitney U test); n = 8.   (Koob, G.F., LeMoal, M.,
and Bloom, F.E., reprinted from Koob, LeMoal and Bloom,
1981 with permission.)

Curiously, large doses of β-endorphin (20 μg/kg as opposed to 2 μg/kg in the other studies) can attenuate acquisition of an avoidance task (Izquierdo, 1980), reflecting a characteristic u-shape dose-response curve seen by others   (Martinez and Rigter, 1980; Rigter, et al., 1981). No really satisfactory explanation for these u-shaped functions has emerged for the opioid peptides although some have speculated as to possible aversive effects at the high doses, or the conversion of the parent compound to a metabolite (Izquierdo, 1980).

In one of the first reports of opioid-learning effects using an appetitively motivated task, [Met] enkephalin and [D-Ala^2, Met^5] enkephalinamide facilitated performance of hungry rats in a complex 12-choice maze (Kastin et al., 1976) at a dose of 80 μg/kg i.p., 15 min before the test. [d-Phe^2, Met^5] Enkephalin, with virtually no opiate activity, also facilitated maze performance, but morphine did the opposite as did [D-Ala^2] β-endorphin (Kastin et al., 1980). We observed similar results using appetitively-motivated tasks and injecting the endorphins during the extinction process (cf. LeMoal, et al., 1979; Koob, et al., 1981). In a continuously reinforced lever press task for food reward, α-endorphin (30 μg/kg, s.c.) delayed extinction and γ-endorphin (30 μg/kg, s.c.) again slightly facilitated extinction (Koob, et al., 1981). However, in a runway task using water as a reward, both α- and γ-endorphin delayed the extinction; naloxone produced no antagonism of either opioid, but did produce a similar delay of extinction (Koob, et al., 1981).

In summary, the effects of opioid peptides administered peripherally are clearly dependent on the time-course and testing protocols. It may be that the immediate effect is a "memory" impairment that is opiate-like in that it is reversed by naloxone, but subsequent break-down products produce opposite effects that are not opiate-like. What is clear is that these peptides do act, at some level, which we infer to be by peripheral mechanisms. If the sites and cellular mechanisms underlying these behavioral effects of peripherally injected peptides could be established with some certainty, such phenomena as these could open a whole new realm of possibilities regarding opioid peptides and brain behavior interactions (see Koob, LeMoal and Bloom, 1981; Koob and Bloom, 1982; 1983).

Behavioral Actions of Naloxone

Recently, a large effect has been direcred at evaluating the effects of naloxone on behavior. Naloxone, at relatively low doses, classi-cally blocks most of the behavioral effects of morphine, such as analgesia (Blumberg et al., 1965), intravenous self-administration (Woods and Schuster, 1971), hyperactivity (Domino et al., 1976), and enhancement of self-stimulation (Wauguier et al., 1974; Holtzman, 1976). However, recent work has demonstrated that naloxone can also antagonize the activation produced by other drugs. Naloxone partially antagonizes the hyperactivity induced by amphetamine (Holtzman, 1974; Dettmar et al., 1978), the feeding produced by chlordiazepoxide (Stapleton et al., 1979), the release of punished responding by benzodiazepines and alcohol (Billingsley and Kubena, 1978; Koob et al., 1980).

Most recently, naloxone has been shown to decrease nondrug-induced behaviors.   Naloxone reduces water intake in water-deprived rats (Holtzman, 1974; Rogers et al., 1978; Frenk and Rogers 1979), food intake in food-deprived rats (Rogers et al., 1979; Frenk and Rogers, 1979), locomotor activity (Segal et al., 1979), exploratory behavior in the open field (Arnsten and Segal, 1979), and intracranial self-stimulation at certain sites (Belluzzi and Stein, 1977), but not at others (Holtzman, 1976; Wauguier et al., 1974; Van der Kooy et al., 1977; Holtzman, 1976; Weibel and Wolf, 1979).   Interestingly, at a somewhat lower dose, 0.5 mg/kg, naloxone actually increased contact time with novel objects in an open field, suggesting an enhanced interaction with environmental stimuli (Arnsten and Segal, 1979).   Another exception to the general sedative action of naloxone on behavior is the facilitation of copulatory behavior in sexually inactive male rats at doses as low as 2-4 mg/kg (Gessa et al., 1979).   Intraventricular injection of β-endorphin or D-ala-met-enkephalin produces the opposite result, consistent with clinical reports of impotence resulting from long-term use of narcotic analgesics (Hollister, 1975).   Interestingly, in view of the above discussion, naloxone has been shown to have anti-amnestic properties in an inhibitory avoidance test (Messing, et al., 1981; Gallagher and Kapp, 1979) and facilitates retention in an active avoidance task (Messing, et al., 1979; see Izauierdo et al., 1981).   This effect may be a central action in that naloxone injected directly into the amygdala has similar effects (Gallagher and Kapp, 1979).

This bewildering variety of effects of naloxone on opiate- and nonopiate-induced events makes it difficult to define a single underlying mechanism.   Perhaps before the endogenous opiate systems are implicated in all these behaviors, studies should be directed at exploring the nature of the general sedative action of naloxone, and at defining the relationship of dose to the specificity of neurochemical action.

THE ROLE OF ENDORPHINS IN PSYCHIATRIC DISORDERS

As already described, the typical behavioral effects of narcotic opiate agonists in man and the behavioral pharmacology of endogenous opiate peptides in animals suggest actions on mood, affect, and the feeling of well-being.   Anatomical data showing that endogenous opiates were localized through the brainstem and the limbic system suggested that they were well positioned for a role not only in sensory-affective interactions, but also in the regulation of emotions and motivations and, by speculative implication, in mental diseases and psychoses.   Moreover, pharmacological studies in correlation with the anatomical results suggested a particular role of these endogenous peptides in altering the metabolism of catecholamines.   Considering the putative role of

DA in schizophrenia (Carlsson, 1978) and the catecholamine theory
of affective disorders (Schildkraut, 1965), it was tempting to
consider this neuropeptide system within the unsolved problems
of psychiatric research. Thus, very soon after their discoveries,
endogenous opiates or opiate antagonists were considered in the
pathophysiology and treatment of schizophrenia (Berger, 1978).
   Animal studies have strongly influenced the subsequent clinical
research resulting in extremely divergent interpretations. In
1976, two groups observed that intracerebral administration of
β-endorphin elicited rigid immobility in rats and both groups
immediately suggested that these findings had mental health implications, but their predictions were in the opposite direction.
Bloom et al. (1976) proposed that an excess in central endorphins,
which induced catalepsy in rats, might have a role in the pathophysiology of schizophrenia, whereas Jacquet and Marks (1976)
suggested that β-endorphin may have therapeutic properties as a
neuroleptic, since it produced extrapyramidal-like rigidity in
rats. Directly or not, most of the clinical studies may be divided
according to these two schools of thought: If an excess of endorphin
or abnormal endorphin at the opiate receptor induces mental disease,
it is logical to give an opiate antagonist as treatment, and, conversely, if endorphin deficiency were posited as producing the
pathology, it is logical to replace the deficiency by injecting
the same product or an artificial agonist. Thus, it is of
interest to evaluate the effects of opiates in mental disease.
Most of the studies of both philosophies have generally led to
negative results, particularly if one considers chronic psychotic
disorders.

Exogenous Opiates in Psychiatric Illness

In addicts or ex-addicts, clinical observations have shown that
some of these patients use opiates to avoid social problems or
as a form of self-medication for their mental disorders (Wikler
and Rasor, 1953; Khantzian, 1974; Verebey et al., 1978).
Opiates appear to produce antidepressant actions by attenuating
dysphoric states involving aggression and rage (Khantzian, 1974).
Patients reportedly use heroin for its tranquilizing and "antipsychotic" properties (Wellisch and Gay, 1972), and schizophrenics
included in a methadone maintenance program appeared to benefit
from the treatment (Salzman and Frosch, 1972). There appears to
be general agreement that maintenance on opiates attenuates the
clinical symptoms of various mental disorders in addicts and ex-
addicts (Verebey et al., 1978) but how these beneficial effects
of opiates relate to the etiology of the drug addiction in the
first place is not clear.
   In nonaddicts, there is little evidence of a beneficial effect
of opiate treatment on mental illness. The clinical investigations
on the effects of exogenous opiates on schizophrenia do not permit
a clear conclusion, at best the effects are weak. The classical
work of Wikler et al., (1952) indicates that addiction to 300 mg/
day of morphine did not produce consistent improvement of the
current schizophrenic symptoms, but the authors noted curiously

that hebephrenic and paranoid patients showed a recession of bizarre mannerisms and sterotyped abnormal patterns of behavior during the morphine-abstinence syndrome.

In summary, the therapeutic effects of opiates in mental illness need to be scientifically re-evaluated, and the safety of long-term opiate treatment remains open to question. Moreover, it is now thought that certain opiate receptors are only weakly responsive to morphine and opiate antagonists. Therefore, in order to explore the opiate-psychosis problem, it will be necessary to demonstrate and characterize other opiate receptors in order to find and use more specific agents.

Endorphin-Deficiency Strategies of Psychiatric Illness

The hypothesis of endorphin deficiency in mental disease could be directly tested by the administration of the peptide itself. However, the cost of β-endorphin as well as the Food and Drug Administration's limitations for its use in psychiatric institutions (little is known at the present time on the toxicity of endorphins) and concerns over the optimal route of administration, have limited such investigations. Kline et al. (1977) administered β-endorphin to a total of 26 patients in an open study with various psychiatric diagnoses (schizophrenia, depression, personality disorders, mental deficiency; one normal subject was also injected) at a dose as high as 10 mg IV. The placebo was injected 30 minutes prior to each injection of β-endorphin. Substantial salutory effects were reported and some of the videotaped nonblinded interviews presented in meetings have been impressive. For the first few minutes post injection, the subjects reported various peripheral effects, such as dry mouth, feeling of warmth; then some presented a state of well-being, became less anxious or more talkative, and finally entered into a period of sedation. Some effects were obtained in schizophrenic and depressed patients with improvement of specific symptom. It is interesting to note that all these sedative and specific effects were noted several hours after the injection and even several days (Kline and Lehmann, 1979) after a single injection. Angst et al. (1979) had treated manic-depressive patients in a nonblind drug trial with the same dose (10 mg β-endorphin, IV) and reported a switch from the depressive state to mania. However, both these groups of studies have been questioned on methodological grounds (Goldstein and Cox, 1979; Berger, 1978); the studies were not carried out in a double blind manner, the placebo was not active, and the patients knew they were receiving active treatment, interviewing physicians were aware of the protocol, and, lastly, no structured evaluation procedures were used. Although these studies may suggest that β-endorphin could have some therapeutic effects in the major psychoses, attempted replications should use more optimal compounds than β-endorphin, which in man has a half-life of about 20 minutes. Synthetic long-lasting materials, such as the D-ala^2-β-endorphin-analogue or the Sandoz FK 33-824, are now available and may be more suitable for studying long-term effects and chronic treatment of opiate agonists. In conclusion, the question of therapeutic effect of the endogenous or exogenous opiates in schizophrenia or affective disorders remains open for future investigation.

Endorphin-Excess in Psychiatric Illness

The alternate hypothesis is based on endorphin excess in schizophrenia, and for this hypothesis the clinical effects of opiate antagonists have been evaluated. The first paper (Terenius et al., 1976) reported that some "endorphins," which have still not been shown to be opiate peptide fragments, were elevated in the cerebrospinal fluid (CSF) in chronic schizophrenics and manic patients, and that these levels decreased after treatment with neuroleptics (schizophrenia) or when mania turned into depression. More recently Domschke et al. (1979) found an increase of radioimmunoassayable β-endorphin concentration in acute but not in chronic schizophrenics, suggesting perhaps the confounding effect of associated stress-induced release (see Guillemin et al., 1977). However, this study has been severely criticized on the grounds of improper statistical analysis and other flaws obviating interpretation of the data. These studies led to the hypothesis that increased CSF endorphin levels may have a role in the pathophysiology of the major psychoses.

Another approach linking the questions of endorphin increase with schizophrenia is the reported therapeutic effects of hemodialysis in chronic schizophrenia patients (Wagemaker and Cade, 1977). Here, there was a remission of psychotic symptoms in a group of drug-resistant schizophrenics after 16 weeks of hemodialysis. In a follow-up study with the same patients, Palmour et al. (1979) analyzed the dialysates from schizophrenic patients. They detected endorphins not present in the dialysates of nonschizophrenic patients and they isolated a peptide that had the general structure of β-endorphin, but contained leucine rather than methionine at position 65. This [Leu$_{65}$]-β-endorphin decreased during the hemodialysis treatment with a time-course that correlated with the clinical improvement. The existence of this specific endorphin molecule had never been previously reported, nor has one been found subsequently (Ross et al., 1979). However, the existence of other endorphinlike peptides with leucine at β-LPH$_{65}$ have been described (Stern et al., 1979). Thus, these results need confirmation with a larger group of subjects and with control studies, if possible, such as sham dialysis in schizophrenia patients, dialysates of normal controls, and also determination of the specific relations of this molecule with schizophrenia and the clinical improvement. A recent study conducted by another group (Lewis et al., 1979) on two schizophrenic and two control patients failed to detect in the dialysates any trace of an increase in [Leu$_{65}$]β-endorphin or even [Met$_{65}$]β-endorphin

in the two groups. At this point it is safe to conclude that
there is no direct proof for an increase or decrease of endorphins nor for the production of any atypical endorphin in
psychoses.

The hypothesis that narcotic antagonists may have some therapeutic indications was also suggested by the paper of Terenius
et al. (1976) that indicated an excess of endorphins in the CSF
of psychotic patients. The first report attempting to test the
hypothesis that increased endorphin production or activity at
the receptor sites may contribute to the pathophysiology of
schizophrenia was published by the same group. The administration of naloxone (0.4 mg, IV) decreased auditory hallucinations in four of the six schizophrenic subjects treated, the
effect lasting several hours after the injections (Gunne et
al., 1977). The use of this briefly active antagonist with
high affinity for the opiate receptor was the obvious way to
test the excess endorphins hypothesis. However, this study
presented some methodological weaknesses. It was single blind
and the basis of subject selection and rating scales were not
clear. Nevertheless, this study prompted many attempted
replications, most of them using naloxone in chronic or acute
schizophrenic patients. The first series of replications
yielded negative results. Volavka et al. (1977), using the
same dose in a carefully selected group of patients and using
coded procedures throughout the experiment, were unable to see
evidence of changes in symptoms. Davis et al. (1977) with
doses of 0.4 to 10 mg found no changes in hallucinations.
Janowsky et al. (1977) used 1.2 mg for one hour in a general
population of schizophrenic subjects with no observable
changes. Kurland et al. (1977) reported no consistent effect
of naloxone at a dose of 1.2 mg in chronically hallucinating,
schizophrenic patients, and finally Gunne et al. (1979) were
unable to replicate their own findings with a double-blind
design. The doses of naloxone injected in these studies ranged
between 0.4 and 10 mg of naloxone but most of the schizophrenic
patients received 1.2 mg or less per injection and the test
periods were generally short.

Emrich et al. (1977) was the first to try systematically
higher doses, up to 4 mg on a variety of hallucinating
patients, and found that the hallucinations decreased some
hours after the injections. A more sophisticated study was
conducted by the Stanford group (Watson, et al., 1978; Berger et
al., 1979) using 10 mg of naloxone in a randomized double-blind
crossover study involved nine schizophrenic subjects. These
patients were chosen after a screening of 1000 psychotic
patients and represented a homogeneous group with the indubi-
table diagnosis of persistent auditory hallucinations, and they
were also good reporters. The study was carried out on two
separate days, at least 48 hours apart. Clinically, six of the
nine subjects reported decreased auditory hallucinations after
naloxone, again suggesting that high doses of naloxone produce

therapeutic effects. Recently van Praag (1979) reviewed 14
published studies including a total of 119 schizophrenic
patients of which 30 presented a reduction of the auditory
hallucinations after naloxone and 16 of 29 manic subjects presented a reduction of their manic symptoms.

Whatever the interpretation of these results, the naloxone-
schizophrenia studies raise several questions: (1) the sample
size is frequently too small for good therapeutic and statistical analysis; (2) in general, low doses of naloxone (under 2
mg IV) do not seem to modify the symptoms in schizophrenic
patients, at least during the short time generally used for the
patient observation; (3) in carefully selected subjects, higher
doses seem to decrease the intensity and frequency of auditory
hallucinations; (4) given the short duration of naloxone's
action, and the necessary coincidence of that short duration
and the production of psychopathological symptoms, the selected
patients must have florid hallucination syndromes, thus repre-
senting a restricted group of the schizophrenic population; (5)
other factors may have influenced the results in almost all of
the studies. The frequent interviews and testing may have influenced the hallucination ratings, especially because the
schizophrenic hallucinations are subjective and transient by
nature (Watson et al., 1979).

For all these reasons, it was more suitable to test a long-
acting antagonist in order to obtain a longer and more stable
opiate receptor blockade. Naltrexone, a long-acting oral
opiate antagonist has been used in schizophrenia in two pilot
studies (Simpson et al., 1977; Mielke and Gallant, 1977) at a
dose of 50 to 800 mg/day without clear beneficial effects,
except that in one study two of five patients became worse
during the treatment (Mielke and Gallant, 1977). Gunne et al.
(1979) reported in a double-blind, placebo-controlled trial
that a dose of 100 mg naltrexone produced no effects in schizophrenia, whereas Watson et al. (2979) also using a double-blind
protocol found slight improvement on the NIMH global illness
scale with a daily dose of 250 mg. Thus, it seems that opiate
antagonists are generally not beneficial (Watson et al., 1979)
but also, except for two patients in the Mielke-Gallant study,
no clear clinical deterioration has emerged from the opiate
receptor blockade, even with doses of naltrexone as high as 800
mg/day. This last observation may also be relevant for the
hypothesis of endorphin deficiency in mental disease. In
effect, if at least a certain portion of opiate receptors must
be occupied by endorphins or other opioid peptides for the
maintenance of a psychophysiological homeostasis, one would
expect that the antagonist would make the patient worse (Verebey
et al., 1978). Thus, the clinical studies with naloxone and
naltrexone do not clearly support either the hypothesis of an
endorphin excess nor the hypothesis of an endorphin deficiency
in mental disease.

Interestingly enough, many commentators, although aware of these negative results, agree that some subjects seemed to benefit by these agents. Some patients may respond to opiate antagonists and thus the reliability of this effect could be established for these individuals. Also, the dose level and repetitive treatment may be critical. These observations might suggest that a particular subgroup of patients may have an important disturbance of the endorphin system. In a critical review of the effects of opiate antagonists in schizophrenia, Davis et al. (1979) proposed alternative strategies, including exploitation of individual differences, use of psychophysiological measures, genetic strategies, and multi-variate statistical techniques with larger sample size. The publication of small sample size studies, the rule of the moment, each failing to reach statistical significance, has increased the danger of premature rejection of effects due to endorphins. The authors also noted that premature rejection is further exacerbated by the likelihood that many biological factors act in the pathophysiology of schizophrenia (Davis et al., 1979). Another hypothesis for the limited efficacy of narcotic antagonists is that there is a therapeutic interaction of narcotic antagonists with more classical antipsychotics (Davis and Bunney [1980]). Indeed, this hypothesis has been explored systematically in a double-blind, World Health Organization collaborative study (Pickar et al., 1982) where there was a significant decrease in schizophrenic symptoms only in patients concurrently treated with neuroleptics.

Finally, some consideration must be given to the hypothesis of de Wied regarding effects of the endorphin-analogue DTγE. Until recently, animal studies had shown that endorphins or enkephalins produced morphine-like actions as assessed by *in vitro* techniques or by intracerebroventricular injections, but not a classical pharmacological neuroleptic profile. However, as the length of the peptide structure decreased, the morphine like potency was markedly reduced, presumably because rapid removal of the N-terminal tyrosine from endorphins destroyed all opiate-like activity. In fact, animal experiments by de Wied's group suggested that γ-endorphin, and in particular the DTγE (β-LPH$_{62-77}$), a nonopiate-like peptide, possessed some "neuroleptic-like" activities (de Wied et al., 1978b,c). This molecule has been suggested as an endogenous neuroleptic with a profile even more specific than the currently used antipsychotic drugs, that is, without the locomotor and the sedative effects of haloperidol (de Wied et al., 1978b, 1978c). Finally, this nonopiate effect was presented as a prt of the general role of the α- and γ-endorphin derivatives in adaptive behavior, and as being independent of opiate receptor sites in the brain (de Wied et al., 1978b,c; vide supra).

The neuroleptic-like activity of the DTγE has been tested in 14 patients with long-lasting, relapsing schizophrenic or schizoaffective psychosis resistant to conventional neuroleptics (Verhoeven et al., 1979). Two studies were conducted, the first for six patients with an open design; all the medications were discontinued and 1 mg of DTγE zinc phosphate (to produce slow release of the peptide) was given daily (IM) for about seven days. The second used a double-blind, crossover design for the other eight patients, of whom six were maintained with additional neuroleptic therapy and two were drug free, all receiving 1 mg of DTγE in saline solution for eight days. Toxic side effects of DTγE have not been observed. The authors reported a transient or semipermanent improvement in both studies in which the psychotic symptoms diminished or even disappeared. With the double-blind crossover design it was possible to show even on the first treatment a slight but significant improvement, which continued so that by day 4 the psychotic symptoms had almost disappeared.

It is interesting to note variabilities in the response to DTγE treatment with some patients having short-term improvement, others a long-lasting one. Moreover, some patients in the first study relapsed after three or four days of treatment and became aggressive or agitated, whereas two improved patients reported euphoria. Verhoeven et al. (1979) concluded that these effects may be a consequence of the normalization of γ-endorphin homeostasis in the brain. van Praag and his collaborators have observed definite improvement in 50% of their patients with DTγE or Des-enkephalin-gamma-endorphin, particularly in a sub-population on acute, non-classical antipsychotic responsive, schizophrenics (van Praag, et al., 1982); several other studies have reported no effect (Tamminga et al., 1981; Casey, et al., 1981; Emrich, et al., 1980; Manchanda and Hirsch, 1981). In two animal studies DTγE failed to produce any characteristics of neuroleptic profile (Weinberger et al., 1979), nor did DTγE bind to dopamine receptors (Pedigo et al., 1979). A possible alternative explanation for these discrepant results would be a presynaptic action for DTγE on dopamine neurons activity (Nickolson and Berendsen, 1980; Davis, et al., 1981). Also, a recent observation that DTγE can markedly increase pineal and plasma levels of melatonin in rats and man (King et al., 1981; Claustrat, 1980) is intriguing. The significance of this finding as regards the clinical relevance of DTγE remains to be determined.

PERSPECTIVES AND CONCLUSIONS

In summary, the endorphins have been shown to produce behavioral effects like those of the opiates, particularly when injected centrally, and some of these naloxone-reversible effects have been localized to the activation of specific neuroanatomical and neurochemical systems (Kelley et al., 1980; Stinus et al., 1980; see Fig. 8-2). On the other hand, small peripheral doses of endorphins can also produce subtle effects on the way animals respond to specific contingencies in their environment, particularly aversive ones. These latter effects appear not to be reversible by naloxone, nor has their specific site of action been identified. The effects of naloxone alone are poorly understood but apparently produce generally mild decreases in stimulated behavior, with the exception of sexual behavior in which naloxone actually increases copulatory behavior and in certain behavioral tests in which rats appear to perseverate in exploration (Arnsten and Segal, 1979) and in extinction (see Fig. 8-7).

These seemingly divergent behavioral results are difficult to integrate into a single hypothesis explaining the functional significance of endorphins. The role of central endorphins in the relief of pain and in producing behavioral activation is relatively clear, particularly since these effects appear opiatelike in that they are naloxone reversible, and since they can be identified with specific CNS pathways. In addition, all the endorphin analogues appear to act in the same direction after central injection, that is, to produce analgesia or hyperactivity, albeit with different potencies (Stinus et al., 1980).

Perhaps, the endorphins from the pituitary have a similar but more subtle role in producing activation and appropriate action during stressful situations. β-Endorphin is released from the pituitary during stress (Rossier et al., 1977) and β-LPH-like material can reverse some of the behavioral deficits observed after hypophysectomy (de Wied, 1977). In normal animals, the endorphins appear most potent in altering the extinction of aversively motivated learning(see de Wied, 1978a,b, and before), although they clearly can influence appetitively motivated learning as well (Kastin et al., 1976; Le Moal et al., 1979). Further, in the periphery, different analogues may be of particular importance in determining the direction of the behavioral effect (de Wied et al., 1978a,b; Le Moal et al., 1979). In this regard it is interesting to note that β-endorphin is a potent stimulator of the release of vasopressin (Weitzman et al., 1977). Also, whether these subtle behavioral effects from peripheral administration of endorphin analogues actually mimic a metabolic degradation of naturally released β-endorphins remains to be determined.

Finally this hypothesis of a two-level (central versus peripheral) mode for the biological activity of endorphins may be related to the already hypothesized multiple opiate receptors. Here, centrally derived endorphins may act at classical opiate receptors (morphine-like action, naloxone-reversible), whereas peripherally derived endorphins may act at nonclassical opiate receptors (ACTH-like action, not naloxone reversible).

No clear conclusions can be drawn from clinical studies of endorphin agonists and antagonists. Although both areas are generally negative, most of the authors interpret these lack of effects cautiously (for review see Verebey et al., 1978; Davis et al., 1979; Watson et al., 1979). Some subjects do seem to benefit from the administration of the antagonists, but even these effects are difficult to characterize for the moment. It appears that the behavioral effects of peripherally injected microdoses of endorphins in specific, highly controlled situations confront the pharmacologists, behaviorists, and clinicians with new challenges because of the unknown mode and site of action of these peripheral endorphins. For example, at this time, it is unclear where peripherally injected naloxone β-endorphin, or even DTγE would be acting to produce therapeutic effects. Indentification and characterization of the site action of the non-naloxone reversible effects of endorphin-like compounds, and of the effects of naloxone itself, will go far toward resolving these questions.

ACKNOWLEDGEMENTS

We thank Robert Strecker and Claudia Gallison for their skilled technical assistance and Nancy Callahan for manuscript assistance. We also thank Dr. Nicholas Ling for providing the endorphin and enkephalin analogues and Drs. Adrian Dunn and Charles Nemeroff for their excellent editorial criticism of the manuscript.

REFERENCES

Angst, J., Antenreith, V., Brem, F., Koukkon, M., Meyer, H., Stassen, H.H., and Storck, U.: Preliminary results of treatment with β-endorphin in depression. In Endorphins in Mental Health Research, E. Usdin, W.E. Bunney, Jr., and N. Kline, eds. MacMillan Press, London, 1979.
Arnsten, A.T., and Segal, D.S.: Naloxone alters locomotion and interaction with environmental stimuli. Life Sci 25:1035-

1042, 1979.

Barker, J.L. Neale, J.H., Smith, T.G., and Macdonald, R.L.: Opiate peptide modulation of amino acid responses suggests novel forms of neuronal communication. Science 199:1451-1453, 1978.

Belcher, G., and Ryall, R.W.: Substance P and Renshaw cells: A new concept of inhibitory synaptic interactions. J Physiol (Lond) 272: 105-119, 1977.

Belluzzi, J.D., and Stein, L.: Enkephalin may mediate euphoria and drive-reduction reward. Nature 266:566-558, 1977.

Belluzzi, J.D., Grant, N., Garsky, V., Sarantakis, D., Wise, C.D., and Stein, L.: Analgesia induced in vitro by central administration of enkephalin in rat. Nature 260:625-626, 1976.

Berger, P.A.: Investigating the role of endogenous opioid peptides in psychiatric disorders. Neurosci Res Program Bull 16:585-599, 1978.

Berger, P.A., Watson, S.J., Akil, H., and Barchas, J.D.: Naloxone administration in chronic hallucinating schizophrenic patients. In Endorphins in Mental Health Research, E. Usdin, W.E. Bunney, Jr., and N. Kline, eds. MacMillan Press, London, 1979.

Billinglsley, M.L., and Kubena, R.K.: The effects of naloxone and picrotoxin on the sedative and anticonflict effects of benzodiazepines. Life Sci 22:897-906, 1978.

Bloom, F.E.: Cyclic nucleotides in central synaptic function. Fed Proc 38:2203-2207, 1979a.

Bloom, F.E.: Chemical integrative processes in the central nervous system. In Neurosciences Fourth Intensive Study Program, F.O. Schmitt and F.G. Worden, eds. MIT Press, Cambridge, Mass. 1979b, pp 51-58.

Bloom, F.E.: Chemical coding: Modulation and level setting. In The Reticular Formation Revisited, J.A. Hobson and M.A.B. Brazier eds. Raven Press, New York, 1980, pp 277-284.

Bloom, F., Segal, D., Ling, N., and Guillemin, R.: Endorphins: Profound behavioral effects in rats suggest new etiological factors in mental illness. Science 194:630-632, 1976.

Bloom, F.E., Battenberg, E.L., Rossier, J., Ling, N., and Guillemin, R.: Neurons containing β-endorphin in rat brain exist separately from those containing enkephalin: Immunocytochemical studies. Proc Natl Acad Sci USA 75:1591-1595, 1978.

Blumberg, H., Wolf, P.S., and Dayton, H.B.: Use of writhing test for evaluating analgesic activity of narcotic antagonists. Proc Soc Exp Biol Med 118:763-766, 1965.

Bohus, B., and de Wied, D.: Inhibitory and facilitatory effect of two related peptides on extinction of avoidance behavior.

Science 153:318-320, 1966.
Borghi, C., Nicosia, S., Giachetti, A., and Said, S.I.: Vaso-
active intestinal peptide (VIP) stimulates adenylate cyclase
in selected areas of rat brain. Life Sci 24:65-70, 1979.
Broekkamp, C.L.E., Phillips, A.G., and Cools, A.R.: Stimulant
effects of enkephalin microinjection into the dopaminergic
AI10 area. Nature 278:560-562, 1979.
Carlsson, A.: Antipsychotic drugs and neurotransmitters. Am J
Psychiatry 135:164-173, 1978.
Casey, D., Korsgaard, S., Gerlach, J., Jorgensen, A. and
Simmelsgaard, H.: Effect of des-tyrosine-gamma-endorphin
in tardive dyskinesia. Archives of General Psychiatry 38(2):
158-168, 1981.
Claustrat, B., Chazot, G. and Brun, J.: Melatonin secretion
in man: Stimulating effects of des-tyrosine-γ-endorphin.
Neuroendocrinology Letters 3(10):35-58, 1980.
Correa, F.M.A., Innis, R.B., Uhl, G.R., and Snyder, S.A.:
Bradykinin-like immunoreactive neuronal systems localized
histochemically in rat brain. Proc Natl Acad Sci USA 76:1489-
1493, 1979.
Davies, J., and Dray, A.: Substance P and opiate receptors.
Nature 268:351-352, 1977.
Davis, G.C., Bunney, W.R., De Fraites, E.G., Kleinman, J.E.,
Van Kammen, D.P., Post, R.M., and Wyatt, R.J.: Intravenous
naloxone administration in schizophrenia and affective
illness. Science 197:74-77, 1977.
Davis, G.C., Buchsbaum, M.D., and Bunney, W.E., Jr.: Research
in endorphins and schizophrenia. Schizoph Bull 5:244-250,
1979.
Davis, G.L. and Bunney, W.E., Jr.: Psychopathology and
endorphins. In: Neural Peptides and Neural Communication,
pp. 455-464, Costa, E. and Trabucchi, M., eds. Raven Press,
New York.
Davis, K.L., Samuel A., Mathe, A.A., and Mohs, R.C. (1981):
Intracerebral des-tyrosine-gamma endorphin inhibits methylphenidate induced locomotor activity. Life Sciences 28(21):
2421-2424.
Dettmar, P.W., Cowan, A., and Walter, D.S.: Naloxone antagonizes
behavioural effects of d-amphetamine in mice and rats. Neuropharmacology 17:1041-1044, 1978.
Dingledine, R., Iversen, L.L., and Breuker, E.: Naloxone as a
GABA antagonist: Evidence from iontophoretic, receptor binding
and convulsant studies. Eur J Pharmacol 47:19-27, 1978.
Domino, E.F., Vasko, M.R., and Wilson, A.E.: Biphasic actions
of selected narcotics on rat locomotor activity and brain
acetylcholine utilization. In Tissue Responses to Addictive

Drugs, D.H. Ford and D.H. Clouet, eds. Spectrum Publications, New York, 1976.

Domschke, W., Dickschas, A., Mitznegg, P.: C.S.F. beta-endorphin in schizophrenia. Lancet 1:1024, 1979.

Emrich, H., Zaudig, M., Kissling, W., Dirlich, G., Zerssen, D. and Herz, A.: Des-tyrosyl-γ-endorphin in schizophrenia. A double blind trial in 13 patients. Pharmopsychiatr. Neuropsychopharmako 13(5):290-298, 1980.

Emrich, H.M., Cording, C., Piree, S., Kolling, A., Von Zerssen, D., and Herz, A.: Indication of an antipsychotic action of the opiate antagonist naloxone. Pharmakopsychiatry Neuropsychopharmalsol 10:265-270, 1977.

Emson, P.C., and Lindvall, O.: Distribution of putative neurotransmitters in the neocortex. Neuroscience 4:1-30, 1979.

Finley, J.C.W., Grossman, G.H., Dimeo, P., and Petrusz, P.: Somatostatin-containing neurons in the rat brain: Widespread distribution revealed by immunocytochemistry after pretreatment by pronase. Am J Anat 153:483-488, 1978.

Frenk, H., and Rogers, G.H.: The suppressant effects of naloxone on food and water intake in the rat. Behav Neural Biol 26:23-40, 1979.

Gale, J.S., Bird, E.D., Spokes, E.G., Iversen, L.L., and Jessell, T.M.: Human brain substance P: Distribution in controls and Huntington's chorea. J. Neurochem:1978.

Gallagher, M., and Kapp, B.S.: Opiate administration into the amygdala: Effects on memory processes. Life Sciences, 23:1978.

Gessa, G.L., Paglietti, E., and Pellegrini Quarantotti, B.: Induction of copulatory behavior in sexually inactive rats by naloxone. Science 204:203-205, 1979.

Goldstein, A., and Cox, B.M.: Studies on endorphin function in animals and man. In Endorphins in Mental Health Research, E. Usdin, W.E. Bunney, Jr., and N. Kline, eds. MacMillan Press, London, 1979.

Guillemin, R.: Peptides in the brain: The new endocrinology of the neuron. Science 202:390-402, 1978.

Guillemin, R., Vargo, T., Rossier, J., Minick, S., Ling, N., Rivier, C., Vale, W., and Bloom, F.: β-Endorphin and adrenocorticotropin are secreted concomitantly by the pituitary gland. Science 197:1367-1369, 1977.

Gunne, L.M., Lindstrom, L., and Terenius, L.: Naloxone-induced reversal of schizophrenic hallucinations. J Neural Transm 40: 13-19, 1977.

Gunne, L.M., Linstrom, L., and Widerlov, E.: Possible role of endorphins in schizophrenia and other psychiatric disorders. In Endorphins in Mental Illness, E. Usdin, W.E. Bunney, Jr., and N.S. Kline, eds. MacMillan Press, London, 1979, pp 547-552.

Guyenet, P., Mroz, E.A., Aghajanian, G.K., and Leeman, S.E.:
Delayed iontophoretic ejection of substance P from glass
micropipettes: Correlation with neuronal excitation in vivo.
Neuropharmacology 18:553-558, 1979.

Henriksen, S.J., Bloom, F.E., McCoy, F., Ling, N., and Guillemin,
R.: β-Endorphin induces nonconvulsive limbic seizures. Proc
Natl Acad Sci USA 75:5221-5225, 1978.

Hökfelt, T., Meyerson, B., Nilsson, G., Pernow, B., and Sachs, C.:
Immunohistochemical evidence for substant P containing nerve
endings in the human cortex. Brain Res 104:181-186, 1976.

Hollister, L.E.: The mystique of social drugs and sex. In Sexual
Behavior; Pharmacology and Biochemistry, M. Sandler and G.L.
Gessa, eds. Raven Press, New York, 1975, pp 81-92.

Holtzman, S.G.: Behavioral effects of separate and combined administration of naloxone and d-amphetamine. J Pharmacol Exp
Ther 189:51-60, 1974.

Holtzman, S.G.: Comparison of the effects of morphine, penta-
zocine, cyclazocine and amphetamine on intracranial self-
stimulation in the rat. Psychopharmacology 46:223-227, 1976.

Houghten, R.A., Swann, R.W., and Li, C.H. (1980): β-endorphin:
Stability, clearance behavior and entry into the central nervous
system after intravenous injection of the tritiated peptide
in rats and rabbits. Proc. Natl. Acad. of Sciences 77:4588-4591.

Hughes, J., Smith, T.W., Kosterlitz, H.W., Fothergill, L.A.,
Morgan, B.A., and Morris, H.R.: Identification of two related
pentapeptides from the brain with potent opioid agonist
activity. Nature 258:577-580, 1975.

Innis, R.B., Correa, F.M.A., Uhl, G.R., Schneider, B., and Snyder,
S.H.: Cholecystokinin octapeptide-like immunoreactivity: Histochemical localization in rat brain. Proc Natl Acad Sci USA
76:521-525, 1979.

Iversen, L.L., Nicholl, R.A., and Vale, W.: Neurobiology of peptides. Neurosci Res Program Bull 16:214-370, 1978.

Izquierdo, I. (1980): Effects of a low and a high dose of B-
endorphin on acquisition and retention in the rat. Behavioral
and Neural Biology, 30:460-464.

Izquierdo, I., and Dias, R.D. (1981): Retrograde amnesia caused
by met-leu- and des-try-met-enkephalin in the rat and its
reversal by naloxone. Neuroscience Lett. 22:189-193.

Izquierdo, I., Dias, R.D., Souza, D.O., Carraso, M.A., Elisabetsky,
E., and Perry, M.D.: The role of opioid peptides in memory and
learning. Beh. Br. Res. 1:451-468, 1980.

Izquierdo, I., Paiva, A.C.M. , and Elisabetsky, E.: Posttraining
intraperitoneal administration of leu-enkephalin and beta-
endorphin causes retrograde amnesia for two different tasks
in rats. Behav. Neural. Biol. 28:246-250, 1980.

Izquierdo, I., Perry, M.L., Dias, R.D., Souza, D.O., Elisabetsky, E., Carrasco, M.A., Orsingher, O.A., and Nelto, C.A.: Endogenous opioids, memory modulation and state dependency. In: Endogenous Peptides and Learning and Memory Processes. J. L. Martinex, Jr., R.A. Jensen, R.B. Messing, H. Rigter, and J.L. McGaugh, eds. Academic Press, New York, 1981.

Jacquet, Y.F., and Marks, N.: The C-fragment of β-lipotropin: An endogenous neuroleptic or antipsychotogen. Science 194: 632-635, 1976.

Janowsky, D.S., Segal, D.S., Bloom, F., Abrams, A., and Buillemin, R.: Lack of effect of naloxone on schizophrenic symptoms. Am J Psychiatry 134:926-927, 1977.

Joyce, E.M., and Iversen, S.D.: The effect of morphine applied locally to mesencephalic dopamine cell bodies on spontaneous activity in the rat. Neurosci Lett 14:207-212, 1979.

Kastin, A.J., Mauk, M.D., Schally, A.V., and Loy, D.H.: Unusual dose-related effect of an endorphin analog in a complex maze. Physiol Behav 25:959-962.

Kastin, A.J., Scollan, E.L., King, M.G., Schally, A.V., and Loy, D.H.: Enkephalin and a potent analog facilitate maze performance after interperitoneal administration in rats. Pharmacol Biochem Behav 5:691-695, 1976.

Katz, R.J., Carroll, B.J., and Baldrighi, G.: Behavioral activation by enkephalins in mice. Pharmacol Biochem Behav 8:493-496, 1978.

Kelley, A.E., Stinus, L., and Iversen, S.D.: Interactions between d-ala-met-enkephalin, and A10 dopaminergic neurones and spontaneous behavior in the rat. Behav Brain Res 1:3-35, 1980.

Khantzian, E.J.: Opiate addiction: A critique of theory and some implications for treatment. Am J Psychother 28:59-70, 1974.

King, M., Geffard, M., Chavreau, J., Gaffori, O., Le Moal, M., and Muyard, J.: Des-tyr-γ-endorphin (DTgammaE) and pineal levels of melatonin and arginine vasopressin (AVP). European J of Pharmacology 76(2-3):271-274, 1981.

Kline, N.S., and Lehman, H.E.: β-Endorphin therapy in psychiatric patients. In Endorphins in Mental Illness, E. Usdin, W.E. Bunney, Jr., and N.S. Kline, eds. MacMillan Press, London, 1979.

Kline, N.S., Li, C.H., Lehmann, H.E., Lajtha, A., Laski, E., and Cooper, T.: Beta-endorphin induced changes in schizophrenic and depressed patients. Arch Gen Psychiatry 34:1111-1113, 1977.

Kobayashi, R.M., Brown, M., and Vale, W.: Regional distribution of neurotensin and somatostatin in rat brain. Brain Res 126:584-588, 1977.

Koob, G.F., and Bloom, F.E.: Behavioral effects of neuropeptides: Endorphins and vasopressin. Ann Rev Physiol 44:571-582, 1982.

Koob, G.F., and Bloom, F.E.: Behavioural effects of opioid peptides. Brit Med Bull 39:89-94, 1983.

Koob, G.F., Strecker, R.E., and Bloom, F.E.: Effects of naloxone on the anticonflict properties of alcohol and chlordiazepoxide. Substance and Alcohol Actions/Misuse 1:447-457, 1980.

Koob, G.F., LeMoal, M., and Bloom, F.E.: Enkephalin and endorphin influences on appetitive and aversive conditioning. In Endogenous Peptides and Learning and Memory Processes. J.L. Martinez, Jr., R.A. Jensen, R.B. Messing, H. Rigter, and J.L. McGaugh, eds. Academic Press, New York, 1981.

Kosterlitz, H.W., Hughes, J., Lord, J.A.H., and Waterfield, A.A.: Enkephalins, endorphins and opiate receptors. In Society Neuroscience Symposium, vol II, W.M. Cowan and J.A. Ferrendelli, eds. 1977, pp 291-301.

Krnjevic, K., and Lekic, D.: Substance P selectively blocks excitation of Renshaw cell by acetylcholine. Can J Physiol Pharmacol 55: 958-961, 1977.

Kurland, A.A., McCabe, O.L., Hanlon, T.E., and Sullivan, D.: The treatment of perceptual disturbances in schizophrenia with naloxone hydrochloride. Am J Psychiatry 134:1408-1410, 1977.

Le Moal, M., Koob, G.F., and Bloom, F.E.: Endorphins and extinction: Differential actions on appetitive and adversive tasks. Life Sci 24: 1631-1636, 1979.

Lewis, R.V., Gerber, L.D., Stein, S., Stephen, R.L., Grosser, B.I., Velick, S.F., and Udenfriend, S.: β-Leu^5-endorphin and schizophrenia. Arch Gen Psychiatry 36:237-239, 1979.

Li, C.H., Barnaft, L., Chretien, M., Chung, D.: Isolation and amino acid sequence of β-LPH from sheep pituitary glands. Nature 208:1092-1094, 1965.

Livett, B.G., Kozousek, V., Mizobe, F., and Dean, D.M.: Substance P inhibits nicotinic activation of chromaffin cells. Nature 278:256-257, 1979.

Lucion, A.B., Rosito, G., Sapper, D., Palmini, A.L., and Izquierdo, I.: Intracerebroventricular administration of nanogram amounts...in rats. Beh Br Res p. 111-114, vol. 4, 1982.

Manchanda, R., and Hirsch, S.: (Des-Tyr)-γ-endorphin in the treatment of Psychological Medicine 11(2):401-404, 1981.

Martinez, J.L., Jr., and Rigter, H. (1980): Endorphins alter acquisition and consolidation of an inhibitory avoidance response in rats. Neuroscience Lett 19:197-20.

Martinez, J.L., Jr., Rigter, H., Jensen, R.A., Messing, R.B., Vasquez, B.J., and McGaugh, J.L.: Endorphin and enkephalin effects on avoidance conditioning: The other side of the pituitary-adrenal axis. In Endogenous Peptides and Learning and Memory Processes. J.L. Martinez, Jr., R.A. Jensen, R.B. Ressing, H. Rigter and J.L. McGaugh, eds. Academic Press, New York, 1981.

Messing, R.B., Jensen, R.A., Martinez, J.L., Jr., Spiehler, V.R., Vasquez, B.J., Soumireu-Mourat, B., Liang, K.C., and McGaugh, J.L.: Naloxone enhancement of memory. Behavioral and Neural Biology, 27:266-275, 1979.

Mason, S.T., and Iversen, S.D.: Effects of selective forebrain noradrenaline loss on behavioral inhibition in the rat. J Comp Physiol Psychol 91:165-173, 1977.

Mielke, D.H., and Gallant, D.M.: An oral opiate antagonist in chronic schizophrenia: A pilot study. Am J Psychiatry 134: 1430-1431, 1977.

Nathanson, J.A.: Cyclic nucleotides and nervous system function. Physiol Rev 57:157-256, 1977.

Nickolson, V., and Berendesen, H.: Effects of the potential neuroleptic peptide des-tyrosine-γ-endorphin and haloperidol on apomorphine-induced behavioral syndromes in rats and mice. Life Sciences 27(15):1377-1385, 1980.

Nicholl, R.A., Siggins, G.R., Ling, N., Bloom, F.E., and Guillemin, R.: Neuronal actions of endorphins and enkephalins among brain regions: A comparative microiontophoretic study. Proc Natl Acad Sci USA 74:2584-2588, 1977.

Noeth, R.A.: Opiates, opioid peptides and single neurones. Life Sci 24:1527-1546, 1979.

Palmour, R.M., Ervin, F.R., Wagemaker, H., and Cade, R.: Characterization of a peptide from the serum of psychiatric patients. In Endorphins in Mental Illness, E. Usdin, W.E. Bunney, Jr., and N.S. Kline, eds. MacMillan Press, London, 1979.

Paxinos, G., Emson, P.C., and Cuello, A.C.: The substance P projections to the frontal cortex and the substantia nigra. Neurosci Lett 7:127-131, 1978.

Pedigo, N.W., Ling, N.C., and Yamamura, H.I.: Examination of des-tyrosine--endorphin activity at ^{3}H-spiroperidol binding sites in rat brain. Life Sci 24:1645-1650, 1979.

Phillis, J.W., and Limacher, J.J.: Substance P excitation of cerebral cortical Betz cells. Brain Res 69:158-163, 1974.

Phillis, J.E., Kirkpatrick, J.R., and Said, S.I.: Vasoactive
intestinal polypeptide excitation of central neurons.
Can J Physiol Pharmacol 56:337-340, 1978.
Pickar, D., Vartanian, F., Bunney, W., Mair, H., Gastpar, M.,
Prakash, R., Sethi, B., Lideman, R., Velyaev, V., Tsutsulkovskaja,
M., Jungkunz, G., Nedopil, N., Verhoeven, W., and van Praag, H.:
Short term naloxone administration in schizophrenic and manic
patients. A World Helath Organization Collaborative Study.
Archives of General Psychiatry 39(3):313-319, 1982.
Pinget, M., Straus, E., and Yalow, R.W.: Localization of
cholecystokinin-like immunoreactivity in isolated nerve ter-
minals. Proc Natl Acad Sci USA 75:6324-6326, 1978.
van Praag, H.M.: Neurotransmitters and neuropeptides in
schizophrenia.   Paper presented at the International Health
Foundation Workshop. Bordeuz, France, October 4-5, 1979.
van Praag, H., Verhoeven, W., van Ree, J., de Wied, D.:
The treatment of schizophrenice psychoses with gamma-type
endorphins. Biological Psychiatry 17(1):83-98, 1982.
Rigter, H.: Attenuation of amnesia in rats by systemically
administered enkephalins. Science 200:83-85, 1978.

Rigter, H., Dekker, I., and Martinez, J.L., Jr.: A comparison
of the ability of opioid peptides and opiates to affect
active avoidance conditioning in rats. Reg Peptides 2:317-
332, 1981.
Rigter, H., Hannan, T.J., Messing, R.B., Martinez, J.L., Jr.,
Vasquez, B.J., Jensen, R.A., Veliquette, J., and McGaugh,
J.L.: Enkephalins interfere with acquisition of an active
avoidance response. Life Sci 26:337-346, 1980a.
Rigter, H., Jensen, R.A., Martinez, J.L., Jr., Messing, R.B.,
Vasquez, B.J., Liang, K.C., and McGaugh. L.J.: Enkephalin
and fear-motivated behavior. Proc Natl Acad Sci USA 77:
3729-3732, 1980b.
Rogers, G.H., Frenk, H., Taylor, A.N., and Liebeskind, J.C.:
Naloxone suppression of food and water intake in deprived
rats. Proc West Pharmacol Soc 31:457-460, 1978.
Ross, M., Berger, P.A., and Goldstein, A.: Plasma β-endorphin
immunoreactivity in schizophrenis. Science 205:163-164,
1979.
Rossier, J., French, E.D., Rivier, C., Ling, N., Guillemin, R.,
and Bloom, F.E.: Foot-shock induced stress increases beta-
endorphin levels in blood but not in brain. Nature 270:618-
620, 1977.
Salzman, B., and Frosch, W.A.: Methodone maintenance for the
psychiatrically disturbed. Proc Natl Conf Methadone Treat
4:117-118, 1972.

Schildkraut, J.J.: The catecholamine hypothesis of affective disorders: A review of supporting evidence. Am J Psychiatry 122:509-522, 1965.

Segal, D.S., Browne, R.G., Arnsten, A., Derrington, D.C., Bloom, F.E., Guillemin, R., and Ling, N.: In Endorphins and Mental Health Research, E. Usdine, W.E., Bunney, Jr. and N.S. Kline, eds. MacMillan Press, London, 1979, pp 307-324.

Simpson, G.M., Branchey, M.H., and Lee, J.H.: A trial of naltrexone in chronic schizophrenia. Curr Therap Res 22: 909-913, 1977.

Stallcup, W.B., and Patrick, J.: Substance P enhances acetylcholine receptor desensitization in a clonal cell line. Proc Natl Acad Sci USA 77: 634-638, 1980.

Stapleton, J.M., Lind, M.D., Merriman, V.J., and Reid, L.D.: Naloxone inhibits diazepam-induced feeding in rats. Life Sci 24:2421-2426, 1979.

Stern, A.S., Lewis, R.V., Kimura, S., Rossier, J., Gerber, L.D., Brink, L., Stein, S., and Udenfriend, S.: Isolation of the opioid heptapeptide Met-enkephalin [Arg^6,phe^7] from bovine adrenal medullary granules and striatum. Proc Natl Acad Sci USA 76:6680-6683, 1979.

Stevens, D.S., and Livermore, A.: Kindling of the mesolimbic dopamine system: Animal model of psychosis. Neurology 28: 36-46, 1978.

Stinus, L., Koob, G.F., Ling, M., Bloom, F.E., and Le Moal, M.: Locomotor activation induced by infusion of endorphins into the ventral tegmental area: Evidence for opiate-dopamine interactions. Proc Natl Acad Sci USA 77:2323-2327, 1980.

Straus, E., and Yalow, R.S.: Species specificity of cholecystokinin in gut and brain of several mammalian species. Proc Natl Acad Sci USA 75:486-489, 1978.

Takagi, H., Satoh, M., Akaike, A., Shibata, T., Yajima, H., and Ogawa, H.: Analgesia by enkephalins injected into the nucleus reticularis gigantocellularis of rat medulla oblongata. Eur J Pharmacol 49:113-116, 1978.

Tamminja, C., Tighe, D., Chase, T., De Fraites, E., Schaffer, M.: Des-tyrosine-γ-endorphin administration in chronic schizophrenics. A preliminary report. Archives of General Psychiatry 38(2):167-168, 1981.

Taylor, D., Hoffer, B., Zieglgänsberger, W., Siggins, G., Ling, N., Seiger, A., and Olson, L.: Opioid peptides excite pyramidal neurons and evoke epileptiform activity in hippocampal transplants in oculo. Brain Res: 1979.

Terenius, L., Wahlstrom, A., Lindstrom, L., and Widerlöv,
E.: Increased CSF levels of endorphins in chronic
psychosis. Neurosci Lett 3:157-162, 1976.

Uhl, G., Kuhar, M.J., and Snyder, S.H.: Neurotensin:
Immunohistochemical localization in rat central nervous
system. Proc Natl Acad Sci USA 74:4059-4063, 1977.

Van der Kooy, D., LePiane, F.G., Phillips, A.G.: Apparent
independence of opiate reinforcement and electrical self-
stimulation systems in rat brain. Life Sci 20:981-986,
1977.

Veith, J.L., Sandman, C.A., Walker, J.M., Loy, D.H., and
Kastin, A.J.: Systemic administration of endorphins
selectively alters open field behavior of rats. Physiol
Behav 20:539-542, 1978.

Verebey, K., Volavka, J., and Clouet, D.: Endorphins in
psychiatry. Arch Gen Psychiatry 35:877-888, 1978.

Verhoeven, W.M.A., van Praag, H.M., van Ree, J.M., and
de Wied. D.: Improvement of schizophrenic patients treated
with [Des-Tyr]-endorphin (DT E). Arch Gen Psychiatry 36:
294-298, 1979.

Volavka, J., Mallya, A., Baig, S., and Perez-Cruet, J.:
Naloxone in chronic schizophrenia. Science 196:1227-1228,
1977.

Volavka, J., Davis, L.G., Ehrlich, Y.H.: Endorphins dopamine
and schizophrenia. Schizophr Bull 5:227-239, 1979.

Wagemaker, H., and Cade, R.: The use of hemodialysis in
chronic schizophrenia. Am J Psychiatry 134:684-685,
1977.

Watson, S.J., Berger, P.A., Akil, H., Hills, M.J., and
Barchas, J.D.: Effects of naloxone on schizophrenia: Re-
duction in hallucinations in a subpopulation of subjects.
Science 201:73-76, 1978.

Watson, S.J., Akil, H., Berger, P.A., and Barchas, J.D.:
Some observations on the opiate peptides and schizophrenia.
Arch Gen Psychiatry 36:35-41, 1979.

Wauguier, A., Niemegeers, C.J.E., and Lal, H.: Differential
antagonism by naloxone of the inhibitory effects of
haloperidol and morphine on brain stimulation. Psychopharmacology 37:303-310, 1974.

Weibel, S.L., and Wolf, H.H.: Opiate modification of intracranial self-stimulation in the rat. Pharmacol Biochem
Behav 10:71-78, 1979.

Weinberger, S.B., Arnsten, A., and Segal, D.C.: Des-
tyrosine-γ-endorphin and haloperidol: Behavioral and
biochemical differentiation. Life Sci 24:1637-1644,
1979.

Weitzman, R., Fisher, D., Minick, S., Ling, N., and
Guillemin, R.: β-Endorphin stimulates secretion of
arginine vasopressin in vivo. Endocrinology 101:1643-
1646, 1977.

Wellisch, D.K., and Gay, G.R.: The walking wounded:
Emergency psychiatric intervention in a heroin addict
population. Drug Forum 1:137-144, 1972.

de Wied, D.: Peptides and behavior. Life Sci 20:195-
204, 1977.

de Wied, D., Witter, A., and Greven, H.M: Behaviorally
active ACTH analogues. Biochem Pharmacol 24:1463-1468,
1975.

de Wied, D., Bohus, B., van Ree, J.M., and Urba, I.: Behavioral and electrophysiological effects of peptides
related to lipotropin (β-LPH). J Pharmacol Exp Ther 204:
570-580, 1978a.

de Wied, D., Kovacs, G.L., Bohus, B., van Ree, J.M., and
Greven, H.M.: Neuroleptic activity of the neuropeptide
βLPH$_{62-77}$[Des-Tyr^1] -endorphin, DT E. Eur J Pharmacol
49:427-436, 1978b.

de Wied, D., Bohus, B., van Ree, J.M., Kovacs, G.L., and
Greven, H.M.: Neuroleptic-like activity of [des-Tyr^1]- -
endorphin in rats. Lancet 1:1046, 1978c.

Wikler, A., and Rasor, R.W.: Psychiatric aspect of drug
addiction. Am J Med 14:566-570, 1953.

Wikler, A., Pescor, M.J., Kalbaugh, E.P., and Angelucci,
R.J.: Effects of frontal lobotomy on the morphine-
abstinence syndrome in man: An experimental study. Arch
Neurol and Psychiat 67:510-521, 1952.

Woods, J.H., and Schuster, C.R.: Opiates as reinforcing
stimuli. In Stimulus Properties of Drugs, T. Thompson
and R. Pickens, eds. Appleton-Century-Crofts, New
York, 1971, pp 163-175.

Zieglgänsberger, W., and Bayerl, H.: The mechanisms of
inhibition of neuronal activity by opiates in the
spinal cord of the rat. Brain Res 115:111-128,
1976.

Zieglgänsberger, W., and Champagnat, J.: Cat spinal
motoneurons exhibit topographic sensitiveity to glutamate and glycine. Brain Res 160:95-104, 1979.

Zieglgänsberger, W., and Tulloch, I.F.: The effects of
methionine- and leucine-enkaphalin on spinal neurons
of the cat. Brain Res 167:53-64, 1979.

Zieglgänsberger, W., Siggins, G.R., French, E.D., and

Bloom, F.E.: Effects of opioids on single units. In
Characteristics and Function of Opioids, J. Van
Ree and L. Terenius, eds. Elsevier, Amsterdame, 1978,
pp 75-86.
Zieglgänsberger, W., French, E.D., Siggins, G.R., and
Bloom, F.E.: Opioid peptides may excite hippocampal
pyramidal neurons by inhibiting adjacent inhibition
neurons. Science 205:415-417, 1979.

Copyright © 1984 by Spectrum Publications, Inc.
Peptides, Hormones and Behavior
Edited bv C.B. Nemeroff and A.J. Dunn

Neurohypophysial Hormones—Their Role in Endocrine Function and Behavioral Homeostasis

TJEERD B. VAN WIMERSMA GREIDANUS
AND DIRK H.G. VERSTEEG

Vasopressin and oxytocin (OXT) (Fig. 9-1) are peptides hormones secreted by the neurohypophysis. They are synthesized in the perikarya of neurons located in the rostral part of the hypothalamus, more particularly in the supraoptic, paraventricular, and suprachiasmatic nuclei (Sachs et al., 1969; Swaab et al., 1975; Scharrer and Scharrer, 1940, 1954; Defendini and Zimmerman, 1978. See also, Rodriguez, Chapter 1). From the supraoptic and paraventricular cell bodies, the neurosecretory material is transported to the posterior pituitary gland, from which it can be released into the general circulation (Scharrer and Scharrer, 1940, 1954; Defendini and Zimmerman, 1978). The transport of the hormones from their hypothalamic sites of synthesis to the nerve terminals occurs by axoplasmic transport. The neurohypophyseal peptides are loosely bound to carrier protein (neurophysin) during their transportation. Most mammalian species probably have two major neurophysins. The hormone-protein complex is packed in special neurosecretory granules and an intragranular association of neurophysin I with OXT and neurophysin II with vasopressin has been suggested (Douglas, 1974; Vandesande et al., 1975; Defendini and Zimmerman, 1978).

Throughout the vertebrates, naturally occurring neurohypophyseal principles are known, which all consist of a ring structure of six amino acids and a C-terminal tail of three amino acids. The ring part is closed by a disulfide link between two cysteine residues in position 1 and 6. Therefore these neurohypophyseal hormones can also be regarded as octapeptides, since the two cysteines can be considered as one cystine residue. Of these nonapeptides arginine[8]-vasotocin (AVT) appears to be the most primitive and may be the ancestor of all neurohypophyseal hormones. It is found in the neuro-hypophyses of all types of vertebrates. OXT is present in all mammals, whereas two types of vasopressin occur among the mammalian species. The most common is arginine[8]-vasopressin (AVP). In the group of *Suinae* the arginine residue is replaced by lysine (lysine[8]-vasopressin, LVP).

The name vasopressin orginates from its effect on blood pressure, the result of constriction of peripheral blood vessels. However, in relatively low amounts vasopressin acts as an antidiuretic principle and for this reason is also known as antidiuretic hormone (ADH). The antidiuretic effect of vasopressin is exerted on the reabsorption of water in the kidney, to conserve free water to the body. Thus ADH increases the osmolality of urine. The regulation of vasopressin secretion into the systemic circulation is complex, involving osmoreceptors and volume receptors in the brain.

Vasopressin also plays a role in the release of adrenocorticotropin (ACTH) from the anterior pituitary, acting as a coticotropin-releasing factor. Nevertheless, despite the recent development in assays for anterior pituitary hormones, the precise role of vasopressin in ACTH release is still unclear.

OXT causes contraction of the myoepithelial elements in the alveolar walls in the mammal, thus inducing milk secretion. This hormone also has an uterotropic action by stimulating contractions of the myometrium. Stimuli from the nipple (suckling) and from the vagina can release OXT (neuroendocrine reflex).

BEHAVIORAL EFFECTS OF SYSTEMIC ADMINISTRATION OF VASOPRESSIN

Active Avoidance Behavior

The first report of a behavioral effect of neurohypophyseal principles described a disturbance in extinction of a learned response in rats after removal of the posterior lobe, including the pars intermedia, of the pituitary (de Wied, 1965). Although acquisition of the avoidance response was not affected and escape behavior of posterior lobectomized rats was not different from that of sham-operated controls, suggesting that motor or sensory systems were not affected, the extinction of a conditioned avoidance response (CAR) was markedly facilitated in these animals. The rapid extinction of the CAR of posterior lobectomized rats appeared to be unrelated to the disturbance in water metabolism of the animals. Subsequent studies revealed that, in posterior lobectomized rats, Pitressin, a crude extract of posterior pituitary tissue containing considerable amounts of vasopressin, was able to restore the disturbed behavior displayed by the operated animals. Long-acting preparations of LVP had the same effect as Pitressin (de Wied, 1965, 1969). The restoration of normal extinction behavior in posterior lobectomized rats was independent of the time of administration of LVP. Treatment with these neurohypophyseal principles, either during acquistion or during extinction, resulted in similar effects on the extinction rate.

Pitressin also induced marked behavioral changes in intact animals. In contrast to control rats, animals injected (SC) with a long-acting Pitressin preparation, either during acquisition or during extinction of a two-way shuttle-box avoidance response, maintained a high level of responding during extinction for several weeks (de Wied and Bohus, 1966; de Wied et al., 1974). These results have recently been confirmed, using an automatically operated shuttle box with a slightly different conditioning schedule (Hagan, personal communication). Similar effects were obtained in one-way active avoidance behavior with synthetic vasopressin. A single SC injection of LVP resulted in a long-term, dose-dependent inhibition of extinction of the CAR, which extended far beyond the presence of the injected peptide in the organism. This suggests that vasopressin triggers a long-term effect on the maintenance of a learned response, probably by facilitating consolidation processes (de Wied, 1971; de Wied at al., 1974, 1976; van Wimersma Greidanus et al., 1973).

Structure-Activity Relationships

Attempts to determine the behaviorally active core of the vasopressin molecule were performed in a one-way active avoidance situation. A number of synthetic analogues and fragments of vasopressin and OXT were tested for their ability to inhibit extinction of a pole-jump avoidance response. Rats trained to jump onto a pole in response to the conditioned stimulus (CS), a light on top of the box, to avoid the unconditioned stimulus (UCS) of electric footshock (FS). Ten trials a day were given with an intertrial interval of 60 seconds, presented in a predetermined sequence for three consecutive days. Peptides were injected SC at the end of the third acquisition session. Extinction sessions in which only the CS was presented were run on days 4, 5, and 8. For each peptide tested, the amount needed to induce six or more positive responses on day 8 (third extinction session) was determined. Saline-treated rats never scored more than two positive responses in this extinction session. If the criterion of six responses was not reached with a dose of 5 μg per rat, the peptide was considered inactive. AVP was the most potent peptide, followed by LVP. Removal of the C-terminal glycinamide decreases the potency to approximately 50 percent. The resulting desglycinamide analogues (DG-AVP and DG-LVP) are partically devoid of the pressor-, antidiuretic-, and ACTH-releasing activities of vasopressin (Lande et al., 1971; de Wied et al., 1972). Oxypressin (OXP) had considerable activity, amounting to 30 percent of that AVP; OXT and vasopressin (VT) were equally effective and possessed about 20 percent of AVP's activity. Pressinamide (PA) retained only 10 percent of the behavioral potency (de Wied and Gispen, 1977; Wang, 1972; Walter et al., 1978; de Wied et al., 1975). These structure-function relationships (Fig. 9-1) indicate that the behavioral action of vasopressin is independent of its more classical endocrine effects.

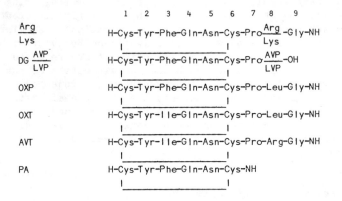

Fig. 9-1.   Structures of neurohypophysial hormones and analogues.

In contrast to these findings, Stark et al., (1978) reported that OXT (SC) had no effect on the extinction of a pole-jump avoidance response, but that a combination of both LVP and OXT affected retention in a way suggesting competition between the hormones. Previous observations by Schultz et al., (1974a, 1974b) indicated that OXT affected avoidance behavior in a direction opposite to that of vasopressin, that is, facilitating extinction of a pole-jump avoidance response after IP administration. However, the most marked effect of oxytocin in this respect was found in thirsty rats, suggesting that oxytocin may be counteracting the behavioral effect of vasopressin. No significant effects were found by these investigators in passive avoidance behavior or on locomotor activity in an open field (Schultz et al., 1974a, 1974b, 1976). However, recently, opposite actions of IP OXT and vasopressin were reported by Kováks et al. (1978) on passive avoidance behavior in rats.

Temporal Aspects

Administration of AVP three hours before or after the behavioral session, markedly decreased the effectiveness of the treatment and administration six hours before or after was ineffective (de Wied, 1971, 1973; Bohus et al., 1972). In general, vasopressin is fully active when given within one hour of the behavior (King and de Wied, 1974).

Vasopressin not only has effects on the maintenance of conditioned avoidance behavior, it also affects acquistion of the CAR. In totally hypophysectomized animals, it normalized the disturbed acquisition of the response (de Wied, 1969; Bohus et al., 1973). In intact rats, vasopressin generally does not alter the rate of acquisition of an avoidance response, but the vasopressin analogue, ornithine-8-vasopressin, improves avoidance learning (de Wied, 1973). Also, a slight improvement of acquisition after vasopressin treatment may occur in situations in which the tendency to respond to relevant stimuli is low (King and de Wied, 1974). However, a single injection of the peptide at the beginning of the training influences extinction of the CAR much more than it affects acquisition (de Wied, 1973; King and de Wied, 1974). Vasopressin-induced resistance to extinction of a CAR is strong when the single injection of the hormone at the first acquisition session was associated with the emission of a correct avoidance response. This suggests that some measure of associative strength must be present before vasopressin is behaviorally effective in an active avoidance situation (King and de Wied, 1974).

In a series of experiments intended to differentiate classical and instrumental components underlying the behavioral effects, it appeared that classical conditioning alone was a sufficient behavioral substrate for the long-term effect of vasopressin on avoidance behavior, but instrumental conditioning was more effective (King and de Wied, 1974). Also, cognitive expectancies may be essential for this effect (Hagan, personal communication). Generally, this agrees with findings on the classical conditioning of a heart-rate response in freely moving rats. This response is of a bradycardic nature, which extinguishes rapidly unless vasopressin has been administered.

The heart rate of non- or pseudoconditioned rats is not affected by vasopressin treatment (Bohus, 1975).

Passive Avoidance Behavior

Vasopressin affects not only active avoidance behavior, but also passive avoidance behavior, as studied in a simple "step-through" situation (Ader et al., 1972). This task uses the innate preference of rats for darkness over light. The apparatus consists of a dark box to which an illuminated platform is attached. Rats are placed on the platform and their latency to enter the dark compartment is recorded. One adaptation trial is given on day 1 and three such trials on day 2. Immediately after the third trial on day 2, the animals receive a FS (0.25 mA for 1 to 3 seconds). The latency to enter is recorded again in the first retention trial, 24 hours after the single acquisition trial. At 48 hours a second retention session is performed. LVP, injected either 1 hour before the first retention trial or immediately after the acquisition trial, considerably increased avoidance latencies. Avoidance latencies were still augmented during the second retention trial at 48 hours. Thus, the long-term preservation of an acquired response was also found in the passive avoidance situation (Ader and de Wied, 1972).

Vasopressin analogues such as desamino-8-D-arginine vasopressin (DDAVP), desamino-6-carba-AVP (DCAVP), and desamino-6-carba-8-ornithine-vasopressin (DCOVP), which are all resistent to enzymic degradation, also induce strong and long-lasting improvement of retention of a passive avoidance response (Krejci and Kupková, 1978a).

A temporal relationship similar to that in active avoidance behavior was found for the retention of a passive avoidance response. The critical time period for the behavioral effect of vasopressin was again found to be approximately 1 hour (Bohus et al., 1972). Administration of vasopressin to rats trained in both active and in a passive avoidance revealed that no generalization or transfer of the effects of vasopressin from one behavioral situation to the other occurred (Bohus et al., 1972). Enhancement of passive avoidance retention of a similar type was also found by Wang (1972), by Lissak and Bohus (1972), and by Bookin and Pfeifer (1977). Vasopressin also induced a long-lasting enhancement of the retention of a passive avoidance response of mice in a situation in which the aversive stimulus was attack by a trained fighter mouse. Vasopressin enhanced retention of this response when administered SC after acquisition or prior to the first retention test. Thus, the effects of vasopressin on avoidance of attack are similar to its effects on shock avoidance (Leshner and Roche, 1977).

Amnesia

Further evidence for the effects of vasopressin and its con-
geners on memory processes was obtained when it appeared that
vasopressin and its analogue antagonize retrograde amnesia.
Lande et al. (1972) reported that the puromycin-induced blockade of expression of maze learning in mice can be prevented by
SC DG-LVP. This suggests a modification of memory consolidation, such that the expression of memory becomes insensitive to
puromycin. Other neurohypophyseal hormone fragments, such as
the tail portions of OXT, Pro-Leu-Gly-NH$_2$ (PLG) and Leu-Gly-
NH$_2$, also attenuate puromycin-induced amnesia (Walter et al.,
1975) and this effect is manifested in a dose-dependent fashion
(Flexner et al., 1977). Vasopressin also antagonized the retrograde amnesia induced by electroconvulsive shock (ECS)
(Pfeifer and Bookin, 1978), by CO$_2$ (Rigter et al., 1974) or by
pentylenetetrazol (Bookin and Pfeifer, 1977).
  Since LVP and DG-LVP diminish amnesia not only when injected
prior to acquisition, but also when given prior to the retrieval test, it appears that vasopressin is able to prevent as well
as to reverse amnesia. These results suggest that vasopressin,
in addition to affecting memory consolidation (de Wied et al.,
1975), also promotes its retrieval (Rigter et al., 1974; Rigter
and van Riezen, 1978; Bookin and Pfeifer, 1977).

Other Behaviors

No effect of vasopressin on open-field behavior or on extinction of a straight runway approach response has been observed
(Garrud et al., 1974). However, vasopressin increased the performance of food-deprived rats trained to hold a lever down in
order to obtain food (Garrud, 1975). Vasopressinlike peptides
have also been shown to facilitate acquisition of an appetitive
task, such as alcohol drinking behavior (Finkelberg et al.,
1978). Relatively complex tasks involving learning or memory
processes also appear to be affected by vasopressin. In an appetitive, black-white discrimination T maze, animals that were
reinforced for choosing the black goal arm demonstrated prolonged extinction if they received vasopressin during extinction, without effects on speed or activity scores. However,
the prolonged extinction was not observed after treatment during acquisition, nor in animals reinforced for choosing the
white arm (Hostetter et al., 1977). These data demonstrate
that vasopressin can effect not only avoidance behavior, but
performance of positively rewarded tasks as well.
  Similar effects of vasopressin were also found in sexually
motivated approach behavior. A higher percentage of male rats

receiving DG-LVP chose the correct arm of the T-maze for reaching a receptive female compared to controls (Bohus, 1977). In this situation, vasopressin exhibited a long-term effect, extending beyond the treatment period (Bohus, 1977). Copulation reward appeared to be essential for this effect; unrewarded rats did not display more correct choices than placebo-treated animals after cessation of the treatment (Bohus, 1977). DG-LVP also delayed the disappearance of intromission and ejaculatory behavior of male rats following castration. Thus vasopressin not only affects the maintenance of learned approach and avoidance behavior, but also the extinction of genetically determined behavioral patterns.

DG-LVP also delayed the approach response to an imprinting stimulus in ducklings, and it is suggested that in this situation also, the peptide interferes with retrieval processes (Martin and van Wimersma Greidanus, 1979). DCOVP was reported by Krejci and Kupková (1978b), to induce sleep, an effect not observed after treatment with vasopressin itself or with other vasopressin analogues.

DG-AVP and PA (SC) have been found to reduce intravenous heroin self-administration in rats, and the effectiveness of these vasopressin analogues appeared to be long term (van Ree and de Wied, 1977). Since memory processes may be involved in heroin control of behavior, the effect of vasopressin fragments on self-administration of heroin agrees with the previously described actions of vasopressin on behavior (van Ree et al., 1978). However, DG-AVP did not decrease morphine self-administration by monkeys physically dependent upon morphine (Mello and Mendelson, 1979).

Finally OXT reduced and vasopressin prolonged the reaction time of rats, that is the time interval between application of an acoustic stimulus and elimination by the animal (Schwarzberg and Unger, 1970).

Development of Tolerance

Development of resistance to morphine analgesia may be regarded as a form of learning or memory (Cohen et al., 1965). This view is supported by the observation that protein-synthesis inhibitors that cause retograde amnesia also prevent the development of tolerance to narcotic analgesics (Cox and Osman, 1970). A further parallel is indicated by the findings of Krivoy et al. (1974) that DG-LVP facilitates the development of resistance to the analgesic action of morphine in mice. Also, it has been reported that DG-AVP, PLG, and OXT facilitate morphine dependence in rats, as measured by the loss of body weight following naloxone administration (van Ree and de Wied,

1976, 1977). However, Schmidt et al. (1978), who measured the effect of peptides on the maximum level of tolerance in the mouse, failed to demonstrate any alteration of the effects of chronic morphine. In fact, they confirmed what had been noted before, that maximal levels of tolerance are not affected. It has therefore been suggested that neurohypophyseal peptides may be endogenous modulators of the chronic responses to morphine and other opiates, albeit that the rate of tolerance induction rather than the end points are affected.

Mice treated with AVP during chronic ethanol administration showed long-term maintenance of tolerance to the hypothermic and sedative effects of ethanol, but OXT was ineffective (Hoffman et al., 1978).

INTRACEREBROVENTRICULAR ADMINISTRATION OF NEUROHYPOPHYSEAL PEPTIDES AND THE BRAIN

Effects on Avoidance Behavior

That vasopressin may be involved physiologically in memory processes was indicated when picogram amounts were found to be active after central administration. As little as 25 pg of AVP injected into the lateral ventricle significantly inhibited ex-tinction of the pole-jump avoidance response (de Wied, 1976). This dose approaches the physiological levels of vasopressin in the cerebrospinal fluid (CSF) (approximately 10 to 30 pg/ml). The vasopressin ring-fragment PA and its C-terminal tail por-tion Pro-Arg-Gly-NH$_2$ (PAG) were behaviorally active in amounts of 40 pg and 100 ng, respectively, given by this route. Thus the ratio of the effective doses of AVP and PA on extinction of a CAR is much smaller after intracerebroventricular (ICV) than after SC administration, suggesting a difference in metabolism of the peptides (de Wied, 1976).

As mentioned previously, systemic OXT had a weak effect in delaying extinction of a CAR compared with vasopressin, but it has also been found to possess behavioral activity opposite to that of vasopressin. Treatment with AVP (ICV) immediately after passive avoidance training improved performance, whereas OXT impaired performance, as judged by the significantly short-er avoidance latencies (Fig. 9-2) (Bohus et al., 1978a, 1978b). Thus OXT may be a naturally occurring amnestic peptide. OXT (1 ng/1 μl ICV) attenuated the passive avoidance response effec-tively when administered up to three hours, but not six hours, after the learning trial. It was also active when given one hour prior to the first retention session. This indicates that OXT not only has an amnestic action, but also exerts inhibitory effects on retrieval processes (Bohus et al., 1978b).

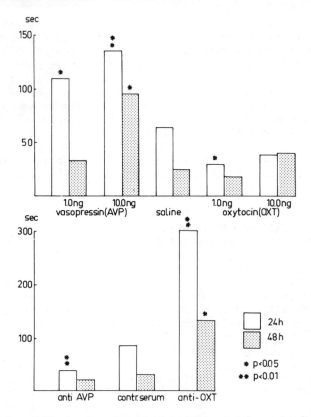

Fig. 9-2. Effect of vasopressin and oxytocin and of
antisera to these hormones on retention of a
passive avoidance response after intracerebroventricular administration immediately after the
learning trial (0.25mA, 2 sec).

Higher doses of OXT (10 ng) were weaker or ineffective after
ICV administration (Bohus et al., 1978a). There is no obvious
explanation for the different results obtained following systemic and ICV injection of OXT. Generally, however, the
results suggest that vasopressin and OXT modulate memory function in an opposite manner by a direct action on the brain.

Other Behavioral Effects of Intracerebroventricular
Neurohypophyseal Peptides

OXT ICV increased the self-stimulation rate of rats and the
animals show a tendency to shorten the reaction time, whereas
vasopressin decreased the rate of self-stimulation and length-
ened reaction time (Table 9-1) (Schwarzberg et al., 1976,
1978a, 1978b). PLG increased the number of responses in a
substantia nigra self-stimulation system, whereas DG-AVP decreased the number of responses (Dorsa and van Ree, 1979).
These observations support the hypothesis of opposite effects
of vasopresssin and OXT on brain function.

Vasopressin increases and OXT decreases the impulse activity
of neurons in the paraventricular and supraoptic nuclei after
ICV administration (Schulz et al., 1971). These effects on the
multiple unit activity of these brain regions were dose dependent (Schwarzberg et al., 1974). Also, ICV OXT interrupted the
high discharge rate of paraventricular and supraoptic neurons
induced by glutamate, and eliminated the typical changes in the
electrocephalogram after this drug in rabbits (Schulz et al.,
1974a). DG-AVP enhanced the high frequencies in hippocampal
theta activity, generated during paradoxical sleep (PS), suggesting a long-lasting increase in excitability induced by the
vasopressin-analogue in midbrain limbic circuits (Urban and de
Wied, 1978). OXT ICV decreased the peak frequency of hippocampal theta rhythm during PS (Bohus er al., 1978a), without affecting the total amount of PS during the eight-hour observation periods. Hippocampal injection of LVP elicited spreading
depression in the hippocampus but not in the neocortex, suggesting that hippocampal neurons are particularly sensitive to
vasopressin (Huston and Jakobartl, 1977).

Vasopressin appears to have other direct effects on the central nervous system (CNS). Relatively large amounts of AVP
(200 ng), AVT (200 ng), or LVP (10 ng) induce barrel rotation
after ICV administration in rats, whereas OXT is practically
ineffective in this respect (Kruse et al., 1977). This barrel
rotation phenomenon may be a toxic effect, since it was accompanied by other toxic symptoms and frequently death. LVP also
caused a dramatic hyperactivity in mice after bilateral ICV
administration of relatively large amounts (5x10 M/5 μl) (Dunn
et al., 1976). Evaluation and extension of these studies with
smaller amounts revealed that LVP, AVP, AVT, and OXT all provoked hyperactivity, extensive foraging, and increased grooming. Administration of higher doses also elicited stereotyped
scratching, squeaking, and occasionally barrel rotation. DG-
LVP was much less potent in this respect (Delanoy et al.,
1978). Extensive structure-activity relationship studies of
these effects are in good agreement with proposals relating the

Table 9.1. Outline of the Effects of Vasopressin and Oxytocin and Their Fragments on Behavior and Related Brain Processes

	Vasopressin[a]	Oxytocin
Avoidance behavior		
Active avoidance: resistance to extinction	+	−
Passive avoidance: consolidation	+	−
retrieval	+	−
Amnesia: prevention	+	+
reversal	+	+
Rewarded behavior		
Operant		
Heroin self-administration: acquisition	−	+
Electrical self-stimulation: at threshold	−	+
Lever pressing to obtain food	+	
Nonoperant		
Sexual reward: T maze choice	+	
Appetitive discrimination task: resistance		
to extinction	+	
Genetically determined behavior (Bohus (1976)		
Intromission and ejaculation:		
delay of disappearance after castration	+	
Approach response to imprinting stimulus	−	
Submissive behavior	+	
Psychophysiology and electrophysiology		
Classical conditioned cardiac response:		
resistance to extinction	+	
Bradycardia during emotional behavior	+	
Hippocampal theta rhythm	+	
Multiple unit discharge:		
N. paraventricularis and N. supraopticus	+	
Development of tolerance and Physical dependence		
Morphine	+	+/−
Ethanol	+	+/0

[a] + improvement, − impairment.

confirmation of neurohypophyseal hormones to their classical
biological activities, with residues in position 2 and 5 forming part of the active sites of the molecule (Delanoy et al.,
1979).

In relation to these effects of ICV vasopressin, the recently
reported data on the effects of ICV-administered vasopressin on
brain water permeability are relevant. Administration of 2
units of vasopressin into the lateral ventricle of anesthetized
rhesus monkeys increased the capillary permeability for water
in the brain without affecting the cerebral blood flow (Raichle
and Grubb, 1978).

Most recently, Pedersen and Prange (1979) have provided evidence that ICV administration of low doses (0.4 μg) of OXT can
induce maternal behavior (grouping, pup licking, crouching,
nest building, and pup retrieval) in virgin female rats. Interestingly, this effect is only manifested in rats in proestrus or estrus. In ovariectomized rats estrogen priming is
necessary and sufficient for appearance of maternal behavior.

EFFECTS OF A REDUCED AMOUNT OF BIOAVAILABLE VASOPRESSIN

Studies on Brattleboro Rats

Evidence for a physiological role of vasopressin in brain processes related to memory function was obtained from experiments
with Brattleboro rats. A homozygous variant of this strain has
hereditary hypothalamic diabetes insipidus because it lacks the
ability of synthesize vasopressin, due to a mutation of a single pair of autosomal loci (Valtin and Schroeder, 1964; Valtin,
1967). By radioimmunoassay (Dogterom et al., 1978), it was
shown that whereas in heterozygous (HE-DI) animals the pituitary vasopressin content was approximately 40 percent of normal, in homozygous diabetes insipidus (HO-DI) rats vasopressin
was virtually absent (van Wimersma Greidanus et al., 1974).
The OXT levels in the pituitaries of HO-DI rats was also lower
(approximately 50 percent) than that in heterozygous or normal
animals (van Wimersma Greidanus and de Wied, 1977). Since in
HO-DI rats elevated plasma levels of OXT have been found
(Dogterom et al., 1978), the decrease in pituitary levels of
this peptide may be due to increased release. A neurohypophyseal activation appears in immunocytochemical studies (Swaab
et al., 1973) and suggests an increase synthesis of OXT to compensate for the absence of vasopressin.

Memory function of HO-DI rats is impaired in one-trial passive avoidance when retention is tested 24 hours or more after
training. Vasopressin (AVP or DG-LVP) administered immediately

after the single acquisition trial restores the disturbed passive avoidance behavior (de Wied et al., 1975). This favors
the hypothesis that memory rather than learning processes are
disturbed in the absence of vasopressin. Indeed, HO-DI rats
are able to acquire fear-motivated responses in multiple-trial
paradigms (shuttle box or pole jump) (Bohus et al., 1975;
Celestian et al., 1975). Furthermore, full retention of passive avoidance behavior is obtained in HO-DI rats when retention is tested shortly after the training trial. The main
disturbance is in the maintenance of the behavior, again suggesting that consolidation of memory is impaired in the absence
of vasopressin. These data were partly confirmed by Bailey and
Weiss (1979), who reported poorer passive avoidance in HO-DI
rats as compared with HE-DI animals. However, under some of
the conditions used by these authors, no differences between
HO-DI and HE-DI animals were found.

Interestingly an inhibition of the development of tolerance
to the antinociceptive action of morphine using a hot plate
test situation was observed in HO-DI rats (de Wied and Gispen,
1976).

Electrophysiological studies of these rats (Urban and de
Wied, 1975) show that the rhythmic slow activity of HO-DI rats
contains substantially lower hippocampal theta frequencies than
that of HE-DI animals during PS, while the other frequencies
remained comparable among the various groups. Injection of
DG-AVP enhanced the generation of higher frequencies in HO-DI
rats, and almost completely restored the distribution of hippocampal theta frequencies (Urban and de Wied, 1978). The
absence of vasopressin in the HO-DI animals did not interfere
with the duration of PS during the observation periods. Therefore it may be that the impaired memory of HO-DI rats is due to
the altered quality of PS in these animals. However, whether
or not the absence of vasopressin is the primary cause of all
the abnormalities of HO-DI rats has to be evaluated. An interaction between bioavailable amounts of vasopressin and OXT or
related neurohypophyseal peptides is an attractive possibility.

Intracerebroventricular Administration
of Antisera to Neurohypophyseal Hormones

Administration of vasopressin antiserum into one of the lateral
ventricles of the brain, immediately after training, induced a
marked deficit in one-trial passive avoidance when tested six
hours or more later. No interference with the behavior was
found when retention was tested less than two hours after

administration of the vasopressin antiserum. Testing at three
and four hours resulted in intermediate responses (van Wimersma
Greidanus and de Wied, 1976a). Intravenous injection of 100
times as much vasopressin antiserum, which effectively removes
the peptide from the circulation, as indicated by the virtual
absence of vasopressin in the urine and by the marked increase
of the urine production, did not affect passive avoidance
behavior. These results indicate the importance of centrally
available vasopressin in relation to avoidance behavior and, in
particular, to memory consolidation (van Wimersma Greidanus et
al., 1975a).

If the injection of antivasopressin serum was postponed for
up to two hours after training, the treatment still resulted in
a marked disturbance of passive avoidance behavior, indicating
that processes of memory consolidation last for several hours
(van Wimersma Greidanus and de Wied, 1976a). Moreover, administration of antivasopressin serum one hour prior to the retention test also resulted in passive avoidance deficits. This
supports the hypothesis that vasopressin is also involved in
retrieval processes (van Wimersma Greidanus and de Wied,
1976a).

Further proof for the role of central vasopressin in avoidance behavior was obtained from experiments using multiple
trial learning paradigms. Treatment with vasopressin antiserum
(ICV) during acquisition of a CAR tended to slow acquisition,
as compared with control rats, but both groups reached the
learning criterion. However, extinction of the avoidance response was faster in the rats injected with vasopressin antiserum, although the treatment was performed during acquisition
and not continued during extinction (van Wimersma Greidanus et
al., 1975c). These results reinforce the notion that learning
can take place in the absence of vasopressin, but the maintenance of the behavior is disturbed.

Neutralization of bioavailable OXT in the brain by ICV administration of specific OXT antiserum, either immediately after
the training trial (Bohus et al., 1978b), or one hour before
the first extinction session, induced longer passive avoidance
latencies than found in controls (Fig. 9-2). Rats that received anti-OXT serum before each acquisition session in the
pole-jump avoidance situation, showed a weak but significant
increase in resistance to extinction of the CAR. These results
again suggest opposing effects of OXT and vasopressin on memory
processes.

In a learned helplessness paradigm, animals display defective
acquisition of escape task after an initial exposure to ines-
capable FS. Mice treated ICV with antiserum to vasopressin
after this shock period did not show the escape deficits.

These data support the hypothesis that endogenous vasopressin may be involved in long-term memory (Leshner et al., 1978).

Treatment of rats with ICV antivasopressin serum lowered the mean and peak frequencies of hippocampal theta (Urban and de Wied, 1978). Conversely ICV OXT antiserum increased the proportion of higher theta frequencies (7.5 to 10 Hz) and decreased the low frequency components. The duration of PS during the observation period was not affected by the anti-OXT serum (Bohus et al., 1978b).

Neutralization of centrally available vasopressin by ICV administration of antivasopressin serum also interferes with the development of tolerence to morphine (van Wimersma Greidanus et al., 1978). Studies with rats in a FS test paradigm revealed that administration of antivasopressin serum inhibited the development of tolerance to morphine. Whereas control rats showed an increasing responsiveness to FS during subsequent morphine injection, presumably reflecting a decrease in the effectiveness of morphine, animals treated with antivasopressin serum showed the same responsiveness to FS in subsequent sessions after repeated morphine treatment. The development of tolerance to morphine was thus inhibited. Time gradient studies suggest that endogenous, centrally available vasopressin may be physiologically involved in the development of tolerance to morphine in a way comparable with its role in memory processes (van Wimersma Greidanus et al., 1978).

SITES OF ACTION OF NEUROHYPOPHYSEAL HORMONES IN RELATION TO AVOIDANCE BEHAVIOR

Vasopressin and the Limbic System

Several attempts have been made to determine the sites of action of neurohypophyseal hormones on avoidance behavior, mostly using one-way active avoidance. Intracerebral injections of LVP in the posterior thalamic area, including the parafascicular nuclei, resulted in a preservation of the pole-jump avoidance response. Other brain areas, including ventromedial and anteromedial parts of the thalamus, posterior hypothalamus, substantia nigra, reticular formation, substantia grisea, putamen, and the dorsal hippocampal complex appeared to be ineffective sites. Thus, the parafascicular area seemed to be involved in this behavioral effect of vasopressin (van Wimersam Greidanus et al., 1973, 1975b). However, ineffective microinjections may not always represent nonsensitive structures of the brain, since a unilateral local microinjection of a small amount of vasopressin into a restricted brain area may

be unable to influence the functions of this area sufficiently
to induce behavioral changes (van Wimersma Greidanus, 1975b).
In fact, intrahippocampal administration of LVP has been found
to prolong retention of a CAR (Stark et al., 1978). Recently,
it was found that bilateral microinjection of small amounts of
AVP (25 pg) in the dentate gyrus, the dorsal septal nuclei, or
the dorsal raphe nuclei improves passive avoidance behavior
(Kovács et al., 1979b).

Further evaluation of the sites of the behavioral effects of
vasopressin used rats bearing lesions in the posterior thalamus; this interfered with both acquisition and extinction of
the pole-jump avoidance response, whereas smaller lesions, restricted to the parafascicular area, only facilitated extinction (van Wimersma Greidanus et al., 1974). The lesions reduced, but did not prevent, the behavioral effect of vasopressin in the pole-jump test. Thus, although the parafascicular
region modulates the action of vasopressin, it does not seem to
be essential for this task. Rats with lesions in this area
only require more vasopressin to increase resistance to extinction (van Wimersma Greidanus et al., 1974, 1975b). Rats with
extensive bilateral lesions, either in the preoptic rostral
septal area, in the dorsal hippocampal complex, or in the amygdala, were resistant to the vasopressin-induced preservation of
avoidance behavior (van Wimersma Greidanus and de Wied, 1976b;
van Wimersma Greidanus et al., 1975b, 1979a). The lesions did
not interfere with the normal rate of extinction of the
avoidance response. Acquisition of the response was retarded
in animals with extensive hippocampal lesions. Smaller lesions
that caused partial damage of the dorsal hippocampal complex
partly inhibited, but did not completely block, the behavioral
effect of vasopressin (van Wimersma Greidanus and de Wied,
1976b). These results point to an important role of midbrain
limbic structures in vasopressin-induced effects on avoidance
behavior, and suggest that vasopressin acts on these structures
to enhance the storage or retrieval of recently acquired information. In addition, the theory that the hippocampus serves as
a "gateway to memory" (Hirsh, 1974) fits the concept that this
brain structure is a site of memory-consolidating action of
vasopressin. The hippocampus has been suggested as a relatively specific target organ for vasopressin (Hutson and Jakobartl,
1979).

To investigate whether the hippocampus, septum, amygdala, and
the parafascicular region are the sites of action of vasopressin, or whether the limbic system needs to be intact in order
to permit vasopressin to exert its behavioral effect, knife
cuts through the fornix were made and the effect of vasopressin
on avoidance extinction was studied. Transection of both pre-
and postcommisural fornix fibers did not block the effect of

vasopressin (van Wimersma Greidanus et al., 1979b), suggesting
that limbic structures may act independently as substrates for
the behavioral effect of vasopressin.

How Do Vasopressin and Oxytocin Reach Their Sites of Action?

Morphological and functional evidence has accumulated for the
release of hypothalamic hormones from neural tissue into the
fluid of the cerebral ventricles. Recent studies indicate that
neurosecretory axons run to the ependyma of the ventricles
(Sterba, 1974a,b; Vorherr et al., 1968; Rodriguez, 1970;
Leonhardt, 1974; Vigh-Tachmann and Vigh, 1974; and Wittkowski,
1968; See Rodriguez, Chapter 1). It has been suggested that
neurohumors are released directly into CSF from these terminals
(Heller and Zaidi, 1974; Sterba 1974a).
   Axons originating in the supraoptic area filled with neurosecretory substances have been shown to terminate in the infundibular recess. Using electron microscopic and immunocytochemical techniques, Goldsmith and Zimmerman (1975) demonstrated
neural processes containing granules with neurophysin and vasopressin close to portal capillary loops also protuding into the
third ventricle along its ventral border, suggesting the possible direct secretion of vasopressin from neuronal terminals
into portal blood and into the CSF (Zimmerman er al., 1977).
Also, vasopressin or VT and OXT have been shown to be present
in rat (Dogterom et al., 1977, 1978; van Wimersma Greidanus et
al., 1979c), rabbit (Heller er al., 1968; Unger et al., 1974),
dog (Dogterom et al., 1977), and human CSF (Gupta, 1969; Pavel,
1970; Coculescu and Pavel, 1973; Dogterom et al., 1978). In
fact, after hypophysectomy, CSF vasopressin in rats is even
higher than controls. This suggests a compensatory release of
this hormone into the ventricular system in association with
undetectable blood levels. By contrast, blood and CSF OXT is
scarcely affected by hypophysectomy (Dogterom et al., 1977).
How the hormones are channeled from CSF to the effector site is
not clear. Perhaps, different pools of ependymal cells or some
circumventricular organs act as mediators in this respect
(Sterba, 1974b).
   Various possibilities for the function of vasopressin in the
CSF have been suggested, such as a role in the feedback of its
own secretion or in the homeostasis of ionic concentration in
the brain (Rodriguez, 1970). Generally, CSF may play an important role in endocrine regulation as a distribution system or
vehicle (Knowles, 1974; Sterba, 1974b; Dunn, 1978). In particular, CSF may be an efficient way of transporting vasopressin
from its hypothalamic sites of synthesis to the loci of action
in the limbic system. Indeed, the fact that ICV administration

of vasopressin analogues affects avoidance behavior at much lower doses than after systemic injection supports this hypothesis (de Wied, 1976).

The dose of vasopressin required to stimulate ACTH release from the anterior pituitary, is approximately the same when administered intravenously or intracerebroventricularly (van Wimersma Greidanus et al., 1977), whereas the effect of vasopressin on blood pressure is initiated at lower doses following IV than ICV administration. These data support the idea that the cerebroventricular fluid is the preferred route of transport for the behavioral effect of vasopressin, the portal vessel system for the hypophysiotropic action, and the general circulation for the systemic effects of vasopressin.

However, release of vasopressin into systemic circulation may be important, since its level in eye-plexus blood as measured by bioassay was elevated in animals with prolonged latencies in passive avoidance behavior (Thompson and de Wied, 1973). This result was also found in trunk blood, but no individual correlation between behavior performance and vasopressin levels in the peripheral blood was determined by radioimmunoassay (van Wimersma Greidanus et al., 1979c). Furthermore, no significant difference was found between the levels of vasopressin in the CSF, collected immediately after retention of a passive avoidance response from the lateral ventricle of rats displaying short or long avoidance latencies. The absence of a direct correlation between plasma or CSF levels of vasopressin and behavioral performance suggests a direct transport of the neurohypophyseal principles from their sites of synthesis to those of their behavioral action. In fact, direct hypothalamic-limbic neurosecretory pathways exist by which neurohypophyseal peptides could affect the limbic system directly. Extrahypothalamic neurosecretory projections to the choroid plexus, circumventricular organs, and limbic system structures have been shown (Kozlowski et al., 1978; Sterba, 1974a, 1978; Buijs, 1978; Buijs et al., 1978; Weindl and Sofroniew, 1976, 1978a). Vasopressin- and OXT-containing pathways were traced from the paraventricular nucleus, among others, toward the dorsal and ventral hippocampus, the amygdaloid nuclei, the substantia nigra, the substantia grisea and other extrahypothalamic structures. (Buijs, 1978; Weindl and Sofroniew, 1978b). Furthermore, vasopressin-neurophysin pathways between the suprachiasmatic nucleus and the medial dorsal thalamus, the lateral septum and the lateral habenula were demonstrated (Weindl and Sofroniew, 1978b; Buijs, 1978). Neurophysin-vasopressin containing terminals have also been found by Kozlowski et al. (1978) in the medial and lateral septal nuclei, in the periventricular and dorsolateral areas of the thalamus, in parts of the amygdaloid complex, and in the habenular region. Neuro-

physin-containing fibers were seen to course within the stria
terminalis (Weindl and Sofroniew, 1976, 1978b; Kozlowski et
al., 1978). It has been suggested that, in addition to its
classical endocrine function, the neurosecretory system affects
systems in the brain by direct neuronal interaction in a specific fashion and thus modulates cerebral functions (Sterba,
1978; Unger, 1977), such as storage and retrieval of information. The observation (Weindl and Sofroniew, 1978b) that axosomatic contacts are formed by vasopressin-containing fibers in
limbic systems regions further supports this assumption.

BIOCHEMICAL ASPECT OF THE CEREBRAL EFFECTS NEUROHYPOPHYSEAL HORMONES

Is Vasopressin a Neuromodulator?

Because vasopressin exerts effects on a variety of physiological processes, of which those on the consolidation of memory
and the retrieval of stored information have been studied most
extensively, and because vasopressin occurs in the CNS in a
neuronal network that projects to various limbic-midbrain and
medullary regions and also to brain structures that have intimate contacts with the ventricles (vide supra), it has been
postulated that vasopressin acts as a neuromodulator, that is,
that it amplifies or dampens neuronal activity (Walter and
Hoffman, 1977; Tanaka et al., 1977b; Versteeg et al., 1978b;
Barchas et al., 1978; van Ree et al., 1978). According to this
postulate, effects of vasopressin on memory processes and other
processes may be the consequence of an interaction with particular neurotransmitter systems.

An abundance of data exists concerning the involvement of
brain catecholamines in memory processes (for references see:
van Ree et al., 1978; Kovács et al., 1979c, 1980). Also, brain
catecholamines have been shown to be involved in other processes that can be influenced by vasopressin and by bioactive
fragments of this and related peptides, such as the development
of tolerance to and the dependence on narcotic drugs, neuroendocrine regulation, and the regulation of arterial blood
pressure (for references see: van Ree et al., 1978). Although
these data suggest the possibility that brain catecholamines
might mediate the effects of vasopressin, only recently has
evidence been presented indicating that vasopressin indeed
interacts with particular catecholamine systems in the brain.

Vasopressin Effects of Cerebral Catecholamines

Whereas no effects were seen of vasopressin on catecholamine
metabolism or on the activity of enzymes involved in the bio-

synthesis of catecholamines in whole brain (Dunn et al., 1976, 1978; Iuvone et al., 1978; Versteeg and Wurtman, unpublished observation), small but significant changes in catecholamine metabolism were observed when catecholamine turnover in regions of brain was measured following systemic (Kovács et al., 1977) or ICV (Tanaka et al., 1977a) administration of vasopressin. Subsequently, vasopressin (ICV), in a dose that was without effect in brain parts, caused marked effects in restricted brain regions obtained by the microdissection technique of Palkovits (Tanaka et al., 1977b). When either vasopressin (30 ng) or vehicle was injected ICV 30 minutes after the IP administration of the tyrosine hydroxylase inhibitor, α-methyl-p-tyrosine (α-MPT), vasopressin was found to increase the α-MPT-induced disappearance of norepinephrine (NE) in, among others, the dorsal septal nucleus, parafascicular nucleus, dorsal raphe nucleus, locus coeruleus, and nucleus tractus solitarii. At the same time, vasopressin reduced the disappearance of NE in the supraoptic nucleus and the nucleus ruber. The most pronounced effects on dopamine (DA) disappearance were seen in the caudate nucleus and median eminence. Interestingly, effects on NE were evident in brain regions that had previously been found to be involved in the expression of the effects of vasopressin on memory processes (see before). These results were interpreted to indicate that the facilitatory effects of vasopressin on memory processes might be due to the effects of the peptide on catecholamine neurotransmission in particular brain regions (Tanaka et al., 1977b; van Ree et al., 1978).

The C-terminal tripeptide of OXT, PLG, affects the retrieval of stored information similarly to vasopressin; it is, however, inactive on the consolidation of memory (Bohus et al., 1978b). PLG accelerated the disappearance of DA from the caudate nucleus but was without effect on that of NE in limbic-midbrain structures (Versteeg et al., 1978a). Based on this finding, it was hypothesized that the effects of vasopressin on memory consolidation are mediated by noradrenergic neurons innervating various limbic-midbrain structures, whereas the effects shared by vasopressin and PLG on the retrieval of stored information might be due to the effects of either peptide on DA-containing neurons of the nigrostriatal system (Versteeg et al., 1978a; van Ree et al., 1978).

Brain Catecholamines in Rats with Reduced Amounts of Available Vasopressin

The results already described were indicative of a correlation between changes induced by neuropeptides on regional catecholamine metabloism and their effects on memory processes, but this evidence is, at best, circumstantial. More direct evidence comes from studies in which regional catecholamine metabolism was measured in rats in which the amount of bioavailable vasopressin in the brain is low.

In agreement with the supposition that brain vasopressin exerts a tonic influence on catecholamines in distinct brain regions, regional brain catecholamine turnover in the HO-DI rat was altered in a direction opposite to that of the changes induced by ICV vasopressin in normal rats (Tanaka et al., 1977b). In most regions in which vasopressin caused an increase in α-MPT-induced catecholamine disappearance, a decreased catecholamine turnover was found in the HO-DI rat (Versteeg et al., 1978b). Interestingly, a markedly lowered epinephrine turnover was found in several regions in the hypothalamus and the medulla oblongata, suggesting that vasopressin modulates neurotransmission in epinephrine-containing neurons as well. Antivasopressin serum, administered ICV to normal Wistar rats, generally induced decreases in the rate of catecholamine disappearance with a pattern similar to that found in HO-DI rats (Versteeg et al., 1979). Taken together, these data support the hypothesis that brain vasopressin modulates catecholamine neurotransmission.

Sites and Mechanism of Action of Vasopressin
as Modulator of Catecholaminergic Neurotransmission

Although studies cited indicate the brain sites where catecholamine neurotransmission is altered as a result of altered amounts of vasopressin, they do not permit conclusion on the site of interaction of the peptide with the catecholamines. The terminal areas of the catecholaminergic systems are obvious candidates.

Recent experiments indicate that the terminal areas of the coeruleo-telencephalic noradrenergic system, which originates in the locus coeruleus and projects via the dorsal noradrenergic bundle to, among others, septum, thalamus, hippocampus, and raphe, may be the sites where vasopressin exerts its facilitatory effects om memory consolidation. Vasopressin, administered SC immediately after the training trial, resulted in a long-term facilitation of a one-trial passive avoidance response; following destruction of the ascending dorsal noradrenergic bundle by bilateral microinjection of 6-hydroxydopamine this effect of vasopressin was abolished (Kovács et al., 1979a). In a second set of experiments, bilateral injection of small amounts of vasopressin (25 pg) in regions containing terminals of the coeruleo-telecephalic noradrenergic system, namely the dorsal septal nucleus, the dentate gyrus of the hippocampus, and the dorsal raphe nucleus, facilitated memory consolidation, whereas microinjections in the central amygdaloid nucleus and the locus coeruleus, which contains the cell bodies of this system, failed to do so (Kovács et al., 1979b).

These results do not exclude the possibility that vasopressin acts via other transmitters. Ramaekers et al. (1977) presented evidence for effects of DG-LVP on serotonin metabolism. Although it is clear that vasopressin affects catecholaminergic systems, little is known concerning the mechanism of its action. Future research should resolve whether this influence is a direct one and, if so, whether the effects of vasopressin are brought about by interaction with pre- or postsynaptic receptors, via processes mediated by cyclic nucleotides, on transmitter release or reuptake, or on transmitter synthesis.

Oxytocin

Whereas there is a large body of evidence supporting a role for brain vasopressin as a modulator of catecholamine neuro-transmission, less is known regarding OXT. Recently Telegdy and Kovács (1979) reported that IP OXT increased DA turnover in the hypothalamus, whereas it decreased NE and DA turnover in the striatum. These authors suggested that the opposite effects of OXT and vasopressin on DA turnover might be related to their opposite effects on various behavioral parameters. However, results of recent experiments, analogous to those performed with ICV vasopressin (Tanaka et al., 1977b), do not support this view (Versteeg et al. unpublished observations). Nevertheless, it seems likely that OXT, like vasopressin, is a neuromodulator. Further studies are needed to substantiate to what extent its effects on catecholamine neurotransmission are opposite to those of vasopressin, and whether other neurotransmitters might be subject to neuromodulation by OXT.

HUMAN STUDIES

Although Pitressin has previously been used in the treatment of schizophrenia (Forizs and Butner, 1952), and a beneficial effect of LVP has been reported in schizophrenic patients (Vranckx et al., 1978), until recently no clear evidence had been obtained for CNS effects of neurohypophyseal hormones in human beings. That OXT release into the CSF might be augmented under pathological conditions had been shown by Unger (1977), who found relatively low levels in the CSF of patients with depressed brain function and high levels in patients suffering from diseases associated with an increased activity of the brain. It is of particular interest that the OXT concentration in the CSF is increased after ECS treatment (Unger et al., 1971), which is amnestic.

It has also been suggested that vasopressin is involved in human affective disorders and that plasma AVP levels are low in depression (Gold et al., 1978). Although ECS elevated plasma AVP, and this elevation might be important in the antidepressant action, the hypothesis that vasopressin function is disturbed in depression remains a hypothesis (Raskind et al., 1979) Interestingly, it has recently been found that plasma vasopressin was elevated in acutely psychotic subjects, but not in acutely anxious nonpsychotic "control" patients. There was also positive correlation between the levels of vasopressin and the degree of psychosis (Raskind et al., 1978). This indicates that the elevation of vasopressin is not just a function of nonspecific anxiety. These results concur with previous reports of water intoxication in acutely psychotic patients, and the occasionally described syndrome of inappropriate ADH secretion in these subjects (Raskind et al., 1974, 1975), and provide evidence for a role of vasopressin in CNS functioning.

There have been several reports on the beneficial effect of vasopressin on memory. A remarkable improvement of memory was found after the use of vasopressin in a patient with Korsakoff's syndrome. After two weeks of vasopressin treatment (nasal spray), a marked increase in the memory scale was observed along with other less circumscribed improvements of memory function (LeBoeuf et al., 1978). Elderly patients given vasopressin (nasal spray) performed better in tests involving attention and concentration and in memory tests than individuals in the placebo group (Legros et al., 1978). Return or improvement of memory or restoration of long-term memory was reported in patients suffering from post-traumatic retrograde or anterograde amnesia stemming from car accidents (Oliveros et al., 1978). Administration of the vasopressin analogue DDAVP (nasal spray) improved the ability to learn a passive avoidance response in Lesch-Nyhan childern (Anderson et al., 1979). This disease is a disorder of the purine metabolism and one of the features is self-mutilation. Skin shock, which is generally very effective in treating self-injurious behavior, is not operative in Lesch-Nyhan disease. This ineffectiveness may be related to the inability to learn a passive avoidance task.

Although only a small number of patients was used in each of these studies, and they need to be extended and carefully evaluated (Poon, 1978), and although improvement of the alcohol-induced amnestic state as part of the Wernicke-Korsakoff syndrome was not observed after vasopressin treatment in two patients (Blake et al., 1978), these clinical data are in good agreement with the results obtained from animal experiments and fit nicely into the hypothesis of the involvement of vasopressin in memory function.

CONCLUDING REMARKS

The finding that neurohypophyseal peptides alter brain processes related to memory may be regarded as a milestone in neuroendocrine research. It is an important aspect of the hypothesis that peptides from the brain or the pituitary act directly on the CNS. This is not only true for neurohypophyseal principles, but also for ACTH, α- and β-MSH, and β-LPH, which all play an important role in cerebral function. Neuropeptides belonging to the ACTH/LPH family, which originate from large prohormones, affect CNS processes related to motivation and attention (de Wied, 1977a,b; Beckwith and Sandman, 1978; Rigter and van Riezen, 1978). The neuropeptides determine the hormonal climate in the brain needed for adaptive functioning of the organism. The fact that vasopressin and OXT exert opposite effects on memory processes again illustrates the endocrinological organization in which processes or functions are stimulated and inhibited by different hormones, thus ensuring homeostasis. Thus, it is tempting to assume that disturbances in the equilibrium between vasopressin and OXT result in alterations of brain function, which may be reflected in behavioral changes. It is conceivable that stimuli triggering memory processes induce an altered activity of the brain systems generating neurohypophyseal principles, thereby affecting the equilibrium between the bioavailability of these entities in the brain. In this way, they might enable the organism to cope with environmental changes when it is confronted with situations previously experienced. Consequently it is likely that some disorders of a psychopathological nature may be the result of disturbances in this adaptive capacity, which in turn may be caused by a dysfunctioning of neuropeptide systems (de Wied, 1978) or, more particularly, when memory processes are involved, by dysfunctioning of the neurohypophyseal system.

REFERENCES

Ader, R., and de Wied, D.: Effects of lysine vasopressin on passive avoidance learning. *Psychon Sci* 29:46-48, 1972.

Ader, R., Weijen, J.A.W.M., and Moleman, P.: Retention of a passive avoidance response as a function of the intensity and duration of electric shock. *Psychon Sci* 26:125-128, 1972.

Anderson, L.T., David, R., Bonnet, K., and Dancis, J.: Passive avoidance learning in Lesch-Nyhan disease: Effect of

1-desamino-8-arginine-vasopressin. *Life Sci* *24*:905-910, 1979.

Bailey, W.H., and Weiss, J.M.: Evaluation of a 'memory dificit' in vasopressin-deficient rats. *Brain Res 162*:174-178, 1979.

Barchas, J.D., Akil, H., Elliott, G.R., Holman, R.B., and Watson, S.J.: Behavioral neurochemistry: Neuroregulators and behavioral states. *Science 200*:964-973, 1978.

Beckwith, B.E., and Sandman, C.A.: Behavioral influences of the neuropeptides ACTH and MSH: A methodological review. *Neurosci Biobehav Rev 2*:311-338, 1978.

Blake, D.R., Dodd, M.J., and Grimley Evans, J.: Vasopressin in amnesia. *Lancet 1*:608, 1978.

Bohus, B.: Pituitary peptides and adaptive autonomic responses. *Prog Brain Res 42*:275-283, 1975.

Bohus, B.: The influence of pituitary peptides on sexual behavior. In *Problèmes Actuels d'Endocrinologie et de Nutrition*, H.P. Klotz, ed. Expansion Scientifique Francaise, Paris, 1976, pp 235-246.

Bohus, B.: Effect of desglycinamide-lysine vasopressin (DG-LVP) on sexually motivated T-maze behavior of the male rat. *Horm Behav 8*:52-61, 1977.

Bohus, B., Ader, R., and de Wied, D.: Effects of vasopressin on active and passive avoidance behavior. *Horm Behav 3*:191-197, 1972.

Bohus, B., Gispen, W.H., and de Wied, D.: Effect of lysine vasopressin and $ACTH_{4-10}$ on conditioned avoidance behavior on hypophysectomized rats. *Neuroendocrinology 11*:137-143, 1973.

Bohus, B., van Wimersma Greidanus, Tj.B., and de Wied, D.: Behavioral and endocrine responses of rats with hereditary hypothalamic diabetes insipidus (Brattleboro strain). *Physiol Behav 14*:607-615, 1975.

Bohus, B., Urban, I., van Wimersma Greidanus, Tj.B., and de Wied, D.: Opposite effects of oxytocin and vasopressin on avoidance behaviour and hippocampal theta rhythm in the rat. *Neuropharmacology 17*:239-247, 1978a.

Bohus, B., Kovács, G.L., and de Wied, D.: Oxytocin, vasopressin and memory: Opposite effects on consolidation and retrieval processes. *Brain Res 157*:414-417, 1978b.

Bookin, H.B., and Pfeifer, W.D.: Effect of lysine vasopressin on pentylenetetrazol-induced retrograde amnesia in rats. *Pharmacol Biochem Behav 7*:51-54, 1977.

Buijs, R.M.: Intra- and extrahypothalamic vasopressin and oxytocin pathways in the rat. *Cell Tissue Res 192*:423-435, 1978.

Buijs, R.M., Swaab, D.F., Dogterom, J., and van Leeuwen, F.W.: Intra- and extrahypothalamic vasopressin and oxytocin pathways in the rat. *Cell Tissue Res 186*:423-433, 1978.

Celestian, J.F., Carey, R.J., and Miller, M.: Unimpaired maintenance of a conditioned avoidance response in the rat with diabetes insipidus. *Physiol Behav 15*:707-711, 1975.

Coculescu, M., and Pavel, S.: Arginine vasotocin-like activity of cerebrospinal fluid in diabetes insipidus. *J Clin Endocrinol Metab 36*:1031–1032, 1973.

Cohen, M., Keats, A.S., Krivoy, W.A., and Ungar, G.: Effect of actinomycin D on morphine tolerance. *Proc Soc Exp Biol Med 119*:381–384, 1965.

Cox, B.M., and Osman, G.H.: Inhibition of the development of tolerance to morphine in rats by drugs which inhibit ribonucleic acid or protein synthesis. *Br J Pharamacol 38*:157–170, 1970.

Defendini, R., and Zimmerman, E.A.: The magnocellar neurosecretory system of the mammalian hypothalamus. In *The Hypothalamus* S. Reichlin, R.J. Baldessarini, and J.B. Martin, ed. Raven Press, New York, 1978, pp 137–154.

Delanoy, R.L., Dunn, A.J., and Tintner, R.: Behavioral responses to intracerebroventricularly administered neurohypophyseal peptides in mice. *Horm Behav 11*:348–362, 1978.

Delanoy, R.L., Dunn, A.J., and Walter, R.: Neurohypophyseal hormones and behavior: Effects of intracerebroventricularly injected hormone analogs in mice. *Life Sci 24*:651–658, 1979.

Dogterom, J., van Wimersma Greidanus, Tj.B., and Swaab, D.F.: Evidence for the release of vasopressin and oxytocin into cerebrospinal fluid: Measurements in plasma and CSF of intact and hypophysectomized rats. *Neuroendocrinology 24*:108–118, 1977.

Dogterom, J., van Wimersma Greidanus, Tj.B., and de Wied, D.: Vasopressin in cerebrospinal fluid and plasma of man, dog, and rat. *Am J Physiol 234*:E463–E467, 1978.

Dorsa, D.M., and van Ree, J.M.: Modulation of substantia nigra self-stimulation by neuropeptides related to neurohypophyseal hormones. *Brain Res 172*:367–371, 1979.

Douglas, W.W.: Mechanism of release of neurohypophysial hormones: Stimulus-secretion coupling. In *Handbook of Physiology*, Section 7, vol IV, part I: *The Pituitary Gland and its Neuroendocrine Control*, E. Knobil and W.H. Sawyer, eds. American Physiological Society, Washington, 1974, pp 191–224.

Dunn, A.J.: Peptides and behavior. *Neurosci Res Program Bull 16*:554–555, 1978.

Dunn, A.J., Iuvone, P.M., and Rees, H.D.: Neurochemical responses of mice to ACTH and lysine vasopressin. *Pharmacol Biochem Behav 5* (Suppl 1):139–145, 1976.

Dunn, A.J., Gildersleeve, N.B., and Gray, H.E.: Mouse brain tyrosine hydroxylase and glutamic acid decarboxylase following treatment with adrenocorticotrophic hormone, vasopressin or corticosterone. *J Neurochem 31*:977–982, 1978.

Finkelberg, F., Kalant, H., and Le Blanc, E.: Effect of vasopressin-like peptides on consumption of ethanol by the rat. *Pharmacol Biochem Behav 9*:453–458, 1978.

Flexner, J.B., Flexner, L.B., Hoffman, P.L., and Walter, R.:
Dose-response relationships in attenuation of puromycin-
induced amnesia by neurohypophyseal peptides. Brain Res
134:139-144, 1977.

Forizs, M.D.L., and Butner, N.C.: The use of pitressin in the
treatment of schizophrenia with deterioration. NC Med J
13:76-80, 1952.

Garrud, P.: Effects of lysine-8-vasopressin on punishment-
induced suppression of a lever-holding response. Prog Brain
Res 42:173-186, 1975.

Garrud, P., Gray, J.A., and de Wied, D.: Pituitary-adrenal hormones and extinction of rewarded behaviour in the rat.
Physiol Behav 12:109-119, 1974.

Gold, P.W., Goodwin, F.K., and Reus, V.I.: Vasopressin in
affective illness. Lancet 1:1233-1235, 1978.

Goldsmith, P.C., and Zimmerman, A.E.: Ultrastructural
localization of neurophysin and vasopressin in the rat median
eminence. In 57th Annual Meeting of the Endocrine Society,
1975, p A-377.

Gupta, K.K.: Antidiuretic hormone in cerebrospinal fluid.
Lancet 1:581, 1969.

Heller, H., and Zaidi, S.M.A.: The problem of neurohypophysial
secretion into the cerebrospinal fluid: Antidiuretic activity
in the liquor and choroid plexus. In Ependyma and Neurohormal
Regulation, A. Mitro, ed. Veda Publishing House, Bratislava,
1974 pp 229-250.

Heller, H., Hasan, S.H., and Saifi, A.Q.: Antidiuretic activity
in the cerebrospinal fluid. J Endocrinol 41:237-280, 1968.

Hirsh, R.: The hippocampus and contextual retrieval of information from memory: A theory. Behav Biol 12:421-444, 1974.

Hoffman, P.L., Ritzmann, R.F., Walter, R., and Tabakoff, B.:
Arginine vasopressin maintains ethanol tolerance. Nature
276:614-616, 1978.

Hostetter, G., Jubb, S.L., and Kozlowski, G.P.: Vasopressin
affects the behavior of rats in a positively-rewarded discrimination task. Life Sci 21:1323-1328, 1977.

Huston, J.P., and Jakobartl, L.: Evidence for selective susceptibility of hippocampus to spreading depression induced by
vasopressin. Neurosci Lett 6:69-72, 1977.

Iuvone, P.M., Morasco, J., Delanoy, R.L., and Dunn, A.J.: Peptides and the conversion of [^{3}H]tyrosine to catecholamines:
Effects of ACTH-analogs, melanocyte-stimulating hormones and
lysine-vasopressin. Brain Res 139:131-139, 1978.

Jones, B.E., and Moore, R.J.: Ascending projections of the
locus coeruleus in the rat. II Autoradiographic studies.
Brain Res 127:23-53, 1977.

King, A.R., and de Wied, D.: Localized behavioral effects of
vasopressin on maintenance of an active avoidance response in

rats. *J Comp Physiol Psychol* 86:1008-1018, 1974.

Knowles, F.: Cerebrospinal fluid and endocrine regulation. In *Ependyma and Neurohormonal Regulation*, A. Mitro, ed. Veda Publishing House, Bratislava, 1974, pp 11-28.

Kovács, G.L., Véscsei, L., Szabó, G., and Telegdy, G.: The involvement of catecholaminergic mechanisms in the behavioural action of vasopressin. *Neurosci Lett* 5:337-344, 1977.

Kovács, G.L., Vécsei, L., and Telegdy, G.: Opposite action of oxytocin to vasopressin in passive avoidance behavior in rats. *Physiol Behav* 20:801-802, 1978.

Kovács, G.L., Bohus, B., and Versteeg, D.H.G.: Facilitation of memory consolidation by vasopressin: Mediation by terminals of the dorsal noradrenergic bundle? *Brain Res* 172:73-85, 1979a.

Kovács, G.L., Bohus, B., Versteeg, D.H.G., de Kloet, E.R., and de Wied, D.: Effects of oxytocin and vasopressin on memory consolidation: Sites of action and catecholaminergic correlates after local microinjection into limbic-midbrain structures. *Brain Res* 175:303-314, 1979b.

Kovács, G.L., Bohus, B., and Versteeg, D.H.G.: The effects of vasopressin on memory processes: The role of noradrenergic neurotransmission. *Neuroscience* 4:1529-1537, 1979c.

Kovács, G.L., Bohus, B., and Versteeg, D.H.G.: The interaction of posterior pituitary neuropeptides with monoaminergic neurotransmission: Significant in learning and memory processes. *Prog Brain Res* 1980.

Kozlowski, G.P., Brownfield, M.S., and Hostetter, G.: Neurosecretory supply to extrahypothalamic structures: Choroid plexus, circumventricular organs, and limbic system. In *Neurosecretion and Neuroendocrine Activity: Evolution, Structure and Function*. W. Bargmann, A. Oksche, A. Polenov, and B. Scharrer. Springer-Verlag, Berlin, 1978 pp 217-227.

Krejci, I., and Kupková, B.: Effects of vasopressin analogues on passive avoidance behavior. *Activ Nerv Sup (Praha)* 20:11-12, 1978a.

Krejci, I., and Kupková, B.: Sleep-inducing effects of a vasopressin analog, deamino-6-carba-ornithine-8-vasopressin (DCOV) in rats. *Activ Nerv Sup (Praha)* 20:60-61, 1978b.

Krivoy, W.A., Zimmerman, E., and Lande, S.: Facilitation of development of resistance to morphine analgesia by desglycinamide[9]-lysine-vasopressin. *Proc Natl Acad Sci USA* 71:1852-1856, 1974.

Kruse, H., van Wimersma Greidanus, Tj.B., and de Wied, D.: Barrel rotation induced by vasopressin and related peptides in rats. *Pharmacol Biochem Behav* 7:311-313, 1977.

Lande, S., Witter, A., and de Wied, D.: Pituitary peptides. An octa-peptide that stimulates conditioned avoidance acquistion in hypophysectomized rats. *Biol Chem* 246:2058-2062, 1971.

Lande, S., Flexner, J.B., and Flexner, L.B.: Effect of corticotropin and desglycinamide[9]-lysine vasopressin on suppression
of memory by puromycin. *Proc Natl Acad Sci USA* 69:558-560,
1972.

Le Boeuf, A., Lodge, J., and Eames, P.G.: Vasopressin and
memory in Korsakoff syndrome. *Lancet* 2:1370, 1978.

Legros, J.J., Gilot, P., Seron, X., Claessens, J., Adam, A.,
Moeglen, J.M., Audibert, A., and Berchier, P.: Influence of
vasopressin on learning and memory. *Lancet* 1:41-42, 1978.

Leonhardt, H.: Ependymstrukturen im Dienst des Stofftransportes
zwischen Ventrikelliquor und Hirnsubstanz. In *Ependyma and
Neurohormonal Regulation.* A. Mitro, ed. Veda Publishing
House, Bratislava, 1974, pp 29-75.

Leshner, A.I., and Roche, K.E.: Comparison of the effects of
ACTH and lysine vasopressin on avoidance-of-attack in mice.
Physiol Behav 18:879-883, 1977.

Leshner, A.I., Hofstein, R., Samuel, D., and van Wimersma
Greidanus, Tj.B.: Intraventricular injection of antivaso-
pressin serum blocks learned helplessness in rats. *Pharmacol
Biochem Behav* 9:889-892, 1978.

Lissák, K., and Bohus, B.: Pituitary hormones and avoidance
behavior of the rat. *Int J Psychobiol* 2:103-115, 1972.

Martin, J.T., and van Wimersma Greidanus, Tj.B.: Imprinting
behavior: Influence of vasopressin and ACTH analogues.
Psychoneuroendocrinology 3:261-269, 1979.

Mello, N.K., and Mendelson, J.H.: Effects of the neuropeptide
DG-AVP on morphine and food self-administration by dependent
rhesus monkey. *Pharmacol Biochem Behav* 10:415-419, 1979.

Oliveros, J.C., Jandali, M.K., Timsit-Berthier, M., Remy, R.,
Benghezal, A., Audibert, A., and Moeglen, J.M.: Vasopressin
amnesia. *Lancet* 1:42, 1978.

Pavel, S.: Tentative identification of arginine vasotocin in
human cerebrospinal fluid. *J Clin Endocrinol Metab* 31:369-
371, 1970.

Pedersen, C.A., and Prange, A.J.: Induction of maternal behavior in virgin rats after intracerebroventricular administration of oxytocin. *Pro Natl Acad Sci USA* 76: 1979.

Pfeifer, W.D., and Bookin, H.B.: Vasopressin antagonizes
retrograde amnesia in rats following electroconvulsive shock.
Pharmacol Biochem Behav 9:261-263, 1978.

Poon, L.W.: Vasopressin and memory. *Lancet* 1:557, 1978.

Raichle, M.E., and Grubb, Jr., R.L.: Regulation of brain water
permeability by centrally-released vasopressin. *Brain Res*
143:191-194, 1978.

Ramaekers, F., Rigter, H., and Leonard, B.E.: Parallel changes
in behaviour and hippocampal serotonin metabolism in rats
following treatment with desglycinamide lysine vasopressin.
Brain Res 120:485-492, 1977.

Raskind, M.A.: Psychosis, polydipsia and water intoxication. *Arch Gen Psychiatry* *30*:112-114, 1974.

Raskind, M.A., Orenstein, H., and Christopher, T.G.: Acute psychosis, increased water ingestion and inappropriate antidiuretic hormone secretion. *Am J Psychiatry* *132*:407-412, 1975.

Raskind, M.A., Weitzman, R.E., Orenstein, H., Fisher, D.A., and Courtney, N.: Is antidiuretic hormone elevated in psychosis? A pilot study. *Biol Psychiatry* *13*:385-390, 1978.

Raskind, M.A., Orenstein, H., and Weitzman, R.E.: Vasopressin in depression. *Lancet* *1*:164, 1979.

van Ree, J.M., and de Wied, D.: Prolyl-leucyl-glycinamide (PLG) facilitates morphine dependence. *Life Sci* *19*:1331-1340, 1976.

van Ree, J.M., and de Wied, D.: Modulation of heroin self-administration by neurohypophyseal principles. *Eur J Pharmacol* *43*:199-202, 1977.

van Ree, J.M., Bohus, B., Versteeg, D.H.G., and de Wied, D.: Neurohypophyseal principles and memory processes. *Biochem Pharamacol* *27*:1793-1800, 1978.

Rigter, H., and van Riezen, H.: Hormones and memory. In *Psychopharmacology: A Generation of Progress*. M.A. Lipton, A. DiMascio, and K.F. Killam, eds. Raven Press, New York, 1978, pp 677-689.

Rigter, H., van Riezen, H., and de Wied, D.: The effects of ACTH- and vasopressin-analogues on CO_2-induced retrograde amnesia in rats. *Physiol Behav* *13*:381-388, 1974.

Rodriguez, E.M.: Morphological and functional relationship between the hypothalamo-neurohypophyseal system and cerebrospinal fluid. In *Aspects in Neuroendocrinology* W. Bargmann and B. Scharrer, eds. Springer-Verlag, Berlin, 1970, pp 352-365.

Sachs, H., Fawcett, P., Takabatake, Y., and Portonova, R.: Biosynthesis and release of vasopressin and neurophysin. *Recent Prog Horm Res* *25*:447-491, 1969.

Scharrer, E., and Scharrer, B.: Secretory cells within the hypothalamus. *Res Publ Assoc Res Nerv Ment Dis* *20*:170-194, 1940.

Scharrer, E., and Scharrer, B.: Hormones produced by neurosecretory cells. *Recent Prog Horm Res* *10*:183-243, 1954.

Schmidt, W.K., Holaday, J.W., Loh, H.H., and Way, E.L.: Failure of vasopressin and oxytocin to antagonize acute morphine antinociception or facilitate narcotic tolerance development. *Life Sci* *23*:151-158, 1978.

Schulz, H., Unger, H., Schwarzberg, H., Pommrich, G., and Stolze, R.: Neuroaktivität hypothalamischer Kerngebiete von Kaninchen nach intraventrikulär Applikation von Vasopressin und Oxytocin. *Experientia* *27*:1482, 1971.

Schulz, H., Schwarzberg, H., and Unger, H.: The effect of

intraventricularly applied oxytocin on seizures of rabbits induced by Na-glutamate. In *Ependyma and Neurohormonal Regulation.* A. Mitro, ed. Veda Publishing House, Bratislava, 1974a, pp 269–280.

Schulz, H., Kovács, G.L., and Telegdy, G.: Effect of physiological doses of vasopressin and oxytocin on avoidance and exploratory behaviour in rats. *Acta Physiol Acad Sci Hung* 45:211–215, 1974b.

Schulz, H., Kovács, G.L., and Telegdy, G.: The effect of vasopressin and oxytocin on avoidance behaviour in rats. In *Cellular and Molecular Bases of Neuroendocrine Processes.* E. Endröczi, ed. Akadémiai Kiadó, Budapest, 1976, pp 555–564.

Schwarzberg, H., and Unger, H.: Anderung der Reaktionszeit von Ratten nach Applikation von Vasopressin, Oxytozin und Nathioglykolat. *Acta Biol Med Ger* 24:507–516, 1970.

Schwarzberg, H., Schulz, H., and Unger, H.: The discharge rate of hypothalamic neurons of rabbits after intraventricular vasopressin and oxytocin application at various concentrations. In *Ependyma and Neurohormonal Regulation.* A. Mitro, ed. Veda Publishing House, Bratislava, 1974, pp 261–267.

Schwarzberg, H., Hartmann, G., Kovács, G.L., and Telegdy, G.: Effect of intraventricular oxytocin and vasopressin on self-stimulation in rats. *Acta Physiol Acad Sci Hung* 47:127–131, 1976.

Schwarzberg, H., Hartmann, G., and Telegdy, G.: Effect of intraventricular administration of oxytocin and vasopressin on self-stimulation and reaction time in rats. In *Neurosecretion and Neuroendocrine Activity: Evolution, Structure and Function.* W. Bargmann, A. Oksche, A. Polenov, and B. Scharrer, eds. Springer Verlag, Berlin, 1978a, p 285.

Schwarzberg, H., Betschen, K., Unger, H., and Schulz, H.: Beeinflussung der hypothalamischen Selbststimulation durch intrazerebroventrikulär verbreichtes Vasopressin und Oxytozin. *Acta Biol Med Germ* 37:1295–1296, 1978b.

Stark, H., Bigl, H., and Sterba, G.: Effects of posterior-pituitary-lobe peptides on the maintenance of a conditioned avoidance response in rats. In *Neurosecretion and Neuroendocrine Activity: Evolution, Structure and Function.* W. Bargmann, A. Oksche, A. Polenov, and B. Scharrer, eds. Springer-Verlag, Berlin, 1978, p 292.

Sterba, G.: Ascending neurosecretory pathways of the peptidergic type. In *Neurosecretion. The Final Neuroendocrine Pathway.* F. Knowles and L. Vollrath, eds. Springer-Verlag, Berlin, 1974a, pp 38–47.

Sterba, G.: Cerebrospinal fluid and hormones. In *Ependyma and Neurohormonal Regulation.* A. Mitro, ed. Veda Publishing House, Bratislava, 1974b, pp 143–179.

Swaab, D.F., Boer, G.J., and Nolten, J.W.L.: The hypothalamoneurohypophyseal system (HNS) of the Brattleboro rat. *Acta Endocrinol (Supp 1) Copenh 80*:177, 1973.

Swaab, D.F., Pool, C.W., and Nijveldt, F.: Immunofluorescence of vasopressin and oxytocin in the rat hypothalamo-neurohypophyseal system. *J Neural Transm 36*:195-215, 1975.

Tanaka, M., Versteeg, D.H.G., and de Wied, D.: Regional effects of vasopressin on rat brain catecholamine metabolism. *Neurosci Lett 4*:321-325, 1977a.

Tanaka, M., de Kloet, E.R., de Wied, D., and Versteeg, D.H.G.: Arginine[8]-vasopressin affects catecholamine metabolism in specific brain nuclei. *Life Sci 20*:1799-1808, 1977b.

Telegdy, G., and Kovács, G.L.: Role of monoamines in mediating the action of hormones on learning and memory. In *Brain Mechanisms in Memory and Learning: From the Single Neuron to Man*. M.A.B. Brazier, ed. Raven Press, New York, 1979, pp 249-268.

Thompson, E.A., and de Wied, D.: The relationship between the antidiuretic activity of rat eye plexus blood and passive avoidance behaviour. *Physiol Behav 11*:377-380, 1973.

Unger, H.: Funktionelle Aspekte der Informationsübermittlung durch die Oligopeptide Vasopressin und Oxytocin bei Säugetieren. In *Hormonale und humorale Informationsmittlung durch Peptide als Mediatoren*. Sitzungsberichte der Akademie Wissenschaft DDR. Akademie-Verlag, Berlin, 1977, pp 62-83.

Unger, H., Pommrich, G., and Beck, R.: Der Oxytocingehalt im menschlichen pathologischen Liquor. *Experientia 27*:1486, 1971.

Unger, H., Schwarzberg, H., and Schulz, H.: The vasopressin and oxytocin content in the cerebrospinal fluid of rabbits under changed conditions. In *Ependyma and Neurohormonal Regulation*, A. Mitro, ed. Veda Publishing House, Bratislava, 1974, pp 251-259.

Urban, I., and de Wied, D.: Inferior quality of RSA during paradoxical sleep in rats with hereditary diabetes insipidus. *Brain Res 97*:362-366, 1975.

Urban, I., and de Wied, D.: Neuropeptides: Effects on paradoxical sleep and theta rhythm in rats. *Pharmacol Biochen Behav 8*:51-59, 1978.

Valtin, H.: Hereditary hypothalamic diabetes insipidus in rats (Brattleboro strain). A useful experimental model. *Am J Med 42*:814-827, 1967.

Valtin, H., and Schroeder, H.A.: Familial hypothalamic diabetes insipidus in rats (Brattleboro strain). *Am J Physiol 206*:425-430, 1964.

Vandesande, F., Dierickx, K., and De Mey, J.: Identification of separate vasopressin-neurophysin II and oxytocin-neurophysin I containing nerve fibers in the external region of the bovine median eminence. *Cell Tissue Res 158*:509-516, 1975.

Versteeg, D.H.G., Tanaka, M., de Kloet, E.R., van Ree, J.M., and de Wied, D.: Prolyl-leucyl-glycinamide (PLG): Regional effects on α-MPT induced catecholamine disappearance in rat brain. *Brain Res* *143*:561-566, 1978a.

Versteeg, D.H.G., Tanaka, M., and de Kloet, E.R.: Catecholamine concentration and turnover in discrete regions of the brain of the homozygous Brattleboro rat deficient in vasopressin. *Endocrinology* *103*:1654-1661, 1978b.

Vigh-Teichmann, I., and Vigh, B.: Correlation of CSF contacting neuronal elements to neurosecretory and ependymosecretory structures. In *Ependyma and Neurohormonal Regulation*, A. Mitro, ed. Veda Publishing House, Bratislava, 1975, pp 281-295.

Vorherr, H., Bradbury, M.W.B., Houghoughi, M., and Kleeman, C.R.: Antidiuretic hormone in cerebrospinal fluid during endogenous and exogenous changes in its blood level. *Endocrinology* *83*:246-250, 1968.

Vranckx, C., Minne, P., Benghezal, A., Moeglen, J.M., and Audibert, A.: Vasopressin et schizophrenie. *Abstract IInd World Congress of Biological Psychiatry*. Barcelona, 1978 p 198.

Walter, R., and Hoffman, P.L.: Proposed mechanisms of action of neurohypophyseal peptides in memory processes and possible routes for the biosynthesis of peptides with a C-terminal carboxamide group. In *Neuropeptide Influences on the Brain and Behavior*, L.H. Miller, C.A. Sandman and A.J. Kastin, eds. Raven Press, New York, 1977, pp 109-126.

Walter, R., Hoffman, P.L., Flexner, J.B., and Flexner, L.B.: Neurohypophyseal hormones, analogs and fragments: Their effect on puromycin-induced amnesia. *Proc Natl Aca Sci USA* *72*:4180-4184, 1975.

Walter, R., van Ree, J.M., and de Wied, D.: Modification of conditioned behavior of rats by neurohypophyseal hormones and analogues. *Proc Natl Acad Sci USA* *75*:2493-2496, 1978.

Wang, S.S.: Synthesis of desglycinamide lysine vasopressin and its behavioral activity in rats. *Biochem Biophys Res Commun* *48*:1511-1515, 1972.

Weindl, A., and Sofroniew, M.V.: Demonstration of extrahypothalamic peptide secreting neurons. A morphologic contribution to the investigation of psychotropic effects of neurohormones. *Pharmakopschiatr Neuropsychopharmakol* *9*:226-234, 1976.

Weindl, A., and Sofroniew, M.V.: The functional morphology of vascular and neuronal efferent connections of neuroendocrine systems. *Drug Res* *28*:1264-1268, 1978a.

Weindl, A., and Sofroniew, M.V.: Neurohormones and circumventricular organs. In *Brain-Endocrine Interation III. Neural Hormones and Reproduction*, D.E. Scott, G.P. Kozlowski, and A. Weindl, eds. 1978b, 117-137.

de Wied, D.: The influence of the posterior and intermediate lobe of the pituitary and pituitary peptides on the maintenance of a conditined avoidance response in rats. *Int J Neuropharmacol* 4:157-167, 1965.

de Wied, D.: Effects of peptide hormones on behavior. In *Frontiers in Neuroendocrinology, 1969*, W.F. Ganong and L. Martini, eds. Oxford University Press, London, 1969, pp 97-140.

de Wied, D.: Long term effect of vasopressin on the maintenance of a conditioned avoidance response in rats. *Nature* 232:58-60, 1971.

de Wied, D.: The role of the posterior pituitary and its peptides on the maintenance of conditioned avoidance behaviour. In *Hormones and Brain Function*, Akadémiai Kiadó, Budapest, 1973, pp 391-397.

de Wied, D.: Behavioral effects of intraventricularly administered vasopressin and vasopressin fragments. *Life Sci* 19:685-690, 1976.

de Wied, D.: Peptides and behavior. *Life Sci* 20:195-204, 1977a.

de Wied, D.: Behavioral effects of neuropeptides related to ACTH, MSH, and βLPH. *Ann NY Acad Sci* 297:264-274, 1977b.

de Wied, D.: Psychopathology as a neuropeptide dysfunction. In *Characteristics and Function of Opioids*, J.M. van Ree and L. Terenius, eds. Elsevier/North Holland Biomedical Press, 1978, pp 113-122.

de Wied, D., and Bohus, B.: Long term and short term effects on retention of a conditioned avoidance response in rats by treatment with long acting pitressin and α-MSH. *Nature* 212:1484-1486, 1966.

de Wied, D., and Gispen, W.H.: Impaired development of tolerance to morphine analgesia in rats with hereditary diabetes insipidus. *Psychopharmacologie* 46:27-29, 1976.

de Wied, D., and Gispen, W.H.: Behavioral effects of peptides. In *Peptides in Neurobiology*, H. Gainer, ed. Plenum Press, New York, 1977, pp 397-448.

de Wied, D., Greven, H.M., Lande, S., and Witter, A.: Dissociation of the behavioural and endocrine effects of lysine vasopressin by tryptic digestion. *Br J Pharmacol* 45:118-122, 1972.

de Wied, D., Bohus, B., and van Wimersma Greidanus, Tj.B.: The hypothalmoneurohypophyseal system and the preservation of conditioned avoidance behavior in rats. *Prog Brain Res* 41:417-428, 1974.

de Wied, D., Bohus, B., and van Wimersma Greidanus, Tj.B.: Memory deficit in rats with hereditary diabetes insipidus. *Brain Res* 85:152-156, 1975.

de Wied, D., van Wimersma Greidanus, Tj.B., Bohus, B., Urban, I., and Gispen, W.H.: Vasopressin and memory consolidation.

Prog Brain Res 45:181-194, 1976.

de Wied, D., Bohus, B., van Ree, J.M., Urban, I., and van Wimersma Greidanus, Tj.B.: Neurohypophyseal hormones and behavior. In *Neurohypophysis*, A.M. Moses and L. Share, eds. S. Karger, Basel, 1977, pp 201-210.

van Wimersma Greidanus, Tj.B., and de Wied, D.: Modulation of passive avoidance behavior of rats by intracerebroventricular administration of antivasopressin in serum. *Behav Biol 18*:325-333, 1976a.

van Wimersma Greidanus, Tj.B., and de Wied, D.: Dorsal hippocampus: A site of action of neuropeptides on avoidance behavior? *Pharmacol Biochem Behav 5 (Suppl 1)*:29-33, 1976b.

van Wimersma Greidanus, Tj.B., and de Wied, D.: The physiology of neurohypophseal system and its relation to memory processes. In *Biochemical Correlates of Brain Structure and Function*. A.N. Davison, ed. Academic Press, London, 1977, pp 215-248.

van Wimersma Greidanus, Tj.B., Bohus, B., and de Wied, D.: Effects of peptide hormones on behavior. In International Congress Series No. 273. *Excerpta Medica*, Amsterdam, 1973, pp 197-201.

van Wimersma Greidanus, Tj.B., Bohus, B., and de Wied, D.: Differential localization of the influence of lysine vasopressin and of $ACTH_{4-10}$ on avoidance behavior: A study in rats bearing lesions in the parafascicular nuclei. *Neuroendocrinology 14*:280-288, 1974.

van Wimersma Greidanus, Tj.B., Dogterom, J., and de Wied, D.: Intraventricular administration of anti-vasopressin serum inhibits memory consolidation in rats. *Life Sci 16*:637-644, 1975a.

van Wimersma Greidanus, Tj.B., Bohus, B., and de Wied, D.: CNS sites of action of ACTH, MSH and vasopressin in relation to avoidance behavior. In *Anatomical Neuroendocrinology*, W.E. Stumpf and L.D. Grant, eds. Karger, Basel, 1975b, pp 284-289.

van Wimersma Greidanus, Tj.B., Bohus, B., and de Wied, D.: The role of vasopressin in memory processes. *Prog Brain Res 42*:135-141, 1975c.

van Wimersma Greidanus, Tj.B., Woutersen, R.A., and de Wied, D.: Stimulation and inhibition of corticotrophin release in rats after intracerebroventricular administration of vasopressin analogues. *J Endocrinol 72*:10p-11p, 1977.

van Wimersma Greidanus, Tj.B., Fat-Bronstein, H.T.K., and van Ree, J.M.: Antisera to pituitary hormones modulate development of tolerance to morphine. In *Characteristics and Function of Opioids*. J.M. van Ree and L. Terenius, eds. Elsevier/North Holland Biomedical Press, Amsterdam, 1978, pp 73-74.

van Wimersma Greidanus, Tj.B., Croiset, G., Bakker, E., and

Bouman, H.: Amygdaloid lesions block the effect of neuropeptides (vasopressin, ACTH $_{4-10}$) on avoidance behavior.
Physiol Behav 22:291-295, 1979a.

van Wimersma Greidanus, Tj.B., Croiset, G., and Schuiling,
G.A.: Fornix transection: Discrimination between neuropeptide
effects on attention and memory. *Brain Res Bull 4*:L625-629,
1979b.

van Wimersma Greidanus, Tj.B., Croiset G., Goedemans, H., and
Dogterom, J.: Vasopressin levels in peripheral blood and in
cerebrospinal fluid during passive and active avoidance
behavior in rats. *Horm Behav 12*:103-111, 1979c.

Wittkowski, W.: Elektronenmikroskopische Studien zur
intraventrikulären Neurosekretion in den Recessus
infundibularis der Maus. *Z Zellforsch 92*:207-216, 1968.

Zimmerman, E.A., Stillman, M.A., Recht, L.D., Antunes, J.L.,
Carmel, P.W., and Goldsmith, P.C.: Vasopressin and
corticotropin-releasing factor: An axonal pathway to portal
capillaries in the zona externa of the median eminence
containing vasopressin and its interaction with corticoids.
Ann NY Acad Sci 297:405-419, 1977.

Angiotensin and Drinking: A Model for the Study of Peptide Action in the Brain

M. Ian Phillips

INTRODUCTION

In the course of evolution, vertebrates have internalized their ancient aquatic environment within the cellular and extracellular fluid compartments of the body. The fluid compositions of these compartments must be constantly maintained against the ever-present challenge of dehydration. So essential is fluid balance that several mechanisms have evolved to maintain it. Skin is an important organ for preventing fluid loss. The kidneys control the amount of water excreted and maintain correct levels of sodium for osmotic balance between the intracellular and extracellular media. The circulatory system maintains a constant volume of fluid by delicately balancing cardiac output with peripheral resistance of the blood vessels. When it is necessary to conserve water, vasopressin is released from the posterior pituitary gland, effecting an antidiuretic action on the kidney tubules. These mechanisms help to retain water and use it efficiently, but eventually the inevitable water loss must be replaced from outside the body by drinking. Although peripheral mechanisms can maintain moment-to-moment regulation of water balance, drinking requires activation of the brain. This is not only for the muscular coordination that is involved in the act of drinking, but for the memory that the brain contains, which can override immediate peripheral demands.

Our current understanding of the brain mechanisms involved in drinking has been accelerated by the discovery that angiotensin II (AII) causes drinking when injected into the brain. Booth (1968) was the first to observe this effect, noted incidentally during a study on cerebral control of feeding. At about the

same time, Fitzsimons and his colleagues, Rolls and Epstein
(Epstein et al., 1969), tested AII as a thirst-inducing hormone
as part of a logical series of experiments in which Fitzsimons
had demonstrated a potential role of the renin-angiotensin system in thirst. The renin-angiotensin system is outlined in
Fig. 10-1. Renin, an enzyme, is produced by the kidneys and
released from the juxtaglomerular cells when renal perfusion
pressure from the renal artery is lowered or when the renal
nerves are stimulated. Renin is released by a number of stimuli, including a state of hypovolemia. This occurs when the
extracellular fluid compartment is reduced in volume without
altering the osmotic balance between intracellular and extracellular fluids. The independence of the two compartments is
maintained because water is lost from inside the cellular compartment only when the extracellular fluid is hyperosmotic. A
reduction in volume can occur, for example, in hemorrhage or
skin burn, when extracellular fluid containing ions and water
is lost. Renin is released into the blood and acts upon angiotensinogen or renin substrate, which is produced in the liver.
Renin cleaves the leucyl-leucyl bond in the substrate to produce the decapeptide angiotensin I (AI). In the lungs, AI is
converted by converting enzyme to AII, an octapeptide of 1000
molecular weight. From the pulmonary circulation, angiotensin
is quickly distributed throughout the body and has a half-life
of one to two minutes in the circulation before aminopeptidases
finally inactivate it. The angiotensin formed in this manner
acts on the adrenal cortex to produce aldosterone secretion,
which effects retention of sodium. Raising Na levels in the
extracellular fluid leads to retention of water in the body.
Angiotensin also causes vasoconstriction by direct action on
the arteriolar smooth muscle and by an increase in sympathetic
activity through an action on the central nervous system and
sympathetic nerve endings (Khairallah et al., 1977). This also
has the effect of offsetting the state of hypovolemia (Fig.
10-2).

It has been found that AII is dipsogenic in may species,
whereas carbachol, a cholinergic agonist (Grossman, 1960),
seems to elicit thirst in only a few species. AII stimulates
drinking in all mammalian species tested, including rat, dog,
cat, sheep, and goat, and in a marsupial, the opossum. Two
avians, the chicken and pigeon, and the reptile iguana are also
susceptible to the dipsogenic activity of angiotensin in the
brain (Fitzsimons, 1978). Doses of the octapeptide lower than
those first reported by Epstein et al. (1969) were shown by
Severs et al. (1970) to be effective in producing thirst. We
now know that when AII is delivered directly into the optic
recess of the third ventricle or the subfornical organ, very
low doses (in the femtogram range) are able to elicit drinking

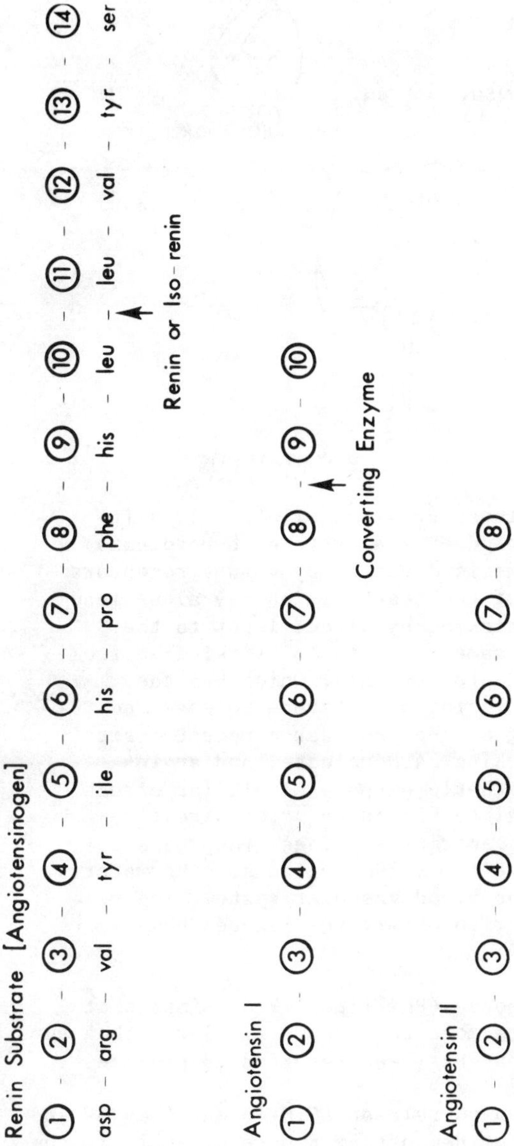

Renin Substrate [Angiotensinogen]

① – ② – ③ – ④ – ⑤ – ⑥ – ⑦ – ⑧ – ⑨ – ⑩ – ⑪ – ⑫ – ⑬ – ⑭
asp – arg – val – tyr – ile – his – pro – phe – his – leu – leu – val – tyr – ser

Renin or Iso-renin

Angiotensin I

① – ② – ③ – ④ – ⑤ – ⑥ – ⑦ – ⑧ – ⑨ – ⑩

Converting Enzyme

Angiotensin II

① – ② – ③ – ④ – ⑤ – ⑥ – ⑦ – ⑧

Fig. 10-1. The renin-angiotensin system.

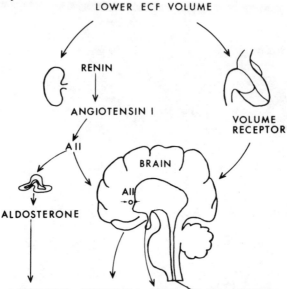

LOWER ECF VOLUME

RENIN

ANGIOTENSIN I

A II

VOLUME RECEPTOR

BRAIN

AII

ALDOSTERONE

RETENTION   DRINKING↑ ADH↑   SYMPATHETIC
of Na⁺↑   OUTPUT ↑

Fig. 10-2.   The effect of lowering extracellular fluid in
volume (i.e., producing a state of hypovolemia).
The lower volume is detected by volume receptors
in the atrium of the heart, which may alone lead
to increased drinking by direct input to the
brain.   At the same time, renin is released from
the kidneys and produces AII, which has the
effect of stimulating aldosterone release and
possibly acting on the brain at a receptor site
to produce drinking, ADH release, and an increase in sympathetic output.   A similar effect
occurs when angiotensin is injected directly
into the brain ventricles.   These responses to
hypovolemia have the effect of conserving water,
constricting the blood vascular system, and re-
plenishing water to offset the reduced body
fluid volume.

behavior and a pressor response (Phillips, 1978; Simpson et
al., 1978).   These doses are more than 2000-fold lower than any
dose given intravenously that has been reported to produce
drinking (Hsaio et al., 1977).
    Angiotensin was also shown to release ACTH in man (Rayyes and
Horton, 1971); this was later demonstrated more directly in the
dog (Maran and Yates, 1977).   AII also releases antidiuretic
hormone (ADH).   This was first shown by Mouw et al. (1971) and

was implied in the earlier studies involving hypophysectomy and urine assay (Severs et al., 1970; Olsson, 1970; Andersson, 1971). Hoffman et al. (1977a) calculated the amount of ADH released by doses of AII and found that ADH was released in quantities sufficiently large to cause a pressor response by 100 ng of AII in the brain. Since hypophysectomy reduced the pressor response to ICV AII by approximately 50 percent, Severs et al. (1970) concluded that ADH contributes about 50 percent to the pressor effect of AII. Gregg and Malvin (1978) demonstrated that AII releases ADH by an action on the hypothalamus and not by direct action on the neurohypophysis. ADH release following ICV AII is easily obtained (Andersson, 1971; Mouw et al., 1971; Keil et al. 1975).

This emphasizes a significant difference between peripheral and central angiotensin actions because it has proved difficult to demonstrate ADH release to angiotensin IV (Claybaugh et al., 1972) and, as stated earlier, direct injection of AII into the brain is far more potent in producing thirst than is peripheral administration.

PERIPHERAL VERSUS CENTRAL ANGIOTENSIN

Thus, not all of the effects of angiotensin can be easily accounted for by the peripheral activation of the renin-angiotensin system. The effects could be due to very high levels of circulating angiotensin or to an angiotensin endogenous to the brain. A difficulty for the peripheral influence of AII on the brain is the lack of evidence that AII can cross the blood-brain barrier. Two studies using radioactively labeled angiotensin claimed evidence of peripheral angiotensin crossing the blood-brain barrier (Volicer and Lowe, 1971; Johnson and Epstein, 1975). The doses used, however, were extremely high (μg range). When high doses of angiotensin are injected intravenously, we have found that the blood-brain barrier opens in various brain sites. This has been shown by using horseradish peroxidase (HRP) as a tracer and also by a dye, Evans blue (Phillips et al., 1977a; Johannsson et al., 1970). The breakdown is temporary and reversible, but it would allow substances to cross the blood-brain barrier during sudden acute hypertension resulting from angiotensin injection. We found that a sudden rise in blood pressure of 35 mm Hg is sufficient for the effect. With lower doses of AII (10 ng), with which the blood pressure increase was less than 35 mm Hg, there was no penetration of the brain by HRP. Johansson et al. (1970) have also shown opening of the blood-brain barrier when blood pressure is high or is suddenly raised by any method. Thus, high blood pressure can open the blood-brain barrier, but whether this occurs by opening tight junctions or by increasing pinocytosis, or both, has yet to be resolved (Westergaard, 1975; Broadwell and Brightman, 1976).

Within the brain, however, there are small regions that do not exhibit a classical blood-brain barrier. These include the circumventricular organs, the organum vasculosum of the lamina terminalis (OVLT), median eminence, area postrema, neurohypophysis, and the pineal and subfornical organ (SFO) (Fig. 10-3). When a tracer, such as HRP, is injected IV, the tracer escapes through the fenestrated capillaries of this area. The extent of spread of the tracer is limited to the organ; it does not spread noticeably into surrounding brain parenchyma. The opposite picture is seen when the tracer is injected ICV. A few minutes after the injection, the parenchymal tissue is darkly stained but the circumventricular organs are unstained (Weindl and Joynt, 1972), except for limited uptake of tracer by tanycytes, which transport proteins from the cerebrospinal fluid to the blood vessels (Fig. 10-4). The differences in distribution are illustrated diagramatically in Fig. 10-5. The mechanisms for this difference of distribution appear to result from the presence of tight junctions between ependymal cells overlying the circumventricular organs, whereas gap junctions occur in the other, ciliated, portions of the ventricles (Phillips et al., 1978a; Brightman and Reese, 1969) (see Fig. 10-5). The limited distribution within the circumventricular organs may be due to rapid uptake of substances into venous blood, since they are characteristically densely vascularized.

The brain therefore has an effective compartmentalization, which is important when considering the acute effects of angiotensin. On the one hand, intravenous AII could reach circumventricular sites, but apparently not spread to other brain sites. On the other hand, intraventricularly injected AII could reach brain parenchymal sites but not spread to the circumventricular sites directly. Unless blood pressure is raised abnormally high, the effects seen using the two routes of injection would be different from one another. This implies that there are two locations for angiotensin receptors in the brain: one on the blood side, which could be reached by peripheral AII; and one in the brain, which could be reached only by endogenous AII.

Is Drinking to Angiotensin Physiological?

Fitzsimons (1972) proposed that the renin-angiotensin system mediates thirst caused by plasma volume or other deficits of water balance in the body. The evidence for this hypothesis is based on three types of experiments. The first was systemic

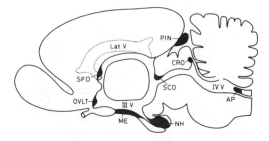

Fig. 10-3. Diagram of the circumventricular organs in the rat
brain shown in sagittal section. The organs lie
close to the brain ventricular system and include:
SFO, subfornical organ; OVLT, organum vasculosum
lamina terminalis; ME, median eminence; NH, neurohypophysis; PIN, pineal; SCO, subcommissural organ;
CRO, collicular recess organ; AP, area postrema.

Fig. 10-4. Photomicrograph of rat brain third ventricle at
level of median eminence after ventricular injection
of HRP. Note that the parenchyma has filled with
reaction product but the median eminence is
comparatively clear. (Data from Phillips et al.,
1978a).

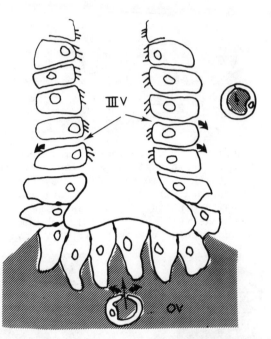

Fig. 10-5. Diagram of the brain ventricle wall and diffusion
         barriers. The figure illustrates that large
         proteinaceous substances injected into the third
         ventricle may diffuse into brain tissue through gaps
         between the ciliated ependyma, but apparently do not
         diffuse to the shaded portion. The shaded portion
         represents a circumventricular organ, such as the
         median eminence or organum vasculosum lamina
         terminalis (OV). Polypeptides arriving here in the
         circulation may enter through the blood vessels,
         which are fenestrated (arrow), and diffuse into the
         organ. The same substances may not enter the organ
         from the ventricle through the ventricular wall
         because of tight junctions between the ependyma. In
         the brain parenchyma, the blood vessel wall is
         impervious to large proteins except under abnormal
         conditions when blood pressure is suddenly raised.

infusion of AII or renin, which elicited water intake in a
dose-dependent manner. The second was by ligation of the vena
cava, which lowers the pressure above the ligation on the
venous side of the blood circulatory system, thus lowering
arterial pressure by decreasing cardiac output. One of the
consequences of this imbalance in pressure is lower renal

arterial pressure. Decreased pressure in the renal artery causes release of renin and activation of the renin-angiotensin system. Another consequence is that there is detection of the low volume in the venous side by volume receptors. The relative contributions of the volume receptors and of the renin-angiotensin system to thirst have never been fully evaluated, but ligation of the inferior vena cava is effective in producing drinking, and the effect is greatly reduced by bilateral nephrectomy (Fitzsimons, 1972). The third method is by administration of the β-adrenergic agonist isoproterenol. This drug stimulates renin secretion and elicits thirst in rats. Again, bilateral nephrectomy virtually abolishes the effect of isoproterenol. However, the question still remains whether any of these treatments produce physiological amounts of angiotensin in the plasma, and whether such amounts will elicit thirst. To investigate this question, Stricker (1977) used these models of thirst and measured plasma renin activity (PRA). This measure is expressed as the amount of AI produced in a given time. He found that the mean PRA must exceed 18 ng AI/ml for 90 minutes before water intake is stimulated by angiotensin during a one-hour drinking test. He determined that increases in PRA of 91 ng AI/ml per 90 minutes beyond that are required for each milliliter of water consumption by the rats tested. According to his calculations, to obtain 6 ml water intake would require a PRA of 700 ng AI/ml per 90 minutes. However, this level is considerably higher than the PRA of 50 to 60 ng AI/ml per 90 minutes found after a 13 percent reduction of plasma volume (Stricker, 1978). Also, circulating levels of AII that are produced by dipsogenic doses of AII infused intravenously in sheep (Abraham et al., 1975) were found to be at supraphysiological levels. Stricker's calculation in an experiment on isoproterenol-elicited drinking showed that angiotensin contributed only 2.1 ml of the total 7.7 ml consumed in the first hour after 0.33 mg/kg isoproterenol. Others have shown that saralasin ([Sar$_1$,Ala$_8$] AII), the specific AII antagonist, does not block thirst after caval ligation or isoproterenol treatment (Tang and Falk, 1974; Rolls and Wood, 1977). Thus, while peripheral AII is clearly involved in the mediation of thirst under some circumstances, its role is not critical, and it is not the major mediator of caval ligation- or isoproterenol-induced thirst. The reason for the lack of drinking when bilateral nephrectomy has been performed after these treatments appears to be the animals' inability to demonstrate behavioral drinking. This deficit may be due to the severe hypotension that is reported to follow removal of the kidneys in combination with caval ligation or isoproterenol treatment (Stricker, 1977). Supporting evidence is that, when the blood pressure is raised by renin injection, hypertonic saline, or intracranial AII injections (Stricker, 1978), drinking is reinstated in nephrectomized rats treated with isoproterenol. Thus a case can be made that peripheral AII does not play a critical role in physiological thirst.

Another approach to the question of whether peripheral AII can normally reach the same central sites as ICV AII has been the use of a competitive angiotensin antagonists such as saralasin. Intravenous infusions of saralasin effectively block the dipsogenic action of IV infusions of AII. Likewise, intracerebral infusions of the peptide and its antagonist effectively cancel out each other (Epstein et al., 1973). Intracranial saralasin reportedly blocked the effects of elevated peripheral angiotensin levels (Ramsay et al., 1978), suggesting that saralasin spread to the same receptor sites as the peptide. However, this study used isoproterenol to raise the level of angiotensin, and Stricker (1977) has contested the assumption that isoproterenol-induced drinking is angiotensin-mediated (vide supra). A direct test of angiotensin IV with saralasin in the third ventricle had no effect in the two dogs tested. In another study, Johnson and Schwob (1975) used extremely high doses of angiotensin (500 μg, SC), which may have opened the blood-brain barrier. Thus the evidence for peripheral AII reaching brain ventricular sites is not compelling. When saralasin was given IV and AII ICV, only a very high dose of saralasin (72 μg bolus injection) could antagonize the central effect of AII (Hoffman and Phillips, 1976a). At this dose, blood pressure is raised another 30 mm Hg because of the agonistic action of the antagonist (Phillips et al., 1977a), a level that would open the blood-brain barrier. If the blood-brain barrier were to open, then angiotensin could reach the same sites as intraventricular saralasin, but it would have little relevance to normal physiology. Thus, before assuming that the peripheral and central angiotensin act on the same receptor sites, the integrity of the blood-brain barrier and the access to circumventricular organs must be taken into account. Thus, peripheral and brain AII may or may not act on the same receptors in the brain. The effects of artificially injecting AII into the brain directly may be more relevant to understanding what brain AII does than how peripheral AII functions.

Dose-response curves showed lower doses of saralasin IV to be ineffective in blocking central angiotensin effects, even though these doses antagonized IV AII-induced drinking (Hoffman and Phillips, 1976a). Therefore, peripheral saralasin, at levels that do not increase blood pressure, does not antagonize central AII. This is evidence that brain angiotensin could act independently of peripheral AII.

These considerations concern the lack of evidence that

circulating AII can reach receptors activated by centrally
injected AII, except when doses are high enough to have
pathophysiological actions on the blood-brain barrier. In
addition to this, there is the evidence of plugging studies.
Ventricular plugging is a technique whereby parts of the
ventricular system can be limited, so that when angiotensin is
injected, for example, its area of influence can be limited to
a specified ventricular region (Phillips et al., 1977b; Buggy
and Fisher, 1976; Hoffman and Phillips, 1976b). First, AII is
injected and the response recorded; then cold cream (2 to 4 μl)
is injected into the desired area. The first cannula is again
tested by injecting AII plus a black dye. A few minutes after
the test, the animal is sacrificed and perfused. The brain is
cut on a freezing
microtome and photographed. When a plug is limited to the
anterior ventral part of the third ventricle over the OVLT,
injections of AII into the lateral ventricles are no longer
effective in producing drinking or pressor responses. Conversely, if all other parts of the ventricular system are plug-
ged except the small region in front of the OVLT, angiotensin
will still be effective when injected into that area (Phillips
et al., 1978b). Peripherally injected angiotensin in rats with
plugs in this area still produced drinking responses (Brody et
al., 1978) (blood pressure and ADH release were not tested).
This difference in results suggests a difference between sites
of action for peripheral and central AII. If blood-borne
angiotensin cannot excite the same receptors as cerebrospinal
fluid (CSF) borne angiotensin, the ventricular AII receptors
may be stimulated by angiotensin produced in the brain.

In summary, AII does not cross the blood-brain barrier, and
there is evidence to indicate that central AII acts upon brain
receptors separate from those activated by peripheral AII. It
appears that the brain has evolved a special receptor system
for brain angiotensin (Fig. 10-6).

Angiotensin and Physiological Thirst

Since it has been shown that angiotensin can be inhibited by
saralasin when both compounds are injected intracerebrally or
when both are injected IV, several experiments have tested whether angiotensin is involved in normal drinking behavior. Rats
deprived of water for 12, 24, and 48 hours were infused with
saralasin (1 μg ICV) and allowed access to water for five minutes later. In no case was there any evidence of a reduction
in the amount of water drunk or the latency to begin drinking
when compared to saline treated rats (Phillips et al., unpublished observations; Fig. 10-7). Similar experiments have been

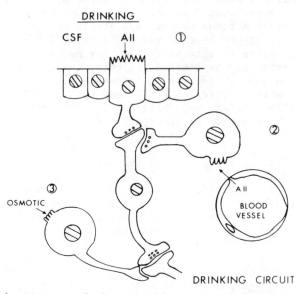

Fig. 10-6. Diagram of the possible angiotensin receptor cells
in the brain. 1. Those that are responsive to
CSF-borne AII and may be on the ependymal wall. 2.
Those that are in a circumventricular organ that
could be reached by angiotensin from the blood. 3.
An interacting osmotic receptor that feeds into the
final common pathway for drinking behavior.

attempted by others. Severs and Summy-Long (1975) induced
drinking in rats by SC polyethylene glycol, which produces
hypovolemia. Rats given 10 μg saralasin ICV 5.5 hours after
polyethylene glycol did not show any reduction in the amount of
water drunk compared with controls. Likewise, a similar dose
of saralasin given 2.5 hours after subcutaneous injection of
hypertonic saline, which produces cellular dehydration and
drinking in rats, failed to alter the water consumption tested
for 60 minutes, three hours after the saline injection. In one
of the few experiments to show an effect of saralasin, Malvin
et al. (1977) infused 66 ng/3 μl/min saralasin into the ventricles of rats that had been water-deprived for 30 hours. Infusions began 75 minutes before rats were allowed access to
water. Rats treated in this way delayed their drinking compared with controls infused with artificial CSF. However, all
the animals in the experiment drank similar total amounts of
water; only the latency to drink was affected by the saralasin.
Thus, in water-deprived rats, where there would be extracellu-

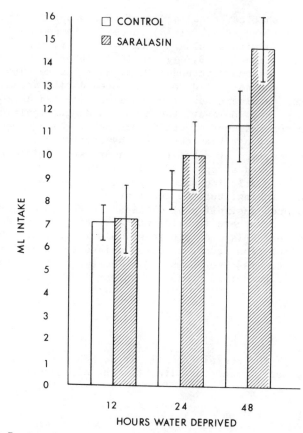

Fig. 10-7. Results of an unpublished study on saralasin
injections ICV (1 μg/μl) in rats deprived of water
for 12, 24, or 48 hours. (N = 6 in each group.) No
significant reduction in drinking was observed.

lar and intracellular dehydration, and in polyethylene glycol-
treated rats, where there would be extracellular dehydration,
and in saline-treated rats, where there would be intracellular
dehydration (Fitzsimons, 1972), saralasin and angiotensin
antagonists have not been shown to inhibit thirst. From this,
one might conclude that angiotensin is not normally involved in
the thirst associated with extracellular or cellular dehydration. There are, however, other factors involved in thirst.
For many years, it has been known that injections of carbachol,a cholinergic agonist, induced thirst (Grossman, 1960).
It is possible that there are thirst-controlling systems that
act in parallel, and if one is inhibited another can take over.

We hypothesized that carbachol and angiotensin act in parallel. To test this, Hoffman et al. (1978) infused rats deprived of water for 48 hours with saralasin (1.4 μg/min ICV) and atropine (a cholinergic antagonist, 0.7 μg/min ICV) either alone or in combination. Infusions were begun 10 minutes before rats had access to water and continued for 30 minutes during access to water. Neither saralasin nor atropine alone exerted an effect on the water intake. However, the combined drugs significantly decreased the total amount of water drunk over the 30-minute period. This study has been confirmed by Severs et al. (1978), who used rats deprived of water for 24 hours and administered saralasin (10 μg) and atropine (2 μg) ICV 30 minutes before the rats had access to drinking water. Again, neither compound alone reduced drinking, but the combination of atropine and saralasin produced a significant decrease in the amount of water drunk. Thus, there is evidence that during water deprivation, angiotensin plays a role in stimulating thirst. However, this stimulation is not revealed unless the parallel cholinergic system is also inhibited. The two systems would appear to work together in producing thirst induced by water deprivation.

RECEPTOR SITES

In recent years, the medial preoptic area, SFO, and OVLT have all been proposed as receptor sites for the dipsogenic effect of AII.

The Medial Preoptic Area

The medial preoptic area is protected by the blood-brain barrier and could not be reached by AII in the blood. When injected directly into this region, higher volumes of AII are more effective than small volumes (Mogenson et al., 1977). This difference may mean that the peptide spreads to the adjacent OVLT region and actually acts there. However, using radioactively labeled AII, Swanson et al. (1978) showed that the label did not spread outside the medial preoptic region and rats drank to angiotensin. Thus the medial preoptic area may be a site that can be stimulated only by endogenous brain angiotensin and not by plasma or CSF AII.

The Subfornical Organ

SFO is a site lying at the entrance to the third ventricle from the lateral ventricles. Simpson et al. (1978) have contributed

to the evidence for the SFO being a receptor site for angiotensin-induced thirst, but it has become apparent that the SFO
cannot be the only such receptor site in the brain. First, the
threshold dose for a drinking response to direct injection of
AII in the SFO is 0.1 to 1.0 pg (Simpson and Routtenberg, 1973;
Simpson et al., 1978). It has never been established, however,
that the injected AII acted only on the SFO. Swanson et al.
(1978) injected ^{3}H-labeled amino acids and AII directly into
the SFO. They found that when the label was confined to the
SFO, no drinking was observed. When the label had spread outside the SFO, drinking was noted. This points to a weakness in
the evidence for the sensitivity of the SFO which needs to be
resolved. Second, lesioning the SFO significantly reduced AII-
induced drinking when the peptide was injected into the
diencephalic preoptic area (Simpson and Routtenberg, 1973), the
cerebral ventricles (Phillips et al., 1974), or intravenously
(Simpson et al., 1978; Abdelaal et al., 1974). Buggy et al.
(1975), however, found recovery of the drinking effect if the
rats were tested 8 to 14 days after lesioning. This implied
that there was either surviving tissue remaining or another
site of action in addition to the SFO. To demonstrate that the
latter was the case, Buggy and Fisher (1976) found they could
obtain responses even after abolishing the SFO when angiotensin
was delivered through a cannula in the anterior third ventricle. Because of such results, attention had been drawn to the
organum vasculosum as a possible site of action, and that
structure lies in the anterior third ventricle (Phillips and
Hoffman, 1974). Third, direct injection of saralasin into the
SFO reduced drinking to angiotensin IV (Simpson et al, 1978).
Again, no measure of spread of this AII antagonist was made,
and it is impossible to know whether or not the saralasin was
limited to the SFO. It could have spread into the third ventricle, for example. Fourth, neurophysiological responses of
SFO cells to AII provided supportive evidence for the presence
of receptor neurons for angiotensin (Felix and Akert, 1974;
Phillips and Felix, 1976). Only the SFO of the cat was tested,
and the role of the SFO in AII-induced drinking in the cat has
not been investigated. This is not to say that the SFO is not
a receptor site for AII in the rat. There is no evidence
against the SFO being a receptor for blood-borne angiotensin,
and this, in fact, may be its role. Upon dehydration, the SFO
undergoes structural changes observed by both scanning and
transmission electron microscopy (Phillips et al., 1978a;
Dellman and Simpson, 1976). It is intriguing that the SFO has
direct connections to the supraoptic nucleus (SON) and could be
involved in the water retention during dehydration by releasing
vasopressin.

The Organum Vasculosum
Lamina Terminalis

The evidence for the OVLT being a dipsogenic receptor site for
AII includes the following: First, it is in the most sensitive
region of brain to angiotensin II detected. As little as 50 fg
AII (5×10^{-17} mol) in 2 μl produce a drinking response (2 ml)
and blood pressure responses (5 mm Hg). However, we cannot be
certain that this effect involves the OVLT, because when dye
injections of the same volume are made, the OVLT and the periventricular area are stained. Also, the same dose of AII in
the OVLT area produces a greater response of drinking compared
with the SFO (Nicolaidis and Fitzsimons, 1975; Phillips et al.,
1977b). Second, cream plugs blocking access of ICV AII to the
ependyma of the OVLT, but not the SFO, prevent the drinking
behavior, blood pressure response, and vasopressin release
(Hoffman and Phillips, 1976b; Phillips et al., 1977b). By contrast, a plug over the SFO, leaving the anterior third ventricle intact, did not prevent drinking to ICV AII (Buggy and
Fisher, 1976). Third, lesions of the ventral anterior third
ventricle prevent responses to AII (ICV or IV) and result in a
temporary adipsia (Buggy et al., 1977). Andersson et al.
(1975) also noted this effect in goats when lesions were made
in the anterior wall of the third ventricle. The hypodipsia
was seen in goats in which the SFO was spared by the lesioning.
Fourth, a knife cut above the OVLT, which would separate the
SFO from efferents to the OVLT, does not interfere with drinking to ICV AII (Phillips et al., 1978b). Finally, direct
recording from OVLT with microiontophoretic application of AII
has shown that cells in that region are specifically sensitive
to it. The majority of AII-sensitive units were excited by the
peptide (see following section).
    Taken together, the data in favor of an area in the anterior
ventral third ventricle being the receptor area for AII are
strong. Since we believe that blood-borne angiotensin could
reach the brain where the blood-brain barrier is not complete,
the OVLT is a likely site. This does not rule out the SFO and,
indeed, it would be foolhardy to suggest that one organ or
another is the only site of angiotensin action in the brain.
Nevertheless, for angiotensin injected into the brain, which
mimics angiotensin released into the CSF endogenously, the anterior ventral third ventricle would seem a likely site. To
piece together the story of the mechanism for the action of the
peptide, we can go beyond arguments for or against particular
brain sites and consider how AII at a brain receptor site would
produce its effects.

ANGIOTENSIN AS A PEPTIDE NEUROTRIGGER

Angiotensin provides a beautiful example of a neuropeptide that acts on the brain and functions to maintain water balance. It can produce a complex behavior, release hormones, and elicit pressor responses. To produce these multiple effects, such a peptide may initiate a primary action that in turn triggers specific neural circuits that motivate an organism to conserve water and drink. One may hypothesize that the primary event is the binding of angiotensin to receptor neurons in specialized parts of the brain. The secondary events would be the mediation by classical neurotransmitters of the complex processes that unfold during searching for water for fluid replacement. These processes involve memory and learning, locomotion, and the physical act of drinking. We may view angiotensin in this case not as a neuromodulator, but as a neurotrigger, with many of the properties of a neurotransmitter that sets in motion a number of neural subroutines resulting in the observed responses. If this view is correct, AII should be released presynaptically and should have postsynaptic effects. It follows that there will not be numerous sites of action of the peptide, but rather a few cells to which it binds. Target neurons may be considered as internal sensory cells of the brain, just as retinal cells in the eye are the starting point for the visual response to light. Once stimulated, the angiotensin-sensitive cells will activate the physiological mechanisms to conserve water balance, such as drinking, ADH release, and a pressor response. By their specialized connections to all parts of the brain that are involved in water seeking and water conservation, angiotensin-sensitive cells give importance to the peptide.

Fig. 10-8 describes a simplified conceptual model for investigating the way in which a peptide, such as angiotensin, could act on the brain. Verification of the model requires that certain types of data be collected. AII serves as a model because some of the data are already available. However, much more remains to be gathered, and it will become increasingly harder technically to do this.

There are six elements required for such a model to be viable.

1. **The peptidergic neuron.** It must be demonstrated that angiotensin is produced in the brain independently of peripheral angiotensin--to show that the octapeptide is endogenous to the brain. The effects of the peptide should also appear upon stimulation of the peptide neurons. The peptidergic terminals must reach areas sensitive to angiotensin as shown by direct injection.

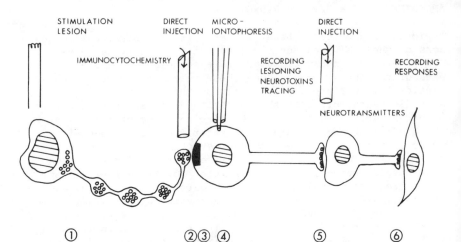

Fig. 10-8. A model for studying peptide action in the brain.
For explanation, see text.

2. **Peptide release.** It will also be necessary to demonstrate that the peptide is actually released in this
   area. A further question is whether such release obeys
   the rules associated with neurotransmitter release, such
   as: Is release quantal? Are its effects graded? Is
   peptide release influenced by the presence of such diva-
   lent ions as calcium?
3. **Receptor binding sites.** The cells found in the same
   areas as the terminals must have receptors of high affinity for the peptide. Binding should be specific to the
   peptide, reversible and saturable.
4. **Neurophysiological responses.** The cells must be shown to
   be neurophysiologically responsive to the peptide. In
   addition, the cell bodies or axonal pathways from the
   neuroeffector neurons must be shown to transmit the message triggered by the peptide into circuits controlling
   specific responses. Interruption or abolition of this
   circuit can be demonstrated by lesioning.
5. **Neurotransmitter release.** The neuroeffector terminals
   must release neurotransmitters to effect synaptic conduction of the pathway; the transmitters may be manipulated by artificial stimulation, or disturbed by neurotoxins and transmitter blockers. At these synapses,
   neuromodulators, such as prostaglandins (PG), may also
   play an active role.

6. **Effector organs.** Finally, an appropriate response must be recordable from the effector organs, whether they be skeletal muscle (as in motor behavior), smooth muscle (as in pressor responses), or hormone secretion (as in ADH- or ACTH-release).

Based on the view that angiotensin triggers a circuit that leads to a response, the model implies that the amount of angiotensin to be found in the brain will be small compared to the commoner neurotransmitters. It is the latter that transmit the message, whereas the peptides prescribe what the message shall be. We should also find for a given response that angiotensin-sensitive neurons will be limited in number and probably localized to limited areas. There may, of course, be more functional roles for angiotensin than are presently known or dreamed of. Receptor-binding studies indicate a wide distribution of AII receptors, and this may reflect multiple functions. In such a case, the number of cells activated by angiotensin will be greater and the sites of angiotensin-containing neurons all the more fascinating.

Angiotensin II-Containing Neurons

Localization of AII has been demonstrated by immunohistofluorescence and immunohistochemical techniques. The distribution of AII as shown by the indirect immunohistofluorescent method (Fuxe et al., 1976; Ganten et al., 1978) indicates regions of high density, including the medial external layer of the median eminence and the amygdala. Those areas with low-to-moderate density include the nucleus of the dorsal medial hypothalamus, locus ceruleus, and caudate nucleus. Scattered terminals were found in the preoptic periventricular gray, septal area, and nucleus tractus solitarius. No immunohistofluorescence was found in the neocortex or cerebellum. Using the same antibody, but the unlabeled antibody enzyme method of Sternberger, we found angiotensinlike material in the same areas of the brain (Phillips et al., 1979; Weyhenmeyer and Phillips, 1979).

The antibody we used had a titer of 1:95,000 at 50 percent AII by radioimmunoassay. It was tested in a dilution of 1:500. There was no cross reactivity with arginine vasopressin (AVP), oxytocin, or AI. The controls included absorption with 2 mM AII, plasma with no antibodies, and goat antirabbit sera, but no angiotensin antisera. None of these controls produced staining.

The distribution within the median eminence lies at the border of the internal and external zones and consists mostly of fiber tracts containing AII-positive material. The SON contain angiotensin in the cell bodies, as revealed by this technique. This is, of course, the region that contains vasopressin. Angiotensinlike material is also detectable in cells of the paraventricular nuclei, which is also a region containing vaso-pressin. For comparison, we used AVP antibody and found a denser distribution of vasopressin in the same regions. There is no cross reaction between the antibody to AII and vasopres-sin. The same cells do not appear to contain the same reactive materials, which leads us to suppose that there is AII in the supraoptic and paraventricular nuclei.

In several places in the brain there are fibers with varico-sities that stain positively and selectively for the angioten-sin antibody (Fig. 10-9). Fibers were found in fiber bundles of the stria medullaris and marked a possible connection from the amygdala and lateral olfactory tract to the septal region. Fibers were most numerous in the anterior hypothalamus, lateral septum and OVLT. Scattered fibers were found in the neocortex and hippocampus. A plexus of fibers innervated a periventri-cular region along the third ventricle. This distribution of AII-like immunoreactivity has now been found in mouse (G. Hoffman, personal communication), rat, monkey, and human brains (Phillips et al., 1979; Weyhenmeyer and Phillips, 1979; Quinlan et al., 1979). For the purposes of the model in Fig. 10-8, we shall consider that the fibers of the peptidergic neu-ron (but not the cell body) are the ones we have located in the OVLT. This region provides answers to several questions that this model of a peptidergic system raises.

Lesioning

Destruction of the tissue in this region has characteristic effects, which have been called the anteroventral third ven-tricle syndrome. Immediately after lesioning there is an acute phase during which rats are adipsic (Buggy et al., 1977). Despite the adipsia, they continue to excrete urine of normal concentration. This inappropriate response to a water deficit implies a disturbance in the release of ADH. Anteroventral third ventricle lesions also prevent drinking elicited by ICV AII. Bealer et al. (1979) have shown that after anteroventral third ventricle lesions, ICV AII was no longer able to release ADH or increase blood pressure. The rats were physically cap-able of releasing ADH, however, because phenylephrine (50 mm ICV) evoked ADH release in the anteroventral third ventricle lesioned rats and produced a pressor response comparable to its effects in sham-lesioned rats. The evidence points to involve-ment of this area in the mediation of central angiotensin responses. The effects of anteroventral third ventricle

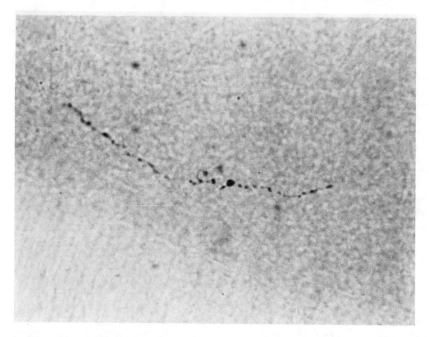

Fig. 10-9. A fiber with varicosities found in the brain, which
has been stained for AII by the peroxidase antiperoxidase (PAP) method. (Data from Weyhenmeyer and
Phillips, 1979).

lesions are not exclusive to the rat. As mentioned previously,
Andersson et al. (1975) made similar lesions in the brain of
goats and observed a permanent hypodipsia and lack of significant release of ADH in response to hypernatremia, plasma hyper-
osmolality, or AII infused via the carotid artery. The anteroventral third ventricle contains the OVLT, which has a distinc-
tive surface facing the third ventricle (Phillips et al.,
1977a; Weindl, 1973). We had earlier shown (Hoffman and
Phillips, 1976b; Buggy and Fisher, 1976) that when access to
this surface by intraventricular AII is blocked with a cream
plug (Hertz et al., 1970), no responses to the octapeptide are
seen. This finding pointed to the OVLT being a critical receptor area for angiotensin (Phillips and Hoffman, 1974; Buggy et
al., 1975).

Release of Angiotensin II

A necessary question yet to be answered is whether angiotensin
is released in the way that other hormones and neurotransmitters are released. If AII is released in the OVLT region, low
doses of exogenously applied AII should produce significant
effects. This is indeed the case. In order to reach the
region accurately, a 30 gauge guide cannula is lowered to the
level of the anterior commissure. Then a longer, 33 gauge
injector is lowered through the guide cannula to end directly
in front of the OVLT in the anterior ventral third ventricle.
This experiment in six rats elicited a dose-response curve with
drinking and pressor responses beginning at doses as low as
$5x10^{-17}$ moles AII (Fig. 10-10). This dose may not be the
absolute threshold for thirst because the subject may sense
thirst but not approach the water and drink, or it may start
drinking and drink more than the dose requires.

Since there are so few AII-containing cells in the adult
brain, this question may not be answerable *in vivo*. Raizada
et al. (1980) reported preliminary success with angiotensin in
cultured brain cells from neonatal rats. They studied the
binding of [^{125}I-Ile]AII ([^{125}I]AII) to substratum-attached
primary cultures of fetal rat brain obtained at 20 days of gestation. Specific binding of [^{125}I]AII to brain cells was
dependent on time, temperature, pH, and cell number. At pH
7.2, 90 to 95 percent specific binding was observed after 30
minutes at room temperature. AII competed for [^{125}I]AII or
[^{3}H]AII binding with 60 percent inhibition by AII at 2 μg/ml.
Peptides structurally related to AII competed for binding in
relation to their biological activity. Unrelated peptides,
including substance P, neurotensin, vasopressin, luteinizing
hormone-releasing hormone (LH-RH), and insulin, did not compete
for [^{125}I]AII binding. Scatchard analysis showed the presence
of about 6000 high-affinity binding sites per cell. These
observations demonstrated the presence of specific high-affinity AII receptors in fetal rat brain cells and provide a model
system for further study of the role of these receptors in the
regulation of neuronal function.

Brain cells from rat fetuses of 20 days gestational age have
also been tested for the presence of AII by the peroxidase/-
antiperoxidase method of Sternberger (Weyhenmeyer et al.,
1980). Cells tested with an antibody to AII demonstrated significant intrinsic staining for AII in the cell body. This
staining was not evident when the primary antibody was previously absorbed with AII. We concluded that there was endogenous AII in fetal rat brain cells. The presence of angiotensin
in cultured brain cells should allow us to test if octapeptide
release occurs and to study the variables that control release.

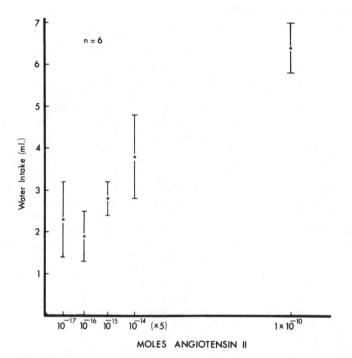

Fig. 10-10. Response to low dose injections of AII when cannulae were placed in the anterior ventral third
ventricle or OVLT.

It may be that, in the neonate, angiotensin plays a role in
growth and development, so that its presence in neonatal neurons relates to other functions of the renin-angiotensin system. There is indirect evidence suggesting that angiotensin
stimulates growth of tumor cells *in vitro* (Schelling et al.,
1979). In the fetal brain cells cultured from whole brains,
angiotensin was found in 2 percent of the cells. This proportion greatly exceeds that found in adult rats. The first two
weeks of life for the neonatal rat are a period of active
growth of brain cells and synaptic proliferation. It would be
interesting to test whether angiotensin blockage could alter
the normal development of the brain.

Angiotensin II Receptors

The next step is the identification of binding sites on effector neurons in the area in which fibers containing the peptide

are found. Binding of a hormone to its receptor leads to chemical and membrane permeability changes in the receptor cell that initiate a neurophysiological response. Structure-activity relationships have been studied on angiotensin analogues to define which part of the molecule is active in binding. The carboxyl terminal hexapeptide Val-Tyr-Ile-His-Pro-Phe- binds to the receptor. This structure is alternately aliphatic and aromatic with Tyr_4, His_6, and Phe_8, and a carboxyl group being essential (Bumpus, 1977; Felix and Phillips, 1979). The characteristics of specific binding of a peptide to its target cells are: reversibility, high affinity, and specificity. In the brain, AII binding studies have been reported using labeled monoiodinated AII. High-affinity binding sites were identified in bovine and rat brain membranes (Bennett and Snyder, 1976). [^{125}I]AII binding was saturable and reversible, with a dissociation constant of 0.2 nM at 37°C. AI was one to two orders of magnitude weaker than AII. The hexapeptide (angiotensin$_{3-8}$) sites were restricted to the cerebellar cortex and dorsal cerebellar nuclei in the calf brain. In rat brain, AII binding was highest in the thalamus, hypothalamus, midbrain, and brainstem. In another study by Sirett et al. (1977), similar results of high affinity and reversibility (with a K_d of 0.9 nM) were reported. The binding was associated with particulate matter in the crude microsomal fraction of the centrifuged membranes. The distribution of receptors was also in the thalamus, septum, hypothalamus, and medulla. By excising smaller pieces of tissue, Sirett et al. (1977) reported very high binding (4.40 fmoles/mg protein) in lateral septum. Cortex (0.33 fmoles/mg protein) and cerebellum (0.52 fmoles/mg protein) were much lower in this study. A finer analysis has been carried out in our laboratory. Stamler et al. (1980) studied angiotensin binding by a similar method, but they used very finely dissected regions and compared spontaneously hypertensive rats of the Okomoto strain (SHR) and their appropriate controls (Wistar Kyoto or WKY). Small areas containing the SFO, OVLT, and cortex were microdissected. Because of the smaller size of each sample (0.1 to 0.5 mg protein), tissue was pooled from 12 SHR and WKY rats matched for age and weight. The specific binding of [^{125}I]AII was determined in supernatants of tissue homogenates. In SHR, the cortical binding was low (0.11 fmoles/mg protein), the SFO was highest (1.33 fmoles/mg protein), and the OVLT medium (0.55 fmoles/mg protein). Compared to the WKY controls, there were no significant differences in binding in cortex or SFO, but in the OVLT (WKY 0.24 fmoles/mg protein), the SHR rats were elevated by 129 percent. Thus, in a region such as the OVLT where AII fibers are found, there is also evidence of AII binding.

Neurophysiological Responses to Angiotensin II

Although there may be several intermediary steps after the peptide has bound to the receptor, the end result will be a characteristic response of the receptor neuron. This may be a graded potential or an all-or-none action potential. The graded potential could be hyperpolarizing or depolarizing. To study the response, the peptide must be applied in quantities low enough to mimic the natural release of the peptide. Neurons receptive to AII have been studied by microiontophoretic application of AII in the SFO of the cat (Felix and Akert, 1974; Phillips and Felix, 1976), the OVLT of the rat (Felix and Phillips, 1979), and in the SON (Nicoll and Barber, 1971; Sakai et al., 1974).

In the study by Nicoll and Barker (1971), the SON in the cat was exposed and cells were identified by antidromic stimulation from the neurohypophysis. The majority of these cells were excited by direct application of AII. A smaller percentage were not responsive and a minority were inhibited. By contrast, cortical neurons were mostly unresponsive to AII; only 1 out of 24 cells was excited. Infusions of AII via the carotids did not stimulate a response in the neurosecretory neurons, but this may reflect the inability of angiotensin to reach supraoptic neurons directly from the blood. In organ-cultured SON cells explanted from neonatal puppies, perfusion of AII immediately elicited a rapid firing activity. The ED_{50} for a response by the cells was 3×10^{-7} M. In retrospect, this dose seems to be very high compared with doses effective in eliciting drinking (Phillips, 1978). Nevertheless, the response was specific because it could be antagonized by AII analogues (Sakai et al., 1974).

Both studies suggest that AII acts directly on specific membrane receptors of cells of the SON. For technical reasons, ADH release was not measured, but Sladek and Joynt (1979), using an organ explant of hypothalamus and neurohypophysis, have shown that 10^{-8} M AII releases ADH in significant amounts compared with unstimulated controls. Saralasin blocks the AII effect, and tetrodotoxin also blocked angiotensin stimulation of vasopressin release, which suggests that Na^+ conductance is important in the response. However, the AII effect need not be direct, and interneurons may be involved.

Iontophoretic application of AII to the SFO of the cat produced an excitatory response that was dose dependent and specifically antagonized by saralasin (Phillips and Felix, 1976). Some of the cells responded to both acetylcholine and angiotensin. In these cases, saralasin inhibited the response to AII but not to acetylcholine. Those units that were responsive only to AII were stimulated by the lowest ejection current values. In the cat SFO, then, there are neurons specifically sensitive to AII. Structure-activity relationships were studied by microiontophoresing purified fragments of angiotensin into the region (Felix and Phillips, 1978; Felix and Schlegel, 1978). Angiotensin$_{2-8}$ was more potent, but angiotensin$_{5-8}$ was less potent than the octapeptide. Angiotensin$_{6-9}$ failed to excite the same neurons that the octapeptide stimulated. Although the phenylalanine at position 8 appears to be important, other amino acids in the chain are necessary for neural response.

Recently we have recorded from the OVLT in rat and have also found cells receptive to angiotensin (Phillips et al., 1979). We used a ventral approach to the OVLT of the rat for direct iontophoresis. Rats were anesthetized and the optic chiasms surgically revealed by microdissection. A 5-barreled pipette was lowered into the OVLT and all excitatory cells encountered were tested with AII± saralasin and LH-RH. Of 87 units tested in the OVLT or lamina terminalis, 55 were excited by AII, 25 were not affected, and 7 were inhibited. Those cells that responded to AII were inhibited by saralasin. When saralasin was applied in the presence of AII, the response to the octapeptide was diminished or abolished (Fig. 10-11). Dose-response curves to AII were established in several of the excited cells. LH-RH produced inhibition in this region. The 22 cells that responded to LH-RH were all inhibited. This recording was made in rats under urethane anesthesia. To avoid the effects of anesthetics, we have also recorded from the OVLT in brain slices (Knowles and Phillips, 1980). The results confirm specific and sensitive responses to AII in the OVLT and lamina terminalis of the rat brain, even in the *in vitro* brain slice preparation.

Thus, the OVLT and adjacent area is sensitive to AII because low doses excite cells there, and the sensitivity is specific because the response can be blocked by a specific antagonist, and another peptide, LH-RH, has opposite effects. A high proportion of angiotensin-sensitive neurons in the preoptic region close to the third ventricle has also been reported (Gronan and York, 1978). This is not to say that other regions of the brain may not contain sensitive AII nerurons, but the proximity to an area we have defined as sensitive for AII-induced drinking supports the hypothesis that these cells play a role in the dipsogenic property of AII.

The Effector Neuron

Once we can establish where the receptor neurons for a peptide are found in the mammalian brain, we can then apply the

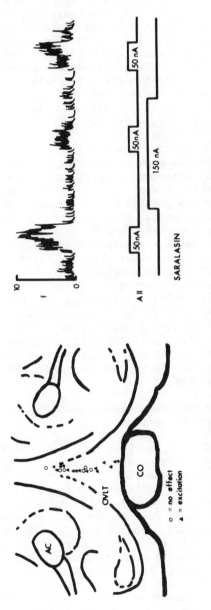

Fig. 10-11. Results of microiontophoretic application of AII in the OVLT of rats. AI

numerous techniques available for studying the connections of
the cells by injection of tracers. Based on clues about
receptor areas, work is already in progress on tracing
pathways with radioactive-labeled amines and HRP (Swanson et
al., 1978; Miselis et al., 1979). This work has indicated that
there are connections between the OVLT and the SFO, and between
the adjacent preoptic area and brainstem. Swanson et al.
(1978) dissolved a [³H]proline-leucine-lysine mixture (0.05
μCi) with 6.6 ng of AII in 20 nl of artificial CSF and injected
the solution into the brains of unanesthetized rats. They
achieved very limited spread by this method and were able to
trace connections from these areas 48 hours after testing for a
drinking response. The area involved in angiotensin-elicited
drinking was defined as a zone in the medial part of the
preoptic region and anterior hypothalamus. Unless the
injection was in this zone or unless it was reached by
ventricular spread, no drinking was observed. This rule
applied to the SFO as well. Only when the cannula had broken
through the ependyma of the SFO and the solution had spread by
way of the ventricles to the anterior ventral third ventricle
region was there a dipsogenic response. The medial preoptic is
adjacent to the OVLT and is connected with it (Miselis et al.,
1979). The efferent connections from the dipsogenic zone were
traced, and two descending pathways were identified. One
passes through the hypothalamus to the midbrain ventral
tegmentum and then courses through a periventricular route to
the central gray. The significance of this dual pathway
remains to be investigated and more precise connections of
cells activated by AII will be necessary to establish a
"thirst" circuit in the brain.

Involvement of Neurotransmitters

In this chapter the case has been made for angiotensin acting
directly on neurons to produce an integrated response. In this
view, the peptide action is similar to other hormonal actions
in which a primary event initiates a sequence of actions that
produce a response. The peptide, as a primary stimulus effecting a neural output, is not necessarily a neuromodulator in
the way suggested for substance P and the enkephalins (Baker,
1976). There is, however, evidence from studies on the sympathetic nervous system that peripheral angiotensin can enhance
norepinephrine (NE) activity. Angiotensin has an excitatory
action on sympathetic ganglion cells (Hughes and Roth, 1971).
It enhances NE synthesis from tyrosine and inhibits NE reuptake
(Khairallah, 1977). Thus, angiotensin has properties that
could modulate neurotransmitter activity presynaptically.

NE appears to be involved in a response initiated by AII. Severs et al. (1970) showed that the pressor response, but not the drinking response, to angiotensin is mediated by sympathetic activation and ADH release. Bilateral adrenalectomy did not interfere with the pressor response, suggesting that circulating NE or epinephrine are not essential for the response to occur. Neither adrenalectomy nor peripheral sympathectomy by 6-hydraxydopamine (6-OHDA) disturbed the drinking response of rats to ICV AII (Falcon et al., 1978). We have also noted that ADH release is not necessary for the drinking response to the peptide (Johnson et al., 1977).

Within the brain, catecholamines are important in mediating the drinking response to AII. Haloperidol and spiroperidol, which are both dopamine (DA) antagonists, significantly inhibited angiotensin-induced drinking without significant effect on carbachol-induced drinking (Fitzsimons and Setler, 1975). However, we have not been able to elicit drinking by injections of DA or the DA agonist apomorphine, or by NE alone. Drinking was not blocked by alpha blockade using phentolamine or phenoxybenzamine. The pressor effect, however, could be blocked by alpha-adrenergic antagonists (Hoffman et al., 1977a; Camacho and Phillips, 1978). The problem with inhibition of drinking as a measure of a dopaminergic mechanism is that drinking is a complex behavior, and any small part of it may be interfered with indirectly. Fitzsimons and Setler used a good control of carbachol-induced drinking that was not affected by haloperidol. The effect might still be due to inhibition of movement because carbachol could act to correct the cholinergic-DA imbalance caused by haloperidol on the function of the basal ganglia.

Therefore the pressor and drinking responses to AII can be separated pharmacologically. This could mean that AII excites receptors that initiate both drinking and pressor responses, but the pressor responses are mediated at some point by a neural circuit that depends on alpha-adrenergic receptors, whereas the drinking circuit does not. Alternatively, there may be independent angiotensin receptors for both circuits. We found that, in spontaneously hypertensive rats, the pressor responses are more sensitive to AII, but the drinking responses are not (Hoffman et al., 1977b). This separation of sensitivity to AII would indicate different receptors.

In summary, the pressor response to ICV AII appears to be mediated by the sympathetic nervous system and ADH release. The sympathetic activation is via an alpha-adrenergic pathway, and ADH may be released directly. The drinking response is not adrenergically mediated. The best clue we have so far concerning drinking is that both angiotensin and cholinergic actions are involved (Hoffman et al., 1978). While neither saralasin nor atropine alone inhibit the amount of water consumed after many hours of water deprivation, there is inhibition when both the angiotensin antagonist and the cholinergic antagonist are given together (vide supra). This implies the existence of parallel pathways. In the course of evolution, several parallel mechanisms and stimuli may have been developed to maintain a function as vital to an organism as water intake and water balance; the peptide angiotensin is but one of the stimuli.

Prostaglandins and Angiotensin II

It is not clear how PGs interact with AII in the brain, but there is evidence that they do. PGE_1 injected directly into the brain was reported by Leksell (1976) to increase drinking in water-replete goats, whereas Kenney and Epstein (1978) reported inhibition of angiotensin-induced drinking in rats. It is unlikely that this discrepancy can be explained by a dose difference, since a wide dose range was used in the rat experiment and there was a dose-response curve showing inhibition of drinking. We found that a prostaglandin inhibitor prolonged the drinking response of rats to AII (Phillips and Hoffman, 1977).

The inhibitory effect of PGs on angiotensin-induced thirst appears to differentiate the drinking response from the pressor response to angiotensin, because PG-synthesis inhibitor did not alter the pressor response. Intraventricular injection of PGE_2 alone produced a slow rise in blood pressure (Hoffman and Schmid, 1979).

An aspect of angiotensin drinking to be explained is why drinking bouts produced by ICV AII rarely last more than a few minutes. If a rat is not allowed to drink after angiotensin injections but is tested 30 minutes later, it drinks very little. One possibility is that the dipsogenic property of the peptide is being counter-inhibited as part of a central homeostatic control. It has been shown recently that angiotensin promotes the synthesis of PGE from endothelial tissue *in vitro* (Gimbrone and Alexander, 1975). Kenney and Epstein have also reported that ICV PGE_1 in doses as low as 10 ng inhibits drinking in rats induced by 5 ng AII injection (ICV).

Meclofenamate (Parke-Davis) was injected ICV at a dose of 1.25 μg in 5 μl two minutes before 50 μg of angiotensin, through the same cannula. Our hypothesis was that if angiotensin increases PG synthesis and PGs inhibit the drinking effect, then meclofenamate, which inhibits the synthesis of PGs, should lead to a prolonged drinking bout of greater amplitude. As a control for the dipsogenic properties of angiotensin, we used carbachol. The results showed that meclofenamate increased

AII-induced drinking, but carbachol-induced drinking was almost totally inhibited. (Fig. 10-12). The angiotensin-induced drinking was increased in both amplitude and duration. The data suggest that the two dipsogenic agents act on different receptors to produce their effect and imply that, whereas PGs inhibit angiotensin-induced drinking, PGs may mediate carbachol-induced drinking.

Since PGE is a vasodilator and angiotensin is a potent vasoconstrictor, it has been proposed that angiotensin has its effect through local vasoconstriction of a highly sensitive central blood vasculature (Nicolaidis and Fitzsimons, 1975). This could explain the opposite effects with carbachol, which is similar in its action to acetylcholine, a vasodilator (Goodman and Gilman, 1970). Vasodilators, such as papaverine (ICV), however, do not block the action of angiotensin (Hoffman et al., 1977a). There is little doubt that angiotensin has an effect on central sympathetic structures in inducing a blood pressure increase in the periphery (Severs et al., 1966), but we do not have any data to support the possibility that the effect is due to local vasoconstrictive effects in the brain.

SUMMARY

In this chapter we have traced the background of the findings and controversies that have led to an understanding of the octapeptide AII in the brain. A model for the mechanism of peptide action in the brain using AII is presented. AII has powerful effects when given centrally; it induces drinking and other responses directed toward fluid replacement, such as cardiovascular changes for preservation of fluid volume, and ADH release to conserve water. Data are reviewed indicating that circulating AII alone cannot account for all the facts. Drinking to AII injection in the brain is much more sensitive and release of vasopressin much more reliable. The model rests on the presence of AII in the brain. Recent evidence from immunocytochemical studies show fibers and cells in limited areas that contain AII-like substance. Whether angiotensin is synthesized in the cell or extracellularly still remains to be resolved. Also open for investigation is how angiotensin is released from the secretory neurons that contain it.

Angiotensin release can be mimicked. Direct injection of AII into an area that contains angiotensin fibers results in responses at very low doses. Receptor sites for AII have been studied by several laboratories, and the OVLT contains binding sites. To show that binding leads to a biological response, it is necessary to demonstrate neural activity in the area of binding. Again, this has been shown for neurons in the OVLT.

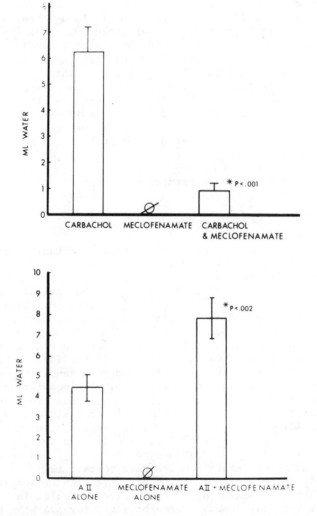

Fig. 10-12. The effect of prostaglandin-synthesis inhibition by
meclofenamate on AII- or carbachol-induced drinking. Meclofenamate alone had no effect on
drinking. (Data from Phillips and Hoffman, 1977.)

Microiontophoretic application of AII produces specific and sensitive responses in these neurons.

The model proposes that a peptide would trigger circuits controlling the responses, and that these effector circuits would be mediated by neurotransmitters but not by the peptide. A consequence of this view is that fewer peptide-containing cells would be found in comparison to other neurotransmitters, and the peptide cells and fibers would be localized to regions sensitive to direct injection of the peptide. Also, lesions in these regions should prevent the responses that the peptide can elicit.

In the case of the OVLT, it has been shown that lesions of the organ and surrounding periventricular tissue abolish the responses to angiotensin. Neurotoxins, such as 6-OHDA, block some of the responses to central angiotensin. Alpha-adrenergic blockers prevent the pressor response but not the drinking response to the octapeptide. PGs interact in an inhibitory fashion with central AII. The model presented here is merely an outline that indicates areas for further study. The information available provides support for the concept of brain angiotensin triggering responses related to water balance, but there are numerous gaps in our knowledge. Angiotensin may also play a role in functions not yet considered or even dreamed of.

NOTE: This chapter was written for publication in 1980.

REFERENCES

Abdelaal, A.E., Assaf, S.Y., Kucharczyk, V., and Mosenson, G.J.: Effect of ablation of the SFO on water intake elicited by systemically injected AII. *Can J Physiol Pharmacol 52*:362-363, 1974.

Abraham, S.F., Baker, R.M., Blaine, E.H., Denton, D.A., and McKinley, M.J.: Water drinking induced in sheep by angiotensin--A physiological or pharmacological effect? *J Comp Physiol Psychol 88*:515, 1975.

Andersson, B.: Thirst and brain control of water balance. *Am Sci 59*:408-415, 1971.

Andersson, B., Leksell, L.G., and Lishajko, F.: Perturbations in fluid balance induced by medially placed forebrain lesions. *Brain Res 99*:261-275, 1975.

Baker, J.L.: Peptides: Roles in neuronal excitability. *Physiol Rev 56*:435-452, 1976.

Bealer, S.L., Phillips, M.I., Johnson, A.K., and Schmid, P.G.: Effect of anteroventral third ventricle lesions on antidiuretic responses to central angiotensin II. *Am J Physiol 236*:E610-E615, 1979.

Bennett, J.P., and Snyder, S.H.: Angiotensin II Binding to mammalian brain membranes. *J Biol Chem 251*:7423-7430,

1976.

Booth, D.: Mechanism of action of norepinephrine in eliciting an eating response on injection into the rat
hypothalamus. *J Pharmacol Exp Ther 160*:336-348, 1968.

Brightman, M.W., and Reese, T.S.: Junctions between intimately opposed cell membranes in the vertebrate brain. *J
Cell Biol 40*:648-677, 1969.

Broadwell, R.D., and Brightman, M.W.: Entry of peroxidase
into neurons of the central and peripheral nervous systems
from extra-cerebral and cerebral blood. *J Comp Neurol
166*:257-284, 1976.

Brody, M.J., Fink, G.D., Bussy, J., Haywood, J.R., Gordon,
F.J., and Johnson, A.K.: The role of anteroventral third
ventricle (AV3V) region in experimental hypertension. *Circ
Res (Suppl 1) 43*:I-2-I-13, 1978.

Buggy, J., and Fisher, A.E.: Anteroventral third ventricle
site of action for angiotensin induced thirst. *Pharmacol
Biochem Behav 4*:651-660, 1976.

Buggy, J., Fisher, A.E., Hoffman, W.E., Johnson, A.K., and
Phillips, M.I.: Ventricular obstruction: Effect on drink-
ing induced by intracranial injection of angiotensin.
Science 190:72-74, 1975.

Buggy, J., Fink, G.D., Johnson, A.K., and Brody, M.J.:
Prevention of the development of renal hypertension by
anteroventral third ventricular tissue lesions. *Cir Res
(Suppl 1) 40*:I-100-I-117, 1977.

Bumpus, F.M.: Angiotensin antagonists: In *Central Actions
of Angiotensin and Related Hormones*, J. Buckley and C.
Ferrario, eds. Pergamon Press, New York, 1977, pp. 17-21.

Camacho, A., and Phillips, M.I.: The role of catecholamines
in central angiotensin II-induced responses. (Abst.),
Physiologist 21 (4):16, 1978.

Claybaugh, J.R., Share, L., Shimizu, K.: The inability of
infusions of angiotensin to elevate the plasma vasopressin
concentration in the anesthetized dog. *Endocrinology
90*:1647-1652, 1972.

Dellman, H-D., and Simpson, J.B.: Regional differences in
the morphology of the rat subfornical organ. *Brain Res
116*:389-400, 1976.

Epstein, A.N., Fitzsimons, J.T., and Rolls, B.J.: Drinking
caused by the intracranial injection of angiotensin into
the rat. *J Physiol (Lond) 200*:98P-100P, 1969.

Epstein, A.N., Fitzsimons, J.T., and Johnson, A.K.: Peptide
antagonists of the renin-angiotensin system and the elucidation of the receptors for angiotensin-induced drinking.
J Physiol (Lond) 238:34-35, 1973.

Falcon, J.C., Phillips, M.I., Hoffman, W.B., and Brody,
M.J.: Effects of intraventricular angiotensin II mediated

by the sympathetic nervous system. *Am J Physiol* *235 (4)*:H392–H399, 1978.

Felix, D., and Akert, K.: The effect of angiotensin II on neurones of the cat subfornical organ. *Brain Res 76*:350–353, 1974.

Felix, D., and Phillips, M.I.: Effects of angiotensin on central neurones: In *Iontophoresis and Transmitter Mechanisms in the Mammalian Central Nervous System*, E. Ryall and J. Kelly, eds. Elsevier/North-Holland Biomedical Press, Amsterdam, 1978, pp 104–106.

Felix, D., and Phillips, M.I.: Inhibitory effects of luteinizing hormone-releasing hormone (LH–RH) on neurons in the organum vasculosum lamina terminalis (OVLT). *Brain Res 169*:204–208, 1979.

Felix, D., and Schlegel, W.: Angiotensin receptive neurones in the subfornical organ: Structure-activity relations. *Brain Res 149*:107–116, 1978.

Fitzsimons, J.T.: Thirst. *Physiol Rev 52*:408–559, 1972.

Fitzsimons, J.T.: Angiotensin, thirst, and sodium appetite: Retrospect and prospect. *Fed Proc 37*:2669–2675, 1978.

Fitzsimons, J.T., and Setler, P.E.: The relative importance of central nervous catecholaminergic and cholinergic mechanisms in drinking in response to angiotensin and other thirst stimuli. *J Physiol (Lond) 250*:613–631, 1975.

Fuxe, K., Ganten, D., Hökfelt, T., and Bolme, P.: Immunohistochemical eveidence for the existence of angiotensin II-containing nerve terminals in the brain and spinal cord in the rat. *Neurosci Lett 2*:229–234, 1976.

Ganten, D., Fuxe, K., Phillips, M.I., Mann, J.F.E., and Ganten, U.: The brain iso-renin-angiotensin system: Biochemistry, localization, and possible role in drinking and blood pressure regulation. In *Frontiers in Neuroendocrinology*, W.F. Ganong and L. Martini, eds. Raven Press, New York, 1978, p 61–99.

Gimbrone, M.A., Jr., and Alexander, R.W.: Angiotensin II stimulation of prostaglandin production in cultured human vascular endothelium. *Science 189*:219–220, 1975.

Goodman, L.S., and Gilman, A.: *The Pharmacological Basis of Therapeutics*. MacMillan, New York, 1970.

Gress, C.M., and Malvin, R.L.: Localization of central sites of action of angiotensin II on ADH release *in vitro*. *Am J Physiol 234*:F135–F140, 1978.

Gronan, R.J., and York, D.H.: Effects of angiotensin II and acetylcholine on neurons in the preoptic area. *Brain Res 154*:172–177, 1978.

Grossman, S.P.: Eating and drinking elicited by direct adrenergic or cholinergic stimulation of hypothalamus. *Science 132*:301–302, 1960.

Hertz, A., Albus, K., Matus, J., Schubart, P., and
Toschenachi, H.J.: On the central sites for the antinociceptive action of morphine and fentanyl. *Neuropharmacology*
9:539-551, 1970.

Hoffman, W.E., and Phillips, M.I.: Evidence for SAR[1]-ALA[8]-
angiotensin II crossing the blood cerebral spinal fluid
barrier to antagonize central effects of angiotensin II.
Brain Res 109:541-552, 1976a.

Hoffman, W.E., and Phillips, M.I.: Regional study of cerebral ventricular sensitive sites to angiotensin II. *Brain
Res 110*:313-330, 1976b.

Hoffman, W.E., and Schmid, P.G.: Cardiovascular and antidiuretic effects of central prostaglandin E_2. *J Physiol
(Lond) 288*:159-169, 1979.

Hoffman, W.E., Phillips, M.I., Schmid, P.G., Falcon, J., and
Weet, J.F.: Antidiuretic hormone release and the pressor
response to central angiotensin II and cholinergic stimulation. *Neuropharmacology 16*:463-472, 1977a.

Hoffman, W.E., Schmid, P.G., and Phillips, M.I.: Evidence
for direct neuronal stimulation by intraventricular angiotensin II. *Brain Res 126*:376-381, 1977b.

Hoffman, W.E., Ganten, U., Phillips, M.I., Schmid, P.,
Schelling, P. and Ganten, D.: Inhibition of drinking in
water-deprived rats by combined central angiotensin II and
cholinergic blockade. *Am J Physiol 234 (1)*:F41-F47, 1978.

Hsiao, S., Epstein, A.N., and Camardo, J.S.: The dipsogenic
potency of intravenous angiotensin. *Horm Behav 8*:129-140,
1977.

Hughes, J., and Roth, R.H.: Evidence that angiotensin
enhances transmitter release during sympathetic nerve
stimulation. *Br J Pharmacol 41*:239-255, 1971.

Johansson, B., Li, C.L., Olsson, Y., and Klatzo, I.: The
effect of acute arterial hypertension on the blood-brain
barrier to protein tracers. *Acta Neuropathol 16*:117-124,
1970.

Johnson, A.K., and Epstein, A.N.: The cerebral ventricles as
the avenue for the dipsogenic action of intracranial angiotensin. *Brain Res 86*:399-418, 1975.

Johnson, A.K., and Schwob, J.E.: Cephalic angiotensin receptors mediating drinking to systemic angiotensin II.
Pharmacol Biochem Behav 3:1077-1081, 1975.

Johnson, A.K., Phillips, M.I., Mohring, J., and Ganten, D.:
Angiotensin-induced drinking in rats with hereditary hypothalamic diabetes insipidus. *Neurosci Lett 4*:327-330,
1977.

Keil, L.C., Summy-Long, J., and Severs, W.B.: Release of
vasopressin by angiotensin II. *Endocrinology 96*:1063-1064,
1975.

Kenny, N.J., and Epstein, A.N.: Antidipsogenic role of the
E-prostaglandins. *J Comp Physiol Psychol 92:*204-219, 1978.

Khairallah, P.A., Moore, A.F., and Gurchinoff, S.: Angioten-
sin receptor sites: In *Central Actions of Angiotensin and
Related Hormones*, J. Buckley and C. Ferrario, eds.
Pergamon Press, New York, 1977, pp 7-16.

Knowles, W.D., and Phillips, M.I.: Hypothalamic brain
slices: Angiotensin II-sensitive cells in OVLT. *Brain Res
(in press)*.

Leksell, L.G.: Influence of prostaglandin E1 on cerebral
mechanisms involved in the control of fluid balance. *Acta
Physiol Scand 96:*1-9, 1976.

Malvin, R.L., Mouw, D., and Vander, A.J.: Angiotensin:
Physiological role in water-deprivation-induced thirst of
rats. *Science 197:*171-173, 1977.

Maran, J.W., and Yates, F.E.: Cortisol secretion during
intrapituitary infusion of angiotensin II in conscious
dogs. *Am J Physiol 233:*E273-E285, 1977.

Miselis, R.R., Shapiro, R.E., and Hand, P.J.: Subfornical
organ efferents to neural systems for control of body
water. *Science 205:*1022-1025, 1979.

Mogenson, G.J., Kucharczyk, J., and Assaf, S.: Evidence for
multiple receptors and neural pathways which subserve
water intake initiated by angiotensin II. In *International
Symposium of the Central Actions of Angiotensin and
Related Hormones*, J.P. Buckley and C. Ferrario, eds.
Pergamon Press, New York, 1977, pp 493-502.

Mouw, D., Bonjour, J.P., Malvin, R.L., and Vander, A.:
Central action of angiotensin in stimulating ADH release.
Am J Physiol 220:239-242, 1971.

Nicolaidis, S., and Fitzsimons, J.T.: La Dependance de la
prise d'eau induite par l'angiotensine II envers la
fonction vasomotrice cerebrale locale chez le rat. *Acad
Sci (Paris) 281D:*1417-1420, 1975.

Nicoll, R.A., and Barker, J.L.: Excitation of supraoptic
neurosecretory cells by angiotensin II. *Nature
233:*172-173, 1971.

Olsson, K.: Effects on water diuresis of infusion of transmitter substances into the third ventricle. *Acta Physiol
Scand 79:*133-135, 1970.

Phillips, M.I.: Angiotensin in the brain. *Neuroendocrinology
25:*354-377, 1978.

Phillips, M.I., and Felix, D.: Specific angiotensin II
receptive neurons in the cat subfornical organ. *Brain Res
109:*531-540, 1976.

Phillips, M.I., and Hoffman, W.E.: Intraventricular organs
as angiotensin II receptor sites for thirst. Proc Int
Symp on Food and Water Intake, XI Internatl Union Physiol,

Jerusalem, 1974, p 161.

Phillips, M.I., and Hoffman, W.E.: Sensitive sites in the brain for the blood pressure and drinking responses to angiotensin II: In *International Symposium on the Central Actions of Angiotensin and Related Hormones*, J.P. Buckley and C. Ferrario, eds. Pergamon Press, New York, 1977, pp 325-356.

Phillips, M.I., Deshmukh, P., and Larsen, L.: Are the central effects of angiotensin due to peripheral angiotensin II crossing the blood brain barrier? *Soc Neurosci Abstr 3:*510, 1977a.

Phillips, M.I., Felix, D., Hoffman, W.E., and Ganten, D.: Angiotensin-sensitive sites in the brain ventricular system: In *Approaches to the Cell Biology of Neurons*, W.M. Cowan and J.A. Ferrendelli, eds. Society for Neuroscience Symp, Bethesda, Md., 1977, pp 308-339.

Phillips, M.I., Mann, J.F.E., Haebara, H., Hoffman, W.E., Dietz, R., Schelling, P., and Ganten, D.: Lowering of hypertension by central saralasin in the absence of plasma renin. *Nature 270:*445-447, 1977c.

Phillips, M.I., Deshmukh, P., and Larsen, L.: Morphological comparisons of the ventricular wall of subfornical organ and organum vasculosum of the lamina terminalis. *Scan Electron Microsc 2:*349-356, 1978a.

Phillips, M.I., Phipps, J., and Bealer, S.: Effect of knife cuts between subfornical organ and organum vasculosum laminae terminalis on the central actions of angiotensin II. *Soc Neurosci Abstr 8:*352, 1978b.

Phillips, M.I., Weyhenmeyer, J., Felix, D., Ganten, D., and Hoffman, W.E.: Evidence for an endogenous brain renin-angiotensin system. *Fed Proc 38:*2260-2266, 1979.

Quinlan, J.T., Phillips, M.I., and Weyhenmeyer, J.A.: Presence of angiotensin II (AII) in the primate brain. *Physiologist 22:*104, 1979.

Raizada, M.K., Yang, J.W., Phillips, M.I., and Fellows, R.E.: Rat brain cells in primary culture: Characterization of angiotensin II binding sites. *Brain Res*, 1980, (in press).

Ramsay, D.J., Reid, I.A., Keil, L.C., and Ganong, W.F.: Evidence that the effects of isoproterenol on water intake and vasopressin secretion are mediated by angiotensin. *Endocrinology 103:*54-59, 1978.

Rayyes, S.S., and Horton, R.: Effect of angiotensin II on adrenal and pituitary function in man. *J Clin Endocrinol 82:*539, 1971.

Rolls, B.J., and Wood, J.R.: Role of angiotensin in thrist. *Pharmacol Biochem Behav 6:*245-250, 1977.

Sakai, K.K., Marks, B.H., George, J., and Koestner, A.:

Specific angiotensin II receptors in organ-cultured canine supraoptic nucleus cells. *Life Sci 14*:1337-1344, 1974.

Schelling, P., Ganten, D., Speck, G., and Fischer, H.: Effects of angiotensin II and angiotensin II antagonist saralasin on cell growth and renin in 3T3 and SV3T3 cells. *J Cell Physiol 98*:503-514, 1979.

Severs, W.B., and Summy-Long, J.Y.: The role of angiotensin in thirst. *Life Sci 17*:1513-1526, 1975.

Severs, W.B., Daniels, A.E., Smookler, H.H., Kinard, W.J., and Buckley, J.P.: Interrelationship between angiotensin II and the sympathetic nervous system. *J Pharmacol Exp Ther 153*:530-537, 1966.

Severs, W.B., Summy-Long, J., Taylor, J.S., and Connor, J.D.: A central effect of angiotensin: Release of pituitary pressor material. *J Pharmac Exp Ther 174*:27-34, 1970.

Severs, W.B., Changaris, D.G., Keil, L.C., Summy-Long, J.Y. Klase, P.A., and Kapsha, J.M.: Pharmacology of angiotensin-induced drinking behavior. *Fed Proc 37*:2699-2703, 1978.

Simpson, B., and Routtenberg, A.: Subfornical organ: Site of drinking elicitation by angiotensin II. *Science 181*:1172-1175, 1973.

Simpson, J.B., Mangiapane, M.L., and Dellman, H-D.: Central receptor sites for angiotensin-induced drinking: A critical review. *Fed Proc 37*:2676-2682, 1978.

Sirett, N.A., McLean, A.S., Bray, J.J., and Hubbard, J.I.: Distribution of angiotensin II receptors in rat brain. *Brain Res 122*:199-312, 1977.

Sladek, C.D., and Joynt, R.J.: Angiotensin stimulation of vasopressin release from the rat hypothalamo-neurohypophyseal system in organ culture. *Endocrinology 104 (1)*:148-153, 1979.

Stamler, J.F., Raizada, M.K., Phillips, M.I., and Fellows, R.E.: Increased specific binding of angiotensin II in the organum vasculosum of the lamina terminalis area of the spontaneously hypertensive rat. *Neurosci Lett 17*:173-177, 1980.

Stricker, E.M.: The renin-angiotensin system and thirst: A reevaluation. II. Drinking elicited in rats by caval ligation or isoproterenol. *J Comp Physiol Psychol 91*:1220-1231, 1977.

Stricker, E.M.: The renin-angiotensin system and thirst: Some unanswered questions. *Fed Proc 37*:2704-2710, 1978.

Swanson, L.W., Kucharczyk, J., and Mogenson, G.J.: Autoradiographic evidence for pathways from the medial preoptic area to the midbrain involved in the drinking response to angiotensin II. *J Comp Neurol 178*:645-660, 1978.

Tang, M., and Falk, J.: SAR[1]-ALA[8]-Angiotensin II blocks renin-angiotensin but not beta-adrenergic dipsogenesis. *Pharmacol Biochem Behav* 2:401-408, 1974.

Volicer, L., and Lowe, C.G.: Penetration of angiotensin II into the brain. *Neuropharmacology* 10:631-636, 1971.

Weindl, A., and Joynt, R.J.: The median eminence as a circumventricular organ: In *Median Eminence, Structure and Function*, K.M. Knigge, D.E. Scott, and A. Weindl, eds. S. Karger, Basel, 1972, pp 280-297.

Weindl, A.: Neuroendocrine aspects of circumventricular organs: In *Frontiers in Neuroendocrinology*, W.F. Ganong adn L. Martini, eds. Oxford University Press, London, 1973, pp 3-32.

Westergaard, E.: The fine structure of nerve fibers and endings in the lateral cerebral ventricles of the rat. *J Comp Neurol* 144:345-354, 1975.

Weyhenmeyer, J.A., and Phillips, M.I.: Immunocytochemical localization of angiotensin II in the CNS of WKY and SH rats. *Soc Neurosci Abst* 5:544, 1979.

Weyhenmeyer, J.A., Raizada, M.K., Phillips, M.I., and Fellows, R.E.: Presence of angiotensin II in neurons cultured from fetal rat brain. *Neurosci Lett* 16:41-46, 1980.

11

Gut Hormones and Feeding Behavior: Intuitions and Experiments

GERARD P. SMITH

INTRODUCTION

The idea that gut hormones affect feeding behavior can be traced back 50 years to clinical reports that insulin promoted weight gain in malnourished patients by increasing appetite and food intake. In the last decade, isolation techniques, made more powerful by immunological procedures, have produced a farrago of gut peptides that clamor for investigation of their effects on feeding behavior. At this time, however, only a small number of them have been analyzed for such effects. Shelley referred to this kind of problem in another context in the last century when he said, "We have eaten more than we can digest." There are two ways to deal with it: One is to give the reader a reflection of all the phenomena in this growing field; the other is to illuminate the problems inherent in the analysis of the effect on feeding behavior of any gut hormone by a critical review of the effect of two that have been studied extensively. The choice is between the mirror and the lamp. I choose the lamp.

ANALYSIS OF FEEDING BEHAVIOR

Phenomenology

Feeding in mammals is a periodic behavior. Even when food is present constantly, bouts of feeding alternate with periods of nonfeeding. A bout of feeding is a meal and the period of

nonfeeding that begins when the meal ends and that ends when
the next meal begins is called the intermeal interval (IMI).
The size and duration of meals and the duration of IMIs are
important measures of feeding behavior. The duration of IMIs
determines meal frequency. Note that the phenomenology of food
intake is exhausted by a description of the 24-hour pattern of
meal size and meal frequency. Thus, the effect of gut hormones
on feeding can be completely described by changes in these two
measures. The effect of gut hormones on these two measures
also determines the effect of gut hormones on body weight, but
this relationship is complicated by other processes, such as
activity and the metabolic utilization of food. A circle
closes here because in some situations changes in activity and
metabolism can change meal size or meal frequency. This opens
up the possibility of indirect effects of gut hormones on
feeding behavior that are mediated by effects of gut hormones
on activity or, more frequently, on metabolism. This issue of
direct and indirect effects is important in the analysis of the
behavioral effects of gut hormones and we shall consider it
again later in this chapter.

Psychobiological Processes

If an accurate description of the behavioral effects of gut
hormones is necessary, it is certainly not sufficient. We want
more. We want to know how the mammal generates its feeding
behavior and how gut hormones alter the psychobiological
processes. In this sense, a gut hormone is a molecular probe
of the mechanisms for fundamental relationships among brain,
mind, and feeding behavior. What would be a satisfactory explanation of feeding behavior? The question needs asking. I
believe that a satisfactory explanation must account for the
onset, duration, and termination of meals, for the length of
IMIs and for the human experiences that correlate with these
behavioral events. It has been traditional to assume that the
explanation for these behavioral changes will be found in the
interaction between the two psychobiological processes of
hunger* and satiety.

---

*The literature frequently attempts to distinguish appetite
from hunger. Usually the distinction is puritanical: hunger
operates when you need food; appetite operates when you want it
but do not need it. Carlson, in his influential monograph of
1916, distinguished appetite from hunger on the basis of
memory: appetite depended much more on memory than hunger did.
Since I am not convinced that we know how to make this
distinction reliably, I do not use it here.

Hunger accounts for the onset and duration of a meal; satiety accounts for the termination of a meal. Both processes determine the IMI: Satiety dominates the early phase and hunger emerges toward the end, finally becoming strong enough to generate the onset of another meal. Hunger is correlated with mild distress until food is found, then it is correlated with the pleasures of feeding. Satiety is correlated with satisfaction, fullness, and tranquilization. Such is the tradition. Unfortunately, we do not have unambiguous measures of hunger or satiety. It is likely that the two processes often operate simultaneously. Thus, it is usually not clear if an increase in meal size is the result of increased hunger, decreased satiety, or both. We (Kraly et al., 1980) recently found a situation in which increased feeding could be ascribed to decreased satiety without a change in hunger, but this kind of analysis is just beginning.

The fact that we will be evaluating the effect of gut hormones in an explanatory framework that is vague and tentative is the reason why explanation in this field is so much more elusive than description. After all, at the descriptive level, a gut hormone can only increase or decrease meal size or meal frequency. But to relate such changes to effects on hunger and satiety is frequently impossible. In fact, we shall see that a gut peptide released by ingested food represents the most promising marker for the initial phase of satiety (see later).

With these ideas in mind, I shall discuss pancreatic insulin as a hormone that increases hunger and intestinal cholecystokinin (CCK) as a hormone that increases satiety. These are paradigmatic cases because the behavioral functions of both hormones have been investigated extensively.

INSULIN AND HUNGER

Early Clinical Studies

In 1923, Pitfield, a practicing pediatrician, reported that insulin improved nutrition in malnourished infants (Pitfield, 1923). Like many other physicians, Pitfield had been impressed by the improvement produced by insulin in the nutrition of diabetic patients. Faced with the frequent and sad clinical problem of malnutrition in infants, Pitfield made a leap of clinical intuition: "Impressed with the stimulating properties of this active principle of the islands of Langherhans, I tried injecting it for a period of time into two young infants under my care, both suffering from malnutrition, with the idea that it would afford a means of combining what blood sugar they had

with the tissues in order to improve their condition (p. 217)."
Both children showed significant improvement.

Pitfield's report was quickly confirmed in the next year in
two larger series of malnourished children by Marriott (1924)
and Barbour (1924). Barbour's report contains the first
explicit reference to increased food intake. In one case,
Barbour wrote that shortly after insulin treatment began (3 to
5 U given IV., 20 minutes prior to a meal, three times a day),
the child complained of hunger until given a daily diet of 3000
calories. Barbour described this child as having a "ravenous
appetite."

Tisdall et al. (1925) could not confirm this initial success,
and Fischer and Rogarty (1926) suggested that insulin treatment
only worked well in those malnourished infants who were
apathetic and indifferent to food.

The phenomenon was extended to undernourished, psychotic
adults by Appel et al. (1928). They noted that appetite and
the amount of food intake were usually increased and that there
was usually a marked increase in weight. These changes appeared to be independent of changes in psychopathology because
the weight gain and improved appetite were usually not
accompanied by psychological improvement.

In the next year, Short reported that in seven adult patients, malnourished as the result of medical diseases, insulin
treatment increased appetite in all and increased body weight
in five (Short, 1929). Short was prescient. He identified two
of the issues that dominated the subsequent study of the effect
of insulin on appetite, food intake, and body weight. The
first is that there is a delay between insulin administration
and increased appetite and, by inference, increased food intake
("Thirty minutes should elapse after insulin administration
before food is taken if the optimum development of appetite is
desired, p. 335").

The second issue Short identified is that the weight gain
could be the result of synergism between two effects of insulin: "Fat is produced in malnutrition under the influence of
insulin. This is due in part, no doubt, to increased food
intake; perhaps also in part to a direct influence of the
insulin in general metabolism (p. 335)."

Over the next couple of years, several reports appeared
confirming the usefulness of insulin treatment of malnutrition.
But in 1935, Freyburg made the first critical study of insulin's effect independent of other aspects of the treatment
program, such as encouragement and expectations of improvement.
The results were interesting: Only 1 of 9 patients who received
insulin without knowing it or knowing what to expect showed a
significant increase in appetite and body weight (Freyburg,
1935). He concluded: "The fact that improvement resulted in

many patients from the suggestion accompanying injections, indicates that much of the benefit commonly attributed to insulin is due to a psychic factor (p. 41)."

Freyburg's conclusion is correct if the behavioral effect of insulin is a strong unconditional response that can be elicited under all conditions. We shall see that this assumption pervades the entire literature concerning the effect of insulin on hunger. In fact it is a common assumption in behavioral neuroendocrinology. I believe that the assumption is too categorical to be likely to have biological validity.

Freyburg drew two other conclusions from his study. The first was that the weight gain that occurred during insulin treatment was not correlated with increased utilization of food, and therefore must reflect increased intake of food. In his view the behavioral effect of insulin was primary.

The second conclusion was concerned with the relationship between the effect of insulin on hunger and the symptoms of hypoglycemia, which are known to be a direct effect of insulin. Freyburg noted that hypoglycemic reactions were frequently observed without an accompanying sensation of hunger. He interpreted this dissociation to mean that the hunger effect does not uniformly occur or that it occurs so briefly that it does not produce a significant increase in food intake.

Early Experimental Studies

The early experimental studies on the effect of insulin took place within the theoretical context of the gastric origin of hunger. This theory had a long history of plausible spec- ulation in the writings of 18th and 19th century physiologists (Sternberg referred to hunger as pruritus stomachi), but it first received experimental support in 1912 in the classic paper of Cannon and Washburn. They reported a temporal correlation between Washburn's report of periodic hunger sensations and contractions recorded by a balloon in his stomach (Cannon and Washburn, 1912). The phenomenon was observed by Carlson and his colleagues in more than 50 men. Carlson, a man known for his devotion to "facts," thought he had found one in this correlation: "When the empty stomach shows strong contractions, the subject invariably signals that he feels hunger, and, on being questioned, he invariably replies that he feels the hunger in his stomach. There is, on the whole, a fairly close correspondence between the duration of stomach contractions and duration of the subjective sensations of hunger (Carlson, 1916)."

Quigley et al. (1929) apparently linked the gastric mechanism of hunger to the hunger produced by insulin in a direct experiment. They studied the balloon-recorded motility of the stomach in three men and one woman after 11 to 44 hours of food deprivation. About one hour after the intravenous injection of 12 to 20 U of insulin, they observed an increase in gastric motility and an increase in gastric sensations. If they blocked the increase in gastric motility by atropine or epinephrine, or by glucose administration into the stomach, the hunger sensations decreased or were abolished. Quigley et al. suggested that the stimulation of gastric motility by insulin and the resultant hunger was a "rationale for the use of insulin for the relief of anorexia (p. 98)."

Thus, before the neutron was discovered in 1932, this behavioral effect of insulin was known and a mechanism for it had been proposed. Subsequent work has been concerned with three aspects of this behavioral effect: (1) psychological and behavioral measures, (2) physiological mechanism, and (3) biological meaning.

Psychological and Behavioral Measures

The behavioral measure most commonly used to demonstrate the stimulation of hunger by insulin is the quantity of food ingested in a test period that lasts one to six hours. It is assumed by those who use it that increased intake is a direct measure of increased hunger. In 1937, Maclagan produced such evidence for the first time. He observed that rabbits ate 10 percent more than the controls in three hours after 10 U of insulin* (Maclagan, 1937). Furthermore, Maclagan measured blood sugar during the tests and demonstrated that this dose of insulin produced hypoglycemia. MacKay et al. (1940) reported that a longer-acting preparation of insulin called protamine zinc insulin produced larger and more reliable increases in food intake. The effect was so robust that repeated administration led to obesity (MacKay et al., 1940; Barnes and Keeton, 1940).

Almost 30 years later, Booth and co-workers made the most detailed study of the phenomenon in rats (Booth and Brookover, 1968; Lovett and Booth, 1970). Insulin increased food intake in rats after 0, 4, or 24 hours of food deprivation. The optimal dose was 16 to 20 U/kg and it produced marked hypoglycemia. The stimulation of hunger by insulin added to that produced by food deprivation. Rats treated with insulin not only ate more, they also performed operant tasks, such as lever pressing and panel pushing, to get food.

---

*Insulin means regular insulin throughout this chapter.

The effect of insulin on hunger, however, could not be obtained when insulin was given the next day. At least four days had to elapse before the effect on hunger was normal.

Insulin also stimulated thirst. In fact, rats drank before they ate. In addition to the effects on feeding and drinking, a low dose of insulin administered with a meal decreased subsequent intake and insulin administered after ingestion of flavored water for 18 days produced a conditioned aversion to the flavored water. Furthermore, Brandes (1977) reported that some neurological impairment was always present when insulin increased feeding.

The satiating, debilitating, and aversive effects, and the stimulation of thirst should be kept in mind. They are characteristics of the response to insulin hypoglycemia that are not observed with food deprivation, the most natural operation that stimulates hunger.

Silverstone and Besser (1971) provided the evidence that human beings feel hungry during insulin hypoglycemia. They administered insulin intravenously to 15 men and 15 women after an overnight fast. At 15-minute intervals throughout the 90-minute test, subjects reported on their state of hunger by marking a visual analogue scale for hunger. Hunger increased after about 30 minutes and remained elevated through 90 minutes (Fig. 11-1). Note that hunger increased most rapidly in the 30 minutes after the nadir of the blood sugar concentration. This confirmed Short's earlier study.

Mechanism Of the Behavioral Effect

Decline of the Gastric Mechanism

We have just seen how subsequent work confirmed the early claim that there was a behavioral effect of insulin. Now we shall see that subsequent work did not support the gastric mechanism of the behavioral effect proposed by Quigley et al. (1929). The gastric mechanism was attacked on general and specific grounds by Christensen in 1931 (Christensen, 1931). The general ground was the failure to find in seven subjects after 10 to 12 hours of food deprivation "any relation between the presence or the strength of the contractions and the presence or the intensity of the hunger sensation."

The specific ground was the failure to observe a correlation between the stimulation of hunger sensation and the activation of gastric contractions by insulin in four patients with anorexia nervosa. Negative results are rarely welcome and this may explain why Christensen's study had so little immediate impact on the field and, this, in turn, would explain why his

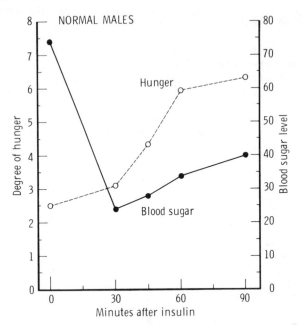

Fig. 11-1. Change in concentration of blood sugar (mg/100 ml) and in visual analogue scale hunger ratings in 15 normal men during the 90 minutes after an IV injection of regular insulin (0.15 U/kg). Very similar results were obtained in normal women. (From Silverstone and Besser, 1971, with the permission of the publisher.)

results are still so rarely quoted. Fortunately, this neglect does not change the fact that his results were very close to the truth as we know it.

In 1957, James reviewed the work on the correlation between gastric contractions and hunger and concluded that there was none (James, 1957)! And in 1971, Stunkard, who had been a proponent of the gastric mechanism, conceded that the gastric factor in hunger was weak and unstable (Stunkard and Fox, 1971).

The failure to confirm the gastric mechanism of hunger did

not exclude the possibility that a gastric mechanism mediated the stimulation of hunger by insulin, at least in subjects without anorexia nervosa (see before). But when it was discovered that vagotomy blocked the stimulation of gastric motility by insulin, it became possible to determine if abdominal vagotomy also abolished insulin's effect on hunger. When this crucial experiment was carried out in rats (Morgan and Morgan, 1940), dogs (Grossman, et al., 1947), and in human beings, vagotomized for peptic ulcer disease (Grossman and Stein, 1948), the results were the same: Vagotomy failed to block the hunger effect of insulin.

If the hunger effect was not due to the gastric motor effect of insulin, investigators reasoned that it might be related to the prominent hypoglycemic effect that was known to be necessary for the hunger effect and the gastric effect.

Rise of the Glucoprivic Mechanism

How would hypoglycemia stimulate hunger? Mayer suggested that hypoglycemia stimulated hunger by decreasing the utilization of glucose by neurons in the brain, especially neurons in the ventromedial area of the hypothalamus (Mayer and Bates, 1952; Mayer, 1955). This is a glucoprivic mechanism because hypoglycemia deprives these central cells of the glucose necessary for normal glucose utilization.

Mayer has been confirmed to the extent that glucoprivation, and not hypoglycemia, has been shown to be the critical stimulus (Smith and Epstein, 1969) and the brain is one of the sites of action (Miselis and Epstein, 1975). The liver is the other (Novin et al., 1973; Stricker et al., 1977).

The mechanisms that are responsible for sensing the intracellular glucoprivation produced by insulin hypoglycemia and transforming this metabolic state into feeding behavior are not known. Hepatic and central glucoreceptors (Niijima, 1969; Oomura et al., 1969) are frequently invoked as mediating mechanisms, but the evidence for this is not convincing (Epstein et al., 1975). It is clear, however, that activation of central noradrenergic synapses are necessary for glucoprivic feeding (Müller et al., 1972; Ritter et al., 1978).

Biological Meaning

At this point I hope the reader is convinced that insulin stimulates the sensation of hunger and increases food intake and that these effects are initiated by glucoprivation of the brain and the liver. What is the biological meaning of this interesting and reliable effect of insulin?

Mayer (1955) argued that the hunger elicited by insulin provided an insight into the normal controls of food intake over
the short-term (minutes to hours). Mayer conceived of glucose
utilization in certain sensitive cells in the brain as a
critical variable for the short-term control of food intake.
He called this the glucostatic hypothesis that related glucose
utilization and food intake inversely: When glucose utilization
decreased, food intake increased; when glucose utilization increased, food intake decreased. This hypothesis has recently
come under withering criticism from Smith (1973) and Epstein et
al. (1975).

Although new evidence may save the glucostatic hypothesis,
the current consensus is that the hunger effect of insulin is a
psychological and behavioral response to an unusual metabolic
state. Grossman (1955) reached this conclusion 25 years ago
and I have not found any recent data that would reverse his
judgment (Smith, 1973). In this view, the hunger and increased
food intake produced by insulin is an example of Richter's
famous insight that behavior can serve metabolic homeostasis.
We (Smith et al., 1972) demonstrated that the feeding initiated
by glucoprivation ameliorates the metabolic challenge of
glucoprivation in two ways. First, ingested food provides an
increased supply of glucose and other nutrients. Second,
feeding restores the normal distribution of glucose to all
tissues by stimulating insulin secretion, which is inhibited
during glucoprivation by epinephrine released from the adrenal
medulla.

SUMMARY

Insulin was the first peptide hormone of the gut to be shown to
have a significant behavioral effect. The behavioral effect is
shown beyond doubt--insulin elicits the sensation of hunger and
an increase in food intake. The mechanism is glucoprivation of
brain and liver tissue. Glucoprivation is a sufficient condition for increased hunger and feeding, but it is not a necessary condition. The feeding response to insulin hypoglycemia
is a capacity of mammals, but this capacity does not appear to
participate in the control of feeding that operates under usual
laboratory conditions or in the lives of human beings when
excess food is constantly available.

The conclusion is supported by three differences between the
hunger produced by insulin hypoglycemia and that produced by
food deprivation (Booth and Brookover, 1968; Lovett and Booth,
1970). First, insulin hypoglycemia stimulates thirst; second,
to be effective, more than one day must elapse between episodes

of insulin hypoglycemia; and third, rats learn to avoid the flavor of water they drink before an episode of insulin hypoglycemia. Since hunger elicited by food deprivation has none of these characteristics, it appears to be clearly different from the hunger stimulated by insulin hypoglycemia. This is a good example of the power of behavioral analysis to distinguish differences in psychological states that elude subjective reports. It also demonstrates how behavioral analysis defines the criteria that a gut hormone must satisfy to be considered a mechanism in the control of feeding behavior.

CHOLECYSTOKININ AND SATIETY

The end of a meal is accompanied by a feeling of fullness in the abdomen. The mechanism of this satisfying and, sometimes, uncomfortable experience has been assumed to be nerve signals arising from a distended stomach. When the gastric mechanism of hunger was widely accepted, this gastric mechanism of satiety was its counterpart. The scheme was a mechanical system for the control of feeding: When the stomach was empty, we were hungry; when the stomach was full, we were satiated. Previously in reviewing insulin and hunger, we noted the demise of the gastric mechanism of hunger. The gastric mechanism of satiety has fared better. It has survived, but it has been so transformed that its early supporters would not recognize it.

Although it is intuitive to equate stomach distention with the quantity of ingested food, the discovery that the entry of different foods into the intestine could slow gastric emptying differentially introduced another factor into the control of gastric fullness (Hunt and Knox, 1978). This appeared to be an explanation for the fact that foods differ in their ability to make us feel full.

The mechanism for the slowing of gastric emptying by food in the small intestine was demonstrated to be humoral by Farrell and Ivy (1926). Kosaka and Lim (1930) called the active factor enterogastrone to designate its origin from the intestine and its inhibitory action on the stomach.

Maclagan, in the same paper in which he obtained the first behavioral evidence for the effect of insulin on hunger, tested the possibility that an impure extract of enterogastrone would decrease food intake in addition to inhibiting the stomach (Maclagan, 1937). He reported that subcutaneous enterogastrone significantly inhibited six hour food intake in rabbits. In discussing this result, he emphasized that the effect was small

and that the enterogastrone had to be administered parenterally.

Janowitz and Grossman (1951) tested the possibility that
endogenous enterogastrone would inhibit food intake. They
stimulated the release of endogenous enterogastrone by
delivering cream to the intestine. They found that cream in
the small intestine significantly inhibited food intake. They
noted that it took a larger volume of cream to inhibit food
intake then to inhibit gastric motility (the visceral effect
that defined enterogastrone activity). They concluded that the
inhibition of food intake by enterogastrone was not a
physiological effect of the hormone. By physiological effect,
they meant that the amount of stimulation required to release
enough enterogastrone to inhibit feeding was unlikely to occur
after the usual meal in the dog. Note that their conclusion
assumes that the potency of enterogastrone for visceral effects
and for behavioral effects must be equivalent and that
important synergistic interactions do not exist for the
behavioral effect. I believe that these assumptions are too
restrictive, and I discuss this issue later.

In 1955, Quigley considered the experimental evidence that
the stomach was inhibited by a humoral intestinal factor
(enterogastrone) to be compelling because the inhibition could
be observed after removal of the vagal and sympathetic innervation of the stomach (Quigley, 1955). Despite the interpretation of Janowitz and Grossman (1951), Quigley suggested
that humoral inhibition of gastric motility was a critical
mechanism for the termination of feeding. This is a clear
hypothesis of a humoral control of a gastric mechanism of
satiety.

The problem lay fallow until the 1960s when Ugolev (1960) and
Glick and Mayer (1968) demonstrated that crude intestinal
extracts inhibited food intake in a dose-related manner. Glick
et al., (1971) failed to replicate the effect of the crude
extract with more pure preparations of CCK and secretin, two
gut hormones known to have enterogastrone activity. Schally's
laboratory, however, reported significant inhibition of food
intake in mice with a preparation of enterogastrone that did
not contain amounts of CCK or secretin that could be detected
by bioassay (Schally et al., 1967; Lucien et al., 1969). It
was against this background of fragmentary results and widespread scepticism that Gibbs and I began to investigate the
satiety effect of gut hormones. Of the five hormones we
tested, CCK, a hormone of the upper small intestine named for
its ability to contract the gall bladder (Ivy and Oldberg,
1928), proved to have the most interesting and potent satiety
effect.

Inhibition of Food Intake in the Rat

A 10 percent pure extract of porcine CCK inhibited food intake
in the rat deprived of food overnight (Gibbs et al., 1973a).
The magnitude of inhibition was a direct function of the dose
of CCK. CCK was administered intraperitoneally so that it was
absorbed primarily into the portal circulation. This mimicked
the vascular route that intestinal CCK follows when it is
released by food stimuli playing upon the mucosal surface of
the duodenum and upper jejunum.

The inhibitory effect was usually over by the end of the
first 30 minutes of the test. Since injections of CCK every
other day for months did not change the initial reaction of the
rats to the test diet, we concluded that CCK did not inhibit
feeding by making food aversive through unconditioned or
conditioned processes. (The fact that under certain conditions
[Deutsch and Hardy, 1977] CCK can elicit conditioned aversions
to food is interesting, but irrelevant [Gibbs and Smith, 1977;
Smith and Gibbs, 1979]). Instead of producing aversion, CCK
appeared to accelerate the satiating process so that the meal
terminated after less food was eaten.

CCK failed to inhibit water intake at a dose that produced a
marked inhibition of liquid food intake. Since the food and
water were both obtained by licking metal spouts, this is
strong evidence that the CCK effect is primarily upon the
central feeding system. This differential effect on feeding
and drinking also diminishes the possibility that CCK produces
a generalized inhibition of behavior by acting as a depressant
or by making rats sick.

Since CCK accounted for only 10 percent of the extract (w/w),
it was possible that a non-CCK factor in the extract was
inhibiting food intake. This possibility was eliminated when
the synthetic C-terminal octapeptide of CCK (CCK-8), a portion
of the molecule that is known to have all the biological activities of the 33 or 39 amino acid molecular forms of CCK
(Ondetti et al., 1970), inhibited food intake in a dose-related
manner. This result demonstrated decisively that the satiety
effect of CCK was a new biological action of the hormone.

This satiety effect had been obtained by administering CCK to
rats under test conditions in which all the natural, endogenous, short-term satiety mechanisms (none of which has been
identified unambiguously, see later) were capable of interacting with the administered CCK. This made it difficult to
evaluate the relative potency of CCK for satiety. To do this,
one must administer CCK to an animal that does not stop feeding
because in that animal the endogenous satiety mechanisms are
either not operating or are activated so weakly that they are

subthreshold for stopping feeding.

Such an assay system became available when we discovered that rats do not stop feeding when they sham feed after a 17-hour deprivation (Young et al., 1974). With this preparation, we showed that impure CCK and CCK-8 inhibited sham feeding (Gibbs et al., 1973b) (Fig. 11-2). This demonstrated that CCK could inhibit feeding when ingested food did not activate other satiety mechanisms by distending the stomach or entering the small intestine.

Figure 11-2 makes another point. Note that desulfated CCK-8 had no significant effect on sham feeding. This demonstrates that the sulfate group in position 7 of the octapeptide is necessary for the satiety effect. This is consistent with earlier work that established that the same sulfate group was necessary for all the known biological actions of CCK (Ondetti et al., 1970).

All of these results were obtained by intraperitoneal in-jection. It was possible that intraperitoneal administration permitted CCK-8 to have abnormal access to abdominal tissue or to flood the liver. These possible artifacts were eliminated by the clear inhibitory effect produced by slow intravenous infusions of CCK-8 (Lorenz et al., 1979).

Inhibition of Food Intake in Other Animals and Man

The satiety effect of impure CCK or CCK-8 has also been observed in lean and genetically obese mice (Koopmans et al., 1972; Strohmayer et al., 1976), rabbits (Houpt et al., 1978), sheep (Grovum, 1977; Della-Fera and Baile, 1979) rhesus monkeys (Gibbs et al., 1976; Falasco et al., 1979), and man (Sturdevant and Goetz, 1976; Kissileff et al., 1979). The inhibition of food intake in man reported by Kissileff et al. (1979) was particularly impressive because it was achieved with a dose within the physiological range, occurred without significant toxic symptoms or signs, and was produced by an acceleration of the satiating process without altering the initial rate of eating.

Inhibition of Food Intake and Satiety

I have been using satiety and inhibition of feeding as equivalent terms. But satiety elicited by ingested food is only one of many possible conditions in which animals and human beings stop feeding. Fright, distress, bad taste, distracting stimuli, and anorective drugs, such as amphetamine, are other conditions for a quicker termination of a meal. This poses an

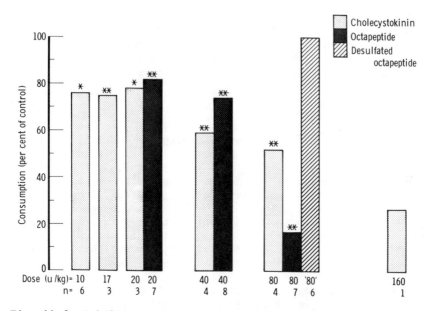

Fig. 11-2. Inhibition of feeding by cholecystokinin (CCK) and
the synthetic C-terminal octapeptide of CCK. After
17 hours of food deprivation, rats were offered
liquid diet. Rats sham fed this liquid diet because
ingested diet drained immediately out of the stomach
through an open gastric cannula. Since the gastric
cannula was closed between tests, rats could eat
normally outside of the test situation and they
remained healthy and maintained their body weight
for months (see Young, et al., 1974 and Antin, et
al., 1977 for details). Sham intake of liquid diet
was measured during the 30 minutes after IP
injections of CCK (stippled bars), synthetic
C-terminal octapeptide of CCK (hatched bar). Sham
intake is expressed as a percentage of the volume
eaten after IP injection of 0.15 M sodium chloride.
Mean control sham intakes for all tests ranged from
50.0 to 73.3 ml. The dose of desulfated octapeptide
was 3.6 μg/kg, a weight equivalant to about 80 Ivy
dog units/kg of the sulfated octapeptide. N is the
number of rats tested. Statistically significant
differences from control, **$\underline{p}$ <.02; *$\underline{p}$ <.05. (From
Gibbs et al., 1973b, with the permission of the
publisher.)

important problem for the interpretation of the inhibitory effect of CCK. Does CCK make animals and human beings eat less by distracting or distressing them, or does it mimic some effect of ingested food that forms part of the endogenous satiating process? The answer to this fundamental question has taken three forms--behavioral, electrophysiological, and subjective.

Behavioral Evidence

The behavioral evidence was obtained first in the rat. We noted that when rats satiated normally, they displayed a specific behavioral sequence (Antin et al., 1975). The sequence began as feeding terminated. Then the rat engaged in nonfeeding activities, such as grooming, sniffing, locomotion, and rearing for about five minutes. The sequence ended when the rat withdrew from the food, went to the rear of the cage, and rested or slept. We decided to use this behavioral index of satiety as a criterion to distinguish between inhibition of feeding that led to satiety and inhibition of feeding that did not.

We pursued the question in the 17-hour food deprived, sham-feeding rat. The sham-feeding rat does not display the satiety sequence because it does not stop feeding. Therefore there are no false positive responses. When CCK was administered to the sham-feeding rat, the rat stopped feeding and displayed the satiety sequence. CCK not only elicited the form of the sequence, it also elicited a normal incidence of the individual behavioral items that occur in the sequence (Antin et al., 1975), (Table 11-1).

The validity of the sequence to index satiety was demonstrated by giving sham-feeding rats quinine-adulterated food or an anorectic dose of amphetamine. Quinine-adulterated food and amphetamine stopped sham feeding, but neither elicited the satiety sequence (Antin et al., 1975).

Identical results have been obtained with CCK in the sham-feeding rhesus monkey. When CCK inhibits sham feeding, it elicits a behavioral sequence of satiety that is characteristic of the monkey feeding in a primate chair (Falasco et al., 1979).

Electrophysiological Evidence

The terminal item of the satiety sequence is resting behavior or sleep. The electrophysiological correlate of this behavioral state is large amplitude, slow waves recorded from

Table 11-1. Individual Behavioral Items Expressed as Percentage of Nonfeeding Behavior Six Minutes Before and After Feeding[a]

Group	Grooming	Sniffing	Rearing	Locomotion	Resting
Unoperated (no injection)	27.9	18.9	9.3	20.9	23.3
Open gastric fistula (Cholecyslokinin injection)	18.9	16.2	5.4	27.0	32.4

[a] Incidence of behavioral items was obtained by time sampling in which the behavioral items were recorded as present or not present every other minute during the time that the satiety sequence occurs, that is, six minutes before and six minutes after feeding stops. Percentage of non-feeding behavior was calculated by the following formula:

$$\frac{\text{Incidence of individual item}}{\text{total incidence of non-feeding items}} \times 100$$

There were five rats in each group. (From Antin et al., 1975, with the permission of the American Psychological Association.)

cortical electrodes (Bernstein, 1974; Rosen et al., 1971). Fara et al. (1969) observed that exogenous CCK could elicit large amplitude, slow waves in the anesthetized cat. Furthermore, they observed the same result when cream was infused into the duodenum. Since cream releases endogenous CCK, they suggested that CCK released by food is a mechanism for the high amplitude, slow waves that occur during postprandial satiety. This is an attractive idea, but no one has tested it in an unanesthetized animal.

A similar phenomenon, however, has been observed in conscious human beings. Stacher et al. (1979) reported that intravenous infusions of CCK increased the spectral power in the theta range and decreased the spectral power in the alpha range of the electroencephalogram (EEG). This increased theta to alpha ratio reflects the shift of the EEG toward high amplitude, slow waves and is characteristic of drowsiness (Gevins et al., 1977). This effect of CCK was obtained with low doses (0.6 or 3.0 U/kg, that is, 0.2–1.0 μg/kg). Since the effect was more pronounced when subjects could smell and see food (their

subjects did not eat the food), Stacher et al. suggested that
it was not a generalized depressant or sedative effect.

Subjective Reports

In their study of the effect of CCK in human beings, Kissileff
et al. (1979) administered an extensive questionnaire about
feelings of postprandial satiety. Subjects reported apparently
normal feelings when CCK decreased food intake.

In the same study that Stacher et al. (1979) reported the EEG
effects of CCK, they also administered self-rating visual analogue scales to assess hunger, satiety, and alertness. Doses
of CCK that increased high amplitude and slow waves also
increased ratings of satiety and feelings of fullness, and
decreased ratings of alertness and behavioral activation.
These effects of CCK were most conspicuous in the presence of
food stimuli.

Finally, Fincham et al. (1977) reported that subjective
ratings on a visual analogue scale for satiety increased
significantly more after ingesting capsules of L-phenylalanine
than after ingesting capsules of lactose. To the extent that
the effect of L-phenylalanine, a potent stimulus for the
release of endogenous CCK, is mediated by endogenous CCK, their
results are consistent with a satiating effect of CCK.

Summary of Evidence for Satiety Effect

The behavioral evidence in animals, the subjective reports in
human beings, and the electrophysiological measurements provide
convergent evidence for a satiating effect of CCK. When this
positive evidence of satiety is complemented by the failure of
human beings to report sickness and the failure of animals to
display toxic signs, I consider the evidence to be compelling
that CCK produces a satiety effect and does not merely inhibit
feeding.

Note the biological implication of CCK's satiety effect.
When animals and human beings receive CCK in a feeding situation, they react to it as if it were food. This is exactly
what one would predict if endogenous CCK formed part of the
normal satiating process. Molecules of CCK are physiological
information for the neural system that controls feeding behavior. The information that CCK carries must be complex. It
includes information for the termination of feeding, for the
elicitation of the behavioral sequence of satiety, and, in
human beings, for the intimate experience of pleasure,
tranquilizaiton, and satisfaction that follows a good meal.

It is important to understand that CCK is not the only mechanism for these effects (see later). It claims our attention because it is a mechanism for which extensive evidence exists.

Mechanism of the Satiety Effect

Synergism with Pregastric Stimuli

In previous sections of this chapter, we noted that the behavioral inhibitory effect of CCK was specific for food and could even be enhanced by the sight and smell of food (see Stacher et al., 1979). This suggests that another mechanism activated by food stimuli must be present for the satiety effect of CCK to be expressed. This other mechanism could function as a cue or as an inhibitory feedback signal to the central feeding system. A cue function would simply permit circulating CCK or, more likely, a visceral effect of CCK to be interpreted as a short-term satiety signal by the central feeding system. An inhibitory feedback signal, on the other hand, would be synergistic with CCK.

To evaluate this question, we investigated the possibility that a pregastric inhibitory mechanism was activated by sham feeding and that it interacts with CCK. We did this by giving a fixed dose of CCK at five different times in relation to the beginning of sham feeding. Thus CCK (40 U/kg) was injected 12 minutes before, 6 minutes before, just before, 6 minutes after or 12 minutes after sham feeding began in 17-hour food-deprived rats. The results were consistent with a potent synergistic interaction (Table 11-2), (Antin et al., 1978).

The pregastric stimuli of sham feeding apparently produce an inhibitory feedback signal whose magnitude is a direct function of the duration of sham feeding. This can be inferred from the significantly larger inhibition produced by CCK when it was administered 12 minutes after sham feeding began than when the same dose of CCK was administered 6 minutes after sham feeding began (Table 11-2). Since the identical dose of CCK was administered at both times, the larger inhibition must be the result of the longer period of sham feeding. Further experiments are necessary to determine the change in magnitude of inhibition produced by periods of sham feeding longer than 12 minutes.

The pregastric inhibitory mechanism also serves a cue function. This can be inferred from the significant increase in inhibition produced by giving CCK 6 minutes before or just before sham feeding began, compared to giving CCK 12 minutes before sham feeding began.

Table 11-2. Comparison of the Inhibitory Effect on Sham Feeding by Duodenal Infusion of Liquid Food and by Injection of CCK as a Function of Time of Administration Relative to the Onset of Sham Feeding[a]

Suppression of sham intake (%) related to time of injection or infusion (min)

	-12	-6	0	+6	+12
Duodenal infusion	-3	39	30	45	76
CCK injections	11	36	54	52	77

[a]Mean percentage inhibition of sham intake was determined by comparing the 30-minute sham intakes that began 12 minutes after the 6-minute infusion of liquid food (6 ml) or intraperitoneal injection of CCK (40 U/kg) to sham intakes after the infusion or injection of 0.15 M NaCl. Times of injection or infusion preceded by a minus sign are times before sham feeding began; times preceded by a plus sign are times after sham feeding began. The -3 percent suppression by duodenal infusions at -12 min means that the mean sham intake after duodenal infusion of food was 3 percent larger than after duodenal infusion of 0.15 M NaCl. The effect of the duodenal infusions in five rats under identical experimental conditions was reported by Antin et al., 1977. (From Antin et al., 1978, with the permission of the American Psychological Association.)

A quantitative analysis is required to determine if the synergism between CCK and the inhibitory mechanism activated by the pregastric stimuli of sham feeding is additive or multiplicative. This will not be possible until the pregastric mechanism is identified and the biological basis of its and CCk's satiety effects are measured precisely.

It is important to note that a nearly identical inhibitory synergism with the same temporal constraints can also be obtained between a duodenal infusion of liquid diet and sham feeding (Antin et al., 1977) (see Table 11-2). This demonstrates that the synergism is not just an artifact of the distribution or biological action of intraperitoneally administered CCK. The fact that the volume of the duodenal infusion is approximately the same as the volume of the same liquid diet that is measured in the small intestine at the onset of spontaneous, postpradial satiety in a rat eating normally (Liebling et al., 1975) is further evidence that pregastric and intestinal inhibitory mechanisms interact synergistically during a normal meal to terminate that meal and elicit satiety.

The close correspondence between the results with CCK and those with duodenal infusion is consistent with the hypothesis that endogenous CCK mediates the satiety effect of food in the small intestine, at least in part. But it is no more than that because biology is notorious for having different mechanisms produce similar effects.

Dependence on Abdominal Vagus

The satiety effect of CCK-8 is abolished or markedly reduced by bilateral abdominal vagotomy (Lorenz and Goldman, 1978; Smith and Cushin, 1978). Although Anika et al. (1977) have not observed this effect, we have now confirmed it in three separate series of vagotomized rats. Furthermore, we have extended the analysis in two ways. First, we have shown that mimicking the motor effects of abdominal vagotomy by peripheral anticholinergic blockade with atropine methyl nitrate does not change the satiety effect of CCK. Since the visceral effects of the loss of vagal motor fibers are apparently not sufficient for the loss of the satiety effect of CCK, the loss of the sensory afferent fibers in the vagus must be necessary for the loss of the satiety effect of CCK.

The second way we have extended the analysis of this phenomenon is by demonstrating that section of only the vagal fibers innervating the stomach is sufficient to reduce markedly the satiety effect of CCK (Smith et al., 1979). Section of the hepatic branch, celiac branch, right vagal trunk, or left vagal trunk alone does not change the satiety effect significantly. These data suggest that the stomach is the major site of action of CCK and that the gastric effect of CCK, for example, muscular contraction, is sensed by vagal receptors and carried up vagal afferents to the medulla oblongata where the neural information enters the central control system for feeding behavior. The reader is warned that this outline is only a working hypothesis at this time.

Although the mechanism is not settled, the fact that abdominal vagotomy markedly reduces the satiety effect of CCK is a clear demonstration of an important peripheral site of action that is probably centered in the stomach. This is the first demonstration of a visceral site for a behavioral action of an endogenous peptide and it suggests that a search for a peripheral site of the behavioral action of other peptides would be heuristic.

CCK may also act directly at brain sites to produce satiety. Three experiments are consistent with this possibility. The first is that of Stern et al. (1976). They injected caerulein, a decapeptide that is very similar in structure to CCK-8 and has all its known biological effects, including satiety (Gibbs et al., 1973a), into the ventromedial (VMH) and lateral (LH) areas of the hypothalamus. Microinjections of caerulein into the VMH inhibited food intake; identical microinjections into the LH had no effect. Furthermore, bilateral VMH lesions blocked the satiety effect of systemic injections of caerulein.

The second experiment was performed by Maddison (1977). He reported that injections of impure CCK into the lateral ventricle increased the duration of the IMI, but did not decrease meal size of rats bar pressing for food.

In the most recent report, Della-Fera and Baile (1979) produced marked inhibition of food intake during a two-hour infusion of CCK-8 into the lateral ventricle of sheep, but identical infusions had no effect in rats.

These positive results have been difficult to replicate. Kulkosky et al. (1976) observed that CCK-8 had its usual satiety effect in rats with bilateral VMH lesions. This contradicts the result of Stern et al. (1976) with caerulein.

We have not been able to replicate Maddison, and Nemeroff et al. (1978) reported that when intraventricular CCK-8 blocked tail-pinch induced feeding, the dose was larger than the threshold dose of peripherally injected CCK-8.

Given these contradictory results, the existence of central sites for a satiety effect of CCK in the rat must be considered hypothetical. This question deserves experimental attention not only because of the interest in finding a central site for the satiety effect of circulating CCK released from the intestine, but also because of the curiosity concerning the function of the large amount of CCK that is stored, synthesized, and released by neurons in numerous regions of the brain. Straus and Yalow (1979) suggested that cerebral CCK also has a satiety effect. Their suggestion was based on their finding that genetically obese mice (ob/ob) have less immunoreactive brain CCK than lean mice. Unfortunately, Schneider et al. (1979) failed to confirm their results. Furthermore, Schneider et al. reported that brain CCK failed to change during prolonged food deprivation or during hyperphagia in rats.

It is safe to say that the function of brain CCK is unknown. Given the diffuse distribution of CCK in the brain (Rehfeld, 1978), its dense concentration in the neocortex (Dockray, 1976; Straus et al., 1977), and its presence in separate cell groups identified at such functionally diverse sites as neocortex, amygdala, and midbrain (Innis et al., 1979), I think it is likely that brain CCK will have many behavioral functions. A satiety effect may be one of them, but it need not be.

Biological Meaning

The discovery of the satiety effect of CCK may mean the identification of the first endogenous, preabsorptive mechanism for
postprandial satiety. I emphasize preabsorptive mechanisms
because of their importance for short-term satiety. Most of
the time, animals stop eating before significant amounts of
ingested nutrients enter metabolic pools. This means that the
chemical and physical stimuli of ingested food must activate
satiety mechanisms before food is absorbed. Thus, at the
mucosal surface of the gut, probably from the tip of the tongue
to the end of the small intestine, food stimuli activate neural
and hormonal preabsorptive satiety mechanisms. Note that these
mechanisms represent a translation of physical and chemical
stimuli of food into biological information coded in neural
impulses and hormonal molecules. This information is processed
by the central control system to command the termination of
feeding and the elicitation of satiety.

The specific importance of a hormonal, preabsorptive satiety
mechanism, such as CCK, is that it can be brought under experimental and therapeutic control more easily than a neural
mechanism. There are two possible ways to manipulate CCK. One
is to administer the exogenous hormone. The other is to
stimulate the release of the endogenous hormone. Since the
release mechanism responds to physical and chemical stimuli, it
should be possible to provide the adequate stimuli for release
with calorically trivial substances (Smith and Gibbs, 1976).
We have modeled this possibility in monkeys (Gibbs et al.,
1976) and Fincham et al. (1977) have done it in human beings.
The selective activation of an endogenous, preabsorptive
satiety mechanism by a calorically trivial, nontoxic substance
is an attractive possibility for the treatment of human
obesity.

Although the satiety effect of exogenous CCK has been established, the satiety effect of endogenous CCK has not been
proved. There is no question that ingested food releases CCK,
but it is not clear if the quantity of CCK released is
sufficient to have a satiety effect. This is because it has
not yet been technically possible to measure the levels or
forms of plasma CCK that circulate in the rat in response to a
meal. It is possible, however, to make an indirect estimate of
the concentration of CCK that circulates after a meal. This
estimate is equivalent to the dose (D_{50}) of CCK required to
stimulate pancreatic enzyme secretion 50 percent of its maximum
response (Grossman, 1977). The D_{50} for pancreatic enzyme
secretion in the rat is 10 to 30 U/kg/hr CCK administered as a
slow intravenous infusion (Fölsch and Wormsley, 1973; Petersen
and Grossman, 1977). When CCK is infused in the same way to

sham-feeding rats, the threshold dose for inhibition of sham feeding is about 20 U/kg/hr (Lorenz et al., 1979). Since this threshold dose is within the range of the D_{50} for pancreatic enzyme secretion, it is possible that the quantity of CCK released by a meal is sufficient to have a satiety effect.

The indirect evidence is encouraging, but the physiological role of intestinal CCK in postprandial satiety will not be settled until the circulating forms of CCK can be reliably measured and the satiety effect of that quantity of CCK can be judged in terms of its synergism with other satiety mechanisms. To estimate synergistic effects, it will be necessary to identify and measure the other satiety mechanisms simultaneously. That will be an immense, experimental enterprise, but nothing short of it will answer the question of the physiological role of CCK in postprandial satiety.

SUMMARY

The synthetic octapeptide of the intestinal hormone CCK has a potent satiety effect. The effect is apparently specific for food, because identical doses of CCK do not inhibit water intake. CCK inhibits the intake of solid and liquid food. Therefore CCK affects the central control system for feeding and not just the motor acts of ingestion.

The satiety effect has been observed in a variety of animals and in man. It does not appear to depend on concomitant toxicity. CCK affects the satiation process primarily; it has little or no effect on the initial rate of feeding.

CCK not only inhibits feeding, it elicits the characteristic behavioral, subjective, and electrophysiological correlates of postprandial satiety. Apparently, animals and people react to CCK as if it were food. This is what one would expect if endogenous CCK were involved in the normal satiating process.

The site of CCK's action in the rat is in a vagally inner-vated organ in the abdomen because bilateral, abdominal vagotomy nearly abolishes the satiety effect. There is pre-liminary evidence that the critical lesion is of afferent vagal fibers from the stomach. This suggests that CCK elicits satiety by altering gastric function, possibly a change of gastric motility, that is sensed by vagal receptors and carried to the medulla oblongata where the vagal nerve impulses enter the central control system for feeding. A direct action on the brain has also been proposed, but the evidence for this is not compelling.

CCK is synergistic with the pregastric satiety mechanisms activated by sham feeding. It is not clear whether the syn-ergism is additive or multiplicative.

Although the satiety effect of exogenous CCK is established, the role of intestinal CCK in the normal physiology of post-prandial satiety is not settled. Proof of such a physiological role will depend on the development of a reliable method for measuring the quantity and molecular forms of CCK released by a meal and a similarly accurate measurement of the synergism between CCK and other mechanism(s) for satiety.

FORMULATION

Insulin and CCK

Intuitions about the control of feeding behavior have come from subjective experience and from recognizing a strong relation-ship between food intake and body weight. Both sources suggest that peripheral visceral events (mechanical and chemical) are important elements of the control system for feeding behavior and, ultimately, for body weight.

The experiments with pancreatic insulin and intestinal CCK confirm these intuitions. Pancreatic insulin increases feeding through the metabolic effects of glucoprivation and the liver is apparently an important site for this action. Intestinal CCK inhibits feeding by producing a change (mechanical or chemical) in a vagally innervated abdominal organ. Both of these peripheral changes are carried to the brain's control system for feeding over vagal afferent fibers. This is part of "what the gut tells the brain about feeding" (Smith and Gibbs, 1975).

In considering the effects of insulin and CCK on feeding behavior, I have emphasized that glucoprivation produced by endogenous insulin is an unusual metabolic situation and that CCK is probably released by every meal. Thus, our primary interest in the stimulation of hunger by insulin is that a gut peptide can elicit subjective and behavioral evidence of hunger through glucoprivation. By analyzing the neurochemical events that connect glucoprivation and increased hunger in this unusual circumstance, one hopes to uncover links in the neurochemical chain that connect the metabolic events of short periods of food deprivation to increased hunger under ordinary circumstances.

The working hypothesis about intestinal CCK and postprandial satiety is quite different. Since intestinal CCK is probably released by every meal, it could be a routine preabsorptive mechanism for postprandial satiety. Intestinal CCK is in the right place (the upper small intestine) and its release begins at the right time (before a meal has ended) for CCK to act with

other mechanisms to terminate feeding. With insulin and
glucoprivation, the question was, "Does it happen?" With CCK
and postprandial satiety, the question shifts to the quantitiative: "Is enough CCK released to have a significant satiety
effect?" Despite a large amount of indirect evidence suggesting that the answer is "Yes," this crucial question must be
considered open at this time.

Growing Points

As I indicated in the Introduction, the effect of gut peptides
on feeding behavior is under very active investigation. Most
of the work concerns possible satiety effects of the peptides.
This is because they, like CCK, are released by ingested food
and thus could serve to terminate feeding or determine the IMI.
There are reports that glucagon (Martin and Novin, 1977),
somatostatin (Lotter et al., 1977), bombesin (Gibbs et al.,
1979), and, even insulin (VanderWeele et al., 1979; Woods and
Porte, 1978) have satiety effects. Gastrin, secretin, and
gastric inhibitory (and insulinotropic) polypeptide, however,
do not under the stringent conditions of sham feeding (Lorenz
et al., 1979). I hope the reader will find the analytical
structure described to test the satiety effect of CCK useful
for testing these recent claims for a satiety effect of other
gut peptides.

Implications for the Psychobiology
of Hunger and Satiety

One of the great promises of the work with gut peptides and
feeding behavior is that we shall discover physiological
mechanisms that can mark hunger and satiety processes unambiguously. Such markers are essential for making a significant
penetration into the current ignorance that surrounds the
control of feeding behavior in animals and in ourselves.

ACKNOWLEDGEMENTS

The writing of this chapter was supported, in part, by Research
Scientist Development Award KO2 MHO0149. I want to thank
Marion Jacobson, Ellen Andrews, and Jeanne Strohmayer for
helping me put this chapter through the typewriter several
times.

REFERENCES

Anika, S.M., Houpt, T.R., and Houpt, K.A.: Satiety elicited by cholecystokinin in intact and vagotomized rats. *Physiol Behav* 19:761-766, 1977.

Antin, J., Gibbs, J., Holt, J., Young, R.C., and Smith, G.P.: Cholecystokinin elicits the complete behavioral sequence of satiety in rats. *J Comp Physiol Psychol 89:* 784-790, 1975.

Antin, J., Gibbs, J., and Smith, G.P.: Intestinal satiety requires pregastric food stimulation. *Physiol Behav 18:* 421-425, 1977.

Antin, J., Gibbs, J., and Smith, G.P.: Cholecystokinin interacts with pregastric food stimulation to elicit satiety in the rat. *Physiol Behav 20:*67-70, 1978.

Appel, K.E., Farr, C.B., and Marshall, H.K.: Insulin therapy in undernourished psychotic patients: Preliminary report. *JAMA 22:*1788-1789, 1928.

Barbour, O.: Use of insulin in undernourished non-diabetic children. *Arch Pediatr 41:*707-711, 1924.

Barnes, B.O., and Keeton, R.W.: Experimental obesity. *Am J Physiol 129:*305-306, 1940.

Bernstein, I.L.: Post-prandial EEG synchronization in normal and hypothalamically lesioned rats. *Physiol Behav 12:*535-545, 1974.

Booth, D.A., and Brookover, T.: Hunger elicited in the rat by a single injection of bovine crystalline insulin. *Physiol Behav 3:*439-446, 1968.

Brandes, J.S.: Insulin induced overeating in the rat. *Physiol Behav 18:*1095-1102, 1977.

Cannon, W.B., and Washburn, A.L.: An explanation of hunger. *Am J Physiol 29:*441-454, 1912.

Carlson, A.J.: *The Control of Hunger in Health and Disease,* University of Chicago Press, Chicago, 1916, p 64.

Christensen, O.: Pathophysiology of hunger pains. *Acta Med Scand (Suppl) 37:*170, 1931.

Della-Fera, M.A., and Baile, C.A.: Cholecystokinin octapeptide: Continuous picomole injections into the cerebral ventricles of sheep suppress feeding. *Science* 206:471-473, 1979.

Deutsch, J.A., and Hardy, W.T.: Cholecystokinin produces bait shyness in rats. *Nature 266:*196, 1977.

Dockray, G.J.: Immunochemical evidence of cholecystokinin-like peptide in brain. *Nature 264:*568-570, 1976.

Epstein, A.N., Nicolaidis, S., and Miselis, R.: The glucoprivic control of food intake and the glucostatic theory of feeding behaviour. In *Neural Integration of Physiological Mechanisms and Behaviour,* G.J. Mogenson and

F.R. Calaresee, eds. University of Toronto Press, Toronto, 1975, pp 148-168.

Falasco, J.D., Smith, G.P., and Gibbs, J.: Cholecystokinin suppresses sham feeding in the rhesus monkey. *Physiol Behav 23*:887-890, 1979.

Fara, J.W., Rubinstein, E.H., and Sonnenschein, R.R.: Visceral and behavioral responses to intraduodenal fat. *Science 166*:110-111, 1969.

Farrell, J.I., and Ivy, A.C.: Studies on the motility of the transplanted gastric pouch. *Am J Physiol 76*:227-228, 1926.

Fincham, J., Silverstone, T., and Saha, B.: The efect of L-phenylalanine on subjective hunger and satiety in man. Program, 6th International Conference on the Physiology of Food and Fluid Intake, 1977.

Fischer, L., and Rogarty, J.L.: Insulin in malnutrition. *Am J Dis Child 31*:363-372, 1926.

Fölsch, U.R., and Wormsley, K.G.: Pancreatic enzyme response to secretin and cholecystokinin-pancreozymin in the rat. *J Physiol (Lond) 234*:79-84, 1973.

Freyburg, R.H.: A study of the value of insulin in under-nutrition. *Am J Med Sci 190*:28-42, 1935.

Gevins, A.S., Zeitlin, G.M., Ancoli, S., and Yeager, C.L.: Computer rejection of EEG artifact. II. Contamination by drowsiness. *Electroencephalogr Clin Neurophysiol 42*:31-42, 1977.

Gibbs, J., and Smith, G.P.: Reply to Deutsch and Hardy. *Nature 266*:196, 1977.

Gibbs, J., Young, R.C., and Smith, G.P.: Cholecystokinin decreases food intake in rats. *J Comp Physiol Psychol 84*:488-495, 1973a.

Gibbs, J., Young, R.C., and Smith, G.P.: Cholecystokinin elicits satiety in rats with open gastric fistulas. *Nature 245*:323-325, 1973b.

Gibbs, J., Falasco, J.D., and McHugh, P.R.: Cholecystokinin-decreased food intake in rhesus monkeys. *Am J Physiol 230*:15-18, 1976.

Gibbs, J., Fauser, D.J., Rowe, E.A., Rolls, B.J., Rolls, E.T., and Maddison, S.P.: Bombesin suppresses feeding in rats. *Nature 282*:208-210, 1979.

Glick, Z., and Mayer, J.: Preliminary observations on the effect of intestinal mucosa extract on food intake of rats. *Fed Proc 27*:485, 1968.

Glick, Z., Thomas, D.W., and Mayer, J.: Absence of effect of injections of the intestinal hormones secretin and cholecystokinin-pancreozymin upon feeding behavior. *Physiol Behav 6*:5-8, 1971.

Grossman, M.I.: Integration of current views on the regulation of hunger and appetite. *Ann NY Acad Sci 63*:76-89,

1955.

Grossman, M.I.: Physiological effects of gastrointestinal hormones. *Fed Proc* 36:1930-1932, 1977.

Grossman, M.I., and Stein, I.F., Jr.: Vagotomy and the hunger producing action of insulin in man. J Appl Physiol 1:263-269, 1948.

Grossman, M.I., Cummins, G.M., and Ivy, A.C.: The effect of insulin on food intake after vagotomy and sympathectomy. *Am J Physiol 149:*100-102, 1947.

Grovum, W.L.: Factors that decrease food intake by sheep. Program, 6th International Conference on the Physiology of Food and Fluid Intake, 1977.

Houpt, T.R., Anika, S.M., and Wolff, N.C.: Satiety effects of cholecystokinin and caerulein in rabbits. *Am J Physiol 235:*R23-28, 1978.

Hunt, J.N., and Knox, M.T.: Regulation of gastric emptying: in *Handbook of Physiology, Section 6: Alimentary Canal, Vol IV, Motility*, C.F. Code, ed., Williams & Wilkins, Baltimore, 1968, pp 1917-1935.

Innis, R.B., Correa, F.M.A., Uhl, G.R., Schneider, B., and Snyder, S.H.: Cholecystokinin octapeptide-like immunoreactivity: Histochemical localization in rat brain. *Proc Natl Acad Sci USA 76:*521-525, 1979.

Ivy, A.C., and Oldberg, E.: A hormone mechanism for gallbladder contraction and evaluation. *Am J Physiol 86:*599-613, 1928.

James, A.H.: *The Physiology of Gastric Digestion*, Edward Arnold, London, 1957, pp 170-176.

Janowitz, H.D., and Grossman, M.I.: Effect of prefeeding, alcohol and bitters on food intake of dogs. *Am J Physiol 164:*182-186, 1951.

Kissileff, H.R., Pi-Sunyer, F.X., Thornton, J., and Smith, G.P.: Cholecystokinin-octapeptide (CCK-8) decreases food intake in man. *Am J Clin Nutr 32:*939, 1979.

Koopmans, H.S., Deutsch, J.A., and Branson, P.J.: The effect of cholecystokinin-pancreozymin on hunger and thirst in mice. *Behav Biol 7:*441-444, 1972.

Kosaka, T., and Lim, R.K.S.: Demonstration of the humoral agent in fat inhibition of gastric secretion. *Proc Soc Exp Biol Med 27:*890-891, 1930.

Kraly, F.S., Cushin, B.J., and Smith, G.P.: Nocturnal hyperphagia in the rat is characterized by decreased postprandial satiety. *J Comp Physiol Psychol 94:*375-387, 1980.

Kulkosky, P.J., Breckenridge, C., Krinsky, R., and Woods, S.C.: Satiety elicited by the C-terminal octapeptide of cholecystokinin-pancreozymin in normal and VMH-lesioned rats. *Behav Biol 18:*227-234, 1976.

Liebling, D.S., Eisner, J.D., Gibbs, J., and Smith, G.P.:
Intestinal satiety in rats. *J Comp Physiol Pyschol*
*89:*955–965, 1975.

Lorenz, D.N., and Goldman, S.A.: Dependence of the
cholecystokinin satiety effect on vagal denervation. *Soc
Neurosci Abstr 4:*178, 1978.

Lorenz, D.N., Kreielsheimer, G., and Smith, G.P.: Effect of
cholecystokinin, gastrin, secretin and GIP on sham feeding
in the rat. *Physiol Behav 23:*1065–1072, 1979.

Lotter, E., Woods, S.C., and Porte, D., Jr.: Somatostatin
decreases food intake. Program, 6th International
Conference on the Physiology of Food and Fluid Intake,
1977.

Lovett, D., and Booth, D.A.: Four effects of exogenous
insulin on food intake. *J Exp Psychol 22:*406–419, 1970.

Lucien, H.W., Itoh, Z., Sun, D.C.H., Meyer, J., Carlton, N.,
and Schally, A.V.: The purification of enterogastrone from
porcine gut. *Arch Biochem Biophys 134:*180–184, 1969.

MacKay, E.M., Callaway, J.W., and Barnes, R.H.: Hyperali-
mentation in normal animals produced by protamine insulin.
*J Nutr 20:*59–66, 1940.

Maclagan, N.F.: The role of appetite in the control of body
weight. *J Physiol (Lond) 90:*385–394, 1937.

Maddison, S.: Intraperitoneal and intracranial cholecystokinin depress operant responding for food. *Physiol Behav
19:*819–824, 1977.

Marriott, W.McK.: The food requirements of malnourished
infants. With a note on the use of insulin. *JAMA
83:*600–603, 1924.

Martin, J.R., and Novin, D.: Decreased feeding in rats
following hepatic-portal infusion of glucagon. *Physiol
Behav 19:*461–466, 1977.

Mayer, J.: Regulation of energy intake and the body weight:
The glucostatic theory and the lipostatic hypothesis. *Ann
NY Acad Sci 63:*15–43, 1955.

Mayer, J., and Bates, M.W.: Blood glucose and food intake in
normal and hypophysectomized, alloxan-treated rats. *Am J
Physiol 168:*812–819, 1952.

Miselis, R.R., and Epstein, A.N.: Feeding induced by
intracerebroventricular 2 deoxy-D-glucose in the rat. *Am J
Physiol 229:*1438–1447, 1975.

Morgan, C.T., and Morgan, J.D.: Studies in hunger II. The
relation of gastric denervation and dietary sugar to the
effect of insulin upon food intake in the rat. *J Gen
Psychol 57:*153–163, 1940.

Müller, E.E., Cocchi, D., and Mantegazza, P.: Brain adrenergic system in the feeding response induced by 2–deoxy-D-
glucose. *Am J Physiol 223:*945–950, 1972.

Nemeroff, C.B., Osbahr, A.J., III, Bissette, G., Jahnke, G., Lipton, M.A., and Prange, A.J., Jr.: Cholecystokinin inhibits tail pinch-induced eating in rats. *Science 200:*793-794, 1978.

Niijima, A.: Afferent impulse discharges from glucoreceptors in the liver of the guinea pig. *Ann NY Acad Sci 157:*690-700, 1969.

Novin, D., VanderWeele, D.A., and Rezek, M.: Infusion of 2-deoxy-D-glucose into the hepatic-portal system causes eating: Evidence for peripheral glucoreceptors. *Science 181:*858-860, 1973.

Ondetti, M.A., Rubin, B., Engel, S.L., Plušcec, J., and Sheehan, J.T.: Cholecystokinin-pancreozymin. *Am J Dig Dis 15:*149-156, 1970.

Oomura, Y., Ono, T., Ooyama, H., and Wayner, M.J.: Glucose and osmosensitive neurons of the rat hypothalamus. *Nature 222:*282-284, 1969.

Petersen, H., and Grossman, M.I.: Pancreatic exocrine secretion in anesthetized and conscious rats. *Am J Physiol 233:*E530-536, 1977.

Pitfield, R.L.: On the use of insulin in infantile inanition. *New York Med J 118:*217-218, 1923.

Quigley, J.P.: The role of the digestive tract in regulating the ingestion of food. *Ann NY Acad Sci 63:*6-14, 1955.

Quigley, J.P., Johnson, V., and Solomon, E.I.: Action of insulin on the motility of the gastrointestinal tract. *Am J Physiol 90:*89-98, 1929.

Rehfeld, J.F.: Immunochemical studies on cholecystokinin. II. Distribution and molecular heterogeneity in the central nervous system and small intestine of man and hog. *J Biol Chem 253:*4022-4030, 1978.

Ritter, S., Pelzer, N.L., and Ritter, R.C.: Absence of glucoprivic feeding after stress suggests impairment of noradrenergic neuron function. *Brain Res. 149:*399-411, 1978.

Rosen, A.J., David, J.D., and Ladove, R.F.: Electrocortical activity: Modification of food ingestion and a humoral satiety factor. *Comm Behav Biol 6:*323-327, 1971.

Schally, A.V., Redding, T.W., Lucien, H.W., and Meyer, J.: Enterogastrone inhibits eating by fasted mice. *Science 157:*210-211, 1967.

Schneider, B.S., Monahan, J.W., and Hirsch, J.: Brain cholecystokinin and nutritional status in rats and mice. *J Clin Invest 64:*1348-1356, 1979.

Short, J.J. Increasing weight with insulin. *J Lab Clin Med 14:*330-335, 1929.

Silverstone, T., and Besser, M.: Insulin, blood sugar and hunger. *Postgrad Med J 47:*427-429, 1971.

Smith, G.P.: Humoral hypotheses for the control of food intake. In *Obesity in Perspective*, Vol. 2, Part 2. G.A. Bray, ed., DHEW Publication No. (NIH) 75-708, Washington, D.C., 1973, pp 19-23.

Smith, G.P., and Cushin, B.J.: Cholecystokinin acts at a vagally innervated abdominal site to elicit satiety. *Soc Neurosci Abstr 4*:180, 1978.

Smith, G.P., and Epstein, A.N.: Increased feeding in response to decreased glucose utilization in the rat and monkey. *Am J Physiol 217*:1083-1087, 1969.

Smith, G.P., and Gibbs, J.: What the gut tells the brain about feeding behavior. In *Appetite and Food Intake*, T. Silverstone, ed., Dahlem Konferenzen, West Berlin, 1975, pp 129-139.

Smith, G.P., and Gibbs, J.: Cholecystokinin and satiety: Theoretic and therapeutic implications. In *Hunger: Basic Mechanisms and Clinical Implications*, D. Novin, W. Wyrwicka, and G. Bray, eds. Raven Press, New York, 1976, pp 349-355.

Smith, G.P., and Gibbs, J.: Postprandial satiety. In *Progress in Psychobiology and Physiological Psychology*, Vol 8, J.M. Sprague and A.N. Epstein, eds., Academic Press, New York, 1979, pp 223-224.

Smith, G.P., Gibbs, J., Strohmayer, A.J., and Stokes, P.E.: Threshold doses of 2-deoxy-D-glucose for hyperglycemia and feeding in rats and monkeys. *Am J Physiol 222*:77-81, 1972.

Smith, G.P., Jerome, C., Eterno, R., and Cushin, B.: Selective gastric vagotomy decreases the satiety effect of cholecystokinin in rats. *Soc Neurosci Abstr 5*:650, 1979.

Stacher, G., Bauer, H., and Steininger, H.: Cholecystokinin decreases appetite and activation evoked by stimuli arising from the preparation of a meal in man. *Physiol Behav 23*:325-332, 1979.

Stern, J.J., Cudillo, C.A., and Kruper, J.: Ventromedial hypothalamus and short-term feeding suppression by caerulein in male rats. *J Comp Physiol Psychol 90*:484-490, 1976.

Straus, E., and Yalow, R.S.: Cholecystokinin in the brains of obese and nonobese mice. *Science 203*:68-69, 1979.

Straus, E.J., Muller, E., Choi, H., Paronetto, F., and Yalow, R.S.: Immunohistochemical localization in rabbit brain of a peptide resembling the COOH-terminal octapeptide of cholecystokinin. *Proc Natl Acad Sci USA 74*:3033-3034, 1977.

Stricker, E.M., Rowland, N., Saller, C.F., and Friedman, M.I.: Homeostasis during hypoglycemia: Central control of adrenal secretion and peripheral control of feeding. *Science 196*:79-81, 1977.

Strohmayer, A., Kreielsheimer, G., and Smith, G.P.: The effects of cholecystokinin on feeding and drinking in genetically obese mice. *Soc Neurosci Abstr 2:*308, 1976.

Stunkard, A.J., and Fox, S.: The relationship of gastric motility and hunger. *Psychosom Med 33:*123-134, 1971.

Sturdevant, R.A.L., and Goetz, H.: Cholecystokinin both stimulates and inhibits human food intake. *Nature 261:*713-715, 1976.

Tisdall, F.F., Brown, A., Drake, T.G.H., and Cody, M.G.: Insulin in treatment of malnourished infants. *Am J Dis Child 30:*10-18, 1925.

Ugolev, A.M.: The influence of duodenal extracts on general appetite. *Dokl Akad Nauk SSSR 133:*632-634, 1960.

VanderWeele, D.A., Haracykiewicz, E., and Van Itallie, T.B.: Insulin and satiety in obese and normal-weight rats. *Soc Neurosci Abstr 5:*225, 1979.

Woods, S.C., and Porte, D., Jr.: The central nervous system, pancreatic hormones, feeding and obesity. *Adv Metab Dis 9:*283-312, 1978.

Young, R.C., Gibbs, J., Antin, J., Holt, J., and Smith, G.P.: Absence of satiety during sham feeding in the rat. *J Comp Physiol Psychol 87:*795-800, 1974.

12

Sleep Peptides:
The Current Status

RENÉ R. DRUCKER-COLIN

INTRODUCTION

The study of sleep rapidly changed course following the land-
mark discovery of Aserinsky and Kleitman (1935) that certain
periods of sleep were accompanied by conjugate rapid eye
movements (REM), while simultaneously the electroencephalogram
(EEG) exhibited an activated pattern consisting of low amplitude (desynchronized) waves with frequencies of 15 to 20 and 5
to 8 Hz. A few years later, Dement and Kleitman (1957) found
that the majority of subjects awakened during REM sleep reported dreams involving visual imagery. These authors also
reported that REM sleep recurred in a cyclic fashion, interspersed with periods of non-REM sleep. These observations led
to the classification of human sleep into five stages, four of
which pertain to non-REM sleep (Rechtschaffen and Kales, 1968).
Two other important observations were made in the late 1950s.
The first was that cats, like human beings, showed cortical
desynchronization during sleep (Dement, 1958), suggesting that
all mammals alternate between at least two distinct phases of
sleep. The second was the loss of muscle tone during REM sleep
in cats (Jouvet and Michel, 1959), an observation later confirmed in human beings (Berger, 1961). This loss of muscle
tone was accompanied by spikes in the pontine reticular formation. Similar spikes were later recorded from the lateral
geniculate body (Mikiten et al., 1961) and the occipital cortex
(Mouret et al., 1963). The anatomical localization of this
activity inspired the name, ponto-geniculo-occipital (PGO)
spikes.
  Today it is universally accepted that mammals and primates
present at least two basic stages of sleep. The state of sleep

is first characterized electroencephalographically by the
appearance of 14 to 18 Hz cortical spindles, which are subsequently replaced by 2 to 4 Hz slow waves. At the same time,
high voltage (500 to 800 μV) sharp waves are recorded from the
hippocampus, while the electromyogram (EMG) decreases slightly.
Usually after some 30 to 40 minutes the electrophysiological
signs of slow wave sleep (SWS) are replaced by low-voltage fast
cortical EEG activity, regular hippocampal theta rhythm (5 to 6
HZ), isoelectric EMG, burst of REMs, and PGO waves. PGO waves
appear in SWS approximately one minute before all REM sleep
periods. These PGO spikes exhibit a fairly constant daily rate
of about 14,000 in the cat (Jouvet, 1969), may exist in man
(Salzarulo et al., 1975), and are made up of simple Type I
spikes and complex Type II spikes (Morrison and Pompeiano,
1966).

The events that occur within REM sleep have been classified
as tonic and phasic. The tonic events refer to those characteristics that more or less define the REM sleep period, such as
EEG activation and EMG suppression. The phasic events are
short-lasting but recurrent activities, including, in addition
to eye movements and PGO spikes, cardiovascular irregularities
(Snyder et al., 1964), respiratory changes (Aserinsky, 1965),
changes in pupil diameter (Berlucchi et al., 1964), fluctuations in penile tumescence (Fisher et al., 1965), muscular
twitchings (Baldridge et al., 1965), middle ear muscle contractions (Dewson et al., 1965), brainstem unit (Pivik et al.,
1973), and multiple unit bursts of activity (Drucker-Colín et
al., 1977).

In animals, the stages of sleep have been variously called
synchronized, non-REM, forebrain or SWS when referring to the
initial phase; and REM, paradoxical, hindbrain, activated,
desynchronized or fast wave sleep when referring to the second
phase. An example of both stages of sleep in the cat as compared to wakefulness is shown in Fig. 12-1.

It is evident that although the electrophysiological signs of
sleep are simple and clear cut, they undoubtedly represent extremely complex central nervous system (CNS) mechanisms, which
are far from being understood. However, partly because of the
exhaustive description of sleep electrographic phenomenology,
experimental results have been obtained that shed some light on
our understanding of some of its mechanisms. Therefore recording of basic electrophysiological parameters is a highly
desirable, if not an indispensable, aspect of any serious sleep
study. Unfortunately, most sleep peptide studies are afflicted
with the absence of such desideratum, rendering their interpretation rather difficult.

THE PROBLEM OF SLEEP-INDUCED SUBSTANCES

The search for sleep-inducing substances has generally taken
one of three different approaches. The first is pure sleep
pharmacology. Historically, it involved the use of herbs and
plants, and more recently the extraction of active compounds
from herbs and plants or the production of synthetic compounds.
It has a very long history, a short memory, and relatively poor
results. Although the discussion of such substances is beyond
the scope of this paper, it should be pointed out that presently, there is no ideal sleep-inducing drug (hypnotic). Most
such drugs are in all probability CNS depressants, which act in
ways unrelated to natural sleep mechanisms (for an excellent
recent review of this problem see Hartmann, 1978). This outcome is not surprising, since researchers trying to develop
hypnotics are more interested in finding out how effective a
particular drug is in inducing sleep, than how it relates to
the mechanisms of sleep.
  The second approach has been championed by Michel Jouvet and
has had a tremendous impact on sleep research. His experiments
explored whether well-known neurotransmitters, such as serotonin (5-HT), norepinephrine (NE), and acetylcholine (ACh),
play an essential role in the regulation and maintenance of the
sleep-wake cycle. The results of these experiments led Jouvet
(1972) to advance the following theory: SWS is initiated by
the release of 5-HT from neurons of the dorsal raphe nuclei of
the brainstem, whereas REM sleep is triggered by the release of
5-HT from neurons of the caudal raphe nuclei. Once initiated,
the "executive" mechanisms of REM sleep involve catecholamines:
the NE-containing neurons of the caudal third of the locus
ceruleus being responsible for the muscle atonia of REM sleep
and those of the medial third being the pacemaker for PGO
activity. On the other hand, wakefulness and cortical arousal
are dependent on NE-containing neurons of the anterior locus
ceruleus, dopamine-containing neurons of the midbrain reticular
formation and cholinergic neurons in the cortex. Unfortunately, a considerable number of experiments in the past few
years has failed to support various aspects of this theory
(Drucker-Colín and Spanis, 1976; Gillin et al., 1978; and Ramm,
1979). This does not mean that neurotransmitters do not play a
role in sleep, but rather that in all probability they do not
play the central role originally proposed.
  The third approach has a longer history than the second but
has been slower to develop, probably because it has had very
few proponents. This approach attempted to identify endogenous
substances from brain or other tissues that are in some way

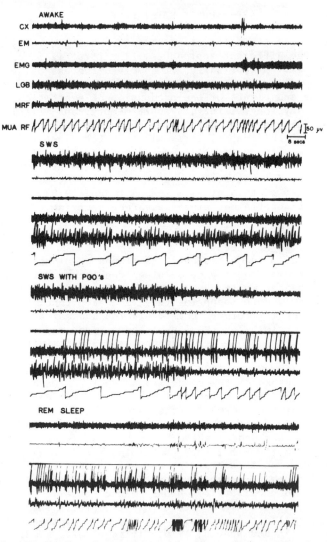

Fig. 12-1.   Typical polygraphic recordings from a cat during
the various stages of the sleep-wake cycle.   The
upper sample shows the cortical (CX) desynchronization, high amplitude electromyogram (EMG), and
isolated eye movements (EM) during wakefulness.
The other three samples are continuous and represent the activity during slow wave sleep (SWS),
during SWS with ponto-geniculo-occipital (PGO)
spikes from the lateral geniculate body, and during
REM sleep.   Note how PGO spikes appear during SWS

related to sleep. This idea was pioneered by Piéron, who reported (Legendre and Piéron, 1910, 1911, 1912) that serum or cerebrospinal fluid (CSF) obtained from sleep-deprived dogs induced a state of somnolence or "sleep" when introduced into the 4th ventricle of normal dogs. If these fluids were ultrafiltered, dialyzed, or heated at 65° C, their sleep-inducing properties were lost. These observations prompted Piéron (1913) to suggest that during wakefulness the CSF accumulated a "hypnotoxin" with properties conducive to sleep, and which dissipated during sleep.

Many years later, Schnedorf and Ivy (1939) repeated Piéron's experiments, observing behavioral sleep in 9 of 24 dogs injected intracisternally with 8 ml of CSF from dogs deprived of sleep for periods of 7 to 16 days. However, "sleep" was also observed in 4 out of 24 dogs receiving CSF from nonsleep-deprived dogs. Similar results were reported by Kroll (1933) with brain extracts of cats and rabbits previously put to sleep with drugs. This brain extract induced "sleep" in a recipient animal. The results of all these studies were complicated by the large volumes of CSF withdrawn and injected into the ventricles, which caused increases in temperature and ventricular pressure. Also, "sleep" was judged by simple behavioral observation, since at the time no objective quantitative measures were available. Without such measures, it is difficult to distinguish sleep from responses to the stressful procedures used.

In the more recent experiments, the volume factors have been controlled. However, other problems relate to the adequacy of the experimental approach. It is not obvious a priori whether sleep-related or sleep-inducing substances will appear during wakefulness, at the onset of sleep, or during sleep. These alternatives can impose important methodological constraints.

No good evidence exists for a substance elaborated during sleep whose levels differ between sleep and wakefulness.

---

a few seconds before the appearance of REM sleep (third sample), and also in the fourth sample that muscle tone has disappeared, eye movements occur in bursts, and that cortical EEG becomes desynchronized, as in wakefulness. The last channel (sixth) of each sample represents the integrated multiple unit activity (MUA) from the midbrain reticular formation (MRF), where we can observe a relatively steady but high frequency activity in the MRF during wakefulness, a decrease of activity during SWS, and an increase prior to and during REM sleep, when bursts of MUA can be observed. Each staircase resetting of MUA MRF activity represents 15 spikes.

Measurement of known neurotransmitters (presumed hypnogenic) following prolonged wakefulness have failed to detect changes in levels of NE and 5-HT (Stern et al., 1971), perhaps because of increased utilization (Hery et al., 1970). This suggests that prolonged wakefulness does not induce the accumulation of "hypnotoxins," be they peptides or other neurotransmitters. Even were there such evidence, there remains the dilemma of determining whether material obtained during sleep is the cause or the result of sleep.

Regardless of whether we search for sleep-inducing substances during wakefulness or during sleep, we face the problem of deciding where to look for them. Shall we look for them in the ventricles or directly in brain tissue? If we choose the latter, where in the brain? The decision is quite difficult. If we decide to look in the brainstem, which seems most reasonable given the available evidence (Moruzzi, 1972), we face the problem of deciding whether to look in the so-called "waking system" or in the "sleep system." These questions can only be answered by careful experimentation.

Very few laboratories have attempted to identify sleep-inducing substances. The reasons are clear. The most important is the absence of a good simple bioassay for determining the sleep-inducing activity of biochemical fractions. Another reason is the length of time that must be invested in this approach, with a relatively poor prognosis. The third reason is that unfortunately, as we shall see later, the results of all the studies that have attempted to identify sleep-inducing substances may merely be artifacts of the procedures.

On the other hand, such studies are laying the groundwork for the discovery of one or more sleep-inducing substances. Perhaps only through them, will the understanding of sleep mechanisms be able to take a great leap forward.

SLEEP PEPTIDES AND POLYPEPTIDES: FACTS AND FANCIES

Sleep Factors from Cerebrospinal Fluid

Pappenheimer and his group have extended Piéron's original observations and apparently subscribe to the view that sleep is the result of substances that accumulate during wakefulness and can be found in the ventricles. In their first study (Pappenheimer et al., 1967), goats chronically implanted with a cisternal tube were deprived of sleep by an automatic device that provided electric shock and acoustic alarm to relaxing animals. At the end of the deprivation period the cisterna magna was punctured and CSF withdrawn at a rate of 0.1 ml/min for five hours.

CSF from sleep-deprived and normal goats was injected into the lateral ventricles of chronically implanted rats at a rate of 3.3 μl/min for 30 minutes (0.1 ml total volume). Injections occurred in the late afternoon prior to the dark period, when rats show their highest level of activity. "Sleep" was measured behaviorally on the basis of locomotor activity monitored constantly by photocells connected to automatic counters. CSF from sleep-deprived goats decreased locomotor activity of rats for several hours, whereas normal CSF had no effect. Since some pilot experiments in cats showed similar effects, the authors wrote: "The fact that fluid from sleep-deprived goats is active in cats and rats suggests that we are dealing with a humoral factor of general and fundamental importance to the sleep mechanism."

Unfortunately this study leaves much to be desired. Firstly, the rats' ventricles were punctured at the time of each infusion, which was not only stressful, but also involved a great deal of handling. It has long been known that handling is aversive to rats (Candland et al., 1962). The procedure may have resulted in periods of inactivity immediately following the treatment, thus decreasing mean locomotor activity. Injection of saline (IP) alone can reduce activity counts (Schnitzer and Ross, 1960). Moreover, the authors only reported standard errors and ranges of activity counts for the pre-infusion control, and not for the periods following treatment. Since the data are presented for 6- or 24-hour periods, it is conceivable that the low mean is the result of very low activities in the periods immediately following injections. It would have been more convincing if the authors had compared hourly trends. On the other hand, since mean activity counts in rats given CSF from sleep-deprived goats were lower than those given normal CSF, the results indicated that the former CSF indeed contained a sleep-inducing humoral factor not present in the latter. However, the only way to ascertain whether this factor induces sleep is to determine it electrophysiologically. Unfortunately, no electrophysiological measurements were made.

In subsequent studies this problem was partially corrected. Locomotor activity and cortical EEG were recorded in a few chronically implanted rats, and a good correlation between low activity counts and increases in EEG synchronization was reported (Fencl et al., 1971). Also, 0.1 ml of CSF from a 48-hour sleep-deprived goat induced the greatest decrease of locomotor activity. It was therefore concluded that the concentration of the sleep-induced factor in CSF peaks after 48 hours of sleep deprivation (Fig. 12-2).

Molecular sieving experiments using calibrated membranes indicated that the sleep-inducing factor had a molecular weight

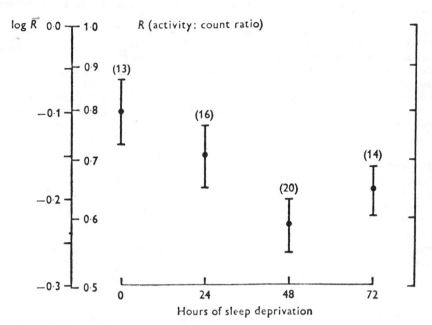

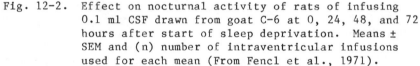

Hours of sleep deprivation

Fig. 12-2. Effect on nocturnal activity of rats of infusing
0.1 ml CSF drawn from goat C-6 at 0, 24, 48, and 72
hours after start of sleep deprivation. Means ±
SEM and (n) number of intraventricular infusions
used for each mean (From Fencl et al., 1971).

of less than 500 and it was named Factor S (Pappenheimer et
al., 1975). During chromatography on Sephadex G-10, Factor S
eluted just before sucrose. Reaction with fluorescamine indicated that 0.5 nmoles of a peptidelike substance in this
region was biologically active, although no correlation was
reported between biological activity and fluorescamine reactivity. Incubation of Factor S with Pronase destroyed its
biological activity, leading Pappenheimer to conclude that it
was a peptide. The sleep-inducing properties of Factor S were
not mimicked by 5-HT, γ-hydroxybutyrate, butyrolactone, γ-
aminobutyric acid (GABA), glutamate, or cyclic 3',5'-AMP. The
presence of Factor S at periventricular sites was suggested.

The ultrafiltrates were tested on locomotor activity in rats
and on the EEG of rabbits bearing screw electrodes over the
frontal and occipital cortices. A 0.1 ml of purified concentrate of CSF from normal and sleep-deprived goats significantly decreased for 12 hours nocturnal locomotor activity of
rats in comparison to those receiving saline (Fig. 12-3a).
Since a 20-fold concentration of CSF from non-sleep-deprived

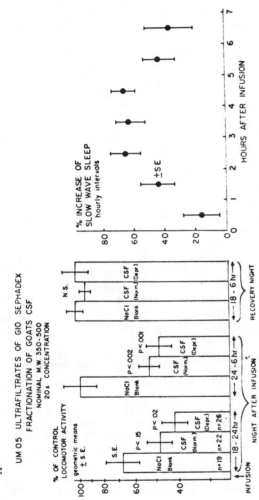

Fig. 12-3. Hourly percentage excess SWS following intraventricular infusion of concentrated Factor S from whole brains of sleep-deprived goats or sheep. Means ± SEM of 13 assays on 5 rabbits. Rabbits normally slept 35 ± 5 percent of any given hour. Effects on nocturnal locomotor activity of rats of infusing product of step 7 from fractionation of goat CSF. Locomotor activity is reduced as a result of infusing concentrates of CSF from both normal and sleep-deprived animals. The depressant effects continue for 12 hours, followed by complete recovery on the second night after infusion (From Pappenheimer et al., 1975).

animals also produced an effect, it was suggested that Factor S is present in small concentrations even in a normal animal. This observation indicated that one could forego the painful process of sleep deprivation. In rabbits, 0.1 ml of purified Factor S increased the relative amount of SWS by approximately 15 to 20 percent for four to five hours (Fig. 12-3b). Also, the mean integrated voltage of the EEG was increased relative to normal and was similar to that observed following sleep deprivation.

A most important observation in this study was that brain extracts, particularly those from brainstem, also appeared to contain Factor S. In the most recent report of this group (Krueger at al., 1978) Factor S was extracted from brain and treated by partition chromatography and high-voltage electrophoresis, to yield an active product purified one million-fold. The concentration of Factor S in brain tissue appears to be of the order of 0.3 nmoles/gm of tissue, the effective sleep-inducing dose being about 0.15 nmoles.

During the course of these studies, it appeared that CSF also contained a factor capable of inducing a long-lasting excitatory behavior. In contrast to Factor S, this excitatory Factor E appeared to exert its effects even if infused into the subarachnoid spaces, thus suggesting receptors in cortical tissue. The presence of so-called "waking factors" had been previously reported in cross circulation experiments in rabbits (Monnier and Hosli, 1964), and in encéphale isolé cats (Purpura, 1956).

Thus, Factor S appears to be a peptide, present throughout the brain, whose concentration increases as a result of sleep deprivation and which has sleep-inducing properties in several species. These observations could be taken to signify that sleep is the result of a delicate balance between the accumulation and utilization of a universal peptide. However, before accepting this postulation, a few additional facts should be considered. First, Ringle and Herndon (1969) did not observe any effects on either locomotor activity, behavior, or EEG patterns with CSF obtained from sleep-deprived rabbits. This might mean that sleep-deprived rabbits do not produce a sleep-inducing factor. In all fairness, however, this study was not strictly comparable to Pappenheimer's since the procedure of extraction of CSF involved prior electrocution of the rabbits.

A serious problem in Factor S studies is the absence of sleep-wake cycle recordings. Granted that the authors have a relatively rapid bioassay, this phase must now be replaced. One looks forward with great anticipation to studies with Factor S that would take into consideration more complete sleep electrophysiology, including REM sleep and its concomitant phasic activities. Until studies become available, the final verdict on Factor S as sleep-inducing peptide must be withheld.

Sleep Factor from Blood

In the oldest modern-day study of sleep factors, Monnier and his group in Switzerland have studied the humoral transmission of sleep via the circulatory system. In these studies the jugular vein of a donor rabbit is connected to the jugular vein of a recipient rabbit. Subsequently the donor's mediocentral intralaminary thalamus is stimulated at low frequencies (6/sec). Such stimulation induces synchronization of the cortical EEG, or what this group calls "orthodox delta sleep," in both the donor and recipient (Monnier et al., 1963). Since the animals are joined by the jugular vein, a sleep substance released into the cerebral venous blood of the donor (as a result of thalamic stimulation) is thought to penetrate the recipient's brain via its heart, lung, and carotid artery, thus showing that sleep can be induced by blood-borne factors. Similar results have been obtained in parabiotic rats (Matsumoto et al., 1972).

Since stimulation of the activating midbrain reticular formation (MRF) induces arousal with EEG low voltage fast activity in the recipient rabbit, Monnier's results imply that the "sleep" humoral factor is produced as a result of being asleep. This is at variance with Pappenheimer's results, although the experimental approaches of these two researchers is not comparable because of the striking differences in methodology.

In subsequent studies, Monnier and Hosli (1964, 1965) dialyzed the cerebral venous blood and showed that the dialysate from a sleeping donor rabbit induced sleep in recipient rabbits. The assessment of the blood-borne hypnogenic factor was originally made by intravenous injections. Later, testing methods were improved by the use of intraventricular infusions (Monnier and Hatt, 1971), and it was shown that a partially purified sleep dialysate induced mild bradycardia but had no effect on respiratory rate and CSF and blood pressures, suggesting that its sleep-inducing properties could not be attributed to visceral factors (Monnier et al., 1973).

The assessment of sleep electrographic data was also improved by expressing an average delta value of 200 mm for a period of 5 minutes or 2360 μV/minutes, chosen as a baseline equivalent to 100 percent. A deflection above 100 percent indicated a tendency to sleep and below 100 percent to arousal. Thereafter, all studies expressed sleep as the average delta value in periods of 5 minutes during 25-minutes samples, and in percentage deviations from 100 percent prior to and following infusions. This method provided a relatively rapid bioassay of "sleep" (Fig. 12-4).

With this, they initiated a series of studies leading to the isolation, purification, and physical-chemical characterization of a substance that turned out to be a nonapeptide, which they

508 Drucker-Colín

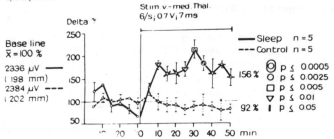

a) Sleep donors

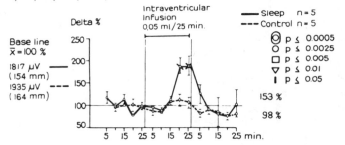

b) Sleep recipients (extracorp. dialysate)

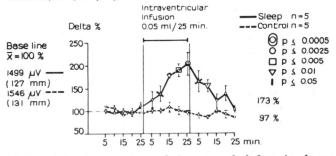

c) Sleep recipients (pool I [G-10])

Fig. 12-4.   (a) Percentage of increased delta in donor-rabbits
during electrical stimulation of the thalamic sleep
area (--) compared with nonstimulated control
animals (---). (b) Transmission of sleep ("delta
sleep") by intraventricular infusion of an extra-
corporal dialysate from donors to recipients (--);
control infusion of dialysate from nonstimulated
donors (---). (c) Increase of delta in recipients
(--) after infusion of Pool I (Sephadex G-10).
Controls (---). (From Schoenenberger et al.,
1972).

called delta sleep inducing peptide (DSIP) with a molecular
weight of 849 and whose amino acid sequence is Trp-Ala-Gly-Asp-
Ala-Ser-Gly-Glu (Monnier et al., 1972, 1977; Schoenenberger et
al., 1972, 1977, 1978; Schoenenberger and Monnier, 1977). In
their more recent experiments, assessment of EEG activity was
carried out using linear and three-dimensional power spectra
analysis, as well as factor analysis, to determine variations
of spectral energies in frequency bands of 2 Hz (Monnier et
al., 1977; Schoenenberger et al., 1978). The experiments leave
little doubt that DSIP induces a higher percentage of slow
waves.

In a recent study, Polc et al. (1978) tested the effects of
intravenous administration of 30 nmol and 300 nmol of DSIP on
the sleep-wake cycle of cats recorded by telemetry for six
hours. Both SWS and REM sleep was increased by the low dose of
DSIP, but the high dose had no effect. However, recently,
Walker and Sandman injected 59 nmol of DSIP into the lateral
ventricle of rats and found no decrease in locomotor activity
(if anything they found an increase). In our laboratory 235
nmoles of DSIP applied to the MRF of rats produced no changes
in sleep (Kastin et al., 1978). In the light of Polc's results, such local application of DSIP may represent a very high
dose, which could explain the absence of effects on sleep and
locomotor activity. Nonetheless, the absence of a dose-effect
is quite strange. Schoenenberger et al. (1978) have justified
the absence of a dose-effect, writing: "DSIP induces EEG
changes similar to those of natural sleep, but differing from
the effects of hypnotic pharmaca. The chief requirement for
classical hypnotic pharmaca (a linear dose-effect curve) was
found unsuitable for the new group of DSIP compound. Indeed,
intravenous administration of DSIP to free moving rabbits
increases only at a definite optimal dose, with narrow dose-
effect relation, the orthodox delta and spindle sleep (slow
wave sleep) the corresponding paradoxical sleep and the
decreases in motor activity."

The Monnier group suggests that DSIP acts more as a modulator
or programmer of sleep that as a neurotransmitter. Whatever
the role of DSIP, we should anticipate studies from this group
on the origin, distribution, and receptor locations of this
substance in the brain, as well as more studies from other
researchers. Although presently one should carefully evaluate
the evidence that DSIP is a true sleep substance, it is the
first compound characterized.

In contrast, however, there are a few isolated studies that
have provided evidence against the presence of blood-borne
sleep factors. Ringle and Herndon (1968) failed to find
dialyzable sleep-promoting factors in plasma of sleep-deprived
rabbits. However, this study was carried out to determine

whether sleep deprivation would promote the appearance of a sleep substance, whereas in Monnier's studies the idea was that sleep itself would promote it. The difference in results may therefore be attributed to a difference in approach. Indeed it may be recalled that Monnier and Hosli (1964) suggested that sustained wakefulness elicited by stimulation produced a "waking factor."

In other studies, it was shown by Lenard and Schulte (1972) that craniopagus twins with a common circulation have independent sleep-wake cycles, and that a dog with an extra head transplanted into its circulation (De Andres et al., 1976) failed to present simultaneous sleep periods. These observations, however, cannot really be taken as a serious criticism of DSIP, because under such abnormal conditions, it is improbable that levels of brain excitability are comparable to those normal subjects.

In summary, on the basis of the available data, we had best be skeptical about DSIP, essentially because the bulk of evidence of its sleep-inducing properties comes from observations on changes in EEG amplitude. The association of a humoral agent with one electrographic sign does not necessarily mean that sleep is being produced.

Sleep Factors from Brain

Recently there have been attempts to extract sleep-promoting material from brain. Here again, the approaches have been based on different hypotheses as to the periods in which this presumed material should be obtained.

Nagasaki et al. (1974), evidently influenced by the Piéron concept that a "sleep substance" accumulates during wakefulness, deprived rats of sleep for a period of 24 hours by applying an electric shock whenever the animals attempted to sleep. At the end of the deprivation period, the rats (N=1000) were sacrificed and the brainstem was separated, homogenized, dialyzed, and lyophylized. This crude brain extract was injected intraperitoneally into rats and was reported to induce a decrease of locomotor activity and an increase of delta activity and SWS time. Unfortunately, this assessment was performed in a very small number of animals (N=3 for locomotor activity; N=4 for sleep time). Nonetheless, the experiment suggests that the brainstem contains a sleep promoting material that accumulates during wakefulness and that appears to cross the blood-brain barrier. In a subsequent study (Nagasaki et al., 1976), this brain extract from sleep-deprived rats was partially purified by Sephadex G-10 and thin-layer chromatography. It has a molecular weight of 200 to 700 and produces inhibition of the spontaneous discharge rates of the crayfish abdominal ganglion. Since this inhibitory effect was 100 times stronger than GABA, while 5-HT increased firing rate, the authors suggest that the brain extract from sleep-deprived rats contains peptide-like substances with strong inhibitory properties. It should be noted that these experiments provide support for Piéron's concept that during wakefulness the CSF accumulates a "hypnotoxin" with inhibitory properties.

Other studies have been performed in our laboratories. The approach was influenced by concepts of synaptic transmission, in the sense that the extraction of a sleep-inducing substance should be made at the time it is released, presumably as a result of a physiological activation during sleep. In these studies, push-pull cannulae were used (see Myers, 1970; Drucker-Colín and Spanis, 1976 for description), which permit the extraction in freely moving animals of "active" substances from specific brain sites. The experiments were based on the simplistic notion that sleep and waking can be explained by a reciprocal interplay of two basically antagonistic neuronal systems, as elegantly described by Moruzzi (1972). Accordingly, experiments were undertaken to extract from the MRF ("waking system") substances released during sleep into the extracellular fluid (by nerve terminals of the "sleep system"), and test the effects of such substances on the homologous MRF neurons of a recipient animal. These experiments showed that the substances (perfusates) obtained from cats during sleep, were capable of inducing sleep when perfused into the homologous MRF of recipient cats. Since the assessment was based on recordings of the entire sleep-wake cycle, it was concluded that perfusates obtained from specific brain sites during sleep contained substances capable of influencing ongoing waking behavior (Drucker-Colín et al., 1970; Drucker-Colín, 1973).

In these types of experiments, isolation of a sleep-inducing factor from perfusates is a serious problem, primarily because extraction through push-pull cannulae yields extremely low quantities and because bioassays based on lengthy sleep-wake cycle recordings makes trial and error assessment at various stages of purification an overwhelming task. For this reason we have approached the problem in a different manner. Although some of the experiments may appear to be unrelated to the isolation of a "sleep factor," we feel they contribute complementary evidence for the concept that a polypeptide factor or factors regulates or triggers REM sleep, and also they provide insights into some aspects of the neurophysiology of sleep.

Macromolecules and REM Sleep

Some 10 years also, Oswald (1969) suggested that the rebound of
REM sleep that occurs during drug withdrawal reflects a phase
of neuronal repair indicated by increased protein synthesis and
further suggested that situations in which protein synthesis
increases should lead to REM sleep. It has been reported
(Takahashi et al., 1968) and confirmed many times (Sassin,
1977) that phases 3 and 4 of sleep in man are associated with a
rise in plasma growth hormone (GH). Since in man the sleep-
related surge of plasma GH occurs in the early phase of the
night, before the late appearing REM sleep stage, it has been
argued (Stern and Morgane, 1977) that GH may play a role in
triggering REM sleep.

The hypothesis has been tested by injecting GH in cats (Stern
et al., 1975) and rats (Drucker-Colín et al., 1975b). Both
studies showed that GH induced a dose-dependent increase of REM
sleep, about three hours following injection.

Since GH is an anabolic hormone that exerts important actions
on protein synthesis (Korner, 1965), it could be argued that
the increase of REM sleep following GH is indirectly produced
by an enhanced formation of macromolecules. However, there are
scanty data on the relationship between protein synthesis and
sleep. Stern et al. (1972) reported an increase of REM sleep
for seven days after the intraventricular administration of
cycloheximide in cats. Since protein synthesis inhibition was
at 75 percent on the first day and 50 percent by the fourth
day, they suggested that the increase of REM sleep was due to
protein synthesis patterns in the brain that were returning
toward a normal state. In this same study, cycloheximide had
no effect on sleep when injected intraperitoneally. However,
Pegram et al. (1973) reported that such injections of cycloheximide in mice produced a specific decrease of REM sleep.
Recently, we also observed a specific decrease of REM sleep in
rats (Rojas-Ramírez et al., 1977) and cats (Drucker-Colín et
al., 1979a) following the administration of the protein synthesis inhibitors, anisomycin and chloramphenicol. Similar
observations with administration of chloramphenicol have been
made by Kitahama and Valatx (1975) in rats and by Petitjean
(1977) in cats. It is important that in all these studies, SWS
was unaffected except when drugs were administered at very high
doses (Petitjean, 1977). It is interesting (Fig. 12-5) that
the decrease of REM is due to a reduction in frequency, since
mean duration is unaffected (Pegram et al., 1973; Rojas-Ramírez
et al., 1977; Drucker-Colín et al., 1979a). These results may
suggest that protein synthesis could be involved in the
mechanisms that trigger REM sleep.

This possibility was tested by determining the effects of

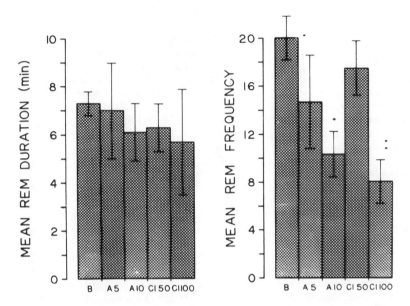

Fig. 12-5. Mean ± SD of duration and frequency of REM sleep periods in cats (N = 13). Note that only frequency was affected. B:baseline; A:anisomycin; Cl:chloramphenicol; 5-10 and 50-100 refers to dose (mg/kg) of these drugs.

chloramphenicol on REM sleep rebound, induced either by REM sleep deprivation or by withdrawal of chronic amphetamine administration. In one such study cats were REM sleep-deprived for 72 hours and the period of recovery recorded for 12 hours on each of two consecutive days with or without chloramphenicol. Chloramphenicol blocked the REM sleep rebound that normally occurs following sleep deprivation, and again it was frequency and not duration of REM that was affected. In another experiment (Drucker-Colín and Benitez, 1977) 12 rats were recorded for one baseline period. They were subsequently administered 10 mg/kg of amphetamine for 15 consecutive days. During this treatment, their sleep-wake cycle was recorded on days 1, 7, and 15. At the end of the 15th day (withdrawal period), the animals were divided into two group of 6, and amphetamine administration was substituted in one group by saline and in the other by 100 mg/kg of chloramphenicol. The results showed that during the withdrawal of amphetamine, the saline group had a significant rebound of both SWS and REM sleep, whereas in the chloramphenicol group the rebound was restricted to SWS only (Fig. 12-6).

Additional experiments have shown that protein synthesis inhibitors also affect some of the components of REM sleep. Observation of any REM sleep period clearly indicates that it is not a homogenous period. This means that eye movements, myoclonic twitches, and high frequency bursts of unit activity constantly oscillate with periods in which these events are absent (Fig. 12-7). Although it has often been reported that the frequency of unit activity, which is a good indicator of phasic events, during REM sleep is higher that that seen in any other phase of the sleep-wake cycle (Steriade and Hobson, 1976), such increase is due solely to bursting periods, since in their absence, unit activity is as low as that seen in other stages (Drucker-Colín et al., 1977). For this reason, REM sleep can be (and should be) analyzed in two distinct periods: phasic (REM_p) and tonic (REM_t).

A sample recording of REM sleep (Fig. 12-7) shows that it can be clearly divided into these two periods. On the right hand graph of this figure, multiple unit activity (MUA) frequency is much higher during REM_p, than during any other period. It is very interesting to note that protein synthesis inhibitors have a very clear and potent effect on MUA frequency during REM_p. In Fig. 12-8, we see that the significant increase in MUA frequency during REM_p is completely abolished by chloramphenicol and anisomycin. A dose-response effect can also be observed.

We have recently calculated that protein synthesis inhibitors decrease to about 50 percent the amount of REM (Drucker-Colín et al., 1979a). A summary of the effects of anisomycin and chloramphenicol on REM latency, REM frequency, and MUA bursts is illustrated in Fig. 12-9. It is interesting to note in this figure that short REM sleep periods seem to be characterized by an absence of phasic bursts of MUA.

Although it is not clear how the decrease of REM_p affects the normal occurrence of REM sleep periods, there is some indication that is may affect its triggering and its duration. Utilizing MUA from MRF neurons as an indicator of phasic activity, the discharge frequency in 30-second periods one minute prior to onset of REM and one minute during REM was calculated. In addition REM periods of more than two minutes duration were compared with those of less than two minutes. An interesting observation emerged. When REM sleep periods were longer than two minutes in duration, spike frequency increased just prior to onset of REM and showed a further increase during the first minute of that state. Similar events have been recorded in pontine reticular structures (Hobson et al., 1974). On the other hand, when REM sleep duration lasted less than two minutes, no such positively inflected rise in discharge

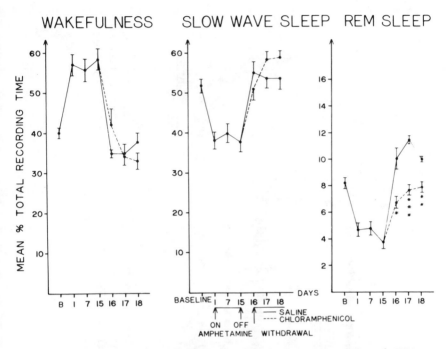

Fig. 12-6. Mean ± SEM percentages of wakefulness, SWS and REM
sleep during the various drug conditions. Note that
chloramphenicol blocks the REM sleep rebound seen
during amphetamine withdrawal. Each point up to day
15 has N=12, each point thereafter has N=6 *P
<0.02, **P < 0.01, ***P < 0.001 (From Drucker-Colín
and Benitez, 1977).

frequency was observed (Drucker-Colín et al., 1979b).

Since, as mentioned previously, it is the phasic discharge of
unit activity that contributes to the rise in frequency observed during REM sleep, these results suggest that in the
absence of phasic discharges REM sleep will be short or
abortive. Most interesting is the observation that in cats
given protein synthesis inhibitors, the discharge rate of MRF
neurons is practically identical to that which occurs in short
duration REM sleep periods (Drucker-Colín et al., 1979b). This
could indicate that by preventing phasic activity, these drugs
reduce the probability of REM sleep periods occurring, and
shorten them when they occur, because of the absence of phasic
avtivity. Unfortunately, the absence of phasic activities with
protein synthesis inhibitors is not always accompanied by an
absence of long REM sleep periods. In fact, as we mentioned

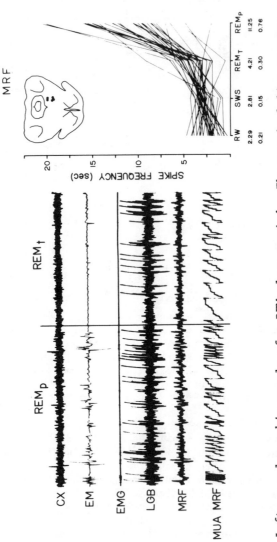

Fig. 12-7. Left, a polygraphic sample of a REM sleep period. The vertical line separates REM$_p$ from REM$_t$. Note how distinct these two periods are. Right: spike frequency per second during relaxed wakefulness, SWS, REM$_t$ and REM$_p$. Note that increased spike frequency during REM is due to activity during REM.

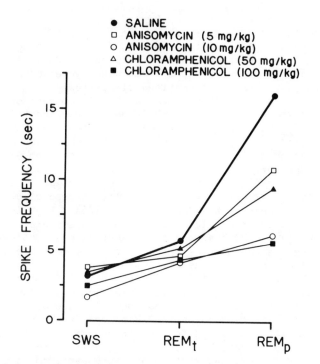

Fig. 12-8. Spike frequency per second of MUA under various drug
conditions during slow wave sleep (SWS), REM$_t$ and
REM$_p$. Note the decrease of spike frequency during
REM$_p$ following administration of protein synthesis
inhibitor, and its dose dependency.

before, these drugs do not seem to alter REM sleep duration.
Although these results give evidence indicating that protein
synthesis inhibitors have a particular effect of REM$_p$, they are
presently difficult to interpret, since there is little information on the meaning of REM sleep periods with little or no
phasic activities. However, in certain pathological states,
such as mental retardation, the amount of REM sleep and
associated phasic activities appears to be significantly
reduced (Clausen et al., 1977; Feinberg, 1968).

Although it is conceivable that certain macromolecules may
participate in the mechanisms that trigger REM sleep, the

evidence so far has been provided by indirect neuropharmacological experiments. We have, therefore, attempted to obtain
support for this hypothesis through more direct experimental
approaches. In one such series of experiments cats implanted
with electrodes for recording the sleep-wake cycle and a push-
pull cannula in the MRF were continuously perfused and recorded
for 12 to 21 hours. At every hour the protein content of the
perfusate and the time spent in wakefulness, SWS, and REM sleep
was determined. We found that protein concentrations in the
perfusate varied in a cyclic fashion and that the peaks of
protein corresponded to those hours in which REM sleep occupied
the greatest proportion of time (Drucker-Colín et al., 1975b)
(Fig. 12-10). Moreover, when insomnia was produced by bilateral lesions of the preoptic area, the cyclic release of proteins
disappeared (Drucker-Colín and Gutiérrez, 1976). Protein
levels also lost their cyclicity following the administration
of chloramphenicol, which specifically decreased REM sleep.
When waking perfusates were compared to those from REM sleep
the latter always contained nearly twice as much protein
(Drucker-Colín and Spanis, 1975). In further studies, separation of released proteins by polyacrylamide sodium dodecyl
sulfate gel electrophoresis revealed the existence of some
large molecular weight proteins in REM sleep perfusates that
were not present in waking perfusates (Spanis et al., 1976).
However, recently we have been unable to confirm this finding.
The differences between two states therefore seem, so far, to
be merely that of concentration, although the possibility
should not be ignored that with more sensitive methods
specificity of the proteins could be detected.

Although these experiments are highly suggestive that proteins may be involved in the regulation of REM sleep, the
evidence is merely correlative. Moreover, it is impossible to
determine whether the protein release is the cause or the
effect of REM sleep. We have therefore tested an approach that
in a more direct form explores whether specific proteins are
indeed involved in the regulation of sleep. This approach
consists in studying the effects of antibodies against proteins
that are obtained from the MRF of cats during REM sleep.

We were led to this approach for three extremely good reasons. Firstly, antibodies can be obtained in relatively large
quantities by immunization with small amounts of perfusate
proteins. Secondly, if a protein is indeed involved in the
regulation of REM sleep, antibodies can amplify the effects on
sleep. Thirdly, antibodies may provide the only available
probe for determining whether proteins really participate in
sleep regulation and for determining receptor regions.

In these experiments Drucker-Colín et al. (*Exp Neurol,
69*:563-575, 1980) some 45 cats were implanted with electrodes

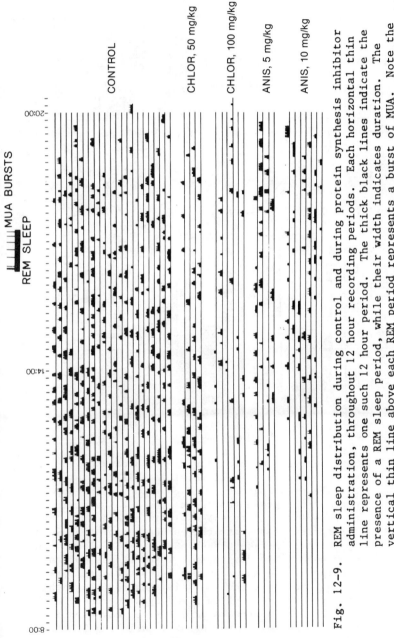

Fig. 12-9. REM sleep distribution during control and during protein synthesis inhibitor administration, throughout 12 hour recording periods. Each horizontal thin line represents one such 12 hour period. The thick black lines indicate the presence of a REM sleep period, while their width indicates duration. The vertical thin line above each REM period represents a burst of MUA. Note the changes in REM sleep, the scarcity of MUA bursts with chloramphenicol (CHLOR) and anisomycin (ANIS) and the usually short duration REM sleep periods in the absence of MUA bursts. (From Drucker-Colin et al., 1979).

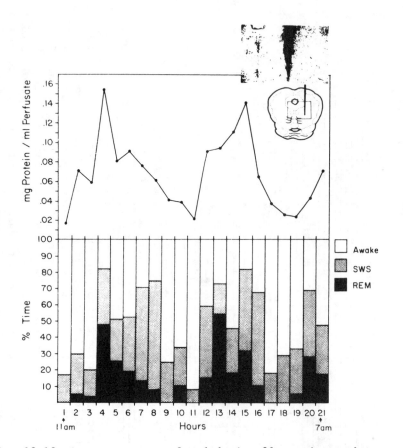

Fig. 12-10. One cat was perfused during 21 continuous hours with continuous electrophysiological monitoring of the sleep-waking cycle. Each point in the upper part of the graph represents the total protein content in the perfusates within each hour. The lower part represents the percentage of time spent by the cat in wakefulness, SWS and REM sleep within each hour. The diagramatic inset of the coronal section of the midbrain indicates the area perfused, and above the inset is a microphotograph of the area squared off in the diagram of the midbrain. Note how peaks of REM sleep correspond to peaks of protein (From Drucker-Colín and Spanis, 1976).

to record the sleep-wake cycle and with a push-pull cannula in the MRF. For a six-month period, perfusates were collected during every REM sleep period, at a flow rate of 20 μl/minutes. Some 70 ml of perfusate was collected, which yielded 1.25 mg of protein. This material was used to obtain antibodies against proteins that appeared in REM sleep. Of 12 bands of proteins in the perfusate, only four were antigenic. Upon purification of the antibodies on a DEAE-cellulose column, they were tested for their effects on the sleep-wake cycle following injection into the MRF of cats who were also implanted with microelectrodes for recording MUA. The effects of the MRF antibodies were compared to gamma globulins from preimmune serum, antialbumin antibodies, anti-cat total serum antibodies, Ringer, and sham injections.

The results showed that the MRF antibodies significantly and specifically decreased REM sleep and, most interestingly, produced effects almost identical to those seen following injections of protein synthesis inhibitors. First of all the antibodies increased the latency of REM sleep to about five hours, in comparison to controls in which mean latency was in order of 50 to 60 minutes. Secondly, antibodies decreased the frequency but not the duration of REM sleep, and only at high doses (500-1000 μg) did the antibodies have any effect on SWS. This effect on SWS which manifested itself as an increase in the latency, was only transient, since total SWS time was unchanged in the 12 hours recorded. In addition to the effects on the sleep-wake cycle, bursts of eye movements and MUA were absent during the first few REM sleep periods but reappeared as the effects of the antibodies started to dissipate.

The significant effects of antibodies provide the most direct evidence for a role of macromolecules in the regulation of sleep. However, a word of caution should be advanced since controls with antibodies from proteins obtained from another area of the brain and antibodies obtained during wakefulness appear to have similar effects. It could therefore be that the antibodies produce nonspecific reactions that manifest themselves in REM sleep. However, antibodies to cat serum proteins failed to have any effects on the sleep-wake cycke. This may suggest that a nonspecific antigen-antibody reaction is not responsible for the effects observed, and that if a polypeptide is indeed involved in the regulation of REM sleep, this molecule is in all probability always present in extracellular fluid. It could then be suggested that this polypeptide is only activated during REM sleep, or else reaches adequate levels only at that particular time. Some support for the latter possibility comes from the fact that during REM sleep there is at least twice the amount of protein present than during wakefulness (Drucker-Colín and Spanis, 1975). It is conceivable therefore that the occurrence of REM sleep depends on a "gating" signal of polypeptide levels, which then initiates all the events that produce REM sleep as we normally see it, that is, atonia, PGO waves, eye movements, and phasic unit activity.

On the other hand, even assuming that some specific macro-molecule is not involved in the regulation of sleep the pro-tein-REM sleep relationship is nonetheless interesting. In a recent article, Adam and Oswald (1977) suggest that sleep may serve the function of tissue restoration. Perhaps during REM sleep proteins subserve the restoration of brain, following waking activity. Or perhaps, protein synthesis during REM sleep aids processes of brain development during the neonatal period, and of information processing in later stages of life. Although there is no substantial evidence supporting these ideas, they may be worth investigating.

ADDENDUM

Since this chapter was originally written, several papers on the subject of sleep peptides have been published. This section will briefly summarize them. Studies with factor S have recently been dropped because extraction from CSF yielded very small quantities. However, on the presumption that Factor S from brain diffuses to the CSF, is then absorbed into the blood and eventually excreted into the urine, a sleep-promoting factor from human urine (SPU) was obtained and shown to have effects similar to those observed with Factor S in rabbits, rats (Krueger et al., *Am J Physiol, 238*:E116-E123, 1980), and cats (García-Arrarás, *Am J Physiol, 241*:E269-E274, 1981). Further studies have recently revealed that SPU is a small glycopeptide, resembling a bacterial peptidoglycan, muramyl dipeptide (Krueger et al., *J Biol Chem, 257*:1664, 1669, 1982). Again the significance of these studies remains to be evaluated by other laboratories, particularly since this glycopeptide could derive from bacterial contamination. Although Krueger et al. (1982) elaborate on several reasons why this couldn't be, it would be of great value to monitor temperature during the course of the sleep studies following SPU administration, since slight increases in temperature may account for the slow waves observed.

In relation to DSIP several recent studies have been unable to replicate the alleged sleep-inducing properties of this nonapeptide (Tobler and Borbely, *Waking and Sleeping,* 4:139-153, 1980; Krueger et al., *Am J Physiol, 238*:E116-E123, 1980; Mendelson et al., in *New Perspectives in Sleep Research,* Spectrum, 1982), despite the fact that DSIP-like immunoreactivity has been found both in developing (Kastin et al., *Brain Res Bull, 7:*687-690, 1981) and adult (Kastin et al., *Brain Res Bull 3:*691-695, 1978) rat brains. One study, however, showed an increase of SWS following 30 nmol/kg of DSIP (Nagasaki et al., *Brain Res, 192:*276-280, 1980). In this same study an increase of both SWS and REM sleep was found following the administration of 4 units of a sleep-promoting factor (SPS) obtained from the brain stem of sleep-deprived rats. In summary, the results obtained with sleep-inducing substances isolated from diverse tissues still remain equivocal.

In relation to growth hormone it has recently been shown that the REM inducing effects observed in rats (Drucker-Colín et al., 1975) and cats (Stern et al., 1975) is also observed in humans (Mendelson et al., *Biol Psychiat, 15:*613-618, 1980). This effect of GH has been suggested to be mediated by protein synthesis (Drucker-Colín, in: *Psychophysiological Aspects of Sleep.* I. Karacan (Ed.), Noyes Medical Publications, pp 80-86, 1981).

It has recently been suggested that arginine vasotocin (AVT) is also a sleep peptide. It has been reported that very small doses of AVT (10^{-6} pg) corresponding to about 600 molecules, administered into the third ventricle of cats induce prolonged periods of SWS, and abolishes REM sleep. This increase in SWS is not observed when AVT is incubated with trypsin and is not mimicked by vasopressin and oxytocin (Pavel et al., *Brain Res Bull, 2:*251-254, 1977). In a subsequent study (Pavel et al., *Brain Res Bull, 4:*731-734, 1979), it was suggested that the effects of AVT are mediated by serotonin, since AVT, but not vasopressin or oxytocin, increases the brain levels of 5-HT. Furthermore, the administration of AVT + fluoxetine, a specific and selective 5-HT uptake inhibitor, doubles the increase in SWS, whereas administration of AVT + methergoline, a 5-HT receptor blocking agent, eliminates SWS. More recently, the administration of a specific vasotocin antiserum to cats induced rapid REM sleep onset periods while increasing REM and decreasing SWS (Pavel and Goldstein, *Brain Res Bull, 7:*453-454, 1981). These effects, being the opposite of those produced by the administration of AVT, are attributed to a neutralization of AVT by the antiserum, which in their words produces "a transient narcoleptic-like disruption of the sleep-wakefulness cycle. Studies with humans have also been done with AVT. Such studies have shown that high amounts of AVT can be detected in lumbar cerebrospinal fluid (CSF) of narcoleptics and Pickwickians coursing with hypersomnia (Popoviciu et al., *Waking and Sleeping, 3:*341-346, 1979) as well as in normal volunteers (Pavel et al., *Waking and Sleeping, 3:*347-352, 1979), when such CSF was removed after awakening from REM sleep. No detectable levels of AVT were observed when awakenings followed non-REM sleep. Subsequently it was shown that 2.5 µg of AVT administered intranasally increased REM sleep and decreased latency for REM in narcoleptic patients (Pavel et al., *Peptides 1:*281-284, 1980). Since in this study melatonin produces similar effects on REM sleep, it is suggested by the authors that AVT is the specific effector of the releasing hormone, pineal melatonin, and that they, respectively, play a role in the induction of REM sleep and in its circadian organization. In contrast to these studies by Pavel, there are two studies which have been unable to find any effects of AVT on sleep of rats (Mendelson et al., *Brain Res, 182:*246-249, 1980; Tobler and Borbely, *Waking and Sleeping,* 4:139-153, 1980). The Mendelson study did however find a decrease of REM sleep at the dose of 50 µg/kg. More recently it has been reported that AVT does decrease REM sleep in rats (Riou et al., *Neuropeptides, 2:* 243-254, 1982).

There are several problems in considering AVT as a sleep peptide. First of all it seems to produce different effects depending on the species. Moreover, it is strange that narcoleptics have high level of AVT, while AVT-antiserum which presumably inactivates AVT, induces narcoleptic-like events in cats. However aside from the difficulties in interpreting the varying results from rats to cats to humans, there are additional problems. The unequivocal presence of AVT in mammalian pineal gland has not been resolved. Originally AVT was detected using a differential bioassay method, though later confirmed by radioimmunoassay and immunochemical staining procedures. Recent reports, however, have failed to confirm such observations (Dogterom et al., *Acta Endocrinol (Copenh) Suppl, 225:*413, 1979; Negro-Vilar et al., *Brain Res Bull,* 4:789-792, 1980). Moreover, by combining high performance liquid chromatography and RIA procedures an almost negligible amount of mammalian pineal AVT has recently been reported (Fisher et al., *66th Ann Meeting of the Endocrine Society,* Washington, pp 103, Abstract 11, 1980). Since AVT has been suggested to have a number of actions in the CNS unrelated to its antidiuretic role (de Wied and Versteeg, *Fed Proc, 38:*2348-2359; Pavel et al., *Neuroendocrinol, 12:*371-375, 1973), it will be important to resolve this controversy because the unequivocal presence of AVT in the pineal gland of mammals is an obvious requirement for accepting its role in physiological events. A related problem is the fact that the pineal gland (origin of AVT) plays a central role in the control of circadian rhythm in birds, but in mammals it plays no such role at all (Kincl et al., *Endocrinol, 87:*38-42, 1970; Menaker, in: *The Neurosciences: Third Study Program,* pp 479-489, MIT Press, 1974).

Finally, three studies have recently reported the effects of various peptides on sleep, and concluded that there are peptides that decrease REM sleep: Angiotensin II, Renin, Substance P and AVT (Riou et al., *Neuropeptides* 2:247-254, 1982); peptides that have no effect on sleep: neurotensin, β-endorphin, leu-enkephalin, met-enkephalin, and choleocystokinin (Riou et al., *Neuropeptides, 2*:255-264, 1982); and one peptide that increases REM which is Vasoactive intestinal Peptide (Riou et al., *Neuropeptides, 2*:265-277, 1982). In this latter study it was also shown that VIP is capable of reversing the insomnia produced by PCPA and the specific decrease of REM sleep induced by choloramphenicol.

CONCLUSIONS

From the foregoing it is evident that the existence of a specific peptide that causes sleep has yet to be conclusively demonstrated, despite the fact that a few are claimed to do so, and one has even been synthesized. It is also evident that the diversity of techniques and hypothesis make the interpretation of the experiments difficult. On the other hand, since sleep is a very complex behavior, it is possible that several sleep factors exist. Regardless, it is quite probable that sleep peptides will turn out to be modulators that will be shown to act at prejunctional or post-junctional sites to regulate or curtail synaptic processes. Therefore in forthcoming studies the interaction of these putative sleep-inducing peptides with neurotransmitters should be studied.

Finally, it should be pointed out that a host of experiments have attempted to explore biochemical parameters of sleep. Two excellent reviews on the biochemistry of sleep have recently been written (Karnovsky and Reich, 1977; Giuditta, 1977). These papers give evidence of the complexity of the sleep process at the cellular level. For this reason, it may be altogether erroneous to believe that a sleep factor exists. It is probably nearer the truth to envisage the sleep process as resulting from a host of factors, among which we could include not only neurotransmitters and peptides, but also such factors as changes in cerebral blood flow and intermediary metabolism. (see McGinty and Drucker-Colín, *Intl Rev Neurobiol, 23*:391-436, 1982). I truly believe that the "one factor-one function" approach to physiological functions usually leads to dead ends. I am equally certain that, although most neurobiologists are aware that only through multifactorial approaches will we understand how sleep occurs, such approaches are not available today. On the other hand, perhaps our hypothetical constructs should be modified and a new way of investigating the nervous system should be sought. Maybe in the future, the long night's sleep of a scientist will provide the answer. Pleasant dreams.

REFERENCES

Adams, K., and Oswald, I.: Sleep is for tissue restoration. *J Roy Coll Physicians Lond 11*:376-388, 1977.

Aserinsky, E.: Periodic respiratory pattern occuring in conjunction with eye movements during sleep. *Science 150*:763-766, 1965.

Aserinsky, E., and Kleitman, N.: Regularly occurring periods of eye motility, and concomitant phenomena during sleep. *Science 118*:273-274, 1953.

Baldridge, B.J., Whitman, R., and Kramer, M.: The concurrence of fine muscle activity and rapid eye movements during sleep. *Psychosom Med 27*:19-26, 1965.

Berger, R.J.: Tonus of extrinsic laryngeal muscles during sleep and dreaming. *Science 134*:840, 1961.

Berlucchi, G., Moruzzi, G., Salvi, G., and Strata, P.: Pupil behavior and ocular movements during synchronized and desynchronized sleep. *Arch Ital Biol 102*:230-244, 1965.

Candland, D.K., Horowitz, S.H., and Culbertson, J.L.: Acquistion and retention of acquired avoidance with gentling as reinforcement. *J Comp Physiol Psychol 61*:50-58, 1962.

Clausen, J., Sersen, E.A., and Lidsky, A.: Sleep patterns in mental retardation: Down's Syndrome. *Electroencephalogr Clin Neurophysiol 43*:183-191, 1977.

DeAndres, I., Gutierrez-Rivas, E., Nava, E., and Reinoso-Suarez, F.: Independence of sleep-wakefulness cycle in an implanted head "encephale isolé." *Neurosci Lett 2*:13-18, 1976.

Dement, W.C.: The occurrence of low voltage fast electroencephalogram patterns during behavioral sleep in the cat. *Electroencephalogr Clin Neurophysical 10*:291-296, 1958.

Dement, W.C., and Kleitman, N.: Cyclic variations in EEG during sleep and their relation to eye movements, body motility and dreaming. *Electroencephalogr Clin Neurophysiol 9*:673-690, 1957.

Dewson, J., Dement, W.C., and Simmons, F.: Middle ear muscle activity in cats during sleep. *Exp Neurol 12*:1-8, 1965.

Drucker-Colín, R.R.: Crossed perfusion of a sleep inducing brain tissue substance in conscious cats. *Brain Res 56*:123-134, 1973.

Drucker-Colín, R.R., and Benitez, J.: REM sleep rebound during withdrawal from chronic amphetamine administration is blocked by chloramphenicol. *Neurosci Lett 6*:267-271, 1977.

Drucker-Colín, R.R., and Gutierrez, M.C.: Effects of forebrain lesions on release of proteins from the midbrain reticular formation during the sleep-wake cycle. *Exp Neurol 52*:339-344, 1976.

Drucker-Colín, R.R., and Spanis, C.W.: Neurohumoral correlates of sleep: Increases of proteins during Rapid Eye Movement sleep. *Experientia 31*:551-552, 1975.

Drucker-Colín, R.R., and Spanis, C.W.: Is there a sleep transmitter? *Progr Neurobiol* 6:1-22, 1976.

Drucker-Colín, R.R., Rojas-Ramirez, J.A., Vera-Trueba, J., Monroy-Ayala, G., and Hernandez-Peon, R.: Effect of cross-perfusion of the midbrain reticular formation upon sleep. *Brain Res* 23:269-273, 1970.

Drucker-Colín, R.R., Spanis, C.W., Hunyadi, J., Sassin, J.F., and McGaugh, J.L.: Growth hormone effects on sleep and wakefulness in the rat. *Neuroendocrinology* 18:1-8, 1975a.

Drucker-Colín, R.R., Spanis, C.W., Cotman, C.W., and McGaugh, J.L.: Changes in protein in perfusates of freely moving cats: Relation to behavioral state. *Science* 187:963-965, 1975b.

Drucker-Colín, R.R., Bernal-Pedraza, J.G., Diaz-Mitoma, F., and Zamora-Quezada, J.: Oscillatory changes in multiple unit activity during rapid eye movement sleep. *Exp Neurol* 57:331-341, 1977.

Drucker-Colín, R.R., Zamora, J., Bernal-Pedraza, J., and Sosa, B.: Modification of REM sleep and associated phasic activities by protein synthesis inhibitors. *Exp Neurol* 63:458-467, 1979a.

Drucker-Colín, R.R., Dreyfus-Cortes, G., and Bernal-Pedraza, J.G.: Differences in multiple unit activity discharge frequency during short and long REM sleep periods: Effects of protein synthesis inhibition. *Behav Neural Biol* 26:123-127, 1979.

Feinberg, I.: Eye movement activity during sleep and intellectual function in mental retardation. *Science* 159:1256, 1968.

Fencl, V., Koski, G., and Pappenheimer, J.R.: Factors in cerebrospinal fluid from goats that affect sleep and activity in rats. *J Physiol (Lond)* 216:565-589, 1971.

Fisher, C., Gross, J., and Zuch, J.: Cycle of penile erection synchronous with dreaming (REM) sleep. *Arch Gen Psychiatry* 12:29-45, 1965.

Gillin, J.C., Mendelson, W.B., Sitaram, N., and Wyatt, R.J.: The neuropharmacology of sleep and wakefulness. *Ann Rev Pharmacol Toxicol* 18:563-579, 1978.

Giuditta, A.: The biochemistry of sleep. In *Biochemical Correlates of Brain Structure and Function*, A.N. Davison, ed. Academic Press, London, 1977, pp 293-337.

Hartmann, E.: *The Sleeping Pill*. Yale University Press, New Haven, 1978.

Hery, F., Pujol, J.F., Lopez, M., Macon, J., and Glowinski, J.: Increased synthesis and utilization of serotonin in the central nervous system of the rat during paradoxical sleep deprivation. *Brain Res* 21:391-403, 1970.

Hobson, J.A., McCarley, R.W., Freeman, R., and Pivik, R.T.: Time course of discharge rate changes by cat pontine brain stem neurons during sleep cycle. *J Neurophysiol* 37:1297-1309, 1974.

Jouvet, M.: Biogenic amines and the states of sleep. *Science* 163:32-41, 1969.

Jouvet, M.: The role of monoamines and acetylcholine in the regulation of the sleep-waking cycle. *Ergeb Physiol* 64:166-307, 1972.

Jouvet, M., and Michel, F.: Correlation electromyographique du sommeil chez le chat decortique et mesencephalique chronique *C R Soc Biol (Paris)* 153:422-425, 1959.

Karnovsky, M.L., and Reich, P.: Biochemistry of sleep. *Adv Neurochem* 2:213-275, 1977.

Kastin, A.J., Nissen, C., Schally, A.V., and Coy, D.H.: Radioimmunoassay of DSIP-like material in rat brain. *Brain Res Bull* 3:691-695, 1978.

Kitahama, K., and Valatx, J.L.: Effect du chloramphenicol et du thiamphenicol sur le sommeil de la souris. *C R Soc Biol (Paris)* 169:1522-1525, 1975.

Korner, A.: Growth hormone control of biosynthesis of protein and ribonucleic acid. *Recent Prog Horm Res* 21:205-238, 1965.

Kroll, F.W.: Veber das Vorkommen von Ubertragbaren schlaferzeugenden stoffen im Hirn schlafender Tiere. *Z Ges Neurol Psychiat* 146:208-218, 1933.

Krueger, J.M., Pappenheimer, J.R., and Karnosky, M.L.: Sleep promoting factor S: Purification and properties. *Proc Natl Acad Sci USA* 75:5235-5238, 1978.

Legendre, R., and Piéron, H.: Le probleme des facteurs du sommeil. Resultats d'injections vasculaires et intra-cerebrales des liquides insomniques. *C R Soc Biol (Paris)* 68:1077-1078, 1910.

Legendre, R., and Piéron, H.: Du developpement au cours de l'insomnie expérimental, des propriétés hypnotoxiques des humeurs en relation avec le besoin croissant de sommeil. *C R Soc Biol (Paris)* 70:190-192, 1911.

Legrendre, R., and Piéron, H.: De la propriété hypnotoxique des humeurs dévelopeés au cours d'une veille prolongée. *C R Soc Biol* 72:210-212, 1912.

Lenard, H.G., and Schulte, F.J.: Polygraphic sleep study in cramopagus twins. *J Neurosurg Psychiatry* 35:756-760, 1972.

Matsumoto, J., Sogabe, K., and Hori-Santiago, Y.: Sleep in parabiosis. *Experientia* 28:1043-1044, 1972.

Mikiten, T., Niebyc, P., and Hendley, C.: EEG desynchronization during behavioral sleep associated with spike discharges from the thalamus of the cat. *Fed Proc* 20:327, 1961.

Monnier, M., and Hatt, A.M.: Humoral transmission of sleep V. New evidence from production of pure sleep hemodialyzate. *Pfluegers Arch* 329:231-234, 1971.

Monnier, M., and Hosli, L.: Dialysis of sleep and waking factors in blood of the rabbit. *Science* 146:796-798, 1964.

Monnier, M., and Hosli, L.: Humoral transmission of sleep and wakefulness II. Hemodialysis of a sleep inducing humor during stimulation of the thalamic somnogenic area. *Pfluegers Arch* 282:60-75, 1965.

Monnier, M., Koller, T., and Graber, S.: Humoral influences of induced sleep and arousal upon electrical brain activity of animals with crossed circulation. *Exp Neurol 8*:264-277, 1963.

Monnier, M., Hatt, A.M., Cueni, L.B., and Schoenenberger, G.A.: Humoral transmission of sleep VI. Purification and assessment of a hypnogenic fraction of "Sleep Dialyzate" (factor delta). *Pfluegers Arch 331*:257-265, 1972.

Monnier, M., Dudler, L., and Schoenenberger, G.A.: Humoral transmission of sleep VIII. Effects of the "Sleep Factor Delta" on cerebral, motor and visceral activities. *Pfluegers Arch 345*:23-35, 1973.

Monnier, M., Dudler, L., Gachter, R., Maier, P.F., Tobler, H.J., and Schoenenberger, G.A.: The delta sleep inducing peptide (DSIP). Comparative properties of the original and synthetic nonapeptide. *Experientia 33*:548-552, 1977.

Morrison, A.R., and Pompeiano, O.: Vestibular influences during sleep IV. Functional relations between vestibular nuclei and lateral geniculate nucleus during desynchronized sleep. *Arch Ital Biol 104*:425-458, 1966.

Moruzzi, G.: The sleep-waking cycle. *Ergeb Physiol 64*:1-165, 1972.

Mouret, J., Jeannerod, M., and Jouvet, M.: L'activité electrique du systeme visual au cours de la phase paradoxale du sommeil chez le chat. *J Physiol (Paris) 55*:305-306, 1963.

Myers, R.D.: An improved push-pull cannula system for perfusing an isolated region of the brain. *Physiol Behav 5*:243-246, 1970.

Nagasaki, H., Iriki, M., Indue, S., and Uchizond, K.: The presence of a sleep promoting material in the brain of sleep-deprived rats. *Proc Jap Acad 50*:241-246, 1974.

Nagasaki, H., Iriki, M., and Uchizono, K.: Inhibitory effect of the brain extract of sleep-deprived rats (BE-SDR) on the spontaneous discharges of crayfish abdominal ganglion. *Brain Res 109*:202-205, 1976.

Pappenheimer, J.R., Miller, T.B., and Goodrich, C.A.: Sleep promoting effects of cerebrospinal fluid from sleep-deprived goats. *Proc Natl Acad Sci USA 58*:513-517, 1967.

Pappenheimer, J.R., Koski, G., Fencl, V., Karnovsky, M.L., and Krueger, J.: Extraction of sleep promoting factor S from cerebrospinal fluid and from brains of sleep deprived animals. *J Neurophysiol 38*:1299-1311, 1975.

Pegram, V., Hammond, D., and Bridgers, W.: The effects of protein synthesis inhibition on sleep in mice. *Behav Biol 9*:377-382, 1973.

Petitjean, F.: Antibiotiques et sommeil. *These se Doctorat*, Lyon, France, 1977.

Piéron, H.: *Le probleme physiologigue du sommeil*. Masson, Paris, 1913.

Pivik, R.T., Hobson, J.A., and McCarley, R.W.: Eye movement associated rate changes in neuronal activity during desynchronized sleep: A comparison of brain stem regions. *Sleep Res 2*:35, 1973.

Polc, P., Schneeberger, J., and Haefely, W.: Effect of the delta sleep-inducing peptide (DSIP) on the sleep-wakefulness cycle of cats. *Neurosci Lett* 9:33-36, 1978.

Purpura, D.: A neurohumoral mechanism of reticular cortical activation. *Am J Physiol* 186:50-54, 1956.

Oswald, I.: Human brain protein, drugs and dreams. *Nature (Lond)* 223:893-897, 1969.

Ramm, P.: The locus coeruleus, catecholamines and REM sleep: A critical review. *Behav Neural Biol* 25:415-448, 1979.

Rechtschaffen, A., and Kales, A.: A manual of standardized terminology, techniques and scoring system for sleep stages of human subjects. Brain Information Service/Brain Research Institute, Los Angeles, 1968.

Ringle, D.A., and Herndon, B.L.: Plasma dialyzates from sleep-deprived rabbits and their effect on the electrocorticogram of rats. *Pfluegers Arch* 303:344-349, 1968.

Ringle, D., and Herndon, B.: Effects on rats of CSF from sleep deprived rabbits. *Pfluegers Arch* 306:320-328, 1969.

Rojas-Ramírez, J.A., Aguilar-Jimenez, E., Posadas-Andrews, A., Bernal-Pedraza, J., and Drucker-Colín, R.R.: The effects of various protein synthesis inhibitors on the sleep-wake cycle of rats. *Psychopharmacology* 53:147-150, 1977.

Salzarulo, P., Lairy, G.C., Bancaud, J., and Munari, C.: Direct depth recording of the striate cortex during REM sleep in man: Are there PGO potentials? *Electroencephalogr Clin Neurophysiol* 38:199-202, 1975.

Sassin, J.: Sleep related hormones. In *Neurobiology of Sleep and Memory*, R.R. Drucker-Colín and J.L. McGaugh, eds. Academic Press, New York, 1977, pp 361-372.

Schnedorf, J.G., and Ivy, A.C.: An examination of the hypnotoxin theory of sleep. *Am J Physiol* 125:191-205, 1939.

Schnitzer, S.B., and Ross, S.: Effects of physiological saline injection on locomotor activity in C57 BL/6 mice. *Psychol Rep* 6:351-354, 1960.

Schoenenberger, G.A., and Monnier, M.: Characterization of a delta-electroencephalogram (sleep) inducing peptide. *Proc Natl Acad Sci USA* 74:1282-1286, 1977.

Schoenenberger, G.A., Cueni, L.B., Monnier, M., and Hatt, A.M.: Humoral transmission of sleep. VII Isolation and physico-chemical characterization of the "Sleep Inducing Factor Delta." *Pfluegers Arch* 338:1-17, 1972.

Schoenenberger, G.A., Maier, P.F., Tobler, H.J., and Monnier, M.: A naturally occurring delta-EEG enhancing nonapeptide in rabbits X. Final isolation, characterization and activity test. *Pfluegers Arch* 369:99-109, 1977.

Schoenenberger, G.A., Maier, P.F., Tobler, H.J., Wilson, K., and Monnier, M.: The delta EEG (sleep) inducing peptide (DSIP) XI. Amino acid analysis, sequence, synthesis and activity of the nonapeptide. *Pfluegers Arch 378*:119-129, 1978.

Snyder, F., Hobson, J., Morrison, D., and Goldfrank, F.: Changes in respiration, heart rate and systolic blood pressure in human sleep. *J Appl Physiol 19*:417-422, 1964.

Spanis, C.W., Gutierrez, M.C., and Drucker-Colín, R.R.: Neurohumoral correlates of sleep: Further biochemical and physiological characterization of sleep purfusates. *Pharmacol Biochem Behav 5*:165-173, 1976.

Steriade, M., and Hobson, J.A.: Neuronal activity during the sleep-waking cycle. *Prog Neurobiol 6*:155-376, 1976.

Stern, W.C., and Morgane, P.J.: Sleep and memory: Effects of growth hormone on sleep, brain biochemistry and behavior. In *Neurobiology of Sleep and Memory*, R.R. Drucker-Colín and J.L. McGaugh, eds. Academic Press, New York, 1977, pp 373-410.

Stern, W.C., Miller, F.P., Cox, R.H., and Maickel, R.P.: Brain norepinephrine and serotonin levels following REM sleep deprivation in the rat. *Psychopharmacology 22*:50-55, 1971.

Stern, W.C., Morgane, P.J., Panksepp, J., Solovick, A.J., and Jalowiec, J.E.: Elevation of REM sleep following inhibition of protein synthesis. *Brain Res 47*:254-258, 1972.

Stern, W.C., Jalowiec, E., Shabshalowitz, H., and Morgane, P.J.: Effects of growth hormone on sleep-waking patterns in cats. *Horm Behav 6*:189-196, 1975.

Takahashi, Y., Kipnis, D.M., and Daughaday, W.H.: Growth hormone secretion during sleep. *J Clin Invest 47*:2079-2090, 1968,

13

Hormones of the Thyroid Axis and Behavior

PETER T. LOOSEN AND ARTHUR J. PRANGE, JR.

The nervous system and the endocrine system are the major
communication systems of the organism. Important aspects of
the mechanisms by which the two systems exchange information
have recently been been described. Thus, some hormones of
peripheral origin (McEwen et al. 1970), including thyroid
hormones (Stumpf and Sar, 1975), are taken up more or less
specifically by brain neurons. Furthermore, groups of hypothalamic cells that receive diverse neural input secrete
peptide hormones that influence the activity of the anterior
pituitary gland. These hypothalamic cells have been called
transducers between the nervous and endocrine systems.

The notion that activity of the thyroid gland influences
human behavior is much older than our knowlede of the exact
nature of thyroid secretions. In 1873 Gull observed that
myxedema could result in psychosis; L-thyroxine (T_4) was
identified in 1926, L-triiodothyronine(T_3) in 1951 (Astwood,
1970). In a similar way, the concept that mental ṣtate, or
life events impinging on mental state, might influence thyroid
activity antedates the description of how this might occur. In
1825 Parry referred to fright as a cause of thyrotoxicosis, but
it has been only 10 years since Guillemin (1978) and Schally
(1978) and their colleagues identified thyrotropin-releasing
hormone (TRH, pGlu-His-Pro-NH$_2$) as an apparent connection
between the nervous system and the hypothalamic-pituitary-
thyroid (HPT) axis.

The two areas of investigation just adumbrated--behavioral
changes in thyroid disease and thyroid changes associated with
altered mental states--provide a useful organizing principle
for considering the mass of data that pertain to brain-thyroid
relationships. Animal investigations have been indispensable

for studying the cellular mechanisms by which thyroid hormones exert their protean tissue effects and their end-point mor-phological results (Sterling, 1979) and for studying syste-matically the many interactions between thyroid state and drug toxicity (Prange and Lipton, 1962). They have been less useful for understanding thyroid-behavioral relationships. For example, it has not yet been possible in an animal model to produce hyperthyroidism by the imposition of stress (Hertzel et al., 1955; Dougier et al., 1956; Tingley et al., 1958; Volpe et al., 1960). This is either a criticism of the concept that in man stress may produce hyperthyroidism or a shortcoming of animal techniques. Tata (1968) pointed out that experimentally induced changes in thyroid state have usually been so gross that secondary effects, for example malnutrition, have com-plicated interpretation of results. This criticism is even more apt when applied to behavioral questions. Elsewhere in this volume, Kuhn and Schanberg review the effects of the thyroid state on brain development and behavior emphasizing animal studies. What follows is concerned only with behavior in man.

BEHAVIORAL DISTURBANCES IN THYROID DISORDERS

Hyperthyroidism

The mental symtoms of patients suffering from hyperthyroidism are well known. Prominent among them are irritability, fatigue, anxiety, tension, and emotional lability (Taylor, 1975; Brenner, 1978, Bursten, 1961; Mandelbrote and Wittkower, 1955; Kleinschmidt and Waxenberg, 1956; Robbins and Vinson, 1960; Wilson et al., 1962; Artunkal and Togrol, 1964; Herman and Quarton, 1965; Whybrow et al., 1969; MacCrimmon, 1979). A very small minority of patients are withdrawn and apathetic, exhibiting so-called "apathetic hyperthyroidism" (Brenner, 1978). Some patients may develop secondary mania (Michael and Gibbons, 1963; Villani and Weitzel, 1979). Profound mental disorder is uncommon in hyperthyroidism, although delirium is a hallmark of thyroid storm, presently a rare condition. Appro-priate antithyroid therapy usually greatly reduces the mental symptoms that may accompany hyperthyroidism. These matters have been throughly reviewed (Relkin, 1969; Michael and Gibbons, 1963; Williams, 1970; Gibson, 1962; Loosen and Prange, 1980a; Whybrow and Hurwitz, 1976).

Table 13-1 summarizes the occurrence of psychiatric mani-festations of hyperthyroidism in eight studies pertaining to unselected hyperthyroid populations (Mandelbrote and Wittkower, e 13-1. Psychiatric Manifestations of Hyperthyroidism

ors	Year	No. of patients	Major psychological disturbance	Psychosis	Cognitive disturbance	Improvement after treatment	Authors' comment
lelbrote and kower	1955	25	Anxiety, depression impulsivity, lack of assertion	0	nc[a]	Not retested	68% showed neurotic difficulties
nschmidt and nberg	1956	17	Anxiety, destructive aggression, and depression	2	nc	8 improved. Pattern inconsistent	Depression usually precedes toxic state
ins and Vinson	1960	10	High "neuroticism" score	0	Impairment	Cognition improved	Significant link between psychologic & clinical assessment of toxicity
on et al.	1962	26	Increased "jittery," decreased "aggressive" Scores on Clyde Mood scale	0	Impairment (14)	Not retested	15 described subjective depression: 2 elation
nkal and Togrol	1964	20	MMPI peaks on paranoia and schizophrenia	Suggested by MMPI	Response to auditory stimuli reduced	Improved. No specific change	MMPI depression scale stable before and after treatment in 10 subjects tested
ann and Quarton	1965	24	Anxiety; irritability	1	Memory impaired	Not retested	
row et al.	1969	10	Fatigue, anxiety, irritability	2	Yes, but milder than in hypo-thyroidism	7 of 9 retested patients improved	In 70% symptoms were enough to "constitute a psychiatric illness"
immon et al.	1979	19	Nervousness, tension irritability, anxiety	0	Impairment	Retesting 3 wk & 10 mo after initial testing. Improvement of psychiatric and cognitive function in the latter	Psychiatric symptoms and cognitive dysfunction significantly correlated to serum T levels prior to treatment

= no specific comment by author.

1955; Kleinschmidt and Waxenberg, 1956; Robbins and Vinson, 1960; Wilson et al., 1962; Artunkal and Togrol, 1964; Herman and Quarton, 1965; Whybrow et al., 1969; MacCrimmon et al., 1979). Whybrow et al. (1969) observed in 7 out of 10 hyperthyroid patients a profound disruption sufficient to "constitute a psychiatric illness." The same authors compared hyperthyroid patients with hypothyroid patients and found hypothyroid patients to be significantly more depressed, with greater psychomotor retardation and reciprocally less motor tension. Cognitive dsyfunction was observed in both conditions, although to a milder degree in hyperthyroid patients. In the latter, cognitive dysfunction returned to normal when euthyrodism was established. In long-standing hypothyroidism, however, there was evidence to suggest that the cognitive disturbance may persist after thyroid replacement therapy. The authors concluded that "in contrast to the cognitive disturbance, where a departure from euthyroidism in either direction resulted in generalized disruption of intellectual function, affective disturbance appeared to be intimately and specifically linked to the prevailing thyroid state."

MacCrimmon et al. (1979) studied 19 consecutive female patients with hyperthyroidism. All patients were studied in the acute hyperthyroid state and one and three weeks thereafter. They were also studied 10 months after the initial testing, when euthyroidism had been achieved by radioiodine treatment. A matched group of controls was studied in parallel. The investigators reported that "group differences on cognitive measures did not reach statistical significance, but cognitive deficits and any symptoms of emotional disorder were significantly associated with the severity of thyroid toxicity previous to treatment."

Taken together the data indicate that a hyperthyroid state is usually associated with significant mental disturbance. However, how, if at all, does a hyperthyroid state developing in a patient with a history of major affective disorder influence the course of the disease? Checkley (1978) studied 267 patients with major affective disorders and found that five patients had eight well-documented episodes of thyrotoxicosis. Only three episodes, however, coincided with an affective illness. In each case an alternative explanation for this association was available.

Hypothyroidism

After Gull's report, previously cited, the Clinical Society of London established a committee on myxedema to study the relationship between this condition and mental disorders. In

1888 they reported that 16 of 45 myxedema cases reviewed had presented with insanity. This "myxematous insanity" was characterized by mania, dementia or melancholia, paranoid delusions, and hallucinations. The committee also observed a general slowing of thought process in their patients (Clinical Society of London, 1888).

Most studies of hypothyroidism have consisted of case reports with no follow-up evaluations (Asher, 1949; Wiesel, 1952; Miller, 1952; Jonas, 1952; Logothetis, 1963; Libow and Durell, 1965; Tonks, 1964; Pomeranze and King, 1966; Pitts and Guze, 1966; Easson, 1966; Nordgren and von Scheele, 1976). Consideration of these studies indicates that frequent mental symptoms in myxedema are depression, pronounced loss of interest, generalized slowing of activity, memory loss, and apathy. Sachar (1975) has stated that depression is present in more than 40 percent of hypothyroid patients. In the study of Whybrow et al. (1969) all but one of the hypothyroid patients reported depression. In addition, delusions and auditory hallucinations may occur, usually persecutory (Pomeranze and King, 1966; Nordgren and von Scheele, 1976; Reed and Bland, 1977; Granet and Kalman, 1978). Dementia may occur in extreme cases (Relkin, 1969; Michael and Gibbons, 1963; Williams, 1970). "Myxedema madness," a term coined by Asher (1949), has been described as an organic psychosis (Asher, 1949; Logothetis, 1963), and the importance of differentiating this organic psychosis from a functional psychosis has been stressed (Nordgren and von Scheele, 1976; Granet and Kalman, 1978). Taken together, the data suggest that there is not a specific psychosis associated with hypothyroidism. As in hyperthyroidism, cognition, mood, and behavior are usually disturbed in varying degrees. As Crammond (1968) put it, "All organic psychoses be they acute or chronic affect all the mental processes: cognition, mood, and behavior. All three aspects of brain function will be affected but the degree of involvement of each component will vary from person to person and also according to the duration and development of the illness."

Table 13-2 summarizes the occurrence of psychiatric manifestations in patients with hypothyroidism. In these five studies (Whybrow et al., 1969; Jain, 1972; Crown, 1949; Reitan, 1963; Schon et al., 1961) an unselected population was used. Whybrow and his co-workers (1969) found depression, fatigue, anxiety, and irritability the most frequent complaints in their patients. Hypothyroid patients were significantly less tense, less agitated and confused, but more depressed than hyperthyroid patients. After euthyroidism was achieved by substitution therapy, there was a reduction of psychopathology in three of four patients evaluated. Improvement was observed in such areas as depressive affect, motor retardation, and

Table 13-2. Psychiatric Manifestations of Hyperthyroidism

Authors	Year	No. of patients	Major psychological disturbance	Psychosis	Cognitive disturbance	Improvement after treatment	Authors' comment
Crown	1949	4	nc[a]	nc	4	4	
Reitan	1963	15	nc	nc	Impaired	nc	On Rorscharch similar to brain damaged group
Schon et al.	1961	24	nc	nc	Impaired	24	Social withdrawal improved with treatment
Whybrow et al.	1969	7	Depression 5 Anxiety 1	1	6	4	nc
Jain	1972	30	Anxiety 10 Depression 13	2[b]	8	Some improvement in all patients but still high depression scores after euthyroidism is achieved	nc

[a] nc = no comment by author.

[b] One had a long history of schizophrenia.

cognitive function. Jain (1972) studied 30 consecutive
patients with a diagnosis of hypothyroidism. Eight patients
were judged to be psychiatrically normal. As indicated in
Table 13-2, 10 patients presented with anxiety, 13 with
symptoms of depression. After euthyroidism was achieved, all
patients were reported to show subjective improvement. However, as the author stated: "even after successful treatment of
the patient's physical condition, abnormally high scores were
obtained on the scales of anxiety and depression [indicating
the importance of] supplementing physical treatment with
psychiatric treatment when there is evidence of affective
disturbance."
   Psychiatric manifestations may be the earliest symptoms and
even precede the onset of recognizable myxedema (Asher, 1949;
Reed and Bland, 1977; Davidoff and Gill, 1977; Logothetis,
1963). Reed and Bland (1977) reported a case of "masked
myxedema madness" and pointed out that "failure to recognize
the endocrinopathy may not only produce recovery difficulties
but also psychiatric and endocrine repercussions if psychotropic medications are given in such masked cases." Replacement therapy of thyroid hormones is reported sometimes to
fail to alleviate the mental symptoms completely (Jain, 1972,
for a review). Pitts and Guze (1966) pointed to the effectiveness of electroconvulsive therapy (ECT) in such cases.

Thyroid Abnormalities in Behavioral Disorders

When describing thyroid changes in behavioral disorders, one is
often unsure whether or not such changes are causes, perpet-
uating influences, or secondary effects. Stress contributes to
illness and illness becomes a stress. As Brown (1951) put it,
"cause is not a chain but a net." In addition, psychiatric
diagnoses, mainly based on phenomenologic techniques, rarely
describe distinct entities and there are often overlapping
symptoms between two or more diagonoses. It is not astonishing
therefore that a specific endocrine finding for a specific
psychiatric entity still awaits description.

General Psychiatric Population

The frequent association between thyroid gland dysfunction and
mental disturbance led several investigators to study the
incidence of thyroid abnormalities in a general psychiatric
population. The early studies on this subject have been ably
reviewed by Michael and Gibbons (1963). Using methods then
available, Reiss and his co-workers (1953) studied thyroid

activity in 1000 psychiatric patients. About 10 percent of these patients showed thyroid values outside normal limits. Nicholson and his co-workers (1976) screened 98 unselected female psychiatric admissions and found a hypothyroid state in three patients. However, if only women older than the age of 40 years were considered, the abnormality was seen in 8 percent of the population. The authors suggested routine screening for thyroid disorder in female psychiatric patients older than the age of 40. Weinberg and Katzell (1977) measured serum T_4 and free $T_4(FT_4)$ index in 50 patients recently admitted to a psychiatric hospital. Three patients were found to have thyrotoxic symptoms with increased serum thyroid hormone levels. Their psychiatric state improved after remission of the thyrotoxicosis. McLarty et al. (1978) measured serum T_4 and T_3 concentration in 1206 psychiatric inpatients. Five men and one woman were hypothyroid (0.5 percent); eight women were hyperthyroid (0.7 percent). The authors found no evidence that treatment with phenothiazines or benzodiazepines had a significant effect on thyroid hormone levels. Cohen and Swigar (1979) studied 480 newly admitted psychiatric patients, measuring total T_4, estimated FT_4 (EFT_4),* and thyroxine binding capacity. Eighteen percent of patients showed abnormal (high or low) EFT_4 levels. In half of these patients, EFT_4 values spontaneously returned to normal within several weeks.

The data suggest that the indidence of thyroid dysfunction in a psychiatric population if higher than in the general population. However, as suggested by Cohen and Swigar (1979), these abnormalities are often of a transient nature.

Depression

Thyroid Hormone Levels

Although most depressed patients appear to be euthyroid by usual criteria, some show thyroid hormone levels in the upper normal range. This was found by Dewhurst et al. (1968) in an early study. A similar finding was reported by Whybrow et al. (1972), and in their patients heightened thyroid activity was positively correlated with a prompt clinical response to imipramine. Our group (Loosen and Prange, 1980b) recently studied baseline thyroid hormone levels in five unipolar

---

*EFT_4 is analogous to FT_4 index. Abnormalities in these parameters have the same significane as changes in the absolute concentrations of free T_4.

depressed women, measuring serum T_3, T_4, and FT_4 index. T_3 and FT_4 index were found to be significantly increased in these depressed patients as compared to sex- and age-matched controls. The resolution of a depressive episode is occasionally followed by hyperthyroidism (Kleinschmidt and Waxenberg, 1956; Lidz, 1949), although this is uncommon.

Some investigators have reported lowered thyroid function in depressed patients. Rybakowski and Sowinski (1973) studied 15 patients with major affective disorders, six showing mania and nine, depression. Thyroid activity (FT_4 index and direct measurement of free T_4) was significantly reduced in both groups. Similar results were reported by Rinieris and his co-workers (1978a), who studied 15 manic patients and 25 depressed patients. Thyroid activity was assessed by measuring serum T_4 and T_3 uptake, and by calculating the FT_4 index. In 14 of 15 manic patients, thyroid function was normal. One manic patient showed clearly subnormal T_4 and FT_4 index values. In contrast, 5 of 25 depressed patients showed subnormal FT_4 index and T_4 values. These five cases, however, did not form a separate group within the depressed population, since plotting the values of all depressed patients revealed a continuous distribution. The mean values of T_4 and FT_4 index in the depressed patients were significantly lower than in a group of normal controls, although within normal limits. Severity of depression or mania was not correlated with any thyroid measures. The same authors (Rinieris et al., 1978b) then compared the FT_4 index values of 25 patients with psychotic depression with sex- and age-matched patients suffering from neurotic depression. The former but not the latter showed a significantly lower FT_4 value when compared to euthyroid controls. The difference was also significant between the two groups, FT_4 index being lower in psychotic than in neurotic depressed patients.

Several authors, however, have not found significant changes in thyroid hormone levels in depressed patients. Leichter et al. (1977) studied thyroid activity in 11 female patients with primary depression before and after treatment with amitriptyline. Before treatment, serum levels of T_4, T_3, T_3 uptake, and thyroid-stimulating hormone (TSH) were found to be normal. Although a satisfactory behavioral improvement occurred during treatment, there was no significant change in any thyroid hormone value. The authors concluded that "the effects of amitriptyline on depression are not related to detectable changes in endogenous thyroid function." Kirkegaard et al. (1978) studied 19 patients with unipolar depression, 12 with bipolar depression, 14 with mania, and five with manic-depressive illness. There was a significantly reduced T_3 level in the manic patients as compared to controls. Serum T_4 was found to be slightly elevated in the mixed manic-depressive and in the unipolar group. However, these differences disappeared when data were corrected for differences in thyroxine-binding proteins. Kolakowska and Swigar (1977) measured the EFT_4 levels in 155 euthyroid patients suffering from depression, depression and alcohol abuse, or alcohol abuse. EFT_4 levels were found to be positively correlated with severity of depression and negatively with alcohol abuse. Nevertheless, in absolute terms the thyroid values of depressed patients without alcohol abuse were in the normal range. Treatment with either tricyclic antidepressants, phenothiazines, or benzodiazepines had no reliable effect on EFT_4 levels. Karlberg et al. (1978) measured FT_4 index values in 12 patients with primary depression. Six received TRH, 80 mg orally for three weeks, six 100 mg amitriptyline daily during the same time interval. FT_4 index values before treatment were normal. After treatment, they increased gradually in the TRH group and remained unchanged in the amitriptyline group. In both groups, behavioral improvement was seen, although in varying degrees. In the TRH group, however, there was a significant negative correlation between pretreatment FT_4 index values and behavioral improvement. The data suggest that a subgroup of depressed patients with low thyroid function may benefit from TRH administration. This concept contrasts with the finding of Whybrow et al. (1972), cited earlier, that thyroid activation correlates positively with imipramine response. Hatotani et al. (1977) studied thyroid activity in 51 depressed patients. Serum T_3, T_4, and the FT_4 index were found to be normal. Gregoire et al. (1977) measured serum T_3 in 19 patients with primary affective disorders before and after clinical improvement. In both conditions T_3 was found to be normal.

The TSH Response to TRH

TSH serum levels are usually found to be undisturbed in depressed patients (Kirkegaard et al. 1975; Loosen et al., 1977,1978a). In a recent study, Weeke and Weeke (1978) measured serum TSH in 19 patients with endogenous depression. Blood samples were taken at 2:00 P.M. and midnight. The degree of diurnal variation of TSH was found to be inversely related to severity of depression. Whereas less depressed patients showed about a twofold increase of TSH at midnight, no such increase occured in severely depressed patients.
    Our group (Prange et al., 1972) first reported that the TSH response to TRH is blunted in some patients with primary depression. Two of 10 unipolar depressed women showed virtually no TSH response to TRH. This finding has been widely

confirmed (Table 13-3) and reviewed (Loosen et al., 1976; Loosen and Prange, 1982).

TSH blunting seems to occur in about 25 percent of patients with primary depression, although these patients have been proved to be euthyroid by all other criteria. The fault does not seem to be related to previous drug intake, diagnosis, sex, age, or severity of illness (Loosen et al., 1976; Loosen and Prange, 1982). There is also no evidence that TSH blunting correlates with mental improvement after TRH injection (Loosen and Prange, 1982). Two recent studies (Gold et al., 1977,1979) reported significantly lower TSH responses in unipolar than in bipolar depressed patients. This difference, however, did not emerge from other studies (Coppen et al., 1974; Loosen et al., 1977; Ehrensing et al., 1974; Takahashi et al., 1974; Maeda et al., 1975). Some reports indicate that the fault usually resolves with remission of depression (Hatatoni et al., 1977, Gregoire et al., 1977; Gold et al., 1977; Van den Burg et al., 1976), although this does not occur in all patients (Coppen et al., 1974; Kirkegaard et al., 1975; Loosen et al., 1977; Maeda et al., 1975). There are several conditions in which blunted TSH responses to TRH may be observed (Table 13-4). However, these have usually been excluded in the depressed patients in whom the response has been studied.

As noted earlier, some degree of thyroid activation is sometimes seen in depression (Dewhurst et al., 1968; Whybrow et al., 1972; Loosen and Prange, 1980b), and it is possible that such increased thyroidal activity could damp the TSH response through enhanced negative feedback. However, there is increasing evidence that depressed patients with blunted TSH responses show low thyroid hormone levels at baseline. Takahashi et al. (1974) identified this phenomenon using the FT_4 index. Hatotani et al. (1977) reported low T_3 levels at baseline in TSH nonresponders. These data, of course, indicate a disturbed feedback inhibition of thyroid hormones on the TSH response in some depressed patients. This concept is further supported by our finding (Loosen et al., 1980) that pretreatment with thyroid hormones inhibits the TSH response in normal subjects but not in those depressed patients, the majority, who do not show blunting.

Our group recently reported that, in some depressed patients, elevated serum cortisol may account for TSH blunting (Loosen et al., 1978a,b). However, TSH blunting sometimes persists into remission (Coppen et al., 1974; Kirkegaard et al., 1975; Loosen et al., 1977; Maeda et al., 1975) when serum cortisol is no longer found to be elevated (Carroll and Mendels, 1976). Thus, in some patients the fault may be related to genetic predisposition rather than to the acute state of depression.

Table 13.3. TRH Induced Response In Psychiatric Disorders

Authors	N	

Depression

Prange et al. (1972)	10	All responses borderline low or absent. thyroid state normal
Kastin et al. (1972)	5	Four showed diminished responses
Shopsin et al. (1973)	2	Responses normal
Coppen et al. (1974)	16	Four showed no response. Thyroid state normal
Chazot et al. (1974)	30	Fifteen showed diminished responses
Ehrensing et al. (1974)	8	Three showed diminished responses
Hutton (1974)	1	Baseline TSH as well as TSH response diminished
Takahashi et al. (1974)	36	Twelve showed diminished responses. TSH nonresponders show low FT_4 index at baseline. High incidence of TSH blunting in chronic depression
Dimitrikoudi et al. (1974)	2	Responses normal
Van den Burg et al. (1975)	10	Mean TSH response diminished. Thyroid state normal
Kirkegaard et al. (1975)	15	Mean response diminished in depression as compared to recovery; diminished response related to early relapse
Widerlov and Sjostrom (1975)	10	Baseline TSH as well as TSH responses lower than in hospitalized, nondepressed controls
Maeda et al. (1975)	13	Mean response diminished as compared to controls
Van den Burg et al. (1976)	10	Responses normal. TRH infused during 4 hour period, possibly masking blunting

Authors	N	
Pecknold and Ban (1977)	6	Diminished in all patients. All patients euthyroid
Loosen et al. (1977)	23	Six showed diminished responses. No correlation with age, severity of illness, clinical subtypes or clinical remission
Gold et al. (1977)	23	TSH responses lower in unipolar than in bipolar patients or controls. Negative correlation of CSF-5 HIAA and TSH response in unipolar patients
Gregoire et al. (1977)	19	TSH response to TRH diminished in depression, normalized after clinical remission
Vogel et al. (1977)	15	Baseline TSH as well as response to TRH diminished
Hatotani et al. (1977)	51	TSH response disturbance in 19 patients: 6 delayed, 6 blunted, and 7 exaggerated. TSH normalization upon recovery. Low T at baseline in TSH nonresponders
Loosen et al. (1978a,b)	7	TSH blunting related to serum cortisol elevation
Kirkegaard et al. (1978)	74	Mean TSH response diminished in manic depressive disease; normal in neurotic and reactive depression
Brambilla et al. (1978)	16	TSH response disturbed in 6 patients: 4 blunted and 2 exaggerated
Naeije et al. (1978)	16	Mean secretory TSH area after TRH significantly lowered in depressed patients, in patients with barbiturate coma due to suicide attempt and after recovery from such coma
Karlberg et al. (1978)	6	No TSH blunting, however blunting was defined as $\Delta TSH < 2.5$ $\mu U/ml$

Authors	N	
Ettigi et al. (1978)	9	Mean TSH response significantly lowered in patients with primary depression (5). No difference between secondary depression (4) and controls
Tsutsui et al. (1979)	11	Mean TSH response significantly lowered in patients with normalization upon recovery
Amsterdam et al. (1979)	34	Clearly blunted TSH response in 7 patients
Gold et al. (1979)	10	ΔTSH in unipolar depression significantly lower than in bipolar depression
Loosen and Prange (1980b)	12	4 clearly blunted responses

Mania

Takahashi (1974)	8	TSH response somewhat diminished
McLarty et al. (1975)	21	Some exaggerated responses to TRH in patients on Lithium maintenance
Kirkegaard (1978)	14	Mean TSH responses somewhat reduced

Alcoholism

Loosen et al. (1979)	26	6 of 12 men in acute alcohol withdrawal and 3 of 10 men after withdrawal (one week later) show TSH blunting. In two men TSH blunting is observed on both occasions

Schizophrenia

Loosen et al. (1977)	7	TSH responses in normal range
Prange et al. (1979a)	17	TSH responses in normal range

Table 13-4. Conditions in Which TSH Blunting May Occur

Thyroid hormone excess
    Administration (Burger and Patel, 1977)
    Hyperthyroidism (Burger and Patel, 1977)
Klinefelter syndrome (Ozawa and Shishiba, 1975)
36-hour starvation (Vinik et al., 1975)
Excess SRIF activity (Vale et al., 1973)
Chronic renal failure (Czernichow et al., 1976)
Glucocortocoid excess
    Administration (Otsuki et al., 1973; Re et al, 1976)
    Cushing's disease (Otsuki et al., 1973)
Dopamine excess (Burrow et al., 1977)
Mental depression (Reviewed in Table 6)
Chronic alcoholism (loosen et al., 1979)
Heroin addiction (Afrasiabi et al., 1977)

It is now generally accepted that biogenic amines may modulate the TRH-induced TSH response (Frohman, 1975). There is evidence that TRH and dopamine are antagonistic in regard to TSH secretion from the anterior pituitary gland, a dopamine excess leading to a blunted TSH response *in vivo* (Burrow et al., 1977). Thus, one may postulate a hyperdopaminergic state in depressed TSH nonresponders. However, data on prolactin (PRL) levels in depressed patients do no support this notion (Maeda et al., 1975; Sachar et al., 1973; Pecknold and Ban, 1977). PRL secretion is considered to be under inhibitory control of dopamine, possibly through the mediation of a PRL inhibitory factor (Schally, 1978). Recently, Gold et al. (1977) demonstrated a significant negative correlation between TSH response and 5-hydroxyindoleacetic acid (5-HIAA) in cerebrospinal fluid of six unipolar depressed men.

Growth hormone (GH) release inhibiting factor (somatostatin, SRIF) inhibits the release of GH from the anterior pituitary gland (Martin, 1973). SRIF also blunts the TSH response to TRH (Vale et al., 1973). Since in some depressed patients, the GH response to insulin hypoglycemia is deficient (Gruen et al., 1975), it appears possible that in some depressed patients excess SRIF activity may account for the observed diminished TSH response to TRH. Several reports, however, do not support this notion. Maeda et al. (1975) and Chazot et al. (1974) recently documented that some depressed patients show a GH hormone release after TRH. This otherwise occurs rarely except in acromegaly (Irie and Tsushima, 1972), a condition in which SRIF activity is more likely to be diminished than increased. Additional study utilizing GH provocation tests in TSH responders and nonresponders is needed to elucidate this problem.

Is the blunted TSH response specific to depression? This question was recently studied by our group (Loosen and Prange, 1980b). TRH was injected in 23 controls and in 17 schizophrenic, 33 alcoholic, and 12 depressed patients. A blunted TSH response was seen in some depressed and some alcoholic patients but not in schizophrenic patients, indicating that the fault is not limited to depression. We concluded that in both conditions, alcoholism and depression, further study is needed to evaluate whether the fault is related to the acute states of these diseases or to some predisposing factor.

Here it is of importance to note that there is increasing clinical and epidemiological evidence for features common in both alcoholism and depression. Patients often show aspects of both disorders and frequently it is impossible to determine which, if either, is primary and which is secondary. Furthermore, both share suicide as a common cause of death and lithium, the drug of choice in the prevention of recurrent depression, may have beneficial effects in chronic alcoholics. Genetic relationships between alcoholism and depression have also been noted (Goodwin and Erickson, 1979; for a review).

Mania

Lithium, the presently preferred treatment for mania (Goodwin and Ebert, 1973), is a potent antithyroid substance (Temple et al., 1972), and manic patients do not escape this effect. It has been suggested that the antithyroid action of lithium is not irrelevant to its antimanic action (Whybrow, 1972). McLarty et al. (1975) have shown that during the course of lithium treatment, manic patients respond to TRH with exaggerated TSH responses, suggesting a persisting shift toward hypothyroidism (Pittman, 1974). Kirkegaard et al. (1978) studied serum T_4, T_3, FT_4 index, FT_3 index, and TSH in 14 manic patients. T_3 and FT_3 index were found to be significantly decreased. The TSH response to TRH tended to be somewhat diminished when compared to values from normal controls.

Schizophrenia

As early as 1890, Kraepelin (1896, 1917) observed a similarity of symptoms between hypothyroidism and schizophrenia. It was suggested that an active or even hyperactive HPT axis may modify beneficially the course of the illness (Bleuler, 1954; Brauchitsch, 1961; Cleghorn, 1950; Freeman, 1935; Gregory,

1956). These notions stimulated psychoendocrine investigations
in schizophrenic patients. Early studies of thyroid function
in schizophrenic patients have been ably reviewed (Bleuler,
1954; Michael and Gibbons, 1963). Here it suffices to state
that several studies have failed to demonstrate a thyroid
impairment in chronic schizophrenia (Brambilla et al., 1976;
Plunkett et al., 1964; Simpson and Cooper, 1966; Suwa and
Yamashita, 1972). Dewhurst et al. (1968) found increased
baseline TSH levels in 5 of 20 patients. Serum TSH elevation
correlated significantly with ratings of paranoid symptoms. In
a recent study Prange et al. (1979a) reported increased FT_4
index and diminished T_3 values in 17 schizophrenic patients as
compared to age- and sex-matched controls. The patients also
showed diminished T_4 binding rather than increased total T_4.
The three findings, of course, suggest the possibility of
diminished enzymatic conversion of T_4 to T_3. They also
suggest, however, a reduction in protein binding as the
initiating fault in thyroid dynamics. MacSweeney and his co-
workers (1978) studied the incidence of thyroid disease in
mothers of 104 schizophrenic patients and found a significantly
higher incidence of thyroid disease than in a carefully matched
control group.

Childhood Autism

Kahn (1970) reported diminished T_3 uptake values in 44 of 62
autistic children. Campbell and her co-workers (1972) studied
14 psychotic children and six severely disturbed nonpsychotic
preschool children. Although all children were reported to be
euthyroid, serum T_4 was somewhat elevated in four children who
responded well behaviorally to pharmacological doses of T_3.
The investigators felt that altered thyroid function and
beneficial behavioral response to T_3 may be related. In a
recent report, however, Abassi et al. (1978) demonstrated
normal serum T_4, T_3, and TSH values in 13 autistic childern.
  The TSH response to TRH was studied by Campbell et al. (1979)
in 10 psychotic childern. Three children showed a diminished
or blunted TSH response; two showed an exaggerated response.

Alcoholism

Earlier reports indicated the occurrence of hypothroidism in
alcoholic patients (Goldberg, 1960;1962), although this was not
confirmed by later studies (Augustine, 1967; Selzer and Van
Housten, 1964). Kolakowska and Swigar (1977) studied EFT_4
values in 23 chronic alcohol abusers and 20 intoxicated

abusers. EFT_4 levels were negatively related to alcohol abuse, intoxicated patients showing the lowest EFT_4 values. Our group (Loosen and Prange, 1977; Loosen et al., 1979) recently studied 21 depressed alcoholic men in acute alcohol withdrawal and 14 men one week after all symptoms of acute withdrawal had shown complete remission. TRH was injected on both occasions and blood was collected for assay of T_3, T_4, FT_4 index, and TSH. There were no reliable baseline differences in TSH between groups. However, both alcoholic groups showed a diminished TSH rise over baseline (ΔTSH) as compared to normal controls. Six of 12 men in acute withdrawal and 3 of 10 men in postwithdrawal showed a blunted TSH response to TRH. Three of the six men with TSH blunting in acute withdrawal were retested after withdrawal. In two men the TSH response remained blunted, whereas it became normal in one, demonstrating that TSH blunting observed in acute withdrawal may or may not be corrected upon recovery. In regard to thyroid hormones, FT_4 index and total T_4, but not binding capacity for thyroid hormones, were found to be significantly increased in patients in acute withdrawal as compared to both patients after withdrawal and to normal controls.

Williams (1976) recently suggested that alcohol may, among other effects, induce increased secretion of TSH and, thus, if ingested chronically, play an important role in the development of thyroid cancer. However, several reports do no support this notion. Lepaeluoto et al. (1975) demonstrated that acute ethanol administration did not alter basal or TRH-induced TSH secretion in nine healthy male volunteers. Ylikahri et al. (1978) studied the effects of ethanol on anterior pituitary secretion in 12 male volunteers during acute alcohol intoxication (4 hours after start of drinking) and during hangover (14 hours after start of drinking). In both conditions baseline levels of T_4, T_3, and TSH were normal, as was the TRH-induced TSH response. The PRL response to TRH, however, was found to be virtually abolished during hangover. As mentioned previously, our group found no evidence for increased TSH secretion in men in acute alcohol withdrawal and one week later when withdrawal symptoms had shown remission (Loosen and Prange, 1979; Loosen et al., 1979).

THE EFFECTS OF HORMONES OF THE HYPOTHALAMIC-PITUITARY-THYROID AXIS ON BEHAVIOR

Thyrotropin-Releasing Hormone

TRH is active in many animal behavior tests and appears to modify the action of several behaviorally active drugs. These

data are discussed by Nemeroff et al. in Chapter 6 of this
volume. Here it may suffice to say that the tripeptide mimics
some effects of amphetamine, although the differences between
TRH and amphetamine are striking (Manberg et al., 1979). It
has recently been reported that TRH selectively depresses
excitation of rat cerebral cortical neurons evoked by L-
glutamate but not by acteylcholine. The data, of course,
suggest a selective interaction between TRH and L-glutamate or,
in broader terms, between a centrally acting peptide and an
excitatory amino acid neurotransmitter (see Renaud, chapter 5).
Neurochemical effects are controversial, but there are data to
suggest an interaction between TRH and certain central nervous
system (CNS) monoamines systems. We have reviewed these data
in detail (Prange et al., 1978a,b;1979b).

Depression

Using a double-blind crossover design, we studied 10 women with
unipolar depression. TRH (0.6 mg IV) produced a rapid, though
brief and partial, improvement (Prange and Wilson, 1972; Prange
et al., 1972). Patients improved within a few hours and tended to relapse to baseline severity within one week. Maximum
improvement as measured by Hamilton's Rating Scale for Depression (Hamilton, 1960) was about 50 percent less than full
remission.

Other investigators have addressed the problem of using TRH
as an efficient remedy for depression. The results are
disappointing. A summary of these studies is given in Table
13-5. Generalizations about the studies performed are difficult; size of dose, frequency and route of administration,
and population characteristics have varied greatly. Pecknold
and Ban (1977) repeated our original experiment and obtained
substantially the same results. Furlong et al. (1976) have
suggested that differences in results might be attributed to
the existence of endocrinologically distinct types of depression. Among the negative trials, the one performed by
Kieley et al. (1976) may be the most instructive. They gave
massive doses of the hormone to patients predicted to be poor
drug responders. Although inferior to placebo, TRH was clearly
active, for it produced intolerable side effects. Some
patients showed evidence of hyperthyroidism. Here it is
important to note that there is increasing evidence that the
behavioral effects of TRH in depression are quite sensitive to
the level of circulating thyroid hormones. In the study by
Karlberg et al. (1978) already cited, the behavioral improvement after TRH in depression was negatively correlated with
pretreatment serum FT_4 index values, patients with low FT_4
values responding best behaviorally to TRH. Loosen et al.

Table 13-5. Behavioral Studies of TRH in Depression and Schizophrenia

Studies	Positive	Negative
Depression		
Oral TRH:		
Single blind	1	0
	(1 patient)	
Double blind	1	4
	(4 patients)	(66 patients)
Total	2	4
	(5 patients)	(66 patients)
Intravenous TRH:		
Single blind	5	6
	(188 patients)	(61 patients)
Double blind	6	8
	(43 patients)	(109 patients)
Total	11	14
	(231 patients)	(170 patients)
Schizophrenia		
Oral TRH:		
Single blind	1	1
	(62 patients)	(9 patients)
Double blind	1	1
	(143 patients)	(12 patients)
Total	2	2
	(205 patients)	(21 patients)
Intravenous TRH:		
Single blind	1	2
	(4 patients)	(6 patients)
Double blind	1	1
	(10 patients)	(10 patients)
Total	2	3
	(14 patients)	(16 patients)

(1980) recently demonstrated that pretreatment with a single
dose of thyroid hormones may virtually abolish the behavioral
response to TRH in depression. Since repeated oral doses of
TRH (but not a single IV dose) often lead to a sustained
thyroid activation (Burger and Patel, 1977), it becomes
plausible that these findings may have contributed to the
negative findings concerning TRH in depression. Table 13-5
contains information relevant to this notion. After oral TRH,
the ratio between positive and negative studies (double-blind)
is one to four. After a single IV injection, however, the same
ratio is six to eight. More study is needed to elucidate this
finding, since its mechanism is presently not well understood.

Mania

The many trials of TRH in depression contrast with the paucity
of such trials in mania. In a double-blind, placebo-controlled
study of five euthyroid manic men, Huey et al. (1975) reported
reliable advantages for TRH (0.5 mg IV) as compared to saline.

Schizophrenia

Most authors have reported that TRH exerts behavioral effects
in schizophrenia, although the effects are not always beneficial. Five studies involving 37 patients did not find TRH to
be active. Drayson (1974) gave repeated injections of TRH, 0.2
mg, to three patients with "cyclical psychosis." None of his
patients showed significant change in psychopathology. In a
double-blind, crossover design, Clark et al. (1975) gave TRH,
300 mg orally daily, for three weeks to 12 schizophrenic
patients. There were no systematic behavioral changes.
Lindstrom and his colleagues (1977) studied 10 chronic
schizophrenic patients in a double-blind crossover design using
TRH (0.6 mg IV) on four consecutive days. The results of this
study were substantially negative. Bigelow et al. (1975) found
slight worsening of depression in two of three treatment-
resistant schizophrenic patients. In an open trial, Davis et
al. (1975) gave nine schizophrenic men TRH, 300 mg orally per
day for 14 days. Seven patients worsened, and this was
especially marked if they showed paranoid symptoms. One
withdrawan schizophrenic patient showed clear improvement.
  Four studies involving 219 schizophrenic patients have revealed benefical effects. Our group described beneficial
effects of TRH in four schizophrenic patients in a preliminary
study (Wilson et al., 1973a). We then studied 10 more patients
in a double-blind crossover design using IV nicotinic acid as

an active placebo to mimic the occasional side effects of the
peptide. TRH (0.5 mg IV) produced beneficial effects in
patients in whom social withdrawal and related symptoms were
prominent. The duration of improvement was variable but
averaged about 10 days. TRH seemed to aggravate the mental
state of a small subgroup of paranoid patients (Prange et al.,
1979a). Inanaga et al. (1975) administered oral TRH (4 mg per
day) in an open study of 62 chronic schizophrenic patients who
concomitantly took standard neuroleptics and in whom reduced
spontaneity and apathy were evident. In about 75 percent of
the patients a favorable response was observed within two
weeks. The same investigators later treated 143 similar
patients with oral TRH, 4 mg per day for 14 days, or with
placebo, in a double-blind procedure (Inanaga et al., 1978).
TRH was significantly more effective in producing over-all
improvement, especially in enhancing motivation and social
contact.

The results outlined suggest that chronic schizophrenic
patients tend to benefit from TRH if they display such symptoms
as social withdrawal, apathy, and anhedonia. Paranoid
patients, however, may be worsened by administration of the
peptide, and the differential effect may be another distinction
between paranoid schzophrenics and nonparanoid schizophrenics
(Tsuang and Winokur, 1974; Buchsbaum and Rieder, 1979). Since
most of the studies cited used TRH (orally or IV) in repeated
doses, more information is needed pertaining to the effects of
thyroid hormones on the behavioral response to TRH in schizophrenic patients.

Childhood Autism

In a preliminary trial Campbell et al. (1978) reported that TRH
(0.4 mg IV) produced beneficial behavioral effects in 8 of 10
autistic children.

Hyperkinetic Syndrome

Tiwary et al. (1975) studied two hyperactive children using a
double-blind crossover design. Both had responded poorly to
methylphenidate administration. For two days after injection
of TRH (0.2 mg), most aspects of their behavior were notably
improved.

Alcoholism

As cited previously, our group has administered TRH in men in
alcohol withdrawal syndrome who showed secondary depression.
In a double-blind study of 33 patients we compared the effects

of TRH (0.5 mg IV) to the effects of nicotinic acid and of
saline, both used as placebos (Loosen et al., 1979; Loosen and
Prange, 1980b). Significant benefits from the hormone were
observed on Factor I of the Hamilton Rating Scale for
Depression, but only three hours after injection. At later
times, all groups showed rapid improvement. Factor I is
concerned mainly with psychomotor retardation and depression.
   Huey et al. (1975) injected TRH (0.5 mg IV) in three patients
in predelirium tremens. They found no beneficial effects. Two
patients in milder stages of withdrawal, however, experienced
an increased sense of well-being and relaxation.

Male Sexual Impotence

Benkert (1975) found oral TRH no more effective in male sexual
impotence than placebo in a double-blind, crossover study of 12
patients.

Parkinsonism

Several investigators have studied the effects of TRH in
Parkinson's disease. Chase et al. (1974) did not find significant changes in neurological symptoms after TRH administration. However, their patients reported an increased sense
of well-being and optimism. Lakke et al. (1974) also found the
neurological symptoms of Parkinson's disease unaltered after
TRH, although the hormone reduced depression scores. McCaul
and his colleagues (1974) gave TRH to three patients already
taking L-DOPA. Two patients experienced a "dramatic
improvement in well-being, including enhanced clarity of
thought." Neurological symptoms were essentially unchanged.

Normal Women

We performed a double-blind, crossover trial of TRH (0.5 mg IV)
compared to saline, in 10 normal women (Wilson et al., 1973b).
After TRH, subjects showed significant relaxation, mild
euphoria, and increased energy. These changes were not related
to frequency of side effects. We performed a similar study in
20 additional women and obtained similar results (Wilson et
al., 1980). Using a double-blind, placebo-controlled crossover
design, Betts et al. (1976) administered TRH (80 mg orally) or
placebo to 10 normal women. For behavioral assessment, they
used subjective and objective rating scales as well as videotaped interview sessions. Using the latter, a blind observer

was asked to analyze the nonverbal components, such as
"smiling, laughing, and looking around." The authors observed
more smiling and laughing after TRH than after placebo and
reported a "significant effect on certain measures of nonverbal
behavior where subjective reports of mood change and conventional objective measures of this were unaffected."

Thyroid Stimulating Hormone

Recently Moldow and Yalow (1978) have reported that TSH occurs
in the hypothalamus as well as the anterior pituitary of man
and is even more widely distributed in rat brain. The pituitary does, however, appear to be the only site of synthesis.

Depression

We performed a double-blind, placebo-controlled study to
discern whether bovine TSH (10 IU IM) would accelerate the
antidepressant action of imipramine in depressed women (Prange
et al., 1970). Patients who received imipramine plus the
hormone improved more rapidly than those who received
imipramine plus saline injection. In this experiment TSH may
have exerted an antidepressant effect independent of its
endocrine effects. However, a sufficient explanation for the
finding is that imipramine potentiation was the consequence of
enhanced thyroid hormone secretion prompted by TSH. This
interpretation is consistent with previous studies demonstrating that oral administration of T_3 accelerates the
therapeutic action of imipramine in depressed women (Prange et
al., 1969; Wilson et al., 1970).

Thyroid Hormones

Early in this century, thyroid extract was a widely used
hormone preparation in psychiatry. It was given to schizophrenic patients by, among others, Kraepelin (1896, 1917), von
Wagner-Jauregg (1914), and M. Bleuler (1954). At that time it
was believed that thyroid hormones might modify the course of
the disease (Bleuler, 1954; Brauchitsch, 1961; Cleghorn, 1950;
Freeman, 1935; Gregory, 1956). However, later studies failed
to establish the efficacy of thyroid hormones in the treatment
of adult psychosis (Michael and Gibbons, 1963; Hollister et
al., 1973, for reviews). The use of thyroid hormones in contemporary psychiatry seems limited to three disorders:
periodic catatonia, childhood autism, and depression, as an
adjunct to tricyclic therapy.

Periodic Catatonia

Kraepelin recognized that some patients with dementia praecox, especially if catatonic, show a periodic course of the disease (Kraepelin, 1896,1917). The syndrome was thoroughly studied and described by Gjessing (1974). It is characterized by periodic phases of catatonic excitement (or, more rarely, stupor) separated by comparatively normal intervals. The syndrome can last for many years or show spontaneous remission, especially in younger patients after brief illness. It was Gjessling's contribution to show that, among other things, basal metabolic rate is low in the free interval and high in the psychotic phase. In addition, like mental states, nitrogen balance shows marked periodicity. Gjessing reported that "the nitrogen balance period is equal to the length of the periodic catatonic syndrome, but does not necessarily coincide in time with the psychotic or vegetative phase." Treatment with thyroid hormones was reported to exert beneficial effects on both metabolic disturbances and mental symptoms (see Gjessling, 1974, for a review).

Recently, patients with periodic catatonia have been treated with lithium. Some studies report beneficial behavioral effects (Annell, 1969; Hanna et al., 1972), although not all agree (Sorner and McHugh, 1974). In the latter study lithium was found to stabilize the cyclic changes in blood urea nitrogen as well as blood pressure and pulse rate, while leaving the behavioral changes and the course of the disease unchanged.

Childhood Autism

Sherwin et al. (1958) first reported the behavioral effects of T_3 in euthyroid autistic children. Two children received 75 μg T_3 per day for an extended time. They showed improvement in "affective contact and attraction to and interest in achievement."

Subsequently, the behavioral effects of T_3 in autistic children were studied by Campbell and her co-workers. In their first uncontrolled study (Campbell et al., 1972) 16 autistic children received T_3 in a daily dose of 75 μg. Marked improvement was noted in 11 children. In a second uncontrolled study (Campbell et al., 1973), T_3 was administered over a period of 11 to 19 weeks to 14 psychotic children and 6 severely disturbed, nonpsychotic, euthyroid, preschool-age children. T_3 was found to be an effective therapeutic agent in both groups: "It had antipsychotic as well as stimulating properties." Campbell and her colleagues (1978) then performed a third study, based on a placebo-controlled crossover design, in 30 euthyroid children with autism. Siginificant behavioral effects were observed in four children. The authors, arriving at a more cautious conclusion, noted that "except for a few symptoms that were reduced on T_3, the drug did not differ from placebo."

Thyroid Hormones as an Adjunct to Tricyclic Antidepressant Therapy

After witnessing in the clinic what appeared to be a transient toxic interaction between a usual dose of imipramine and an excessive replacement dose of thyroid hormones (Prange, 1963), and after confirming, in mice, that gross hyperthyroidism enhances imipramine lethality, we proposed that the drug-hormone interaction appearing as toxicity might be rendered therapeutically useful, if the dose of hormone were reduced. As reviewed elsewhere (Prange et al., 1976), this appears to be the case. Two studies by our group (Prange et al., 1969; Wilson et al., 1970), and two studies by English groups (Wheatley, 1972; Coppen et al., 1972), together involving 127 patients, have shown that when imipramine or amitriptyline is given in usual doses and accompanied by as little as 20 μg T_3 per day, the antidepressant effect of the regimen occurs much faster that with tricyclic antidepressants alone. This acceleration of the therapeutic response has been more marked in women that in men. Indeed, in a third study, in men (Prange, 1971), we could find no effect of T_3 whatsoever. We did find, however, that men responded faster to imipramine alone than did women. Thus, men have less to gain from T_3 and, plausibly for this reason, show less benefit.

Feighner and his colleagues (1972) found only a trend for T_3 to hasten the imipramine response. They used somewhat more imipramine than employed in the positive studies and used T_3 for a shorter time. Using a very small number of patients, Steiner et al. (1978) found no differences in response rates between three groups of four depressed women, one group receiving imipramine alone, and another imipramine plus T_3, the third ECT alone.

In all seven of the double-blind, controlled studies (Table 13-6), patients were euthyroid at the outset and remained euthyroid throughout the course of treatment. T_3 has not increased toxicity; in the study by Coppen et al. (1972) the hormone appeared to reduce toxicity.

Although extended latency of therapeutic action has been a complaint about tricyclic antidepressants since their introduction, the T_3 phenomenon has been addressed more often,

-6. Thyroid Hormones in Combination With Tricyclics in Untreated Depressed Patients

rs	Dose[a]	Patients	N	Blind	Control	Type	Results and Comment
t al.	150 mg IMI q D 25 µg T_3 q D or P. Days 4–28	Women with primary depression, mostly unipolar. Motor retardation	20	Double	Placebo against T_3	Parallel comparison	T_3-treated patients achieved remission twice as fast as controls
t al.	150 mg IMI q D 25 µg T_3 q D or P. Days 4–28	Women with primary depression, mostly unipolar. No motor retardation. Some agitated	20	Double	Placebo against T_3	Parallel comparison	T_3-treated patients achieved remission twice as fast as controls
	150 mg IMI q D 25 µg T_3 q D or P. Days 4–28	Men with primary depression, mostly unipolar. Mixed retarded and non-retarded	10	Double	Placebo against T_3	Parallel comparison	No T_3 advantage. Patients showed a prompter IMI response than women in preceding studies
et al.	200 mg IMI q D 25 µg T_3 q D or P. Days 2–11	Consecutive patients with primary depression	49	Double	Placebo against T_3	Parallel comparison	Nonsignificant trend for T_3-treated patients to improve more rapidly than control patients
	100 mg AMI q D 20 or 40 µg T_3 q D or P. Days 4–28	Men and women. Depressed out-patients	57	Double	Placebo against T_3	Parallel comparison	Both doses of T_3 were significantly more effective than P. Larger dose of T_3 tended to be more effective. Women profited more from T_3 than men
t al.	150 mg IMI q D 25 µg T_3 q D or P. Days 1–14; or 9 g L-trypto-phan q D 25 µg T_3 q D or P. Days 1–14	Men and women with severe unipolar depression	30	Double	Placebo against T_3	Parallel comparison	T_3 potentiated response to IMI but not to L-tryptophan. T_3 effect limited to women, all of whom achieved depression scores of 0
et al.	150 mg IMI q D 25 g T_3 or P. Days 1–35 or ECT alone	Women with endo-genous depression unipolar or bipolar	12	Double	Placebo against T_3	Parallel comparison	All treatments equally effective, 3 of 4 patients being "responders" in each group

ramine; q D: every day; P: placebo; AMI: amitriptyline.

although less systematically, to a related problem, the problem of inefficacy. During the past decade, six published studies (Table 13-7) and one unpublished study (F.K. Goodwin et al., 1982), have been concerned with T_3 as an adjunctive means of treating patients who have failed to show an adequate response to tricyclic antidepressants alone. All studies have been positive. They have involved a variety of tricyclic antidepressants in a variety of doses with various amounts of T_3. The conversion rate--therapeutic failure to therapeutic success--is in the order to two-thirds. The conversion rate, quite interestingly, appears about as high in men as in women. Here it is important to remember that depressed men, like depressed women, who have failed to respond to tricyclic antidepressants are, at least by virture of failure, a special case. Goodwin et al. (1982) showed that T_3 can convert "failures" to "successes" even when blood levels of tricyclic antidepressants have been in the range that is usually effective.

How does T_3 accelerate the tricyclic antidepressant response in pharmacologically naive women and convert failures to successes in both sexes? It is unlikely that T_3 exerts an effect on the metabolism of the drugs. First, tricyclic antidepressant toxicity is unchanged or reduced; second, T_3 can exert its effect even when drug levels have been apparently adequate; third, a preliminary study of T_3 on imipramine (and desmethylimipramine) blood levels showed no effect whatever (Garbutt et al., 1979). This latter finding was consistent with an earlier study in animals by Breese et al. (1974).

It remains likely that T_3 in the presence of imipramine enhances the activity of central noradrenergic receptors. The work of Frazer et al. (1974) tends to support this notion. They found that chronic administration of T_3 did not affect adrenergic responsiveness in brain as measured by the ability of norepinephrine (NE) to stimulate cyclic AMP in brain slices. Chronic imipramine treatment, however, reduced cyclic AMP formation caused by NE stimulation. When T_3 was added, the imipramine effect was partly offset. The data suggest that T_3 may offset an antiadrenergic effect of chronic imipramine.

DISCUSSION

Many data support the concept that hormones of the HPT axis play a role in mental disorders. When the function of the HPT axis is grossly disturbed, mental disorder usually occurs; when TRH or a thyroid hormone is given to a patient with mental disorder, a behavioral change is usually effected. Mason (1968) pointed out that both the nervous system and the HPT e 13-7: Thyroid Hormones In Combination With Tricyclics In Depressed Patients Who Did Not Respond to Tricyclics Alone

thors	Dose[a]	Patients	N	Blind	Control	Type	Comment
e 970)	IMI, AMI and protrypt In various doses 25 μg T_3	Men and women with retarded depression	25	Single	None	Before-after comparison	70% showed improvement under T_3 and tricyclics
a et al. 974)	Various tricyclics, 20-30 μg T_3	Men and women old and young, unipolar and bipolar depressed	44	Single	None	Before-after comparison	66% showed a good to excellent response after T_3 was added, usually within four days
ca et al. 974)	Various doses of AMI or chlor IMI. Various doses of T_4-T_3 preparation	Men and women depressed patients, 42-52 years old	16	Single	None	Before-after comparison	Thyroid preparations effective, especially with chlor imipramine
l 975)	75-200 mg AMI q D, 100-300 mg trl IMI q D, 20-40 μg T_3 after day 10	Men and women hospitalized depressed patients all of whom had shown a poor tricyclic antidepressant response after 10 days	96	Single	Increased dose of AMI to 300 mg of D	Parallel com- parison, after 10 days, between patient receiv- increased tri- cyclic & patient receiving T_3	T_3 effective In 39 of 52 patients, additional tricyclic effective in 10 of 44 patients
l 977)	75-200 AMI q D 20-40 μg T_3 q D after day 14	Women with primary depres- sion, unipolar and bipolar, all of whom had shown poor AMI response after 14 days	49	Single	Increased dose of AMI to 300 mg of D	Parallel com- parsion, after 14 days, between patient receiving increased AMI and patient receiving T_3	T_3 effective In 23 of 33 patients; additional AMI effective in 4 of 16 patients
sui et al. 979)	Various doses of any of six TCA; 10-25 μg T_3 q D	Men with protrac- ted primary depression who showed diminished TSH response to TRH challenge	11	Single	None	Before-after comparison	10 of 11 rated (globally) Improved by T_3

imipramine; AMI:amitriptyline; q D:everyday; TCA:

axis participate in adaptations to environmental events. Although this review has been limited to hormones of the HPT axis, we wish specifically to disclaim the notion that these hormones are unique in influencing mental processes. Probably most hormones play a role. Some, like the HPT hormones, may exert global effects. Others, angiotensin for example, may influence behaviors as discrete as fluid ingestion (see Phillips, chapter 10).

However discrete the behavioral actions of some hormones may be in animal models, there has yet to be described a precise correlation between an endocrinopathy and a human mental illness. If one considers that most hormones probably affect many brain processes (which in turn are affected by many hormones, among other factors) and that the symptoms of any diagnostic entity are not only varied but overlap with those of other entities, then this degree of specificity is probably not to be expected. It still might be possible, however, to identify the mental effects characteristic of a given endocrinopathy. Here we think that careful attention to the concomitants of slight degrees of an endocrinopathy are most likely to yield useful information. As an endocrinopathy becomes more severe or more chronic, its mental effects tend increasingly to resemble the mental effects of other endocrinopathies. The end result is usually an organic brain syndrome. Longitudinal studies, difficult as they are to perform, would allow consideration of mild disturbances and would provide the opportunity to distinguish cause from effect in many cases. A related problem has been evident in most work on mental changes in patients with abnormal endocrine function. This is the lack of a suitable comparison population to control for morbidity. Although depression is common in hypothyroid patients, one cannot presently be sure that it would not be equally common in a population experiencing equal morbidity from, say, diabetes mellitus.

Despite the limitations of the available data, we think that evidence converging from several points suggest that thyroid hormones have more to do with affective state than with other aspects of mentation, save possibly recent memory. First, hypothyroidism does seem to be accompanied by a high incidence of depression. Second, it now has been widely confirmed that a small dose of a thyroid hormone will accelerate, or in other cases permit, the antidepressant effect of tricyclic antidepressants. On the other hand, although hypothyroidism may produce depression, depressed patients as a group clearly are not hypothyroid. Some may even show thyroid activation, but this may represent an adaptive change. Thyroid activation correlates with a favorable response to the use of tricyclic antidepressants alone, and it subsides with remission from illness. In another place we have formulated these findings more extensively (Prange et al., 1976).

TRH provides an example of the broad spectrum of activity of some hormones. It appears to exert behavioral effects in patients in several diagnostic categories and in normal subjects and may benefit all of them, except paranoid schizophrenics, at least briefly. TRH seems to increase the sense of well-being, motivation, and coping capacity. In all instances of mental disorder these functions are probably impaired to some extent. However, variability in these functions is probably more descriptive of the fundamental set of the organism than of any specific disease state. Furthermore, they are not the specific foci of rating scales designed for assessing one or another mental illness. In this regard it is noteworthy that Betts et al. (1976) could identify TRH effects from watching, in a distributive way, videotapes, but not from the discrete information contained in rating scales.

TRH has also been used in a way that illustrates another stragegy of modern psychoendocrinology, the strategy of administering a hormone (or some other substance) that elicits an endocrine response or responses, noting these responses and comparing them between a diagnostic group and normal controls and between different diagnostic groups. Such a strategy might be expected to clarify the relationships between groups and also to help describe the nature of the fundamental alteration of the organism whether neural or endocrinological. TRH has been useful in the first instance in revealing that some depressed and some alcoholic patients, but not schizophrenic patients, have in common a blunted TSH response. In both conditions, further study is needed to evaluate whether the fault is related to the acute disease state or its genetic predisposition. The latter possibility is especially tantalizing in view of the postulated genetic relationship between the two disorders (Winokur et al., 1971).

TRH has also been helpful in elucidating the state of the nervous system in some patients. The implication of a blunted TSH response, of course, depends upon its total context. In alcoholic patients a nexus of other endocrine findings permits the generalization that a hyperdopaminergic state exists. The endocrine context is less clear in depression and interpretation awaits addition research.

REFERENCES

Abassi, V., Linscheid, T., and Coleman, M.: Triiodothyronine (T$_3$) concentration and therapy in autistic children. *J Autism Child Schizophr 8*:383-387, 1978.

Afrasiabi, M.A., Flom, M.S., and Valenta, L.Y.: Hormonal

studies in heroin addicts with special attention to the tests of the anterior pituitary reserve. *Clin Res* 25:228A, 1977.

Amsterdam, J.D., Winokur, A., Mendels, J., Brunswick, D., and Snyder, P.: Multiple hormonal response to TRH in depressed patients and normal controls. Tenth Annual Meeting of the International Society for Psychoneuro-endocrinology, Park City, Utah, August 8-10, 1979.

Annell, A.L.: Lithium in the treatment of children and adolescents. *Acta Psychiat Scand (Suppl)* 207:19-33, 1969.

Artunkal, S., and Togrol, B.: Psychological studies in hyperthyroidism: In *Brain Thyroid Relationships*, Little Brown, Boston, 1964.

Asher, R.: Myxoedematous madness. *Bri Med J* 2:555-562, 1949.

Astwood, E.B.: Hormones and agonists: In *The Pharmacological Basis of Therapeutics*, L.S. Goodman and A. Gilman, eds. Macmillan, London, 1970, pp 1464-1500.

Augustine, J.R.: Laboratory studies in acute alcoholics. *Can Med Assoc J* 96:1367, 1967.

Banki, C.M.: Triiodythyronine in the treatment of depression. *Orv Hetil* 116:2543-2546, 1975.

Banki, C.M.: Cerebrospinal fluid amine metabolites after combined amitriphyline-triiodothyronine treatment of depressed women. *Eur J Pharmacol* 41:311-315, 1977.

Benkert, O.: Studies on pituitary hormones and releasing hormones in depression and sexual impotence: In *Hormones, Homeostasis and the Brain*. W.H. Gispen, Tj.B. van Wimersma Greidanus, B. Bohus, and D. de Wied, eds. *Progress in Brain Research* 42, Elsevier, Amsterdam, 1975, pp 25-36.

Betts, T.A., Smith, J., Pidd, S., Machintosh, J., Harvey, P., and Funicane, J.: The effects of thyrotropin-releasing hormone on measures of mood in normal women. *Br J Clin Pharmacol* 3:469-473, 1976.

Bigelow, L.G., Gillin, J.C., Semal, C., and Wyatt, R.J.: Thyrotropin releasing hormone in chronic schizophrenia. *Lancet* 2:869-870, 1975.

Bleuler, M.: *Endokrinologische Psychiatrie*. Georg Thieme, Stuttgart, 1954.

Brambilla, F., Guastalla, A., Guerrini, A., Rovere, C., Legnani, G., Sarno, M., and Riggi, F.: Prolactin secretion in chronic schizophrenia. *Acta Psychiatr Scand* 54:275-286, 1976.

Brambilla, F., Emeraldi, E., Sacchetti, E., Flammetta, N., Cocchi, D., and Muller, E.: Deranged anterior pituitary responsiveness to hypothalamic hormones in depressed patients. *Arch Gen Psychiatry* 35:1231-1238, 1978.

Brauchitsch, H.V.: Endokrinologische Aspekte des

Wirkungsmechanismus neuroplegischer Medikamente. *Psychopharmacologia* 2:1-21, 1961.

Breese, G.R., Prange, A.J., Jr., and Lipton, M.A.: Pharmacological studies of thyroid-imipramine interactions in animals. In *The Thyroid Axis, Drugs, and Behavior*, A.J. Prange, Jr., ed. Raven Press, New York, 1974, pp 29-48.

Brenner, I.: Apathetic hyperthyroidism. *J Clin Psychol* 39:479-480, 1978.

Brown, J.S.L.: Discussion. Hormones and metabolism. Parameters of metabolic problems. *Recent Prog Horm Res* 5:159-194, 1951.

Buchsbaum, M.S., and Rieder, R.O.: Biological heterogeneity and psychiatric research. *Arch Gen Psychiatry* 36:1063-1169, 1979.

Burger, H.G., and Patel, J.C.: TSH and TRH: Their physiologic regulation and the clinical application of TRH. In *Clinical Neuroendocrinology*, L. Martin and G.M. Bessers, eds. Academic Press, New York, 1977.

Burrow, G.N., May, P.B., Spaulding, S.W., and Donabedian, R.K.: TRH and dopamine interaction affecting pituitary hormone secretion. *J Clin Endocrinol Metab* 45(1):65-72, 1977.

Bursten, B.: Psychoses associated with thyrotoxicosis. *Arch Gen Psychiatry* 4:267-273, 1961.

Campbell, M., Fish, B., David, R., Shapiro, T., Collins, P., and Koh, C.: Response to T_3 and dextroamphetamine: A study of preschool schizophrenic children. *J Autism Child Schizophr* 2:343-358, 1972.

Campbell, M., Fish, B., David, R., Shapiro, T., Collins, P., and Koh, C.: Liothyronine treatment in psychotic and nonpsychotic children under 6 years. *Arch Gen Psychiatry* 29:602-608, 1973.

Campbell, M., Small, A.M., Hollander, C.S., Korein, J., Cohen, I.L., Kalmijn, M., and Ferris, S.: A controlled crossover study of T3 in autistic children. *J Autism Child Schizophr* 8:371-381, 1978.

Campbell, M., Hollander, C.S., Ferris, S., and Greene, L.W.: Response to TRH stimulation in young psychotic children: A pilot study. *Psychoneuroendocrinology* 3:195-201, 1979.

Carroll, B.J., and Mendels, J.: Neuroendocrine regulation in affective disorders. In *Hormones, Behavior, and Psychopathology*, E.J. Sachar, ed. Raven Press, New York, 1976, pp 41-68.

Cavalca, G.G., Covezzi, E., and Boncinelli, A.: Clinical experiences with the combination of thyroid extract and tricyclics in the treatment of depressed patients. *Riv Sper Freniat* 98:271-300, 1974.

Chase, T.N., Woods, A.C., Lipton, M.A., and Morris, C.E.:

Hypothalamic releasing factors and Parkinson's disease. *Arch Neurol* 31:55-56, 1974.

Chazot, G., Chalumeau, A., Aimard, G., Mornex, R., Garde, A., Schott, B., and Girard, P.F.: Thyrotropin releasing hormone and depressive states: From agroagonines to TRH. *Lyon Med* 231:831-836, 1974.

Checkley, S.A.: Thyrotoxicosis and the course of manic-depressive illness. *Am J Psychiatry* 133:219-223, 1978.

Clark, M.S., Parades, A., Costiloe, J.P., and Wood, F.: Synthetic thyroid releasing hormone (TRH) administered orally to chronic schizophrenic patients. *Psychopharmacol Commun* 1:191-200, 1975.

Cleghorn, R.A.: Endocrine influences on personality and behavior. In *Biology of Mental Health and Disease*. Hoeber Publishing, New York, 1950, pp 265-272.

Clinical Society of London: Report on myxedema. *Transactions of the Clinical Society of London* 21 *(Suppl)*. Longmans, Green, London, 1888.

Cohen, K.L., Swigar, M.E.: Thyroid function screening in psychiatric patients. *JAMA* 242:254-257, 1979.

Coppen, A., Whybrow, P.C., Noguera, R., Maggs, R., and Prange, A.J., Jr.: The comparative antidepressant value of L-tryptophan and imipramine with and without attempted potentiation by liothyronine. *Arch Gen Psychiatry* 26:234-241, 1972.

Coppen, A., Montgomery, S., Peet, M., and Bailey, J.: Thyrotropin-releasing hormone in the treatment of depression. *Lancet* 2:433-434, 1974.

Crammond, W.A.: Organic psychosis. *Br Med J* 4:497-500, 1968.

Crown, S.: Notes on an experimental study of intellectual deterioration. *Br Med J* 2:684-685, 1949.

Czernichow P., Dauzet, M.C., Broyer, M., and Rappoport R.: Abnormal TSH, PRL, and GH response to TSH releasing factor in chronic renal failure. *J Clin Endocrinol Metab* 43:630-637, 1976.

Davidoff, F., and Gill, J.: Myxedema madness: Psychosis as an early manifestation of hypothyroidism. *Conn Med* 41:618-621, 1977.

Davis, K.L., Hollister, L.E., and Berger, P.A.: Thyrotropin-releasing hormones in schizophrenia. *Am J Psychiatry* 132(9):951-953, 1975.

Dewhurst, K.E., Elkabir, D.J., Exley, D., Harris, G.W., and Mandelbrote, B.M.: Blood levels of thyrotropic hormone, protein bound iodine, and cortisol in schizophrenia and affective states. *Lancet* 2:1160-1162, 1968.

Dimitrikoudi, M., Hanson-Norty, E., and Jenner, F.A.: T.R.H. in psychoses. *Lancet* 1:456, 1974.

Dougier, M., Wittkower, E.D., Stephens-Newsham, L., Hoffman,

M.M.: Psychophysiological studies in thyroid function. *Psychoso Med* 18:310-323, 1956.

Drayson, A.M.: TRH in cyclical psychoses. *Lancet* 2:312, 1974.

Earle, B.V.: thyroid hormone and tricyclic antidepressants in resistant depressions. *Am J Psychiatry* 126:1667-1669, 1970.

Easson, W.: Myxedema with psychosis. *Arch Gen Psychiatry* 14:277-283, 1966.

Ehrensing, R.H., Kastin, A.J., Schalch, D.S., Friesen, H.G., Vargas, J.R., and Schally, A.V.: Affective state and thyrotropin and prolactin responses after repeated injections of thyrotropin-releasing hormone in depressed patients. *Am J Psychiatry* 131(6):714-718, 1974.

Ettigi, P.G., and Brown, G.: TSH and LH responses in subtypes of depression. Annual Meeting, American Psychosomatic Society, Washington, D.C., March 1978.

Feighner, J.P., King, L.J., Schuckit, M.A., Croughan, J., and Briscoe, W.: Hormonal potentiation of imipramine and ECT in primary depression. *Am J Psychiatry* 128:1230-1238, 1972.

Frazer, A., Pandey, G., Mendels, J., Neeley, S., Kane, M., and Hess, M.E.: The effect of triiodothyronine in combination with imipramine on (^{3}H)-cyclic AMP production in slices of rat cerebral cortex. *Neuropharmacology* 13:1131-1140, 1974.

Freeman, W.: Personality and endocrines. *Ann Intern Med* 9:444, 1935.

Frohman, L.A.: Neurotransmitters as regulators of endocrine function. *Hosp Pract* 10:54-67, 1975.

Furlong, F.W., Brown, G.M., and Beeching, M.F.: Thyrotropin-releasing hormone: Differential antidepressant and endocrinological effects. *Am J Psychiatry* 133:1187-1190, 1976.

Garbutt, J., Malekpour, B., Brunswick, D., Jonnalagadda, M.R., Jolliff, L., Podolak, R., Wilson, I., and Prange, A.J., Jr.: Effects of triiodothyronine on drug levels and cardiac function in depressed patients treated with imipramine. *Am J Psychiatry* 136(7):980-982, 1979.

Gibson, J.G.: Emotions and the thyroid gland: A critical appraisal. *J Psychosom Res* 6:93-116, 1962.

Gjessing, L.R.: A review of periodic catatonia. *Biol Psychiatry* 8:23-45, 1974.

Gold, P.W., Goodwin, F.K., Wehr, T., and Rebar, R.: Pituitary thyrotropin response to thyrotropin-releasing hormone in affective illness: Relationship to spinal fluid amine metabolites. *Am J Psychiatry* 134(9):1028-1031, 1977.

Gold, M.S., Pottash, A.L.L., Davies, R.K., Ryan, N.,

Sweeney, D.R., and Martin, D.M.: Distinguishing unipolar and bipolar depression by thyrotropin release test. *Lancet* 2:411-412, 1979.

Goldberg, M.: The occurrence and treatment of hypothyroidism among alcoholics. *J Clin Endocrinol Metab* 20:609-621, 1960.

Goldberg, M.: thyroid function in chronic alcoholism. *Lancet* 2:746, 1962.

Goodwin, D.W., and Erickson, C.K.: *Alcoholism and Affective Disorders*. Spectrum Publications, New York, 1979.

Goodwin, F.K., and Ebert, M.H.: Lithium in mania: Clinical trials and controlled studies. In *Lithium: Its Role in Psychiatric Research and Treatment*, S. Gershon and B. Shopsin, eds. Plenum Press, New York, 1973, pp 237-252.

Goodwin, F.K., Prange, A.J., Post, R.M., Muscettola, G., and Lipton, M.A.: Potentiation of antidepressant effects by L-triiodothyroninde in tricyclic nonresponders. *Am J Psychiatry* 139:34-38, 1982.

Granet, R.B., and Kalman, T.P.: Hypothyroidism and psychosis: A case illustration of the diagnostic dilemma in psychiatry. *J Clin Psychiatry* 39:260-263, 1978.

Gregoire, F., Brauman, H., de Buck, R., and Corvilain, J.: Hormone release in depressed patients before and after recovery. *Psychoneuroendocrinology* 2:303-312, 1977.

Gregory, Y.: Mental disorder associated with thyroid dysfunction. *Can Med Assoc J* 75:489, 1956.

Gruen, P.H., Sachar, E.J., Altman, N., and Sassin, J.: Growth hormone responses to hypoglycemia in postmenopausal depressed women. *Arch Gen Psychiatry* 32:31-33, 1975.

Guillemin, R.: Peptides in the brain: The new endocrinology of the neuron. *Science* 202:390-402, 1978.

Gull, W.: On a cretinoid state supervening in the adult. *Br Med J* 2:528-529, 1873.

Hamilton, M.: A rating scale for depression. *J Neurol Neurosurg Psychiatry* 23:56-62, 1960.

Hanna, S.M., Jenner, F.A., Pearson, I.B., Sampson, G.A., and Thompson, E.A.: The therapeutic effect of Lithium carbonate on a patient with a 48 hour periodic psychosis. *Br J Psychiatry* 121:271-280, 1972.

Hatotani, N., Nomura, J., Yamaguchi, T., and Kitayama, I.: Clinical and experimental studies on the pathogenesis of depression. *Psychoneuroendocrinology* 2:115-130, 1977.

Hermann, H.T., and Quarton, G.C.: Psychological changes and psychogenesis in thyroid hormone disorders. *J Clin Endocrinol* 25:327-338, 1965.

Hetzel, B.S., Schottstaedt, W.W., Grace, W.J., and Woeff, H.G.: Changes in urinary 17-hydroxycortico-steroid excretion during stressful life experiences in man. *J Clin*

Endocrinol 15:1057-1068, 1955.

Hollister, L.E., Davis, K.L., and Davis, B.M.: Hormones in the treatment of psychiatric disorders. *Hosp Pract* 9:103-110, 1973.

Huey, L.Y., Janowsky, D.S., Mandell, A.J., Judd, L.L., and Pendery, M.: Preliminary studies on the use of thyrotropin releasing hormone in manic states, depression, and the dysphoria of alcohol withdrawal. *Psychopharmacol Bull* 11:24-27, 1975.

Hutton, W.N.: Thyrotropin-releasing hormone in depression. *Lancet* 2:53, 1974.

Inanaga, K., Nakano, T., and Nagato, T.: Effects of thyrotropin releasing hormone in schizophrenia. *Kurume Med J* 22:159-168, 1975.

Inanaga, K., Nakano, T., Nagato, T., Tanaka, M., and Ogawa, N.: Behavioral effects of protirelin in schizophrenia. *Arch Gen Psychiatry* 35:1011-1014, 1978.

Irie, M., and Tsushima, T.: Increase of serum growth hormone concentration following thyrotropin releasing hormone injection in patients with acromegaly or gigantism. *J Clin Endocrinol Metab* 35:97-100, 1972.

Jain, V.K.: A psychiatric study of hypothyroidism. *Psychiatr Clin* 5:121-130, 1972.

Jonas, A.D.: Hypothyroidism and neurotic depression. *Am Practit* 3:103-105, 1952.

Karlberg, B.E., Kjellman, B.F., and Kagedol, B.: Treatment of endogenous depression with oral thyrotropin. *Acta Psychiatr Scand* 58:389-400, 1978.

Kastin, A.J., Ehrensing, R.H., Schalch, D.S., and Anderson, M.S.: Improvement in mental depression with decreased thyrotropin response after administration of thyrotropin-releasing hormone. *Lancet* 2:740-742, 1972.

Khan, A.A.: *Thyroid dysfunction* Br Med J 4:495, 1970.

Kieley, W.F., Adrian, A.D., Lee, J.H., and Nicoloff, J.T.: Therapeutic failure of oral TRH in depression. *Psychosom Med* 38:233-241, 1976.

Kirkegaard, C., Norlem, N., Lauridsen, U.B., Bjorum, N., and Christiansen, C.: Protirelin stimulation test and thyroid function during treatment of depression. *Arch Gen Psychiatry* 32:1115-1118, 1975.

Kirkegaard, C., Bjorum, N., Cohn, D., and Lauridsen, U.B.: TRH stimulation test in manic depressive disease. *Arch Gen Psychiatry* 35:1017-1023, 1978.

Kleinschmidt, H.J., and Waxenberg, S.E.: Psychophysiology and psychiatric management of thyrotoxicosis: A two year follow-up study. *J Mt Sinai Hosp NY* 23:131-153, 1956.

Kolakowska, T., and Swigar, M.E.: Thyroid function in depression and alcohol abuse. *Arch Gen Psychiatry*

34: 984-988, 1977.

Kraepelin, E.: *Psychiatrie* 5th ed. Barth, Leipzig, 1896.

Kraepelin, E.: *Lectures on Clinical Psychiatry* . William Wood and Co., New York, 1917.

Lakke, J.P.W.F., van Praag, H.M., van Twisk, R., Doorenbos, H., and Witt, F.G.J.: Effects of administration of thyrotropin releasing hormone in Parkinsonism. *Clin Neurol Neurosurg 3/4:* 1-5, 1974.

Leichter, S.B., Kirstein, L., and Martin, N.D.: Thyroid function and growth hormone secretion in amitriphyline-treated depression. *Am J Psychiatry 134:* 1270-1272, 1977.

Lepaeluoto, J., Papeli, M., Varis, R., Ranta, T.: Secretion of anterior pituitary hormones in man: Effects of ethyl alcohol. *Acta physiol Scand 95:* 400-406, 1975.

Libow, L.S., and Durell, J.: Clinical studies on the relationship between psychosis and the regulation of thyroid gland activity. *Psychosom Med 28:* 377-383, 1965.

Lidz, T.: Emotional factors in the etiology of hyperthyroidism. *Psychosom Med 2:* 2-8, 1949.

Lindstrom, L.H., Gunne, L.M., Oest, L.G., and Person, E.: Thyrotropin-releasing hormone (TRH) in chronic schizophrenia. *Acta Psychiat Scand 55:* 74-80, 1977.

Logothetis, J.: Psychotic behavior as the initial indicator of adult myxedema. *J. Nerv Ment Dis 136:* 561-568, 1963.

Loosen, P.T., and Prange, A.J., Jr.: Alcohol and anterior pituitary secretion. *Lancet 2:* 985, 1977.

Loosen, P.T., and Prange, A.J., Jr.: Psychoendocrinology: A Brief Survey. In *Psychiatric Complications of Medical Drugs*, R.I. Shader, ed. Raven Press, New York, 1980a.

Loosen, P.T., and Prange, A.J., Jr.: Thyrotropin releasing hormone (TRH): A useful tool for psychoneuroendocrine investigation. *Psychoneuroendocrinology*, 1980b.

Loosen, P.T., and Prange, A.J., Jr.: The serum BH response to TRH in psychiatric patients: A review. *Am J Psychiatry 139:* 405-416, 1982.

Loosen, P.T., Wilson, I.C., Lara, P.P., Prange, A.J., Jr., and Pettus, C.: thyrotropin releasing hormone in depressed patients: A review. *Pharmacol Biochem Behav 5:* 95-101, 1976.

Loosen, P.T., Prange, A.J., Jr., Wilson, I.C., Lara, P.P., and Pettus, C.: Thyroid stimulating hormone response after thyrotropin releasing hormone in depressed, schizophrenic and normal women. *Psychoneuroendocrinology 2:* 137-148, 1977.

Loosen, P.T., Prange, A.J., Jr., and Wilson, I.C.: Influence of cortisol on TRH induced TSH response in depression. *Am J Psychiatry 135:* 244-246, 1978a.

Loosen, P.T., Prange, A.J., Jr., and Wilson, I.C.: The TSH

response to TRH in psychiatric patients: Relation to serum cortisol. *Prog Neuropsychopharmacol* 2:479–486, 1978b.

Loosen, P.T., Prange, A.J., Jr., and Wilson, I.C.: TRH (Protirelin) in depressed alcoholic men: Behavioral changes and endocrine responses. *Arch Gen Psychiatry* 36:540–547, 1979.

Loosen, P.T., Wilson, I.C., and Prange, A.J., Jr.: Endocrine and behavioral changes in depression after TRH: Alteration by pretreatment with thyroid hormones. *J Affective Disorders* 2:267–278, 1980.

McCaul, J.A., Cassell, K.J., and Stern, G.M.: Intravenous thyrotropin releasing hormone in Parkinson's disease. *Lancet* 2:735, 1974.

MacCrimmon, D.J., Wallace, J.E., Goldberg, W.M., and Streiner, D.L.: Emotional disturbance and cognitive deficits in hyperthyroidism. *Psychosom Med* 41:331–340, 1979.

McEwen, B.S., Zigmond, R.E., Azmitia, E.C., Jr., and Weiss, J.M.: Steroid hormone interaction with specific brain regions. In *Biochemistry of Brain and Behavior*, R.E. Bowman and S.P. Datta, eds. Plenum Publishing, New York, 1970, pp 123–167.

McLarty, D.G., O'Boyle, J.H., Spencer, C.A., and Ratcliffe, J.G.: Effects of lithium on hypothalamic-pituitary-thyroid function in patients with affective disorders. *Br Med J* 3:623–626, 1975.

McLarty, D.G., Ratcliffe, W.A., Ratcliffe, J.G., Shiminins, J.G., and Goldberg, A.: A study of thyroid function in psychiatric inpatients. *Br J Psychiatry* 133:211–218, 1978.

MacSweeney, D., Timms, P., and Johnson, A.: Thyro-endocrine pathology, obstetric mobility and schizophrenia: Survey of a hundred families with a schizophrenic proband. *Psychol Med* 8:151–155, 1978.

Maeda, K., Kato, Y., Ohgo, S., Chihard, K., Yoshimoto, Y., Yamaguchi, N., Kuromaru, S., and Imura, H.: Growth hormone and prolactin release after injection of thyrotropin releasing hormone in patients with depression. *J Clin Endocrinol Metab* 40:501–505, 1975.

Manberg, P.J., Nemeroff, C.B., and Prange, A.J., Jr.: TRH and amphetamine: A comparison of pharmacological profiles in animals. *Prog Neuropsychopharmacol* 3(4):303–314, 1979.

Mandelbrote, B.M., and Wittkower, E.D.: Emotional factors in Graves' disease. *Psychosom Med* 17:109–123, 1955.

Martin, J.B.: Neural regulation of growth hormone secretion. *N Engl J Med* 288:1384–1393, 1973.

Mason, J.W.: A review of psychoendocrine research on the pituitary-thyroid system. *Psychosom Med* 30:666–681, 1968.

Michael, R.P., and Gibbons, J.L.: Interrelationships between the endocrine system and neuropsychiatry. *Int Rev Neurobiol* 5:243-302, 1963.

Miller, R.: Mental symptoms from myxedema. *J Lab Clin Med* 40:267-270, 1952.

Moldow, R.L., and Yalow, R.S.: Extrahypophyseal distribution of thyrotropin as a function of brain size. *Life Sci* 22:1859-1864, 1978.

Naeije, R., Golstein, J., Zegers de Beyl, D., Linkowski, P., Mendlewicz, G., Copinschi, G., Badawi, M., Leclercq, R., L'Hermite, M., and Vanhaelst, L.: Thyrotrophin, prolactin and growth hormone responses to TRH in barbiturate coma and in depression. *Clin Endocrinol* 9:49-58, 1978.

Nicholson, G., Liebling, L.I., and Hall, R.A.: Thyroid disfunction in female psychiatric patients. *Br J Psychiatry* 129:236-238, 1976.

Nordgren, L., and Scheele, C. von.: Myxedema madness without myxedema. *Acta Med Scand* 199:233-236, 1976.

Ogura, C., Okuma, T., Uchida, Y., Imai, S., Yogi, H., and Sunami, Y.: Combined thyroid (triiodothyronine)-tricyclic antidepressant treatment in depressive states. *Folia Psychiat Neurol Jpn* 28:179-186, 1974.

Otsuki, M., Dakoda, M., and Baba, S.: Influence of glucocorticoids on TRH mediated TSH response in man. *J Clin Endocrinol Metab* 36:95-102, 1973.

Ozawa, J., and Shishiba, J.: Lack of TRH induced TSH secretion in a patient with Klinefelter's syndrome: A case report. *Endocrinol Jpn* 22:269-275, 1975.

Parry, C.H., ed.: *Collections from the Unpublished Writings of the Late Caleb Hillier Parry*, Vol. I. Underwoods, London, 1825.

Pecknold, L.C., and Ban, T.A.: TRH in depressive illness. *Pharmakopsychiatr Neuropsychopharmacol* 12:166-173, 1977.

Pittman, J.A., Jr.: Thyrotropin releasing hormone. *Adv Intern Med* 19:303, 325, 1974.

Pitts, F.N., and Guze, S.B.: Psychiatric disorders and myxedema. *Am J Psychiatry* 118:142-147, 1966.

Plunkett, E.R., Rangecroft, G., and Heagy, P.C.: Thyroid function in patients with sex chromosomal abnormalities. *J Ment Defic Res* 8:25-34, 1964.

Pomeranze, J., and King, E.: Psychosis as first sign of thyroid dysfunction. *Geriatrics* 21:211-213, 1966.

Prange, A.J., Jr.: Paroxysmal auricular tachycardia apparently resulting from combined thyroid-imipramine treatment. *Am J Psychiatry* 119:994-995, 1963.

Prange, A.J., Jr.: Therapeutic and theoretical implications of imipramine-hormone interactions in depressive disorders. In *Proceedings of the V World Congress of Psychiatry*. Excerpta Medica, Amsterdam, 1971.

Prange, A.J., Jr., and Lipton, M.A.: Enhancement of imipramine instability in hypothyroid mice. *Nature* *196:*588-589, 1962.

Prange, A.J., Jr., and Wilson, I.C.: Thyrotropin releasing hormone (TRH) for the immediate relief of depression: A preliminary report. *Psychopharmacologia 26:*82, 1972.

Prange, A.J., Jr., Wilson, I.C., Rabon, A.M., and Lipton, M.A.: Enhancement of imipramine antidepressant activity by thyroid hormone. *Am J Psychiatry 126:*457-469, 1969.

Prange, A.J., Jr., Wilson, I.C., Knox, A., McClane, T.K., and Lipton, M.A.: Enhancement of imipramine by thyroid stimulating hormone: Clinical and theoretical implications. *Am J Psychiatry 127:*191-199, 1970.

Prange, A.J., Jr., Wilson, I.C., Lara, P.P., Alltop, L.B., and Breese, G.R.: Effects of thyrotropin-releasing hormone in depression. *Lancet 2:*999-1002, 1972.

Prange, A.J., Jr., Wilson, I.C., Breese, G.R., and Lipton, M.A.: Hormonal alteration of imipramine response: A review. In *Hormones, Behavior and Psychopathology*, E.J. Sachar, ed. Raven Press, New York, 1976, pp 41-67.

Prange, A.J., Jr., Nemeroff, C.B., Lipton, M.A., Breese, G.R., and Wilson, I.C.: Peptides and the central nervous system. In *Handbook of Psychopharmacology*, Vol. 13, L.L. Iversen, S.D. Iversen, and S.H. Snyder, eds. Plenum Publishing, New York, 1978a, pp 1-107.

Prange, A.J., Jr., Nemeroff, C.B., and Loosen, P.T.: Behavioral effects of hypothalamic peptides. In *Centrally Acting Peptides*, J. Hughes, ed. MacMillan Press, Basingstoke, England, 1978b, pp 99-118.

Prange, A.J., Jr., Loosen, P.T., Wilson, I.C., Meltzer, H.Y., and Fang, V.S.: Behavioral and endocrine responses of schizophrenic patients to TRH (Protirelin). *Arch Gen Psychiatry 36:*1086-1093, 1979a.

Prange, A.J., Jr., Loosen, P.T., and Nemeroff, C.B.: Peptides: Application to research in nervous and mental disorders. In *New Frontiers of Psychotropic Drug Research*, S. Fielding, ed. Futura Publishing Co., Mt. Kisco, New York, 1979b, pp 117-189.

Re, R.N., Kourides, I.A., Ridgway, E.C., Weintraub, B.D., and Maloof, F.: The effect of glucocorticoid administration on human pituitary secretion of thyrotropin and prolactin. *J Clin Endocrinol Metab 43:*338-346, 1976.

Reed, K., and Bland, R.D.: Masked "myxedema madness." *Acta Psychiatr Scand 56:*421-426, 1977.

Reiss, M., Hemphill, R.E., Maggs, R., Haigh, C.P., and Reiss, J.M.: The significance of the thyroid in psychiatric illness and treatment. *Br Med J 1:*906-910, 1953.

Reitan, R.M.: Intellectual functions in myxedema. *Arch Neurol Psychiatry* 69:436-449, 1963.

Relkin, R.: Effect of endocrines on central nervous system. *NY State J Med* 69:2133-2145, 2247-2265, 1969.

Rinieris, P.M., Christodoulou, G.N., Sourateoglou, A., Koutras, D.A., and Stefanis, C.N.: Free-thyroxine index in mania and depression. *Compr Psychiatry* 19:561-564, 1978a.

Rinieris, P.M., Christodoulou, G.N., Sourateoglou, A., Koutras, D.A., and Stefanis, C.N.: Free-thyroxine index in psychotic and neurotic depression. *Acta Psychiatr Scand* 58:56-60, 1978b.

Robbins, L.R., and Vinson, D.B.: Objective psychologic assessment of the thyrotoxic patient and the response to treatment: Preliminary report. *J Clin Endocrinol* 20:120-129, 1960.

Rybakowski, J., and Sowinski, J.: Free-thyroxine index and absolute free-thyroxine in affective disorders. *Lancet* 1:889, 1973.

Sachar, E.J.: Psychiatric disturbances associated with endocrine disorders. In *American Handbook of Psychiatry*, D.X. Freedman and J.E. Dyrud, eds. Basic Books, New York, 1975.

Sachar, E.J., Frantz, A.G., Altman, N., and Sassin, J.: Growth hormone and prolactin in unipolar and bipolar depressed patients: Responses to hypoglycemia and L-DOPA. *Am J Psychiatry* 130:1362-1367, 1973.

Schally, A.V.: Aspects to hypothalamic regulation of the pituitary gland. *Science* 202:18-28, 1978.

Schon, M., Sutherland, A.M., and Rawson, R.W.: Hormones and neuroses: The psychological effects of thyroid deficiency. In *Proceedings III World Congress in Psychiatry*. McGill University Press, Montreal, 1961.

Selzer, M.L., and VanHouten, W.H.: Normal thyroid function in chronic alcoholism. *J Clin Endocrinol* 24:380-382, 1964.

Sherwin, A.C., Flach, F.F., and Stokes, P.E.: Treatment of psychoses in early childhood with triiodothyronine. *Am J Psychiatry* 115:166-167, 1958.

Shopsin, B., Shenkman, L., Blum, M., and Hollander, C.: T and TSH response to TRH: Newer aspects of lithium-induced thyroid disturbances in men. *Psychopharm Bull* 9:29, 1973.

Simpson, G.M., and Cooper, T.B.: Thyroid indices in chronic schizophrenia. *J Nerv Ment Dis* 142:58-61, 1966.

Sorner, R.D., and McHugh, D.R.: Lithium in the treatment of periodic catatonia: A case report. *J Nerv Ment Dis* 158:214-221, 1974.

Steiner, M., Radwan, M., Elizur, A., Blum, I., Atsomon, A., and Davidson, S.: Failure of L-triiodothyronine (T_3) to potentiate tricyclic antidepressant response. *Curr Ther*

Res 23: 655–659, 1978.
Sterling, K.: Thyroid hormone action at the Cch level. N Engl J Med 300: 117–123, 1979.
Stumpf, W.E., and Sar, M.: Localization of thyroid hormone in the mature rat brain and pituitary. In Anatomical Neuroendocrinology, W.E. Stumpf and L.D. Grant, eds. S. Karger, Basel, 1975, pp 318–327.
Suwa, N., and Yamashita, I.: Psychological Studies of Emotion and Mental Disorders. Hokkaido University School of Medicine, Japan, 1972, pp 59–60.
Takahashi, S., Kondo, H., Yoshimura, M., and Ochi, Y.: Thyrotropin responses to TRH in depressive illness: Relation to clinical subtypes and prolonged duration of depressive episode. Folia Psychiats Neurol Jpn 28: 355–365, 1974.
Tata, J.R.: Biological action of thyroid hormones at the cellular and molecular levels. In Actions of Hormones on Molecular Processes, E. Litwack and D. Kritchevsky, eds. John Wiley & Sons, New York, 1968, pp 58–131.
Taylor, J.W.: Depression in thyrotoxicosis. Am J Psychiatry 132: 552–553, 1975.
Temple, R., Berman, M., Carlson, H.E., Robbins, J., and Woeff, J.: The use of lithium in Graves' disease. Mayo Clin Proc 47: 872–878, 1972.
Tingley, J.O., Morris, A.W., and Hill, S.R.: Studies on the diurnal variation and response to emotional stress of the thyroid gland. Clin Res Proc 6: 134, 1958.
Tiwary, G.M., Rosenbloom, A.L., Robertson, M.F., and Parker, J.C.: Effects of thyrotropin releasing hormone in minimal brain dysfunction. Pediatrics 56: 119–121, 1975.
Tonks, C.M.: Mental illnesses in hypothyroid patients. Br J Psychiatry 110: 706–710, 1964.
Tsuang, M.T., and Winokur, G.: Criteria for subtyping schizophrenia. Arch Gen Psychiatry 31: 43–47, 1974.
Tsutsui, S., Yamazaki, Y., Namba, T., and Tsushima, M.: Combined therapy of T3 and antidepressants in depression. J Int Med Res 7: 138–146, 1979.
Vale, W., Brazeau, P., Rivier, C., Rivier, J., Grant, G., Burgus, R., and Guillemin, R.: Inhibitory hypophysiotropic activity of hypothalamic somatostatin. Fed Proc 32: 211, 1973.
Van den Burg, W., van Praag, H.M., Box, E.R.H., Piers, D.A., van Zanton, A.K., and Doorenbos, H.: TRH as a possible quick-acting but short-lasting antidepressant. Psychol Med 5: 404–412, 1975.
Van den Burg, W., van Praag, H.M., Box, E.R.H., Piers, D.A., van Zanten, A.K., and Doorenbos, H.: TRH by slow, continuous infusion: An antidepressant? Psychol Med

576 Loosen and Prange

6:393-397, 1976.

Villani, S., and Weitzel, W.D.: Secondary mania (Letter to the Editor). *Arch Gen Psychiatry* 36:1031, 1979.

Vinik, A.I., Kalk, W.J., McLaren, H., Hendricks, S., and Pristone, B.L.: Fasting blunts the TSH response to synthetic thyrotropin releasing hormone (TRH). *J Clin Endocrinol Metab* 40:509-511, 1975.

Vogel, H.P., Benkert, O., Illig, R., Mueller-Oerlinghausen, B., and Poppenberg, A.: Psychoendocrinological and therapeutic effects of TRH in depression. *Acta Psychiatr Scand* 56:223-232, 1977.

Volpe, R., Vale, J., Johnston, M.W.: The effects of certain physical emotional tensions and strains on fluctuations in the level of serum protein-bound iodine. *J Clin Endocrinol* 20:415-428, 1960.

von Wagner-Jauregg, J.: *Lehrbuch der Organotherapie*. Georg Thieme, Leipzig, 1914.

Weeke, A., and Weeke, J.: Disturbed circadian variation of serum thyrotropin in patients with endogenous depression. *Acta Psychiatr Scand* 57:281-289, 1978.

Weinberg, A.D., and Katzell, T.D.: Thyroid and adrenal function among psychiatric patients. *Lancet* 1:1104-1105, 1977.

Wheatley, D.: Potentiation of amitriptyline by thyroid hormone. *Arch Gen Psychiatry* 26:229-233, 1972.

Whybrow, P.C.: Synergistic action between iodine and lithium. *JAMA* 221:506, 1972.

Whybrow, P., and Hurwitz, T.: Psychological disturbances associated with endocrine disease and hormone therapy. In *Hormones, Behavior, and Psychopathology*, E.J. Sachar, ed. Raven Press, New York, 1976, pp 125-143.

Whybrow, P.C., Prange, A.J., Jr., and Treadway, C.R.: The mental changes accompanying thyroid gland dysfunction. *Arch Gen Psychiatry* 2 :48-63, 1969.

Whybrow, P.C., Coppen, A., Prange, A.J., Jr., Noguera, R., and Bailey, J.E.: Thyroid function and the response to L-triiodothyronine in depression. *Arch Gen Psychiatry* 26:242-245, 1972.

Widerlov, E., and Sjostrom, R.: Effects of thyrotropin releasing hormone on endogenous depression. *Nord Psychiatr Tidskr* 29:503-512, 1975.

Wiesel, C.: Psychosis with myxedema. *J Kentucky Med Assoc* 50:395-397, 1952.

Williams, R.H.: Metabolism and mentation. *J Clin Endocrinol* 31:461-479, 1970.

Williams, R.: Breast and thyroid cancer and malignant melanoma promoted by alcohol induced pituitary secretion of prolactin, T.S.H., and M.S.H. *Lancet* 1:996-999, 1976.

Wilson, W.P., Johnson, J.E., and Smith, R.B.: Affective change in thyrotoxicosis and experimental hypermetabolism. In *Recent Advances in Biological Psychiatry*, Plenum Publishing, New York, 1962.

Wilson, I.C., Prange, A.J., Jr., McClane, T.K., Rabon, A.M., and Lipton, M.A.: Thyroid hormone enhancement of imipramine in non retarded depression. *N Engl J Med* *282*:1063-1067, 1970.

Wilson, I.C., Lara, P.P., and Prange, A.J., Jr.: Thyrotropin releasing hormone in schizophrenia. *Lancet* 2:43-44, 1973a.

Wilson, E.C., Prange, A.J., Jr., Laura, P.P., Alltop, L.B., Stikeleather, R.A., and Lipton, M.A.: TRH (lopremone): Psychobiological responses to normal women: (I) subject experience. *Arch Gen Psychiatry* 29:15-32, 1973b.

Wilson, I.C., Prange, A.J., Jr., and Loosen, P.T.: Behavioral changes after TRH in normal women. Antagonism by pretreatment with thyroid hormones. *J Psychiatr Res* 2:211-222, 1980.

Winokur, G., Cadoret, R., Dorzab, T., and Baker, M.: Depressive disease, a genetic study. *Arch Gen Psychiatry* 24:135-144, 1971.

Ylikahri, R.H., Huttersen, M.O., Haerkoenen, M., Leior, T., Helenins, T., Liewendahl, K., and Karonen, S.: Acute effects of anterior pituitary secretion of the tropic hormones. *J Clin Endocrinol Metab* 46:715-720, 1978.

14

Glucocorticoids and Mineralocorticoids: Actions on Brain and Behavior

HOWARD D. REES AND HARRY E. GRAY

INTRODUCTION

The adrenal cortex produces more than 40 steroids, but only
a few of these are secreted in biologically significant
quantities. The major active secretions are classified as
glucocorticoids, such as cortisol, corticosterone, and 11-
deoxycortisol, or mineralocorticoids with 21 carbon atoms,
such as aldosterone and deoxycorticosterone (DOC), and weak
androgens with 19 carbon atoms, such as dehydroepiandrosterone. Minor amounts of progestational and estrogenic steroids are also secreted. The predominant glucocorticoid is
cortisol in some species, for instance, hamster, sheep, and
primates, but is corticosterone in others, for instance rat,
mouse, and rabbit (Seth, 1969). Although cortisol is the
predominant circulating glucocorticoid in primates, corticosterone may comprise a significant or even major fraction
of the glucocorticoid bound in the brain (Turner et al.,
1979b).

Biosynthesis and Secretion

The enzymatic pathways for synthesis of adrenal steroids
from cholesterol, which the adrenals may take up from blood
or synthesize from acetate, is shown in simplified form in
Fig. 14-1. The adrenal cortex does not store its secretions
but does store an abundance of the precursor cholesterol in
lipid droplets. The adrenal stimulators adrenocorticotropic
hormone (ACTH) and angiotensin II, by processes involving
cyclic AMP, accelerate the conversion of cholesterol to

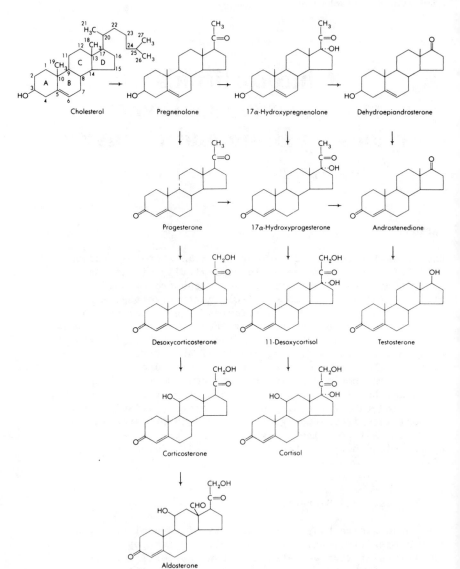

Fig. 14-1. Principal pathways for biosynthesis of adrenocorti-
costeroids and adrenal androgens. (From Haynes and
Larner, 1975, with permission.)

pregnenolone, the rate-limiting step in adrenal steroidogenesis. This conversion is rapidly followed by the synthesis and secretion into the circulation of the active corticosteroids.

Metabolic Effects of Glucocorticoids

Glucocorticoids are so named for their role in regulating glucose metabolism. They act directly on most tissues and indirectly influence all tissues. An early effect of glucocorticoids is the inhibition of glucose uptake by adipose tissue, skin, fibroblasts, and lymphoid tissue. There is a decrease in macromolecular (protein, lipid, and nucleic acid) synthesis and an increase in protein degradation in these tissues and in muscle. There is an increase in lipolysis in fat cells, and a depletion of glycogen in muscle. These catabolic actions result in the release of amino acids, free fatty acids, glycerol, and nucleotides into the circulation. In contrast, the actions of glucocorticoids in the liver are primarily anabolic, resulting in a general increase in RNA and protein synthesis, as well as the specific induction of a number of enzymes. The amino acids derived from peripheral catabolism are substrates for increased glucose formation, or gluconeogeneis, in liver, and to a lesser extent in the kidney. Glycogen accumulates in liver, and blood glucose levels tend to rise. The latter change induces a compensatory increase in insulin, which counteracts many of the glucocorticoid effects.
  Certain tissues are spared from the catabolic actions of glucocorticoids—brain, red blood cells, heart, liver, kidney. These tissues, perhaps more essential than others, may rather enjoy the additional circulating glucose diverted from elsewhere or produced by gluconeogenesis. The diverse actions of glucocorticoids may thus make teleological sense as a coordinated mechanism for making glucose maximally available to certain essential tissues during and immediately following periods of environmental challenge or stress.
  Glucocorticoids also have many diverse effects not related to glucose metabolism. They act at multiple sites to suppress inflammatory and allergic reactions, including inhibition of extravasation and migration of leukocytes, edema, and phagocytosis, decrease in the circulating lymphocytes and eosinophils, involution of the thymus, lymph nodes, and spleen, and decrease in antibody production. The lungs respond to glucocorticoids with enhanced catecholamine sensitivity, bronchodilation, and decreased vascular resistance. There may be anabolic effects in the development of some organs, for example, the induction of surfactant secretion in the fetal lung. Glucocorticoid actions have been comprehensively reviewed by Baxter (1976).

Metabolic Effects of Mineralocorticoids

Mineralocorticoids regulate water and electrolyte metabolism by promoting the retention of sodium and loss of potassium (Forman and Mulrow, 1975). Their major target is the epithelium of the renal distal tubules, although they also exert similar effects on epithelial cells of the salivary and sweat glands and gastrointestinal tract. The most potent natural mineralocorticoids are aldosterone and 11-DOC. The absence of mineralocorticoids, following adrenalectomy, results in decreased tubular reabsorption of sodium chloride, and water, and increased reabsorption of potassium, eventually leading to extracellular dehydration, hemoconcentration, acidosis, hypotension, decreased glomerular filtration rate, extrarenal uremia, and shock.

Regulation of Glucocorticoid Secretion

Glucocorticoids are released from the adrenals in response to ACTH, which is in turn secreted by cells of the anterior pituitary in response to as yet unidentified corticotropin-releasing factors (CRF) of hypothalamic and perhaps extra-hypothalamic origin (Sayers and Portanova, 1975; Vale et al., 1981). Three factors are known to control glucocorticoid secretion: stress, rhythms, and corticosteroid feedback.

Stress

The pituitary-adrenal system responds within a few minutes to a wide variety of noxious stimuli, termed "stressors." The "general adaptation syndrome" (Selye, 1950) produced by this pituitary-adrenal stress response is both nonspecific with respect to a variety of stimuli and relatively slow to develop, in contrast to the autonomic responses that can produce relatively stimulus-specific and extremely rapid adaptive changes in many organ systems. Although a distinction is frequently drawn between "physiological" stressors, such as trauma, hemorrhage, hypoxia, infection, ether, cold, heat, and fasting, and less noxious "psychological" stressors, such as immobilization, handling, mild electric shock, loud noise, and situations that produce fear, guilt, anxiety, or frustration, this distinction is often blurred in practice. When pain, discomfort, and emotional reactions were carefully avoided, several "physiological stressors," such as fasting, heat, and exercise, no longer elevated corticosteroid levels. This finding led Mason (1971) to suggest that the essential property of all stressors may be the ability to elicit a behavioral response of emotional arousal or hyperalerting, which prepares the organism for flight, struggle, or strenuous exertion in a threatening situation.

Circadian and Ultradian Rhythms

The circadian rhythm in glucocorticoid secretion appears to be entrained by the organism's rest-activity cycle. The secretory peak occurs just before the active phase, even when the relation of activity to the lighting cycle is reversed, as in human beings working on night shifts or in rats fed only during the day (Morimoto et al., 1977). In addition to the circadian rhythm, higher frequency oscillations in corticosteroid secretion have been revealed by frequent sampling, every 20 minutes. Recent evidence (Holaday et al., 1977) has shown that this pulsatile secretion, previously regarded as "episodic," actually follows an ultradian rhythm, or frequency greater than one cycle in 20 hours, having a predominant periodicity of about 90 minutes and other components harmonically related to the circadian rhythm. These rhythmic fluctuations in plasma cortisol were synchronized among eight isolated, restrained, undisturbed monkeys, indicating their entrainment by environmental factors, such as the feeding or lighting schedule. The unexpected finding that ultradian cortisol rhythms were not disrupted by infusion of supramaximal ACTH challenges the classic concerpt that periodic bursts of corticosteroid output depend entirely on the immediately preceding release of ACTH. The physiological function of corticosteroid rhythms is unknown.

Corticosteroid Feedback

ACTH secretion is regulated by two temporally distinct negative feedback mechanisms, a rate-sensitive fast feedback (FFB), which occurs 5 to 30 minutes after steroid administration, and a proportional, delayed feedback (DFB), which appears after one or more hours (Dallman and Yates, 1969; Jones et al., 1977). These two phases are separated by a "silent" period, during which negative feedback is not observed. Differences in the steroid structure-activity relationship for FFB and DFB indicate that different receptor mechanisms may be involved.

There is evidence that corticosteroid feedback actions may be exerted at multiple sites. The sensitivity of ACTH-secreting pituitary cells to inhibition by physiological doses of natural and synthetic corticosteroids has been clearly established by studies in which the possibility of hypothalamic involvement was circumvented and by studies of pituitary cells *in vitro* (Kendall, 1971; Jones et al., 1977). Hypothalamic tissue *in vitro* also showed a FFB effect of corticosterone, due to decreased release of CRF, and a DFB effect, due to both decreased synthesis and release of CRF (Jones et al., 1977). A series of studies (see Feldman and Conforti, 1980) demonstrated that posterior hypothalamic deafferentation, dorsal fornix section, or dorsal hippocampectomy reduced the inhibitory DFB effect of dexamethasone on both basal and other stress-induced corticosterone secretion in the rat, suggesting that the dorsal hippocampus also participates in the feedback regulation of pituitary-adrenal function. Furthermore, several studies (see Carsia and Malamed, 1979) have indicated a direct inhibitory effect of corticosterone and cortisol on ACTH-induced corticosteroidogenesis, suggesting that the self-suppression of adrenocortical cells by end products may provide an additional fine adjustment of steroidogenesis. What remains to be determined is the relative physiological importance of glucocorticoid feedback at the various sites—anterior pituitary, hypothalamus, extrahypothalamic structures, and the adrenal cortex itself. Although it appears that the synthetic glucocorticoid dexamethasone may act primarily at the pituitary level (Kendall, 1971; de Kloet et al., 1974), the pituitary may be less responsive to natural glucocorticoids. The explanation for this difference is that anterior pituitary cytosol contains a transcortin-like macromolecule, which like plasma transcortin binds corticosterone but not dexamethasone and has negligible affinity for DNA-associated acceptor sites in the nucleus (de Kloet et al., 1977). Thus, while the transcortin-like binders cannot interfere with the action of dexamethasone, they can, by competing with the "true" cytoplasmic glucocorticoid receptors, reduce the amount of corticosterone able to interact with these receptors. Such a mechanism might insure that under nonstress conditions the pituitary glucocorticoid receptors would not be occupied, thus allowing them to function only in response to much higher, stress-induced levels of corticosterone.

Glucocorticoids may also act directly upon the hypothalamus and pituitary to modulate the production or secretion of hormones other than CRF and ACTH, such as thyrotropin-releasing hormone (TRH), thyroid-stimulating hormone (TSH), and growth hormone (GH) (see Burger and Patel, 1977).

The "compensatory" hypertrophy of the remaining adrenal following unilateral adrenalectomy was long considered a result of decreased corticosteroid feedback. However, compensatory adrenal growth was recently shown to be neurally rather than hormonally mediated, dependent on reciprocal neural connections between the hypothalamus and adrenal (Dallman et al., 1976).

Although much progress has been made in elucidating individual factors influencing pituitary-adrenal activity (neural input, feedback, stress, rhythms), we are still far from gaining a clear understanding of how these isolated components function together in the intact living organism.

Regulation of Mineralocorticoid Secretion

Three factors have been identified in the control of aldosterone secretion: ACTH, the renin-angiotensin system, and the
plasma concentrations of sodium and potassium (Williams and
Dluhy, 1972; Davis, 1975). Although ACTH may stimulate
aldosterone under some circumstances, its role appears to be
relatively minor. Vascular baroreceptors in the juxtaglomerular apparatus of the kidney respond to decreased renal
perfusion pressure, resulting, for instance, from exsangui-
nation, dehydration, sodium depletion, or hypoalbuminemia, by
stimulating the secretion of renin. This enzyme cleaves the
circulating inactive peptide angiotensinogen to release the
decapeptide angiotensin I, which is further hydrolyzed by a
converting enzyme in the lung to the active octapeptide
angiotensin II. Angiotensin II stimulates the production of
aldosterone by cells in the zona glomerulosa of the adrenal.
By acting on the kidney to increase sodium reabsorption, aldosterone causes expansion of the extracellular fluid, increased effective blood volume, and elevation of renal perfusion pressure. Thereby the stimulus for renin production is
removed and the negative feedback loop completed. The other
identified control mechanism is the direct effect of high
potassium and low sodium in the plasma to stimulate aldosterone
secretion. The possibility that the nervous system may exert
significant regulatory effects on the secretion of aldosterone
has not been adequately explored and cannot be dismissed.

MECHANISMS OF ACTION

Overview of Corticosteroid Mechanisms

Many of the effects of corticosteroids are believed to be
mediated by interactions of the steroid molecules with steroid-
specific cytoplasmic macromolecular receptors that "trans-
locate" as hormone-receptor complexes to the target cell
nuclei, where they initiate the alterations in specific RNA and
protein metabolism that then lead to the ultimate physiological, neuroendocrine, and behavioral effects. Other less well-
understood steroid effects may result from direct interactions
of the steroid molecule with components of target cell
membranes (see later). Some established and hypothetical
events in corticosteroid action are represented in Fig. 14-2.

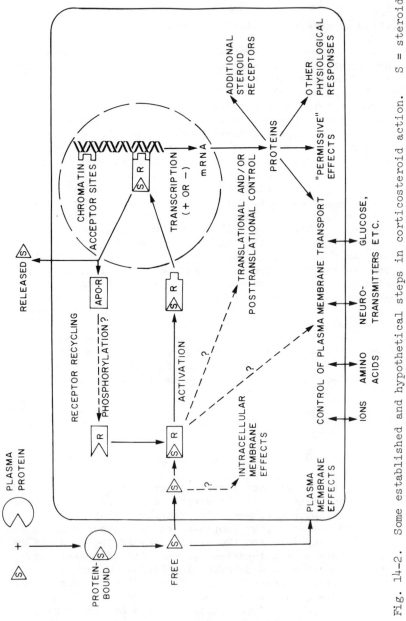

Fig. 14-2. Some established and hypothetical steps in corticosteroid action. S = steroid molecule, R = receptor, APO.R = aporeceptor. Different shapes of R represent different conformations or covalent modifications. The broken lines indicate

Although the natural adrenal steroids are soluble enough to be transported unassisted in plasma, about 75 percent of the circulating glucocorticoid, cortisol or corticosterone, is bound to an α-globulin called transcortin or corticosteroid-binding globulin, about 15 percent is bound to serum albumin, and only about 10 percent is free (Westphal, 1975). Transcortin does not seem to be necessary in any way for the biological activity of the steroids; current evidence supports the dogma that only the free steroid in the plasma can exert physiological action. Transcortin does reduce the amplitude of free glucocorticoid variations in response to large rhythmic or stress-induced changes in adrenal output. The potent synthetic fluorinated glucocorticioids, dexamethasone and triamcinolone, as well as the natural mineralocorticoid aldosterone, are only very weakly bound by transcortin. Since the natural gluco-corticoids can significantly occupy mineralcorticoid receptors when present in high concentrations and since the total concentration of plasma glucocorticoids is much greater than the normal concentration of aldosterone, the "buffering" effect of glucocorticoid binding by transcortin is apparently neces-sary to prevent the saturation of mineralocorticoid receptors by glucocorticoids (Funder et al., 1973). It is not known why the glucocorticoid to mineralocorticoid ratio is so large, requiring this rather peculiar mechanism to confer specificity of hormone action.

It is often assumed that target cell membranes do not present a barrier to free lipophilic steroids and that their passage into the target cell is governed solely by simple diffusion. Recent studies have, however, demonstrated for at least several different cell types (isolated rat liver cells and ACTH-secreting mouse pituitary tumor cells) that glucocorticoid passage through the membrane may involve carrier-mediated transport in addition to simple diffusion (see Harrison et al., 1977). It is not yet clear how general this phenomenon may be in terms of other target cells and hormones. It has been observed that some steroids may have several different actions in the nervous system that are mediated independently by their different metabolites, but there is no evidence that the metabolites of the natural glucocorticoids, corticosterone and cortisol, are functionally important and possess their own nonenzymatic high-affinity binding sites in brain or pituitary. Indeed, following *in vivo* injections of [³H]corticosterone, the radioactivity extracted from the nuclear fraction of rat brain was found to consist of approximately 90 percent authentic cor-ticosterone (McEwen et al., 1972).

The cytoplasmic steroid receptors are thermolabile proteins with stereo-specific binding sites. Before the steroid can bind to the receptor with high affinity ($K_d \approx 10^{-9}$ M), the

corticosteroid receptor protein ("aporeceptor") may be required
to undergo an energy-dependent transformation (possibly a
phosphorylation) in order that the potential binding site may
be "switched on" to the appropriate conformation for interaction with the steroid. A rapid "switching-off" or down-
regulation of the steroid-binding sites, possibly mediated by a
phosphatase, has also been observed, suggesting that cells may
utilize an internal phosphorylation-dephosphorylation feedback
cycle to modulate physiological responses by regulating the
amount of receptor capable of interacting with free steroid.
Thus, under many circumstances a substantial pool of latent or
"cryptic" aporeceptors may be present in many target cells.
This dynamic regulation of the hormone binding site itself has
only recently been explored in cells derived from a few
peripheral tissues (see Sando et al, 1979), and is an
intriguing area for brain studies. The number of receptors
capable of interacting with free steroid may be subject to
additional regulation by certain proteolytic enzymes that can
disconnect the steroid binding site from the region of the
receptor molecule that contains the nuclear binding site,
resulting in nonfunctional steroid-binding fragments termed
"mero-receptors" (see Sherman et al., 1978).

After binding, the noncovalent cytoplasmic steroid-receptor
complex must next undergo a transformation that results in the
development of a high affinity for certain nuclear components
associated with the genome. This process of "activation"
probably involves a steroid-induced conformational change in
the acidic receptor protein that brings a positively charged
"acceptor" binding site to the surface of the molecule. The
nuclear "acceptors" to which the activated receptor complexes
now bind are unidentified components of chromatin, possibly
basic non-histone proteins, that possess high affinity and, to
varying extents, tissue- and receptor-specific binding domains.
All nuclear and chromatin preparations that have been found to
"accept" or bind corticosteroid-receptor complexes contain DNA.
Although the acceptors are probably not merely specific DNA
nucleotide sequences, they do appear to regulate the interactions of the steroid-receptor complexes with the DNA.

Following the formation of the ternary steroid-receptor-
acceptor complex, a process often described as "nuclear
translocation," the chromatin structure becomes altered in
subtle ways that lead to changes in the rates of transcription
of specific mRNA species. These specific mRNA molecules are
then translated to produce the proteins that mediate the
hormone-induced physiological responses. Figure 14-2 indicates
that the proteins whose rates of synthesis are modulated by the
steroid may encompass a broad spectrum of cellular functions:
additional steroid receptors; components or modulators of

membrane transport mechanisms; enzymes of intermediary
metabolism; protein kinases, components of peptide hormone- or
neurotransmitter-sensitive receptor-adenylate cyclase complexes, and other modulators whose altered synthesis may
contribute to the so called "permissive" effects of steroids;
and even specific proteins required for some catabolic steroid
effects, such as thymus involution, are all examples of
proteins that may be regulated to produce the ultimate steroid
response.

After exerting their genomic effect, the receptors are either
degraded or recycled back to their unbound, nonactivated cytoplasmic form by a process that may be linked to cell metabolism
by requirement for adenosine triphosphate (ATP). The nuclear
"processing" of the receptors, the "off-reaction," and receptor
recycling are very poorly understood; it is possible that some
steroid dissociation may occur before the receptors are
released from their chromatin acceptor sites, and there are
hints that the process may be coupled to the proposed cyclic
transformations of the steroid binding sites. The released
steroid molecules may either re-enter the receptor cycle or
diffuse out of the cell into the circulation to enter another
cell or to be metabolized and excreted.

Figure 14-2 also indicates several largely unexplored
potential mechanisms of steroid influence on cellular function
that do not directly involve events at the genome. The
suggestions that corticosteroid-receptor complexes may directly
exert translational or post-translational control over specific
protein synthesis, a reported effect of androgen-receptor
complexes in the prostate, or that they may directly regulate
membrane transport mechanisms are hypothetical at present. The
suggestion that some glucocorticoid effects may result from the
interaction of free steroids with intracellular membrane
systems is also hypothetical; glucocorticoids are known to
modify some membrane properties, but no functional consequences
of such changes are yet well established. Free steroids may
also exert important effects at the cell plasma membrane.

Since a steroid's affinity for the cytoplasmic receptors does
not adequately predict the magnitude of the physiological
response, it is necessary to classify all steroids into one of
four categories on the basis of their physiological effectiveness. Optimal inducers are steroids that all produce the same
maximal response when present in saturating amounts. For
example, aldosterone will produce as great a glucocorticoid
response as dexamethasone (in many tissues) when present in
very high concentrations. Suboptimal inducers elicit smaller,
less-than-maximal responses even when present in saturating
concentrations; 11-DOC is an example of a suboptimal glucocorticoid. Anti-inducers or antihormones produce no typical

physiological responses by themselves, but rather behave as
competitive inhibitiors of the active hormones; progesterone
and cortexolone (11-deoxycortisol) are antiglucocorticoids.
Finally, inactive steroids do not bind to the specific steroid
receptors at all. It should be stressed that the classification of a particular steroid must refer to a specific,
measurable response and may vary among species and from one
tissue to another.

It is believed that different ligands can promote different
degrees of conformational change, leading to the formation of
steroid-receptor complexes with different states of "partial
activation" (different affinities for nuclear acceptor components). Munck and Leung (1977) have proposed a model in
which each relevant steroid or class of steroids binds to the
receptor and promotes a subsequent conformational change that
differs in degree from that produced by other steroids.
Optimal inducers are those that produce the highest degrees of
activation, and anti-inducers either do not promote activation
or promote minimal, ineffective increases in affinity for
nuclear acceptors. Suboptimal inducers produce intermediate
states of activation. Other models of agonist and antagonist
interactions with the glucocorticoid receptor are also under
active consideration (Rousseau and Baxter, 1979: Sherman 1979).

Anatomical Distribution of Corticosteroid Binding

Neuronal nuclear concentration of [³H]corticosterone has been
demonstrated by autoradiography in structures of the limbic
system, brainstem, and spinal cord, but not in the hypothalamus
(McEwen et al., 1975; Stumpf and Sar, 1975; Warembourg, 1973).
In adrenalectomized rats, nuclear accumulation of [³H]cortico-
sterone was most intense in structures related to the hippocampus, including the postcommissural hippocampus, dentate
gyrus, induseum griseum (supracallosal hippocampus), anterior
(precommissural) hippocampus, and subiculum. Strong nuclear
labeling was also seen in the lateral septum, amygdala
(cortical, central, and basomedial nuclei), and the piriform,
entorhinal, suprarhinal, and cingulate cortices. Additional
labeling, although weaker and less frequent, was present in the
anterior olfactory nucleus, medial amygdaloid nucleus,
habenula, red nucleus, and subfornical organ. Motor neurons in
cranial nerve nuclei and spinal cord were strongly labeled, and
some glial cells were weakly labeled (Stumpf and Sar, 1979).
The pattern of *in vivo* uptake of [³H]corticosterone determined
by autoradiography (highest in hippocampus and septum, followed
by amygdala, cortex, and hypothalamus) agrees well with the
anatomical distribution of cytoplasmic [³H]corticosterone

binding sites (McEwen et al., 1972; Grosser et al., 1973); and with the patterns of [³H]corticosterone binding found in purified nuclei both following [³H]hormone injections *in vivo* (McEwen et al., 1970) and after incubation of brain slices with [³H]corticosterone *in vivo* (McEwen and Wallach, 1973; de Kloet et al, 1975). Corticosterone target cells have been observed in the anterior pituitary of the rat (Warembourg, 1973) and Pekin duck (Rhees et al., 1972), but not the rhesus monkey (Pfaff et al., 1976).

The distribution of target cells for [³H]cortisol in the brains of adrenalectomized rats was identical to that for [³H]corticosterone (Stumpf and Sar, 1973). Nuclear binding sites for cortisol appeared to be saturated by endogenous corticosteroids in adrenally intact mice (Schwartz et al., 1972) and guinea pigs (Warembourg, 1973).

The synthetic glucocorticoid dexamethasone displayed a surprisingly different pattern of uptake from that of natural glucocorticoids. Whereas corticosterone and cortisol were concentrated strongly by neurons, [³H]dexamethasone was accumulated weakly by all types of cells in the brain (Rees et al., 1975; Rhees et al, 1975; Warembourg, 1975). The labeling was heaviest in epithelial cells of the choroid plexus and ventricular lumen and was also observed in vascular endothelial cells, glia, meninges, ependyma, circumventricular organs, and in neurons in areas near the third ventricle (preoptic area, hypothalamus, thalamus) and lateral ventricle (septum, caudate, amygdala). In contrast to [³H]corticosterone, [³H]dexamethasone was concentrated only very weakly by hippocampal neurons. Furthermore, the presence of endogenous adrenal hormones in intact rats did not affect the pattern of [³H]dexamethasone localization (Rhees et al., 1975). In the pituitary, dexamethasone was heavily concentrated by cells in the pars distalis (particularly corticotrophs) and pars nervosa, but not pars intermedia (Rees et al, 1977). The autoradiographic data were consistent with the pattern of *in vivo* uptake of [³H]-dexamethasone revealed by direct measurements of tissue radioactivity (de Kloet et al., 1974). No detailed *in vitro* comparison of the concentrations of cytosol [³H]dexamethasone and [³H]corticosterone binding sites across different brain regions has yet been reported.

The strikingly different patterns of distribution of these natural and synthetic glucocorticoids have been interpreted as evidence for the existence of at least two classes of gluco-corticoid receptors differing in their distribution and steroid specificity. However, some of the findings may be explained without reference to the concept of receptor heterogeneity. There are large differences in the permeability of the blood-brain barrier to different steroids (Pardridge and Mietus,

1979). Dexamethasone appears to enter the brain more slowly than corticosterone; time course studies showed that maximal binding in hippocampal cell nuclei occurred one hour after injection of [^{3}H]corticosterone, but two hours after [^{3}H]-dexamethasone (de Kloet et al., 1975). Similarly, the cellular accumulation of [^{3}H]dexamethasone in the hippocampus seen autoradiographically three hours after injection was not yet evident at 30 minutes (Rees et al., 1975). The greater blood-brain barrier to dexamethasone may also explain some discrepancies between the patterns of nuclear binding of glucocorticoids obtained *in vivo* and in slices incubated *in vitro*. Although the *in vivo* nuclear binding of [^{3}H]corticosterone in hippocampus was more than 10 times that of [^{3}H]dexamethasone, this difference was dramatically reduced (corticosterone: dexamethasone ratio of 1.2 to 1.5) when slices of hippocampus were incubated with the steroids *in vitro* (de Kloet et al., 1975; McEwen et al., 1976). It is possible that the small but significant remaining differences in nuclear binding observed in the *in vitro* slice experiments, that is, greater binding of corticosterone in hippocampus and of dexamethasone in hypothalamus and pituitary, may have resulted from factors other than glucocorticoid receptor heterogeneity, for example, differences in the rates of cellular penetration of the two steroids, which may persist in the slice experiments, and differences in the relative abilities of the two steroids to promote activation and nuclear binding of the steroid-receptor complexes (see Svec and Harrison, 1979). The question of multiple glucocorticoid receptors will be further discussed in the next section.

The mineralocorticoid aldosterone displayed an autoradiographic pattern of distribution nearly identical to that of corticosterone, that is, greatest labeling in hippocampus and related structures, motor nuclei of cranial nerves, and the reticular formation (Ermisch and Rühle, 1978; Birmingham et al., 1979). The only differences were in the relative intensity of labeling; for example, with [^{3}H]aldosterone the induseum griseum was labeled more strongly than the postcommissural hippocampus, whereas with [^{3}H]corticosterone the reverse was true. In only one respect did aldosterone resemble dexamethasone rather than corticosterone—its accumulation by cell nuclei of the arachnoid. The high proportion of cells labeled with [^{3}H]aldosterone or [^{3}H]corticosterone in certain regions, such as hippocampus, makes it probable that at least some cells concentrate both hormones. Clearly, future autoradiographic studies should assess the extent to which aldosterone and corticosterone may compete for common binding sites. The pattern of *in vivo* nuclear concentration of [^{3}H]-aldosterone revealed by autoradiography also agrees well with the anatomical distribution (highest in hippocampus, followed by septum, amygdala, cortex, and hypothalamus; no binding sites in preoptic area) of specific high-affinity cytoplasmic mineralocorticoid binding sites recently described by Moguilewsky and Raynaud (1980). Biochemical studies of the nuclear binding of [^{3}H]aldosterone in discrete brain regions have not yet been reported.

Characterization of Soluble Corticosteroid
Receptors in Brain

The natural glucocorticoids, corticosterone and cortisol, the synthetic glucocorticoids, triamcinolone acetonide (TA) and dexamethasone, and the natural mineralocorticoids, aldosterone and 11-DOC, bind to specific, saturable brain cytosol components believed to be the physiological transducer molecules or "receptors." The goal of receptor research is to identify the different classes or categories of adrenal steriod action in the brain and to study individually the receptors mediating these actions. The categories "glucocorticoid" and "mineralocorticoid" are defined by distinguishable peripheral physiological effects and steroid specificities; this distinction may be meaningful in the nervous system, but it should not be assumed a priori that brain steroid effects and specificities closely correspond to those of other organ systems.

Basic Physicochemical Properties of the Brain Receptors

In comparison with the corticosteroid receptors found in other tissues, the properties of brain receptors are generally unremarkable. They have been distinguished from those of transcortin by a number of physicochemical criteria; unlike transcortin, the brain cytoplasmic glucocorticoid-binding protein was found to bind the synthetic steroids dexamethasone and TA with high affinity and to possess sulfhydryl groups whose modification led to the loss of functional steroid binding sites (see Chytil and Toft, 1972; McEwen and Wallach, 1973). The nonactivated (untransformed) mouse brain [^{3}H]TA-receptor complex (protected *in vitro* by sodium molybdate) is a large, asymmetric, possibly multimeric species having the following well-resolved properties: Stokes radius a = 7.7 nm; sedimentation coefficient $S_{20,w}$ = 9.5 S; molecular weight, 315,000 daltons; frictional ratio (f/fo), 1.7; and prolate axial ratio, 13:1 (Gray and Luttge, unpublished). These properties are shared by several different classes of steroid-receptor complexes extracted from a variety of tissues. The physicochemical properties, other than steroid specificity, of specific mineralocorticoid binding sites in the brain have not yet been investigated.

Resolution of Brain Corticosteroid Receptor Classes on the Basis of Differential Steroid Affinity and Binding Capacity

The resolution of the different classes of brain corticosteroid receptors is confusing because it involves two distinct but related issues: the possible existence of separate binding sites for natural and synthetic glucocorticoids, and the distinction between glucocorticoid and mineralocorticoid binding sites. The principal technique for defining distinct binding sites is the *in vitro* measurement of steroid specificity in competition experiments. Oddly, the actual affinities of different competing steroids for brain cytosol [³H]steroid binding sites have seldom been reported. For example, the extent to which progesterone competes for [³H]corticosterone binding sites in the brain has not yet been measured precisely. Specificity data have typically been reported only as a rank ordering of the abilities of different steroids to compete for the binding of a given [³H]steroid.

Although both natural and synthetic agonists were bound with similar high affinities by brain cytosol (Chytil and Toft, 1972), dexamethasone did not reduce the binding of [³H]corticosterone in whole brain cytosol to the extent that was predicted from its physiological potency as a peripheral glucocorticoid (Grosser et al., 1973; McEwen and Wallach, 1973). In most studies, corticosterone and dexamethasone were equally effective in competition for [³H]dexamethasone and [³H]TA binding (see Chytil and Toft, 1972; Stevens et al., 1975; de Kloet and McEwen, 1976).

Comparative measurements of the total cytosol binding capacity for corticosterone and dexamethasone have been reported. Because a number of poorly understood variables, such as the ionic composition of the incubation buffer, the presence or absence of phosphatase inhibitors, the time elapsed between tissue homogenization, and cytosol labeling with [³H]steroid, have not yet been fully explored or controlled, published estimates of apparent binding capacity (B_{max}) are sometimes in conflict. The discovery by de Kloet et al. (1975) that the spontaneous loss of [³H]dexamethasone binding capacity from unlabeled cystosol was more rapid than the loss of [³H]corticosterone binding sites stimulated experiments in which tissues were homogenized in the presence of the [³H]steroids. With this important improvement in methodology, binding capacities for [³H]dexamethasone and [³H]corticosterone in hippocampal cytosol were found to be equal, and hypothalamic cytosol had an even slightly higher capacity for [³H]dexamethasone than for [³H]corticosterone (e.g., de Kloet et al., 1975; Turner and McEwen, 1980).

High-affinity mineralocorticoid (type I) binding sites, which occur in significant concentrations principally in the hippocampus and associated structures, have been studied in rat brain cytosol (Anderson and Fanestil, 1976; Moguilewsky and Raynaud, 1980). The mineralocorticoid receptors have high affinity for aldosterone and DOC and, surprisingly, an almost equally high affinity for progesterone. Since [³H]aldosterone, [³H]corticosterone and [³H]dexamethasone all bind, with different affinities, to both glucocorticoid and mineralocorticoid brain receptor sites, it has been most productive to study the binding of [³H]aldosterone in the presence of an excess of the "pure" glucocorticoid R26988 (Moguilewsky and Raynaud, 1980). Although concentrations of glucocorticoid and mineralocorticoid receptor sites were comparable in hippocampal cytosol, the concentration of high-affinity mineralocorticoid binding sites was much lower than the concentration of glucocorticoid binding sites in whole brain. Very little cell nuclear retention of radioactivity was found in any brain following *in vivo* injections of [³H]DOC (McEwen et al., 1976), but potential nuclear binding may have been reduced by the extensive metabolism of injected [³H]DOC (seen in eviscerated rats, Kraulis et al., 1975). The awareness that glucocorticoids can bind to mineralocorticoid receptor sites, coupled with the current realization that the hippocampus contains a significant concentration of mineralocorticoid binding sites, will require the reassessment of a number of studies previously thought to be concerned only with glucocorticoid receptors.

Glucocorticoid Receptor Heterogeneity

Although the discrepancies between *in vivo* and *in vitro* (brain slice) nuclear binding of [³H]corticosterone and [³H]dexamethasone can be largely explained without reference to the possible heterogeneity of unbound glucocorticoid receptors, several observations, such as the relatively poor ability of dexamethasone to compete for [³H]corticosterone binding sites, have suggested that natural and synthetic glucocorticoids may bind to somewhat different receptor populations. When cytosol samples were chromatographed on DEAE-cellulose ion exchange columns, complexes formed with [³H]corticosterone and [³H]dexamethasone were eluted as multiple peaks, and the proportion of bound [³H]steroid in each of the two major peaks differed for the two glucocorticoids (de Kloet and McEwen, 1976). Although it is unlikely that the two major peaks of both hippocampal [^{3}H]dexamethasone and [^{3}H]corticosterone binding merely represent different pools of activated and nonactivated receptors, it is quite possible that they are distinct proteolytic fragments, created *in vitro*, of a single larger intact glucocorticoid receptor. Affinity chromatography of rat brain cytosol on columns of immobilized DOC hemisuccinate, subsequently eluted with [^{3}H]corticosterone, selectively purified one of the two major [^{3}H]corticosterone binding peaks resolved by ion exhange chromatography (de Kloet and Burbach, 1978).

Apparent receptor heterogeneity was also observed in rat brain (and pituitary) cytosol following isoelectric focusing of labeled samples on polyacrylamide gels (Maclusky et al., 1977). In brain cytosol three major specific [^{3}H]corticoterone binding peaks were resolved; these had isoelectric points (pI) of approximately 6.8, 5.9, and 4.3. When [^{3}H]dexamethasone was the ligand, only two peaks were found (the peak at pI 4.3 was absent). Furthermore, the relative sizes of the two remaining peaks were different for the two ligands. Wrange (1979) has suggested that the apparent receptor heterogeneity observed by Maclusky et al. (1977) may have resulted from proteolytic artifacts and that the brain cytosol [^{3}H]corticosterone binding peak at pI 4.3 (reported by the same workers) can probably be attributed to residual transcortin remaining in the tissue following incomplete perfusion. Wrange found only a single peak of radioactivity (at pI 6.1-6.2) when rats were extensively perfused and hippocampal cytosol samples labeled with either [^{3}H]corticosterone or [^{3}H]dexamethasone were analyzed. Wrange also found that limited tryptic digestion of hippocampal cytosol labeled with either [^{3}H]corticosterone or [^{3}H]dexamethasone produced two peaks of bound radioactivity having pI values of 5.9 to 6.1 and 6.3 to 6.5. These pI values are close enough to those reported by Maclusky et al. (1977) to suggest that proteolytic fragments of a single molecule may have been responsible for the observed heterogeneity. Furthermore, the relative sizes of the two trypsin-induced peaks were different when [^{3}H]dexamethasone was substituted for [^{3}H]corticosterone.

The autoradiographic data reviewed in the previous section have led to the suggestion that neurons contain glucocorticoid receptors that preferentially bind corticosterone and cortisol, whereas glial cells contain glucocorticoid receptors having higher affinity for dexamethasone and TA (see McEwen et al., 1979). There is, however, very little evidence to support this dichotomy. Although dexamethasone is a potent inducer of glycerol phosphate dehydrogenase (GPDH) in cultured glial tumor cells (see later), and cytosol prepared from these tumor cells and from optic nerve oligodendrocytes contains high-affinity

[^{3}H]dexamethasone binding sites, glial cells also respond to natural glucocorticoids (see Breen et al., 1978; Cotman et al., 1978). Furthermore, there is no evidence (e.g., Clayton et al., 1977) that brain [^{3}H]dexamethasone binding capacity increases faster than [^{3}H]corticosterone binding capacity during the period of rapid glial growth associated with myelination. Glial tumor cells were found to contain only one glucocorticoid receptor with a pI of 5.9, corresponding to the molecular species that bound [^{3}H]dexamethasone preferentially in the rat brain (Maclusky, unpublished, cited by McEwen et al., 1979). This observation cannot, however, be considered strong evidence for a neuronal-glial receptor dichotomy, since Wrange (1979) reported a similar isoelectric binding profile containing only a single radioactive peak in hippocampal cytosol from throughly perfused animals. It is possible that different concentrations of proteolytic enzymes in the cytosol samples prepared from brain tissue and from cultured glial cells could explain the differences between the [^{3}H]gluco-corticoid binding profiles observed by Maclusky and colleagues in these different preparations. Even the demonstration that [^{3}H]dexamethasone binding sites disappear from hippocampal cytosol, in the absence of steroid, more rapidly than [^{3}H]corticosterone binding sites (de Kloet et al., 1975) may result from the gradual alteration of a single initial population of steroid binding sites by enzymatic processes that are triggered upon cell disruption and that proceed rapidly in the absence of protective steroid ligands. Clarification of this complex issue must await the purification and comparison of both intact unbound receptors and steroid-receptor complexes.

Physiological Control of Corticosteroid Receptors

Changes During Development

The ontogeny of the capacity of rat brain cytosol to bind both natural and synthetic glucocorticoids has been studied. Binding of [^{3}H]dexamethasone was very low immediately after birth, but it seemed to reach the adult level sooner than [^{3}H]corticosterone binding, which was higher than that of [^{3}H]dexamethasone immediately after birth. Adult levels of [^{3}H]corticosterone binding were similar to those of [^{3}H]dexamethasone in both hippocampal and hypothalamic cytosols (Olpe and McEwen, 1976). Turner (1978) found that the amount of [^{3}H]corticosterone bound by hippocampal nuclei in adrenalectomized rat pups injected with steroid *in vivo* was very small in comparison with adult levels. Furthermore, the nuclear binding of [^{3}H] corticosterone by hippocampal pyramidal and dentate granule cells as determined by autoradiography was directly correlated with neuronal age; in the neonatal hippocampus the oldest cells revealed the heaviest labeling, whereas newly arrived cells showed little nuclear retention of steroid. Thus, functional receptor production may occur relatively late in the differentiation of these neruons.

An age-related decline in corticosterone receptors has been reported in mouse hippocampus (Finch and Latham, 1974) and in rat cerebral cortex (Roth, 1976). Evidence suggests that senescent intracellular biochemical changes rather than cellular losses are responsible for the decline in cortical receptors (Roth, 1976).

Effects of Endogenous Corticosteroids

The concentration of intracellular corticosteroid binding sites rises in response to steroid deprivation. Adrenalectomy caused a two-stage increase in the nuclear binding of [^{3}H]corticosterone by hippocampus *in vivo* and *in vitro* (McEwen et al., 1974) and increased glucocorticoid cytosol receptor concentrations (Stevens et al., 1975; Olpe and McEwen, 1976). The apparent receptor content increased rapidly for the first two hours after adrenalectomy and then remained at a plateau for about 12 hours; the second, slower increase began between 12 and 18 hours after adrenalectomy and approached a new plateau after about three days. The first, rapid change, which parallels the decline in plasma corticosterone, certainly represents the disappearance of endogenous corticosterone from brain binding sites and may also reflect the "switching-on" of the steroid binding sites of receptors. The interesting long-term increase results from either the synthesis of new receptors or the "switching-on" of previously unobservable "cryptic" aporeceptors.

Both the concentration of endogenous corticosteroids and the occupancy of corticosteroid receptors in the brain vary with changes in plasma steroid levels. Brain glucocorticoid concentrations, which were intermediate between free plasma and total plasma glucocorticoid concentrations (Carroll et al., 1975) were found to fluctuate in parallel with both basal circadian and stress-induced changes in the plasma steroid concentrations (Butte et al., 1976; Carroll et al., 1975). Furthermore, the diurnal and stress-induced increases in plasma corticosterone decreased the *in vitro* cytosol binding of [^{3}H]-corticosterone (Stevens et al., 1973). In all brain regions examined in unstressed animals, glucocorticoid receptor occupancy varies between about 50 percent at the diurnal trough and approximately 80 percent at the peak (Stevens et al.,

1973; McEwen et al., 1974; Turner et al., 1978a,b). In contrast to other brain regions, the preoptic and septal areas exhibited a high level of receptor occupancy during the morning corticosterone minimum, and no increase at the evening peak (Turner et al., 1978a,b). However, all brain regions showed a circadian variation in the total concentration of cytosol [^{3}H] corticosterone binding sites. Furthermore, Angelucci et al. (1980) found that the same dose of [^{3}H]-corticosterone injected into adrenalectomized mice produced higher hippocampal steroid concentrations at different times of the day. The peak brain concentrations varied as the normal circadian rhythm, suggesting that a steroid-independent rhythm of receptor concentration may persist in the adrenalectomized animals.

An investigation of the temporal relationship between glucocorticoid nuclear binding and the availability of cytosol binding sites led to the unexpected finding that there was no net depletion of total hippocampal cytosol binding capacity as a result of nuclear translocation 15 to 60 minutes after the injection of fully saturating doses of either [^{3}H]corticosterone or [^{3}H]dexamethasone (Turner and McEwen, 1980). The predicted cytosol receptor depletion was based on hippocampal nuclear uptake measured following the *in vivo* [^{3}H]steroid injections. Cytosol samples from rats injected with [^{3}H]-steroids were incubated *in vitro* with additional [^{3}H]steroids to determine the maximal cytosol steroid binding capactity, but no depletion of this total capacity as a result of nuclear translocation was ever observed. This investigation also revealed that [^{3}H]corticosterone injected *in vivo* could occupy no more than 40 percent of the total cytosol binding sites measured *in vitro*. These results suggest the existence of a reserve pool of aporeceptors or "cryptic" receptors that can be rapidly converted to the form capable of binding steroids.

Effects of Neuropeptides

The increase in [^{3}H]corticosterone binding capacity of rat hippocampal cytosol observed after hypophysectomy combined with adrenalectomy was greater than that after adrenalectomy alone. Both ACTH$_{1-24}$ (steroidogenic) and ACTH$_{4-10}$ (devoid of cortico-trophic activity) eliminated the additional increase attributed to hypophysectomy. Furthermore, vasopressin-deficient (Brattleboro strain) rats were found to have abnormally low hippocampal cytosol [^{3}H]corticosterone receptor levels that could be restored by physiological doses of vasopressin or des-glycinamide^9-arg^8-vasopressin (a behaviorally potent analogue having low antidiuretic activity) (de Kloet et al., 1980).

Corticosterone "Membrane Effects" and Receptors

Some corticosteroid effects on the nervous system are not mediated by the mobile cytoplasmic receptors that affect gene expression; these effects may derive from the alteration of membrane properties by the free steroids themselves or may be mediated by specific membrane-associated receptors. Proposed mechanisms for such effects have included the stabilization of lysosomal membranes, which could delay the release of hydrolytic enzymes, alteration of ribosomal attachment to the endoplasmic reticulum, which could modify protein synthesis, and alteration of the binding of calcium to intracellular membranes, which could influence synaptic function.

Dexamethasone appeared to elevate tyrosine hydroxylase (TH) activity in the superior cervical ganglion of adrenally intact rats by exerting an excitatory pharmacological influence directly on preganglionic cholinergic nerve terminals; very large doses of corticosterone were completely ineffective, and the slowly developing effect of dexamethasone was abolished by a cholinergic receptor antagonist (Sze and Hedrick, 1979). The effects of synthetic glucocorticoids on cholinergic neurotransmission have been both excitatory and inhibitory. The excitability of cat somatic motoneurons was increased (Riker et al., 1975), and the contraction of guinea pig ileum in response to nerve stimulation was decreased (Cheng and Araki, 1978) by the steroids; in both cases the evidence suggested a direct steroid action on cholinergic terminals.

The membrane effects of glucocorticoids probably also include the physiologically important FFB suppression of the release of CRF by hypothalamic neurons. The addition of corticosterone to the *in vitro* incubation medium blocked the release of CRF produced by the electrical stimulation of sheep hypothalamic synaptosomes (Edwardson and Bennett, 1974), and there is evidence that the FFB action of corticosterone, which is unaffected by a number of pharmacological agents, may be mediated by a direct stabilizing interaction of the steroid with the membrance of the CRF-secreting cell, which decreases the flux of calcium into its terminals (Jones et al., 1977).

Glucocorticoids regulate the uptake of [³H]tryptophan by isolated brain synaptosomes incubated *in vitro* with the steroids; corticosterone and dexamethasone at concentrations above 10^{-7} M elevated the maximal rate (V_{max}) of tryptophan transport by a high-affinity synaptosomal uptake system from mouse brain (see Sze, 1976). This effect contrasts sharply with the reversal by glucocorticoids of the increase in the V_{max} of high-affinity gamma aminobutyric acid (GABA) uptake into rat hippocampal synaptosomes observed after adrenalectomy (Miller et al., 1978). The latter effect required hormone pretreatment *in vivo* and was not observed when the synaptosomes were incubated *in vitro* with glucocorticoids, suggesting that in this case the steroid action was probably mediated by the "classical" mobile receptor pathway.

Synaptic plasma membranes prepared from osmotically shocked rat brain synaptosomes contain specific binding sites for glucocorticoids (Towle and Sze, 1978). Specific, saturable binding of [^{3}H]corticosterone by synaptic membranes was greatest in the hypothalamus, and lower, but approximately equal, levels were found in hippocampus and cerebral cortex. The affinity of corticosterone for the binding sites ($K_d \approx 10^{-7}$ M) was similar in three brain regions, and both corticosterone and synthetic glucocorticoids had similar affinities for the membrane sites. Soluble cytoplasmic receptors and synaptic membrane binding sites in brain are characterized by somewhat different properties of thermal stability and resistance to hydrolytic enzyme attack, making unlikely the possibility of artifactual contamination of the membranes by cytoplasmic receptors. Since the affinity of corticosterone for the membrane binding sites (Towle and Sze, 1978) agrees well with the concentration-response relation for the *in vitro* stimulation of synaptosomal tryptophan uptake by corticosterone (see Sze, 1976), it is possible that the membrane binding sites are involved in the regulation of tryptophan uptake in the brain.

EFFECTS OF CORTICOSTEROIDS ON THE NERVOUS SYSTEM

Brain Structure and Metabolism

Inhibitory effects of corticosteroids on the growth and development of the nervous system are well documented. A single injection of the glucocorticoid methylprednisolone in neonatal rats caused long-lasting impairments of brain weight, myelination, and dendritic branching (Gumbinas et al., 1973; Oda and Huttenlocher, 1974). Neonatal administration of cortisol similarly caused irreversibe reductions in cell proliferation in the cerebrum and cerebellum (Balázs and Cotterrell, 1972; Ardeleanu and Sterescu, 1978; Bohn and Lauder, 1978). That physiological levels of adrenal hormones may modulate brain growth was demonstrated by the finding that one month after young (50-day-old) rats were adrenalectomized, their brains were heavier than those of sham-operated rats, whether measured as wet or dry weight, absolute or relative to body weight (Devenport, 1979). Mineralocorticoids rather than glucocorticoids were implicated in this effect.

Similarities between the pathologies of stress and aging have often been noted, and elevated corticosterone levels have been found in aging rats (Riegle and Hess, 1972; Tang and Phillips, 1978). The hypertrophy of hippocampal astrocytes, and age-related pathological change, was positively correlated with adrenal activity in intact rats (Landfield et al., 1978), and was increased by exogenous corticosterone in adrenalectomized rats (Scheff et al., 1979). The amount of axonal sprouting in the dentate gyrus in response to a lesion in the entorhinal cortex was reduced both by normal aging and by corticosterone treatment in adrenalectomized rats (Scheff et al., 1979). Furthermore, two other morphological indices of aging, the number of hippocampal microglia and the depth of the hippocampus, were directly correlated with the dose of cortisol given to adrenalectomized rats (Landfield et al., 1979). In view of these correlations between corticosteroids and aging processes, Landfield et al. (1978) have suggested that hormones may gradually damage their neural target cells and contribute to some aspects of aging.

The effects of corticosteroids on brain electrolytes and excitability have been reviewed by Woodbury (1972). Adrenalectomy increased the ratio of intracellular to extracellular Na^+ concentration, decreased that for K^+, and increased brain excitability (lower threshold to electroshock seizures), while mineralocorticoids (DOC and aldosterone) reversed these effects. Acute administration of cortisol increased both intracellular Na^+ concentration and excitability (Woodbury, 1972), whereas long-term treatment increased the activity of Na^+, K^+-stimulated ATPase in the cerebral cortex, brainstem, and cerebellum, which would be expected to reduce excitability by hyperpolarizing cell membranes (Sadasivudu et al., 1977). Injection of cortisol increased the excitability (seizure susceptibility) of the hippocampus, but not of the septum, hypothalamus, or reticular formation in intact rats, while slowing the spontaneous electroencephalographic (EEG) activity in all four regions (Conforti and Feldman, 1975).

Cerebral blood flow and oxygen consumption in rat brain were decreased by adrenalectomy and then normalized by replacement therapy with cortisone but not DOC. In contrast, acute cortisol administration inhibited cerebral oxygen consumption in neonatal rats but was ineffective or stimulatory in older rats (Woodbury, 1972). Glucocorticoids inhibited the oxidation of glucose and increased the deposition of glycogen in the brain (Woodbury, 1972), but Wallis and Printz (1980) found no effect of adrenalectomy on local cerebral glucose utilization in 10 brain regions.

Adrenalectomy reduced the concentrations of free amino acids in brain, and cortisol (acute and chronic treatment) increased the concentrations of most amino acids, except for glutamine and GABA (Woodbury, 1972; Diez et al., 1977). Although acute

administration of corticosterone *in vivo* did not affect [^{3}H]ly-
sine uptake or incorporation into protein in whole brain (Rees
and Dunn, 1977), glucocorticoids have exhibited specific
effects on RNA and protein synthesis in hippocampal tissue *in
vitro*. Corticosterone increased the incorporation of [^{3}H]uri-
dine into hippocampal nuclear RNA (Dokas, 1979), and increased
the incorporation of [^{3}H]-leucine into specific cytoplasmic
proteins similar in molecular weight to tubulin (Etgen et al.,
1979). Glucocorticoids also induced GPDH, a glial-specific
enzyme that plays a role in cerebral phospholipid synthesis, in
adrenalectomized rats (Meyer et al., 1979), and in rat glial
cell cultures (see Breen et al., 1978).

Sensitivity to opiates, and to some other drugs, is modulated
by adrenal steroids. Adrenalectomized mice displayed enhanced
analgesia and straub tail responses to administration of morphine (4 mg/kg) and β-endorphin (12 mg/kg) (Holaday et al,
1979). Opiate toxicity and brain uptake of [^{3}H]-morphine were
also increased by adrenalectomy, and all apparent effects of
adrenalectomy were reversible by pretreatment with dexamethasone. Although these results were interpreted as indirect
effects of decreased drug metabolism and the consequent increased bioavailability of opiates, glucocorticoids may alter
opiate sensitivity by modulating brain opiate receptor concentrations. Adrenalectomy produced a significant and dexametha-
sone-reversible increase in the concentration of rat hypothalamic opiate binding sites (Roosevelt et al., 1979).

Adrenal effects on regional concentrations of a few neuropeptides have been studied. Although adrenalectomy did not
affect the level of endogenous "brain" ACTH in the rat hypothalamus, where concentrations of the peptide are highest, it
did produce a transient but significant decrease of the ACTH
content of the hippocampus (van Dijk et al., 1979). This
effect was not reversed by corticosterone replacement, and hypophysectomy alone did not decrease hippocampal ACTH levels.
The effect of adrenalectomy was apparent even after hypophysectomy, suggesting that hippocampal ACTH concentrations may be
modulated by some adrenal secretion other than corticosterone.
Furthermore, experiments in which pituitary ACTH was incubated
with a hippocampal synaptosomal membrane fraction suggested
that adrenalectomy may profoundly alter the metabolism of ACTH
in the brain. In contrast, Gibson et al. (1979) found that intact rats treated with dexamethasone for one week had elevated
levels of immunoreactive ACTH-like material in the hypothalamic
arcuate nucleus, whereas concentrations of ACTH in other brain
regions studied were not affected. Adrenalectomy or chronic
administration of low doses of dexamethasone had no effect on
the levels of met-enkephalin and leu-enkephalin in rat hypothalamus or striatum, but chronic treatment with high (1 mg/kg

twice per day for seven days) doses of dexamethasone profoundly increased the levels of both enkephalins in the striatum, but not hypothalamus (Rossier et al., 1979). The recent finding (Wallis and Printz, 1980) that adrenalectomy decreased and corticosterone replacement restored brain intracellular concentrations of the peptide angiotensinogen, the renin substrate, is of interest because of the possible role of an endogenous brain renin-angiotensin system in the control of blood pressure.

Synthetic glucocorticoids are widely used clinically to reduce vasogenic brain edema (Reulen, 1976). The nuclear concentration of [³H]-dexamethasone by cerebral capillary endothelial cells and by choroid plexus epithelial cells (Rees et al., 1975) suggests that dexamethasone could act on these cells to reduce the extravasation of protein into the brain through the damaged blood-brain barrier or to reduce the rate of formation of cerebrospinal fluid. The mechanism by which dexamethasone prevents the dramatic abnormal increase in cerebrospinal fluid Ca^{++} concentration that follows intravenous calcium infusion in adrenalectomized animals (Fukuda and Ui, 1967) may also involve these cell types. Direct effects on the transport of electrolytes across brain cell membranes might also contribute to the clinical efficacy of steroids. Aldosterone has also been clinically effective in the treatment of brain edema in man (Schmiedek et al., 1974). Since high concentrations of dexamethasone can occupy the high-affinity (type I) brain mineralocorticoid binding sites (Anderson et al., 1976) it is not yet known whether the clinically useful steroid effects are mediated by specific glucocorticoid or mineralocorticoid receptors (or, indeed, if they are receptor-mediated at all).

Neurotransmitter Metabolism

Corticosteroids directly modulate the synthesis and degradation of enzymes involved in catecholamine metabolism in the peripheral nervous system (Thoenen and Otten, 1978). In the superior cervical ganglion, glucocorticoids greatly amplify the acetylcholine-mediated transsynaptic induction of TH and dopamine β-hydroxylase (DBH). In the adrenal medulla, however, glucocorticoids inhibit the inducibility of TH, and also inhibit the rates of degradation of DBH and phenylethanolamine-N-methyl transferase (PNMT).

The effects of corticosteroids on catecholamines in the central nervous system (CNS), summarized in Table 14-1, are not so well understood. In some cases, TH activity was reduced by adrenalectomy and restored by exogenous glucocorticoids. Dexamethasone apparently elevates epinephrine levels by increasing the activity of PNMT.

Table 14-1. Effects of Manipulations of the Pituitary-Adrenal System on Neurotransmitter Metabolism in the Brain

Treatment	Observations[a]	Authors
Catecholamines		
Hypophysectomy	↓NE, DA turnover; no change in steady state levels	Versteeg et al., 1972
Adrenalectomy	↑NE turnover	Javoy et al., 1968
Adrenalectomy	↓TH activity in median eminence	Kizer et al., 1974
Adrenalectomy	↓DBH activity in hypothalamus, brainstem; ↑NE turnover; no change in NE or DA content	Shen and Ganong, 1976
Adrenalectomy	↓NE content in hypothalamus, striatum; ↓DA content in brainstem, striatum; ↑TH activity in striatum; ↑MAO activity in cerebral cortex	Rastogi and Singhal, 1978b
Corticosterone	↑NE turnover; no change in NE content	Iuvone et al., 1977
Corticosterone	↑TH activity in hypothalamus	Dunn et al., 1978
Corticosterone	↑NE content in supraoptic nucleus, median eminence; ↑DA content in median eminence	Fekete et al., 1978
Cortisol **in vitro**	↑NE uptake, ↓NE turnover in brain slices	Maas and Mednieks, 1971
Cortisol	↑TH activity in neonatal but not adult mice	Diez et al., 1977
Dexamethasone	Partially restored lowered TH activity in median eminence in adrenalectomized rats	Kizer et al., 1974

Treatment	Observations	Authors
Dexamethasone	No change in DBH activity	Kizer et al., 1974
Dexamethasone	↑E content in hypothalamus	Moore and Phillipson, 1975
Dexamethasone	↑DBH activity in hypothalamus	van Loon and Mascardo, 1975
Dexamethasone	↓MAO activity	Veals et al., 1977
Dexamethasone	↑PNMT activity in brainstem	Turner et al., 1979a
Serotonin		
Adrenalectomy	↓TPH activity in midbrain	Azmitia and McEwen, 1969
Adrenalectomy	↓5-HT content in hypothalamus	Vermes et al., 1973
Adrenalectomy	Blocked developmental rise in TPH activity; blocked reserpine-induced rise in TPH activity; no effect on TPH activity in adult mice	Sze et al., 1976
Adrenalectomy	↓Tryptophan content and TPH activity in brainstem; ↓5-HT content, ↑5-HIAA content in brainstem, striatum	Rastogi and Singhal, 1978a
Corticosterone	Normalized low TPH activity in midbrain in adrenalectomized rats	Azmitia and McEwen, 1969
Corticosterone	↑Rate of synthesis of 5-HT	Millard et al., 1972
Corticosterone	Normalized low 5-HT content in hypothalamus in adrenalectomized rats; ↑5-HT content in hypothalamus in intact rats	Vermes et al., 1973
Corticosterone	↑TPH activity in midbrain, forebrain	Azmitia and McEwen, 1974,1976
Corticosterone	Restored developmental and	

Treatment	Observations	Authors
	reserpine-induced increases in TPH activity in adrenalectomized animals; ↑TPH activity in neonatal rats but not in adult mice	Sze et al., 1976
Corticosterone	↑5-HT content and turnover in hypothalamus and midbrain 30 minutes but not 2 hours after injection of low dose; opposite effects after high dose	Kovacs et al., 1977a
Corticosterone	↑Tryptophan, 5-HT content, ↑TPH activity, ↓5-HIAA content in brainstem in adrenalectomized rats	Rastogi and Singhal, 1978a
Cortisol	↑5-HT, 5-HIAA content in young but not old rats	Green and Curzon, 1975
Cortisol	↑Tryptophan content, ↑5-HT synthesis in pargyline-treated rats	Neckers and Sze, 1975
Cortisol **in vitro**	↑Uptake of tryptophan by synaptosomes	Neckers and Sze, 1975
Cortisol	↑5-HT content in hypo-thalamus after intracranial implantation; ↑5-HT content in hypothalamus after in-jection in neonatal rats	Ulrich et al., 1975
Betamethasone	↑5-HT, 5-HIAA content in hippocampus, not hypothalamus	Balfour and Benwell, 1979

GABA

Andrenalectomy	↓Glutamate, GABA content, GAD, GABA transaminase activity in cortex	Pandolfo and Macaione, 1964

Treatment	Observations	Authors
Adrenalectomy	↑Maximal velocity for GABA uptake by hippocampal synaptomes	Miller et al., 1978
Costicosterone	No effect on GAD activity	Dunn et al., 1978
Corticosterone **in vivo**	↓Maximal velocity for GABA uptake by hippocampal synaptosomes in adrenalectomized rats	Miller et al., 1978
Corticosterone	No effect on GAD activity in hippocampus in adrenalectomized rats	Meyer et al., 1979
Acetylcholine		
Methylprednisolone, triamcinolone diacetate, cortisol acetate, and deoxy corticosterone acetate **in vivo**	↑Maximal velocity for choline uptake by cat caudate-putamen synaptosomes	Riker et al., 1979

[a]↑:increased; ↓:decreased; DA:dopamine; DBH:dopamine β-hydroxylase;
E:epinephrine; GABA:γ-aminobutyric acid; GAD: glutamic acid decarboxylase;
MAO:monoamine oxidase; NE:norepinephrine; PNMT:phenylethanolamine-N-methyl
transferase; TH:tyrosine hydroxylase; TPH:tryptophan hydroxylase;
5-HIAA:5-hydroxyindoleacetic acid; 5-HT:5-hydroxytryptamine (serotonin)

Much of the evidence cited in Table 14-1 indicates a stimulatory action of corticosteroids on serotonin (5-HT) metabolism. Adrenalectomy decreased the synthesis of 5-HT by reducing the activity of tryptophan hydroxylase (TPH), and glucocorticoids reversed this deficit. In intact animals, exogenous glucocorticoids acted rapidly (within one hour) to increase TPH activity and 5-HT turnover. The natural glucocorticoid-induced increase in uptake of the 5-HT precursor, tryptophan, by synaptosomes *in vitro* suggests that at least some of these effects may have a membrane site of action.

Information about glucocorticoid effects on amino acid neurotransmitters is fragmentary and sometimes contradictory (Table 14-1).

Electrical Activity

Glucocorticoid effects on the firing rates of neurons have been studied by a number of investigators. The use of different routes of administration (systemic vs. iontophoretic) and different hormones (cortisol and corticosterone versus dexamethasone) makes direct comparison between these studies difficult. Since systemically administered hormones may affect the cells indirectly, there is no a priori reason to expect similar effects of systemic or local application of these substances. Studies in which either cortisol or dexamethasone were administered peripherally may not be directly comparable, because these hormones enter the brain at different rates. Furthermore, systemic injection of glucocorticoids into animals with intact pituitaries confounds the steroid effect with possible effects mediated by the feedback suppression of ACTH release.

Systemic or local injection of glucocorticoids has been observed to affect the spontaneous activity of neurons in many brain loci. Systemic injection of cortisol in intact, freely moving rats rapidly increased spontaneous activity of units in the anterior hypothalamus and mesencephalic reticular formation, and decreased unit activity in the ventromedial and basal hypothalamus (Phillips and Dafny, 1971a,b). Spontaneous activity of basal hypothalamic neurons in completely deafferented hypothalamic islands was also rapidly decreased following systemic injection of either cortisol (Feldman and Sarne, 1970) or dexamethasone in intact rats (Ondo and Kitay, 1972). Similarly, iontophoretic application of dexamethasone onto medial basal hypothalamic neurons in intact rats produced an immediate depression of cell firing (Steiner et al., 1969). Mesencephalic neurons also responded to direct application of dexamethasone with a decrease in firing rate (Steiner et al., 1969), in contrast to the increase in firing rate after systemic injection of cortisol reported by Phillips and Dafny (1971a,b).

The responses of hippocampal neurons to exogenous cortisol or corticosterone have been studied in adrenally intact rats. Following systemic injection of cortisol, 56 percent of the hippocampal neurons observed responded with either an increase (50 percent) or decrease (40 percent) in spontaneous activity (Phillips and Dafny, 1971a). However, none of the 500 hippocampal neurons tested by Barak et al. (1977) were responsive to iontophoretic application of either cortisol or corticosterone. In contrast, injection of dexamethasone into the vicinity of the recording electrode rapidly produced a dramatic decrease in hippocampal multiple unit activity (Michal, 1974). In hypophysectomized rats, Pfaff et al. (1971) found many neurons in the dorsal hippocampus that responded to systemic injection of corticosterone, predominantly by a delayed decrease in spontaneous activity (latency 10-40 minutes). Although it is not known which of the reported electrophysiological effects of corticosteroids are mediated by specific receptors, the discrepancy between some findings in intact and adrenalectomized or hypophysectomized rats is not surprising, since a majority of the corticosterone binding sites in the hippocampus are normally occupied in intact rats (McEwen et al., 1974). A direct comparison of the latencies and effects of dexamethasone and corticosterone on hippocampal neurons in adrenalectomized and hypophysectomized rats would aid in resolving some of these apparent discrepancies.

Effects of corticosteroids on evoked neural activity have also been reported. In intact control rats, stimulation of the sciatic nerve increased firing rates of single neurons in the anterior hypothalamus, whereas in cortisol-treated rats such stimulation decreased the firing rates of these neurons (Feldman and Dafny, 1970). A decrease in the amplitude of evoked sciatic potentials was observed in brainstem sites soon after the injection of DOC (Kraulis et al., 1975). In adrenalectomized rats, injection of corticosterone slowly increased (for at least two hours following the injection) the amplitudes of hippocampal potentials evoked by visual or somatosensory stimulation (McGowan-Sass and Timiras, 1975). These data suggest the possible alteration of neural responses to sensory input by corticosteroids (see the following section).

EFFECTS OF CORTICOSTEROIDS ON BEHAVIOR

Sensory Perception

Both hypo- and hypersecretion of adrenocortical hormones disrupt sensory processes (reviewed by Henkin, 1975). Human beings with adrenocortial insufficiency exhibited dramatic increases in their ability to detect gustatory, olfactory, and auditory stimuli. These increases in detection sensitivity were at least 100-fold for each of the four taste qualities, 1000-fold for various odors, and 13 db for auditory stimuli. Such patients also showed an increased sensitivity to pain. Treatment with glucocorticoids, such as cortisol, cortisone, and dexamethasone, but not mineralocorticoids, returned detection sensitivity to within normal limits in 24 to 48 hours. In contrast, patients with Cushing's syndrome (excessive glucocorticoid secretion) exhibited decreased detection sensitivity for taste, odor, and sound. Variations in glucocorticoid secretion within the normal range may also influence sensory processing, as evidenced by the fact that the circadian pattern of detection sensitivity in normal subjects closely follows the pattern of cortisol secretion.

Whereas detection sensitivity appears to be affected in opposite ways by increases or decreases in glucocorticoid levels, conditions of glucocorticoid insufficiency and excess both result in decreased recognition sensitivity. For example, both types of patient suffer exceptional difficulty in identifying specific taste qualities, such as salty, bitter, at low concentrations, and in understanding speech in the absence of extraneous cues. Thus, there is ample evidence that the ability of human beings to integrate sensory information is impaired by deviations of glucocorticoid levels above or below optimal limits. Whether natural fluctuations in glucocorticoid levels produce physiologically significant changes in sensory capacities, and whether such changes are due to central or peripheral actions, remain open questions.

Feeding and Drinking

The data regarding adrenal effects on food and water intake are sketchy and confusing. Whereas treatment of intact rats with dexamethasone reduced food intake (Beatty et al., 1970) and body weight (Kendall, 1970; Beatty et al., 1971; Katz and Carroll, 1978), treatment of intact rabbits with physiological doses of cortisol and corticosterone increased food intake, but not body weight (Blaine et al., 1975). An extra-adrenal inhibitory effect of ACTH was suggested, since ACTH (4 U/day) increased food intake in intact rabbits but reduced it in adrenalectomized rabbits. However, administration of ACTH (5 U/day) caused weight loss in intact but not adrenalectomized rats (Weisinger et al., 1978). The authors speculated that ACTH-induced elevations of blood glucose and free fatty acids might affect glucostats and lipostats believed to regulate food intake.

Recent evidence (Yukimura et al., 1978) suggests that overeating in genetically obese (Zucker) rats may be attributable to a hypersensitivity to glucocorticoids. Zucker rats dramatically decreased their food intake and weight gain following adrenalectomy. Replacement therapy with corticosterone (2 or 10 mg/day) caused a much greater increase in food intake and weight gain in these animals than in normal controls. It is not known whether corticosterone acts directly on the brain to increase feeding behavior or indirectly, for example, by elevating insulin secretion.

It is well known that adrenalectomized animals voluntarily drink more sodium chloride solution than do intact animals. A U-shaped dose-response relationship was shown between the dose of aldosterone or DOC and the sodium chloride intake of adrenalectomized rats, with physiological doses causing the greatest reduction in salt appetite (Fregly and Waters, 1966). The mechanism for this mineralocorticoid effect is unknown, but the authors suggested a peripheral action on the Na^+ and K^+ composition of saliva, therby altering the bias of the taste receptors in the mouth. Evidence for a central mineralocorticoid action was provided by a study of [^{3}H]-deoxycorticosterone binding to brain cytosol in the Long-Evans strain of rat, which is resistant to induction of increased salt appetite by deoxycorticosterone (Lassman and Mulrow, 1974). Compared to behaviorally responsive Sprague-Dawley rats, the deoxycorticosterone-resistant rats had a deficiency of [^{3}H]-deoxycorticosterone binding proteins in the hypothalamus, while binding in other areas was similar in both strains. In view of the confusion surrounding the properties of the DOC-receptor complex itself (the failure to observe significant nuclear binding of [^{3}H]DOC *in vivo* and the possible lack of stringent glucocorticoid or mineralocorticoid binding specificity), the molecular mechanism of the possible behavioral effect of DOC on salt appetite remains obscure.

Physiological doses of glucocorticoids also stimulated sodium chloride appetite in intact and adrenalectomized rabbits. In addition, ACTH increased salt appetite by an unexpected extra-adrenal pathway (Blaine et al., 1975). In the rat, ACTH also stimulated salt appetite, but its effect was entirely dependent on adrenal hormones (Weisinger et al., 1978).

Motor Activity

Several studies have investigated the relationship between the circadian rhythms of locomotor activity and corticosteroid secretion. Corticosteroids have been shown to influence both running wheel activity and general activity.

Running wheel activity of rats was markedly depressed by adrenalectomy (Leshner, 1971) or hypophysectomy (Richter and Wislocki, 1930) and was restored to normal by injection of corticosterone (5 mg/kg, Leshner, 1971). Surgical isolation of the basal hypothalamus abolished the rhythms of both corticosteroid secretion and running activity (Greer et al., 1972). However, these authors suggested that the dramatic decrease in running was not due to adrenal insufficiency per se, since plasma corticosterone levels in the operated rats were intermediate between the preoperative A.M. low and P.M. high, yet running was suppressed even below the normally low daytime level.

In adrenally intact male rats, exogenous corticosterone (5 mg/kg) decreased running wheel activity (Leshner, 1971). In contrast, addition of dexamethasone-21-phosphate to the drinking water (0.1-20 μg/ml) caused a dose-dependent increase in running behavior of intact male rats (Kendall, 1970), and injections of dexamethasone (1 mg/kg, SC) had a similar effect in females (Beatty et al., 1971). In Kendall's study, the increase in activity, which occurred entirely in the nocturnal phase, began one to three days after starting dexamethasone treatment, and reached a maximum of up to 10 times pretreatment levels after five to seven days. The stimulatory effect of dexamethasone was not due to suppression of ACTH secretion, since implantation of cortisol acetate in the median eminence, which suppressed ACTH secretion, reduced activity. Kendall suggested that the enhanced activity may be related to starvation, since dexamethasone reduced eating and body weight (see previous section), and food restriction also increased running activity (Kendall, 1970; Beatty et al., 1971). A reciprocal relationship is suggested by the observation that rats starved for five days displayed high corticosterone levels without circadian rhythmicity (Inoue et al., 1976). Although spontaneous running activity is reduced by adrenalectomy and enhanced by supraphysiological doses of dexamethasone, the causal relationships between corticosteroids, food intake, and activity are poorly understood.

Glucocorticoid administration had an opposite (inhibitory) effect on general activity, operationally defined as any gross body movements that produce vibrations of the home cage mounted on an activity monitoring platform. Dexamethasone in the drinking water (5 μg/ml) markedly reduced daily activity of intact male rats (Katz and Carroll, 1978). Most of the activity still occurred nocturnally. The steroid-induced decrease in body weight was not responsible for this change, since food-restricted controls were more active than normal. The opposing effects of dexamethasone on general and running wheel activity support a distinction between the mechanisms controlling these two types of spontaneous activity.

Corticosteroids also affect activity in an open-field situation (a novel empty enclosed area). Adrenalectomized rats were more "emotional" than controls in the open-field test, that is, they moved about less (Joffe et al., 1972; McIntyre, 1976) and defecated more (Joffe et al., 1972). Injections of cortisone acetate (25 mg/kg daily) in intact mice stimulated open-field activity within two days of starting treatment (Fuller et al., 1956). Similarly, both cortisol and cortisol-21-sulfate (3 mg every other day) enhanced the exploratory activity of weanling rats (Miyabo et al., 1972). These authors suggested that the effect was centrally mediated, since the sulfate conjugate of cortisol was shown to have none of the systemic catabolic, gluconeogenic, lympholytic, or ACTH-suppressive effects of glucocorticoids. Since cortisol-21-sulfate did not compete for *in vitro* binding of [^{3}H] corticosterone by adult rat brain cytosol, Ueda (1978) proposed that its behavioral action may be independent of glucocorticoid receptor binding. This hypothesis is intriguing, but alternative explanations exist, for example, that some of the conjugate could be converted to cortisol by a sulfohydrolase in brain or elsewhere. Despite the consistent finding that large doses of glucocorticoids increase open-field activity, Stern et al. (1973) found no reliable correlation between physiological levels of corticosteroids and open-field behavior.

Sexual Behavior

There is little evidence for a consistent effect of the adrenals on male sexual behavior. The adrenals played no role in the sexual behavior of male rats following castration (Bloch and Davidson, 1968). Although adrenalectomy eliminated the estrogen stimulation of ejaculatory behavior in castrated male rats (Gorzalka et al., 1975), it had no significant effect on the copulatory behavior of estrogen-treated castrated male mice (Wallis and Luttge, 1975). Adrenalectomy reduced the sexual behavior of individually housed, but not group housed, gonadally intact male mice, but corticosterone did not reverse this effect (Gorzalka and de Catanzaro, 1979).

There is evidence for a progestin-like effect of adrenal hormones in stimulating female sexual receptivity. In ovariectomized, estrogen-primed guinea pigs, corticosterone was less effective than progesterone, but as effective as pregnanedione in facilitating lordosis (Wade and Feder, 1972). In ovariectomized, estrogen-primed rats, DOC was as effective as progesterone in facilitating lordosis, whereas aldosterone and corticosterone were ineffective (Gorzalka and Whalen, 1977). More work is needed to evaluate the steroid specificity and species differences of this behavioral effect.

The adrenals may also influence maternal behavior. Adrenalectomized mother rats took longer to initiate retrieval of their pups and retrieved a smaller percentage than did sham-operated controls (Hennessy et al., 1977).

Aggression

Glucocorticoids facilitate some types of intermale aggression in rodents, although less consistently than do androgens. Adrenalectomy reduces aggressiveness and corticosterone or dexamethasone restores it in adrenalectomized animals. There is also evidence for a direct, extra-adrenal suppressive effect of ACTH on aggressive behavior (see review by Brain, 1972; Leshner, 1975). Single injections of cortisol or a low dose of ACTH in intact rats increased shock-induced attack behavior, whereas adrenalectomy, hypophysectomy, or a high dose of ACTH suppressed this behavior (Rogers and Semple, 1978). However, treatment with corticosterone for several days had no effect on aggression in intact male mice, but reduced the aggression-promoting effect of testosterone in castrated males, suggesting a modulatory action of corticosterone on the responsiveness to androgen (Simon and Gandelman, 1978).

Submissiveness to attack, which can vary independently of aggressiveness, may also be facilitated by corticosteroids. Long-term treatment with ACTH increased submissiveness in intact mice, but not in adrenalectomized mice (Leshner and Politch, 1979).

Aversive Conditioning

Because corticosteroids are released in response to aversive stimuli, their influence on conditioned avoidance behavior has been studied extensively.

Acquisition of Active Avoidance

Adrenalectomy increased and ACTH or corticosterone acetate (10 to 20 mg/kg) decreased the number of intertrial responses during acquisition of one-way active avoidance, but the number of avoidance responses was unaffected (Bohus and Lisśak, 1968). Adrenalectomy, when combined with a 6-hydroxydopamine lesion of the dorsal noradrenergic bundle, permanently impaired acquisition of both one-way (Ogren and Fuxe, 1977) and two-way active avoidance (Ogren and Fuxe, 1974), although neither treatment alone had any effect.

Extinction of Active Avoidance

During extinction of avoidance, the aversive stimulus is no longer presented, making responding unnecessary. Adrenalec-

tomized rats continued responding longer than intact rats
during active avoidance extinction (Bohus et al., 1968; Weiss
et al., 1970). Conversely, rats treated with corticosteroids
emitted fewer responses during extinction of active avoidance
behavior, that is, they stopped responding earlier than untreated animals. This effect of corticosteroids was opposite
to that of ACTH, which delayed avoidance extinction (de Wied,
1974; see Dunn, chapter 7). However, the effect of corticosteroids was evidently not solely due to their suppression of
ACTH secretion, since corticosteroids facilitated active avoidance extinction in hypophysectomized as well as intact rats (de
Wied, 1974). A facilitation of active avoidance extinction has
been observed after systemic administration of a variety of
steroids, including corticosterone, cortisone acetate, dexamethasone, aldosterone, progesterone, pregnenalone and 17β-
ethyl-androst-4-ene-3-one, whereas cholesterol, testosterone,
and estradiol were ineffective (de Wied, 1967; Bohus and
Lisśak, 1968; van Wimersma Greidanus, 1970).

A biphasic dose-response effect of corticosterone on extinction has been correlated with its effect on 5-HT content in the
hypothalamus and the mesencephalon. Low doses of corticosterone (1 and 5 mg/kg) facilitated active avoidance extinction
and elevated brain 5-HT levels, whereas a higher dose (10
mg/kg) delayed extinction and reduced 5-HT levels (Kovács et
al., 1976). Intracerebral implantation of corticosteroids also
alters the rate of extinction of one-way active avoidance.
Cortisol acetate facilitated extinction when implanted in the
mesencephalic reticular formation, median eminence parafascicular region of the medial thalamus, rostral septum, anterior
hypothalamus, and amygdala, but delayed extinction when implanted in the lateral and anterior thalamus, and posterior and
lateral hypothalamus (Bohus, 1970). Extinction was also facilitated by implants of corticosterone in the parafascicular
thalamus, or of dexamethasone phosphate in several parts of the
medial thalamus, and the lateral ventricle, but not in the
hippocampus, septum, or reticular formation (van Wimersma
Greidanus and de Wied, 1969).

Maintenance of Free-Operant Avoidance

Adrenal hormones affect performance on free-operant (unsignal-
led or Sidman) active avoidance tasks, in which each response,
such as pressing a lever, postpones the regularly scheduled
delivery of shock for some specified interval, so that by
responding at a sufficient rate, the subject may avoid shocks
indefinitely. Both ACTH and dexamethasone (200 μg per rat)
increased the efficiency of free-operant avoidance responding,

as evidenced by longer inter-response times and fewer shocks (Wertheim et al., 1967).

Rats made more avoidance responses in a free-operant avoidance task during sessions conducted 30 minutes after a brief period of "prestimulation" (footshock, air blast, or only handling) than during control sessions not preceded by prestimulation (Gray, 1976). The augmentation of avoidance responding was manifested only after the first 15 minutes of the session. The prestimulation effect appeared to depend on mineralocorticoids, but not on other adrenal or pituitary hormones, since it was abolished by adrenalectomy but not by adrenal demedullation or hypophysectomy. Furthermore, the effect was also absent in intact rats depleted of endogenous mineralocorticoid by substitution of hypertonic saline for their drinking water. In either adrenalectomized or intact, sodium-loaded rats, daily injections of aldosterone acetate or DOC acetate, but not of corticosterone acetate, restored the effect of prestimulation while leaving the nonprestimulated response rate unaffected. The role of mineralocorticoids was "permissive," since it was the presence of a basal level of the hormone, not its release in response to the prestimulation, that seemed important. Gray (1976) attributed the effect of prestimulation to a temporary arousal of the subject, which predisposes it to respond actively in the avoidance apparatus rather than to freeze.

Acquisition of Passive Avoidance

Several studies have indicated that pretraining administration of glucocorticoids to intact animals shortened passive avoidance response latencies immediately after training, indicating an impairment of acquisition. Thus, rats given cortisone (20 mg/kg, but not 10 mg/kg) three hours before training had shorter latencies to re-enter a chamber where they had just been shocked (Bohus et al., 1970). Dexamethasone (20 μg/rat) given three hours before training impaired the acquisition of spatially discriminated passive avoidance behavior (Bohus et al., 1970). Water-deprived rats treated with cortisol (1 mg), corticosterone (5 mg), or 6-dehydro-16-methylene-cortisol (0.5 mg), but not DOC (5 mg), resumed drinking in a chamber where they had been shocked more quickly than did controls (Bohus, 1973). In a similar task, exogenous ACTH had the opposite effect, that is, it facilitated passive avoidance behavior, whereas corticosterone (7.5 mg/kg) or dexamethasone (150 μg/kg) had no significant effect (Pappas and Gray, 1971). Intracranial implantation studies showed that glucocorticoids may act directly on certain brain structures to impair the acquisition of passive avoidance behavior. Impaired acquistion in rats was observed one day after implantation of cortisone acetate in the anterior hypothalamus, medial thalamus, rostral septum, dorsal hippocampus, amygdala, and cerebral ventricles (Bohus, 1971). Cortisol impaired acquisition when implanted in the anterior hypothalamus, medial thalamus, septum, and dorsal hippocampus, but not in the frontal cortex; corticosterone had a similar effect when implanted in the anterior hypothalamus or septum, but not in the medial thalamus, dorsal hippocampus, or frontal cortex (Bohus, 1973).

In contrast to these observations of impairment of passive avoidance acquisition by corticosteroids, Kovács et al (1977b) found that the effect of corticosterone was either facilitatory or inhibitory, depending on the dose. Rats injected with low doses of corticosterone (1 to 5 mg/kg) 30 minutes before training made fewer attempts to drink water from an electrified tube than did vehicle-injected controls, whereas those given a high dose of corticosterone (10 mg/kg) made more drinking attempts than controls. The biphasic effect of corticosterone on behavior was paralleled by that on 5-HT content in the hypothalamus and mesencephalon; 5-HT levels were increased by the 1 mg/kg dose, unaffected by the 5 mg/kg dose, and decreased by the 10 mg/kg dose. The mineralocorticoid DOC (25 mg/kg) had no significant action on either passive avoidance or brain 5-HT. These findings suggest that the effect of glucocorticoids on acquistion of passive avoidance may be facilitatory in physiological doses and suppressive in pharmacological doses, and they emphasize the need for the determination of dose-response relationships in behavioral studies.

Retention of Passive Avoidance

Adrenalectomized rats performed better than intact rats in passive avoidance retention tests, demonstrating longer response latencies as well as increased defecation (Weiss et al., 1970). Hypophysectomy caused the opposite effects, that is, shorter latencies and decreased defecation, suggesting that these changes may reflect a facilitation of passive avoidance by ACTH, rather than an inhibition by corticosteroids.

There is evidence for interactions of adrenal hormones and noradrenergic systems in passive avoidance behavior. 6-Hydroxydopamine lesions impaired acquistion and retention of passive avoidance in rats only when combined with adrenalectomy; neither treatment alone significantly altered retention (Roberts and Fibiger, 1977; Wendlandt and File, 1979).

There are several reports that the administration of corticosteroids to intact animals before passive avoidance training caused poorer retention one day later. Retention was impaired by the injection of cortisone acetate (10 or 20 mg/kg) three hours before training (Bohus et al., 1970), or by the injection of cortisol (1 mg/rat), corticosterone (1 to 5 mg), or 6-de-hydro-16-methylene-cortisol (0.5 mg), but not DOC (5 mg) two hours before training (Bohus, 1973). A similar retention defi-cit was seen in rats that on the day before training received intracranial implants of cortisol or corticosterone in the anterior hypothalamus, medial thalamus, septum, or dorsal hippocampus, or of 6-dehydro-16-methylene-cortisol in the septum (Bohus, 1973).

The nature of the effect of corticosterone on passive avoid-ance retention, like that on acquistion, appears to depend on the dosage. Rats that received a low dose of corticosterone (1 or 5 mg/kg) 30 minutes before training and testing exhibited better retention than controls, whereas a higher dose (10 mg/kg) produced a deficit in retention (Kovács et al., 1977b). As noted in the preceding section, the biphasic behavioral effect was correlated with a biphasic effect on brain 5-HT content.

Extinction of Passive Avoidance

Since animals treated with high doses of glucocorticoids ac-quire and retain passive avoidance behavior less readily than normal, it is not surprising that they stop avoiding earlier in extinction. Injection of dexamethasone (20 μg/rat) three hours before the training session and each daily test session facilitated the extinction of passive avoidance behavior (Bohus et al., 1970).

Corticosterone appears to be necessary for the behavioral effect of forced extinction (Bohus and de Kloet, 1977). Rats were given one-trial passive avoidance training, in which foot-shock was delivered when the rats stepped from a well-lighted compartment into a dark compartment. A day later they display-ed long latencies to re-enter the dark (shock) compartment. However, this passive avoidance behavior was absent in the rats subjected to a forced extinction procedure three hours after the training trial, that is, confinement in the dark compart-ment for five minutes. Adrenalectomy, but not sham-adrenalec-tomy, one hour before the forced extinction procedure prevented the extinction of the avoidance behavior. A physiological dose of corticosterone (0.3 mg/kg) at the time of adrenalectomy normalized the extinction behavior of adrenalectomized rats. The same dose of either dexamethasone or progesterone did not normalize the behavior of adrenalectomized rats, and both steroids attenuated the effect of corticosterone when injected one hour before corticosterone administration. These authors suggest that the interference by dexamethasone and progesterone with the behavioral action of corticosterone is related to their competition for the binding of corticosterone to hippocampal cytosol receptors, but the similarities in nuclear binding of [^{3}H]corticosterone and [^{3}H]dexamethasone by incubated hippocampal slices (see earlier discussion) are difficult to reconcile with this proposed "antiglucocorticoid" action of dexamethasone.

Fear Conditioning

This experimental paradigm is also known as the Estes-Skinner, or conditioned emotional response (CER), procedure. During the Pavlovian conditioning phase of the paradigm, presentations of a previously neutral conditioned stimulus (CS) are followed by unavoidable shock. Later, the experimenter tests the effect of presenting the CS alone on the performance of a well-established operant response. Typically the CS suppresses appetitive responding and enhances avoidance responding.

Although pituitary-adrenal activation is a normal correlate of fear conditioning, the blockade of this system by hypothalamic implantation of cortisol had no effect on the acquisition of a CER (Davis et al., 1978). The retardation of CER acquisition caused by adrenalectomy (Randich et al., 1979) was apparently not due to the lack of adrenal hormones, since hypophysectomized rats acquired the response normally.

Fear conditioning produced a greater suppression of appetitive responding in rats that received the same treatment before both the shocking and the testing sessions, whether the treatment was saline or dexamethasone (50 μg/rat), than in rats that received different treatments on the two occasions (Pappas and Gray, 1971). The simplest interpretation of this result was in terms of state dependence, that is, that injection of dexamethasone and saline produce discriminably different internal states and that retrieval is best when the state at testing most closely resembles that at original learning. However, no evidence for state dependence was found in a similar study with pigs (Mormède and Dantzer, 1977). Treatment with dexamethasone (0.2 mg/kg) before Pavlovian fear conditioning alone decreased the suppressive effect of a CS imposed on a baseline of food-reinforced behavior, whereas treatment before conditioning or testing increased the facilitatory effect of a CS imposed on a baseline of unsignaled active avoidance (Mormède and Dantzer, 1977; 1978). These authors also rejected an interpretation of the results in terms of fear, since dexamethasone decreased an index of fear in the appetitive task and increased it in the avoidance task.

Kamin Effect

In 1957 Kamin reported the unexpected observation that retention of an incompletely acquired active avoidance response was
a U-shaped function of the time between original training and
testing, that is, rats displayed much poorer retention at
intermediate intervals of one to four hours than immediately or
24 hours after training. Later studies replicated and extended
this observation, demonstrating a Kamin-like effect in passive
avoidance (Pinel and Cooper, 1966) and appetitive training as
well (Seybert et al., 1979). Furthermore, the use of a wider
range of testing intervals revealed an oscillating function,
whereby retention was good immediately after training and after
multiples of 12 hours, but poor between these peaks (Wansley
and Holloway, 1975; Elson et al., 1977). Although the mechanism underlying the Kamin effect is not understood, the predominant hypothesis accounts for the effect in terms of state-
dependent learning, that is, that retention is optimal when
both the external and internal stimuli during testing most
closely approximate those during the initial learning. Oscillations of the pituitary-adrenal system have frequently been
suggested as a candidate for a biological rhythm mediating the
Kamin effect. Indeed, a decline in plasma corticosterone after
avoidance training was correlated with the declining arm of the
U-shaped behavioral function (Brush and Levine, 1966). Furthermore, stimulation of adrenal activity before testing by stress
(Klein, 1972) or injection of ACTH or cortisol (Levine and
Brush, 1967) prevented the retention deficit normally seen at
intermediate intervals. However, the hypothesis that pituitary-adrenal activity is the primary mediator of the Kamin
effect appears to be untenable, because a Kamin effect has been
demonstrated in hypophysectomized (Marquis and Suboski, 1969),
adrenalectomized (Suboski et al., 1970), and dexamethasone-
injected animals (Snider et al., 1971). The biological rhythm
responsible for Kamin-like phenomena remains to be identified.

Appetitive Conditioning

Adrenal influences have been studied much less extensively in
appetitive than in aversive conditioning. Injections of corticosterone (10 mg/kg) in intact rats delayed the acquisition of
a spatial discrimination for water reward, but improved performance on a discrimination reversal (Bohus, 1973).
   Appetitive extinction (withholding a positive reinforcer) is
known to cause elevation of plasma corticosteroid levels (Davis
et al., 1976), but the effect of the hormone on performance is
not clear. In one study (Garrud et al., 1974), administration

of corticosterone (1 mg daily) to intact rats facilitated the
extinction of a food-reinforced response. In contrast, however, adrenalectomy also facilitated the extinction of a food-
reinforced response, and corticosterone (2 mg daily) normalized
the behavior of adrenalectomized rats but was without effect in
intact rats (Micco et al., 1979).

Consolidation of Memory

A variety of treatments, including electroconvulsive shock
(ECS) and protein synthesis inhibitors, administered shortly
after training, have been found to produce retrograde amnesia
for training experiences, especially with passive avoidance
tasks (for review see Dunn, 1980). In a retention test of passive avoidance, usually given 24 hours or more after training,
the effectiveness of an amnestic agent is evidenced by decreased response latency. This paradigm thus insures that a
possible debilitating effect of a treatment on locomotor activity will not be interpreted as a deficiency of memory.
   Studies in rats and mice have found that prior adrenalectomy
prevented or attenuated the amnestic effect of several agents,
including intracranial puromycin (Flexner and Flexner, 1970),
subcutaneous cycloheximide (Nakajima, 1975), kindled convul-
sions from amygdaloid stimulation (McIntyre, 1976; McIntyre and
Wann, 1978), and ECS (Bookin and Pfeifer, 1978). Various explanations for this effect have been proposed. The finding
that cycloheximide was not amnestic in adrenalectomized mice,
despite its inhibition of brain protein synthesis by more than
90 percent, led Nakajima (1975) to propose that the amnestic
effect of peripherally administered cycloheximide is mediated
by its action on the adrenal glands to suppress corticosteroidogenesis rather than by its inhibition of cerebral protein
synthesis. This hypothesis seems unlikely, since emetine, a
peripheral protein synthesis inhibitor, also blocked ACTH-
induced corticosteroid secretion but was not amnestic (Dunn et
al., 1977), and since aminoglutethimide, which suppressed
plasma corticosterone to the same extent as a cycloheximide
without inhibiting protein synthesis, was not amnestic (Squire
et al., 1976; Dunn and Liebman, 1977). An alternative explanation is that the longer latency exhibited by adrenalectomized
animals in passive avoidance retention tests is not an "anti-
amnestic" effect but is simply due to a reduction in locomotor
activity (Squire et al., 1976). This hypothesis is partially
supported by reports that adrenalectomized animals were less
active than intact ones in an open field (see previous discus-

sion). The hypothesis should be tested in an active avoidance paradigm, where reduced activity and an antiamnestic effect would be expected to produce opposite results. The elevation of ACTH levels in response to long-term adrenalectomy may play a mediating role, since ACTH analogues administered before retention testing attenuated CO_2-induced amnesia for passive avoidance (Rigter et al., 1976).

The administration of exogenous corticosteroids in intact animals can also antagonize experimentally induced amnesia. Cortisol (30 mg/kg, SC), given immediately after training, improved the retention performance of cycloheximide-treated mice (Nakajima, 1975) and of ECS-treated mice (Nakajima, 1978). Post-training injection of dexamethasone (4 mg/kg, SC) similarly blocked the amnestic effect of anisomycin for both active and passive avoidance tasks (Flood et al., 1978). These effects may be pharmacological, however, since Squire et al. (1976) determined that a 30 mg/kg dose of corticosterone elevated the plasma corticosterone levels of mice to about seven times normal (140 μg/100 ml). A lower dose (1.2 mg/kg), which produced more physiological levels (about 35 μg/100 ml), did not block the amnestic effect of cycloheximide for passive avoidance training, and actually enhanced the amnestic effect of cycloheximide for visual discrimination escape training (Squire et al., 1976). Thus, glucocorticoids may promote amnesia at low doses and attenuate amnesia at high doses.

To localize the brain site for the anti-amnestic action of glucocorticoids, Cottrell and Nakajima (1977) gave cycloheximide-treated mice intracerebral injections of cortisol succinate immediately after passive avoidance training and tested their retention six days later. Avoidance responding was restored by bilateral injections in the hippocampus, but not in the septum or hypothalamus. The effective hippocampal dose (13 μg per side) was approximately one thousandth the effective subcutaneous dose.

Gibbs and Ng (1977) have advanced a three-phase model of memory, based upon the use of various drugs to inhibit memory for a one-trial avoidance task in day-old chicks. Injection of corticosterone at the time of training was found (1) to delay the oubain-induced decay of the second phase ("labile" memory), presumably by antagonizing the inhibition of sodium pump activity by oubain, and (2) to overcome the aminoisobutyric acid induced inhibition of amino acid uptake required for the formation of long-term memory, possibly by stimulating Na^+,K^+-ATPase activity (Gibbs, personal communication). There is some evidence that these corticosterone effects are mediated by norepinephrine release.

Sleep

A reduction in the proportion of rapid eye movement (REM) sleep
in human beings occurred following administration of the glucocorticoid prednisone (Gillin et al., 1972) or ACTH (Gillin et
al., 1974). The suppression of REM sleep induced by ACTH appeared to be adrenally mediated, since it was greater in normal
volunteers than in patients with Addison's disease. This
effect did not appear until at least four hours after starting
the ACTH infusion, and it persisted for at least 12 hours after
the infusion. Patients with Cushing's syndrome exhibited a
similar alteration of sleep patterns (Krieger and Glick, 1972).

Mood and Affect

Many authors, including Addison himself, have noted psychiatric
symptoms in the vast majority of patients with Addison's
disease (adrenal insufficiency). These symptoms, which include
apathy, depression, negativism, lack of concentration, lack of
initiative, seclusivenes, irritability, lability of mood, reduction of appetite, and impairment of memory, are usually improved by glucocorticoid treatment (Michael and Gibbons, 1963;
Carpenter et al., 1972; von Zerssen, 1976). A wide variety of
psychiatric symptoms have also been observed in Cushing's
syndrome (adrenal hypersecretion). Depression is present in
well over half of the cases and has been reported in patients
with adrenal tumors as well as those with ACTH-dependent
Cushing's syndrome (Haskett et al., 1979). Other reported
symptoms include fatigue, drowsiness, insomnia, excitement,
apathy, anxiety, irritability, gross overreaction to emotional
stimuli, increased appetite, delusions, hallucinations, and
acute psychotic episodes (Michael and Gibbons, 1963; Carpenter
et al., 1972; von Zerssen, 1976).
  The possibility of a link between corticosteroids and depression is suggested by observations of adrenal hypersecretion in
many depressed patients. Frequent blood sampling (every 20
minutes) revealed that the cortisol secretory episodes of
depressed patients were more numerous and of greater magnitude
than normal, occurring during both sleeping and waking hours.
The cortisol levels were normalized after recovery (Sachar et
al., 1973). Furthermore, about half of a sample of patients
with a diagnosis of endogenous depression showed abnormally
early recovery from dexamethasone-induced suppression of
cortisol secretion (Carroll et al., 1976). The proportion of
depressed patients who were resistant to the suppression of
cortisol secretion by dexamethasone was higher among patients
having a first-degree relative also diagnosed as depressive (82

percent) than among those with no familial history of depression (24 percent), suggesting a hereditary factor in this type of unipolar depressive illness (Schlesser et al., 1979). These studies raise some intriguing questions: Do the reported abnormalities in pituitary-adrenal function reside in the brain, pituitary, or peripheral metabolism, and do they contribute directly to certain psychiatric symptoms, or are they independent manifestations of an underlying neural pathology, perhaps in the limbic system.

SUMMARY AND CONCLUSIONS

A rapid increase in the secretion of glucocorticoids is a normal concomitant of behavioral arousal in response to various stressors. In addition to their widespread effects in peripheral tissues, these hormones act on the nervous system to modify a variety of behaviors, including the perception of sensory stimuli, locomotion, food and water intake, sleep, aggression, learning, and memory. Although certain hormone-induced changes may have adaptive value to animals in threatening situations, there are indications that chronic adrenocortical excess may eventually promote pathological or senescent changes in the brain. The diversity of the behavioral actions of glucocorticoid hormones has defied attempts to characterize them in terms of a unified global hypothesis. Although the investigation of mineralocorticoid effects on brain and behavior is in its infancy, there is evidence that aldosterone and DOC directly influence both brain metabolism, such as electrolyte and water balance, and behavior, such as salt ingestion.

Our current understanding of the mechanisms mediating adrenocortical actions on brain and behavior is far from satisfying. Some of these effects, such as FFB, some electrophysiological changes, and the prolongation of labile memory in chicks, occur too rapidly to involve the modulation of gene transcription. Such actions may be mediated by the membrane-associated glucocorticoid receptors recently demonstrated in several regions of brain, or perhaps directly by the free steroids themselves. The delayed onset of other, both organizational and activational, effects suggests that they may be consequences of gene regulation mediated by the mobile cytoplasmic steroid receptors that have been demonstrated in the CNS. The complete resolution and characterization of the different classes of receptors in the brain mediating the actions of mineralocorticoids and glucocorticoids will require additional effort. The evidence for distinct subpopulations of glucocorticoid receptors having preferences for either natural or synthetic glucocorticoids is not compelling. Striking differences in the *in vivo* pattern of distribution of dexamethasone compared to corticosterone and cortisol may reflect differences in the rates of penetration of these steroids into brain cells. Apparent physicochemical differences between receptors that bind synthetic and natural glucocorticoids may result from differences in the properties of the steroid-receptor complexes formed between different steroids and single initial populations of receptors. These issues are now under active investigation in a number of laboratories.

ACKNOWLEDGMENTS

The authors gratefully thank Dr. J.A. Weisberg and Dr. N.R. Kramarcy for their participation in a number of helpful discussions and their direct assistance in the organization of this material.

REFERENCES

Anderson, N.S., and Fanestil, D.D.: Corticoid receptors in rat brain: Evidence for an aldosterone receptor. *Endocrinology* *98*:676-684, 1976.

Angelucci, L., Valeri, P., Palmery, M., Patacchioli, F.R., and Catalani, A.: Brain glucocorticoid receptor: Correlation of *in vivo* uptake of corticosterone with behavioral, endocrine, and neuropharmacological events. *Adv Biochem Psychopharmacol* *21*:391-406, 1980.

Ardeleanu, A., and Sterescu, N.: RNA and DNA synthesis in developing rat brain: Hormonal influences. *Psychoneuroendocrinology* *3*:93-101, 1978.

Azmitia, E.C., and McEwen, B.S.: Corticosterone regulation of tryptophan hydoxylase in midbrain of the rat. *Science* *166*:1274-1276, 1969.

Azmitia, E.C., and McEwen, B.S.: Adrenalcortical influence on rat brain tryptophan hydroxylase activity. *Brain Res 78*:291-302, 1974.

Azmitia, E.C., and McEwen, B.S.: Early response of rat brain tryptophan hydroxylase activity to cycloheximide, puromycin, and corticosterone. *J Neurochem 27*:773-778, 1976.

Balázs, R., and Cotterrell, M.: Effect of hormonal state on cell number and functional maturation of the brain. *Nature* *236*:348-350, 1972.

Balfour, D.J.K., and Benwell, M.E.M.: Betamethasone-induced pituitary-adrenocortical suppression and brain 5-hydroxytryptamine in the rat. *Psychoneuroendocrinology* *4*:83-86, 1979.

Barak, Y.B., Gutnick, M.J., and Feldman, S.: Iontophoretically applied corticosteriods do not affect the firing of hippocampal neurons. Neuroendocrinology 23:248-256, 1977.

Baxter, J.D.: Glucocorticoid hormone action. Pharmacol Ther (B) 2:605-659, 1976.

Beatty, P.A., Beatty, W.W., Bowman, R.E., and Gilchrist, J.C.: The effects of ACTH, adrenalectomy, and dexamethasone on the acquistion of an avoidance response in rats. Physiol Behav 5:939-944, 1970.

Beatty, W.W., Scouten, C.W., and Beatty, P.A.: Differential effects of dexamethasone and body weight loss on two measures of activity. Physiol Behav 7:869-871, 1971.

Birmingham, M.K., Stumpf, W.E., and Sar, M.: Nuclear localization of aldosterone in rat brain cells assessed by autoradiography. Experientia 35:1240-1241, 1979.

Blaine, E.H., Covelli, M.D., Denton, D.A., Nelson, J.F., and Shulkes, A.A.: The role of ACTH and adrenal glucocorticoids in the salt appetite of wild rabbits [Oryctolagus cuniculus (L)]. Endocrinology 97:793-801, 1975.

Bloch, G.J., and Davidson, J.M.: Effects of adrenalectomy and prior experience on postcastrational sex behavior in the male rat. Physiol Behav 3:461-465, 1968.

Bohn, M.C., and Lauder, J.M.: The effects of neonatal hydrocortisone on rat cerebellar development: An autoradiographic and light-microscopic study. Dev Neurosci 1:250-266, 1978.

Bohus, B.: Central nervous structures and the effect of ACTH and corticosteroids on avoidance behavior: A study with intracerebral implantation of corticosteroids in the rat. Prog Brain Res 32:171-183, 1970.

Bohus, B.: Adrenocortical hormones and central nervous functions: The site and mode of their behavior in the rat. In V.H.T. James and L. Martini eds. Proceedings of the Third International Congress on Hormonal Steroids, Hamburg, 7-12 September 1970, Excerpta Medica, Amsterdam, 1971, pp 752-758.

Bohus, B.: Pituitary-adrenal influences on avoidance and approach behavior of the rat. Prog Brain Res 39:407-419, 1973.

Bohus, B., and de Kloet, E.R.: Behavioral effects of corticosterone related to putative glucocorticoid receptor properties in the rat brain. J Endocrinol 72:64P, 1977.

Bohus, B., and Lissák, K.: Adrenocortical hormones and avoidance behavior in rats. Int J Neuropharmacol 7:301-306, 1968.

Bohus, B., Nyakas, C., and Endröczi, E.: Effects of adrenocorticotrophic hormone on avoidance behavior of intact and adrenalectomized rats. Int J Neuropharmacol 7:307-314, 1968.

Bohus, B., Grubits, J., Kovács, G., and Lissák, K.: Effect of corticosteroids on passive avoidance behavior of rats. Acta Physiol Acad Sci Hung 38:381-391, 1970.

Bookin, H.B., and Pfeifer, W.D.: Adrenalectomy attenuates

electroconvulsive shock-induced retrograde amnesia in rats. *Behav Biol* 24:527-532, 1978.

Brain, P.F.: Mammalian behavior and the adrenal cortex--a review. *Behav Biol* 7:453-477, 1972.

Breen, G.A., McGinnis, J.F., and de Vellis, J.: Modulation of the hydrocortisone induction of glycerol phosphate dehydrogenase by N6, 02'-dibutyryl cyclic AMP, norepinephrine, and isobutylmethylxanthine in rat brain cell cultures. *J Biol Chem* 253:2254-2262, 1978.

Brush, F.R., and Levine, S.: Adrenocortical activity and avoidance learning as a function of time after fear conditioning. *Physiol Behav* 1:309-311, 1966.

Burger, H.G., and Patel, Y.C.: TSH and TRH: Their physiological regulation and the clinical applications of TRH. In *Clinical Neuroendocrinology*, L. Martini and G.M. Besser, eds. Academic Press, New York, 1977, pp 67-131.

Butte, J.C., Kakihana, R., and Noble, E.P.: Circadian rhythm of corticosterone levels in rat brain. *J Endocrinol* 68:235-239, 1976.

Carroll, B.J., Heath, B., and Jarrett, D.B.: Corticosteroids in brain tissue. *Endocrinology* 97:290-300, 1975.

Carroll, B.J., Curtis, G.C., and Mendels, J.: Neuroendocrine regulation in depression. II. Discrimination of depressed from nondepressed patients. *Arch Gen Psychiat* 33:1051-1058, 1976.

Carpenter, W.T., Jr., Strauss, J.S., and Bunney, W.E., Jr.: The psychobiology of cortisol metabolism: Clinical and theoretical implications. In *Psychiatric Complications of Medical Drugs*, R.I. Shader, ed. Raven Press, New York, 1972, pp 48-72.

Carsia, R.V., and Malamed, S.: Acute self-suppression of corticosteroidogenesis in isolated adrenocortical cells. *Endocrinology* 105:911-914, 1979.

Cheng, J.-T., and Araki, H.: Inhibitory mechanisms of dexamethasone on contractions induced by drugs and by transmural stimulation in isolated guinea pig ileum. *Jpn J Pharmacol* 28:755-762, 1978.

Chytil, F., and Toft, D.: Corticoid binding component in rat brain. *J Neurochem* 19:2877-2880, 1972.

Clayton, C.J., Grosser, B.I., and Stevens, W.: The ontogeny of corticosterone and dexamethasone receptors in rat brain. *Brain Res* 134:445-453, 1977.

Conforti, N., and Feldman, S.: Effect of cortisol on the excitability of limbic structures of the brain in freely moving rats. *J Neurol Sci* 26:29-38, 1975.

Cotman, C.W., Scheff, S.W., and Benardo, L.S.: Glucocorticoids decrease reactive synaptogenesis in the rat dentate gyrus. *Soc Neurosci Abstr* 4:469, 1978.

Cottrell, G.A., and Nakajima, S.: Effect of corticosteroids in the hippocampus on passive avoidance behavior in the rat. *Pharmacol Biochem Behav 7*:277-280, 1977.

Dallman, M.F., and Yates, F.E.: Dynamic asymmetries in the corticosteroid feedback path and distribution – metabolism-binding elements of the adrenocortical system. *Ann NY Acad Sci 156*:696-721, 1969.

Dallman, M.F., Engeland, W.C., and Shinsako, J.: Compensatory adrenal growth: A neurally mediated reflex. *Am J Physiol 231*: 408-414, 1976.

Davis, H., Memmott, J., MacFadden, L., and Levine, S.: Pituitary-adrenal activity under different appetitive extinction procedures. *Physiol Behav 17*:687-690, 1976.

Davis, H., Green, B., Herrmann, T., and Levine, S.: Blockage of pituitary-adrenal activity does not affect conditioned suppression. *Physiol Behav 20(4)*:423-425, 1978.

Davis, H., Memmott, J., MacFadden, L., and Levine, S.: Pituitary-adrenal activity under different appetitive extinction procedures. *Physiol Behav 17*:687-690, 1976.

Davis, J.O.: Regulation of aldosterone secretion. In *Handbook of Physiology Sec 7, Endocrinology*, vol VI, R.O. Greep and E.B. Astwood, eds. American Physiological Society, Washington, D.C., 1975, pp 77-106.

Devenport, L.D.: Adrenal modulation of brain size in adult rats. *Behav Neural Biol 27*:218-221, 1979.

Diez, J.A., Sze, P.Y., and Ginsburg, B.E.: Effects of hydrocortisone and electric footshock on mouse brain tyrosine hydroxylase activity and tyrosine levels. *Neurochem Res 2:* 161-170, 1977.

van Dijk, A.M.A., van Wimersma Greidanus, Tj., de Kloet, E.R., and de Wied, D.: Adrenocorticotrophin concentration in the brain after adrenalectomy. *J Endocrinol 80*:60P, 1979.

Dokas, L.A.: Corticosterone and RNA metabolism in the rat hippocampus. *Soc Neurosci Abstr 5*:443, 1979.

Dunn, A.J.: Neurochemistry of learning and memory: An evaluation of recent data. *Annu Rev Psychol 31*:343-390, 1980.

Dunn, A.J., and Liebmann, S.: The amnestic effect of protein synthesis inhibitors is not due to the inhibition of adrenal corticosteroidogenesis. *Behav Biol 19*:411-416, 1977.

Dunn, A.J., Gray, H.E., and Iuvone, P.M.: Protein synthesis and amnesia: Studies with emetine and pactamycin. *Pharmacol Biochem Behav 6*:1-4, 1977.

Dunn, A.J., Gildersleeve, N.B., and Gray, H.E.: Mouse brain tyrosine hydroxylase and glutamic acid decarboxylase following treatment with ACTH, vasopressin or corticosterone. *J Neurochem 31*:977-982, 1978.

Edwardson, J.A., and Bennett, G.W.: Modulation of corticotrophin-releasing factor from hypothalamic synaptosomes.

Nature 251:425–427, 1974.

Elson, I.J., Seybert, J.A., and Ghiselli, W.B.: Retention of aversively motivated behavior: Effects of time of training and associative versus nonassociative processes. *Behav Biol 20*:337–353, 1977.

Ermisch, A., and Rühle, H.-J.: Autoradiographic demonstration of aldosterone-concentrating neuron populations in rat brain. *Brain Res 147*:154–158, 1978.

Etgen, A.M., Lee, K.S., and Lynch, G.: Glucocorticoid modulation of specific portein metabolism in hippocampal slices maintained in vitro. *Brain Res 165*:37–46, 1979.

Fekete, M.I., Stark, E., Herman, J.P., Palkovits, M., and Kanyicska, B.: Catecholamine concentration of various brain nuclei of the rat as affected by ACTH and corticosterone. *Neurosci Lett 10*:153–158, 1978.

Feldman, S., and Conforti, N.: Participation of the dorsal hippocampus in the glucocorticoid feedback effect on adrenocortical activity. *Neuroendocrinology 30*:52–55, 1980.

Feldman, S., and Dafny, N.: Effects of cortisol on unit activity in the hypothalamus of the rat. *Exp Neurol 27*:375–387, 1970.

Feldman, S. and Sarne, Y.: Effect of cortisol on single cell activity in hypothalamic islands. *Brain Res 23*:67–75, 1970.

Finch, C., and Latham, K.: Corticosterone receptors and aging in mouse hippocampus. *56th Annual Meeting of the Endocrine Society*, Philadelphia, p A-110, 1974.

Flexner, J.B., and Flexner, L.B.: Adrenalectomy and the supression of memory by puromycin. *Proc Nat Acad Sci USA 66*:48–52, 1970.

Flood, J.F., Vidal, D., Bennett, E.L., Orme, A.E., Vasquez, S., and Jarvik, M.E.: Memory facilitating and antiamnestic effects of corticosteroids. *Pharmacol Biochem Behav 8*:81–88, 1978.

Forman, B.H., and Mulrow, P.J.: Effect of corticosteroids on water and electrolyte metabolism. In *Handbook of Physiology, section 7: Endocrinology,* vol VI, R.O. Greep and E.B. Astwood, eds. Washington, American Physiological Society, 1975, pp 179–190.

Fregly, M.J., and Waters, I.W.: Effects of mineralocorticoids on spontaneous sodium chloride appetite of adrenalectomized rats. *Physiol Behav 1*:65–74, 1966.

Fukuda, T., and Ui, J.: Breakdown of blood-cerebrospinal fluids barrier for calcium ions in the absence of glucosteroids. *Nature 214*:598–599, 1967.

Fuller, J.L., Chambers, R.M., and Fuller, R.P.: Effects of cortisone and of adrenalectomy on activity and emotional behavior of mice. *Psychosom Med 18*:234–242, 1956.

Funder, J.W., Feldman, D., and Edelman, I.S.: The roles of

plasma binding and receptor specificity in the mineralocorticoid action of aldosterone. *Endocrinology 92*:994-1004, 1973.

Garrud, P., Gray, J.A., and de Wied, D.: Pituitary-adrenal hormones and extinction of rewarded behaviour in the rat. *Physiol Behav 12*:109-119, 1974.

Gibbs, M.E., and Ng, K.: Psychobiology of memory: Towards a model of memory formation. *Biobehav Rev 1*:113-136, 1977.

Gibson, M.J., Krieger, D.T., Liotta, A.S., Brownstein, M.J., and McEwen, B.S.: Chronic dexamethasone alters content of immunoreactive ACTH-like material in rat arcuate nucleus. *Soc Neurosci Abstr 5*:446, 1979.

Gillin, J.C., Jacobs, L.S., Fram, D.H., and Snyder, F.: Acute effect of a glucocorticoid on normal human sleep. *Nature 237*: 398-399, 1972.

Gillin, J.C., Jacobs, L.S., Snyder, F., and Henkin, R.I.: Effects of ACTH on the sleep of normal subjects and patients with Addison's disease. *Neuroendocrinology 15*:21-31, 1974.

Gorzalka, B.B., and de Cantanzaro, D.: Pituitary-adrenal effects on sexual behavior in isolated and group-housed mice. *Physiol Behav 22*(5):939-945, 1979.

Gorzalka, B.B., and Whalen, R.E.: The effects of progestins, mineralocorticoids, glucocorticoids, and steroid solubility on the induction of sexual receptivity in rats. *Horm Behav 8*: 94-99, 1977.

Gorzalka, B.B., Rezek, D.L., and Whalen, R.E.: Adrenal mediation of estrogen-induced ejaculatory behavior in the male rat. *Physiol Behav 14*:373-376, 1975.

Gray, P.: Effects of prestimulation on avoidance responding in rats, and hormonal dependence of the effect. *J Comp Physiol Psychol 90*:1-17, 1976.

Green, A.R., and Curzon, G.: Effects of hydrocortisone and immobilization on tryptophan metabolism in brain and liver of rats of different ages. *Biochem Pharmacol 24*:713-716, 1975.

Greer, M.A., Panton, P., and Allen, C.F.: Relationship of nyctohemeral cycles of running activity and plasma corticosterone concentration following basal hypothalamic isolation. *Horm Behav 3*:289-295, 1972.

Grosser, B.I., Stevens, W., and Reed, D.J.: Properties of corticosterone-binding marcomolecules from rat brain cytosol. *Brain Res 57*:387-395, 1973.

Gumbinas, M., Oda, M., and Huttenlocher, P.: The effects of corticosteroids on myelination of the developing rat brain. *Biol Neonate 22*:355-366, 1973.

Harrison, R.W., Fairfield, S., and Orth, D.M.: Effect of cell membrance alteration on glucocorticoid uptake by the AtT-20/D-1 target cell. *Biochim Biophys Acta 466*:357-365, 1977.

Haskett, R.F., Schteingart, D.E., Starkman, M.N., and Carroll,

B.J.: Use of psychiatric research diagnostic criteria in Cushing's syndrome. *Proceedings of the Tenth International Congress of the International Society of Psychoneuroendocrinology*, 1979, p 12.

Haynes, R.C., Jr., and Larner, J.: Adrenocorticotropic hormone; adrenocortical steroids and their synthetic analogs; inhibitors of adrenocortical steroid biosynthesis. In *The Pharmacological Basis of Therapeutics*, 5th ed. L.S. Goodman and A. Gilman, eds. Macmillan, New York, 1975, pp 1472-1506.

Henkin, R.I.: The role of adrenal corticosteroids in sensory processes. In *Handbook of Physiology-Endocrinology*, vol. VI. H. Blaschko, G. Sayers, and A.D. Smith, eds. Williams & Williams, Baltimore, 1975, pp 209-230.

Hennessy, M.B., Harney, K.S., Smotherman, W.P., Coyle, S., and Levine, S.: Adrenalectomy-induced deficits in maternal retrieval in the rat. *Horm Behav 9*:222-227, 1977.

Holaday, J.W., Law, P.-Y., Loh, H.H., and Li, C.H.: Adrenal steroids indirectly modulate morphine and β-endorphin effects. *J Pharm Exp Therap 208*:176-183, 1979.

Holaday, J.W., Martinez, H.M., and Natelson, B.H.: Synchronized ultradian cortisol rhythms in monkeys: Persistence during corticotropin infusion. *Science 198*:56-58, 1977.

Inoue, K., Takahashi, K., and Takahashi, Y.: Influence of change in feeding regimen and food deprivation on circadian rhythm of adrenal cortical activity in rats. *Folia Endocrinol Jpn 52*:898-907, 1976.

Iuvone, P.M., Morasco, J., and Dunn, A.J.: Effect of corticosterone on the synthesis of ^{3}H-catecholamines in the brains of CD-1 mice. *Brain Res 120*:571-576, 1977.

Javoy, F., Glowinski, J., and Kordon, C.: Effects of adrenalectomy on the turnover of norepinephrine in rat brain. *Eur J Pharmacol 4*:103-104, 1968.

Joffe, J.M., Mulick, J.A., and Rawson, R.A.: Effects of adrenalectomy on open-field behavior in rats. *Horm Behav 3*:87-96, 1972.

Jones, M.T., Hillhouse, E.W., and Burden, J.: Dynamics and mechanics of corticosteroid feedback at the hypothalamus and anterior pituitary gland. *J Endocrinol 73*:405-417, 1977.

Kamin, L.J.: Retention of an incompletely learned avoidance response. *J Comp Physiol Psychol 50*:457-460, 1957.

Katz, R.J., and Carroll, B.J.: Endocrine control of psychomotor activity in the rat: Effects of chronic dexamethasone upon general activity. *Physiol Behav 20*:25-30, 1978.

Kendall, J.W.: Dexamethasone stimulation of running activity in the male rat. *Horm Behav 1*:327-336, 1970.

Kendall, J.W.: Feedback control of adrenocorticotropic hormone secretion. In *Frontiers in Neuroendocrinology*, Vol. 2, L. Martini and W.F. Ganong, eds. Oxford University Press, New

York, 1971, pp 177-207.

Kizer, J.S., Palkovits, M., Zivin, J., Brownstein, M., Saavedra, J., and Kopin, I.J.: The effect of endocrinological manipulations on tyrosine hydroxylase and dopamine-β-hydroxylase activities in individual hypothalamic nuclei of the adult male rat. *Endocrinology 85*:799-812, 1974.

Klein, S.B.: Adrenal-pituitary influence in reactivation of avoidance-learning memory in the rat after intermediate intervals. *J Comp Physiol Psychol 79*:341-359, 1972.

de Kloet, E.R., and Burbach, P.: Selective purification of a single population of glucocorticoid receptors from rat brain. *J Neurochem 30*:1505-1508, 1978.

de Kloet, E.R., and McEwen, B.S.: Differences between cytosol receptor complexes with corticosterone and dexamethasone in hippocampal tissue from rat brain. *Biochim Biophys Acta 421*: 124-132, 1976.

de Kloet, E.R., van der Vies, J., and de Wied, D.: The site of the suppressive action of dexamethasone on pituitary-adrenal activity. *Endocrinology 94*:61-73, 1974.

de Kloet, E.R., Wallach, G., and McEwen, B.S.: Differences in corticosterone and dexamethasone binding to rat brain and pituitary. *Endocrinology 96*:598-609, 1975.

de Kloet, E.R., Burbach, P., and Mulder, G.H.: Localization and role of transcortin-like molecules in the anterior pituitary. *Mol Cell Endocrinol 7*:261-273, 1977.

de Kloet, R., Veldhuis, D., and Bohus, B.: Significance of neuropeptides in the control of corticosterone receptor activity in rat brain. *Adv Biochem Psychopharmacol 21*:373-382, 1980.

Kovács, G.L., Telegdy, G., and Lissák, K.: 5-Hydroxytryptamine and the mediation of pituitaryadrenocortical hormones in the extinction of active avoidance behaviour. *Psychoneuroendocrinology 1*:219-230, 1976.

Kovács, G.L., Telegdy, G., and Lissák, K.: Dose-dependent action of corticosteroids on brain serotonin content and passive avoidance behavior. *Horm Behav 8*:155-165, 1977b.

Kovács, G.L., Kishonti, J., Lissák, K., and Telegdy, G.: Dose-dependent dual effect of corticosterone on cerebral 5-HT metabolism. *Neurochem Res 2*:311-322, 1977a.

Kraulis, I., Foldes, G., Traikov, H., Dubrovsky, B., and Birmingham, M.K.: Distribution, metabolism and biological activity of deoxycorticosterone in the central nervous system. *Brain Res 88*:1-14, 1975.

Krieger, D.T., and Glick, S.M.: Growth hormone and cortisol responsiveness in Cushing's syndrome. *Am J Med 52*:25-40, 1972.

Landfield, P.W., Waymire, J.C., and Lynch, G.: Hippocampal aging and adrenocorticoids: Quantitative correlations.

Science 202:1098-1102, 1978.

Landfield, P.W., Wurtz, C., Lindsey, J.D., and Lynch, G.: Long-term adrenalectomy reduces some morphological correlates of brain aging. *Soc Neurosci Abstr* 5:7, 1979.

Lassman, M.N., and Mulrow, P.J.: Deficiency of deoxycorticosterone-binding protein in the hypothalamus of rats resistant to DOC-induced hypertension. *Endocrinology* 94:1541-1546, 1974.

Leshner, A.I.: The adrenals and the regulatory nature of running wheel activity. *Physiol Behav* 6(5):551-558, 1971.

Leshner, A.I.: A model of hormones and agonistic behavior. *Physiol Behav* 15(2):225-235, 1975.

Leshner, A.I., and Politch, J.A.: Hormonal control of submissiveness in mice: Irrelevance of the androgens and relevance of the pituitary-adrenal hormones. *Physiol Behav* 22(3):531-534, 1979.

Levine, S., and Brush, F.R.: Adrenocortical activity and avoidance learning as a function of time after avoidance training. *Physiol Behav* 2:385-388, 1967.

Maas, J.W., and Mednieks, M.: Hydrocortisone-mediated increase of norepinephrine uptake by brain slices. *Science* 171:178-179, 1971.

Maclusky, N.J., Turner, B.B., and McEwen, B.S.: Corticosteroid binding in rat brain and pituitary cytosols: Resolution of multiple binding components by polyacrylamide gel-based isoelectric focusing. *Brain Res* 130:564-571, 1977.

McEwen, B.S., and Wallach, G.: Corticosterone binding to hippocampus: Nuclear and cytosol binding *in vitro*. *Brain Res* 57:373-386, 1973.

McEwen, B.S., Weiss, J.M., and Schwartz, L.S.: Retention of corticosterone by cell nuclei from brain regions of adrenalectomized rats. *Brain Res* 17:471-482, 1970.

McEwen, B.S., Magnus, C., and Wallach, G.: Soluble corticosterone-binding macromolecules extracted from rat brain. *Endocrinology* 90:217-226, 1972.

McEwen, B.S., Wallach, G., and Magnus, C.: Corticosterone binding to hippocampus: Immediate and delayed influences of the absence of adrenal secretion. *Brain Res* 70:321-334, 1974.

McEwen, B.S., Gerlach, J.L., and Micco, D.J.: Putative glucocorticoid receptors in hippocampus and other regions of the rat brain. In *The Hippocampus, Vol. 1: Structure and Development,* R.L. Isaacson and K.H. Pribram, eds. Plenum, New York, 1975, pp 285-322.

McEwen, B.S., de Kloet, E.R., and Wallach, G.: Interactions *in vivo* and *in vitro* of corticoids and progesterone with cell nuclei and soluble macromolecules from rat brain regions and pituitary. *Brain Res* 105:129-136, 1976.

McEwen, B.S., David, P.G., Parsons, B., and Pfaff, D. W.: The brain as a target for steroid hormone action. *Annu Rev Neurosci* 2:65-112, 1979.

McGowan-Sass, B.K., and Timiras, P.S.: The hippocampus and hormonal cyclicity. In *The Hippocampus, Vol. 1: Structure and Development*, R.L. Isaacson and K.H. Pribram, eds. Plenum, New York, 1975, pp 355-374.

McIntyre, D.C.: Adrenalectomy: Protection from kindled convulsion induced amnesia in rats. *Physiol Behav* 17(5):789-795, 1976.

McIntyre, D.C., and Wann, P.D.: Adrenalectomy: Protection from kindled convulsion induced dissociation in rats. *Physiol Behav* 20(4):469-474, 1978.

Marquis, H.A., and Suboski, M.D.: Hypophysectomy and ACTH replacement in the incubation of passive and shuttle box avoidance responses. *Proc 77th Ann Conv Am Psychol Assoc* 4:207-208, 1969.

Mason, J.W.: A re-evaluation of the concept of "nonspecificity" in stress theory. *J Psychiat Res* 8:323-333, 1971.

Meyer, J.S., Luine, V.N., Khylchevskaya, R.I., and McEwen, B.S.: Glucocorticoids and hippocampal enzyme activity. *Brain Res* 166:172-175, 1979.

Micco, D.J.., Jr., McEwen, B.S., and Shein, W.: Modulation of behavioral inhibition in appetitive extinction following manipulation of adrenal steroids in rats: Implications for involvement of the hippocampus. *J Comp Physiol Psychol* 93:323-329, 1979.

Michael, R.P., and Gibbons, J.L.: Interrelationships between the endocrine system and neuropsychiatry. *Int Rev Neurobiol* 5:243-302, 1963.

Michal, E.K.: Dexamethasone inhibits multi-unit activity in the rat hippocampus. *Brain Res* 65:180-183, 1974.

Millard, S.A., Costa, E., and Gal, E.M.: On the control of brain serotonin turnover rate by end product inhibition. *Brain Res* 40:545-551, 1972.

Miller, A.L., Chaptal, C., McEwen, B.S., and Peck, E.J.: Modulation of high-affinity GABA uptake into hippocampal synaptosomes by glucocorticoids. *Psychoneuroendocrinology* 3:155-164, 1978.

Miyabo, S., Hisada, T., Ueno, K., Kishida, S., and Kitanaka, I.: Behavioral and other systemic effects of cortisol-21-sulfate. *Horm Behav* 3:227-236, 1972.

Moguilewsky, M., and Raynaud, J.P.: Evidence for a specific mineralocorticoid receptor in rat pituitary and brain. *J Steroid Biochem* 12:309-314, 1980.

Moore, K.E., and Phillipson, O. T.: Effects of dexamethasone on phenylethanolamine N-methyltransferase and adrenaline in the brains and superior cervical ganglia of adult and neonatal rats. *J Neurochem* 25:289-294, 1975.

Morimoto, Y., Arisue, K., and Yamamura, Y.: Relationship between circadian rhythm of food intake and that of plasma corticosterone and effect of food restriction on circadian adrenocortical rhythm in the rat. *Neuroendocrinology 23*: 212–222, 1977.

Mormède, P., and Dantzer, R.: Effects of dexamethasone on fear conditioning in pigs. *Behav Biol 21*: 225–235, 1977.

Mormède, P., and Dantzer, R.: Pituitary-adrenal influences on avoidance behavior of pigs. *Horm Behav 10*:285–297, 1978.

Munck, A., and Leung, K.: Glucocorticoid receptors and mechanisms of action. In *Receptors and Mechanism of Action of Steroid Hormones. Modern Pharmacology-Toxicology*, Vol. 8, Part 2, J.R. Pasqualini, ed. Marcel Dekker, New York, 1977, pp 311–397.

Nakajima, S.: Amnesic effect of cycloheximide in the mouse mediated by adrenocortical hormones. *J Comp Physiol Psychol 88*:378–385, 1975.

Nakajima, S.: Attenuation of amnesia by hydrocortisone in the mouse. *Physiol Behav 20*:607–611, 1978.

Neckers, L., and Sze, P.Y.: Regulation of 5-hydroxytryptamine metabolism in mouse brain by adrenal glucocorticoids. *Brain Res 93*:123–132, 1975.

Oda, M.A.S., and Huttenlocher, P.R.: The effect of corticosteroids on dendritic development in rat brain. *Yale J Biol Med 3*:155–165, 1974.

Ogren, S.-O., and Fuxe, K.: Learning, brain noradrenaline and the pituitary-adrenal axis. *Med Biol 52*:399–405, 1974.

Ogren, S.-O., and Fuxe, K.: On the role of brain noradrenaline and the pituitary-adrenal axis in avoidance learning. I. Studies with corticosterone. *Neurosci Lett 5*:291–296, 1977.

Olpe, H.-R., and McEwen, B.S.: Glucocorticoid binding to receptor-like proteins in rat brain and pituitary: Ontogenetic and experimentally-induced changes. *Brain Res 105*:121–128, 1976.

Ondo, J.G., and Kitay, J.I.: Effects of dexamethasone and stressful stimuli on hypothalamic electrical activity in rats with diencephalic islands. *Neuroendocrinology 9*:215–227, 1972.

Pandolfo, L., and Macaione, S.: Effect of adrenalectomy on activity of GABA transaminase and glutamic acid decarboxylase from rat brain cortex. *Ital J Biochem 13*:247–252, 1964.

Pappas, B.A., and Gray, P.: Cue value of dexamethasone for fear-motivated behavior. *Physiol Behav 6*:127–130, 1971.

Pardridge, W.M., and Mietus, L.J.: Transport of steroid hormones through the rat blood-brain barrier. *J Clin Invest 64*: 145–154, 1979.

Pfaff, D.W., Gerlach, J.L., McEwen, B.S., Ferin, M., Carmel, P., and Zimmerman, E.A.: Autoradiographic localization of hormone-concentrating cells in the brain of the female rhesus monkey. J Comp Neurol 170:279-294, 1976.

Pfaff, D.W., Gregory, E., and Silva, M.T.A.: Testosterone and corticosterone effects on single unit activity in the rat brain. In Influence of Hormones on the Nervous System, D.H. Ford, ed. S. Karger, Basel, 1971, pp 269-281.

Phillips, M.I., and Dafny, N.: Effect of corisol on unit activity in freely-moving rats. Brain Res 25:651-655, 1971a.

Phillips, M.I., and Dafny, N.: Effect of a dose range of cortisol on brain unit activity. Fed Proc 30:203, 1971b.

Pinel, J.P.J., and Cooper, R.M.: Demonstration of the Kamin effect after one-trial avoidance learning. Psychonom Sci 4: 17-18, 1966.

Randich, A., Smith, I.M., and LoLordo, V.M.: Pituitaryadrenal influences on the acquisition of excitatory and inhibitory conditioned responses in rats. Physiol Psychol 7:75-83, 1979.

Rastogi, R.B., and Singhal, R.L.: Adrenocorticoids control 5-hydroxytryptamine metabolism in rat brain. J Neural Transm 42:63-71, 1978a.

Rastogi, R.B., and Singhal, R.L.: Evidence for the role of adrenocortical hormones in the regulation of noradrenaline and dopamine metabolism in certain brain areas. Br J Pharmacol 62:131-136, 1978b.

Rees, H.D., and Dunn, A.J.: Role of the pituitary-adrenal system in the footshock-induced increase of ^{3}H-lysine incorporation into mouse brain and liver proteins. Brain Res 120: 317-325, 1977.

Rees, H.D., Stumpf, W.E., and Sar, M.: Autoradiographic studies with ^{3}H-dexamethasone in the rat brain and pituitary. In Anatomical Neuroendocrinology, W.E. Stumpf and L.D. Grant, eds. S. Karger, Basel, 1975, pp 262-269.

Rees, H.D., Stumpf, W.E., Sar, M., and Petrusz, P.: Autoradiographic studies of ^{3}H-dexamethasone uptake in immunocytochemically characterized cells of the rat pituitary. Cell Tissue Res 182:347-356, 1977.

Reulin, H.J.: Vasogenic brain oedema: New aspects in its formation, resolution and therapy. Br J Anaesth 48:741-752, 1976.

Rhees, R.W., Abel, J.H., and Haack, D.W.: Uptake of tritiated steroids in the brain of the duck (Anas platyrhynchos): An autoradiographic study. Gen Comp Endocrinol 18:292-300, 1972.

Rhees, R.W., Grosser, B.I., and Stevens, W.: The autoradiographic localization of [^{3}H]dexamethasone in the brain and pituitary of the rat. Brain Res 100:151-156, 1975.

Richter, C.P., and Wislocki, G.B.: Anatomical and behavioral changes produced in the rat by complete and partial extirpation of the pituitary gland. Am J Physiol 95:481-492, 1930.

Riegle, G.D., and Hess, G.D.: Chronic and acute dexamethasone suppression of stress activation of adrenal cortex in young and aged rats. *Neuroendocrinology 9*:175-187, 1972.

Rigter, H., Janssens-Elbertse, R., and van Riezen, H.: Reversal of amnesia by an orally active ACTH 4-9 analog (Org. 2766). *Pharmacol Biochem Behav 5* (Suppl. 1):53-58, 1976.

Riker, D.K., Sastre, A., Baker, T., Roth, R.H., and Riker, W.F.: Regional high-affinity [^{3}H]choline accumulation in cat forebrain: Selective increase in the caudate-putamen after corticosteroid pretreatment. *Mol Pharmacol 16*:886-899, 1979.

Riker, W.F., Baker, T., and Okamoto, M.: Glucocorticoids and mammalian motor nerve excitability. *Arch Neurol 32*:688-694, 1975.

Roberts, D.C.S., and Fibiger, H.C.: Evidence for interactions between central noradrenergic neurons and adrenal hormones in learning and memory. *Pharmacol Biochem Behav 7*:191-194, 1977.

Rodgers, R.J., and Semple, J.M.: Pituitary-adrenocortical axis and shock-induced fighting in rats. *Physiol Behav 20*:533-537, 1978.

Roosevelt, S., Wolfsen, A.R., and Odell, W.D.: Modulation of brain endorphin-opiate receptor by glucocorticoids. *Endocrinology 104*[Suppl.]:A125, 1979.

Rossier, J., French, E., Gros, C., Minick, S., Guillemin, R., and Bloom, F.E.: Adrenalectomy, dexamethasone or stress alters opioid peptide levels in rat anterior pituitary but not intermediate lobe or brain. *Life Sci 25*:2105-2112, 1979.

Roth, G.S.: Reduced glucocorticoid binding site concentration in cortical neuronal perikarya from senescent rats. *Brain Res 107*:345-354, 1976.

Rousseau, G.G., and Baxter, J.D.: Glucocorticoid receptors. In *Glucocorticoid Hormone Action,* J.D. Baxter and G.G. Rousseau, eds. Springer-Verlag, Berlin, 1979, pp 49-77.

Sachar, E.J., Hellman, L., Roffwarg, H.P., Halperm, F.S., Fukushima, D.K., and Gallagher, T.F.: Disrupted 24-hour patterns of cortisol secretion in psychotic depression. *Arch Gen Psychiat 28*:19-24, 1973.

Sadasivudu, B., Rao, T.I., and Murthy, R.K.: Metabolic effects of hydrocortisone in mouse brain. *Neurochem Res 2*:521-532, 1977.

Sando, J.J., Hammond, N.D., Stratford, C.A., and Pratt, W.B.: Activation of thymocyte glucocorticoid receptors to the steroid binding form. *J Biol Chem 254*:4779-4789, 1979.

Sayers, G., and Portanova, R.: Regulation of the secretory activity of the adrenal cortex: Cortisol and corticosterone. In *Handbook of Physiology. Section 7: Endocrinolgy,* Vol. VI, R.O. Greep and E.B. Astwood, eds. American Physiological Society, Washington, D.C., 1975, pp 41-53.

Scheff, S.W., Thorne, D.R., Sasvary, G., Benardo, L.S., and

Cotman, C.W.: Chronic glucocorticoid administration alters axon sprouting in the rat hippocampal formation. *Soc Neurosci Abstr 5*:1555, 1979.

Schlesser, M.A., Winokur, G., and Sherman, B.M.: Genetic subtypes of unipolar primary depressive illness distinguished by hypothalamic-pituitary-adrenal axis activity. *Lancet 1*:739-741, 1979.

Schmiedek, P., Aettinger, W., Baethmann, A., Enzenback, R., and Marguth, F.: Aldosterone—a new therapeutic principle for the treatment of brain oedema in man. *Acta Neurochir 30*:59-68, 1974.

Schwartz, M.L., Tator, C.H., and Hoffman, H.J.: The uptake of hydrocortisone in mouse brain and ependymoblastoma. *J Neurosurg 36*:178-183, 1972.

Selye, H.: *Stress: The Physiology and Pathology of Exposure to Stress.* Acta Medica, Montreal, 1950.

Seth, P.: Occurrence and function of corticosteroids in some selected mammalian species. *Gen Comp Endocrinol (Suppl)2*: 317-324, 1969.

Seybert, J.A., Vandenberg, G.L., Harvey, R.J., Budd, J.R., and McClanahan, L.G.: Retention of appetitive instrumental behavior: The Kamin effect. *Behav Neural Biol 26*:266-286, 1979.

Shen, J.T., and Ganong, W.F.: Effect of variations in adrenocortical function on dopamine β-hydroxylase and norepinephrine in the brain of the rat. *J Pharm Exp Therap 199*:639-648, 1976.

Sherman, M.R.: Allosteric and competitive steroid-receptor interaction. In *Glucocorticoid Hormone Action,* J.D. Baxter and G.G. Rousseau, eds. Springer-Verlag, Berlin, 1979, pp 123-133.

Sherman M.R., Pickering, L.A., Rollwagen, F.M., and Miller, L.K.: Meroreceptors: Proteolytic fragments of receptors containing the steroid-binding site. *Fed Proc 37*:167-173, 1978.

Simon, G., and Gandelman, R.: Influence of corticosterone on the development and display of androgen dependent aggressive behavior in mice. *Physiol Behav 20*:391-396, 1978.

Snider, N., Marquis, H.A., Black, M., and Suboski, M.D.: Adrenal corticosteroids and the Kamin effect. *Psychonom Sci 22*: 309-310, 1971.

Squire, L.R., St. John, S., and Davis, H.P.: Inhibitors of protein synthesis and memory: Dissociation of amnestic effects and effects of adrenal steroidogenesis. *Brain Res 112*:200-206, 1976.

Steiner, F.A., Ruf, K., and Akert, K.: Steroid-sensitive neurons in rat brain: Anatomical localization and responses to neurohumors and ACTH. *Brain Res 12*:74-85, 1969.

Stern, J.M., Erskine, M.S., and Levine, S.: Dissociation of

open-field behavior and pituitary-adrenal function. *Horm Behav* 4:149-162, 1973.

Stevens, W., Reed, D.J., Erickson, S., and Grosser, B.I.: The binding of corticosterone to brain proteins: Diurnal variation. *Endocrinology* 93:1152-1156, 1973.

Stevens, W., Reed, D.J., and Grosser, B.I.: Binding of natural and synthetic glucocorticoids in rat brain. *J Steroid Biochem* 6:521-527, 1975.

Stumpf, W.E., and Sar, M.: Hormonal inputs to releasing factor cells, feedback sites. *Prog Brain Res* 38:54-70, 1973.

Stumpf, W.E., and Sar, M.: Anatomical distribution of corticosterone-concentrating neurons in rat brain. In *Anatomical Neuroendocrinology,* W.E. Stumpf and L.D. Grant, eds. S. Karger, Basel, 1975, pp 82-103.

Stumpf, W.E., and Sar, M.: Steroid hormone target cells in the extrahypothalamic brain stem and cervical spinal cord: Neuroendocrine significance. *J Steroid Biochem* 11:801-807, 1979.

Suboski, M.D., Marquis, H.A., Black, M., and Plutenius, P.: Adrenal and amygdala function in the incubation of aversively conditioned responses. *Physiol Behav* 5:283-289, 1970.

Svec, F., and Harrison, R.W.: The intracellular distribution of natural and synthetic glucocorticoids in the AtT-20 cell. *Endocrinology* 104:1563-1568, 1979.

Sze, P.Y.: Glucocorticoid regulation of the serotonergic system of the brain. *Adv. Biochem Psychopharmacol* 15:251-265, 1976.

Sze, P.Y., and Hedrick, B.J.: Dexamethasone effect on ganglionic tyrosine hydroxylase activity. *Trans Am Soc Neurochem* 10:163, 1979.

Sze, P.Y., Neckers, L., and Towle, A.C.: Glucocorticoids as a regulatory factor for brain tryptophan hydroxylase. *J Neurochem* 26:169-173, 1976.

Tang, F., and Phillips, J.G.: Some age-related changes in pituitary-adrenal function in male laboratory rat. *J Gerontol* 33:377-382, 1978.

Thoenen, H., and Otten, U.: Role of adrenocorticoid hormones in the modulation of synthesis and degradation of enzymes involved in the formation of catecholamines. In *Frontiers in Neuroendocrinology,* vol. 5, W.G. Ganong and L. Martini, eds. Raven Press, New York, 1978, pp 163-184.

Towle, A.C., and Sze, P.Y.: Binding of corticosterone to synaptic plasma membrane from rat brain. *Soc Neurosci Abstr* 4:356, 1978.

Turner, B.B.: Ontogeny of glucocorticoid binding in rodent brain. *Am Zoologist* 18:461-475, 1978.

Turner, B.B., and McEwen, B.S.: Hippocampal cytosol binding capacity of corticosterone (b): No depletion with nuclear receptor loading. *Brain Res* 189:169-182, 1980.

Turner, B.B., Smith, E.M., and Carroll, B.J.: Diurnal pattern of corticosterone binding in the brain. *American Physiological Society* Fall Meeting, (Abstr), 1978a.

Turner, B.B., Smith, E., and Carroll, B.J.: Regional patterns of endogenous corticosterone binding in rat brain: Basal vs. stressed states. *Soc Neurosci Abstr* 4:357, 1978b.

Turner, B.B., Katz, R.J., and Carroll, B.J.: Neonatal corticosteroid permanently alters brain activity of epinephrine-synthesizing enzyme in stressed rats. *Brain Res* 166:426-430, 1979a.

Turner, B.B., Smith, E.M., and Carroll, B.J.: Baboon corticosterone: Substantial brain binding of a "minor" adrenal glucocoricoid. *Soc Neurosci Abstr* 5:462, 1979b.

Ueda, M.: Failure to interpret the behavioral effect of cortisol-21-sulfate by steroid-receptor interaction. *J Steroid Biochem* 9:1261-1262, 1978.

Ulrich, R., Yuwiler, A., and Geller, E.: Effects of hydrocortisone on biogenic amine levels in the hypothalamus. *Neuroendocrinology* 19:259-268, 1975.

Vale, W., Spiess, J., Rivier, C., and Rivier, J.: Characterization of a 41-residue ovine hypothalamic peptide that stimulates secretion of corticotropin and β-endorphin. *Science* 213:1394-1397, 1981.

Van Loon, G.R., and Mascardo, R.N.: Effect of hypophysectomy, ACTH and dexamethasone on brain dopamine-β-hydroxylase activity in rats. *Soc Neurosci Abstr* 1:444, 1975.

Veals, J.W., Korduba, C.A., and Symchowicz, S.: Effect of dexamethasone on monoamine oxidase inhibition by iproniazid in rat brain. *Eur J Pharmacol* 41:291-299, 1977.

Vermes, I., Telegdy, G., and Lissák, K.: Correlation between hypothalamic serotonin content and adrenal function during acute stress: Effect of adrenal corticosteroids on hypothalamic serotonin content. *Acta Physiol Acad Sci Hung* 43:33-42, 1973.

Versteeg, D.H.G., Gispen, W.H., Schotman, P., Witter, A., and de Wied, D.: Hypophysectomy and rat brain metabolism: Effects of synthetic ACTH analogs. *Adv Biochem Psychopharmacol* 6:219-239, 1972.

Wade, G.N., and Feder, H.H.: Effects of several pregnane and pregnene steroids on estrous behavior in ovariectomized estrogen-primed guinea pigs. *Physiol Behav* 9:773-775, 1972.

Wallis, C.J., and Luttge, W.G.: Maintenance of male sexual behavior by combined treatment with oestrogen and dihydrotestosterone in CD-1 mice. *J Endocrinol* 66:257-262, 1975.

Wallis, C.J., and Printz, M.P.: Adrenal regulation of regional brain angiotensinogen content. *Endocrinology* 106:337-342, 1980.

Wansley, R.A., and Holloway, F.A.: Multiple retention deficits

642 Rees and Gray

following one-trial appetitive training. *Behav Biol 14*:135–149, 1975.

Warembourg, M.: Etude radiographique des retroactions centrales des corticosteroides ³H chez le rat et le cobaye. In *Neuroendocrinologie de l'Axe Coricotrope, Brain-Adrenal Interactions*. INSERM, September, 11, 12, 1973. Vol. 22, pp 41–66.

Warembourg, M.: Radioautographic study of the rat brain and pituitary after injection of ³H-dexamethasone. *Cell Tissue Res 161*:183–194, 1975.

Weisinger, R.S., Denton, D.A., McKinley, M.J., and Nelson, J.F.: ACTH induced sodium appetite in the rat. *Pharmacol Biochem Behav 8*:339–342, 1978.

Weiss, J.M., McEwen, B.S., Silva, M.T., and Kalkut, M.: Pituitary-adrenal alterations and fear responding. *Am J Physiol 218*:864–868, 1970.

Wendlandt, S., and File, S.E.: Behavioral effects of lesions of the locus coeruleus noradrenaline system combined with adrenalectomy. *Behav Neural Biol 26*:189–201, 1979.

Wertheim, G.A., Conner, R.L., and Levine, S.: Adrenocortical influences on free-operant avoidance behavior. *J Exp Anal Behav 10*:555–563, 1967.

Westphal, U.: Binding of corticosteroids by plasma proteins. In *Handbook of Physiology, Section 7: Endocrinology*, Vol. VI, R.O. Greep and E.B. Astwood, eds. American Physiological Society, Washington, D.C., 1975, pp 117–125.

de Wied, D.: Opposite effects of ACTH and glucocorticoids on extinction of conditioned avoidance behavior. In *Proceedings of the Second International Congress Hormonal Steroids, Milan, May 1966, International Congress Series 132*: L. Martini, F. Fraschini, and M. Motta, eds. Excerpta Medica Foundation, amsterdam, 1967, pp 945–951.

de Wied, D.: Pituitary-adrenal system hormones and behavior. In *The Neurosciences: Third Study Program*. F.O. Schmitt and F.G. Worden, eds. MIT Press, Cambridge, Mass, 1974, pp 653–666.

van Wimersma Greidanus, Tj.B.: Effects of steroids on extinction of an avoidance response in rats: A structure-activity relationship study. *Prog Brain Res 32*:185–191, 1970.

van Wimersma Greidanus, Tj.B., and de Wied, D.: Effects of intracerebral implantation of corticosteroids on extinction of an avoidance response in rats. *Physiol Behav 4*:365–370, 1969.

Williams, G.H., and Dluhy, R.G.: Aldosterone biosynthesis: Interrelationship of regulatory factors. *Am J Med 53*:595–605, 1972.

Woodbury, D.M.: Biochemical effects of adrenocortical steroids on the central nervous system. In *Handbook of Neurochemistry*, Vol. VII, A. Lajtha, ed. Plenum, New York, 1972, pp 225–287.

Wrange, O.: A comparison of the glucocorticoid receptor in cytosol from rat liver and hippocampus. *Biochim Biophys Acta* *582*:346-357, 1979.

Yukimura, Y., Bray, G.A., and Wolfsen, A.R.: Some effects of adrenalectomy in the fatty rat. *Endocrinology* *103*:1924-1928, 1978.

von Zerssen, D.: Mood and behavioral changes under cortico-steroid therapy. In *Psychotropic Action of Hormones,* T.M. Itil, G. Laudahn, and W.M. Herrmann, eds. Spectrum Publications, 1976, pp 195-222.

15

Cerebral Effects of Gonadal Steroid Hormones

WILLIAM G. LUTTGE

The concept that steroid hormones secreted by the testis, ovary, corpus luteum, and adrenal cortex can have direct actions upon neurons in specific regions of the brain and spinal cord, which in turn precipitate certain specific behavioral or physiological responses, is accepted universally today. The first direct demonstration of the endocrine involvement in behavior is credited to Berthold (1849), who reported observations on the reversible inhibition of sexual, agonistic, and crowing behaviors following castration and subsequent testis transplantation in roosters. Although it generally was assumed by early researchers that the behavioral actions of gonadal hormones were attributable to their biochemical actions in the brain (Beach, 1948), it was the provocative demonstration that intracerebral administration of progestins, androgens, and estrogens can facilitate the display of reproductive behavior (Fisher, 1956; Kent and Liberman, 1949; Lisk, 1962) that focused behavioral endocrinologists' attention clearly on the potential mechanisms of hormone action in the brain. Early neuroendocrine studies (Moore and Price 1932; Pfeiffer, 1936) strongly implied a direct role for gonadal steroids in gonadotropin regulation in adults and in the sexual differentiation of that feedback regulation during the perinatal period. However, these and other contemporary studies ascribed the primary site of this hormone action to the pituitary rather than to the brain. It was not until the pioneering work of Harris (1955) and others that researchers became enamored with the concept that steroid hormones may have direct intracerebral effects, other than on behavior, that are critically relevant to the regulation of adenohypophysial hormone secretion.

The present chapter reviews the literature up to July, 1980, on the molecular actions of gonadal steroid hormones in the mammalian brain. The discussion also includes brief summaries of the potential relationships of the molecular mechanisms to current theories of the role of gonadal steroid hormones in the regulation of reproductive behavior in small laboratory mammals, such as rat, mouse, hamster, and guinea pig. No attempt is made to review this latter material in detail. This chapter also excludes detailed discussions of the involvement of gonadal steroids in the regulation of other behaviors or the secretion of adenohypophysial hormones, since these subjects are reviewed elsewhere in this book (see chapter 4 by Frohman and Berelowitz; chapter 5 by Renaud; chapter 17 by Simpkins and Estes) and in books by Beyer (1979), and Adler and Pfaff (1983), and Svare (1983).

OVERVIEW OF STEROID MECHANISMS OF ACTION

Research on the molecular mechanisms of gonadal hormone actions in the brain has clearly been influenced by the much longer history of the actions of these steroids in peripheral tissues. Since one of the primary roles of gonadal hormones in these tissues is the maintenance of cell mass, that is, hyperplasia and hypertrophy, and such secretory activity as the estrogen-progestin controlled endometrial cycle of the uterus, it is not surprising that the genomic actions of steroids have received so much attention. These mechanisms (Fig. 15-1) require that steroids physically enter their target cells in order to initiate the required sequences of biochemical reactions. This uptake process is not limited to target cells, since "nonspecific" steroid uptake by most animal cells is a common observation. Neuronal uptake of steroid hormones is modulated primarily by the relative solubility of free steroids in lipid-rich membranes, the relative availability of vascular and cerebrospinal fluid (CSF) sources, and by the extra-intracellular steroid diffusion gradients. Recent evidence has suggested that carrier-mediated transport may also be important in some steroid target cell systems (Allera et al., 1980; Baulieu et al., 1978.) Steroids bound to albumin can readily be disassociated, thus facilitating their transport across the blood-brain barrier and neuronal plasma membranes, whereas steroids bound to a variety of globular transport proteins, for example, testosterone-binding globulin, may be effectively excluded from the brain (Pardridge and Mietus, 1979). This exclusion can at times afford the brain a degree of protection from the undesirable actions of certain endogenous or exogenous hormones. For example, in neonatal female rats the blood-borne estrogen-binding α-fetoprotein is thought to be essential for preventing the sexual differentiating actions of the female's own adrenal estrogens (McEwen et al., 1976; Weisz and Gunsalus, 1973).

Once in the neuronal cytoplasm, the hormones may undergo specific metabolic transformations (Figs. 15-3 and 15-4), which either increase or decrease the efficacy of their molecular actions and which may provide a mechanism for further regulating the various possible actions of steroids. The original steroid, or one of its metabolites, may then form a noncovalent high-affinity ($K_d \simeq 0.1$ to 1.0 nM) attachment to specific receptor sites on soluble protein-rich cytoplasmic macromolecules. Although cellular uptake may not provide target cell or steroid structure specificity, the cytoplasmic steroid receptors can provide both qualitative and quantitative specificity, that is differential binding affinities and capacities for different steroids in different neuronal populations. The transient hormone-receptor complexes can freely diffuse through pores in the nuclear membrane and undergo an, as yet poorly understood, "activation" process that facilitates the binding of the complex to its nuclear "acceptors." The interaction of the hormone-receptor complex with this second set of protein-rich macromolecules in turn facilitates the synthesis of DNA preceding hormone-induced karyokinesis (such as cell division during perinatal neural sexual differentiation) or the synthesis of various species of RNA, including those that code for specific proteins required for the maintenance of stimulatory actions in target tissues.

The steroid or its receptor may then be catabolized, conjugated, or in some other way modified in the nucleoplasm. Alternatively, the steroid receptor complex may also be returned intact or dissociated to the cytoplasm wherein it can be similarly modified or reutilized in another receptor-binding cycle. Gonadal steroid receptor reutilization in the brain may require a "capacitation" step prior to hormone binding similar to that described for glucocorticoid receptors in non-neural target tissues (Sando et al., 1979a,b). Cytoplasmic receptor depletion (by translocation to the nucleus) and replenishment (by translocation from the nucleus or *de novo* synthesis in the cytoplasm), following this cyclic sequence of gonadal hormone action, has been demonstrated in numerous studies with brain tissue (Cidlowski and Muldoon, 1976; Jungblut et al., 1979; Morris, 1976; White et al., 1978).

Although the genomic mechanisms of steroid hormone action are by far the most widely studied, there are viable alternatives that may be important in the brain. The most obvious of these

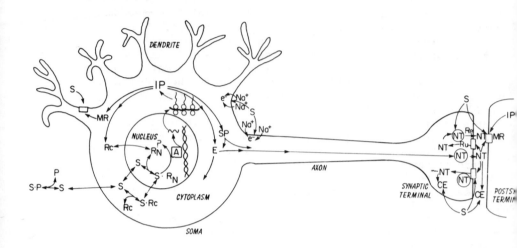

Fig. 15-1. Diagrammatic representation of various possible
molecular and electrophysiological actions of
gonadal steroid hormones in CNS neurons. In the
extracellular space and blood the steroid exists
in both the free (S) and protein-bound (S-P)
states. The free steroid enters the neuron
either by diffusion or carrier-mediated
transport after which it may bind to a cytosolic
receptor macromolecule (Rc) and diffuse as a
steroid-receptor complex (S-Rc) to the nucleoplasm or it may directly enter the nucleoplasm
and then bind to a nuclear receptor (Rn). In
either case, the intranuclear steroid-receptor
complex (S-Rn) may then undergo a poorly understood activation process that facilitates its
interaction with basic nonhistone acceptor proteins (A) in the chromatin, which in turn may
precipitate the production of specific mRNA's
that code for the production of specific "induced proteins" (IP). Examples of these IP's
include the 39-45K Dalton E_2IP, structural proteins (SP), such as tubulin, membrane receptor
proteins (MR), such as MR's for DA, β-adrenergic
agents, and catechol estrogens), cytoplasmic re-

either by diffusion or carrier-mediated transport after which it may bind to a cytosolic receptor macromolecule (Rc) and diffuse as a steroid-receptor complex (S-Rc) to the nucleoplasm or it may directly enter the nucleoplasm and then bind to a nuclear receptor (Rn). In either case, the intranuclear steroid-receptor complex (S-Rn) may then undergo a poorly understood activation process that facilitates its interaction with basic nonhistone acceptor proteins (A) in the chromatin, which in turn may precipitate the production of specific mRNA's that code for the production of specific "induced proteins" (IP). Examples of these IP's include the 39-45K Dalton E_2IP, structural proteins (SP), such as tubulin, membrane receptor proteins (MR), such as MR's for DA, β-adrenergic agents, and catechol estrogens), cytoplasmic receptor proteins for steroids (Rc), such as E_2 and P_4 receptors, and various anabolic or catabolic enzymes (E), such as choline acetyltransferase and L-Cys-arylmidase. Release (Re) and reuptake (Ru) of neurotransmitters may also be modulated by the direct actions of steroids. Lastly steroids may interact directly with the neuronal plasma membrane to modulate the transmembrane diffusion of specific ions, for example, Na+, and thereby produce an electrophysiological response (e-). Further details and other examples of these various types of responses to gonadal steroid hormone actions can be found in the text.

---

is that steroids may interact with receptors on the neuronal membrane in much the same manner as peptidergic hormones and neurotransmitters. Although there is little direct evidence supporting the existence of membrane-associated steroid receptors in the brain, there is a wealth of evidence demonstrating that gonadal hormones can exert regionally, and structurally, specific rapid modifications of postsynaptic neuronal firing and presynaptic neurotransmitter/hormone release rates. Pre- and postsynaptic modulation of the efficacy of chemical transmission can also result from modifying the activity of neurotransmitter metabolism enzymes.

NEUROANATOMICAL DISTRIBUTION OF RECEPTORS

Steroid Autoradiography

The development of histological techniques for the high resolution autoradiographic localization of diffusible substances,
coupled with the ready availability of high specific activity
tritiated steroids, has facilitated detailed investigations of
the neuroanatomical distribution of gonadal steroid-retaining
cells in a wide variety of vertebrates, including fish, reptiles, amphibians, birds, and mammals (reviewed by Kim et al.,
1978; Morrell and Pfaff, 1978; Morrell et al., 1975; Stumpf and
Grant, 1975; Stumpf and Sar, 1978). As might be expected with
any analysis of such a wide range of animal species, clear examples of species or hormone specific differences in the regional pattern and density of labeled neurons do exist, but
these differences are minor in comparison with the abundant
interspecies similarities. Some of these differences appear to
be due to the occurrence of relatively unique groupings of neurons involved in specific hormone-mediated behaviors, whereas
others may be attributed to a less extensive development of
certain allocortical structures. For example, in the hyper-
striatum ventrale, pars caudale of the canary, there is a distinct grouping of androgen-concentrating neurons that other
evidence has indicated are intimately involved in androgen-
sensitive song production (reviewed by Kelly, 1978). Similarly, although estrogen-concentrating neurons are found in the
amygdala of mammals, reptiles, amphibians, and birds, they
appear to be absent in the pallium region of the less enceph-
alized brains of teleost fish (reviewed by Kim et al., 1978;
Morrell et al., 1975).

Thus following the systemic administration of [^{3}H]estra-diol
(E_2), [^{3}H]testosterone (T) or [^{3}H]dihydrotestosterone (DHT)
(see Fig. 15-4 for structural formulas and Table 15-1 for International Union for Pure and Applied Chemistry [IUPAC] names
and abbreviations) silver grains in the autoradiographs of vertebrates concentrate over neuronal nuclei primarily in certain
hypothalamic (medial basal and anterior), preoptic (medial),
septal (ventral lateral and bed nucleus of stria terminalis
[NIST]), and amygdaloid (medial and cortical) regions of the
brain or the homologous structures in nonmammalian species (see
Table 15-2 for neuroanatomical abbreviations and Table 15-3 for
autoradiography examples). These areas appear to belong to
phylogenetically older periventricular brain regions and they
further appear to be interconnected by distinct fiber tracts,
such as the stria terminalis in mammals and the occipitomes-
encephalic in birds (Martinez-Vargas et al., 1976; Pfaff and

Keiner, 1973; Stumpf, 1970). Gonadal steroid accumulating cells have also been found in lower brainstem regions and in the spinal cord. For example, although both androgen- and estrogen-labeled neurons can be found in the midbrain central gray of rats, estrogen-concentrating cells predominate in certain sensory cranial nerves and spinal cord regions, whereas androgen-concentrating cells predominate in certain somatomotor cranial nerves and spinal cord regions (Keefer et al., 1973; Stumpf and Sar, 1979). Similar gonadal steroid-concentrating cells have not been observed in primate spinal cord, but this may be due to the markedly reduced cell labeling seen in these larger animals (Keefer and Stumpf, 1975; Pfaff et al., 1976).

In contrast to the wealth of information on the neuroanatomical distribution of estrogen and androgen receptors, there have been comparatively few successful reports on progestin-labeled neurons. This is because early attempts to characterize progestin receptor neuroanatomical distribution patterns often used unprimed ovariectomized animals, which resulted in an unnecessarily low cell labeling density. Recent biochemical studies (*vida infra*) have revealed that progestin receptor concentrations are greatly magnified with estrogen priming and that noncovalent binding interaction between progestin receptors and the naturally occurring progestins, for example, progesterone (P_4) and dihydroprogesterone (5α-DHP) (see Table 15-1 and Fig. 15-3), is much more labile than similar interactions between estrogens and androgens and their receptors (*vide infra*). The latter problem has recently been, at least partially, circumvented through the use of an extremely potent progestin analogue, R5020. Thus, in recent studies with rats and guinea pigs using either $[^3H]P_4$ or $[^3H]R5020$, estrogensensitive, limited capacity, progestin-labeled neuronal nuclei were found to be localized primarily in the preoptic area (POA) and in the medial basal hypothalamus (MBH)(see Table 15-3).

Intracerebral Steroid Cannulation and Behavior

The relationship of the autoradiographic distribution patterns of hormone receptors to the potential sites of gonadal steroid actions in reproductive behavior has been most studied actively with stereotaxic implantation techniques. In addition to drug implantation, electrical stimulation, lesion and electrophysiological recording data (*vida infra*), considerable information has been derived from examining the behavioral effects of direct exposure of various brain regions to hormones and receptor-blocking antihormones using cannulas. Early studies with this technique in female rats often concluded that estrogen im-

Table 15-1. Gonadal Steroid Hormone Names and Abbreviations

IUPAC names	Abbreviations	Trivial names
5α-Androstan-3,17-dione	5α-AA	5α-Androstanedione
5α-Androstan-3α,17β-diol	3α-Adiol	3α-Androstanediol
5α-Androstan-3β,17β-diol	3β-Adiol	3β-Androstanediol
5α-Androstan-3α-ol-17-one	A	Androsterone
5α-Androstan-17β-ol-3-one	DHT	Dihydrotestosterone
5β-Androstan-3,17-dione	5β-AA	5β-Androstanedione
4-androsten-17β-ol-3-one	T	Testosterone
4-Androsten-17α-ol-3-one	Epi-T	Epitestosterone
4-Androsten-17β,19-diol-3-one	19-OH-T	19-Hydroxytestosterone
4-Androsten-17β-ol-3,19-dione	19-OXO-T	19-Oxotestosterone
4-Androsten-2β,17β-diol-3,19-dione	2-OH-19-OXO-T	2-Hydroxy-19-oxotestosterone
4-Androsten-3,17-dione	AE	Androstenedione
4-Androsten-19-ol-3,17-dione	19-OH-AE	19-Hydroxyandrostenedione
4-Androsten-3,17,19-trione	19-OXO-AE	19-Oxoandrostanedione
4-Androsten-2β-ol-3,17,19-trione	2-OH-19-OXO-AE	2-Hydroxy-19-oxoandrostenedione
5-Androsten-3β-ol-17-one	DHEA	Dehydroepiandrosterone
5-Androsten-3β,17β-diol	AEDiol	5-Androstenediol

Systematic name	Abbreviation	Trivial name
1,3,5(10)-Estratrien-3-ol-17-one	E_1	Estrone
1,3,5(10)-Estratrien-3,17α-diol	$17α\text{-}E_2$	17α-Estradiol
1,3,5(10)-Estratrien-3,17β-diol	E_2	Estradiol
1,3,5(10)-Estratrien-3,16α-17β-triol	E_3	Estriol
1,3,5(10)-Estratrien-2,3-diol-17-one	$2\text{-OH-}E_1$	2-Catechol Estrone
1,3,5(10)-Estratrien-3,4-diol-17-one	$4\text{-OH-}E_1$	4-Catechol Estrone
1,3,5(10)-Estratrien-2,3,17β-triol	$2\text{-OH-}E_2$	2-Catechol Estradiol
1,3,5(10)-Estratrien-3,4,17β-triol	$4\text{-OH-}E_2$	4-Catechol Estradiol
5α-Pregnan-3,20-dione	5α-DHP	5α-Dihydroprogesterone
5α-Pregnan-3α-ol-20-one	3α-OH-5α-DHP	3α-Dihydroxydihydroprogesterone
5α-Pregnan-3α,20α-diol	3α,20α-OH-5α-DHP·	3α,20α-Dihydroxydihydroprogesterone
5α-Pregnan-20α-ol-3-one	20α-OH-5α-DHP	20α-Hydroxydihydroprogesterone
5β-Preganan-3,20-dione	5β-DHP	5β-Dihydroprogesterone
4-Pregnen-3,20-dione	P_4	Progesterone
4-Pregnen-11β,17α,21-triol-3,20-dione	F	Cortisol
4-Pregnen-11β,21-diol-3,20-dione	B	Corticosterone
4-Pregnen-17α-ol-3,20-dione	$17α\text{-OH-}P_4$	17α-Hydroxyprogesterone
4-Pregnen-20α-ol-3-one	$20α\text{-OH-}P_4$	20α-Hydroxyprogesterone
5-Pregnen-3β-ol-20-one	Pg	Pregnenolone
5-Pregnen-3β,17α-diol-20-one	17α-OH-Pg	17α-Hydroxypregnenolone
19-nor-4-9-Pregnadien-17,21-dimethyl-3,20-dione	R5020	

Table 15-2. Abbreviations for Neuroanatomical Structures

Abbreviation	Complete name	Abbreviation	Complete name
AH	Anterior hypothalamus	Mes	mesencephalon
Amg	Amygdala	MH	middle hypothalamus
Arc	Arcuate nucleus (hypothalamus)	MPOA	medial preoptic area
CAmg	Cortical amygdaloid nucleus	MVM	medial ventromedial nucleus (hypothalamus)
CG	Central gray	NA	Nucleus accumbens
CMAmg	Corticomedial amygdaloid nucleus	NISM	Bed nucleus of the stria medullaris
CN	Caudate nucleus	NIST	Bed nucleus of the stria terminalis
CX	Cerebral cortex	OB	Olfactory bulb
DB	Diagonal band nucleus	PAVN	Paraventricular nucleus (hypothalamus)
DHpc	Dorsal hippocampus	PeVN	Periventricular nucleus (hypothalamus)
DM	Dorsomedial nucleus (hypothalamus)	PH	Posterior hypothalamus
FCx	Frontal cortex	PM	Premamillary nucleus (hypothalamus)
HTh	Hypothalamus	POA	Preoptic area
LH	Lateral hypothalamus	SCN	Suprachiasmatic nucleus (hypothalamus)
LPOA	Lateral preoptic area	SM	Supramamillary nucleus (hypothalamus)
LS	Lateral septal nucleus	SN	Substantia nigra
LVM	Lateral ventromedial nucleus (hypothalamus)	SON	Supraoptic nucleus (hypothalamus)
MAmg	Medial amygdaloid nucleus	Str	Striatum
MBH	Medial basal hypothalamus	VM	Ventromedial nucleus (hypothalamus)
ME	Median eminence	VS	Ventral subiculum
MFB	Medial forebrain bundle	VTA	Ventral tegmental area

Steroid:	Hypothalamus							Preoptic Area				Septal Area			Amygdala	Reference
	PeVN	VM	Arc	PM	AH	LH	PH	MPOA	POA-SCN	POA-PeVN	LS	NIST	DB	CAmg	MAmg	
[³H]E₂	++	++	++	++	++	+	+	++	++	++	+	++	+	++	++	Pfaff and Keiner (1973)
[³H]DHT	++	+	+	++[b]	+[c]	+[c]	+[b]	++	+	+	+[b]	++	+	+	++	Sar and Stumpf (1977a,b)
[α-³H]T	++	++	++	++	++	+	+	++	++	++	+	++	+	+	++	Sheridan (1979)
[β-³H]T	-	-	-	+[d]	+[e]	-	-	-	-	-	-	++	-	-	+	Sheridan (1979)
[³H]R5020	++	++	++	++	+	-	-	+	++	++	-	++	+	-	-	Warembourg (1978)

[a] See Tables 15-1 and 15-2 for steroid and neuroanatomical abbreviation definitions. The dramatically different autoradiographic patterns derived with [α-³H]T and [β-³H]T occurs because the β-³H is lost during the aromatization of T to E₂. Thus, the autoradiographic pattern with [β-³H]T should be more representative of true androgen binding (similar to [³H]DHT) than [α-³H]T, which is confounded with the aromatization-produced [³H]E₂. Note, however, that [³H]DHT and [β-³H]T labeling patterns are very different in several nuclei, such as PeVN, VM, Arc, POA-SCN, POA-PeVN, DB, and CMAmg, whereas in other nuclei the reduction in the [³H]DHT labeling with E₂ pretreatment, such as PM, MPOA, and LS, suggests that some of the binding may be due to association with E₂ receptors.

[b] Labeling reduced with E₂.

[c] Dispersed cells.

[d] Labeled cells in only three out of eight rats.

[e] Labeled cells only in dorsolateral AH.

plants in many brain regions effectively facilitates the induction of sexual receptivity (testing usually performed four to six hours after the systemic administration of P_4). This apparent lack of anatomical specificity is now thought to be due to diffusion artifacts (for review see Davidson, 1972). Recent work has shown that the ventromedial hypothalamus (VM) and, to a lesser extent, the medial POA (MPOA) are the most sensitive sites of estrogenic action in female-typical reproductive behavior (Davis and Barfield, 1979b). The autoradiographic data indicate that both of these regions are densely populated with E_2-concentrating neurons (see Table 15-3).

In a study designed to test the necessary versus sufficient sites for estrogen action, induction of receptivity in ovariectomized rats following systemic E_2-benzoate (E_2B) injections was blocked by implanting cannulas containing the potent receptor blocking antiestrogen, CN-69,725-27, into the MPOA, but not by similar cannulas placed in the VM or ventral tegmental area (VTA) (Luttge, 1976). This lack of concordance between the effects of estrogen and antiestrogen implants in the VM and MPOA is just one example of the data (electrical stimulation, lesion and recording findings to be discussed later) that have led to the theory that the MPOA exerts a tonic, but estrogen-reversible, inhibitory influence on receptivity, whereas the VM exerts an estrogen-dependent facilitatory influence. The failure of antiestrogen implants in the VM to inhibit receptivity may relate to the recent observation with $[^3H]E_2$ implants, that only 4 percent of the VM E_2 receptors need to be occupied in order to facilitate the induction of receptivity (Davis et al., 1979). Since the MPOA is less sensitive than the VM to the facilitatory actions of estrogen on receptivity, it is entirely possible that antiestrogen implants were effective in the former, but not in the latter region because insufficient antiestrogen was released to inhibit the requisite 96 percent of VM estrogen receptor sites. This hypothesis is supported by the observation that when cholesterol was not used to dilute the antiestrogen, implants placed in the VM were effective in reversibly inhibiting the induction of receptivity (Luttge, 1976).

In rats, mice, hamster, and guinea pigs, the induction of sexual receptivity during the ovarian cycle requires the synergistic facilitatory actions of endogenous progestins. Studies with gonadectomized females have shown that progestins can often both facilitate and inhibit the induction of receptivity with exogenous estrogen. For example, in female rats primed 48 to 72 hours earlier with systemic injections of E_2B, a single injection of P_4 increases both the qualitative and quantitative assessments of receptivity four to six hours later. However, if a subsequent injection of P_4 is given 24 hours later (at a time when receptivity levels have returned to oil control levels), a second increase in receptivity is not seen even though high levels of receptivity can be achieved in control females treated with oil instead of P_4 on the previous day (Nadler, 1970). The P_4-induced facilitation of receptivity normally seen during the first test can also be inhibited if a comparatively large dose of P_4 is given at the same time as the original injection of E_2B (Blaustein and Wade, 1977). The former example is now known as "sequential" inhibition, and the latter is referred to as "concurrent" inhibition.

Intracerebral cannulation studies in female rats have implicated the VTA as one of the principal sites for the facilitatory actions of P_4 (Luttge and Hughes, 1976), whereas the MBH has been implicated as a site for the sequential and concurrent inhibitory actions of P_4 (Marrone et al., 1979). Facilitatory and concurrent inhibitory actions of P_4 have also been demonstrated with cannulas placed in the caudate-putamen region (Yanase and Gorski, 1976a,b). Dense concentrations of estrogen-dependent P_4 nuclear receptors are found in the MBH (see Table 15-3), but little evidence for similar receptors has been reported for the VTA, even though this region does accumulate considerable concentrations of the hormone (Whalen and Luttge, 1971a). Antiestrogen implantation studies, however, have suggested that P_4 actions in the VTA do require estrogen-priming (Yanase and Gorski, 1976a). Cannulation studies in mice and guinea pigs have further implicated the VTA and MBH as the primary sites of P_4 action in female sexual behavior (Hall and Luttge, 1974; Morin and Feder, 1974a,b).

The induction of male sexual behavior in castrated rats can also be facilitated with gonadal hormones implanted into the MPOA and VM; however, in contrast to the results obtained for female sexual behavior, the MPOA appears to be more sensitive than the VM (and posterior hypothalamus) to androgen (T or TP) and estrogen (E_2 or E_2B) induced male reproductive behavior (Christensen and Clemens, 1974; Davis and Barfield, 1979a). As already described (see Table 15-3), both of these regions contain high densities of androgen- and estrogen-concentrating neurons. Electrical stimulation and lesion data (*vida infra*) have further suggested that, in contrast to female sexual behavior, the MPOA exerts a preferential facilitation of male sexual behavior, whereas the VM may act to inhibit male copulation.

SUBCELLULAR LOCALIZATION, PHYSIOCHEMICAL PROPERTIES, AND MOLECULAR/PHYSIOLOGICAL REGULATION OF RECEPTORS AND ACCEPTORS

Estrogens
Receptor Physico- and Biochemical Properties

In early studies (Eisenfeld, 1970; Kahwanago et al., 1970), [^{3}H]E_2 was incubated at 4°C for several hours with buffered cytosol, that is 100,000 x g supernatant of brain homogenates. A peak of radioactivity was excluded from gel chromatography columns and sedimented in the 7S range using either no- or low-salt sucrose gradients. Treatment of cytosol with sulfhydryl blocking agents or with proteolytic enzymes, but not with RNase or DNase, dramatically reduced its binding capacity, suggesting that estrogen receptors were large proteins. The highest concentrations of these receptors were found in the hypothalamus, POA, and amygdala with much lower amounts in the cerebral and cerebellar cortices. The regional distribution of estrogen receptors in brain cytosol thus closely parallels that determined autoradiographically (see Table 15-3). The cytosolic receptors were shown to be saturable and apparently specific for estrogens. For example, unlike E_2, estrone (E_1), diethylstilbestrol (DES), and other natural and synthetic estrogens and antiestrogens, various androgens (such as T, DHT), progestins (such as P_4, 5α-DHP) and glucocorticoids (such as corticosterone [B], cortisol [F]), all failed to reduce [^{3}H]E_2 binding during competition studies.

It now appears that at physiological salt concentrations the estrogen receptor found in brain and uterine cytosol sediments as a single 4S peak, rather than as the 8S aggregate suggested in earlier work done at hypophysiological salt concentrations (Fox and Johnston, 1974). Numerous studies with uterine material have shown that the 4S estrogen receptor complex does not readily associate with nuclear acceptors, such as DNA, partially purified chromatin, oligo(dT)-cellulose, and that in order for this association to take place the 4S monomer must be "activated." This activation process involves combining of the 4S estrogen receptor complex (about 80K Daltons: S-R in Fig. 15-2) with a second subunit (about 50K Daltons: component X in Fig. 15-2) to form the 5S estrogen receptor complex (about 130K Daltons: S-R-X in Fig. 15-2), which is typically isolated from estrogen target tissue nuclei and which readily associates with DNA, chromatin, and other components of the nucleus (Weichman and Notides, 1979).

The 4S and 5S estrogen receptor complexes from hypothalamus, uterus, and other target tissues have similar sedimentation coefficients, similar estrogen binding affinities, and similar antigenic properties (Fox, 1977; Greene et al., 1979). Antibodies generated in rabbit or goat to the calf uterine estrogen receptor cross react with nuclear and cytosolic estrogen receptors obtained from a wide range of species (Radanyi et al., 1979). It is now possible, however, to select rat or mouse myeloma hybridoma clones that produce antibodies demonstrating species specificity or nuclear versus cytosolic estrogen receptor preference (Greene et al., 1980a,b).

The 4S to 5S transformation of estrogen receptor complexes found in brain, pituitary, and uterine samples is facilitated by DNA (Fox, 1977; Yamamoto, 1974). It occurs in nucleoplasm as well as in cytosol (Jungblut et al., 1978; Linkie and Siiteri et al., 1978) (see Fig. 15-2), at different rates in different tissues (Fox, 1977; Linkie, 1977), and is both temperature and time dependent (Fox, 1977; Gorski and Gannon, 1976). Kinetic studies suggest that the activation ($4S \rightarrow 5S$) process produces a conformational change in the 4S portion of the complex, which in turn increases the binding affinity of the active region of the receptor for the E_2 ligand (Weichman and Notides, 1979). The rate of estrogen dissociation from the activated receptor complex, and from the nucleus, closely correlates with the hormone's biological activity (Jungblut et al., 1979; Weichman and Notides, 1980). Thus, although the rate constants for estrogen association with the 4S receptor and for 4S to 5S transformation were similar for E_1, E_2, and estriol (E_3), the dissociation rate constants between these hormones were clearly different (Weichman and Notides, 1980). There is a similar dose-response relationship between the biological effectiveness of various estrogens in the uterus with those seen in the brain.

Receptor Utilization and Replenishment

The possibility of hormone-induced fluctuations in estrogen receptors during the estrous cycle of rodents has been a major topic of interest during the last decade. Early *in vivo* and *in vitro* studies with hypothalamic, pituitary, and uterine systems revealed that the total number of available, that is unbound, binding sites fluctuate during the estrous cycle in inverse proportions with endogenous concentrations of circulating estrogens, that is receptor availability was maximal during diestrus and metestrus and minimal during proestrus and estrus (Ginsberg et al., 1975; Luttge 1972). In more recent studies utilizing exchange assays to determine the total binding capacity, that is, unbound plus bound receptors, the content of nuclear receptors for E_2 was shown to fluctuate in concert with the known cyclic variations in plasma estrogen titers, that is

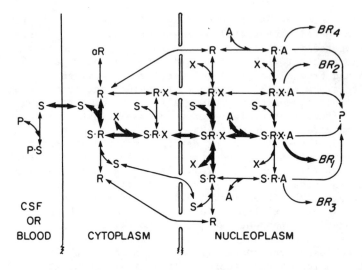

Fig. 15-2. A diagram of the various possible steroid-receptor-
acceptor-effector interactions. In the CSF and
blood the steroid may exist in either the free or
protein-bound state (S or P•S). Only the free
steroid is transported across the plasma membrane
(diffusion or carrier-mediated transport). In the
cytoplasm or nucleoplasm it may form a high-affinity noncovalent bond with a receptor macromolecule
(R). The unbound receptor may exist either in the
cytoplasm or in the nucleoplasm. It may also exist
in an apo-form (aR), requiring "capacitation" prior
to its interaction with the steroid. The steroid-
receptor complex (S•R) may then undergo an "activation" process, involving an interaction with a
second macromolecule (X) in the cytoplasm or nucleoplasm (to form S•R•X) before interacting with
the intranuclear acceptor macromolecules (A). This
final interaction may in turn lead to a biological
response (BR$_1$) typified by the increased production
of specific mRNA's and their associated proteins
(see Fig. 15-1). Recent evidence also suggests the
possibility that other combinations of steroid, receptor, and acceptor molecules may be able to
elicit biological responses (BR$_2$, BR$_3$, and BR$_4$),
which may or may not be similar to those elicited
by the traditional steroid-receptor-acceptor complex (note that this latter pathway is highlighted
with bold arrows). Following, or as a necessary
consequence of, the elicitation of the biological

hypothalamic and uterine nuclear estrogen receptor complex concentrations were maximal during proestrus as were endogenous estrogen blood titers, whereas the opposite was true during diestrus (White et al., 1978). In these same studies, the uterine cytosol estrogen receptor concentrations did not vary during the estrous cycle, but the content of hypothalamic cytosol estrogen receptors showed marked fluctuations. Receptor numbers were at a minimum during proestrus and at a maximum during diestrus. Therefore in the uterus, the translocation of estrogen receptor complexes into the nucleus did not appreciably deplete the total cytosolic estrogen receptor pool, whereas in the hypothalamus there was such a depletion. These findings suggest that in the uterus there is a replenishment or synthesis system that homeostatically maintains cytosol estrogen receptor concentration, whereas in the hypothalamus, cytosol receptor depletion exceeds the rate of replenishment. This situation contrasts sharply with recent findings in the hippocampus, where cytosol glucocorticoid receptor concentration was not diminished by pretreatment with high doses of glucocorticoids (see Turner and McEwen, 1980; also Rees and Gray, chapter 14).

The cytosolic or nuclear estrogen receptor depletion or replenishment cycle in the hypothalamus has also been studied in experiments with ovariectomized rats subjected to exogenous estrogen challenge (Cidlowsky and Muldoon, 1974, 1976). These studies have revealed that as the dose of E_2 increases from 1 to 10 μg/animal, the magnitude of cytosolic estrogen receptor depletion also increases. This depletion typically reaches its maximum by one hour following estrogen, whereas the replenishment requires 15 hours or more. The protein synthesis inhibitor, cycloheximide, does not effect receptor depletion, but it does have a marked dose-dependent effect on receptor replenishment. As the dose of estrogen increases, the relative proportion of receptor replenishment *not affected* by the cycloheximide administration also increases. These results, together with those from other laboratories (Jungblut et al., 1979; Little et al., 1975), suggest that the bulk of cytosolic receptor replenishment in the hypothalamus and uterus that takes place as a result of interactions with endogenous estrogens, is due to the de novo synthesis of new receptors, or enzymes

---

response, the steroid-receptor-acceptor complex may undergo a poorly understood "processing" reaction (?). It is also possible that the complex may not be involved in all stages of the genomic response including processing, and thus the steroid, receptor or acceptor molecules may be avialable for recycling.

required for the capacitation of aporeceptors, rather than to
the recycling of old receptors from the nucleoplasm back into
the cytoplasm. This conclusion is also consistent with the
intranuclear degradation or "processing" of estrogen receptors
recently demonstrated in brain (White and Lim, 1978), uterine
(Jungblut et al., 1979), and breast cancer cell studies (Horwitz and McGuire, 1978a). The processing of nuclear receptors
appears to be required for the genomic actions of estrogen
(Horwitz and McGuire, 1978b), and it may serve as a stimulant
for the de novo synthesis of new cytoplasmic estrogen receptors
(Little et al., 1975).

Several laboratories have suggested that newly synthesized
estrogen receptors are first detectable in the microsomal fraction of target cells. These microsomal estrogen receptors appear to have the same binding affinity and steroid-specificity
as those found in cytosol. Upon solubilization, the microsomal
receptors form dimers and higher order aggregates similar to
those seen in studies with the cytosol estrogen receptors. De
novo estrogen receptor synthesis has been hypothesized to proceed via the following pathway: 3.5S "basic" microsomal receptor → 3.5S "acidic" microsomal receptor → 4S "acidic" cytosol
receptor → 5S "acidic" cytosol and nuclear receptor (Jungblut
et al., 1978; Little et al., 1975). Since estrogen receptors
have already been detected in the microsomal fraction from several other estrogen target tissues, including anterior pituitary (Watson and Muldoon, 1977), and microsomal high-affinity
receptors have been reported for other gonadal hormones, such
as P_4 and DHT (Little et al., 1973), microsomal receptors also
may play a role in estrogen receptor synthesis in the brain.

Unoccupied Nuclear Receptors

Isolation of target cell nuclei from ovariectomized and ovari-
ectomized-adrenalectomized animals, has revealed the existence
of unoccupied estrogen receptors within the nucleus (Jungblut
et al., 1979; Levy et al., 1980; Zava and McGuire, 1977). It
now appears possible in some target cell systems, most notably
those with high nucleoplasm to cytoplasm ratios, such as many
neurons, that a majority of the unbound estrogen receptors may
be located in the nucleoplasm rather than the cytoplasm. This
intranuclear accumulation of hydrophilic unoccupied estrogen
receptors may be the result of their relative exclusion from
the cytoplasm, where unbound water space is comparatively limited, to the nucleoplasm, where unoccupied water space is considered to be more abundant (Horowitz and Moore, 1974).

At present, the biological significance of unoccupied nuclear
estrogen receptors is unclear. The fact that some of these re-

ceptors can associate with DNA and chromatin (Fox, 1977) suggests, but certainly does not prove, that unbound 4S and 5S nuclear estrogen receptors may cause genomic responses similar to
those documented for bound 5S nuclear estrogen receptor (Lippmann et al., 1976; Little et al., 1975; Zava and McGuire, 1977)
(see Fig. 15-2). Estrogen binding and estrogen receptor activation may function primarily to increase the avidity of the
receptor for its nuclear acceptors. This concept is consistent
with that described earlier for uterus, wherein the biological
potency of E_1, E_2, and E_3 correlated better with their relative
abilities to stabilize the 5S estrogen receptor complex than it
did with their relative rates of association and dissociation
with the 4S receptor or with the rates of activation of the 4S
estrogen receptor complex (Weichman and Notides, 1980).

Nuclear Acceptors

In vitro experiments have shown that the 5S and, to a lesser
extent, the 4S estrogen receptor complex from hypothalamus
binds to DNA- and oligo(dT)-cellulose (Fox, 1977; Fox and
Johnston, 1974). *In vivo* experiments have shown that this complex also binds to chromatin isolated from hypothalamic, but
not cerebral cortical samples (Perry and Lopez, 1978; Whalen
and Olsen, 1978). Work on other steroid target tissue systems
strongly suggests that the intracerebral estrogen receptor complex must interact with both DNA and chromatin in order to decrease or increase selectively the synthesis of mRNAs coding
for specific proteins. Several groups have argued that this is
achieved by interaction of different regions or subunits of the
receptors, for example, mero-receptors, with each of the various ligands and acceptors (Jungblut et al., 1979; Sherman et
al., 1978).
    Following the early observations that not all E_2 receptors
can be extracted with hypertonic potassium chloride (KCl) from
uterine cell nuclei, several investigators have suggested that
the "salt-resistant" fraction represents a specific class of
nuclear estrogen receptors tightly bound to acceptors and are
essential for the physiological response to estrogens (Clark
and Peck, 1976). Although there is evidence supporting the
distinction between salt-resistant and salt-extractable estrogen-receptor-acceptor nuclear complexes, several recent studies
have argued strongly that this dichotomy has been over emphasized (Muller et al., 1977; Traish et al., 1977). For example,
even though salt-resistant complexes can be readily demonstrated in hypothalamic and uterine nuclei with *in vitro* [^{3}H]E_2 exchange assays, it is difficult to demonstrate these complexes
with *in vivo* [^{3}H]E_2 injections (Barrack and Coffey, 1979;

Barrack et al., 1977). Furthermore, by slightly modifying the standard salt extraction procedures, virtually all of the nuclear estrogen receptor complexes found in hypothalamic and uterine samples can be extracted with 0.4-0.6 M KCl (Roy and McEwen, 1979; Traish et al., 1977). Lower concentrations of salt yield lower recoveries. Hence salt extractability is a relative and not an absolute classification. The observation that steroid nuclear receptor (and acceptor?) dissociation rates vary inversely with the salt concentrations required for receptor extraction (Roy et al., 1979) could indicate that it is the unbound rather than the bound receptor that is salt extractable. This seems unlikely because it fails to explain how the unbound receptor reacquires its (radiolabeled) ligand in the extraction buffer. Alternatively, the differential dissociation rates of various ligands could reflect a differential ability of the antagonists and agonists to induce conformational changes in the estrogen receptor which alter its binding affinity to the intranuclear acceptors. This possibility is consistent with the widespread observation that steroid receptor lability is markedly reduced after steroid binding.

The saturability and specificity of the intranuclear acceptors is still under considerable debate, even in non-neural estrogen target tissues. *In vivo* it is possible to demonstrate a clear dose-dependent increase in nuclear steroid binding, which levels off at high doses of estrogen. This does not necessarily indicate that nuclear acceptor capacity is limited, the limiting factor could be the capacity of the cytosolic receptors. Support for this interpretation is provided by the observation that when uterine nuclei, previously subjected to saturating levels of E_2 *in vivo*, are subsequently incubated with fresh $[^3H]E_2$-labeled cytosol, there is no apparent reduction in the nuclear uptake of the $[^3H]E_2$ receptor complexes (Shepherd et al., 1974). Thus, with crude nuclear preparations there is little evidence for acceptor saturability and, as other data have suggested, tissue or steroid specificity. With more highly purified acceptor preparations, however, the situation changes markedly.

Using nonhistone nuclear proteins coupled to Sepharose, several laboratories have used affinity chromatography to characterize the acceptors (Mainwaring et al., 1976; Puca et al., 1975). Most of this work suggests that the nuclear acceptors for estrogen, progestin, androgen, and glucocorticoid receptors are basic (pI $\simeq$ 8.6-9.3), nonhistone proteins (70-85K Daltons), with high affinity (K_d = 0.2 nM for E_2), saturable and, to varying extents, tissue- and hormone-specific binding for steroid-receptor complexes. For example, uterine nonhistone basic proteins coupled to Sepharose retained nearly four times more $[^3H]E_2$-labeled uterine receptor complexes than were similarly

prepared acceptor proteins extracted from rat liver or prostate. Uterine receptor-acceptor-Sepharose binding was greatly
improved if the receptors were first exposed to E_2, but not P_4,
T, or F. Addition of native or heat-denatured DNA reduces the
total binding capacity, but not the affinity of the acceptor-
Sepharose columns for uterine E_2 receptors. This undoubtedly
reflects the strong binding affinity of the nonhistone basic
protein acceptors for DNA. Using DNA-Sepharose it is possible
to isolate estrogen receptor-acceptor-DNA-Sepharose complexes
by binding sequentially the acceptors to the DNA followed by
the receptors to the DNA-bound acceptors.

In these same studies, saturation of Sepharose-coupled acceptors was demonstrated at total protein concentrations 100 to
1000 times lower than those found in uterine cytosol (Puca et
al., 1975). This finding obviously vitiates the argument that
acceptor saturation is an *in vitro* artifact produced by highly
concentrated cytosol preparations (Chamness et al., 1974) known
to contain low and high molecular weight inhibitors of receptor
activation and receptor-acceptor binding (Asai et al., 1979;
Sato et al., 1980; Shen et al., 1979). It is important to
note, however, that the binding capacity of nonhistone basic
protein acceptors extracted from uterine nuclei may be as much
as 5 to 10 times greater than the estrogen receptor concentration of the same cells (Puca et al., 1974). Thus, although the
nuclear acceptor capacity may not be limiting *in vivo*, the hormone and tissue specificity of these nonhistone basic proteins,
as well as their high affinity for both the receptors and DNA,
clearly suggests that these proteins have the appropriate characteristics required for the selective expression of hormone-
induced genomic responses. The marked physiocochemical similarity of the acceptors for estrogen (uterine), progestin
(uterine), androgen (prostate), and glucocorticoid (liver)
receptors (Mainwaring et al., 1976) lends support to the notion
that the intracerebral acceptors for these steroid receptor
complexes will also be similar.

Behavioral Correlates of Steroid-Receptor Interactions

With the exception of the direct electrophysiological and other
membrane actions of steroids (see Fig. 15-1), very few of the
molecular actions are thought to be mediated by mechanisms
other than those involving the high-affinity intracellular
receptors. Thus, it is not surprising that these receptors
have received considerable attention in studies of the behavioral actions of steroids. In a preceding section, it was
noted that most, but not all, of the behaviorally effective
implantation sites for steroid-containing cannulas were localized in brain regions containing high densities of receptors

for that steroid. Intracerebral and systemic administration of receptor-blocking antiestrogens have implicated that nuclear receptor binding is critical in the initiation of female (Etgen, 1979) and, to a lesser extent, male (Luttge, 1975) sexual behavior in the rat. This conclusion is also supported by observations derived from streptozotocin-diabetic rats in which ovariectomized females exhibit dramatic reductions in hypothalmic intranuclear estrogen binding capacity and in the ability of exogenous estrogen to induce sexual receptivity (Gentry et al., 1977; Siegel and Wade, 1979). These effects are reversed with insulin replacement.

In spite of these findings, attempts to correlate the clear individual differences in behavioral sensitivity to estrogen with differences in cytosolic and nuclear binding capacity for estrogens have consistently met with failure (Gentry et al., 1976). Furthermore, although it is a common observation that the effectiveness of estrogen to facilitate the induction of receptivity declines with time after ovariectomy, and that this loss can be reversed by estrogen priming (Beach and Orndoff, 1974; Whalen and Nakayama, 1965), there does not appear to be a correlation between the capacity or affinity of hypothalmic estrogen receptors and the increased behavioral sensitivity following estrogen priming (Parsons et al., 1979). This lack of concordance between receptor density and behavior may occur because circulating levels of estrogen often exceed those required for the induction of receptivity (Johnston and Davidson, 1979) and, as discussed before, $[^3H]E_2$ binding with as little as 4 percent of the VM estrogen receptors is sufficient to facilitate the induction of receptivity in ovariectomized rats (Davis et al., 1979). It is also significant that these hormones and their receptors are involved in a number of neuroendocrine and behavioral actions in addition to their presumed roles in sexual behavior.

An incubation period of at least 17 hours after a single IV injection of E_2 is required in ovariectomized female rats for the first significant increase in sexual receptivity, that is, in tests performed three to four hours after an SC injection of P_4 (Green et al., 1970). Maximal receptivity (and P_4 receptor binding capacity, *vide infra*) is not achieved until 24 to 48 hours after the E_2 injection (Moguilewsky and Raynaud, 1979a). This comparatively delayed action raises the question of whether the estrogen must remain bound to its target cell nuclear receptors during the entire incubation period or only during the first few hours, that is, a maintenance or a triggering action. Both the correlative and interventive approaches have been used in studies designed to answer this question. For example, the administration of $[^3H]E_2$ in doses sufficient to induce receptivity in rats has revealed that very

little, if any, of the radioactivity is retained within the hypothalmic target cell nuclei beyond 12 hours (McEwen et al.,
1975). In other studies, the administration of receptor blocking antiestrogens to rats and hamsters within the first 8 to 12
hours after E_2 or E_2B injection, and not later, effectively reduced the induction of receptivity (Morin et al., 1976; Whalen
and Gorzalka, 1973). In guinea pigs, however, antiestrogen administration as late as one to two days after E_2B was often as
effective in inhibiting receptivity as that given during the
first few hours after estrogen (Feder and Morin, 1974; Walker
and Feder, 1979). The only reductions in receptivity achieved
in female rats and hamsters with similary delayed injection
schedules appeared to be those confounded by nonspecific depressive side-effects of antiestrogen (Blaustein et al., 1979;
Morin et al., 1976). These data suggest that in female guinea
pigs, E_2 has both a triggering and a maintenance action in the
induction of sexual receptivity, whereas in female rats and
hamsters only the triggering action of E_2 appears to be essential.

Progestins

Receptor Physico- and Biochemical Properties

The intracerebral receptor systems for progestins are in many
ways similar to those described for estrogens; however, it
proved exceptionally difficult to establish this. Numerous
early *in vivo* and *in vitro* studies purported to demonstrate
unique, and functionally significant, regional distribution
patterns for progestin uptake sites within the brain (Luttge et
al., 1974; Marrone and Feder, 1977); however, the uptake demonstrated in these studies was often neither progestin-specific,
progestin-saturable, nor estrogen-dependent. The regional differences probably reflected either nonspecific association with
brain lipids or nonspecific binding of the progestins to either
glucocorticoid or mineralocorticoid receptors (Moguilewsky and
Raynaud, 1980).

A major problem with these early studies was the nearly universal failure to allow for the estrogen-dependence and the extreme lability and short half-life of the progestin receptor
and the progestin-receptor complex. Seiki et al. (1977) were
among the first groups to demonstrate macromolecular receptor
binding for P_4 in MBH cytosol from estrogen-primed ovariectomized rats. Scatchard plots yielded a K_d of 6 nM and a maximum
binding capacity of 20 fmol/mg protein, whereas low-salt sucrose gradient separations yielded sedimentation coefficients
of approximately 7S. Later work revealed that the 7S species
is probably an aggregate of the 4S species present at higher

and, presumably, more physiological salt conditions (Seiki et al., 1979). Thus, the progestin-receptor complex in hypothalmic cytosol appears to have similar physicochemical properties to those described earlier for the estrogen-receptor complex. Other laboratories have confirmed and extended the findings of Seiki et al. by demonstrating the susceptibility of the receptor to destruction by proteases, the specificity of the binding to progestins (that is P_4, 5α-DHP, and R5020, but not B, F, T, and E_2, effectively competed with $[^3H]P_4$ for binding to the cytosol, and a half-life of progestin-receptor association that is on the order of minutes (37 to 160 minutes) rather than hours to days as demonstrated for other steroid-receptor complexes (Lee et al., 1979; Thrower and Lim, 1980).

Many of the problems associated with progestin-receptor complex instability have been reduced dramatically with the introduction of the synthetic progestin R5020. This steroid forms high-affinity associations with progestin receptors throughout the body, is much more potent in bioassays than P_4, and does not bind to the plasma glucocorticoid binding globulin as will P_4 (Raynaud, 1977). Hypothalmic cytosol receptor studies with R5020 have produced similar physicochemical findings to those described earlier for P_4 (K_d = 0.3 to 4.0 nM, B_{max} = 10 fmol/mg protein, 6 to 7S in low-salt and 3.5 to 4.5S in high-salt sucrose gradients), as well as some new ones not previously described (pI = 5.8, electrophoretic separability from the estrogen (R2858)- and androgen (R1881)-receptor complexes) (Kato and Onouchi, 1979; MacLusky and McEwen, 1980; Moguilevesky and Raynaud, 1979b). Similar R5020-binding macromolecular receptors have also been found in hypothalmic cytosol from male rats and female rhesus monkeys (MacLusky et al., 1980; Naess and Attramadal, 1978). The R5020-receptor complex is theromlabile, sensitive to proteolytic enzymes, and specific for progestins. However, the fact that as much as 50 percent of the ^3H bound to hypothalmic cytosol macromolecules incubated with $[^3H]R5020$ is nonsupressible following co- or preincubation with P_4, has prompted Thrower and Lim (1980) to advise caution in interpreting the data obtained with this synthetic progestin.

The *in vivo* administration of $[^3H]R5020$ to estrogen-primed ovariectomized rats is reported to result in progestin-saturable nuclear binding in purified nuclei obtained from hypothalamus and, to a lesser extent, POA and septal area, cerebral cortex, and midbrain (Blaustein and Wade, 1978). Working with 1-mm thick slices of hypothalamic and other brain regions from estrogen-primed female rats, Seiki et al. (1979) demonstrated the *in vitro* association of $[^3H]R5020$ with salt-extractable nuclear proteins that sedimented as 9S aggregates in sucrose gradients. By first stimulating the nuclear accumulation of progestin receptors with unlabeled progestin injec-

tions in estrogen-primed ovariectomized rats and guinea pigs,
it is possible to isolate the nuclei and exchange the unlabeled
progestin with [^{3}H]R5020 (a competitive exchange reaction based
on the reversibility of the nuclear progestin-receptor binding). These [^{3}H]R5020-labeled nuclear receptor complexes have
a K_d of 0.16 to 1.4 nM, a binding maximum of at least 466
fmol/mg DNA (the actual magnitude depends upon the extent of
both estrogen and progestin priming), and a sedimentation coefficient of 5 to 6S (Blaustein and Feder, 1980; Kato and
Onouchi, 1979). These findings suggest that the 9S species
described by Seiki et al. (1979) is an aggregate, whereas the 5
to 6S species described in the exchange studies may represent a
dimer of the 4S cytosol receptor (or a heterologous aggregate)
similar to that described earlier for the nuclear estrogen-
receptor complex.

Receptor Utilization and Replenishment

The intracerebral progestin receptor is an estrogen-dependent
protein: hypothalamic and preoptic (but not cerebral cortical,
amygdaloid, or midbrain) progestin receptor concentrations decrease after gonadectomy and increase in a dose- and time-dependent manner after exogenous estrogen replacement therapy in
female and, to a lesser extent, male rats, guinea pigs, and
bonnet monkeys (Blaustein and Feder, 1979a,b; MacLusky and
McEwen, 1980; MacLusky et al., 1980; Moguilewsky and Raynaud,
1979b; Schwartz et al., 1979). Removal of an exogenous estrogen sources, such as a Silastic capsule, from ovariectomized
rats, leads to a rapid depletion of MBH progestin receptors,
suggesting a half-life of approximately one day in the absence
of estrogen (Parsons et al., 1980). Since antiestrogen (CI-
628) administration can inhibit this estrogen-dependent process
(Roy et al., 1979b), and maximal stimulation is not achieved
until 24 to 48 hours after the injection of E_2B (SC in oil)
(Blaustein and Feder, 1979a; Moguilewsky and Raynaud, 1979a;
Schwartz et al., 1979), these data suggest that the increase in
progetin receptor concentration is mediated by an extrogen-re-
ceptor-acceptor interaction resulting in *de novo* synthesis of
progestin receptor proteins. Future work with protein synthesis inhibitors, similar to that described earlier for the estrogen receptor, should help resolve this possibility.
    Since the appearance of progestin-receptor complexes in the
nucleoplasm is probably dependent upon their prior presence in
the cytoplasm, except for the possible contribution by unoccupied nuclear progestin-receptors, (Saffran and Loesser,
1979), the appearance of the nuclear progestin-receptor complex
is also an estrogen-dependent process (Blaustein and Feder,

1980; Blaustein and Wade, 1978). A positive interaction exists between the dose of estrogen used to stimulate the appearance of the cytoplasmic receptor and the dose of progestin used to stimulate the redistribution of that receptor to the nucleus. Coincident with the progestin-induced nuclear accumulation of progestin-receptor complexes, there is a dose-dependent, long-duration (at least 24 hours) reduction in cytosolic progestin receptor concentrations (Blaustein and Feder, 1979b, 1980; Moguilewsky and Raynaud, 1979a). The failure of this system to homeostatically regulate the level of progestin receptors is similar to that described for estrogen receptors and suggests that receptor "processing" may be required for the complete expression of progestin-receptor action. When progestins are administered concurrently with the priming injection of estrogen, there is a dose-dependent negative interaction between the two steroids resulting in a progestin-induced reduction in estrogen-stimulated progestin receptor appearance in the cytosol fraction (Blaustein and Feder, 1979b; Moguilewsky and Raynaud, 1979b). The mechanism of this inhibition is at present unclear, since it does not appear to involve a direct interaction between progestins and the estrogen-receptor.

Behavioral Correlates of Steroid-Receptor Interactions

The potential role of high-affinity receptors in the intracerebral actions of progestins in female sexual behavior has received considerable support from correlational studies, but the lack of specific receptor blocking antiprogestins has prevented the appropriate interventive studies. For example, recent studies with rats and guinea pigs have shown that the effectiveness of exogenous progestins to induce receptivity closely parallels the time- and dose-response characteristics of estrogen-stimulated increases in hypothalamic progestin receptor binding capacity (Blaustein and Feder, 1979a, 1980; Moguilewsky and Raynaud, 1979a; Schwartz et al., 1979). Also, antiestrogen administration produces a parallel inhibition of both responses (Roy et al., 1979a). The sequential and concurrent inhibitory actions of progestin pretreatment on progestin-stimulated receptivity (*vide supra*) also correlates with similar reductions in hypothalamic progestin receptor binding capacity (Blaustein and Feder, 1979b, 1980; Moguilewsky and Raynaud, 1979a; Schwartz et al., 1979).

In spite of these positive findings, there are others that do not appear to support a direct relationship between progestin receptor binding capacity and behavioral responsiveness. For example, as discussed earlier, intracerebral cannulation studies in the rat, mouse, and guinea pig have suggested that in addition to the MBH the VTA is a major site of progestin in female reproductive behavior (Hall and Luttge, 1974; Luttge and Hughes, 1976; Morin and Feder, 1974a). However, in receptor studies estrogen stimulation has not been shown to increase the cytosolic or nuclear progestin binding capacity of this mesencephalic region (MacLusky and McEwen, 1978). Since antiestrogen pretreatment of the VTA blocks the effectiveness of P_4 implants in this region to facilitate receptivity (Yanase and Gorski, 1976a), it is clear that even though mesencephalic progestin receptor binding capacity may not be increased with estrogen pretreatment, mesencephalic behavioral responsiveness to progestin stimulation is increased.

Androgens

Early studies on androgen receptors concentrated on establishing brain regional uptake patterns for various androgens given systemically (Blaquier et al., 1970; Stern and Eisenfeld, 1971; Perez-Palacios et al., 1970; Whalen and Luttge, 1971b). In general these studies found that hypothalamic, POA and septal area concentrate more androgen-derived radioactivity than other brain regions, thus confirming the results obtained with the anatomically more precise, but quantitatively less precise autoradiographic procedures described earlier. *In vitro* studies with hypothalamic and POA cytosol preparations from rats, mice, and other mammals find that [^{3}H]T and [^{3}H]DHT associate with high affinity (K_d = 0.3 - 8.0 nM), limited capacity (B_{max} = 2 to 14 fmol/mg protein), negatively charged (pI = 5.8), protease- and sulfhydryl reagent-sensitive macromolecules sedimenting as 5-8S aggregates in low-salt linear sucrose gradients and as 4S (monomer?) species in gradients with higher salt concentrations (Clark and Nowell, 1979; Gustafsson et al., 1976; Hannouche et al., 1977; Kato, 1975; Naess, 1976). As is the case for the estrogen and progestin receptors, the hypothalamic receptors for androgens display considerable hormone specificity (that is T, DHT, and certain antiandrogens, such as flutamide and cyproterone acetate, bind avidly to the receptor, whereas estrogens, progestins, glucocorticoids, and weaker androgens and other antihormones do not bind to the high affinity receptor sites), but they display little species specificity (that is, high-affinity androgen receptors have been found in hypothalamic tissues from all mammals tested, inluding rats, mice, hamsters, guinea pigs, rabbits, cats, dogs, cattle, and primates).

Salt-extractable and salt-resistant nuclear binding in hypothalamic, septal and other limbic brain structures has also been reported for androgens (Kato, 1975; Lieberburg and McEwen,

1977; Lieberburg, 1977; Monbon et al., 1974), and for estrogens
derived from the intracerebral metabolic aromatization of androgens (Lieberburg and McEwen, 1975; *vide infra*). *In vivo*
studies with male rats indicate that [^{3}H]DHT binding nuclear
macromolecules are saturated in the gonadally intact animal and
only become apparent following castration, often combined with
adrenalectomy (Lieberburg et al., 1977). In agreement with the
results recently obtained with the more intensively studied
peripheral androgen target tissues, such as rat prostate,
testis, epididymis and seminal vesicle (Wilson and French,
1979), the nuclear-extracted androgen receptor from brain
appears to sediment in sucrose gradients with the same rate,
that is 3 to 4.5S, as the cytosol receptor (Lieberburg et al.,
1977). Thus, in contrast to the situation described earlier
for the estrogen and progestin receptors, the intracerebral
androgen receptor does not appear to undergo a size change, for
example, aggregation, as a consequence of, or precondition to,
its redistribution from the cytoplasm to the nucleoplasm.

Although there are clear individual, and even strain, differences in the reproductive behavioral sensitivity to androgens,
these differences have not been found to correlate with similar
differences in cerebral androgen receptor binding capacity
(Harding and Feder, 1976; Luttge et al., 1976b). As with estrogen receptors, failure to establish an unambiguous relationship between androgen receptor binding capacity and behavioral
sensitivity may be a consequence of the relative insensitivity,
and nonspecificity, of the receptor assays, thus precluding the
detection of potentially minor differences in the binding capacity of a small fraction of androgen-concentrating neurons.
Furthermore, while binding and biochemical and morphological
studies in peripheral target tissues have shown that antiandrogens can inhibit the actions of T and DHT, similar studies
on cerebral receptors and reproductive behavior have often led
to conflicting, or at least controversial, findings (for
reviews see Luttge, 1979,1983; Whalen et al., 1983).

INTRACEREBRAL STEROID METABOLISM

The gonadal steroid hormones, estrogens, progestins, and androgens can all be metabolized to varying extents within the brain
to form numerous steroids that may have quantitatively or qualitatively different biological activities from their parent
steroids. In some cases these metabolic pathways have previously been studied extensively in peripheral tissues as part of
a biosynthetic sequence, such as the aromatization of androgens
to estrogens in the ovary, or as part of a target-tissue specific metabolic activation sequence, such as the 5α-reduction

of T to DHT in the ventral prostate, whereas in other cases the metabolites have only recently been examined in peripheral as well as intracerebral tissue studies, as in the 2- and 4-hydroxylation of estrogens to catechol estrogens in liver and brain (for general review see Fregly and Luttge, 1982). The following sections briefly review the major intracerebral steroid metabolites and, when possible, the regional or subcellular distribution of the relevant enzymes, steroid, precursors, and final products (see also Luttge, 1979,1983; Whalen et al., 1983). The biological significance of the various metabolites will be reviewed later, and their structural formulas and IUPAC names and abbreviations are given in Figs. 15-3 and 15-4 and Table 15-1.

Estrogens

E_2 has long been considered to be the principal estrogen secreted by the ovary and active in peripheral and neural target tissues. However, recent studies have shown that E_1 and E_2, and the catechol estrogens, $2\text{-OH-}E_2$, $2\text{-OH-}E_1$, $4\text{-OH-}E_2$, and $4\text{-OH-}E_1$ can also be found in peripheral plasma or intracerebral extracts and that these metabolites may actively participate in mediating certain estrogenic functions.

Estradiol, Estrone, and Estriol

In early *in vivo* studies $[^3H]E_1$ was detected in certain brain regions of male and female rats administered $[^3H]E_2$, whereas $[^3H]E_2$ was detected in the brains of animals injected with $[^3H]E_1$ (Kato and Villee, 1967; Luttge, 1971; Presl, 1973). Metabolically-produced E_1 was comparatively enriched in posterior hypothalamic and anterior mesencephalic samples, whereas metabolically produced E_2 was most prelevant in anterior hypothalamic and POA samples. Very little of the more polar estrogen, E_3 was found in these studies; but Reddy (1979a) has shown conclusively that $[^3H]E_3$ can be synthesized *in vitro* by rat brain from $[^3H]E_2$ and, to a much lesser extent, $[^3H]E_1$. A similar preferential *in vivo* conversion of $[^3H]E_1$ to $[^3H]E_2$ by hypothalamic, and to a lesser extent septal, amygdaloid, and cerebral cortical samples has recently been demonstrated in adult male and female guinea pigs (Landau and Feder, 1980). The soluble enzyme responsible for the interconversion of E_2 and E_1 (17β-oxidoreductase) has been isolated from rat, rabbit, and other mammalian brains (Kazama and Longcope, 1974; Reddy, 1979b).

After systemic injections of $[^3H]E_2$, isolated nuclei from rat brain tissue pools containing hypothalamic, preoptic, septal, amygdaloid, and olfactory bulb regions retain mostly E_2 (86

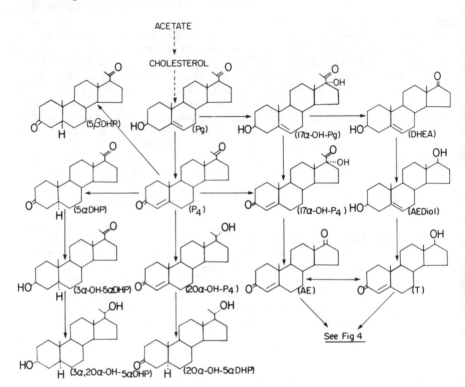

Fig. 15-3. Structural formulas and metabolic pathways for
progestin metabolism in the brain, and androgen
biosynthesis in peripheral such organs as the
testis. Abbreviations are defined in Table
15-1.

percent), although significant amounts of E_1 (4 percent) and E_3
(2 percent) can also be identified (Zigmond and McEwen, 1970).
Nuclei from tissue pools containing hippocampal and cerebral
cortical regions retain comparatively less E_2 (75 percent),
more E_3 (8 percent) and similar amounts of E_1 (4 percent) to
that detected in the pooled limbic brain regions. E_3 retention
by hypothalamic nuclei following systemic injection of $[^3H]E_3$
in ovariectomized guinea pigs has also been demonstrated
(Landau and Feder, 1977). Even though the $[^3H]E_3$ was apparently not metabolized to E_1 or E_2, the free estrogens usually
accounted for less than 1 percent of the total radioactivity,
suggesting that most of the retained ^3H was associated with
water-soluble, conjugated estrogens. It should be noted,

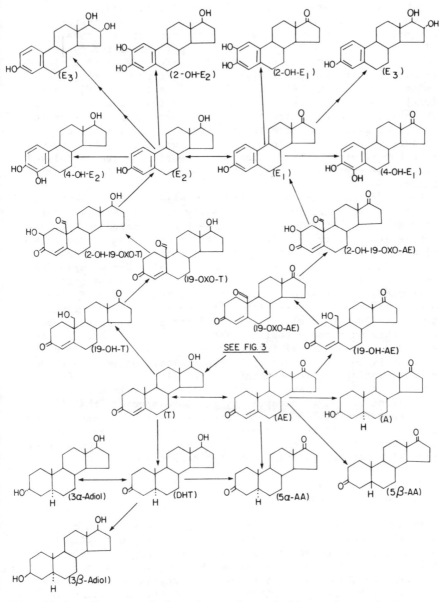

Fig. 15-4. Structural formulas and metabolic pathways for androgen and estrogen metabolism in the brain. Abbreviations are defined in Table 15-1.

however, that conjugated estrogens can also be metabolized by the brain back into free steroids (Kishimoto, 1973).

The bulk of the available evidence suggests that intracerebral estrogen metabolism is fairly consistent across species, with E_2 accounting for 76 to 90 percent of the hypothalamic radioactivity retained after systemic injections of $[^3H]E_2$ in mice, hamsters, guinea pigs, and rats (Feder et al., 1974; Luttge, 1972; Luttge et al., 1976a). In each case, E_1 accounted for the bulk of the remaining radioactivity. Since $[^3H]E_2B$ is thought to be metabolized to $[^3H]E_2$ before entering the brain (Eaton et al., 1975), the intracerebral metabolism of this frequently used synthetic estrogen, is essentially the same as that for E_2 with the exception that the time course is delayed because of the slow release rate from peripheral depot storage sites.

The principal ovarian estrogen, E_2, effectively facilitates female sexual receptivity in virtually all mammalian species studied, with several prominant exceptions, including the human female (Luttge, 1971). Other naturally occurring estrogens are also effective (for further details and specific references, see Luttge, 1979,1983; Whalen et al., 1983). For example, in behavioral tests with ovariectomized rats and guinea pigs, systemic administration of the free alcohol forms of E_1, E_2, and E_3 has revealed only minor differences in their relative potencies (Beyer et al., 1971; Feder and Silver, 1974). However, this does not establish that each of these estrogens is required in the natural induction of receptivity. Because E_1 is readily converted into E_2 by hypothalamic enzymes, although E_3 does not appear to be metabolied to either E_1 or E_2 (*vide supra*), and E_3 may have a specific role in the induction or regulation of sexual receptivity in these species, that is, different from that of E_2, the effectiveness of E_1 may well be due to its intracerebral conversion to E_2. Until specific inhibitors of these metabolic interconversions are found, their necessity for reproductive behavior cannot be ascertained.

Catechol Estrogens

Although once thought to be of importance only in the catabolism of estrogens in the liver, the recent demonstration that E_2 and E_1 can also be metabolized within the brain to their 2- and 4-hydroxy- or catechol-derivatives has sparked considerable interest in these novel estrogens (for reviews see Fishman et al., 1976; Paul et al., 1980). Catechol estrogen formation is generally greater in hypothalamus than in cerebral cortex (Ball and Knuppen, 1978; Fishman et al., 1976; Paul and Axelrod, 1977), is equivalent in male and female brains (Barbieri et al., 1978; Reddy and Rajan, 1978), is equally prevalent at the 2 and 4 positions in brain (Ball et al., 1978), increases dramatically in brain (but not liver) during proestrus in rats (Fishman et al., 1980), and is more effective with E than E as the enzyme substrate (Barbieri et al., 1978; Fishman and Norton, 1975; Reddy and Rajan, 1978). The enzyme itself appears to be a cytochrome P-450 mixed function oxidase localized in the microsomal fraction of brain cells (Paul et al., 1977; Sasame et al., 1977). The early report by Paul and Axelrod (1977) that the brain contains more than 10 times as much catechol estrogen as E_2 and E_1 has been strongly disputed by two recent independent studies (Ball and Knuppen, 1978; Fishman and Martucci, 1979). It now appears that catechol estrogens formed intracerebrally are rapidly turned over and that they form highly reactive superoxide intermediates leading to the covalent binding of catechol estrogen derivatives to brain microsomal proteins (Fishman and Martucci, 1979; Sasame et al., 1977). Catechol estrogens can also compete with $[^3H]E_2$ for high-affinity binding to brain cytosol estrogen receptors (Davies et al., 1975) and with norepinephrine (NE) and other biogenic amines for high-affinity interactions with brain tyrosine hydroxylase (TH), catechol-O-methyltransferase, and other enzymes (Breuer and Koster 1974; Lloyd and Ebersole, 1980).

Extensive testing of the effectiveness of 2-OH-E_2 and 2-OH-E_1 in ovariectomized rats and guinea pigs to induce sexual receptivity indicates that these highly labile compounds are considerably less potent than E_2 when tested after IV, SC, or intracerebral administration (Luttge and Jasper, 1977; Marrone et al., 1977). Preliminary work with 4-OH-E_2 has suggested that this catechol estrogen may be effective in stimulating receptivity in female rats when infused directly into the brain with an Alzet minipump (Jellinck et al., 1979). However, more work with enzyme inhibitors, to block the formation and catabolism of the various catechol estrogens, and other manipulations must be done before any strong conclusions can be drawn regarding the role of catechol estrogens in reproduction behavior.

Progestins

Progesterone, Dihydroprogesterone, and Other Metabolites

Although P_4 is commonly assumed to be the major circulating progestin in mammals, 20α-hydrozy-P_4 and, to a lesser extent, 5α-DHP, 5α-pregnan-3α,20αdiol and 5α-pregnan-20α-ol-3-one also contribute, in a major way, to the progestins available to the brain (for review see Fregly and Luttge, 1982; Whalen et al., 1983). All of these progestins should readily cross the blood-brain barrier (Pardridge and Mietus, 1980).

Early studies indicatd that one to four hours after the systemic administration of [^{3}H]P$_4$ only 20 to 30 percent of the extractable radioactivity from ovariectomized (estrogen-primed and nonestrogen-primed) rat and mouse brains were isopolar with authentic P$_4$, whereas in similarly prepared guinea pigs, P$_4$ often accounted for up to 70 percent of the intracerebrally retained radioactivity (Luttge et al., 1974; Wade and Feder, 1972a). Further analysis indicated that the principal metabolite retained in hypothalamic, mesencephalic, and cerebral and cerebellar cortical samples was 5α-DHP, the 5α-reductase product of P$_4$ (Johnson et al., 1976; Karavolas et al., 1979). By contrast with results from the pituitary and other peripheral target tissues, very little 5α-pregnan-3α-ol-20-one, 5α-pregnan-20α-ol-3-one and 5α-pregnan-3α,20α-diol was detected. After [^{3}H]5α-DHP injections, there was no evidence for significant retention of P$_4$ or 5α-pregnan-20α-ol-3-one by rat and rabbit brain tissues; most of the radioactivity was associated with 5α-DHP (Johnson et al., 1976; Karavolas et al., 1979). Although [^{3}H]5α-DHP-associated radioactivity was comparatively high in hypothalamic and mesencephalic brain samples after systemic injections of either [^{3}H]P$_4$ or [^{3}H]5α-DHP, it is not known whether any of the ^{3}H was retained by high-affinity progestin receptors, nor just how much of the 5α-DHP was produced within the brain as opposed to the periphery.

With *in vitro* studies a much wider range of progestin metabolites has been detected than with the *in vivo* work. Following incubations with either [^{3}H]P$_4$ or [^{14}C]P$_4$, the principle metabolite formed in hypothalamic and, especially, in mesencephalic tissue samples in most, but not all mammalian species, is 5α-DHP. In rats, 5α-reduction can account for 10 to 50 percent of the radioactivity extracted from hypothalamic tissues after [^{3}H]P$_4$ incubations, whereas in baboons 20α-reductase activity greatly exceeds (2 to 10 times) that of 5α-reductase. The further reduction of 5α-DHP to 5α-pregnan-3α-ol-20-one is especially prelevant in rat brains, often accounting for 2 to 10 percent of the extracted radioactivity (Cheng and Karavolas, 1973; Hanukoglu et al., 1977; Krause and Karavolas, 1978; Tabei and Heinrichs, 1974). Other metabolites have been detected, but their concentrations are often very low and highly variable.

Subcellular studies have indicated that the 3α-oxidoreductase enzyme predominates in the cytosol fraction, whereas the 5α-reductase enzyme activity is maximal in the microsomal fractions (Cheng and Karavolas, 1975; Rommerts and van der Molen,

1971). More detailed investigations of the hypothalamic 5α-reductase enzyme have shown that P_4 ($K_m = 0.48$ μM) and 20α-OH-P_4 (K = 0.86 M) are both more effective substrates for the enzyme than T ($K_m = 16$ μM) (Cheng and Karavolas, 1975). The latter difference is potentially very important in males in whom the 5α-reduction of T to DHT is of major concern (*vide infra*). 17α-OH-P_4 also proved to be a very effective competitive inhibitor of P_4 5α-reduction ($K_m = 0.29$ μM) in the hypothalamus, whereas E_2 inhibited noncompetitively (Cheng and Karavolas, 1975; Krause and Karavolas, 1978). The comparatively extensive accumulation of 5α-DHP suggest that it and other metabolites may play a significant role in mediating the neurochemical, neurobehavioral, and neuroendocrine actions of progestins.

Behavioral Correlates of Intracerebral
Progestin Metabolites

In estrogen-primed ovariectomized female rats, the comparative potency of P_4 and 5α-DHP in the induction of sexual receptivity depends to a large extent on the carrier vehicle used in the injections of 5α-DHP. For example, when dissolved in Tween-80 and injected SC, 5α-DHP is equipotent with P_4 in the induction of receptivity (Gorzalka and Whalen, 1977); however, when the progestins are injected SC in oil, 5α-DHP is only 60 percent as potent as P_4 (Whalen and Gorzalka, 1972) and when injected IV in propylene glycol, 5α-DHP loses all potency (Kubli-Garfias and Whalen, 1977). P_4 remains active under each of these injection routes or vehicle conditions. Similar data are obtained from guinea pigs, hamsters, and mice (Czaja et al., 1974; Gorzalka and Whalen, 1974; Johnson et al., 1976a; Luttge and Hall, 1976; Wade and Feder, 1972b). The relative behavioral potencies of P_4 and 5α-DHP have also been shown to be dependent upon as yet unknown genetically determined factors (Gorzalka and Whalen, 1976). Further metabolites of 5α-DHP, such as 5α-pregnan-3α-ol-20-one and 5α-pregnan-3α,20α-diol, are even less active in facilitating the induction of receptivity in estrogen-primed ovariectomized rats and guinea pigs (Czaja et al., 1974; Kubli-Garfias and Whalen, 1977; Whalen and Gorzalka, 1972). Thus, although the administration of 5α-DHP and its metabolites can, in some cases, mimic the behavioral effects of exogenous P_4, the comparatively low potency of these 5α-reduced progestins casts doubt on their physiological significance.

The second major pathway of cerebral P_4 metabolism involves the formation of 20α-OH-P_4 and its 5α-reductase product, 5α-pregnan-20α-ol-3-one (see Fig. 15-3). Both progestins display considerable potency in the induction of receptivity in estrogen-primed ovariectomized rats when they are administered IV in propylene glycol. In fact, when tested five minutes after injection, 20α-OH-P_4 is even more effective than similarly injected P (Kubli-Garfias and Whalen, 1977). 20α-OH-P_4-treated females remain sexually receptive for at least two hours under these conditions. Thus, the possibility that the 20α-hydroxylation of P_4 may be essential for the early onset of receptivity clearly needs to be tested when specific enzyme inhibitors are developed.

Androgens

Testosterone

T is the predominant circulating androgen in most adult male mammals but androstenedione (AE), dehydroepiandrosterone (DHEA), DHT, 3α- and 3β-androstanediol (3α- and 3β-Adiol), and epitestosterone (Epi-T) (see Table 15-1 and Fig. 15-4) can also be found in blood samples taken from a large number of mammalian species (for review see Fregly and Luttge, 1982; Whalen et al., 1983). Peripheral target and nontarget tissue metabolism, coupled with the release of these metabolites into blood, and their metabolism by blood-born enzymes combine to increase substantially the total number of androgen metabolites available for uptake and still further metabolism by the brain. The sheer magnitude of this non-neural metabolism of androgens complicates the interpretation of studies using the systemic administration of androgens.

17α-Oxidoreductase and Androstenedione

Intracerebral interconversion of T and AE by 17β-oxidoreductase has been demonstrated *in vivo* and *in vitro* in a number of mammalian species, including rats, mice, guinea pigs, rabbits, dogs, rhesus monkeys, and human beings (for review see Luttge, 1979,1983; Whalen et al., 1983). The cytosol localized enzyme is uniformly distributed in relatively low concentrations throughout the brain (Rommerts and van der Molen, 1971).

5α-Reductase and Dihydrotestosterone

The intracerebral conversions of T to DHT by 5α-reductase has also been demonstrated *in vivo* and *in vitro* in a wide range of mammalian species (for review see Luttge, 1979,1983; Whalen et

al., 1983). Although the microsomal- or nuclear-localized
enzyme is distributed (unevenly) throughout the brain (Denef et
al., 1973; Rommerts and van der Molen, 1971; Selmanoff et al.,
1977), the metabolically derived DHT is found in the hypothalamus to a much greater extent than in other brain regions
(Lieberburg and McEwen, 1977; Loras et al., 1974; Rezek, 1977).
The intracerebral 5α-reductions of both T and P₄ appear to be
essentially irreversible and they also appear to be mediated by
the same enzyme, with P₄ acting as the preferred substrate in
direct competition studies (Cheng and Karavolas, 1975;
Karavolas et al., 1979; Whalen and Rezek, 1972).

3α-Oxidoreductase and Other Enzymes
and Androgen Metabolites

In vivo and *in vitro* studies with [¹⁴C]- and [³H]-labeled
precursors (T, DHT, and AE) have indicated that the mammalian
brain can form still other androgen metabolites, including 3α-
and 3β-Adiol, 5α- and 5β-androstandedione (5α- and 5β-AA), and
androsterone (A) (see Table 15-1 and Fig. 15-3). The interconversion of DHT and 3α-Adiol by 3α-oxidoreductase, a soluble
enzyme distributed widely throughout the brain, has undoubtedly
been the most widely studied of these additional metabolic
pathways (Celotti et al., 1979; see also Luttge, 1979,1983;
Whalen et al., 1983). Compared to the relatively abundant formation of 3α-Adiol, the formation of its 3β-epimer is markedly
smaller and although 3α-Adiol can be rapidly converted back to DHT,
3β-Adiol does not appear to display this reaction readily in the
brain (Kao et al., 1977; Noma et al., 1975; Sholl et al., 1975;
Van Doorn et al., 1975). The 5α-reduction of AE to 5α-AA in
hypothalamic and other brain regions is probably mediated by
the same microsomal or nuclear enzyme that acts on T and P₄
(Cheng and Karavolas, 1975; Perez et al., 1975; School et al.,
1975). In male and female hamsters, preoptic-hypothalamic,
anterior limbic cortex and frontal cerebral cortex tissue
homogenates readily convert [³H]AE to 5α-AA, but in contrast to
most other mammalian species, an even larger percentage of the
precursor is converted to 5β-AA (Callard et al., 1979). The *in
vitro* conversion of [³H]AE to [³H]A has also been observed with
tissue minces of hypothalamic and hippocampal brain material
from gonadectomized and intact male rats (Perez et al., 1975).

Aromatization and Estrogens

The most extensively studied of the intracerebral metabolic
pathways for androgens is the aromatization of these C-19

steroids to C-18 estrogens. *In vivo* and *in vitro* evidence for the conversion of T to E_2 and AE to E_1 within the brain have been obtained for a wide range of mammalian species, including rats, mice, rabbits, hamsters, cats, cows, rhesus monkeys, and human beings (Callard et al., 1978; Weisz and Gibbs, 1974; see also Luttge, 1979,1983; Whalen et al., 1983). Callard et al. (1978) have shown that the aromatase enzyme system is found in the brains of representative species from every major vertebrate group with the exception of the Agnatha. A prominant exception to this generalization is the guinea pig in which no *in vivo* or *in vitro* evidence for the aromization of [^{3}H]T by hypothalamic tissues has been detected (Sholl et al., 1975). The highest activities are concentrated in the MPOA, periventricular POA, and medial amygadaloid nucleus, whereas much lower levels (6- to 30-fold lower) are found in the lateral POA, MBH, and lateral hypothalamus of male rats (Kobayashi and Reed, 1977; Selmanoff et al., 1977). Cerebral cortex and other nonlimbic brain regions usually are reported to lack aromatase activity in the adult (Callard et al., 1979; Naftolin et al., 1975; Reddy et al., 1973). These findings are in excellent agreement with recent *in vivo* studies in which estrogens were found to account for a comparatively high percentage of the total radioactivity retained in purified nuclei from hypothalamic, preoptic, amygdaloid, and septal tissue pools obtained from male and female rats injected systemically with [^{3}H]T (Lieberberg and McEwen, 1975, 1977; Rezek, 1977). Little if any evidence was found for estrogens produced metabolically in cerebral cortical nuclei.

The human placental microsome *in vitro* system has provided the most detailed information on the actual sequence of metabolic conversions involved in the aromatization of androgens (Goto and Fishman, 1977; Selmanoff et al., 1977). It involves a series of three hydroxylations (see Fig. 15-4) catalyzed by a single cytochrome P-450 requiring mixed function oxidase (yielding the temporary intermediates 19-OH-T and 19-OH-AE). The product of this multistep hydroxylation then undergoes a nonenzymatic collapse within the A-ring to form the final estrogen product. Although this sequence is well established for the human placenta, there is little direct evidence for it in the brain (Goto and Fishman, 1977; Selmanoff et al., 1977). In fact, since there is clear evidence for interspecies differences in placental aromatization (Ainsworth and Ryan, 1966), and for the aromatization of some rather unusual androgens such as 19-nor-T, (Thompson and Siiteri, 1973), intracerebral studies that tacitly accept this mechanism must be viewed with caution. A by-product of the acceptance of the human placental aromatase system is the classification of steroids into two classes: aromatizable and nonaromatizable. The former class includes those androgens with an unsaturated A-ring, such as T, AE, DHEA, 19-OH-T, 19-OHAE, whereas the latter includes those with saturated A-rings, such as DHT, 5α- and 5β-AA, 3α- and 3β-Adiol and A. Although DHT and 5α-AA have been clearly shown to be nonaromatizable in the placental systems (Gual et al., 1962; Ryan, 1960), this relationship has not been established unequivocally for the brain aromatase system. However, the irreversibility of the intracerebral conversion of T to DHT supports the assumption that at least this androgen, and presumably its 3α-, 3β-, and 17β-oxidoreductase products, is probably not readily aromatized by brain enzymes. This classification does not rule out the possibility that estrogens may be produced *in vivo* from DHT, since the bacterial flora of the human gut, and presumably other mammals can perform this synthesis (Hill et al., 1971).

Behavioral Correlates of Intracerebral Androgen Metabolites

The possibility that the intracerebral production of specific metabolites may be required for the complete expression of the behavioral responses to gonadal steriods has been examined most extensively in studies of the involvement of T metabolites in male sexual behavior (for detailed reviews and specific references see Luttge, 1979, 1983; Whalen et al., 1983). The complexity of this problem is considerable in that it is compounded by a large number of metabolites (see Fig. 15-4), numerous strain and species differences, and by qualitatively and quantitatively different results obtained with various doses, treatment durations and routes of hormone administration. For example, studies with systemic injections of AE suggest that this metabolite may be as potent (rat), less potent (mouse), or more potent (hamster) than T in either the induction or maintenance of male copulatory behavior following castration. The extensive formation, intranuclear localization, and high potency of DHT in many sex accessory organs, such as the prostate, seminal vesicles, and penis, suggests that this metabolite plays an important, if not essential, role in these tissues. The importance of the intracerebral formation of DHT is, however, not nearly as clear. In hamsters and in certain strains of rats and mice, DHT administration has little effect on male sexual behavior, whereas in other strains of rats and mice and in guinea pig, it may be nearly as potent as T. Some of these differences can be eliminated by beginning the replacement therapy shortly after castration rather than waiting weeks to months for all remnants of the presumed hormone-dependent behaviors to vanish, that is, maintenance versus restoration design, while other differences can be markedly reduced by changing the dose, duration, or route of DHT administration.

Other androgenic metabolites of T (and DHT) have also been tested, such as 3α- and 3β-Adiol, but none of these hormones, which sometimes show considerable potency in peripheral target tissues, appears to be more potent than their parent steroids, and thus they have questionable physiological significance. Use of a 5α-reductase blocker to test the importance of the intrahypothalamic production of DHT, similar to the studies with "17βC" in peripheral target tissues (Luttge et al., 1978), will be very helpful in future studies on male sexual behavior.

The potential role of hypothalamic aromatization of androgens to estrogens in the regulation of male sexual behavior has also been aggressively studied in many mammalian species. For example, although there are some notable exceptions, such as rabbits, guinea pigs, and rhesus monkeys, the systemic or intrahypothalamic administration of E_2 (or E_2B), or one of its immediate precursors such as 19-OH-T, or metabolites, for example E_1 and E_3 (the catechol estrogens have yet to be examined in this context) have been shown to facilitate the induction or restoration of male sexual behavior in many laboratory species (rats, mice, hamsters). The observation that this sometimes incomplete response, for instance reduced frequency of intromissions or ejaculations, can often be greatly facilitated by additional treatment with DHT, which by itself may be behaviorally ineffective, has led to the theory that male sexual behavior in these species may actually be dependent upon the dual metabolism of T to E_2 in the brain and T to DHT in the brain and in certain peripheral target tissues, such as the penis (Luttge et al., 1975; Wallis and Luttge, 1975). Definitive rather than correlative support for the importance of aromatization to androgen action in male sexual behavior has been obtained in the rat by demonstrating a reduction in copulatory performance following either the systemic or intrahypothalamic administration of receptor blocking anti-estrogens or aromatization inhibitors in T-treated castrates. The demonstration, however, that the purportedly nonaromatizable androgen, DHT (for critique see Yahr, 1979), can facilitate copulatory performance in a number of species, including rats, combined with the failure of estrogens to stimulate male sexual behavior in some of these species, such as guinea pigs, clearly indicates that the intrahypothalamic aromatization of androgens to estrogens is not universally required for androgenic behavioral activity.

NEUROCHEMICAL ACTIONS OF GONADAL STEROID HORMONES

The genomic model of steroid hormone action (see Figs. 15-1 and 15-2) predicts that *de novo* synthesis of specific proteins should directly follow the steroid-receptor-acceptor activation of transcription in the nucleus. In peripheral target tissues these effects have been studied in great detail and have resulted in the isolation of specific mRNAs and their respective proteins (Chan et al., 1978; Gorski and Ganon, 1976). However, in the brain, progress in this area has been severly hampered by the limited availability of steroid-specific target tissue, for unlike the uterine endometrium, the brain is a very heterogenous tissue containing more nontarget than target cells, and the lack of a major protein product, that is, the brain does not produce a steriod-induced protein with a similar selective magnification as ovalbumin, avidin, vitellogenin, casein, and other unique proteins found in peripheral target tissues. The recent demonstration of estrogen-specific induction of P_4 receptors in the brain (*vide supra*) and various enzymes (*vide infra*) may change this.

Deoxyribonucleic Acid, Karyokinesis, and Synaptogenesis

Since neurogenesis within the brain is believed to be resticted to the early perinatal period in mammals (Creps, 1974), it is not surprising that there are no definitive reports of gonadal steriod-induced increases in DNA synthesis in the adult brain. There is, however, indirect evidence for such an effect in the neonatal brain. DNA synthesis inhibitors ameliorate androgen- and estrogen-induced perinatal neuroendocrine sexual differentiation (Salaman, 1974; see also chapter 16 by Kuhn and Schanberg). Since these gonadal hormones can produce clear neuroanatomical changes in preoptic, hypothalamic, and septal regions (Gorski et al., 1978; Greenough et al., 1977; Matsumoto and Arai, 1980; Raisman and Field, 1973), it is possible that the DNA synthesis inhibitor-sensitive effects and the neuroanatomical changes reflect the same mechanisms. Alternatively, the neuroanatomical response may reflect dendritic or axonal hypertrophy or sprouting, that is, RNA and protein synthesis changes, rather than selective neuronal karyokinesis. The former possibility is supported by the recent observation that estrogen therapy in ovariectomized adult rats facilitates axodendritic synapse formation in the arcuate nucleus following hypothalamic deafferentation (Matsumoto and Arai, 1979). Because this same surgical procedure reduces estrogen receptor density within the deafferented tissue (Carrillo, 1980), the maintenance of hyperphysiological estrogen titers may be required for the maintenance of target neuron stimulation and increased synaptogenesis. This direct effect on neuronal circuitry appears to be on a dose- and age-related continuum, since the maintenance of high estrogen titers in adulthood can lead to the production of multifocal lesions within the arcuate region (Brawer and Naftolin, 1979; Brawer et al., 1980).

Ribonucleic Acids

Correlative support for gonadal steroid-induced increases in neuronal RNA synthesis has been obtained from *in vivo* and *in vitro* studies purporting to demonstrate an increase in either total RNA content or [^{3}H]uridine labeling of RNA in adult female rat hypothalamus during proestrus or estrus (Foreman et al., 1977; Novakova et al., 1971; Salaman, 1970). Although these changes coincide temporally with increased E_2 and P_4 blood titers, they also synchronize with estrous cycle fluctuations in other hormones that may themselves directly produce these effects, such as luteinizing hormone-releasing hormone (LH-RH) (Biro et al., 1978). Furthermore, other studies have failed to find estrous cycle-related changes in hypothalamic RNA content or synthesis (Litteria and Schapiro, 1972; Luck, 1975). This confusion may be a consequence of the comparatively low sensitivity of the crude RNA measurements, compounded by the low density of estrogen target cells in the hypothalamus (Stumpf and Grant, 1975).

Direct support for an effect of estrogen on brain RNA synthesis has been derived from *in vitro* studies demonstrating close temporal correlations between a transient estrogen-induced increase in hypothalamic DNA-dependent RNA polymerase II activity, that associated with mRNA synthesis, and estrogen nuclear receptor occupancy (Peck et al., 1979). Also, intracerebral applications of RNA synthesis inhibitors can block behavioral and neuroendocrine actions of estrogens. For example, in one study implantation of actinomycin D into the POA at 12, but not at 21, hours after an SC injection of E_2B inhibited the display of sexual receptivity for at least 68 hours (Whalen et al., 1974). This inhibition was not permanent, however, since the hormone administration one week later produced normal levels of receptivity. The transient loss and recovery of estrogen behavioral sensitivity following actinomycin D administration parallels the transient appearance of intraneuronal nucleolar segregation in the POA of similarly treated ovariectomized and intact female rats (Hough et al., 1974; Quadagno et al., 1980).

Proteins, Peptides, and Peptidases

General Effects on Protein and Peptide Synthesis

There have been numerous studies examining cerebral protein synthetic responses to endogenous and exogenous steroids. Somewhat surprisingly, the early autoradiographic work on this subject indicated that gonadectomy of adult male and female rats resulted in an increase of amino acid incorporation into proteins in the arcuate nucleus and many other regions of the hypothalamus (Litteria, 1973a; Nikai et al., 1971; Yaginuma et al., 1969). A notable exception was the MPOA in which ovariectomy reduced the incorporation, determined autoradiographically (Litteria and Thorner, 1974). These effects were interpreted to reflect the gonadectomy-induced hypersecretion of gonadotropins from the adenohypophysis and the presumed hyperstimulation of hypothalamic LH-RH synthesis and release. This explanation is consistent with the recent demonstration that LH-RH concentrations in the MBH are more responsive to the direct feedback actions of gonadal steroids (T, DHT, and E_2) than are those in the POA (Kalra and Kalra, 1980).

In a rather unusual study with adult male rats, it was discovered that the *in vitro* incorporation of [^{3}H]Phe into proteins in hypothalamic slices was decreased three weeks after castration, but it was significantly increased, compared to gonadally intact controls, six weeks after castration (Moguilevsky et al., 1971a). In each case, the systemic administration of T prevented the effects of castration. Neither castration nor exogenous T had any effect on protein synthesis in the cerebral cortex. Since LH secretion also responded biphasically to castration in the adult male rat (Badger et al., 1978) and since gonadal steroids are thought to activate the *de novo* synthesis of LH-RH (Kalra and Kalra, 1980), the fluctuations in hypothalamic protein synthesis observed following castration and androgen replacment therapy may have reflected the differential rates of LH-RH synthesis.

In adult female rats, the incorporation of amino acids into proteins peaked at proestrus in the anterior hypothalamus (AH) and at estrus in the posterior hypothalamus (PH); the sample included the bulk of the MBH (Moguilevsky and Christot, 1972; Moguilevsky et al., 1971b). When exogenous E_2 was added to the incubation media containing brain fragments from immature female rats, [^{3}H]Leu incorporation into proteins was increased in the POA-AH, but not in other brain regions (Faigon and Moguilevsky, 1976). The *in vitro* studies clearly suggest a direct stimulatory effect of ovarian steroids on hypothalamic protein synthesis. This is further supported by semiquantita-

tive autoradiographic work in which the MPOA, arcuate, paraventricular, and supraoptic areas displayed an increase in the
density of silver grains after [^{3}H]Lys injections during proestrus with progressive diminution during estrus and diestrus,
thus paralleling the endogenous E_2 blood titers (Litteria,
1973b).

An extensive series of studies on [^{35}S]Met incorporation into
rat brain proteins by MacKinnon and co-workers (Burnet and
MacKinnon, 1975; terHaar and MacKinnon, 1973, 1975, 1977) has
added certain complexities to this system. The *in vivo* incorporation of this precursor into proteins in the median eminence
(ME) was variable with a peak each day at the onset of the dark
phase of the lighting cycle. The largest dark-onset peak was
found on the evening of proestrus when endogenous estrogen
levels were also highest. To normalize the diurnal fluctuations in protein synthesis seen in all brain regions, the
[^{35}S]Met incorporation values for the putamen, which lacks
nuclear E_2 receptors, were divided into those for the other
brain regions. Using these normalized values, the amygdala and
POA were revealed to show diurnal fluctuations in [^{35}S]Met
incorporation, with the biggest peak occurring 15 to 18 hours
prior to the proestrus LH surge. The administration of an
anti-E_2 antibody preparation on diestrus II at 1600 hours
blocked the [^{35}S]-Met incorporation peak usually seen in the ME
and pituitary, at the onset of darkness during proestrus. This
antibody injection also blocked ovulation as well as the diurnal fluctuations and proestrus peaks in [^{35}S]-Met incorporation, relative to putamen, in the amygdala and POA. The administration of exogenous E_2 and P_4 on the morning of proestrus
tended to advance the preovulatory surge of LH secretion and
[^{35}S]Met incorporation in the ME and pituitary, by three to six
hours. This effect was more pronounced with P_4 than with E_2,
suggesting a synergistic interaction of the exogenous P_4 with
endogenous E_2 in a fashion similar to that normally observed
during the evening of proestrus. Similar injections of 5α-DHP
and 20α-OH-P_4 tended to reduce the proestrus surge in LH secretion and [^{35}S]Met incorporation in the ME and amygdala, but
not in the POA, and pituitary. Ovariectomy blocked both the
diurnal fluctuations in LH secretion and [^{35}S]-Met incorporation into brain proteins. A single 20 μg injection of E_2B
restored both these diurnal rhythms for at least 12 days.

The consistency of the positive correlation between increased
LH secretion, presumably due to the positive feedback of E_2 and
P_4 and the increased incorporation of labeled amino acids into
proteins in the rat brain, and pituitary, strongly suggests a
functional relationship. This hypothesis is strengthened by
the numerous studies (Jackson, 1975) in which protein or RNA
synthesis inhibitors blocked the positive feedback actions of

gonadal hormones on LH secretion. It is relevant that the induction of a proestruslike LH surge requires virtually the same gonadal steroid blood titer as does the induction of sexual receptivity (*vide infra*). Furthermore, protein or RNA synthesis inhibitors block the onset of sexual receptivity in female rats in response to either endogenous or exogenous ovarian steroid stimulation (Quadagno et al., 1980; Watts et al., 1979; *vide infra*). A generalized increase in [^{3}H]Leu incorporation into brain proteins following E_2B administration has also been reported in ovariectomized guinea pigs (Wade and Feder, 1974).

Specific Hormone-Induced Brain Proteins

Although these demonstrations of steroid-induced increases in total protein synthesis are important, they provide no evidence for the induction of specific proteins related to the neuroendocrine or behavioral actions of gonadal hormones. Studies with rat uterus using a double-labeling technique and sodium dodecyl sulfate (SDS) polyacrylamide gels led to the discovery of a soluble estrogen-induced protein (E_2IP) with an approximate molecular weight of 39 to 45K Daltons and a pI of 4.5. The synthesis of E_2IP was maximal during proestrus in intact adult females and at one hour after exogenous E_2 administration in immature or ovariectomized females. E_2IP synthesis in the uterus was blocked in estrogen-primed females by the administration of estrogen antibodies, or antiestrogens or protein or RNA synthesis inhibitors (King et al., 1974; Manak et al., 1980).

Remarkably, the use of similar techniques has revealed the existence of a hypothalamic E_2IP, which is apparently identical to uterine E_2IP! The kinetics of production, mobility on polyacrylamide gels, pI, peptide maps, and immunological cross reactivity all support the contention that uterine and hypothalamic E_2IP are the same protein (Beinfeld and Packman, 1976; Reiss and Kay, 1979; Vertes et al., 1978; Walker et al., 1979). E_2IP synthesis has not been observed in nontarget regions of the brain, including the cerebral cortex. The function of E_2IP is still a mystery, but the intertissue similarity strongly suggests that it plays a fundamental regulatory role in estrogen-receptor-acceptor interactions. This role could include: (1) the synthesis of the 4S estrogen receptor, note that the *de novo* synthesis of both the E_2IP and 4S receptor proteins are stimulated by estrogen, but they continue to be synthesized, at markedly reduced rates, even in the absence of estrogen (Little et al., 1975; Walker et al., 1979); (2) the second component of the 5S receptor complex, note that both

E_2IP and component X have been shown to be similar if not identical in all target tissues (Reiss and Kaye, 1979; Yamamoto, 1974); (3) the nuclear nonhistone acceptor, note, however, that the E_2IP and acceptor proteins have markedly different physicochemical properties (Mainwaring et al., 1976; Reiss and Kaye, 1979; Walker et al., 1979); or (4) the cytosol factors that inhibit estrogen receptor activation or aggregation and receptor-acceptor binding (Asai et al., 1979; Sato et al., 1979, 1980; Shen et al., 1979).

There is a considerable body of biochemical, physiological, and behavioral data on another intracerebral or uterine E_2IP, namely, the high-affinity progestin receptor. As described previously, the hypothalamic, preoptic, and uterine progestin receptor concentrations decrease following ovariectomy and increase in an estrogen-specific, dose-dependent fashion after hormone replacement therapy. The time course of synthesis (24 to 48 hours versus 1 hour), the pI (5.8 versus 4.5), and other factors make it unlikely that the progestin receptor and the E_2IP are the same protein. The gonadal steroid induced activation or synthesis of brain choline acetytransferase (Ch AcT) and other enzymes involved in neurotransmitter and carbohydrate metabolism will be discussed later.

A final example is a specific class of hypothalamic peptidases (Griffiths et al., 1975; Kuhl et al., 1979). The activity of these soluble L-Cys arylamidase enzymes are restricted to certain oligopeptides, such as LH-RH, decrease following gonadectomy in adult male and female rats and rabbits, and increase after hormone replacement therapy, with T, DHT, and E_2, with a latency of approximately 16 to 18 hours. In gonadally intact adult female rats, the activity of these enzymes fluctuate during the estrous cycle (maximal during diestrus) and during the daily lighting cycle (maximal during darkness). Although these variations in LH-RH degradative activity of hypothalamic extracts clearly correlate negatively with LH secretion rates, they also correlate negatively with the induction of sexual behaviors (*vide infra*). Furthermore, since LH-RH administration facilitates the display of both male and female sexual behavior (for reviews, see Mauk et al., 1980; Moss et al., 1975), it is possible that gonadal hormones regulate these behaviors partially through a modulation of the synthesis and degradation of hypothalamic (and mesencephalic) LH-RH.

Behavioral Correlates

POA implants of the protein synthesis inhibitor cycloheximide, six hours before or up to 12 hours after an SC injection of E_2B

in ovariectomized rats (Quadagno and Ho, 1975) or during the late afternoon of diestrus day 2 in intact rats (Watts et al., 1979), inhibits the induction of sexual receptivity. Dose-response studies have further shown that the VM is as sensitive as the POA to the inhibitory effects of cycloheximide on female sexual behavior (Meinkoth et al., 1979). Similar implants in the lateral septum, caudate nucleus, or cortical and medial amygdaloid nuclei failed to inhibit receptivity (Leehan et al., 1979). These data therefore clearly support the notion that estrogen-dependent cerebral protein and RNA synthesis is required for the induction of sexual activity in the rat. The apparent localization of these necessary estrogenic actions to the VM and POA is in agreement with the findings from estrogen and antiestrogen intracerebral cannulation studies (*vide supra*) and from intracerebral electrical stimulation and lesion studies (*vide infra*).

Limited work has suggested that cerebral protein synthesis may also be involved in the sequential inhibitory action of P_4 on sexual receptivity in estrogen-primed female guinea pigs (Wallen et al., 1972), the inhibitory actions of DHT on estrogen-induced receptivity in female mice (Gray and Luttge, 1978), and in the facilitatory actions of endogenous hormones (androgens ?) on copulatory behavior in mice (Quadagno et al., 1976).

Polyamines and Gamma-Aminobutyric Acid

The diamine putrescine (PUT) and the polyamines spermidine (SPD) and spermine (SPM) are ubiquitously distributed in mammalian tissues, whereas the neurotransmitter gamma aminobutyric (GABA) is confined almost exclusively to the nervous system (for reviews, see Roberts et al., 1976; Russell, 1980; Shaw, 1979; Williams-Ashman and Canellakis, 1979).

Synthesis and Interconversions

Studies on the biosynthesis and turnover of the polyamines and GABA are confounded by the fact that many of these compounds are interconvertible and by the fact that they display a variety of end-product inhibition- and facilitation-modulations of the synthetic enzymes. The traditional biosynthetic pathway for the polyamines (see Fig. 15-5) has L-ornithine (ORN) and L-methionine (Met) as the primary precursors with the rate limiting enzyme for PUT synthesis being ornithine decarboxylase (ODC), whereas propylamine group addition to PUT and SPD to form SPD and SPM, respectively, are catalyzed by spermidine and spermine synthase. This ostensively straight-through biosynthetic scheme of ORN → PUT → SPD → SPM has been challenged by the recent demonstration that PUT, SPD, and SPM are all readily interconvertible in mammalian brain (Antrup and Seiler, 1980). Similarly, although GABA can be synthesized (see Fig. 15-5) from glutamic acid by the action of glutamic acid decarboxylase (GAD), it can also be formed in the hypothalamus and other brain regions by acetylation of PUT followed by deamination (Seiler et al., 1979a,b). Ornithine can also act as an effective precursor for GABA in synaptosomal preparations from hypothalamus and other brain regions as well as in tissue slices from striatum (Murrin, 1980). The pathway for this latter synthesis could be via PUT, but a more likely route is via production of glutamic acid semialdehyde followed by oxidation to glutamic acid and finally decarboxylation to GABA (Jung and Seiler, 1978; Peraino and Pitot, 1963; Ragland and Pitot, 1971) (see Fig. 15-5). A further complication of these intertwined biosynthetic schemes was provided by the recent demonstration that in addition to PUT, SPD, and SPM controlling their own biosynthesis (Williams-Ashman and Cannelakis, 1979), GABA can stablize ODC and thereby increase the over-all production of PUT (McCann et al., 1979; Seiler et al., 1979a,b). This latter effect is undoubtedly magnified by the fact that ODC has by far the shortest half-life of any known mammalian enzyme (Russell, 1980).

Polyamine Actions and Hormone-Induced Synthesis Modulations

Although our understanding of polyamine actions is incomplete, there is strong support for a regulatory role in nucleic acid and protein synthesis and membrane function. In their non-neural target tissues, trophic hormones characteristically stimulate a rapid increase in ODC activity, polyamine concentrations, and, in some cases, tissue hypertrophy or hyperplasia. This includes the actions of the gonadal hormones in the uterus (Kaye et al., 1971), prostate (Danzin et al., 1979), kidney (Nawata et al., 1980), and pituitary (Gray et al., 1980). Even though the intracerebral actions of polyamines are probably much more pronounced during perinatal neurogenesis and myelination (see chapter 16 by Kuhn and Schanberg), they have also been implicated in various aspects of DNA and protein synthesis in the adult brain (Fleisher-Lambropoulos et al., 1975; Goertz, 1979; Stipek et al., 1978). A single 20 μg injection of E_2B in ovariectomized adult rats has been shown to increase the concentrations of PUT in the MBH and POA-AH 12 hours after the E_2B, whereas the concentration of PUT in the cerebral cortex, as well as that of SPD and SPM in all three brain regions, did not change (Gray et al., 1980). The elevated

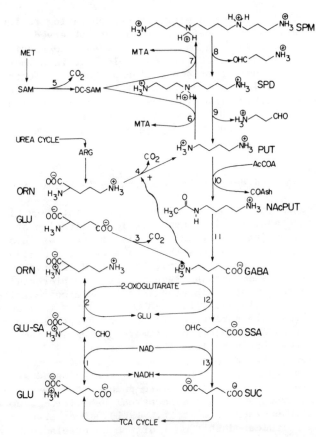

Fig. 15-5.   Structural formulas and metabolic pathways for
polyamine and GABA biosynthesis and catabolism.
Abbreviations: GLU:glutamic acid; GLU-SA:glu-
tamic acid semialdehyde; PUT:putrescine;
NAcPUT:N-acetylputrescine; ORN:ornithine; SPM:
spermine; SPD:spermidine; SUC:succinic acid;
SSA:succinic acid semialdehyde; AcCOA;acetyl
coenzyme A; COAsh:reduced coenzyme A; NAD:
nicotine adenine dinucleotide; SAM:S-adeno-
sylmethionine; DC-SAM:decarboxylated S-adenosylmethionine; MTA:methylthioadenosine.
Enzymes: (1) GLU-SA dehydrogenase, (2) ORN
aminotransferase, (3) GAD, (4) ODC, (5) SAM
decarboxylase, (6) SPD synthase, (7) SPD
synthase, (8) SPM oxidase, (9) SPD oxidase, (10)
PUT N-acetyltransferase, (11) MAO, (12) GABA
transaminase, (13) SSA dehydrogenase.

levels of PUT in these two estrogen-concentrating regions were
only temporary, since baseline values were observed in the 24,
36, and 48 hour samples. There are at least two possible mechanisms for this effect: PUT values could have been increased
consequent to induction or activation of ODC or an estrogen-
induced increase in the conversion of SPD to PUT (Antrup and
Seiler, 1980). It will be interesting to correlate the effects
of various hormones, antihormones, and RNA and protein synthesis inhibitors on the production of brain polyamines and
hormone-induced proteins.

GABA Actions and Hormone-Induced Synthesis Modulations

GABA is present in the hypothalamus in comparatively high concentrations (Fahn, 1976); iontophoretic application of GABA can
inhibit the ongoing electrical activity of neurons in several
hypothalamic nuclei, such as the paraventricular nucleus
(PaVN), supraoptic nucleus (SON), VM, and LH (Curtis and
Johnston, 1974) and the GABA receptor blocker picrotoxin can
abolish the recurrent inhibition seen after arcuate stimulation
(Yagi and Sawaki, 1975). Although GAD activity is distributed
unevenly across the various nuclei of the hypothalamic region,
the rates of GABA synthesis often markedly exceed those observed in other brain regions (Tappaz et al., 1977; Wallis and
Luttge, 1980). Exceptions to this generalization include the
substantia nigra (SN), which has a well-established striatal
GABAergic input to link the caudate-putamen and amygdala with
this highly important (to female sexual behavior - *vide infra*)
mesencephalic nucleus (Kim et al., 1971). Deafferentation
studies have indicated that most of the hypothalamic GAD is
localized within intrinsic interneurons (Tappaz and Brownstein,
1977; Wallis and Luttge, 1977). Numerous neuroendocrine and
behavioral studies have suggested a facilitatory role of
GABAergic neurons in the release of LH (Vijayan and McCann,
1978; Wallis, 1976) and prolactin (Mioduszewski et al., 1976),
and in the induction of sexual receptivity in female rats
(McGinnis et al., 1980b). Since the gonadal steroids are well
known to be directly involved in these same neuroendocrine (see
chapter 4 by Frohman and Berelowitz) and behavioral (*vide
infra*) events, it is likely that they are also involved in
modulating the relevant GABAergic neurons (see Table 15-4).
    The data on gonadal hormone effects on cerebral GABA concen-
trations are summarized in Table 15-4. Early work on the
biosynthetic enzyme, GAD, suggested that chronic estrogen
treatment reduced whole-brain activity in rats (Pandolfo and
Macione, 1963). Recently it has been shown that treatment of
ovariectomized rats with as little as 1 μg E_2B/day for two days

Table 15-4. Representative Effects of Steroid Hormones on Brain GABA Neurons

Treatment	Effect(s)	Reference(s)
Content (GABA)[a]		
♂ → ♂̸	↓ in Amg & Septum, ↑ in Midbrain	Early and Leonard (1978)
♂̸ + TD*	↑ in Amg & Septum, – in Midbrain	Early and Leonard (1978)
♂̸ + E_2D	↑ in Amg, Septum & Midbrain	Early and Leonard (1978)
♂̸ + DHTD	– in Amg & Septum, ↓ in Midbrain	Early and Leonard (1978)
♀ → ♀̸	↑ in FCx	Saad (1970)
Synthesis (GAD activity)		
♀̸ + E_2B†	↓ in CMAmg, AH, Arc, NIST, NISM, VM, SM, VTA & SN	Gordon et al. (1977, 1979), Wallis and Luttge (1980)
♀̸ + P_4	↓ in AH, Arc, DHpc, NIST, NISM, OB, MPOA, VS, VM & VTA	Wallis and Luttge (1980)
♀̸ + E_2B + P_4	↓ in LS, – in AH, Arc, Cx, DHpc, OB, MPOA SM & VTA	Wallis and Luttge (1980)
♀̸ + E_2 in CN or Amg	↓ in SN, – in VTA	McGinnis et al. (1979)
♀̸ + E_2 in NA	– in SN, ↓ in VTA	McGinnis et al. (1979)
♀̸ + E_2 in POA or VM	– in SN & VTA	McGinnis et al. (1979)

[a] ♂: intact male; ♂̸: castrated ♂; ♀: intact female; ♀̸: ovariectomized ♀; ↑: increase; ↓: decrease; –: no change;
*: decanoate; †: benzoate.
All other abbreviations are defined in the text or in Tables 15-1 and 15-2.

significantly reduces GAD activity in a variety of known estrogen target regions (see Table 15-4) (Wallis and Luttge, 1980). The addition of a single 500 μg injection of P_4 on day three, instead of E_2B or oil, blocked the reduction of GAD activity in most of these brain regions, whereas it stimulated the reduction in GAD in the lateral septal nucleus. None of these manipulations affected GAD activity in cerebral cortex, dorsal hippocampus, olfactory bulb, or, surprisingly, MPOA. In later work it was shown that with long-term ovariectomy, E_2B replacement therapy also significantly reduces GAD in the VTA (McGinnis et al., 1980a).

Enzyme kinetic experiments were performed revealing that estrogen reduced the V_{max} of GAD for glutamate, without changing the K_m for this substrate or for the required cofactor, pyridoxyl phosphate (Wallis and Luttge, 1980). These data suggest that estrogen reduces GAD activity by direct or indirect action on its synthesis in the soma rather than by modulating the efficacy of its utilization of substrate or cofactor within the presynaptic terminal (Fonnum, 1968). It is also possible that estrogen reduces the V_{max} values by selectively increasing the rate of GAD catabolism.

The inhibitory effects of estrogen on GAD need not be direct; they could be mediated transsynaptically. In support of this, electrolytic lesions of the septal region in both male and female gonadectomized rats or suction aspirations of the olfactory bulbs in ovariectomized rats eliminates the inhibitory effects of E_2B replacement therapy on GAD activity in the VTA and SN (Gordon et al., 1977, 1979; Tyler et al., 1979). E_2-containing cannulas in the caudate and amygdala significantly reduced GAD activity in the SN, but not in the VTA, whereas implants in the nucleus accumbens reduced GAD in the VTA, but not in the SN (McGinnis et al., 1980a). Unilateral implantation of E_2 in the POA and VM of the hypothalamus failed to influence GAD in either the VTA or SN. It is particularly interesting to note that estrogen-induced reductions in GAD activity in the VTA and SN were not produced by implants in the two regions richest in estradiol receptors, the POA and VM. GAD activity was reduced in the VTA and SN with E_2-implants in brain regions known to be rich in dopaminergic afferents as well as GABAergic efferents to either the VTA or SN (Fuxe et al., 1975; Kim et al., 1971). A further differentiation of the striatal-nigral and mesolimbic-VTA estrogen-sensitive GABAergic systems was provided by the demonstration that a single systemic injection of 8 μg E_2B in long-term (60 days) ovariectomized rats resulted in a significant reduction in VTA GAD activity within 12 hours, whereas a reduction in SN GAD activity was not observed before 29 hours after E_2B injection (McGinnis et al., 1980a). Furthermore, whereas GAD activity in the VTA returned to control levels by 29 hours, GAD in the SN remained depressed for at least another 24 hours.

Several lines of evidence suggest that the estrogen-induced effects on brain polyamines and GAD activity are mediated via different mechanisms and possibly in different neurons. The fact that the biosynthesis of PUT and GABA are interrelated should not be overinterpreted, since GAD is localized within the synaptosomal fraction, whereas ODC is not (Fonnum, 1968; Seiler and Sarhan, 1980). Thus, although an increase in the cerebral levels of GABA, for instance, via an inhibition of the degradative enzyme, GABA-T, leads to an increase in the intra-cerebral concentrations of PUT (Seiler et al., 1979a,b), these effects probably do not occur in the immediate presynaptic region. That the estrogen-induced effects on hypothalamic PUT concentrations and VTA GAD activity are both detected within the first 12 hours of estrogen exposure is probably coinci-dental, since they are in opposite directions (Gray et al., 1980; McGinnis et al., 1980a). It should also be noted that, whereas E_2B exposure stimulated an increase in PUT in the POA, this brain region failed to display an E_2B-induced reduction in GAD activity and when directly exposed to E_2B it failed to alter GAD in either the VTA or SN (Gray et al., 1980; Wallis and Luttge, 1980; McGinnis et al., 1980a).

GABA Actions and Gonadal Hormone-Induced Sexual Behavior

A recent series of studies by Gorski and co-workers has led to the theory that ovarian hormone-induced variations in the ac-tivity of the GABAergic input to the SN and VTA dopaminergic neurons may significantly influence the induction of sexual receptivity in female rats. For example, infusion of the GABA-T inhibitor, hydrazinopropionic acid, directly into the SN in-creased receptivity scores in E_2B-primed ovariectomized females (2 μg/day for three days; note that no P_4 was given prior to testing) (McGinnis et al., 1980b). Since this treatment pre-sumably increased the endogenous levels of GABA, as well as PUT via the GABA-mediated stabilization of ODC, it was suggested that increased GABAergic activity in the SN facilitates recep-tivity, possibly via an inhibition of the inhibitory dopamin-ergic influences on receptivity. This hypothesis was supported by the observation that infusions of exceptionally low doses of the GABA agonist, muscimol, into the SN also increased recep-tivity, but in only three out of five females, whereas infu-sions of the GABA receptor blocker picrotoxin inhibited the induction of receptivity. This latter effect, however, was only apparent in females previously subjected to electrolytic lesions in the septal area. These lesions have been shown in earlier work to produce a dramatic increase in estrogen-induced receptivity in both female and male rats (Gordon et al., 1977, 1979). Septal, as well as olfactory bulb lesions, which also increases estrogen-induced receptivity, have been further shown to inhibit the reductions in GAD activity in the SN and VTA, which normally result from either systemic or intracerebral administration of estrogen (see Table 15-4) (Gordon et al., 1977; Tyler et al., 1979). Thus, by preventing the normal estrogen-induced reductions of GABAergic activity in the SN and VTA, these regions remain under the inhibitory influence of GABA and are therefore less involved in the dopamine-mediated inhibition of sexual receptivity.

Catecholamines

The catecholamine neurotransmitters, dopamine (DA), NE and, to a lesser extent, epinephrine (E) have been widely implicated in the mediation of certain aspects of gonadal steroid hormone actions in neural development and aging, gonadotropin and prolactin secretion regulation, and sexual, agonistic, parental, and other behaviors. Since many of these topics are extensively reviewed elsewhere in this volume (see Frohman and Berelowitz, chapter 4; Kuhn and Schanberg, chapter 16; Renaud, chapter 5; Simpkins and Estes, chapter 17), the present discussion will focus on selected examples related to the gonadal hormone-induced variations in catecholamine content, turnover, synthesis, catabolism, release, reuptake and receptor binding in the brains of mammals (see Fig. 15-6 for synthetic and catabolic pathways and structural formulas).

Content

Gonadectomy, exogenous hormone therapy, and estrous cycle-related hormone fluctuations have been shown to modify the concentrations of NE and DA in numerous hypothalamic and limbic regions. Examples of these effects are presented in Table 15-5. Although it is tempting to correlate catecholamine content changes with presumed variations in catecholaminergic neural activity, it is important to remember that changes in concentration may be the result of several factors, including transmitter turnover, synthesis, and catabolism.

Turnover

Catecholamine turnover can be measured by preventing synthesis with a drug, normally α-methyl-p-tyrosine (α-mpt) that inhibits TH and by measuring the rates of NE and DA depletion. An elec-

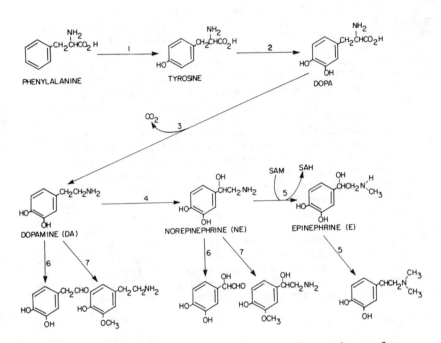

Fig. 15-6. Structural formulas and metabolic pathways for catecholamine biosynthesis and catabolism. Enzymes: (1) phenylalanine hydroxylase or tyrosine hydroxylase (TH), (2) TH, (3) aromatic amino acid decarboxylase, (4) dopamine-β-hydroxylase (DBH), (5) phenylalanine N-methyltransferase (PNMT), (6) monoamine oxidase (MAO), (7) catechol-0-methyltransferase (COMT).

trophysiologically active system is believed to display a greater degree of depletion, and hence a higher turnover rate, than a less active system. Using this technique, it has been reported that 10 days following castration there is a significant increase in DA, but not NE turnover in the ME of male rats (Kizer et al., 1978). In contrast, it has been reported very recently that systemic administration of T decreases NE, but has no effect on DA turnover in the slightly larger MBH region (Simpkins et al., 1980). In the latter study, the systemic T treatment reduced DA turnover in the POA, whereas NE turnover in this region was unchanged. When the MBH was stimulated directly with T, NE turnover in this region was again depressed (Simpkins et al., 1980), whereas DA turnover in the POA was unchanged.

In a very recent study of the female rat estrous cycle, NE turnover rates increased dramatically in the MPOA during the proestrus estrogen/progestin surge (13-fold difference in pro-estrus/diestrus turnover rates) (Honma and Wuttke, 1980). Other techniques have yielded qualitatively similar findings (Lofstrom, 1977). In ovariectomized female rats, the systemic administration of 5 μg E_2B decreased NE turnover rates 54 hours later in the lateral septal area, NIST and in the central gray catecholamine-rich regions, but it increased NE turnover in the anterior hypothalamic nuclei and decreased DA turnover in the diagonal band nucleus (Crowley et al., 1978b). All of these estrogenic effects were prevented by a single injection of 1.5 mg P_4 given six hours before the catecholamine assays. P_4 treatment of E_2B-primed females actually increased DA turnover in the diagonal band nucleus and in the NIST, and it either increased or decreased NE turnover in several hypothalamic and limbic area nuclei (see Table 15-5). Similar findings have been reported by other laboratories (Lofstrom et al., 1977; Munaro, 1977; Simpkins et al., 1979). If turnover rates do in fact reflect changes in neuronal synaptic activity, then both estrogen- and progestin-induced increases and decreases in NE and DA synaptic activity are suggested by these studies.

Synthesis

TH activity in the male rat ME has been reported to increase by more than 50 percent seven days after castration, an effect that was prevented by replacement therapy with 100 μg TP/day (Kizer et al., 1974). TH activity in several other hypothal-amic nuclei did not change (see Table 15-5). Lesions of the ventral noradrenergic afferents to the hypothalamus markedly reduced TH activity in the ME (by 85 percent), but failed to alter the castration-induced effects, suggesting that the latter are directed primarily at the DA neurons of the arcuato-ME pathway (Kizer et al., 1976a). This is consistent with the lack of effects of castration on hypothalamic dopamine-β-hydro-xylase (DBH) (Kizer et al., 1974) and with the reported in-crease in DA, but not NE turnover in the ME in castrated male rats (Kizer et al., 1978). They are not, however, consistent with the reported decrease in NE turnover in the larger MBH sample (Simpkins et al., 1980). Furthermore, not all labora-tories agree with the reported increase in TH activity in the ME following castration (e.g., Nakahara et al., 1979). Also, although it originally was claimed that long-term castration increased DA synthesis in pooled samples of limbic tissue, paralleling the loss in male sexual behavior (Engel et al., 1979), these findings were not replicated in a subsequent study (Alderson et al., 1979).

Estrous cycle changes, ovariectomy, and subsequent estrogen/progestin replacement therapy have also been reported to modify catecholamine synthesis in female brains. [^{3}H]Tyrosine conversion into [^{3}H]NE increases in the hypothalamus during proestrus (Donoso and Moyano, 1970), whereas it decreases in the anterior hypothalamus following E$_2$B and P$_4$ treatment of ovariectomized females (Bapna et al., 1971). Short-term or low-dose estrogen replacement therapy appears to increase TH activity in hypothalamus, but the addition of P$_4$ or increasing the estrogen duration or dose reduces it in the MBH (Beattie et al., 1972; Luine et al., 1977; Tobias et al., 1979). P$_4$ and other progestational steroids reduce TH activity *in vitro* by competitive inhibition with the pterin cofactor (DMPH$_4$), which is similar to NE and DA end-product inhibition (Beattie and Soyka, 1973). In contrast, when E$_2$ is incubated with samples of MBH, TH is not altered (Luine et al., 1977). The latter finding is surprising, because the catechol estrogens, 2-OH-E$_2$ and 2-OH-E$_1$ are very effective in competitively inhibiting hypothalamic, preoptic, amygdaloid, striatal, and hippocampal TH *in vitro* by a mechanism like that for P$_4$ (Foreman and Porter, 1980; Lloyd and Weisz, 1978; Lloyd and Eversole, 1980). There are few reports on the effects of ovarian hormones on DBH, although drug studies have frequently suggested that this enzyme is required for certain gonadal hormone actions on behavior and adenohypophysial hormone secretion regulation (Everitt et al., 1975; Kalra and McCann, 1975).

Catabolism

The activity of monoamine oxidase (MAO) in the anterior hypothalamus, amygdala, and other estrogen target regions increases during proestrus and estrus in the rat (Kurachi and Hirota, 1969; Zolovick et al., 1966). Following ovariectomy, MAO increases in hypothalamic regions, but not in cerebral cortex. Replacement therapy with E$_2$ or E$_2$B, but not TP or P$_4$ reduces MAO activity in MBH and amygdala (Kato and Minaguchi, 1964; Kurachi and Hirota, 1969; Luine et al., 1975b). The significance of these findings for catecholamine catabolism in female rats has been challenged recently by the demonstration that it is MAO type A, preferential for indoleamines, rather than MAO type B, preferential for catecholamines, which is modified by estrogen replacement therapy in females (Luine and McEwen, 1977a). In castrated male rats and mice, MAO activity was increased in medial preoptic (rats) and POA, hypothalamic and amygdaloid area tissue pools (mice) following T or TP, but not E$_2$B replacement therapy (Luine et al., 1975a,b,1979); types A and B were not distinguished.

The catechol estrogens are extraordinarily effective inhibitors of methylation by catechol-O-methyltransferase (COMT) in brain (Breuer and Koster, 1974; Lloyd et al., 1978). In brain a molar ratio for 2-OH-E_2:NE of as little as 0.3 inhibits NE methylation by more than 97 percent and a molar ratio for 2-OH-E_2:DA of 0.03 still inhibits the DA methylation by more than 50 percent. With a K_i of 6-9 μM, and an endogenous concentration in brain that may exceed these values in certain regions (however, see Fishman and Martucci, 1979; *vide supra*), these data are consistent with the notion that the catechol estrogens may function as endogenous modulators of catecholamine catabolism.

Release

As described earlier, ovariectomy and estrogen replacement therapy, respectively, increase and decrease the NE concentrations of several hypothalamic nuclei, including the anterior hypothalamus. This suggests changes in catecholamine release from presynaptic terminals. Recently, the efflux rate of both NE and DA from short-term organ culture explants of whole hypothalamus from immature female rats was shown to be increased in a dose-dependent fashion by E_2 and the nonsteroidal estrogen, DES (Paul et al., 1979). The biologically inactive enantiomer of E_2, 17α-estradiol, and the comparatively weak estrogens E_1 and E_3 and the glucocorticoid B were ineffective. The latency for the estrogen-induced catecholamine efflux was between 40 and 60 minutes, suggesting that the effect may not be directed at the presynaptic terminals but may instead involve protein synthesis in the neuronal somas. Future tests with RNA and protein synthesis inhibitors should help resolve this possibility. Surprisingly similar *in vivo* results have recently been reported (Nagle and Rosner, 1980). Jugular vein concentrations of NE were consistently increased in ovariectomized estrogen-primed rats following the acute administration of P_4. Since the concentrations of NE in the carotid artery were much lower than those in the jugular vein during the peak of NE release (at four hours after the SC injection of 1.5 mg P_4), it is reasonable to conclude that the NE was derived mainly from brain and not the adrenals. NE concentrations in the anterior hypothalamus displayed a coincident decrease, further suggesting hypothalamic origin of the NE. Preliminary results with estrogen suggest that this steroid alone can increase NE release from the hypothalamus.

Experiments designed specifically to examine the release of DA from the tuberoinfundibular neurons into the hypothalamo-pituitary portal circulation have revealed dramatically different results from those previously discussed (Cramer et al.,

1979a,b). Portal blood concentrations of DA decreased following systemic estrogen administration, whereas they increased with subsequent P_4 administration. The fact that prolactin secretion rates in similarly treated females (both groups subjected to acute ovariectomy during proestrus) were inversely proportional to the DA portal blood concentrations clearly suggests that these two estrogen-induced effects are causally related. Thus, apparently ovarian hormones can affect differentially the release of catecholamines from different sets of hypothalamic neurons.

Reuptake

The principal mechanism for the termination of catecholamine's synaptic actions is the reuptake of the neurotransmitter by the presynaptic terminal. *In vitro* uptake studies with synaptosomal preparations or tissue minces from "whole" brain indicate that high doses of estrogens or progestins *in vitro* typically reduce [^{3}H]DA or [^{3}H]NE accumulation (Janowsky and Davis, 1970; Kendall and Tonge, 1977; Nixon et al., 1974; Wirz-Justice et al., 1974). The significance of these findings has been disputed by other studies in which the *in vivo* administration of much lower, but still physiologically effective, doses of E_2 or P_4 to ovariectomized rats increased [^{3}H]NE uptake by synaptosomal preparations from MBH and PH (Cardinali and Gomez, 1977) or by tissue minces from whole hypothalamus (Endersby and Wilson, 1974). [^{3}H]NE uptake by synaptosomes from AH was not increased by the *in vivo* E -treatment, whereas [^{3}H]DA uptake was increased by E_2 only in AH synaptosomes. Somewhat paradoxically, the uptake of [^{3}H]DA by the hypothalamic tissue minces was not affected by systemic E_2, but [^{3}H]DA uptake was decreased in those females receiving concurrent injections of E_2 and P_4. The apparent differences between the [^{3}H]DA uptake results obtained with the synaptosomes and tissue minces remains to be resolved.

Receptors

The postsynaptic receptors for the catecholamines are also influenced by gonadal steroids. In adult rats the daily injection of E_2 (10 μg) to ovariectomized females or a single injection of E_2-valerate (125 μg SC) to intact males results six to seven days later in a 20 percent increase in the density of DA receptors, that is [^{3}H]spiroperidol binding, in the striatum, nucleus accumbens plus olfactory bulb, caudate and frontal cortex (Hruska and Silbergeld, 1979; Paola et al., 1979). This increase was not eliminated after 6-hydroxydopamine (6-OHDA) destruction of the catecholamine input to these regions, supporting the contention that the binding changes occurred on the postsynaptic membranes. Scatchard analyses indicated that the increased binding was due to an increase in the number of receptor sites, that is B_{max} increased, and not to an increase in the affinity of the receptors for the catecholamine ligand, that is, no change in K_d. An increase in the binding of [^{3}H]DA to striatal membranes has also been reported in ovariectomized female rats six days after the cessation of chronic treatment with E_2B (10 μg/kg/day for 16 days) (Gordon and Diamond, 1979).

Two recent studies have used [^{3}H]dihydroergokryptine (DHE) and [^{3}H]dihydroalprenolol (DHA) binding to the P_2 membrane fraction from brain homogenates to study α- and β-receptors, respectively, (Vacas and Cardineli, 1980; Wilkinson et al., 1979). In intact adult female rats they found that the density (B_{max}) of both receptor types was unchanged during the estrous cycle in tissue blocks that included the hypothalamus (minus the mamillary bodies) and POA. However, there was a reduction in the K_d for [^{3}H]DHE (but not [^{3}H]DHA) binding from 0.6 nM in metestrus, diestrus, and estrus to 0.2 nM during proestrus. Following ovariectomy, there was no effect on α-receptors (B_{max} or K_d), but there was a significant increase in the B_{max} for the binding of [^{3}H]DHA to the β-receptors. A three-day exposure to exogenous E_2 replacement therapy (0.5 and 50 μg E_2/day on days 1 and 2 or 25 μg E_2B on day 1 only) significantly increased [^{3}H]DHA binding in both the large hypothalamic-POA tissue block and in a much smaller sample restricted to the MBH. Neither ovariectomy nor estrogen replacement had any effect on the K_d of the β-receptors, nor on either the B_{max} or the K_d of [^{3}H]DHE binding. The subsequent administration of P_4, five hours prior to killing, had no effect on the E_2-induced increase in [^{3}H]DHA binding. To date, there is no information on the minimal latency or dose to produce this increase in β-receptor density; however, chronic treatment with high doses of 17α-ethynyl-E_2 (50 μg/kg/day for 12-14 days) reduced the B_{max} for [^{3}H]DHA binding to membranes isolated from the cerebral cortex, striatum, olfactory bulb, and hypothalamus without affecting the [^{3}H]DHA binding to membranes isolated from the hippocampus, thalamus, cerebellum, medulla, and pons (Wagner et al., 1979). Since none of these treatments influenced the K_d, both the increase and decrease in apparent β-receptor density probably resulted from estrogen-induced alterations in the de novo synthesis of β-receptor proteins (or in their half-life), rather than to estrogen-induced competitive interactions between [^{3}H]DHA and existing β-receptors.

Direct interactions between estrogen and catecholamine receptors have been demonstrated in vitro. High concentrations of either E_2 or E_1 $(10^{-5}-10^{-6} M)$ reduced both the apparent B_{max}, and the apparent K_d of [^3H]DA and [^3H]NE binding to P_2 membrane fragments from whole rat brain (Inaba and Kamata, 1979). These results contrast with the increase in [^3H]DA (but not [^3H]NE) binding following incubation with T or DHT, and the lack of effect of P_4, B, or F on either [^3H]DA or [^3H]NE. Estrogens and androgens also interact, by some as yet unknown mechanism, to nullify the effect of either class of steroid administered by itself. Although the molecular mechanisms for these gonadal steroid induced effects is unknown, their occurrence in the absence of intact cells excludes the genomic mechanism from serious consideration.

SUMMARY

A summary of the major effects of gonadal hormones on the content, turnover, synthesis, catabolism, release, reuptake, and receptors for brain catecholamines is presented in Table 15-5 (see also Fig. 15-1). Because each hormone may have facilitatory or inhibitory effects at any step of catecholaminergic neurotransmission, because the gonadal hormones may interact with each other in a facilitatory or inhibitory fashion, and finally because there is extensive regional specificity to the effects, it makes summary of these various relationships difficult and the explanation of the catecholamine-related behavioral effects of gonadal steroid hormones speculative at best.

For example, following castration there are transient increases in NE concentrations in the AH and ME accompanied by transient decreases in DA concentration in the AH without similar effects in the ME. These changes in catecholamine content are paralleled by a castration-induced increase in DA and no effect on NE turnover in the ME and an increase in TH activity in the ME without a similar effect in several other hypothalamic nuclei. Replacement therapy with T, and in some cases DHT and AE, following castration restores or prevents castration-induced catecholamine content changes, but it reduces TH activity in the ME as well as turnover of NE in the MBH and of DA in the POA. DBH activity in these same brain regions was uneffected by castration and exogenous androgen treatment. These results suggest that endogenous androgens have a differential negative feedback effect on DA- and NE-containing neurons in different regions of the hypothalamus and limbic system. Although some of the effects may be mediated by direct actions of the androgen on the nerve terminals in the hypothalamus, such as suppressing NE turnover in the MBH following implantation of T into the MBH, other effects may be

Table 15-5. Representative Effects of Steroid Hormones on Brain Catecholamine Neurons

Treatment	Effects	References
Content		
♂ → ♀ (5 days)	[DA] – In AH, FCx, MH & PH [NE] – In AH, FCx, MH & PH	Donoso et al. (1967)
♂ → ♀ (10–20 days)	[DA] ↓ In AH; – In FCx, ME, MH & PH [NE] ↑ In AH; – in FCx, ME, MH & PH	Donoso et al. (1967); Kizer et al. (1978)
♂ → ♀ (4–8 hr)	[DA] – In ME^a [NE] ↑ In ME	Chiocchio et al. (1976)
♂ → ♀ (12 hr)	[DA] – In ME^a [NE] – In ME	Chiocchio et al. (1979)
♂ → ♀ (8–12 hr)	[DA] – In ME [NE] – In ME	Chiocchio et al. (1976)
♂ + TP, DHT or AE	[DA] ↓ In Amg & HTh; – Mes & Str [NE] – In Amg, HTh, Mes & Str	Vermes et al. (1979)
Met ♀ → Di^b ♀	[DA] ↑ In LS; ↓ In DB; – In AH, CG, CMAmg, CN, ME, MPOA, NA, NIST, OB, PaVN, PeVN, SON & VM [NE] ↓ In CG; – In AH, CMAmg, CN, DB, LS, ME, MPOA, NA, NIST, PaVN, PeVN, SON & VM	Crowley et al. (1978a)

Di ♀ → Pro ♀	[DA] ↓ in LS; – in AH, CG, CMAmg, CN, DB, ME, MPOA, NA, NIST, OB, PaVN, PeVN, SON & VM [NE] ↓ in PaVN; – in AH, CG, CMAmg, CN, DB, LS, ME, MPOA, NA, NIST, PeVN, SON & VM	Crowley et al. (1978a)
Pro ♀ → Est ♀	[DA] ↑ in ME; ↓ in CN; – in AH, CG, CMAmg, DB, LS, MPOA, NA, NIST, OB, PaVN, PeVN, SON & VM [NE] – in AH, CG, CMAmg, CN, DB, LS, ME, MPOA, NA, NIST, PaVN, PeVN, SON & VM	Crowley et al. (1978a)
Est ♀ → Met ♀	[DA] ↑ in DB; – in AH, CG, CMAmg, CN, LS, ME, MPOA, NA, NIST, OB, PaVN, PeVN, SON & VM [NE] ↑ in CG, LS, MPOA & PaVN; – in AH, CMAmg, CN, DB, ME, NA, NIST, PeVN, SON & VM	Crowley et al. (1978a)
Met ♀ → Pro ♀[c]	[DA] ↑ in CN; ↓ in MPOA [NE] ↓ in MPOA	Crowley et al. (1978a)
Est ♀ → Di ♀[c]	[DA] ↑ in CN	Crowley et al. (1978a)
⌀ + E₂B	[NE] ↓ in AH	Nagle and Rosner (1980)
E₂B ⌀ + P₄	[NE] ↓ in AH	Nagle and Rosner (1980)
Turnover[d]		
♂ → ⌀ (10 days)	[DA] ↓ in ME[d] [NE] – in ME	Kizer et al. (1978)

(continued)

Table 15-5. Representative Effects of Steroid Hormones on Brain Catecholamine Neurons (continued)

Treatment	Effects	References
♀ + T	[DA] ↑ in POA; - in MBH [NE] ↑ in MBH; - in POA	Simpkins et al. (1980)
♀ + T in MBH	[DA] - in MBH & POA [NE] ↑ in MBH; - in POA	Simpkins et al. (1980)
Di ♀ → Pro ♀	[NE] ↓ in MPOA	Honma and Wuttke (1980)
♀ + E₂B	[DA] ↑ in CN, LS, ME, NA, NIST, OB & PeVN [NE] ↑ in CG, LS & NIST; ↓ in Arc, DB, LVM, MAmg, ME, MPOA, PaVN, PeVN & SON	Crowley et al. (1978b)
E₂B ♀ + P₄	[DA] ↓ in DB & NIST; - in CN, LS, ME, NA, OB & PeVN [NE] ↑ in AH, LVM, PaVN & PeVN; ↓ in Arc, CG, LS, NIST & SON; - in DB, MAmg, ME & MPOA	Crowley et al. (1978b)

Synthesis

Treatment	Effects	References
♂ → ♀ (7 days)	TH ↑ in ME; - in Arc, DM, MFB, PaVN, PeVN & POA DBH - in Arc, DM, ME, MFB, PaVN, PeVN & POA	Kizer et al. (1974)
♂ → ♀ + TP (7 days)	TH - - in Arc, DM, ME, MFB, PaVN, PeVN & POA DBH - in Arc, DM, ME, MFB, PaVN, PeVN & POA	Kizer et al. (1974)

Di ♀→Pro♀	[³H]Tyr → [³H]NE ↑ in HTh (in vivo)	Donoso and Moyano (1970)
♂ + E$_2$[Low]	TH ↑ in HTh	Tobias et al. (1979)
♂ + E$_2$[High]	TH → in MBH	Luine et al. (1977)
E$_2$B ♂ + P$_4$	TH ↓ in HTh	Beattie et al. (1972)
♂ + E$_2$B + P$_4$	[³H]Tyr → [³H]NE ↓ in AH; – in PH (in vivo)	Bapna et al. (1971)
Media + P$_4$, 17α–OH–P$_4$ or 20α–OH–P$_4$	TH ↓ in HTh	Beattie and Soyka (1973)
Media + 2–OH–E$_2$ or 2–OH–E$_1$	TH ↓ in Amg, Hpc, HTh, POA & Str	Foreman and Porter (1980); Lloyd and Ebersole (1980)
Media + E$_2$ or E$_1$	TH – in Amg, Hpc, HTh, POA & Str	Foreman and Porter (1980); Lloyd and Ebersole (1980); Luine et al. (1977)

Catabolism

♂ + TP	MAO ↑ in MPOA & Amg + HTh + POA[e]	Luine et al. (1975b, 1979)
Di ♀ → Est♀	MAO ↑ in Amg, FCx & HTh[e]	Zolovick et al. (1966)

(continued)

Table 15-5. Representative Effects of Steroid Hormones on Brain Catecholamine Neurons (continued)

Treatment	Effects	References
♀ → ♂	MAO ↑ in HTh[e]	Kato and Minaguchi (1964)
♂ + E_2 or E_2B	MAO ↓ in Amg & MBH[e]	Luine et al. (1975b, 1979)
Media + 2-OH-E_2	COMT ↓ in whole brain & neuroblastoma culture	Breurer and Koster (1974); Lloyd et al. (1978)
Release		
E_2B ♂ + P_4	[NE] ↑ in jugular vein	Nagle and Rosner (1980)
♂ + E_2	[DA] ↓ in tuberohypophysial vessels	Cramer et al. (1979a)
E_2 ♂ + P_4	[DA] ↑ in tuberohypophysial vessels	Cramer et al. (1979b)
Media + E_2	[DA] & [NE] ↑ in efflux from HTh cultures	Paul et al. (1979)
Media + E_1, E_3 or B	[DA] & [NE] – in efflux from HTh cultures	Paul et al. (1979)

Reuptake

♀ + E_2	[DA] – in HTh minces [NE] ↑ in HTh minces	Endersby and Wilson (1974)
♀ + E_2 + P_4	[DA] ↓ in HTh minces [NE] ↑ in HTh minces	Endersby and Wilson (1974)
♀ + E_2	[DA] ↑ in AH; – in MBH & PH synaptosomes [NE] ↑ in MBH & PH; – in AH synaptosomes	Cardinali and Gomez (1977)

Receptors

♂ + E_2V	[DA-receptors] ↑ in CN	Hruska and Silvergeld (1979)
♀ + E_2	[DA-receptors] ↑ in FCx, Str & NA + OB	Paola et al. (1979)
♀ + E_2 or E_2B	[α-receptors] – in HTh & MBH [β-receptors] ↑ in HTh & MBH	Vacas and Cardinali (1980); Wilkinson et al. (1979)
Media + T or DHT	[DA & NE] ↑ in brain P_2 membranes	Inaba and Kamata (1979)
Media + E_2 or E_1	[DA & NE] ↓ in brain P_2 membranes	Inaba and Kamata (1979)
Media + B, F or P_4	[DA & NE] – in brain P_2 membranes	Inaba and Kamata (1979)

(continued)

712 Luttge

Table 15-5. Representative Effects of Steroid Hormones on Brain Catecholamine Neurons (continued)

a Comparison made with sham ♀, which is itself decreased from ♂ controls.

b Met: metestrous; Di: diestrous; Pro: proestrus; and Est: estrous. All other abbreviations are defined in the text or in Tables 15-1, 15-2, or 15-4.

c Note that intermediate day's values were not different from either group in this special comparison. Null effect values have been ommitted.

d Turnover rates are expressed as the rate of loss of the catecholamine after synthesis inhibition with α-MPT. Thus, an increase in the rate of loss (that is, [DA] or [NE]↓) is interpreted as an increase in the turnover rate.

e Controversial results--see text for explanation.

mediated by (genomic) actions on the cell bodies both within and outside the hypothalamus (see Heritage and Grant, 1979; Simpkins and Estes, chapter 17).

In the AH of ovariectomized female rats, NE concentrations are decreased and turnover increased by exogenous E_2 replacement therapy. NE turnover in the central gray, lateral septal nucleus and NIST and DA turnover in the diagonal band nucleus (DB) are decreased with the same treatment. The addition of P_4 to the E_2-primed ovariectomized rat further decreases NE concentrations in the AH and increases NE turnover in the structures showing NE turnover decreases after E_2 priming alone and decreases NE turnover in structures previously displaying NE turnover increases or no effect with E_2 alone. P_4 treatment also reverses the E_2-induced decrease in DA turnover in the DB and NIST. The effects of the ovarian hormones on TH activity are further complicated by dose-related differential responses, such that with low doses of E_2, TH activity increases in the hypothalamus, whereas with higher doses it decreases in the MBH as it does in the hypothalamus with the combination of E_2 and P_4. NE release and reuptake by the hypothalamus are both increased with ovarian hormone replacement therapy, and although DA release is also increased from hypothalamic organ cultures with E_2 treatment, it is decreased when measured in the hypophysial portal vessels in *in vitro* studies. DA release into the portal vessels increases following P_4 injections in the E_2-primed ovariectomized rat, and although the reuptake of this catecholamine decreases in hypothalamic minces treated with these steroids, it is increased in AH synaptosomes. The increased density of DA- and β-receptors in many brain regions, such as striatum, frontal cortex, nucleus acumbens plus olfactory bulb for DA- and hypothalamus and MBH for β-receptors following estrogen treatment combined with the direct competition of E_2 and E_1 for NE and DA binding to P_2 membrane fragments adds still further levels of complexity.

In attempting to interpret these sometimes seemingly conflicting results, it should be noted that some of the effects may only represent pharmacological anomalies. Although it may seem more physiological to only consider the catecholamine system during natural endocrine fluctuations, such as the estrous cycle, it must also be noted that these endogenous hormone fluctuations are not mono-hormonal, but multi-hormonal, thus adding to their own interpretational complexities. Although it may be difficult to sort out the multiple possibilities, it is likely that at least some of the changes in the catecholamine system result from direct actions of these hormones on the catecholamine-containing neurons themselves, or on neurons that synaptically interact with catecholamine-containing neurons. This conclusion is supported by histochemical data illustrating a close correspondence between the estrogen- and androgen-concentrating neurons and the catecholamine nerve terminals and cell bodies (Grant and Stumpf, 1975; Heritage and Grant, 1979; Heritage et al., 1977). This conclusion is challenged, however, by the observations that the gonadotropins may also influence catecholamine activity, presumably reflecting their short feedback loop regulatory mechanisms (Anton-Tay et al., 1969). In speculating about the possible genomic versus nongenomic mechanisms of gonadal hormone action on catecholamine systems, it appears likely that both mechanisms are potentially operational. For example, the estrogen-induced increase in DA- and β-receptor density could well reflect a genomic mechanism, whereas the competitive interaction between P_4 and 2-OH-E_2 on TH, as well as the competitive interaction between 2-OH-E_2 and NE and DA on COMT, are clear examples of a nongenomic mechanism of gonadal steroid action.

Catecholamine Actions and Hormone-Induced Sexual Behavior

The potential role of catecholamines in gonadal hormone-induced sexual behavior is extremely complicated and controversial. This is because of both the complex interactions between the steroid hormones and the neurotransmitters (*vide supra*) and the incomplete specificity of the neuropharmacological agents, such as receptor agonists and antagonists, and other neurobiological techniques, such as electrolytic, microknife, and neurotoxin-induced lesions. Despite these limitations, a general, although not universal, concensus has arisen for the rat holds that DA neurotransmission inhibits display of lordosis behavior, (that is, an index of female sexual receptivity), while it facilitates to the display of mounts, intromissions, and ejaculatory behavior, but apparently not the interval between ejaculations, in the male. The data on NE and E are far less convincing, but both catecholamines may facilitate both female and male sexual behavior (for reviews and conflicting opinions, see Barfield et al., 1975; Caggiula et al., 1973; Carter and Davis, 1977; Clark et al., 1975; Foreman and Moss, 1978; Malmnas, 1973; Meyerson et al., 1979; Nock and Feder, 1979; Sandler and Gessa, 1975; Yanase, 1977). A number of theories have emerged to explain these effects. For example, Gordon and co-workers have recently championed the notion that sexual receptivity in female rats is facilitated via an estrogen-dependent inhibition of the postsynaptic efficacy of DA neurotransmission (e.g., Gordon et al., 1977, 1979, 1980a,b; McGinnis et al., 1980b), and other groups have argued for the importance of hormone-induced variations in presynaptic events (e.g., Crowley et al., 1978a,b). The details of these and other theories of gonadal

hormone and catecholamine interactions involved in sexual behavior are beyond the scope of the present review; thus, the
reader is referred to the aforementioned reviews for comprehensive discussions.

Indolamines

The principal mammalian cerebral indolamines serotonin (5-HT)
and melatonin have been implicated in the regulation of gonadal
hormone-influenced gonadotropin secretion or behaviors (see
Fig. 15-7 for synthetic and catabolic pathways and structural
formulas). 5-HT is synthesized mainly, although not exclusively, within the midbrain raphe nuclei whose axons reach the
hypothalamus and limbic system via the medial forebrain bundle
and other long tracts (see chapter 17 by Simpkins and Estes).
Melatonin, by contrast, is synthesized exclusively within the
pineal gland where it can be released into CSF and blood to act
as a neurohormone (for review see Reiter, 1980). (It should be
noted that extremely high levels of 5-HT are also found within
the pineal body, where its conversion into melatonin is catalyzed by enzymes whose activity is strongly influenced by
changes in environmental lighting.) Since the involvement of
these two indolamines in gonadotropin secretion regulation has
been the subject of extensive reviews in recent years (Reiter,
1980; Weiner and Ganong, 1978; see chapter 4 by Frohman and
Berelowitz, and chapter 17 by Simpkins and Estes) and since
melatonin does not appear to be directly involved in mediating
the effects of gonadal hormones on behavior, the following section will be restricted to a brief review of the effects of
gonadal steroids on cerebral 5-HT.

Content

Castration of adult male rats causes a time-dependent, that is,
six weeks more than three weeks, increase in the concentration
of 5-HT in a "whole-brain" sample that excludes hypothalamus,
but not in the hypothalamus itself (Bernard and Paolino, 1974).
However, in another study one week following castration a four-
to-fivefold increase in 5-HT concentration was observed in the
VM and medial forebrain bundle (van de Kar et al., 1978). Replacement therapy with 5 or 10 mg TP every other day, but not 1
mg TP per day, decreased 5-HT content significantly in several
hypothalamic nuclei (see Table 15-6). Treatment of ovariectomized female rats with oil or 5 μg E_2B followed 48 hours later
by an injection of oil or 1.5 mg P_4 produced mixed effects in
hypothalamic and limbic nuclei microdissected six hours after

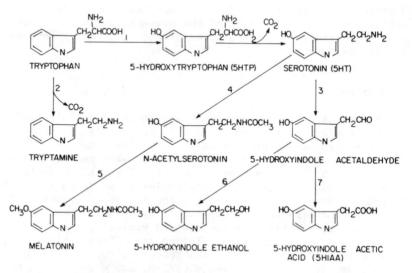

Fig. 15-7. Structural formulas and metabolic pathways for
indolamine biosynthesis and catabolism. Enzymes:
(1) tryptophan hydroxylase, (2) 5-hydroxytryptophan
decarboxylase, (3) MAO, (4) serotonin acetylase,
(5) 5-hydroxyindole O-methyltransferase, (6)
aldehyde reductase, (7) aldehyde dehydrogenase.

the last set of injections (Crowley et al., 1979). E_2B injections alone did not alter 5-HT concentrations, but P_4 injections alone increased 5-HT concentrations in the diagonal band
and VTA by 40 to 50 percent. The combination of E_2B and P_4
increased 5-HT concentration in the ME by 140 percent and nul-
lified the P_4 effects in the diagonal band and VTA. Plasma LH
determinations revealed the expected decrease in secretion
following E_2B alone, no effect with P_4 alone, and a marked
surge in LH in females receiving both hormones. Although not
measured in these females, this latter hormone treatment
schedule should also have induced sexual receptivity.

The concentrations of 5-HT and its catabolite, 5-hydroxy-
indoleacetic acid (5-HIAA), are higher in the morning than in
the late afternoon (see chapter 14 by Rees and Gray). Three
days after a single 20 μg E_2B injection in ovariectomized
female rats the anterior hypothalamic content of 5-HT and 5-
HIAA is reduced significantly in the morning, but not in the
afternoon (Munaro, 1978). However, if these same females also
received a single 2 mg P_4 injection three hours before killing,
all parameters were restored to control levels.

Turnover and Synthesis

A single injection of 20 μg E_2B reduces 5-HT turnover in the
anterior hypothalamus (morning samples only) of ovariectomized
female rats, as determined by the decrease in 5-HIAA concentration following inhibition of MAO by pargyline) (Munaro, 1978).
Treatment of ovariectomized female rhesus monkeys with either
E_2B (15 μg/day for 10 days) or TP (250 or 400 μg/day for 10
days) lowered whole brain 5-HT turnover, as determined by the
increase in 5-HIAA concentrations in CSF before and after
treatment with probenecid (Gradwell et al., 1975). In both of
these studies combined treatment with P_4 and E_2B eliminated all
changes in 5-HT turnover. The combined treatment of E_2B and P_4
has also been reported not to affect the whole brain rate of
conversion of [^{3}H or ^{14}C]Try $\rightarrow$ [^{3}H or ^{14}C]5-HT in ovariectomized
rats (Bapna et al., 1971; Hyyppa et al., 1973), whereas long-
term (50 days) (Engle et al., 1979), but not short-term (9
days) (Kizer et al., 1976b), castration of adult male rats has
been reported to produce a small but significant increase in
diencephalic and limbic tryptophan hydroxylase activity. Androgen replacement therapy in the latter, but not in the
former, study was further shown to reduce the rate of 5-HT
synthesis. However, the magnitude of these changes were comparatively small, and still controversial, thus caution should
be exercised in interpreting their significance.

Catabolism

The principal catabolic enzyme in the 5-HT system is MAO Type
A. The activity of this enzyme was increased significantly in
the MPOA of castrated adult male rats following TP treatment
(28 μg/day for 7 days), but not following treatment with an
equimolar amount of E_2B (30 μg/day for 7 days) (Luine et al.,
1975b). MAO activity in the hypothalamus and amygdala were
unaffected by these treatments. In similar experiments with
castrated adult mice, MAO activity in a pooled tissue sample
from hypothalamus, POA, and amygdala increased following androgen, whereas no effects were seen in cerebral cortical
samples or in Tfm mutant males known to have diminished or
defective intracerebral androgen receptors (Luine et al.,
1979). In female rats MAO activity was reported to increase in
the hypothalamus, amygdala, and frontal cortex during the
transition from diestrus to estrus (Zolovick et al., 1966).
MAO activity in the MBH and centromedial amygdaloid nucleus
(CMAmg), but not in the MPOA, of ovariectomized female rats was
reduced following treatment with E_2B, but not TP (30 and 28
μg/day for 7 days, respectively)(Luine et al., 1975b). The

Table 15-6. Representative Effects of Steroid Hormones on Brain Serotonin Neurons.

Treatment	Effects	References
Content		
♂ → ♀ (6 wk)	[5HT] ↑ in brain (minus HTh); – in HTh	Bernard and Paolino (1974)
♂ → ♀ (1 wk)	[5HT] ↑ in MFB & VM; – in AH, Arc, DM, MPOA & SCN	Van der Kar et al. (1978)
♀ + TP (5-10 mg/2 days)	[5HT] ↓ in AH, Arc, MFB, SCN & VM; – in DM & MPOA	Van der Kar et al. (1978)
♀ + TP (1 mg/day)	[5HT] – in AH, Amg, Arc, CN, DM, Hpc, MFB, MPOA, SCN & VM	Van der Kar (1978)
♀ + E$_2$B	[5HT] ↓ in AH[a]; – in DB, DM, ME, MPOA, NIST, PaVN, SCN, SON & VTA	Crowley et al. (1979)
	[5HIAA] ↓ in AH[a]	
♀ + P$_4$	[5HT] ↑ in DB & VTA; – in AH, DM, ME, MPOA, NIST, PaVN, SCN & SON	Crowley et al. (1979)
E$_2$B ♀ + P$_4$	[5HT] ↑ in ME; – in AH, DB, DM, MPOA, NIST, PaVN, SCN, SON & VTA	Crowley et al. (1979)
	[5HIAA] – in AH	Munaro (1978)
Turnover [b]		
♀ + E$_2$B	[5HIAA] ↑ in AH[a,b]	Munaro (1978)
♀ + E$_2$B + P$_4$	[5HIAA] – in AH	Munaro (1978)
Synthesis		
♂ → ♀ (9 days)	TPH[c] – in Amg, Arc, Hpc, ME, MFB, MPOA, PeVN, SCN, SON & VM	Kizer et al. (1976b)
♂ → ♀ (50 days)	TPH ↑ in diencephalon & limbic forebrain[d]	Engel et al (1979)
♀ (50 days) + TP	TPH ↓ in diencephalon & limbic forebrain[d]	Engel et al. (1971)

∅ + E₂B + P₄ — [³H or ¹⁴C]Try → [³H or ¹⁴C]5HT – In brain (In vivo) — Bapna et al. (1971)

Catabolism

∅ + TP	MAO ↑ in MPOA; – in Amg & HTh	Luine et al. (1975b)
∅ + E₂B	MAO – in Amg, HTh & MPOA	Luine et al. (1975b)
∅ + TP	MAO – in CMAmg, Cx, MBH & MPOA	Luine et al. (1975b)
∅ + E₂B	MAO ↓ in CMAmg & MBH; – in Cx & MPOA	Luine et al. (1975b)
DI♀ → Est♀	MAO ↑ in Amg, FCx & HTh	Zolovick et al. (1966)

Reuptake

∅ + E₂	[5HT] ↑ in AH[a] & PH; – in MBH synaptosomes	Cardinali and Gomez (1977)
∅ + E₂ ∓ P₄	[5HT] – in HTh minces	Zolovick et al. (1966)
Media + T (10⁻⁴ M)	[5HT] ↓ in whole brain synaptosomes[d]	Nixon et al. (1974)
Media + E₂ or P₄	[5HT] – in whole brain synaptosomes	Nixon et al. (1974)

[a]Morning samples only.
[b]Turnover rates are expressed as the rate of loss of 5-HIAA after MAO inhibition with pargyline. Thus, a decrease in the rate of loss (i.e., [5HIAA] ↑) is interpreted as a decrease in the rate of 5-HT turnover.
[c]TPH: tryptophan hydroxylase; Try: tryptophan. All other abbreviations are defined in the text or in Tables 15-1, 15-2, and 15-4.
[d]Controversial results--see text for explanation.

effects of E_2B on MAO in both the MBH and amygdala were inhibited by antiestrogen treatment (MER-25), but hypophysectomy
blocked the effects of E_2B on MAO only in the MBH (Luine et
al., 1975b). These latter results suggest that although the
estrogen-receptor system may be involved, the effects of estrogen on MAO, at least in the MBH, may not be direct and may
in fact be mediated via the short feedback actions of the
gonadotropins or prolactin. The effects of hypophysectomy on
the MAO responses to androgen therapy in the castrated male rat
have not been reported. In a further analysis of the inhibitory effects of estrogen (indirect or direct) on intracerebral
MAO activity, one to four weeks of treatment with either E_2 or
E_2B in ovariectomized female rats reduced MAO activity (30
percent) in the MBH and CMAmg, but not in cerebral cortex.
This effect appears in large part to be due to an increase in
the rate of degradation of the enzyme, since turnover was
decreased from 12.7 to 7.8 days in the CMAmg and from 9.8 to
7.6 days in the MBH (Luine and McEwen, 1977a).

Reuptake

In whole brain synaptosomes, extremely high concentrations of T
(10^{-4} M), but not E_2 (10^{-5} M) or P_4 (10^{-4} M) were found to reduce
[^{3}H]5-HT uptake (Neckers and Sze, 1975; Nixon et al., 1974).
However, with *in vivo* treatment, high doses of E_2 (50 μg/day)
increased the *in vitro* uptake of [^{3}H]5-HT in synaptosomes from
PH, both morning and late afternoon, and AH, morning only, but
not MBH (Cardinali and Gomez, 1977). When reuptake is assessed
using hypothalamic minces, no effects on [^{3}H]5-HT uptake were
found following *in vivo* treatment with either E_2, or P_4 or both
steroids (Endersby and Wilson, 1974). Therefore until additional experiments are reported, the significance of these
scattered, and pharmacological, findings must be questioned.

Serotonin Actions and Hormone-Induced Sexual Behavior

As in the case of the catecholamines, the details of the role
of 5-HT in sexual behavior are somewhat controversial, due
primarily to methodological problems. However, for the rat it
appears that serotonergic neurotransmission tonically inhibits
both female and male sexual behavior (for reviews, see Carter
and Davis, 1977; Malmnas, 1973; Meyerson et al., 1979; Sandler
and Gessa, 1975; Sodersten et al., 1980). For example, drugs
that reduce the efficacy of serotonergic neurotransmission,
such as via synthesis inhibition and receptor blockade, facilitate receptivity in estrogen-primed females, and mounts,

intromissions, and ejaculations in testosterone- or estrogen-primed males. These drugs do not induce sexual behavior in gonadectomized animals deprived of hormone replacement therapy. The data from neurotoxin and electrolytic lesions of serotonergic cell bodies or pathways are often inconsistent or in disagreement with the neuropharmacological data (see Meyerson et al., 1979). Drugs that mimic serotonergic activity or increase the endogenous serotonergic activity, in general, depress copulatory activity, but there are some exceptions. Observations, such as the demonstration that the 5-HT agonist alpha-methyltryptamine can inhibit receptivity in ovariectomized female rats primed with both E_2 and P_4, but not in females receiving chronic treatment with E_2B alone (Espino et al., 1975), have prompted the theory that P_4 stimulation of receptivity involves a selective suppression of 5-HT-mediated inhibition. However, although both P_4 and the 5-HT receptor blocker methysergide can produce a similar facilitation and subsequent inhibition, that is, sequential and concurrent inhibition, of receptivity in estrogen-primed female rats, cross tolerance can only be demonstrated for the concurrent inhibitory actions (Davis and Kohl, 1978; Rodriquez-Sierra and Davis, 1978,1979). These results suggest that the mechanism of progestin actions in the facilitation and inhibition of receptivity is not simply a modulation of serotonergic neurotransmission. Until the hormone-induced variations in serotonin neurochemistry are better understood, it will be difficult to resolve the role of serotonergic neurotransmission in sexual behavior.

Acetylcholine

Acetylcholine (ACh) is a predominantly excitatory neurotransmitter that has been implicated in the mediation of gonadal steroid hormone effects on reproductive behavior and adenohypophysial hormone secretion (for reviews, see chapter 4 by Frohman and Berelowitz; Ganong, 1974; Meyerson and Eliasson, 1977). Several recent studies have indicated that in the rat, gonadal hormones may directly influence the synthesis and catabolism of this important neurotransmitter (see Fig. 15-8 for synthetic and catabolic pathways and structural formulas). For example, in ovariectomized, ovariectomized and adrenalectomized, and hypophysectomized adult female rats three days of treatment with E_2B, but not TP, results in an increase in ChAcT activity in the estrogen concentrating MPOA and CMAmg (Luine and McEwen, 1977b; Luine et al., 1975b). Interestingly, ChAcT activity was not increased in the MBH, which also avidly concentrates estradiol, but it was increased by higher doses of

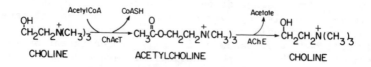

Fig. 15-8. Structural formulas and metabolic pathways for
acetylcholine synthesis and catabolism. Enzymes:
choline acetyltransferase (ChAcT),
acetylcholinesterase (AChE).

E_2B (30 μg/day versus 3 μg/day) in the hippocampus, an area
that concentrates little [^{3}H]E_2. The latter effect may have
been due to the actions of E_2 in the septal region, an estrogen-concentrating nuclear group with strong cholinergic projections to the hippocampus (Kuhar et al., 1973; Pfaff and Keiner,
1973), rather than to the direct actions of this steroid in the
hippocampus.

Similar treatment of castrated adult male rats with E_2B did
not alter ChAcT activity, whereas treatment with TP (28 μg/day
for 7 days) increased ChAcT activity in the MPOA (Luine et al.,
1975b). The failure of E_2B to increase ChAcT activity in the
MPOA of males and of TP to increase the activity of this enzyme
in females is interesting in light of the finding that the MPOA
has exceptionally high aromatase activity (*vide supra*), suggesting that the conversion of T to E_2 is not required for the
increase in ChAcT in males. Since antiestrogen (MER-25) administration blocked the effects of E_2B on ChAcT in females
(Luine and McEwen, 1977b; Luine et al., 1975b), it is probable
that ChAcT activity increases are mediated via the steroid-
receptor-acceptor mechanism. The possibility that the mechanism is genomic has been strengthened by the observations that
there is no direct cell-free effect of E_2 on ChAcT activity
(Luine et al., 1975b) and no increase in ChAcT activity within
three hours after a single 100 μg injection of E_2 (Luine et
al., 1980), but there is a clear increase in immunoprecipitable
ChAcT, indicating an increase in the number of ChAcT molecules,
following one or three days of E_2B treatment (Luine et al.,
1980).

The termination of the synaptic actions of ACh is mediated
primarily by degradation by acetylcholinesterase (AChE). A
single injection of E_2 (10 μg) can result four hours later in a
significant increase in AChE activity in the cerebral and
cerebellar cortices of immature and adult rats (Moudgil and
Kanungo, 1973). Since this effect can be blocked by actinomycin D, administered one hour before E_2, the observed increase
in AChE activity may represent yet another example of E_2-stim-

ulated *de novo* protein synthesis. However, since there are few E_2-concentrating cells in these two brain regions (*vide supra*), further work must be done before this possibility can be considered seriously.

Recent evidence suggests that cholinergic systems have a facilitatory role in the induction of male and female sexual behavior in rats (for reviews see Meyerson and Eliasson, 1977; Sandler and Gessa, 1975). For example, bilateral stimulation of MPOA or mesencephalic reticular formation (MRF) (just dorsal to the interpeduncular nucleus) with cholinergic muscarinic agonists, carbachol or bethanechol, facilitated the induction of sexual receptivity in ovariectomized, and adrenalectomized or dexamethasone-primed, female rats pretreated with E_2B (1 $\mu g/$ rat/day for 3 days, SC) Clemens et al., 1980). These effects were detected within 30 minutes following intracerebral drug administration, were inhibited by atropine administered systemically, and were not achieved with muscarinic stimulation of the neocortex. In a similar experiment bilateral infusion of the ACh biosynthesis inhibitor hemicholinium, which reportedly inhibits the presynaptic reuptake of choline, into the NIST or MPOA inhibited the display of receptivity in ovariectomized female rats primed with both E_2B and P_4 (Clemens and Dohanich, 1980). These effects were detectable within 15 minutes following intracerebral drug administration, were inhibited by concurrent infusion of choline chloride, and were not achieved with hemicholinium infusions into the neocortex, but a similar inhibition of receptivity was observed with atropine sulfate infusions into the MPOA. The previously described high uptake of estrogen in the NIST and MPOA and of progestins in the MRF-IP, induction of receptivity following E_2 and P_4 implantation into these brain regions and increased ChAcT activity in the MPOA following systemic E_2B administration are all consistent with these behavioral data, suggesting a steroid-induced cholinergic involvement in the induction of sexual receptivity.

Histamine

The possible role of histamine (HA) as a neurotransmitter in the mammalian brain is still controversial (see Schwartz et al., 1980). However, the administration of this amine, or of drugs that block or mimic its actions at membrane receptors, that is, the H_1 and H_2 receptor sites, do influence gonadotropin secretion (Donoso, 1978). Furthermore, since HA powerfully stimulates cerebral adenylate cyclase activity (Schwartz, 1977) and since the activation of this enzyme has been linked with the behavioral actions of gonadal steroids (*vide infra*), it is undoubtedly important that gonadal steroids

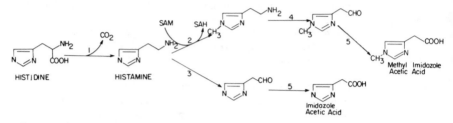

Fig. 15-9.   Structural formulas and metabolic pathways for
histamine biosynthesis and catabolism. Enzymes:
(1) histidine decarboxylase, (2) histamine
methyltransferase, (3) histaminase, (4) MAO, (5)
aldehyde dehydrogenase.

can influence both the synthesis and actions of HA in the mammalian brain (see Fig. 15-9 for synthetic and catabolic pathways and structural formulas).

The activity of histadine decarboxylase in rabbit hypothalamus is increased significantly following ovariectomy, and
following thyroidectomy--the effect being pronounced especially
in those females subjected to both operations (Bjorklund et
al., 1972). A similar potentiation in HA synthesis is observed
in the male rat hypothalamus following castration (Orr and
Quay, 1975). Castration also shifts the diurnal rhythm of HA
synthesis such that the peak in synthesis was much sharper and
occurred several hours earlier following the transition from
darkness to light. Lastly, it has been shown recently that HA
stimulation of adenylate cyclase activity in hypothalamic membrane fragments from gonadectomized adult male and female rats
is inhibited by both the *in vivo* and *in vitro* administration of
E_2 (Portaleone et al., 1980). Methyltestosterone did not affect HA-stimulated adenylate cyclase activity under either *in
vivo* or *in vitro* conditions, whereas 2-OH-E_2 inhibited and P_4
had no effect *in vitro* on HA-facilitated cyclic AMP synthesis.
E_2 and 2-OH-E_2 *in vitro* inhibit the response to HA, clearly
suggesting a nongenomic competitive interaction between the
steroids and HA for the H_2 binding sites in hypothalamus
(Portaleone et al., 1978). This hypothesis is supported by the
observation that incubation of E_2 and 2-OH-E_2 without HA results in a moderate increase in hypothalamic membrane adenylate
cyclase activity (Portaleone et al., 1980). (Further information on the effects of gonadal hormones on cerebral cAMP is
discussed later.) The relationship of these provocative

findings to the behavioral and neuroendocrine actions of the gonadal steroid hormones clearly deserves further investigation.

Cyclic AMP

Although cAMP is generally considered to play a central role in the mediation of the actions of many protein and polypeptide hormones (Sundberg et al., 1976) and in the mediation of the actions and synthesis of several neurotransmitters (Debus and Kehr, 1979; Williams, 1979), it is not often considered in discussions of the potential mechanisms of gonadal hormone actions in the brain. However, several recent studies have presented direct and indirect biochemical and behavioral data suggesting that the actions of steroids may also be mediated, at least in part, via a modulation of cAMP synthesis. For example, the systemic administration of 5 μg E_2B to immature female rats has been shown to increase, and pretreatment with antiestrogens, to decrease the hypothalamic concentration of cAMP (Gunaga et al., 1974). This effect can be blocked by *in vivo* pretreatment with either α- or β-blockers, suggesting catecholaminergic mediation. *In vitro* studies have confirmed and extended these findings, establishing that they may be restricted to estrogen-concentrating regions of the hypothalamus, are mediated probably via an initial interaction with cytosolic (or nuclear) estrogen receptors, inhibited by DA-receptor blockers, such as pimozide and haloperidol, as well as by α- and β-blockers, require a longer time to increase hypothalamic cAMP than does direct exposure to catecholamines (50 versus 5 minutes), and that they can facilitate NE and DA efflux with a similar time course to the estrogen-induced increase in cAMP concentrations (Gunaga and Menon, 1973; Paul et al., 1979; Wiessman and Johnson, 1976; Weissman and Skolnick, 1975; Weissman et al., 1975). Taken together these *in vitro* results clearly support the notion that the estrogen-induced increases in hypothalamic cAMP concentrations observed the *in vivo* are mediated probably via an estrogen-induced increase in DA or NE synaptic activity (Williams, 1979).

The correlation between estrogen-induced catecholamine efflux from hypothalamic tissue fragments and increased hypothalamic cAMP concentrations has been extended by the observation that unlike E_2, neither E_1 nor E_3 exhibits physiologically relevant potency in either catecholamine efflux or cAMP accumulation (Paul and Skolnick, 1977; Paul et al., 1979). In these studies, incubation with 2-OH-E_2 inhibited E_2- and DES-induced increases in hypothalamic cAMP. 2-OH-E_2 alone had no effect on cAMP concentrations, and it failed to inhibit the DES effect if added 40 minutes after the estrogen, or if the 2-methoxy derivative was used instead of the free catechol. Despite the very recent demonstration that $[^3H]2$-OH-E_2 binds with high affinity to membrane receptors in the pituitary, cerebral cortex, hypothalamus, striatum, and other brain regions that also demonstrate high-affinity binding for DA (Schaeffer and Hseuh, 1979; Schaeffer et al., 1980), it is unlikely that these receptors are involved in the 2-OH-E_2-induced anti-estrogenic activity. This conclusion is based on several inconsistencies, including the fact that, whereas DA effectively competes with 2-OH-E_2 for binding to membrane-associated receptors, E_2 does not compete; yet 2-OH-E_2 inhibits E_2-induced increases in cAMP only when administered prior to the expected release of catecholamines (Paul and Sklonick, 1977; Paul et al., 1979). Furthermore, as discussed earlier, E_2 effectively competes with 2-OH-E_2 for binding to hypothalamic cytosol receptors (Davies et al., 1975), clearly suggesting it is the inhibition of this (genomic ?) mechanism by 2-OH-E_2 that produces in the loss of estrogen-induced catecholamine-dependent adenylate cyclase.

The possibility that estrogens may also directly affect hypothalamic cAMP synthesis was supported by the recent demonstration with hypothalamic membrane fragments in which E_2 and, in contrast to the data discussed already, 2-OH-E_2, but not T or P_4, increased adenylate cyclase (Portaleone et al., 1980). Although the mechanism of this facilitatory effect remains to be discovered, it cannot involve genomic interactions.

Other examples of the involvement of gonadal steroid hormones in cerebral cAMP systems include the previously discussed estrogen inhibition of HA-induced increases in hypothalamic cAMP (Portaleone et al., 1978, 1980), and the estrogen-dependent maintenance of adenylate cyclase-sensitive DA- and β-receptors in the hypothalamus and other brain regions (Kumakura et al., 1979; for an opposing view, see Paola et al., 1979). The potential involvement of gonadal hormone-induced alterations in cAMP in sexual behavior has received comparatively little attention. The administration of the cAMP phosphodiesterase inhibitor, theophylline, which should increase endogenous cAMP, was shown to potentiate E_2B- and TP-induced male, but not female, sexual behavior (Christensen and Clemens, 1974; Clemens and Christensen, 1977). Although it is not known whether this represents a direct potentiation of a steroid-induced effect, it could represent a potentiation of an androgen- or estrogen-stimulated DA-dependent facilitation of adenylate cyclase and copulation (*vide supra*).

Carbohydrate Metabolism

Changes in glucose utilization and metabolism, as occur in natural and experimental diabetes, have been shown to influence the receptor binding and the behavioral and neuroendocrine actions of the gonadal hormones (Denari and Rosenr, 1972; Florez-Lozano et al., 1978; Gentry et al., 1977; Siegel and Wade, 1979; Schiaffini et al., 1970). Thus, predictably, direct and indirect actions of gonadal steroids on carbohydrate oxidation and tricarboxylic acid cycle enzyme activities have been reported (see Table 15-7). Furthermore, although there have been no positive findings with exogenous hormone treatment, the activity of the glycolytic enzymes has also been reported to display regular fluctuations in certain regions of the hypothalamus during the ovarian estrous cycle of the rat (see Tables 15-7, 15-8).

In the male rat, castration decreases oxygen consumption using either glucose, citrate or succinate as the carbon source, in slices of anterior and posterior hypothalamus, whereas it increases oxidative activity in the amygdala, cerebral cortex, and pituitary and has no effect in the middle hypothalamus, which includes the ME and the arcuate, VM, and dorsomedial nuclei, and in the hippocampus (Moguilevsky et al., 1971a; Schiaffini and Martini, 1972). Similar results were obtained by assaying $^{14}CO_2$ production from [U-^{14}C]glucose, or by direct measurement of cytochrome oxidase activity, but there were no significant effects of castration on glutamate and pyruvate oxidative metabolism or on succinic dehydrogenase activity (Moguilevsky et al., 1971a; Scacchi et al., 1971). *In vivo*, but not *in vitro*, except in cerebral cortex and pituitary, replacement therapy with T (or TP) has been found to reverse these effects (Moguilevsky et al., 1971a; Scacchi et al., 1971; Schiaffini and Martini, 1972) and produce significant increases in isocitric and glucose-6-phosphate dehydrogenase activities in the MBH and malic dehydrogenase activity in the CMAmg (Luine et al., 1975a). *In vivo* replacement therapy with E_2B produces a different pattern of effects, characterized by an increase in isocitric dehydrogenase activity in the amygdala and malic dehydrogenase in the MBH, although it did not influence glucose-6-phosphate dehydrogenase activity in the MBH (Luine et al., 1975a).

The lack of effect of T *in vitro* has prompted the suggestion that the effects of castration on oxidative metabolism in the hypothalamus and amygdala are due to the short feedback actions of the gonadotropins, which increase and decrease their secretion rates following castration and exogenous T and E_2 replace-

Table 15-7. Representative Effects of Steroid Hormones on Brain Carbohydrate Metabolism

Treatment[a]	Enzymatic analyses or substrates for oxidative metabolism	Effects									
		AH	MH	PH	HTh	MBH	MPOA	Amg	Hpc	Cx	Pit
♂ → ♀	Glucose	↓7[b]	-7	↓7				↑12	-12	↑7	↑7
	[U-¹⁴C]Glucose				↓9					-9	
	Citrate or Succinate	↓7	-7	↓7							
	Glutamate or Pyruvate										
	CytOx	↓7	-7	↑7							
	SDH	-7	-7	-7							
	Glucose	↑7	-7	↑7				↓12	-12	↓7	↓7
	[U-¹⁴C]Glucose				↑9					-9	
	Citrate	↑7	-7	↑7							
♀ + T or P	Glutamate or Pyruvate										
	CytOx	↑7	-7	↑7							
	SDH	-7	-7	-7							
	MDH					-2		↑2			
	ICDH					↑2		-2			
	G6PDH					↑2					
	MDH					↑2		-2			
♀ + E₂B	ICDH					-2		↑2			
	G6PDH					-2					
♂ → Hypox♂	Glucose	↑6	-6	↑6				↓12	-12	-12	-12
♀ → Hypox♀	Glucose	-6	-6	-6				↓12	-12	-12	-12
Hypox♀ + TP	Glucose	↓5	-5	-5					-12	-12	
♀ + T (in vitro) Glucose		-5	-5	-5				-12	-12	↓5	↓5

Treatment	Substrate / Enzyme[a]						
♂ → PancreX♂	Glucose	↓13			-13		-13
Di ♀ → Es†♀	Glucose	↑4	-4	↑4	↑11	-11	↑3
	[U-14C]Glucose				↓11		-10
	Succinate	↑8	-8	↑8			-1
	CytOx	-8	-8	-8			↑1
	SDH	↑8	-8	↑8			↑1
							-1
Di ♀ → ♂	Glucose	↑11	-11		↑11	-11	↑3
	[U-14C]Glucose	↑10	-10				-10
♂ + E2B	MDH & ICDH	↑1	↑1	-1	↑1		-1
	G6PDH	-1	↑1	-1	-1		-1
	6PGDH & LDH	-1	-1	-1	-1		-1
	HK	-1	-1	-1	-1		-1
♂ + TP	MDH, ICDH & G6PDH	-2		-2	-2		
Hypox ♂ + E2B	MDH & ICDH	↑1		↑1	↑1		
	G6PDH	↑1		↑1			
Adrex ♂ + E2B	MDH & ICDH	↑1				-1	
♂ + E2 (in vitro)	MDH & ICDH	-1					
Di♀ > PancreX♀	Glucose	↓14		↑14	↑14	-14	

[a] Cy+Ox: cytochrome oxidase; SDH: succinic dehydrogenase; MDH: malic dehydrogenase; ICDH: isocitric dehydrogenase; G6PDH: glucose-6-phosphate dehydrogenase; 6PGDH: 6-phosphogluconate dehydrogenase; LDH: lactic dehydrogenase; HK: hexokinase; hypox: hypophysectomy; adrex: adrenalectomy; pancreX: pancreatectomy. All other abbreviations are defined in the text or in Tables 15-1, 15-2, and 15-4.

[b] 1. Luine et al. (1974), 2. Luine et al. (1975a), 3. Menendez-Patterson et al. (1979), 4. Moguilevsky (1965), 5. Moguilevsky et al. (1970a), 6. Moguilevsky et al. (1970b), 7. Moguilevsky et al. (1971a), 8. Moguilevsky et al. (1971b), 9. Scacchi et al. (1971), 10. Scacchi et al. (1973), 11. Schiaffini and Marin (1971), 12. Schiaffini and Martini (1972), 13. Schiaffini et al. (1968), 14. Schiaffini et al. (1970).

Table 15-8. Relative Activities of Hypothalamic Carbohydrate Metabolizing Enzymes During Rat Estrous Cycle[a]

Tissue:	MPOA	LPOA	PaVN	AH	SCN	SON	MVM	LVM	Arc	PH
Enzyme										
PK[b]	=[c]	E>P,D₁,D₂,D₃[d]	E>P,D₁; D₂>D₁	≈	E>D₂	≈	E>P,D₁,D₂,D₃	E>P,D₁,D₂,D₃	E,D₁>P,D₂,D₃	E>P,D₁,D₂,D₃
HK	D₂>E	E>D₃	E,D₁,D₂>D₃	E,D₂,D₃>P	E,D₁,D₂,D₃>P	D₁>P	E,D₁,D₂>P,D₁	E>D₃; D₁,D₂>P,D₃	E,D₁,D₂>D₃; D₁,D₂>P	D₂>E
PFK	≈	≈	≈	≈	≈	≈	≈	≈	≈	P,D₂>D₃

[a] Results taken from Packman et al. (1977).

[b] PK = pyruvate kinase, HK = hexokinase, PFK = phosphofructokinase, D₁ = diestrus day 1, D₂ = diestrus day 2, D₃ = diestrus day 3, P = proestrus, E = estrus. All other abbreviations are defined in the text and/or in Tables 15-1 and 15-2.

[c] = no difference.

[d] Days to right of > and separated by commas are equivalent.

ment therapy, respectively, rather than to their direct actions on the enzymes (Moguilevsky et al., 1970a). This hypothesis received considerable support from the observations that hypophysectomy of gonadally intact, but not castrated, males increases oxidative activity in the AH and PH, decreases it in the amygdala, and has no effect on oxygen consumption in the middle hypothalamus, cerebral cortex, and hippocampus (Moguilevsky et al., 1970b; Schiaffini and Martini, 1972). The lack of effect in cerebral cortex is consistent with the lack of an *in vitro* effect of T on oxidative metabolism in this brain region (Moguilevsky et al., 1970a). The mechanism of this effect is unclear, however, since the cerebral cortex has few high-affinity receptors for androgens in the adult animal (*vide supra*). Additional support for the short feedback hypothesis was provided by the demonstrations that *in vivo* administration of gonadotropins to hypophysectomized castrated male rats reduces hypothalamic oxidative activity, that is LH decreases oxygen consumption in the AH, but not in the PH, or middle hypothalamus, whereas follicle–stimulating hormone was effective only in the posterior hypothalamus (Moguilevsky et al., 1970a). *In vitro* administration of gonadotropins increase amygdaloid, but not cortical or hippocampal oxidative activity (Schiaffini and Martini, 1972).

In spite of these provocative demonstrations of direct gonadotropin effects on intracerebral carbohydrate metabolism, the administration of TP to hypophysectomized castrated male rats also reduces oxygen consumption in the AH but not in the PH or middle hypothalamus (Moguilevsky et al., 1970a). These latter findings clearly suggest that at least in the AH androgens can influence carbohydrate metabolism independent of their negative feedback actions on gonadotropin secretion. Therefore it is particularly interesting that in pancreatectomy–induced diabetes the AH was the only hypothalamic region in which oxidative carbohydrate metabolism was reduced (Schiaffini et al., 1968), whereas in alloxan–induced diabetes it had a dramatically reduced ability to accumulate [^{3}H]T (Denari and Rosner, 1972). It is entirely possible that these two findings are related and that diabetes produces a castrationlike reduction in anterior hypothalamic oxidative activity via a localized reduction in androgen binding.

In the female rat, the transition from the diestrus to estrus is associated with an increase in anterior and posterior, but not middle, hypothalamic oxygen consumption in *in vitro* studies using either glucose or succinate as the carbon source (Moguilevsky, 1965; Moguilevsky et al., 1971b,c). Similar findings have recently been reported for the female hamster (Menendez-Patterson et al., 1978). In limbic and cortical structures, the transition to diestrus is accompanied by an

increase in oxidative activity in the amygdala, a decrease in
the hippocampus, and no consistent effects in the cerebral
cortex (Sacchi et al., 1973; Schiaffini and Marin, 1971). As
in the male rat, gonadectomy in female rats and hamsters reduces oxygen consumption in the hypothalamus, increases it in
the amygdala and cerebral cortex, and has no consistent effect
in the hippocampus (Menendez-Patterson et al., 1979; Schiaffini
and Marin, 1971). In contrast to the results with male rats,
however, the endocrine-induced changes in hypothalamic oxygen
consumption in the female were not accompanied by parallel
changes in $^{14}CO_2$ production or cytochrome oxidase activity,
although they were accompanied by changes in succinic dehydrogenase activity, which in the male was insensitive to gonadal
hormone actions (Moguilevsky et al., 1971a,b,c; Sacchi et al.,
1973). Estrogen replacement therapy in the gonadectomized
female rat also increased activity of both malic and isocitric
dehydrogenase in both the hypothalamus and amygdala and of
glucose-6-phosphate dehydrogenase in the hypothalamus; a
pattern again distinctly different from that observed in the
male rat (Luine et al., 1974, 1975a). Glucose-6-phosphate
dehydrogenase activity in the amygdala was not influenced in
the female by either E_2B or TP replacement therapy. TP treatment also had no effect in the female on either malic or isocitric dehydrogenase activity in the MBH or in the CMAmg.

The apparent lack of effectiveness of E_2B replacement therapy
in the ovariectomized female rat on hexokinase activity in hypothalamic and limbic brain regions (Luine et al., 1974) contrasts with the significant fluctuations in the activity of
this glycolytic enzyme in many of the individually assayed hypothalamic nuclei during the estrous cycle of the intact female
(see Table 15-8 and Packman et al., 1977). That proestrous
values were frequently lower than those observed during diestrus in many of the hypothalamic nuclei, such as arcuate, VM,
suprachiasmatic, and anterior, further supports the notion that
the estrous cycle-associated changes in hexokinase activity are
not due to the direct actions of estrogen within the hypothalamus. It is quite possible that changes in progestin or gonadotropin secretion may be involved in this regulation. Furthermore, although the pattern of changes in enzymatic activity
were different for hexokinase and pyruvate kinase (see Table
15-8), neither of these key glycolytic enzymes displayed maximal activity during proestrus when the circulating levels of
E_2 are at their highest point of the cycle.

As in the male, the possiblity that ovarian hormone-induced
changes in cerebral carbohydrate metabolism are due to short
feedback actions of gonadotropins has received some attention
and positive support. Thus, although E_2 did not influence
either malic or isocitric dehydrogenase activities in hypotha-

lamic tissues incubated directly with the steroid, *in vivo* replacement therapy with E_2B increases the activity of both enzymes in MBH and CMAmg tissues obtained from hypophysecto- mized ovariectomized animals (Luine et al., 1974). These results therefore suggest that estrogens can influence hypo- thalamic and amygdaloid carbohydrate metabolism independent of the confounding influence of gonadotropin. The fact that the estrogen-induced increases in the activity of these enzymes was often greater in ovariectomized females who were not subjected to the additional loss of either their pituitary or adrenal glands (Luine et al., 1974) is, however, consistent with the notion that the gonadotropin and glucocorticoid hormones have a permissive role. It remains to be discovered whether these seemingly direct effects of estrogen are due to a genomic, for example, *de novo* synthesis of new enzyme, or to a nongenomic, such as allosteric activation of pre-existing enzyme, action of the steroid.

In summary, the gonadal steroids have been shown to influence oxidative carbohydrate metabolism directly or indirectly in various regions of the hypothalamus and limbic system. Al- though these effects appear to result primarily from modifica- tions in the activities of various tricarboxylic acid cycle enzymes, changes in glycolytic enzymatic activities have also been reported. The observation that the pentose phosphate cycle enzyme, 6-phosphogluconate dehydrogenase was not altered by gonadal hormones in the hypothalamus (Luine et al., 1974, 1975a) is curious in light of the recently established effects of these steroids on hypothalamic protein and nucleic acid synthesis (*vide supra*), a response that obviously requires the ribose sugar intermediates and NADPH reducing equivalents formed by the pentose phosphate cycle. The lack of a signi- ficant effect on this alternative pathway for glucose meta- bolism, may be compensated for, at least partially, by an increased production of reducing equivalents by malic and isocitric dehydrogenases following gonadal hormone stimulation.

NEUROELECTROPHYSIOLOGICAL ACTIONS OF GONADAL STEROID HORMONES

Although it may be customary to think of the actions of gonadal steroids in terms of biochemical effects in peripheral non- neural tissues or even in terms of the modulation of neurot- ransmitters, it is important to consider that these hormones also affect the electrophysiological activity of neurons in the brain, and in so doing presumably modulate the expression of certain behaviors and neuroendocrine responses. The following section briefly reviews these neuroelectrophysiological ac- tions, which have recently been extensively reviewed elsewhere

(Komisaruk, 1971; Moss, 1976; Pfaff and Modianos, 1980). The potential importance of these responses to the behavioral actions of gonadal steroids is evident from the observations that the hormonal activation or display of certain hormone-dependent behaviors results in alterations in the ongoing electrophysiological activity in specific brain regions whose electrophysiological stimulation or lesion can activate or inhibit these same behaviors (Arendash and Gorski, 1979; Kurtz, 1975; Malsbury et al., 1980; Merari and Ginton, 1975; Pfaff and Modianos, 1980).

EEG and Evoked Potentials

The major changes in electroencephalographic (EEG) recordings from the high amplitude slow wave or synchronous activity characteristic of sleep, to the low amplitude fast wave or desynchronous activity characteristic of arousal, and of paradoxical sleep, can readily be modified by variations in gonadal hormones, either endogenous or exogenous (Branchey et al., 1973; Colvin et al., 1969; Vogel et al., 1971). The site of action of these effects is poorly understood and certainly confounded by possible variations in numerous nonspecific factors, including sensory input, the effects of anesthestics and a general inability to localize changes in the EEG. The problem is compounded by the gonadal hormone-induced variations in the general arousal state of the animal (Endroczi, 1967), peripheral somatosensory receptive fields (Bereiter and Barker, 1980; Bereiter et al., 1980), and other sensory systems (Ward et al., 1977). Thus, even though gonadal hormone administration can modulate the amplitude or latency of evoked potentials (Cartas-Heredia et al., 1978), it is not possible to localize precisely the site or mechanism of this effect. Thus, for the purposes of this chapter, the EEG and evoked potentials do not contribute greatly to our understanding of the molecular mechanisms of gonadal hormone action.

Neuronal Activity

Both single and multiple unit activity (SUA and MUA, respectively) recordings can provide quantitatively meaningful assessments of increases and decreases in the frequency of action potentials for those neurons in close proximity to the recording electrode. There is, however, still the problem of determining the exact site of hormonal action, since neural activity in the hypothalamus and other "target" regions for gonadal hormones, as determined by the presence of high-affinity receptors, can be influenced by nonspecific changes in arousal or sensory input, which in turn may or may not be a direct consequence of gonadal hormones. For example, it is well known that systemic or intracerebral P_4 administration can result in the development of a sleeplike cortical EEG and a general suppression in the responsiveness of hypothalamic neurons to exteroceptive stimuli (Arai et al., 1967; Komisaruk et al., 1967). Unfortunately, it is difficult to record from the same neuron both before and after the administration of a hormone, and it can only be assumed that electrodes placed similarly in control and hormone-treated animals will record from similar neurons. In spite of these very real limitations, important contributions to our understanding of hormone actions in the brain have been made with SUA and MUA recording techniques.

For example, SUA and MUA recordings from intact female rats have revealed estrous cycle-related changes in neuronal activity in certain hypothalamic and limbic recording sites. In one study, unit firing rates in the POA and AH were much faster during proestrus than during diestrus or estrus, whereas in the cingulate cortex MUA was greater during diestrus, and in the lateral septum there were no differences throughout the cycle (Moss and Law, 1971). In another study, a 12 to 25 minute burst of increased MUA was recorded during the 2 to 4 PM proestrus "critical period" for LH release in the MBH, MPOA, VM, arcuate nucleus, septum, amygdala and NIST (Kawakami et al., 1970). Since ovulation can be elicited by electrical stimulation of these same sites in a proestrus female rat, these findings clearly suggest a causal relationship. A decrease in the SUA of 60 percent of the POA neurons that project to the MBH, as determined by antidromic stimulation, was also reported following the IV administration of E_2; a short-lived increase in SUA was found in 30 percent and no effect in the remaining 10 percent (Whitehead and Ruf, 1974). Similar results were obtained with IV E_2 in rabbits (Dufy et al., 1976) and following long-term SC E_2B (10 days) in ovariectomized rats (Bueno and Pfaff, 1976). In the latter study an increase in SUA was reported in the MBH following E_2B.

These electrophysiological findings may be directly relevant to the hormone-induced female sexual behavior, since electrical stimulation of the VM facilitates the induction of sexual receptivity in rats by systemic E_2B and P_4 (Pfaff and Sakuma, 1979b), whereas stimulation of the MPOA inhibits the induction of receptivity (Moss et al., 1974). Lesions of the VM can also totally eliminate female sexual behavior (Pfaff and Sakuma, 1979a), but lesions of the POA can actually facilitate the induction of receptivity (Powers and Valenstein, 1972). Similar, although less extensive, data have been obtained with hamsters and guinea pigs (Malsbury et al., 1980; Rodriguez-

Sierra and Terasawa, 1979). The inhibitory effect of VM
lesions on sexual receptivity in female rats has recently been
shown to be temporary (Okada et al., 1980). The fact that this
recovery of receptivity was observed only after prior treatment
with estrogen suggests that it may represent yet another example of estrogen-stimulated neuronal plasticity following hypothalamic lesion or deafferentation (Matsumota and Arai, 1979).
These lesions, electrical stimulation, and SUA recording data
are also consistent with the previously discussed theory that
in rats the POA exerts a tonic inhibition on the induction of
female sexual receptivity, whereas the VM facilitates it. Neuronal activity in the POA appears to be inhibited by estrogen,
thus removing the inhibitory tone, but neuronal activity in the
VM is stimulated by estrogen, thus increasing its facilitatory
influence on female sexual behavior.

Although these SUA and MUA data are provocative, they still
cannot be ascribed to the direct action of the hormone on that
neuron. This problem has been largely circumvented through the
use of microelectrophoretic, or microelectro-osmotic, application of hormones directly onto, or in the immediate vicinity
of, the neurons from which the SUA is recorded (for review of
techniques, see Kelly et al., 1977b; Yamada and Veshima, 1978).
For example, in one large series of studies (Kelly et al.,
1976, 1977a,b, 1978) a majority of POA-septal area neurons,
which antidromic stimulation data indicate do not project to
the MBH, and are therefore potentially more likely to be involved in the regulation of behavior rather than gonadotropin
secretion, increase their SUA during diestrus day 1 in response
to the microelectrophoretic application of E_2-hemisuccinate
(E_2S), but they decrease their SUA in response to E_2S challenge
during diestrus day 2, proestrus, and estrus. The specificity
of this response was demonstrated by the observations that
$17\alpha-E_2S$, unlike $17\beta-E_2S$, failed to elicit these responses.
Since the latency of these acute alterations in neuronal activity were generally on the order of seconds or less, it is highly unlikely that they are mediated by a cytosolic or nuclear
receptor mechanism of action. The possibility of a direct
electrophysiological interaction of the steroid with the neuronal membrane was futher strengthened by the observation that
$E_2-7\alpha$-butyric acid can also influence the SUA of POA neurons,
yet it fails to compete with E_2 for binding to the uterine cytosol receptor or stimulate ODC in the chick oviduct (Carrette
et al., 1979). The responsiveness of the POA-septal area neurons to E_2S was reduced markedly following ovariectomy, although they remained normally responsive to the microelectrophoretic application of ACh (Kelly et al., 1978). The latter
strongly suggests that while the genomic mechanism of steroid

action may not be involved in eliciting the electrophysiological response, it may be involved in the maintenance of the receptors required for that response.

In comparison to these generally inhibitory actions of estrogens on the SUA of POA-septal area neurons, the microelectrophoretic application of testosterone sulfate (TS) facilitated the firing rate of 13 of 200 units examined in the AH-POA-septal area of intact male rats (Yamada, 1979). The latency for these responses ranged between 2 and 30 seconds, an inhibition of activity was not reported for any of the 200 units, and a concurrent microelectrophoretic application of E_1-sulfate did not alter the facilitatory response to TS. The latter observation clearly suggests that TS and E_1S interact with different neurons, presumably reflecting the specificity of the membrane receptors for the steroids.

These electrophysiological data may also reflect the mechanisms of androgen- and estrogen-stimulated male sexual behavior in the rat. Electrical stimulation of the MPOA clearly facilitates the display of male copulatory behavior (Malsbury, 1971; Merari and Ginton, 1975), whereas lesions in this brain region block the induction of this behavior with exogenous androgens (Ginton and Merari, 1977). Lesions in the VM, however, may facilitate male sexual behavior (Christensen et al., 1977). Thus, the electrical stimulation and lesion data are essentially opposite for male and female sexual behavior, although they are consistent with the opposite effects of androgens and estrogens on the SUA in the MPOA. The fact that estrogen as well as androgen implants in the MPOA can facilitate male sexual behavior in rats (*vide supra*) clearly indicates, however, that the electrophysiological reactions of hypothalamic neurons to gonadal steroids cannot totally explain the effect of these hormones on sexual behavior. There appears to be no simple answer to this problem, since the gonadal steroids elicit a wide range of genomic and nongenomic responses in the brain (see Fig. 15-1).

ACKNOWLEDGMENTS

The author is grateful for the support received from an Alfred P. Sloan Research Fellowship. The expert clerical assistance of Radara Grover and the typing assistance of Debe Martin and Vicki Durrance is also acknowledged and appreciated.

Adler, N.T., and Pfaff, D.: Neurobiology of Reproduction. Plenum Press, New York, 1983.

REFERENCES

Ainsworth, L., and Ryan, K.J.: Steroid hormone transformation by endocrine organs from pregnant mammals. I. Estrogen biosynthesis by mammalian placental preparations *in vitro*. *Endocrinology 79:*875-883, 1966.

Alderson, L.M., Starr, M.S., and Baum, M.J.: Effects of castration on dopamine metabolism in rat striatum and limbic forebrain. *Soc Neurosci Abstr 5:*1471, 1979.

Allera, A., Rao, G.S., and Breuer, H.: Specific interaction of corticosteroids with components of the cell membrane which are involved in the translocation of the hormone into the intravesicular space of purified rat liver plasma membrane vesicles. *J Steroid Biochem 12:*259-266, 1980.

Anton-Tay, F., Pelham, R.W., and Wurtman, R.J.: Increased turnover of ^{3}H-norepinephrine in rat brain following castration or treatment with ovine follicle-stimulating hormone. *Endocrinology 84:*1489-1492, 1969.

Antrup, H., and Seiler, N.: On the turnover of polyamines spermidine and spermine in mouse brain and other organs. *Neurochem Res 5:*123-143, 1980.

Arai, Y., Hiroi, M., Mitra, J., and Gorski, R.A.: Influence of intravenous progesterone administration on the cortical electroencephalogram of the female rat. *Neuroendocrinology 2:* 275-282, 1967.

Arendash, G.W., and Gorski, R.A.: Suppression of lordosis behavior in the female rat during mesencephalic electrical stimulation. *Soc Neurosci Abstr 5:*1473, 1979.

Asai, M., Yu, W., and Leung, B.S.: Cytoplasmic competitiors of estrogen receptor at chromatin binding site. *Endocr Soc Abstr 781:* 1979.

Badger, T.M., Wilcox, C.E., Meyer, E.R., Bell, R.D., and Cicero, T.J.: Simultaneous changes in tissue and serum levels of luteinizing hormone, follicle-stimulating hormone and luteinizing hormone/follicle-stimulating hormone-releasing factor after castration in the male rat. *Endocrinology 102:* 136-141, 1978.

Ball, P., and Knuppen, R.: Formation of 2- and 4-hydroxyestrogens by brain, pituitary, and liver of the human fetus. *J Clin Endocrinol Metab 47:*732-737, 1978.

Ball, P., Haupt, M., and Knuppen, R.: Comparative studies on the metabolism of oestradiol in the brain, the pituitary and the liver of the rat. *Acta Endocrinol (Copenh) 87:* 1-11, 1978.

Bapna, J., Neff, N.H., and Costa, E.: A method for studying norepinephrine and serotonin metabolism in small regions of rat brain: Effect of ovariectomy on amine metabolism in anterior and posterior hypothalamus. *Endocrinology 89:*

1345-1349, 1971.

Barbieri, R.L., Canick, J.A., and Ryan, K.J.: Estrogen 2-hydroxylase: Activity in rat tissues. *Steroids 32*:529-538, 1978.

Barfield, R.J., Wilson, C., and McDonald, P.G.: Sexual behavior: Extreme reduction of postejaculatory refractory period by midbrain lesions in male rats. *Science 189*:147-149, 1975.

Barrack, E.R., and Coffey, D.S.: The specific binding of estradiol to the nuclear matrix of estrogen target tissues. *Endocr Soc Abstr 783:* 1979.

Barrack, E.R., Hawkins, E.F., Allen, S.L., Hicks, L.L., and Coffey, D.S.: Concepts related to salt resistant estradiol receptors in rat uterine nuclei: Nuclear matrix. *Biochem Biophys Res Commun 79*:829-836, 1977.

Baulieu, E.E., Godeau, F., Schorderet, M., and Schorderet-Slatkin, S.: Steroid-induced meiotic division of Xenopus laevis oocytes: Surface and calcium. *Nature 275*:593-598, 1978.

Beach, F.A.: *Hormones and Behavior*, Hoeber-Harper, New York, 1948.

Beach, F.A., and Orndoff, R.K.: Variation in the responsiveness of female rats to ovarian hormones as a function of preceding hormonal deprivation. *Horm Behav 5*:201-206, 1974.

Beattie, C.W., and Soyka, L.F.: Influence of progestational steroids in hypothalamic tyrosine hydroxylase activity *in vitro*. *Endocrinology 93*:1453-1455, 1973.

Beattie, C.W., Rodgers, C.H., and Soyka, L.F.: Influence of ovariectomy and ovarian steroids on hypothalamic tyrosine hydroxylase activity in the rat. *Endocrinology 91*:276-279, 1972.

Beinfeld, M.C., and Packman, P.M.: Estrogen induction of specific soluble proteins in the hypothalamus of the immature rat. *Biochem Biophys Res Commun 73*:646-652, 1976.

Bereiter, D.A., and Barker, D.J.: Hormone-induced enlargement of receptive fields in trigeminal mechanoreceptive neurons. I. Time course, hormone, sex and modality specificity. *Brain Res 184*:395-410, 1980.

Bereiter, D.A., Stanford, L.R., and Barker, D.J.: Hormone-induced enlargement of receptive fields in trigeminal mechanoreceptive neurons. II. Possible mechanisms. *Brain Res 184:* 411-423, 1980.

Bernard, B., and Paolino, R.: Time dependent changes in brain biogenic amine dynamics following castration in male rats. *J Neurochem 22*:951-956, 1974.

Berthold, A.A.: Transplantation der Hoden. *Arch Anat Physiol Wiss Med 16*:42-46, 1849.

Beyer, C.: *Endocrine Control of Sexual Behavior*. Raven Press, New York, 1979.

Beyer, C., Morali, G., and Vargas, R.: Effect of diverse estrogens on estrous behavior and genital tract development in
ovariectomized rats. *Horm Behav 2*:273-277, 1971.

Biro, J.: The effect of luteinizing hormone releasing hormone
(LH-RH) and oestrogen on RNA synthesis in anterior pituitary
and different brain regions of rats. *Endokrinologie 72*:285-
290, 1978.

Bjorklund, A., Hakamson, R., Nobin, A., and Sjoberg, N.O.: Increase in rabbit hypothalamic histidine decarboxylase activity after oophorectomy and thyroidectomy. *Experientia 28*:
1232-1233, 1972.

Blaquier, J.A., Cameo, M.S., and Charreau, E.H.: Comparative
uptake of androstenediol, testosterone and dihydrotestosterone by tissues of the male rat. *J Steroid Biochem 1*:
327-334, 1970.

Blaustein, J.D., and Feder, H.H.: Cytoplasmic progestin receptors in guinea pig brain: Characteristics and relationship to
the induction of sexual behavior. *Brain Res 169*:481-497,
1979a.

Blaustein, J.D., and Feder, H.H.: Cytoplasmic progestin receptors in female guinea pig brain and their relationship to re-
fractoriness in expression of female sexual behavior.
Brain Res 177:489-498, 1979b.

Blaustein, J.D., and Feder, H.H.: Nuclear progestin receptors
in guinea pig brain measured by an *in vitro* exchange assay
after hormonal treatments that affect lordosis.
Endocrinology 106:1061-1069, 1980.

Blaustein, J.D., and Wade, G.N.: Concurrent inhibition of sexual behavior, but not brain (^{3}H)estradiol uptake, by progesterone in female rats. *J Comp Physiol Psychol 91*: 742-751,
1977.

Blaustein, J.D., and Wade, G.N.: Progestin binding by brain and
pituitary cell nuclei and female rat sexual behavior.
Brain Res 140:360-367, 1978.

Blaustein, J.D., Dudley, S.D., Gray, J.M., and Wade, G.N.:
Long-term retention of estradiol by brain cell nuclei and
female rat sexual behavior. *Brain Res 173*:355-359, 1979.

Branchey, L., Branchey, M., and Nadler, R.D.: Effects of sex
hormones on sleep patterns of male rats gonadectomized in
adulthood and in the neonatal period. *Physiol Behav 11*:
609-611, 1973.

Brawer, J.R., and Naftolin, F.: The effects of oestrogen on
hypothalamic tissue. *Sex Horm Behav 62*:19-40, 1979.

Brawer, J.R., Ruf, K.B., and Naftolin, F.: Effects of estradiol-induced lesions of the arcuate nucleus on gonadotropin
release in response to preoptic stimulation of the rat.
Neuroendocrinology 30:144-149, 1980.

Breuer, H., and Koster, G.: Interaction between oestrogens and

neurotransmitters at the hypophysial-hypothalamic level.
*J Steroid Biochem 5:*961-968, 1974.

Bueno, J., and Pfaff, D.W.: Single unit recording in hypothalamus and preoptic area of estrogen-treated and untreated ovariectomized female rats. *Brain Res 101:*67-78, 1976.

Burnet, F.R., and MacKinnon, P.C.B.: Restoration by oestradiol benzoate of a neural and hormonal rhythm in the ovariectomized rat. *J Endocrinol 64:*27-35, 1975.

Caggiula, A.R., Antelman, S.M., and Zigmond, M.J.: Disruption of copulation in male rats after hypothalamic lesions: A behavioral, anatomical and neurochemical analysis.
*Brain Res 59:*273-287, 1973.

Callard, G.V., Petro, Z., and Ryan, K.: Conversion of androgen to estrogen and other steroids in the vertebrate brain.
*Am Zoologist 18:*511-523, 1978.

Callard, G.V., Hoffman, R.A., Petro, Z., and Ryan, K.J.: *In vitro* aromatization and other androgen transformations in the brain of the hamster (Mesocricetus auratus). *Biol Reprod 21:* 33-38, 1979.

Cardinali, D.P., and Gomez, E.: Changes in hypothalamic noradrenaline, dopamine and serotonin uptake after estradiol administration to rats. *J Endocrinol 73:*181-182, 1977.

Carrette, B., Barry, J., Linkie, D., Ferin, M., Mester, J., and Balieu, E.-E.: Effets de l' «oestradiol-7α-acide butyrique» au niveau de cellules hypothalamiques. *CR Acad Sci (Paris) 288D:*631-634, 1979.

Carrillo, A.J.: Estrogen receptors in the medial basal hypothalamus of the rat following complete hypothalamic deafferentation. *Brain Res 186:*157-164, 1980.

Cartas-Heredia, L., Guevara-Aguilar, R., and Aguilar-Baturoni, H.U.: Oestrogenic influences on the electrical activity of the olfactory pathway. *Brain Res Bull 3:*623-630, 1978.

Carter, C.S., and Davis, J.M.: Biogenic amines, reproductive hormones and female sexual behavior: A review.
*Biobehav Rev 1:*213-224, 1977.

Celotti, F., Farina, J.M.S., Santaniello, E., Martini, L., and Motta, M.: Effect of testosterone, its 5α-reduced metabolites and the corresponding propionates on testosterone metabolism. I. In the hypothalamus and in the anterior pituitary.
*J Steroid Biochem 11:*215-219, 1979.

Chamness, G.C., Jennings, A.W., and McGuire, W.L.: Estrogen receptor binding to isolated nuclei. A nonsaturable process.
*Biochemistry 13:*327-331, 1974.

Chan, L., Means, A.R., and O'Malley, B.W.: Steroid hormone regulation of specific gene expression. *Vitam Horm 36:*259-295, 1978.

Cheng, Y.-J., and Karavolas, H.J.: Conversion of progesterone to 5α-pregnane-3,20-dione and 3α-hydroxy-5α-pregnan-20-one by rat medial basal hypothalami and the effects of estradiol and stage of estrous cycle on the conversion. *Endocrinology 93:* 1157-1162, 1973.

Cheng, Y.-J., and Karavolas, H.J.: Subcellular distribution and properties of progesterone (Δ^4-steroid)5α-reductase in rat medial basal hypothalamus. *J Biol Chem 250:*7997-8003, 1975.

Chiocchio, S.R., Negro-Vilar, A., and Tramezami, J.H.: Acute changes in norepinephrine content in the median eminence induced by orchidectomy or testosterone replacement. *Endocrinology 99:*629-635, 1976.

Christensen, L.W., and Clemens, L.G.: Possible involvement of cyclic AMP in regulation of masculine sexual behaviour by testosterone. *J Endocrinol 61:*153-161, 1974.

Christensen, L.W., Nance, D.M., and Gorski, R.A.: Effects of hypothalamic and preoptic lesions on reproductive behavior in male rats. *Brain Res Bull 2:*137-141, 1977.

Cidlowski, J.A., and Muldoon, T.G.: Estrogenic regulation of cytoplasmic receptor populations in estrogen-responsive tissue of the rat. *Endocrinology 95:*1621-1629, 1974.

Cidlowski, J.A., and Muldoon, T.G.: Sex-related differences in the regulation of cytoplasmic estrogen receptor levels in responsive tissues of the rat. *Endocrinology 98:*833-841, 1976.

Clark, C.R., and Nowell, N.W.: Binding properties of testosterone receptors in the hypothalamic-preoptic area of the adult male mouse brain. *Steroids 33:*407-426, 1979.

Clark, J.H., and Peck, E.J., Jr.: Nuclear retention of receptor oestrogen complex and nuclear acceptor sites. *Nature 260:*635-637, 1976.

Clark, T.K., Cagiula, A.R., McConnell, R.A., and Antelman, S.M.: Sexual inhibition is reduced by rostral midbrain lesions in the male rat. *Science 190:*169-171, 1975.

Clemens, L.G., and Christensen, L.W.: Theophylline, potentiation of testosterone propionate and of estradiol benzoate in inducing male but not female sexual behavior in the female rat. *Horm Behav 9:*170-177, 1977.

Clemens, L.G., and Dohanich, G.P.: Inhibition of lordotic behavior in female rats following intracerebral infusion of anticholinergicc agents. *Pharmacol Biochem Behav 13:*89-95, 1980.

Clemens, L.G., Humphrys, R.R., and Dohanich, G.P.: Cholinergic brain mechanisms and the hormonal regulation of female sexual behavior in the rat. *Pharmacol Biochem Behav 13:* 81-88, 1980.

Colvin, G.B., Whitmoyer, D.T., and Sawyer, C.H.: Circadian sleep-wakefulness patterns in rats after ovariectomy and treatment with estrogen. *Exp Neurol 25:*616-625, 1969.

Cramer, O.M., Parker, C.R., and Porter, J.C.: Estrogen inhibition of dopamine release into hypophysial portal blood.

Endocrinology 104:419-422, 1979a.

Cramer, O.M., Parker, C.R., and Porter, J.C.: Stimulation of dopamine release into hypophysial portal blood by administration of progesterone. *Endocrinology 105*:929-933, 1979b.

Crowley, W.R., O'Donohue, T.L., and Jacobowitz, D.M.: Changes in catecholamine content in discrete brain nuclei during the estrous cycle of the rat. *Brain Res 147*:315-326, 1978a.

Crowley, W.R., O'Donohue, T.L., Wachslicht, H., and Jacobowitz, D.M.: Effects of estrogen and progesterone on plasma gonadotropins and on catecholamine levels and turnover in discrete brain regions of ovariectomized rats. *Brain Res 154*:345-357, 1978b.

Crowley, W.R., O'Donohue, T.L., Muth, E.A., and Jacobowitz, D.M.: Effects of ovarian hormones on levels of luteinizing hormone in plasma and on serotonin concentrations in discrete brain nuclei. *Brain Res Bull 4*:571-574, 1979.

Curtis, D.R., and Johnston, G.A.R.: Amino acid transmitters in the mammalian central nervous system. *Ergeb Physiol 69*:97-188, 1974.

Czaja, J.A., Goldfoot, D.A., and Karavolas, H.: Comparative facilitation and inhibition of lordosis in the guinea pig with progesterone, 5α-pregnane-3,20-dione or 3α-hydroxy-5α-pregnan-20-one. *Horm Behav 5*:261-274, 1974.

Danzin, C., Jung, M.J., Claverie, N., Grove, J., Sjoerdsma, A., and Koch-Weser, J.: Effects of α-difluoromethyl-ornithine, and enzyme-activated irreversible inhibitor of ornithine decarboxylase, on testosterone-induced regeneration of prostate and seminal vesicle in castrated rats. *Biochem J 180*:507-513, 1979.

Davidson, J.M.: Hormones and reproductive behavior. In *Hormones and Behavior*, S. Levine, ed. Academic Press, New York, 1972, p 64-103.

Davies, I.J., Naftolin, F., Ryan, K.J., Fishman, J., and Siu, J.: The affinity of catechol estrogens for estrogen receptors in the pituitary and anterior hypothalamus of the rat. *Endocrinology 97*:554-557, 1975.

Davis, G.A., and Kohl, R.L.: Biphasic effects of the antiserotonergic methysergide on lordosis in rats. *Pharmacol Biochem Behav 9*:487-491, 1978.

Davis, P.G., and Barfield, R.J.: Activation of masculine sexual behavior by intracranial estradiol benzoate implants in male rats. *Neuroendocrinology 28*:217-227, 1979a.

Davis, P.G., and Barfield, R.J.: Activation of feminine sexual behavior in castrated male rats by intrahypothalamic implants of estradiol benzoate. *Neuroendocrinology 28*:228-233, 1979b.

Davis, P.G., McEwen, B.S., and Pfaff, D.W.: Localized behavioral effects of tritiated estradiol implants in the ventromedial hypothalamus of female rats. *Endocrinology 104*:898-

903, 1979.

Debus, G., and Kehr, W.: Catecholamine and 5-hydroxytryptamine synthesis and metabolism following intraventricular injection of dibutyryl cyclic AMP. J Neural Transm 45:195-206, 1979.

Denari, J.H., and Rosner, J.M.: Sexual steroid uptake in the alloxanized diabetic rat. Steroids Lipids Res 3:151-155, 1972.

Denef, C., Magnus, C., and McEwen, B.S.: Sex differences and hormonal control of testosterone metabolism in rat pituitary and brain. J Endocrinol 59:605-621, 1973.

Donoso, A.O.: Induction of prolactin and luteinizing hormone release by histamine in male and female rats and the influence of brain transmitter antagonists. J Endocrinol 76: 193-202, 1978.

Donoso, A.O., and Moyano, M.B.G.: Adrenergic activity in hypothalamus and ovulation. Proc Soc Exp Bio Med 135:633-635, 1970.

Donoso, A.O., Stefano, F.J.E., Biscardi, A.M., and Cukier, J.: Effects of castration on hypothalamic catecholamines. Am J Physiol 212:737-739, 1967.

Dufy, B., Partouche, C., Poulain, D., Dufy-Barbe, L., and Vincent, J.D.: Effects of estrogen on the electrical activity of identified and unidentified hypothalamic units. Neuroendocrinology 22:38-47, 1976.

Early, C.J., and Leonard, B.E.: GABA and gonadal hormones. Brain Res 155:27-34, 1978.

Eaton, G.G., Goy, R.W., and Resko, J.A.: Brain uptake and metabolism of estradiol benzoate and estrous behavior in ovariectomized guinea pigs. Horm Behav 6:81-97, 1975.

Eisenfeld, A.J.: ^{3}H-Estradiol: In vitro binding to macromolecules from rat hypothalamus, anterior pituitary and uterus. Endocrinology 86:1313-1318, 1970.

Endersby, C.A., and Wilson, C.A.: The effect of ovarian steroids on the accumulation of ^{3}H-labelled monamines by hypothalamic tissue in vitro. Brain Res 73:321-331, 1974.

Endroczi, E.: Neural and hormonal regulation of the patterns of sexual behavior. In Symposium On Reproduction, K. Lissak, ed. Akadademiai Kiado., Budapest, 1967, pp 39-63.

Engel, J., Ahlenius, S., Almgren, O., Carlsson, A., Larsson, K., and Sodersten, P.: Effects of gonadectomy and hormone replacement on brain monoamine synthesis in male rats. Pharmacol Biochem Behav 10:149-154, 1979.

Espino, C., Sano, M., and Wade, G.N.: Alpha-methyltryptamine blocks facilitation of lordosis by progesterone in spayed, estrogen-primed rats. Pharmacol Biochem Behav 3:557-559, 1975.

Etgen, A.M.: Antiestrogens: Effects of taxoxifen, nafoxidine, and CI-628 on sexual behavior, cytoplasmic receptors, and nuclear binding of estrogen. *Horm Behav 13:* 97–112, 1979.

Everitt, B.J., Fuxe, K., Hökfelt, T., and Jonsson, G.: Role of monoamines in the control by hormones of sexual receptivity in the female rat. *J Comp Physiol Psychol 89:* 556–572, 1975.

Fahn, S.: Regional distribution studies of GABA and other putative neurotransmitters and their enzymes. In *GABA in Nervous System Function.* E. Roberts, T.N. Chase, and D.B. Tower, eds. Raven Press, New York, 1976, pp 169–186.

Faigon, M.R., and Moguilevsky, J.A.: Effect of estradiol on amino acid incorporation into proteins of different hypothalamic areas in prepubertal rats. *Experientia 32:* 392–394, 1976.

Feder, H.H., and Morin, L.P.: Suppression of lordosis in guinea pigs by ethamoxy-triphetol (MER-25) given at long intervals (34–36 hr) after estradiol benzoate treatment. *Horm Behav 5:* 63–72, 1974.

Feder, H.H., and Silver, R.: Activation of lordosis in ovariectomized guinea pigs by free and esterified forms of estrone, estradiol-17β and estriol. *Physiol Behav 13:* 251–255, 1974.

Feder, H.H., Siegel, H., and Wade, G.N.: Uptake of [6,7-^{3}H] estradiol-17β in ovariectomized rats, guinea pigs, and hamsters: Correlation with species differences in behavioral responsiveness to estradiol. *Brain Res 71:* 93–103, 1974.

Fisher, A.E.: Maternal and sexual behavior induced by intracranial chemical stimulation. *Science 124:* 228–229, 1956.

Fishman, J., and Martucci, C.: Absence of measurable 2-hydroxyestrone in the rat brain--evidence of rapid turnover. *J Clin Endocrinol Metab 49:* 940–942, 1979.

Fishman, J., and Norton, B.: Catechol estrogen formation in the central nervous system of the rat. *Endocrinology 96:* 1054–1058, 1975.

Fishman, J., Naftolin, F., Davies, I.J., Ryan, K.J., and Petro, Z.: Catechol estrogen formation by the human fetal brain and pituitary. *J Clin Endocrinol Metab 42:* 177–180, 1976.

Fishman, J., Norton, B.I., and Krey, L.: 2-Hydroxylation of estrogens in the brain participates in the initiation of the preovulatory LH surge in the rat. *Biochem Biophys Res Commun 93:* 471–477, 1980.

Fleischer-Lambropoulos, H., Sarkander, H.-I., and Brade, W.P.: Effects of polyamines on amino acid incorporation into protein by cerebral and cerebellar as well as "neural" and "glial" nuclei of rat brain. *Biochem Biophys Res Commun 63:* 792–800, 1975.

Florez-Lozano, J.A., Menendez-Patterson, A., Marin, B.: Sexual behavior of the pancreatectomized (95%) male hamster (Mesocricetus auratus). *Physiol Behav 20:* 465–468, 1978.

Fonnum, F.: The distribution of glutamate decarboxylase and asparate transamininase in subcellular fractions of rat and guinea pig brain. *Biochem J 106:*401-412, 1968.

Foreman, M.M., and Moss, R.L.: Role of hypothalamic alpha and beta andrenergic receptors in the control of lordotic behavior in the ovariectomized estrogen primed rat. *Pharmacol Biochem Behav 9:*235-241, 1978.

Foreman, M.M., and Porter, J.C.: Effects of catechol estrogens and catecholamines on hypothalamic and corpus striatal tyrosine hydroxylase activity. *J Neurochem 34:*1175-1183, 1980.

Foreman, M.M., Wickersham, E.W., and Anthony, A.: Cytophotometric analysis of hypothalamic RNA fluctuations during the rat estrous cycle. *Brain Res 119:*471-475, 1977.

Fox, T.O.: Conversion of the hypothalamic estradiol receptor to the 'nuclear' form. *Brain Res 120:*580-583, 1977.

Fox, T.O., and Johnston, C.: Estradiol receptors from mouse brain and uterus: Binding to DNA. *Brain Res 77:*330-336, 1974.

Fregly, M.J., and Luttge, W.G.: *Human Endocrinology--An Interactive Text,* Elsevier Biomedical, New York, 1982.

Fuxe, K., Hökfelt, T., Ljungdahl, A., Agnati, L., Johansson, O., and Perez de la Mora, M.: Evidence for an inhibitory gabaergic control of the mesolimbic dopamine neurons: Possibility of improving treatment of schizophrenia by combined treatment with neuroleptics and gabaergic drugs. *Med Biol 53:* 177-183, 1975.

Ganong, W.F.: The role of catecholamines and acetylcholine in the regulation of endocrine function. *Life Sci 15:*1401-1414, 1974.

Gentry, R.T., Wade, G.N., and Roy, R.T.: Individual differences in estradiol-induced behaviors and in neural ^{3}H-estradiol uptake in rats. *Physiol Behav 17:*195-200, 1976.

Gentry, R.T., Wade, G.N., and Blaustein, J.D.: Binding of [^{3}H]estradiol by brain cell nuclei and female rat sexual behavior: Inhibition by experimental diabetes. *Brain Res 135:*135-146, 1977.

Ginsburg, M., MacLusky, N.J., Morris, J.D., and Thomas, P.J.: Physiological variation in abundance of oestrogen specific high-affinity binding sites in hypothalamus, pituitary and uterus of the rat. *J Endocrinol 64:*443-449, 1975.

Ginton, A., and Merari, A.: Long range effects of MPOA lesion on mating behavior in the male rat. *Brain Res 120:*158-163, 1977.

Goertz, B.: Effect of polyamines on cell-free protein synthesizing systems from rat cerebral cortex, cerebellum and liver. *Brain Res 173:*125-135, 1979.

Gordon, J.H., and Diamond, B.I.: The modulation of tardive dyskinesia by estrogen: Neurochemical studies in an animal model. *Ann Neurol 6:*152, 1979.

Gordon, J.H., Nance, D.M., Wallis, C.J., and Gorski, R.A.: Effects of estrogen on dopamine turnover, glutamic acid decarboxylase activity and lordosis behavior in septal lesioned female rats. *Brain Res Bull* 2:341-346, 1977.

Gordon, J.H., Nance, D.M., Wallis, C.J., and Gorski, R.A.: Effect of septal lesions and chronic estrogen treatment on dopamine, GABA and lordosis behavior in male rats. *Brain Res Bull* 4:85-89, 1979.

Gordon, J.H., Borison, R.L., and Diamond, B.I.: Modulation of dopamine receptor sensitivity by estrogen. *Biol. Psychiatry* 15:389-396, 1980a.

Gordon, J.H., Gorski, R.A., Borison, R.L., and Diamond, B.I.: Postsynaptic efficacy of dopamine: Possible suppression by estrogen. *Pharmacol Biochem Behav* 12:515-518, 1980b.

Gorski, J., and Gannon, F.: Current models of steroid hormone action: A critique. *Annu Rev Physiol* 38:425-450, 1976.

Gorski, R.A., Gordon, J.H., Shryne, J.E., and Southam, A.M.: Evidence for a morphological sex difference within the medial preoptic area of the rat brain. *Brain Res* 148:333-346, 1978.

Gorzalka, B.B., and Whalen, R.E.: Genetic regulation of hormone action: Selective effects of progesterone and dihydroprogesterone (5α-pregnane-3,20-dione) on sexual receptivity in mice. *Steroids* 23:499-505, 1974.

Gorzalka, B.B., and Whalen, R.E.: Effects of genotype on differential behavioral responsiveness to progesterone and 5α-dihydroprogesterone in mice. *Behav Genet* 6:7-15, 1976.

Gorzalka, B.B., and Whalen, R.E.: The effects of progestins, mineralocorticoids, glucocortiocids and steroid sobulity on the induction of sexual receptivity in rats. *Horm Behav* 8:94-99, 1977.

Goto, J., and Fishman, J.: Participation of a nonenzymatic transformation in the biosynthesis of estrogens from androgens. *Science* 195:80-81, 1977.

Gradwell, P., Everitt, B.J., and Herbert, J.: 5-Hydroxytryptamine in the central nervous system and sexual receptivity of female rhesus monkeys. *Brain Res* 88:281-293, 1975.

Grant, L.D., and Stumpf, W.E.: Hormone uptake sites in relation to CNS biogenic amine systems. In *Anatomical Neuroendocrinology*, W.E. Stumpf and L.D. Grant, eds. S. Karger, Basel, 1975, pp 445-463.

Gray, H.E., and Luttge, W.G.: Apparent role of protein synthesis and LH-RH in the inhibition by dihydrotestosterone of estrogen-induced sexual receptivity in mice. *Physiol Behav* 21:973-977, 1978.

Gray, H.E., Jasper, T.W., Luttge, W.G., Shukla, J.B., and Rennert, O.M.: Estrogen increases hypothalamic and pituitary polyamine levels in ovariectomized rats. *J Neurochem* 34:753-755, 1980.

Green, R., Luttge, W.G., and Whalen, R.E.: Induction of receptivity in ovariectomized female rats by a single intravenous injection of estradiol-17β. *Physiol Behav 5*:137-142, 1970.

Greene, G.L., Closs, L.E., DeSombre, E.R., and Jensen, E.V.: Antibodies to estrophilin: Comparison between rabbit and goat antisera. *J Steroid Biochem 11*:333-341, 1979.

Greene, G.L., Closs, L.E., DeSombre, E.R., and Jensen, E.V.: Estrophilin: Pro and anti. *J Steroid Biochem 12*:159-167, 1980a.

Greene, G.L., Fitch, F.W., and Jensen, E.V.: Monoclonal antibodies to estrophilin: Probes for the study of estrogen receptors. *Proc Natl Acad Sci USA 77*:157-161, 1980b.

Greenough, W.T., Carter, C.S., Steerman, C., and DeVoogel, T.J.: Sex differences in dendritic patterns in hamster preoptic area. *Brain Res 126*:63-72, 1977.

Griffiths, E.C., Hooper, K.C., Jeffcoate, S.L., and Holland, D.T.: The effects of gonadectomy and gonadal steroids on the activity of hypothalamic peptidases inactiviating luteinizing hormone-releasing hormone (LH-RH). *Brain Res 88*:384-388, 1975.

Gual, C., Morato, T., Hayano, M., Gut, M., and Dorfman, R.I.: Biosynthesis of estrogens. *Endocrinology 71*:920-925, 1962.

Gunaga, K.P., and Menon, K.M.J.: Effect of catecholamines and ovarian hormones on cyclic AMP accumulation in rat hypothalamus. *Biochem Biophys Res Commun 54*:440-448, 1973.

Gunaga, K.P., Kawano, A., and Menon, K.M.J.: *In vivo* effect of estradiol benzoate on the accumulation of adenosine 3', 5'-cyclic monophosphate in the rat hypothalamus. *Neuroendocrinology 16*:273-281, 1974.

Gustafsson, J.-A., Pousette, A., and Svensson, E.: Sex-specific occurrence of androgen receptors in rat brain. *J Biol Chem 251*:4047-4054, 1976.

terHaar, M.B., and MacKinnon, P.C.B.: Changes in serum gonadotrophin levels, and in protein levels and *in vivo* incorporation of [35S]methionine into protein of discrete brain areas and the anterior pituitary of the rat during the oestrous cycle. *J Endocrinol 58*:563-576, 1973.

terHaar, M.B., and MacKinnon, P.C.B.: Effects of antibody to oestrogen or of ovariectomy on the incorporation of [35S]-methionine into brain protein and on gonadotropin levels during the oestrous cycle in the rat. *J Endocrinol 65*:399-410, 1975.

terHaar, M.B., and MacKinnon, P.C.B.: Ovarian steroid effects on gonadotrophin output, and on the incorporation of 35S from methionine into protein of discrete brain areas and the anterior pituitary. *Acta Endocrinol (Copenh) 85*:279-290, 1977.

Hall, N.R., and Luttge, W.G.: Intracerebral progesterone: Effects on sexual behavior in female mice. *Soc Neurosci Abstr*

270: 1974.
Hannouche, N., Thieulant, M.-L., Samperez, S., and Jouan, P.:
Androgen binding proteins in the cytosol from prepubertal
male rat hypothalamus, preoptic area and brain cortex.
J Steroid Biochem 9:147-151, 1977.
Hanukoglu, I., Karavolas, H.J., and Goy, R.W.: Progesterone
metabolism in the pineal, brain stem, thalamus and corpus
callosum of the female rat. Brain Res 125:313-324, 1977.
Harding, C.F., and Feder, H.H.: Relation of uptake and metabolism of (1,2,6,7³H)testosterone to individual differences
in sexual behavior in male guinea pigs. Brain Res 105:137-
149, 1976.
Harris, G.W.: Neural Control of the Pituitary Gland, Arnold,
London, 1955.
Heritage, A.S., and Grant, L.D.: ³H-Dihydrotestosterone in
catecholamine neurons of rat brain stem. Anat Rec 193:564,
1979.
Heritage, A.S., Grant, L.D., and Stumpf, W.E.: ³H-Estradiol in
catecholamine neurons of rat brain stem: Combined localization by autoradiography and formaldehyde-induced fluorescence. J Comp Neurol 176:607-630, 1977.
Hill, M.J., Goddard, P., and Williams, R.E.O.: Gut bacteria and
aetiology of cancer of the breast. Lancet 2:472-473, 1971.
Honma, K., and Wuttke, W.: Norepinephrine and dopamine turnover
rates in the medial area and medial-basal hypothalmaus of the
rat brain after various endocrinological manipulations.
Endocrinology 106:1848-1853, 1980.
Horowitz, S.B., and Moore, L.C.: The nuclear permeability,
intracellular distribution and diffusion of insulin in the
amphibian oocyte. J Cell Biol 60:405-415, 1974.
Horwitz, K.B., and McGuire, W.L.: Estrogen control of progesterone receptor in human breast cancer. Correlation with nuclear processing of estrogen receptor. J Biol Chem 253:2223-
2228, 1978a.
Horwitz, K.B., and McGuire, W.L.: Actinomycin D prevents nuclear processing of estrogen receptor. J Biol Chem 253:6319-
6322, 1978b.
Hough, J.C., Ho, K.-W., Cooke, P.H., and Quadagno, D.M.: Actinomycin-D: Reversible inhibition of lordosis behavior and
correlated changes in nucleolar morphology. Horm Behav 5:367-
376, 1974.
Hruska, R.E., and Silbergeld, E.K.: Increased dopamine receptor
sensitivity after estrogen treatment. Soc Neurosci Abstr 5:
236, 1979.
Hyyppa, M.T., Cardinali, D.P., Baumgarten, H.G., and Wurtman,
R.J.: Rapid accumulation of H³-serotonin in brains of rats
receiving intraperitoneal H³-tryptophan: Effects of 5,6-
dihydroxytroptamine or female sex hormones. J Neural Transm

*34:*111-124, 1973.

Inaba, M., and Kamata, K.: Effect of estradiol-17β and other steroids on noradrenaline and dopamine binding to synaptic membrane fragments of rat brain. *J Steroid Biochem 11:* 1491-1497, 1979.

Jackson, G.L.: Blockage of progesterone-induced release of LH by intrabrain implants of actinomycin D. *Neuroendocrinology 17:*236-244, 1975.

Janowsky, D.S., and Davis, J.M.: Progesterone-estrogen effects on uptake and release of norepinephrine by synaptosomes. *Life Sci 9:*525-531, 1970.

Jellink, P.H., Davis, P.G., Krey, L.C., Luine, V.N., Roy, E.J., and McEwen, B.S.: Central and peripheral action of estradiol and catchol estrogens administered by continuous infusion. *Endocrine Soc Abstr 838:* 1979.

Johnson, W.A., Billiar, R.B., and Little, B.: Progesterone and 5α-reduced metabolites: Facilitation or lordosis behavior and brain uptake in female hamsters. *Behav Biol 18:*489-497, 1976a.

Johnson, W.A., Billiar, R.B., Rahman, S.S., and Little, B.: The head extractions brain and pituitary uptake and metabolism of progesterone and 5α-dihydroprogesterone in anesthetized, female rabbits. *Brain Res 111:*147-155, 1976b.

Johnston, P.G., and Davidson, J.M.: Priming action of estrogen: Minimum duration of exposure for feedback and behavioral effects. *Neuroendocrinology 28:* 155-159, 1979.

Jung, M.J., and Seiler, N.: Enzyme-activated irreversible inhibitors of L-ornithine: 2-Oxoacid aminotransferase. Demonstration of mechanistic features of the inhibition of ornithine aminotransferase by 4-aminohex-5-ynoic acid and gabaculine and correlation with *in vivo* activity. *J Biol Chem 253:* 7431-7439, 1978.

Jungblut, P.W., Kallweit, E., Sierralta, W., Truitt, A.J., and Wagner, R.K.: The occurrence of steroid-free, "activated" estrogen receptor in target cell nuclei. *Hoppe Seylers Z Physiol Chem 359:*1259-1268, 1978.

Jungblut, P.W., Hughes, A., Gaues, J., Kallweit, E., Marschler, I., Parl, F., Sierralta, W., Szendro, P.I., and Wagner, R.K.: Mechanisms involved in the regulation of steroid receptor levels. *J Steroid Biochem 11:*273-278, 1979.

Kahwanago, I., Heinrichs, W.L.R., and Herrmann, W.L.: Estradiol receptors in hypothalamus and anterior pituitary gland: Inhibition of estradiol binding by SH-group blocking agents and clomiphene citrate. *Endocrinology 86:*1319-1326, 1970.

Kalra, P.S., and Kalra, S.P.: Modulation of hypothalamic luteinizing hormone-releasing hormone levels by intracranial and subcutaneous implants of gonadal steroids in castrated rats: Effects of androgen and estrogen antagonists.

Endocrinology 106:390-397, 1980.

Kalra, S.P., and McCann, S.M.: Effects of drugs modifying catecholamine synthesis on plasma LH and ovulation in the rat. *Neuroendocrinology* 15:79-91, 1975.

Kao, L.W.L., Lloret, A.P., and Weisz, J.: Metabolism *in vitro* of dihydrotestosterone, 5α-androstane-3α,17β-diol and its 3β-epimer, three metabolites of testosterone by three of its target tissues: The anterior pituitary, the medial basal hypothalamus and the seminiferous tubules. *J Steroid Biochem 8:* 1109-1115, 1977.

van de Kar, L., Levine, J., and Van Orden, L.S.: Serotonin in hypothalamic nuclei: Increased content after castration of male rats. *Neuroendocrinology* 27:186-192, 1978.

Karavolas, H.J., Hodges, D., and O'Brien, D.: Uptake of [³H]-progesterone and [³H]5α-dihydroprogesterone by rat tissues *in vivo* and analysis of accumulated radio-activity: Accumulation of 5α-dihydroprogesterone by pituitary and hypothalamic tissues. *Endocrinology* 98:164-175, 1976.

Karavolas, H.J., Hodges, D.R., O'Brien, D.J. and MacKenzie, K.M.: *In vivo* uptake of [³H]progesterone and [³H]-5α-dihydroprogesterone by rat brain and pituitary and effects of estradiol and time: Tissue concentration of progesterone itself or specific metabolites. *Endocrinology* 104:1418-1425, 1979.

Kato, J.: The role of hypothalamic and hypophyseal 5α-dihydrotestosterone, estradiol and progesterone receptors in the mechanism of feedback action. *J Steroid Biochem* 6:979-987, 1975.

Kato, J., and Minaguchi, H.: Cholinergic and adrenergic mechanisms in the female rat hypothalamus with special reference to reproductive functions. *Gunma Sympos Endocrinol* 1:269-281, 1964.

Kato, J., and Onouchi, T.: Nuclear progesterone receptors and characterization of cytosol receptors in the rat hypothalamus and anterior hypophysis. *J Steroid Biochem* 11:845-854, 1979.

Kato, J., and Villee, C.A.: Preferential uptake of estradiol by the anterior hypothalamus of the rat. *Endocrinology* 80:567-575, 1967.

Kawakami, M., Teresawa, E., and Ibuki, T.: Changes in multiple unit activity of the brain during the estrous cycle. *Neuroendocrinology* 6:30-48, 1970.

Kaye, A.M., Icekson, I., and Lindner, H.R.: Stimulation by estrogens of ornithin and S-adenosylmethionine decarboxylase in the immature uterus. *Biochem Biophys Acta* 252:150-159, 1971.

Kazama, N., and Longcope, C.: *In vivo* studies of the metabolism of estrone and estradiol-17β by the brain. *Steroids* 23:469-481, 1974.

Keefer, D.A., and Stumpf, W.E.: Atlas of estrogen-concentrating cells in the central nervous system of the squirrel monkey.

*J Comp Neurol 160:*419-442, 1975.

Keefer, D.A., Stumpf, W.E., and Sar, M.: Topographical localization of estrogen-concentrating cells in the rat spinal cord
following ^{3}H-estradiol administration. *Proc Soc Exp Biol Med*
*143:*414-417, 1973.

Kelly, D.B.: Neuroanatomical correlates of hormone sensitive
behaviors in frogs and birds. *Am Zoologist 18:*477-488, 1978.

Kelly, M.J., Moss, R.L., and Dudley, C.A.: Differential sensitivity of preoptic-septal neurons to microelectrophoresed estrogen during the estrous cycle. *Brain Res 114:*152-157, 1976.

Kelly, M.J., Moss, R.L., Dudley, C.A., and Fawcett, C.P.: The
specificity of the response of preoptic-septal neurons to estrogen: 17α-estradiol versus 17β-estradiol and the response
of extrahypothalamic neurons. *Exp Brain Res 30:*43-52, 1977a.

Kelly, M.J., Moss, R.L., and Dudley, C.A.: The effects of microelectrophoretically applied estrogen, cortisol and acetylcholine on medial preoptic-septal unit activity throughout
the estrous cycle of the female rat. *Exp Brain Res 30:*53-64,
1977b.

Kelly, M.J., Moss, R.L., and Dudley, C.A.: The effects of
ovariectomy on the responsiveness of preoptic-septal neurons
to microelectrophoresed estrogen. *Neuroendocrinology 25:*204-
211, 1978.

Kendall, D.A., and Tonge, S.R.: Effects of testosterone and
ethinyloestradiol on the synthesis and uptake of noradrenaline and 5-hydroxytryptamine in rat hindbrain: Evidence
for a presynaptic regulation of monoamine synthesis.
*Br J Pharmacol 60:*310P-311P, 1977.

Kent, G.C., Jr., and Liberman, M.L.: Induction of psychic
estrus in the hamster with progesterone administered via the
lateral brain ventricle. *Endocrinology 45:*29-32, 1949.

Kim, J.S., Bak, I.J., Hanler, R., and Okada, Y.: Role of γ-
aminobutyric acid (GABA) in the extrapyramidal motor system.
2. Some evidence for the existence of a type of GABA-rich
strionigral neurons. *Exp Brain Res 14:*95-104, 1971.

Kim, Y.S., Stumpf, W., Sar, M., and Martinez-Vargas, M.C.:
Estrogen and androgen target cells in the brain of fishes,
reptiles and birds: Phylogeny and ontogeny. *Am Zoologist 18:*
425-434, 1978.

King, R.J.B., Somjen, D., Kaye, A.M., and Lindner, H.R.:
Stimulation by oestradiol-17β of specific cytoplasmic and
chromosomal protein synthesis in immature rat uterus.
*Mol Cell Endocrinol 1:*21-36, 1974.

Kishimoto, Y.: Estrone sulphate in rat brain: Uptake from blood
and metabolism *in vivo. J Neurochem 20:*1489-1492, 1973.

Kizer, J.S., Palkovits, M., Zivin, J., Brownstein, M.,
Saavedra, J.M., and Kopin, J.J.: The effect of endocrinological manipulations of tyrosine hydroxylase and dopamine-β-

hydroxylase activities in individual hypothalamic nuclei of the adult male rat. *Endocrinology 95:*799-812, 1974.

Kizer, J.S., Muth, E., and Jacobowitz, D.M.: The effect of bilateral lesions of the ventral noradrenergic bundle on endocrine-induced changes of tyrosine hydroxylase in the rat median eminence. *Endocrinology 98:*886-893, 1976a.

Kizer, J.S., Palkovits, M., Kopin, I.J., Saavedra, J.M., and Brownstein, M.J.: Lack of effect of various endocrine manipulations on tryptophan hydroxylase activity of individual nuclei of the hypothalamus, limbic system and midbrain of the rat. *Endocrinology 98:*743-747, 1976b.

Kizer, J.S., Humm, J., Nicholson, G., Greeley, G., and Youngblood, W.: The effect of castration, thyroidectomy and haloperidol upon the turnover rates of dopamine and norepinephrine and the kinetic properties of tyrosine hydroxylase in discrete hypothalamic nuclei of the male rat. *Brain Res 146:*95-107, 1978.

Kobayashi, R.M., and Reed, K.C.: Conversion of androgens to estrogens (aromatization) in discrete regions of the rat brain: Sexual differences and effects of castration. *Soc Neurosci Abstr 3:*1115, 1977.

Komisaruk, B.R.: Strategies in neuroendocrine neurophysiology. *Am Zoologist 11:*741-754, 1971.

Komisaruk, B.R., McDonald, P.G., Whitmoyer, D.I., and Sawyer, C.H.: Effects of progesterone and sensory stimulation on EEG and neuronal activity in the rat. *Exp Neurol 19:* 494-507, 1967.

Krause, J.E., and Karavolas, H.J.: The effect of progesterone analogues, naturally occurring steroids, and contraceptive progestins on hypothalamic and anterior pituitary Δ^4-steroid (Progesterone) 5α-reductase. *Steroids 31:* 823-839, 1978.

Kubli-Garfias, C., and Whalen, R.E.: Induction of lordosis behavior in female rats by intravenous administration of progestins. *Horm Behav 9:* 380-386, 1977.

Kuhar, M.J., Stethy, V.H., Roth, R.H., and Aghajanian, G.D.: Choline: Selective acumulation by central cholinergic neurons. *J Neurochem 20:* 581-593, 1973.

Kuhl, H., and Tubert, H.-D.: Inactivation of luteinizing hormone releasing hormone by rat hypothalamic L-cystine arylamidase. *Acta Endocrinol (Copenh) 78:*634-648, 1975.

Kuhl, H., Rosniatowski, C., and Taubert, H.-D.: Effect of sex hormones on LH-RH-degrading hypothalamic enzyme system during estrus cycle in rats. *Endocrinol Exp (Bratisl) 13:*29-38, 1979.

Kumakura, K., Hoffman, M., Cocchi, D., Trabucchi, M., Speno, P.F., and Muller, E.E.: Long term effect of ovariectomy on dopamine stimulated adenylate cyclase in rat striatum and nucleus accumbens. *Psychopharmacology 61:*13-16, 1979.

Kurachi, K., and Hirota, K.: Catecholamine metabolism in rat's brain related with sexual cycle. *Endocrinol Jpn Suppl 1*:69-73, 1969.

Kurtz, R.G.: Hippocampal and cortical activity during sexual behavior in the female rat. *J Comp Physiol Psychol 89*:158-169, 1975.

Landau, I.T., and Feder, H.H.: Whole cell and nuclear uptake of [³H]estriol in neural and peripheral tissues of the ovariectomized guinea pig. *Brain Res 121*:190-195, 1977.

Landau, I.T., and Feder, H.H.: Uptake and metabolism of ³H-estrone in neural and peripheral tissue of gonadectomized adult and neonatal guinea pigs. *Psychoneuroendocrinology 5*: 25-32, 1980.

Lee, H., Davies, I.J., and Ryan, K.J.: Progesterone receptor in the hypothalamic cytosol of female rats. *Endocrinology 104*: 791-800, 1979.

Leehan, S.W., Quadagno, D.M., and Bast, J.D.: The effects of extrahypothalamic cycloheximide on sexual receptivity in the rat. *Horm Behav 12*:264-268, 1979.

Levy, C., Mortel, R., Eychenne, B., Robel, P., and Baulieu, E.E.: Unoccupied nuclear oestradiol-receptor sites in normal human endometrium. *Biochem J 185*:733-738, 1980.

Lieberburg, I., and McEwen, B.S.: Estradiol-17β a metabolite of testosterone recovered in cell nuclei from limbic areas of adult male rat brains. *Brain Res 91*:171-174, 1975.

Lieberburg, I., and McEwen, B.S.: Brain cell nuclear retention of testosterone metabolites, 5α-dihydrotestosterone and estradiol-17β in adult rats. *Endocrinology 100*:588-597, 1977.

Lieberburg, I., MacLusky, N.J., and McEwen, B.S.: 5α-Dihydrotestosterone (DHT) receptors in rat brain and pituitary cell nuclei. *Endocrinology 100*:598-607, 1977.

Linkie, D.M.: Estrogen receptors in different target tissues: Similarities of form—dissimilarities of transformation. *Endocrinology 101*:1862-1870, 1977.

Linkie, D.M., and Siiteri, P.K.: A re-examination of the interaction of estradiol with target cell receptors. *J Steroid Biochem 9*:1071-1078, 1978.

Lippmann, M., Bolan, G., Monaco, M., Pinkus, L., and Engel, L.: Model systems for the study of estrogen action in tissue culture. *J Steroid Biochem 7*:1045-1051, 1976.

Lisk, R.D.: Diencephalic placement of estradiol and sexual receptivity in the female rat. *Am J Physiol 203*:493-496, 1962.

Litteria, M.: Increased incorporation of ³H-lysine in specific hypothalamic nuclei following castration in the male rat. *Exp Neurol 40*:309-315, 1973a.

Litteria, M.: *In vivo* alterations in the incorporation of [³H]lysine into the medial preoptic nucleus and specific hypothalamic nuclei during the estrous cycle of the rat.

Brain Res 55:234-237, 1973b.

Litteria, M., and Schapiro, S.: Brain nucleic acid content during the estrous cycle in the rat. *Experientia* 28:898, 1972.

Litteria, M., and Thorner, M.W.: Alterations in the incorporation of [³H]lysine into proteins of the medial preoptic area and specific hypothalamic nuclei after ovariectomy in the adult female rat. *J Endocrinol* 60:377-378, 1974.

Little, M., Szendro, P.I., and Jungblut, P.W.: Hormone-mediated dimerization of microsomal estradiol receptor. *Hoppe Seylers Z Physiol Chem* 354:1599-1610, 1973.

Little, M., Szendro, P., Hughes, A., and Jungblut, P.W.: Biosynthesis and transformation of microsomal and cytosol estradiol receptors. *J Steroid Biochem* 6:493-500, 1975.

Lloyd, T., and Ebersole, B.J.: Feedback inhibition of tyrosine hydroxylase from five regions of rat brain by 2-hydroxyestradiol and dihydroxyphenylalanine. *J Neurochem* 34:726-731, 1980.

Lloyd, T., and Weisz, J.: Direct inhibition of tyrosine hydroxylase activity by catechol estrogens. *J Biol Chem* 253:4841-4843, 1978.

Lloyd, T., Weisz, J., and Breakfield, X.O.: The catechol estrogen, 2-hydroxyestradiol, inhibits catechol-O-methyltransferase activity in neuroblastoma cells. *J Neurochem* 31:245-250, 1978.

Lofstrom, A.: Catecholamine turnover alterations in discrete areas of the median eminence of the 4- and 5-day cyclic rat. *Brain Res* 120:113-131, 1977.

Lofstrom, A., Eneroth, P., Gustafsson, J.-A., and Skett, P.: Effects of estradiol benzoate on catecholamine levels and turnover in discrete areas of the median eminence and the limbic forebrain, and on serum luteinizing hormone, follicle stimulating hormone and prolactin concentrations in the ovariectomized female rat. *Endocrinology* 101:1559-1569, 1977.

Loras, B., Genot, A., Monbon, M., Nuecher, F., Reboud, J.P., and Bertrand, J.: Binding and metabolism of testosterone in the rat brain during sexual maturation-II. Testosterone metabolism. *J Steroid Biochem* 5:425-432, 1974.

Luck, D.N.: Comparison of the effects of oestrogen on macromolecular synthesis in the uterus and brain of the immature mouse. *J Reprod Fertil* 43:359-362, 1975.

Luine, V.N., and McEwen, B.S.: Effect of oestradiol on turnover of type A monoamine oxidase in brain. *J Neurochem* 28:1221-1227, 1977a.

Luine, V.N., and McEwen, B.S.: Effects of an estrogen antagonist on enzyme activities and [³H]estradiol nuclear binding in uterus, pituitary and brain. *Endocrinology* 100:903-910, 1977b.

Luine, V.N., Khylcheveskaya, R.I., and McEwen, B.S.: Oestrogen effects on brain and pituitary enzyme activities. *J Neurochem* *23*:925-934, 1974.

Luine, V.N., Khylcheveskaya, R.I., and McEwen, B.S.: Effect of gonadal hormones on enzymes activities in brain and pituitary of male and female rats. *Brain Res 86*:283-292, 1975a.

Luine, V.N., Khylcheveskaya, R.I., and McEwen, B.S.: Effect of gonadal steroids on activities on monoamine oxidase and choline acetylase in rat brain. *Brain Res 86*:293-306, 1975b.

Luine, V.N., McEwen, B.S., and Black, I.B.: Effect of 17β--estradiol on hypothalamic tyrosine hydroxylase activity. *Brain Res 120*:188-192, 1977.

Luine, V.N., MacLusky, N.J., and McEwen, B.S.: Testosterone effects on enzymes in central and peripheral target sites in Tfm mutant mice. *Soc Neurosci Abstr 5*:1529, 1979.

Luine, V., Park, D., Joh, T., Reis, D., and McEwen, B.S.: Immunochemical demonstration of increased choline acetyl-transferase concentration in rat preoptic area after estradiol administration. *Brain Res 191*:273-277, 1980.

Luttge, W.G.: The role of gonadal hormones in the sexual behavior of the rhesus monkey and human: A literature survey. *Arch Sex Behav 1*:61-88, 1971.

Luttge, W.G.: The estrous cycle of the rat: Effects on the accumulation of estrogenic metabolites in brain and peripheral tissues. *Brain Res 38*:315-325, 1972.

Luttge, W.G.: Intracerebral implantation of the anti-estrogen CN-69,725-27: Effects on female sexual behavior in rats. *Pharmacol Biochem Behav 4*:685-688, 1976.

Luttge, W.G.: Effects of anti-estrogens on testosterone stimulated male sexual behavior and peripheral target tissues in the castrate male rat. *Physiol Behav 14*:839-846, 1975.

Luttge, W.G.: Endocrine control of mammalian male sexual behavior: An analysis of the potential role of testosterone metabolites. In *Endocrine Control of Sexual Behavior*. C. Beyer, ed. Raven Press, New York, 1979, pp 341-363.

Luttge, W.G.: Molecular mechanisms of steroid hormone actions in the brain. In *Hormones and Agressive Behavior*. B.B. Svare, ed. Plenum Press, New York, 1983, pp. 247-312.

Luttge, W.G., and Hall, N.R.: Interactions of progesterone and dihydroprogesterone with dihydrotestosterone on estrogen activated sexual receptivity in female mice. *Horm Behav 7*: 253-257, 1976.

Luttge, W.G., and Hughes, J.R.: Intracerebral implantation of progesterone: Re-examination of the brain sites responsible for facilitation of sexual receptivity in estrogen-primed ovariectomized rats. *Physiol Behav 17*:771-775, 1976.

Luttge, W.G., and Jasper, T.W.: Studies on the possible role of 2-OH-estradiol in the control of sexual behavior in female rats. *Life Sci 20*:419-426, 1977.

Luttge, W.G., and Whalen, R.E.: The accumulation, retention and interaction of oestradiol and oestrone in central neural and peripheral tissues of gonadectomized female rats. *J Endocrinol* 52:379-395, 1972.

Luttge, W.G., Wallis, C.J., and Hall, N.R.: Effects of pre- and post-treatment with unlabeled steroids on the *in vivo* uptake of ³H-progestins in selected brain regions, uterus and plasma of the female mouse. *Brain Res* 71:105-115, 1974.

Luttge, W.G., Hall, N.R., Wallis, C.J., and Campbell, J.C.: Stimulation of male and female sexual behavior in gonadectomized rats with estrogen and androgen therapy and its inhibition with concurrent anti-hormone therapy. *Physiol Behav* 14:65-73, 1975.

Luttge, W.G., Gray, H.E., and Hughes, J.R.: Regional and subcellular ³H-estradiol localization in selected brain regions and pituitary of female mice: Effects of unlabeled estradiol and various anti-hormones. *Brain Res* 104:273-281, 1976a.

Luttge, W.G., Grant, E.G., and Gray, H.E.: Regional and subcellular localization of ³H-dihydrotestosterone in brain and pituitary of male mice. *Brain Res* 109:426-429, 1976b.

Luttge, W.G., Jasper, T.W., Sheets, C.S., and Gray, H.E.: Inhibition of 5α-reductase activity and growth of accessory sex tissues in castrated male mice. *J Reprod Fertil* 53:45-50, 1978.

McCann, P.P., Hornsperger, J.-M., and Seiler, N.: Regulatory interrelationships between GABA and polyamines. II. Effect of GABA on ornithine decarboxylase and putrescine levels in cell culture. *Neurochem Res* 4:437-447, 1979.

McEwen, B.S., Pfaff, D.W., Chaptal, C., and Luine, V.N.: Brain cell nuclear retention of [³H]estradiol doses able to promote lordosis: Temporal and regional aspects. *Brain Res* 86:155-161, 1975.

McEwen, B.S., Lieberburg, I., MacLusky, N., and Plapinger, L.: Interactions of testosterone and estradiol with the neonatal rat brain: Protective mechanism and possible relationship to sexual differentiation. *Ann Biol Anim Biochim Biophys* 16:471-478, 1976.

McGinnis, M.Y., Gordon, J.H., and Gorski, R.A.: Time course and localization of the effects of estrogen on glutamic acid decarboxylase activity. *J Neurochem* 34:785-792, 1980a.

McGinnis, M.Y., Gordon, J.H., and Gorski, R.A.: Influence of γ-aminobutyric acid on lordosis behavior and dopamine activity in estrogen primed spayed female rats. *Brain Res* 184:179-191, 1980b.

MacLusky, N.J., and McEwen, B.S.: Oestrogen modulates progestin receptor concentrations in some brain regions, but not in others. *Nature* 274:276-278, 1978.

MacLusky, N.J., and McEwen, B.S.: Progestin receptors in rat brain; Distribution and properties of cytoplasmic progestin-binding sites. *Endocrinology 106:*192-202, 1980.

MacLusky, N.J., Lieberburg, I., Krey, L.C., and McEwen, B.S.: Progestin receptors in the brain and pituitary of the bonnet monkey (macaca radiata): Differences between the monkey and the rat in the distribution of progestin receptors. *Endocrinology 106:*185-191, 1890.

Mainwaring, W.I.P., Symes, E.K., and Higgins, S.J.: Nuclear components responsible for the retention of steroid-receptor complexes, especially from the stand-point of the specificity of hormonal responses. *Biochem J 156:*129-141, 1976.

Malmnas, C.O.: Monoaminergic influence on testosterone-activated copulatory behavior in the castrated male rat. *Acta Physiol Scan [Suppl] 395:*1-128, 1973.

Malsbury, C.W.: Facilitation of male rat copulatory behavior by electrical stimulation of the medial preoptic area. *Physiol Behav 7:*797-805, 1971.

Malsbury, C.W., Pfaff, D.W., and Malsbury, A.M.: Suppression of sexual receptivity in the female hamster: Neuroanatomical projections from the preoptic and anterior hypothalamic electrode sites. *Brain Res 181:*267-284, 1980.

Manak, R., Wertz, N., Slabaugh, M., Denari, H., Wang, J.-T., and Gorski, J.: Purification and characterization of the estrogen-induced protein (IP) of the rat uterus. *Mol Cell Endocrinol 17:*119-132, 1980.

Marrone, B.L., and Feder, H.H.: Characteristics of (^{3}H)estrogen and (^{3}H)progestin uptake and effects of progesterone on (^{3}H)-estrogen uptake in brain, anterior pituitary and peripheral tissues of male and female guinea pigs. *Biol Reprod 17:*42-57, 1977.

Marrone, B.L., Rodriguez-Sierra, J.F., and Feder, H.H.: The role of catechol estrogens in activation of lordosis in female rats and guinea pigs. *Pharmacol Biochem Behav 7:*13-17, 1977.

Marrone, B.L., Rodriguez-Sierra, J.F., and Feder, H.H.: Intrahypothalamic implants of progesterone inhibit lordosis behavior in ovariectomized, estrogen-treated rats. *Neuroendocrinology 28:*92-102, 1979.

Martinez-Vargas, M.C., Stumpf, W.E., and Sar, M.: Anatomical distribution of estrogen target cells in the avian CNS: A comparison with the mammalian CNS. *J Comp Neurol 167:*83-104, 1976.

Matsumoto, A., and Arai, Y.: Synaptogenic effect of estrogen on the hypothalamic arcuate nucleus of the adult female rat. *Cell Tiss Res 198:*427-433, 1979.

Matsumoto, A., and Arai, Y.: Sexual dimorphism in wiring pattern in the hypothalamic arcuate nucleus and its modification by neonatal hormone environment. *Brain Res 190:*238-242, 1980.

Mauk, M.D., Olsen, G., Kastin, A.J., and Olson, R.D.: Behavioral effects of LH-RH. *Neurosci Biobehav Rev 4:* 1-8, 1980.

Meinkoth, J., Quadagno, D.M., and Bast, J.D.: Depression of steroid-induced sex behavior in the ovariectomized rat by intracranial injection of cycloheximide: Preoptic area compared to the ventromedial hypothalamus. *Horm Behav 12:* 199-204, 1979.

Menendez-Patterson, A., Florez-Lozano, J.A., and Marin, B.: Oxidative activity during the sexual cycle of the central nervous system, adrenal glands and ovaries in the hamster (Mesocricetus auratus). *Experientia 34:* 190-191, 1978.

Menendez-Patterson, A., Florez-Lozano, J.F., and Marin, B.: Effects of ovariectomy on the oxidative metabolism of the central nervous system and adrenal glands in female hamster (Mesocricetus auratus). *Experientia 35:* 349-350, 1979.

Merari, A., and Ginton, A.: Characteristics of exaggerated sexual behavior induced by electrical stimulation of the medial preoptic area in male rats. *Brain Res 86:* 97-108, 1975.

Meyerson, B.J., and Eliasson, M.: Pharmacological and hormonal control of reproductive behavior. In *Drugs, Transmitters, and Behavior*, Vol. 8, *Handbook of Psychopharmacology*, L.L. Iverson, S.D. Iversen, and S.H. Snyder, eds. Raven Press, New York, 1977, pp 159-232.

Meyerson, B.J., Palis, A., and Sietnieks, A.: Hormone-monoamine interactions and sexual behavior. In *Endocrine Control of Sexual Behavior*, C. Beyer, ed. Raven Press, New York, 1979, pp 389-404.

Mioduszewski, R., Grandison, L., and Meites, J.: Stimulation of prolactin release in rats by GABA. *Proc Soc Exp Biol Med 151:* 44-46, 1976.

Moguilevsky, J.A.: Oxidative activity of different hypothalamic areas during sexual cycle in rats. *Acta Physiol Lat Am 15:* 423-424, 1965.

Moguilevsky, J.A., and Christot, J.: Protein synthesis in different hypothalamic areas during the sexual cycle in rats: Influence of castration. *J Endocrinol 55:* 147-152, 1972.

Moguilevsky, J.A., Libertum, C., and Foglia, V.G.: Metabolic sensitivity of different hypothalamic areas to luteinizing hormone (LH), follicle stimulating hormone (FSH), and testosterone. *Neuroendocrinology 6:* 153-159, 1970a.

Moguilevsky, J.A., Libertun, C., and Foglia, V.G.: Oxidative metabolism of the hypothalamus in hypophysectomized-castrated rats. *Experientia 26:* 421-422, 1970b.

Mogilevsky, J.A., Kalbermann, L.E., Libertun, C., and Gomez, C.J.: Effect of ovariectomy on the amino acid incorporation into proteins of anterior pituitary and hypothalamus of rats. *Proc Soc Exp Biol Med 136:* 1115-1118, 1971a.

Moguilevsky, J.A., Schiaffini, O., Szwarcfarb, B., and

Libertun, C.: Metabolic evidences of the short feed back mechanism controlling LH and FSH secretion. *Acta Endocrinol Panam* 2:177-186, 1971b.

Moguilevsky, J.A., Szwarcfarb, B., and Schiaffini, O.: Succinic-dehydrogenase and cytochrome-oxidase activity in the hypothalamus during the sexual cycle in rats. *Neuroendocrinology* 8:334-339, 1971c.

Moguilewsky, M., and Raynaud, J.-P.: The relevance of hypothalamic and hypophyseal progesin receptor regulation in the induction and inhibition of sexual behavior in the female rat. *Endocrinology* 104:516-522, 1979a.

Moguilewsky, M., and Raynaud, J.-P.: Estrogen-sensitive progestin-binding sites in the female rat brain and pituitary. *Brain Res* 164:165-175, 1979b.

Moguilewsky, M., and Raynaud, J.-P.: Evidence for a specific mineralocorticoid receptor in rat pituitary and brain. *J Steroid Biochem* 12:309-314, 1980.

Monbon, M., Loras, B., Reboud, J.P., and Bertrand, J.: Binding and metabolism of testosterone in the rat brain during sexual maturation--I. Macromolecular binding of androgens. *J Steroid Biochem* 5:417-424, 1974.

Moore, C.R., and Price, D.: Gonad hormone functions, and the reciprocal influence between gonads and hypophysis with its bearing on the problem of sex hormone antagonism. *Am J Anat* 50:13-71, 1932.

Morin, L.P., and Feder, H.H.: Inhibition of lordosis behavior in ovariectomized guinea pigs by mesencephalic implants of progesterone. *Brain Res* 70:71-80, 1974a.

Morin, L.P., and Feder, H.H.: Hypothalamic progesterone implants and facilitations of lordosis behavior in estrogen-primed ovariectomized guinea pigs. *Brain Res* 70:81-93, 1974b.

Morin, L.P., Powers, J.B., and White, M.: Effects of the anti-estrogens, MER-25 and CI-628, on rat and hamster lordosis. *Horm Behav* 7:283-292, 1976.

Morrell, J.I., and Pfaff, D.W.: A neuroendocrine approach to brain function: Localization of sex steroid concentrating cells in vertebrate brains. *Am Zoologist* 18:447-460, 1978.

Morrell, J.I., Kelly, D.B., and Pfaff, D.W.: Autoradiographic localization of hormone-concentrating cells in the brain of an amphibian, xenopus laevis. II. Estradiol. *J Comp Neurol* 164:63-78, 1975.

Morris, I.D.: Changes in brain, pituitary and uterine cytoplasmic oestrogen receptors induced by oestradiol-17β in the ovariectomized rat. *J Endocrinol* 71:343-349, 1976.

Moss, R.L.: Unit responses in preoptic and arcuate neurons related to anterior pituitary function. In *Frontiers In Neuroendocrinology*, Vol 4, L. Martini and W.F. Ganong, eds. Raven Press, New York, 1976, pp 95-128.

Moss, R.L., and Law, O.T.: The estrous cycle: Its influence on single unit activity in the forebrain. *Brain Res 30:* 435-438, 1971.

Moss, R.L., McCann, S.M., and Dudley, C.A.: Releasing hormones and sexual behavior. *Prog Brain Res 42:* 37-46, 1975.

Moss, R.L., Paloutzian, R.F., and Law, O.T.: Electrical stimulation of forebrain structures and its effect on copulatory as well as stimulus-bound behavior in ovariectomized hormone-primed rats. *Physiol Behav 12:* 997-1004, 1974.

Moudgil, V.K., and Kanungo, M.S.: Effect of age of the rat on induction of acetylcholinesterase of the brain by 17β-estradiol. *Biochem Biophys Acta 329:* 211-220, 1973.

Muller, R.E., Traish, A.M., and Wotiz, H.H.: Interaction of receptor-estrogen complex (R-E) with uterine nuclei. *J Biol Chem 252:* 8206-8211, 1977.

Munaro, N.I.: The effect of ovarian steroids on hypothalamic norepinephrine neuronal activity. *Acta Endocrinol (Copenh) 86:* 235-242, 1977.

Munaro, N.I.: The effect of ovarian steroids on hypothalamic 5-hydroxytryptamine neuronal activity. *Neuroendocrinology 26:* 270-276, 1978.

Murrin, L.C.: Ornithine as a precursor for γ-aminobutyric acid in mammalian brain. *J Neurochem 34:* 1779-1781, 1980.

Nadler, R.D.: A biphasic influence of progesterone on sexual receptivity of spayed female rats. *Physiol Behav 5:* 95-97, 1970.

Naess, O.: Characterization of the androgen receptors in the hypothalamus, preoptic area and brain cortex of the rat. *Steroids 27:* 167-185, 1976.

Naess, O., and Attramadal, A.: Progestin receptors in the anterior pituitary and hypothalamus of male rats. *Brain Res Suppl 2:* 175-183, 1978.

Naftolin, F., Ryan, K.J., Davies, I.J., Reddy, V.V., Flores, F., Petro, Z., Kuhn, M., White, R.J., Takaoka, Y., and Wolin, L.: The formation of estrogens by central neuroendocrine tissues. *Recent Prog Horm Res 31:* 295-319, 1975.

Nagle, C.A., and Rosner, J.M.: Rat brain norepinephrine release during progesterone-induced LH secretion. *Neuroendocrinology 30:* 33-37, 1980.

Nakahara, T., Uchimura, H., Hirano, M., Saito, M., Kim, J.S., and Matsumoto, T.: Effects of gonadectomy and thyroidectomy on tyrosine hydroxylase in discrete areas of the rat median eminence. *Brain Res 179:* 396-400, 1979.

Nakai, T., Kagawa, T., and Sokamoto, S.: ^{3}H-Leucine uptake of hypothalamic nuclei in female male rats and its fluctuation after castration. *Endocrinol Jpn 18:* 353-357, 1971.

Nawata, H., Yamamoto, R.S., and Poirier, L.A.: Ornithine decarboxylase induction and polyamine levels in the kidney of estradiol-treated castrated male rats. *Life Sci* 26: 689–698, 1980.

Neckers, L., and Sze, P.Y.: Regulation of 5-hydroxytryptamine metabolism in mouse brain by adrenal glucocorticoids. *Brain Res* 93: 123–132, 1975.

Nixon, R.L., Janowsky, D.S., and Davis, J.M.: Effects of progesterone, β-estradiol, and testosterone on the uptake and metabolism of ^{3}H-norepinephrine, ^{3}H-dopamine and ^{3}H-serotonin in rat brain synaptosomes. *Res Commun Chem Pathol Pharmacol* 7: 233–236, 1974.

Nock, B., and Feder, H.H.: Noradrenergic transmission and female sexual behavior in guinea pigs. *Brain Res* 16: 369–380, 1979.

Noma, K., Sato, B., Yano, S., Yamamura, Y., and Seki, T.: Metabolism of testosterone in hypothalamus of male rat. *J Steroid Biochem* 6: 1261–1266, 1975.

Novakova, V., Sandritter, W., Krecek, J., and Ostadalova, I.: Sexual differences in total RNA content of rat brain cells of medial hypothalamus and hippocampus. *Beitr Pathol Bd* 143: 295–300, 1971.

Okada, R., Watanabe, H., Yamanouchi, K., and Arai, Y.: Recovery of sexual receptivity in female rats with lesions of the ventromedial hypothalamus. *Exp Neurol* 68: 595–600, 1980.

Orr, E., and Quay, W.: The effects of castration on histamine levels and 24-hour rhythm in the male rat hypothalamic. *Endocrinology* 97: 481–484, 1975.

Packman, P.M., Bragdon, M.J., and Boshans, R.L.: Quantitative histochemical studies of the hypothalamus. Control point enzymes during the estrous cycle. *Neuroendocrinology* 23: 76–87, 1977.

Pandolfo, L., and Macione, S.: Influenza della surrenectoma sufa attivita della GABA transaminase e della glutamico de carbossilasi di corteca cerebrale di ratto. *G Biochem* 13: 256–261, 1963.

Paolo, T.D., Labrie, F., Dupont, A., Barden, N., and Langelier, P.: Effects of estrogen treatment on normal and sensitized rat striatal dopamine (DA) receptors and DA-sensitive adenylyl cyclase. *Soc Neurosci Abstr* 5: 1495, 1979.

Pardridge, W.M., and Mietus, L.J.: Transport of steroid hormones through the rat blood-brain barrier. Primary role of albumin-bound hormone. *J Clin Invest* 64: 145–154, 1979.

Pardridge, W.M., and Mietus, L.J.: Effects of progesterone-binding globulin versus a progesterone antiserum on steroid hormone transport through the blood-barrier. *Endocrinology* 106: 1137–1141, 1980.

Parsons, B., MacLusky, N.J., Krieger, M.S., McEwen, B.S., and Pfaff, D.W.: The effects of long-term estrogen exposure on the induction of sexual behavior and measurements of brain estrogen and progestin receptors in the female rat. *Horm Behav* 13:301-313, 1979.

Parsons, B., MacLusky, N.J., Krey, L., Pfaff, D.W., and McEwen, B.S.: The temporal relationship between estrogen-inducible progestin receptors in the female rat brain and the time course of estrogen activation of mating behavior. *Endocrinology* 107:774-779, 1980.

Paul, S.M., and Axelrod, J.: Catechol estrogens: Presence in brain and endocrine tissues. *Science* 197:657-659, 1977.

Paul, S.M., and Skolnick, P.: Catechol oestrogens inhibit oestrogen elicited accumulation of hypothalamic cyclic AMP suggesting role as endogenous antioestrogens. *Nature* 266: 559-561, 1977.

Paul, S.M., Axelrod, J., and Dilibertor, E.J., Jr.: Catechol estrogen-forming enzyme of brain: Demonstration of a cytochrome P450 monooxygenase. *Endocrinology* 101:1604-1610, 1977.

Paul, S.M., Axelrod, J., Saavedra, J.M., and Skolnick, P.: Estrogen-induced efflux of endogenous catecholamines from the hypothalamus *in vitro*. *Brain Res* 178:499-505, 1979.

Paul, S.M., Hoffman, A.R., and Axelrod, J.: Catechol estrogens: Synthesis and metabolism in brain and other tissues. In *Frontiers in Neuroendocrinolgy*, Vol. 6. L. Martini and W.F. Ganong, eds. Raven Press, New York, 1980, pp 203-217.

Peck, E.J., Jr., Miller, A.L., and Kelner, K.L.: Estrogen receptors and the activation of RNA polymerase by estrogens in the central nervous system. In *Ontogeny of Receptors and Reproductive Hormone Action*, T.H. Hamilton, J.H. Clark, and W.A. Sandler, eds. Raven Press, New York, 1969, pp 403-410.

Peraino, C., and Pitot, H.C.: Ornithine-γ-transaminase in the rat. I. Assay and some general properties. *Biochim Biophys Acta* 73:223-231, 1963.

Perez, A.E., Ortiz, A., Cabeza, M., Beyer, C., and Perez-Palacios, G.: *In vitro* metabolism of ^{3}H-androstenedione by the male pituitary, hypothalamus, and hippocampus. *Steroids* 25:53-62, 1975.

Perez-Palacios, G., Castaneda, E., Gomez-Perez, F., Perez, A.E., and Gaul, C.: *In vitro* metabolism of androgens in dog hypothalamus, pituitary, and limbic system. *Biol Reprod 3*: 205-213, 1970.

Perry, B.N., and Lopez, T.A.: The binding of ^{3}H-labelled oestradiol- and progesterone-receptor complexes to hypothalamic chromatin of male and female sheep. *Biochem J 176*: 873-883, 1978.

Pfaff, D.W., and Keiner, M.: Atlas of estradiol-concentrating cells in the central nervous system of the female rat. *J Comp Neurol* 151:121-158, 1973.

Pfaff, D.W., and Modianos, D.: Neural mechanisms of female reproductive behavior. In *Neurobiology of Reproduction,*

R. Goy and D. Pfaff, eds. Plenum Press, New York, 1980.

Pfaff, D.W., and Sakuma, Y.: Deficit in the lordosis reflex of female rats caused by lesions in the ventromedial nucleus of the hypothalamus. J Physiol (Lond) 288:203-210, 1979a.

Pfaff, D.W., and Sakuma, Y.: Facilitation of the lordosis reflex of female rats from the ventromedial nucleus of the hypothalamus. J Physiol (Lond) 288:189-202, 1979b.

Pfaff, D.W., Gerlach, J.L., McEwen, B.S., Ferin, M., Carmel, P., and Zimmerman, E.A.: Autoradiographic localization of hormone-concentrating cells in the brain of the female rhesus monkey. J Comp Neurol 170:279-294, 1976.

Pfeiffer, C.A.: Sexual differences of the hypophyses and their determination by the gonads. Am J Anat 58:195-225, 1936.

Portaleone, P., Genazzani, E., Pagnini, G., Crispino, A., and DiCarlo, F: Interaction of estradiol and 2-hydroxy-estradiol with histamine receptors at hypothalamic level. Brain Res 187:216-220, 1980.

Portaleone, P., Pagnini, G., Crispino, A., and Genazzani, E.: Histamine-sensitive adeylate cyclase in hypothalamus of rat brain: H_1 and H_2 receptors. J Neurochem 31:1371-1374, 1978.

Powers, B., and Valenstein, E.S.: Sexual receptivity: Facilitation by medial preoptic lesions in female rats. Science 170:1003-1005, 1972.

Presl, J., Herzmann, J., Rohling, S., and Horsky, J.: Regional distribution of estrogenic metabolites in the female rat hypothalamus. Endocrinol Exp (Bratisl) 7:119-123, 1973.

Puca, G.A., Sica, V., and Nola, E.: Identification of a high affinity nuclear acceptor site for estrogen receptor of calf uterus. Proc Natl Acad Sci USA 71:979-983, 1974.

Puca, G.A., Nola, E., Hibner, U., Cicala, G., and Sica, V.: Interaction of the estradiol receptor from calf uterus with its nuclear acceptor sites. J Biol Chem 250:6452-6459, 1975.

Quadagno, D.M., and Ho, G.K.W.: The reversible inhibition of steroid-induced sexual behavior by intracranial cyclohexi- mide. Horm Behav 6:19-26, 1975.

Quadagno, D.M., Albelda, S.M., McGill, T.E., and Kaplan, L.J.: Intracranial cycloheximide: Effect on male mouse sexual be- havior and plasma testosterone. Pharmacol Biochem Behav 4: 185-189, 1976.

Quadagno, D.M., Hough, J.C., Ochs, R.L., Renner, K.J., and Bast, J.D.: Intrahypothalamic actinomycin-D: Sexual behavior and nucleolar ultrastructure in the steroid-primed ovari- ectomized rat and the intact cyclic rat. Physiol Behav 24: 169-172, 1980.

Radanyi, C., Redeuilh, G., Eigenmann, E., Lebau, M.C., Massol, N., Secco, C., Baulieu, E.E., and Richard-Foy, H.: Production et detection d'anticorps anti-recepteur de l'oestradiol d'uterus de Veau. Interaction avec le recepteur d'oviducte de

Poule. *CR Acad Sci (Paris)* *288D*:255-258, 1979.

Ragland, W.L., and Pitot, H.C.: Enzymes of carbohydrate and amino acid metabolism in liver, brain and kidney of *Macaca mulatta*. *Experientia* *27*:1023, 1971.

Raisman, G., and Field, P.M.: Sexual dimorphism in the neuropil of the preoptic area of the rat and its dependence of neonatal androgen. *Brain Res* *54*:1-29, 1973.

Raynaud, J.P.: R5020, a tag for the progestin receptor. In *Progesterone Receptors in Normal and Neoplastic Tissues*. W.L. McGuire, J.P. Raynaud, and E.E. Baulieu, eds. Raven Press, New York, 1977, pp 9-21.

Reddy, V.V.R.: Estriol synthesis in rat brain and pituitary. *Brain Res* *175*:165-168, 1979a.

Reddy, V.V.R.: Estrogen metabolism in neural tissues of rabbits: 17β-hydroxysteroid oxidoreductase activity. *Steroids* *34*:207-215, 1979b.

Reddy, V.V.R., and Rajan, R.: Catecholestrogens formation in male and female rats' brain. *J Steroid Biochem* *9*:881, 1978.

Reddy, V.V.R.. Naftolin, F., and Ryan, K.J.: Aromatization in the central nervous system of rabbits: effects of castration and hormone treatment. *Endocrinology* *92*:589-594, 1973.

Reiss, N., and Kaye, A.M.: Separation of a protein from rat brain resembling the uterine "estrogen-induced protein" in electrophoretic and immunologic properties. *Isr J Med Sci* *15*: 545, 1979.

Reiter, R.J.: The pineal and its hormones in the control of reproduction in mammals. *Endocrine Rev* *1*:109-131, 1980.

Rezek, D.L.: Nuclear localization of testosterone, dihydrotestosterone and estradiol-17β in basal rat brain. *Psychoneuroendocrinology* *2*:173-178, 1977.

Roberts, E., Chase, T.N., and Tower, D.B.: *GABA in Nervous System Function*, Raven Press, New York, 1976.

Rodriquez-Sierra, J.F., and Davis, G.A.: Progesterone does not inhibit lordosis through interference with estrogen priming. *Life Sci* *22*:373-378, 1978.

Rodriguez-Sierra, J.F., and Davis, G.A.: Tolerance to the lordosis-facilitating effects of progesterone or methysergide. *Neuropharmacology* *18*:335-339, 1979.

Rodriguez-Sierra, J.F., and Terasawa, E.: Lesions of the preoptic area facilitate lordosis behavior in male and female guinea pigs. *Brain Res Bull* *4*:513-517, 1979.

Rommerts, F.F.G., and van der Molen, H.J.: Occurrence and localization of 5α-steroid reductase, 3α- and 17β-hydroxysteroid dehydrogenase in hypothalamus and other brain tissues of the male rat. *Biochem Biophys Acta* *248*:489-502, 1971.

Roy, E.J., and McEwen, B.S.: Oestrogen receptors in cell nuclei of the hypothalamus--preoptic area--amygdala following an injection of oestradiol or the antioestrogen CI-628.

J Endocrinol 83:285-293, 1979.

Roy, E.J., MacLusky, N.J., and McEwen, B.S.: Antiestrogen inhibits the induction of progestin receptors by estradiol in the hypothalamus preoptic area and pituitary. *Endocrinology* 104:1333-1336, 1979a.

Roy, E.J., Schmit, E., McEwen, B.S., and Wade, G.N.: Antiestrogens in the central nervous system. In *Antihormones*, M.K. Agarwal, ed. Elsevier/North Holland Biomedical Press, 1979b, pp 181-197.

Russell, D.H.: Ornithine decarboxylase as a biological and pharmacological tool. *Pharmacology* 20:117-129, 1980.

Ryan, K.J.: Estrogen formation by the human placenta: Studies on the mechanisms of steroid aromatization by mammalian tissue. *Acta Endocrinol (Copenh) 35 (Suppl)* 51:697-698, 1960.

Saad, S.F.: The effect of ovariectomy on the γ-aminobutyric acid content in the cerebral hemispheres of young rats. *J Pharm Pharmacol* 22:307, 1970.

Saffran, J., and Loeser, B.K.: Nuclear binding of guinea pig uterine progesterone receptor in cell-free preparations. *J Steroid Biochem* 10:43-51, 1979.

Salaman, D.F.: RNA synthesis in the rat anterior hypothalamus and pituitary: Relation to neonatal androgen and oestrous cycle. *J Endocrinol* 48:125-137, 1970.

Salaman, D.F.: The role of DNA, RNA and protein synthesis in sexual differentiation of the brain. *Prog Brain Res* 41:349-362, 1974.

Sandler, M., and Gessa, G.L.: *Sexual Behavior--Pharmacology and Biochemistry*. Raven Press, New York, 1975.

Sando, J.J., LaForest, A.C., and Pratt, W.B.: ATP-dependent activation of L cell glucocorticoid receptors to the steroid binding form. *J Biol Chem* 254:4772-4778, 1979a.

Sando, J.J., Hammond, N.D., Stratford, C.A., and Pratt, W.B.: Activation of thymocyte glucocorticoid receptors to the steroid binding form. *J Biol Chem* 254:4779-4789, 1979b.

Sar, M., and Stumpf, W.E.: Androgen concentration in motor neurons of cranial nerves and spinal cord. *Science* 197:77-79, 1977a.

Sar, M., and Stumpf, W.E.: Distribution of androgen target cells in rat forebrain and pituitary after [^{3}H]-dihydrotestosterone administration. *J Steroid Biochem* 8:1131-1135, 1977b.

Sasame, H.A., Ames, M.M., and Nelson, S.D.: Cytochrome P-450 and NADPH cytochrome C reductase in rat brain. *Biochem Biophys Res Commun* 78:919-926, 1977.

Sato, B., Nishizawa, Y., Noma, K., Matsumoto, K., and Yamamura, Y.: Estrogen-independent nuclear binding of receptor protein of rat uterine cytosol by removal of low molecular weight inhibitor. *Endocrinology* 104:1474-1479, 1979.

Sato, B., Noma, K., Nishizawa, Y., Nakao, K., Matsumoto, K., and Yamamura, Y.: Mechanism of activation of steroid receptors: Involvement of low molecular weight inhibitor in activation of androgen, glucocorticoid, and estrogen receptor systems. Endocrinology 106:1142-1148, 1980.

Scacchi, P., Moguilevsky, J.A., and Szwarcfarb, B.: Effect of castration on $C^{14}O_2$ production from glucose-U-C^{14} in the hypothalamus and cerebral cortex. Proc Soc Exp Biol Med 136: 1068-1071, 1971.

Scacchi, P., Moguilevsky, J.A., and Schiaffini, O.: Glucose oxidation in the hypothalamus during the sexual cycle in rats: Influence of castration. Neuroendocrinology 11: 321-327, 1973.

Schaeffer, J.M., and Hseuh, A.J.W.: 2-Hydroxyestradiol interaction with dopamine receptor binding in rat anterior pituitary. J Biol Chem 254:5606-5608, 1979.

Schaeffer, J.M., Stevens, S.R., and Smith, R.G., and Hseuh, A.J.: Binding of 2-hydroxyestradiol to rat anterior pituitary cell membranes. J Biol Chem 255:9838-9843, 1980.

Schiaffini, O., and Marin, B.: Effect of ovariectomy on the oxidative activity of the hypothalamus and of the limbic system of the rat. Neuroendocrinology 7:302-307, 1971.

Schiaffini, O., and Martini, L.: The amygdala and the control of gonadotrophin secretion. Acta Endocrinol (Copenh) 70: 209-219, 1972.

Schiaffini, O., Moguilevsky, J.A., Libertun, C., and Foglia, V.G.: Oxidative and glycolytic metabolism of different hypothalamic areas in diabetic rat. Acta Physiol Lat Am 18: 257-262, 1968.

Schiaffini, O., Marin, B., and Foglia, V.G.: Metabolic alterations in hypothalamus and limbic structures in female diabetic rats. Experientia 26:610-611, 1970.

Schwartz, J.C.: Histaminergic mechanism in brain. Ann Rev Pharmacol Toxicol 17:325-339, 1977.

Schwartz, J.C., Pollard, H., and Quach, T.T.: Histamine as a neurotransmitter in mammalian brain: Neurochemical evidence. J Neurochem 35:26-33, 1980.

Schwartz, S.M., Blaustein, J.D., and Wade, G.N.: Inhibition of estrous behavior by progesterone in rats: Role of neural estrogen and progestin receptors. Endocrinology 105:1078-1082, 1979.

Seiki, K., Haruki, Y., Imanishi, Y., and Enomoto, T.: Further evidence of the presence of progesterone-binding proteins in female rat hypothalamus. Endocrinol Jpn 24:233-238, 1977.

Seiki, K., Haruki, Y., Imanishi, Y., and Enomoto, T.: Progestin binding in vitro by brain cell nuclei of ovariectomized oestrogen-primed rats. J Endocrinol 82:347-360, 1979.

Seiler, N., and Sarhan, S.: On the nonoccurrence of ornithine decarboxylase in nerve endings. *Neurochem Res* 5:97-100, 1980.

Seiler, N., Bink, G., and Grove, J.: Regulatory interrelations between GABA and polyamines. I. Brain GABA and polyamine metabolism. *Neurochem Res* 4:425-435, 1979a.

Seiler, N., Schmidt-Glenewinkel, T., and Sarhan, S.: On the formation of γ-amino butyric acid from putrescine in brain. *J Biochem (Tokyo)* 86:277-278, 1979b.

Selmanoff, M.K., Brodkin, L.D., Weiner, R.J., and Siiteri, P.: Aromatization and 5α-reduction of androgens in discrete hypothalamic and limbic regions of the male and female rat. *Endocrinology* 101:841-848, 1977.

Shaw, G.G.: The polyamines in the central nervous system. *Biochem Pharmacol* 28:1-6, 1979.

Shen, G., Thrower, S., and Lim, L.: Uterine oestrogen-receptor binding to oligo(dT)-cellulose. An inhibitor from hypothalamic cytosol. *Biochem J* 182:241-243, 1979.

Shepherd, R.E., Huff, K., and McGuire, W.L.: Non-interaction between *in vivo* and cell free nuclear binding of estrogen receptor. *Endocr Res Commun* 1:73-85, 1974.

Sheridan, P.J.: The nucleus interstitialis striae terminalis and the nucleus amygdaloideus medialis: Prime targets for androgen in the rat forebrain. *Endocrinology* 104:130-136, 1979.

Sherman, M.R., Pickering, L.A., Rollwagen, F.M., and Miller, L.K.: Mero-receptors: Proteolytic fragments of receptors containing the steroid-binding site. *Fed Proc* 37:167-173, 1978.

Sholl, S.A., Robinson, J.A., and Goy, R.W.: Neural uptake and metabolism of testosterone and dihydrotestosterone in the guinea pig. *Steroids* 25:203-216, 1975.

Siegel, L.I., and Wade, G.N.: Insulin withdrawal impairs sexual receptivity and retention of brain cell nuclear estrogen receptors in diabetic rats. *Neuroendocrinology* 29:200-206, 1979.

Simpkins, J.W., Huang, H.H., Advis, J.P., and Meites, J.: Changes in hypothalamic NE and DA turnover resulting from steroid-induced LH and prolactin surges in ovariectomized rats. *Biol Reprod* 20:625-632, 1979.

Simpkins, J.W., Kalra, P.S., and Kalra, S.P.: Effects of testosterone on catecholamine turnover and LHRH contents in the basal hypothalamus and preoptic area. *Neuroendocrinology* 30:94-100, 1980.

Slotkin, T.A.: Ornithine decarboxylase as a tool in developmental neurobiology. *Life Sci* 24:1623-1630, 1979.

Sodersten, P., Hansen, S., Eneroth, P., Wilson, C.A., and Gustafsson, J.-A.: Testosterone in the control of rat sexual behavior. *J Steroid Biochem* 12:337-346, 1980.

Stern, J.M., and Eisenfeld, A.J.: Distribution and metabolism of ³H-testosterone in castrated male rats; effects of cyproterone, progesterone and unlabeled testosterone. *Endocrinology* 88:1117-1125, 1971.

Stipek, S., Crkovska, J., Trojan, S., and Prokes, J.: The effects of polyamines on RNA synthesis in cell nuclei isolated from the rat brain. *Physiol Bohemoslov* 27:280-281, 1978.

Stumpf, W.E.: Estrogen-neurons and estrogen-neuron systems in the periventricular brain. *Am J Anat* 129:207-218, 1970.

Stumpf, W.E., and Grant, L.D.: *Anatomical Neuroendocrinology*, S. Karger, Basel, 1975.

Stumpf, W.E., and Sar, M.: Anatomical distribution of estrogen, androgen, progestin, corticosteroid and thyroid hormone target sites in the brain of mammals: Phylogeny and ontogeny. *Am Zoologiol* 18:435-445, 1978.

Sundberg, D.K., Fawcett, C.P., and McCann, S.M.: The involvement of cyclic-3'5'AMP in the release of hormones from the anterior pituitary *in vitro*. *Proc Soc Exp Biol Med* 151:149-154, 1976.

Tabei, T., and Heinrichs, W.L.: Metabolism of progesterone by the brain and pituitary gland of subhuman primates. *Neuroendocrinology* 15:281-289, 1974.

Tappaz, M.L., and Brownstein, M.J.: Origin of glutamate-decarboxylase (GAD) containing cells in discrete hypothalamic nuclei. *Brain Res* 132:95-106, 1977.

Tappaz, M.L., Brownstein, M.J., and Kopin, I.J.: Glutamate decarboxylase (GAD) and γ-aminobutyric acid (GABA) in discrete nuclei of hypothalamus and substantia nigra. *Brain Res* 125:109-121, 1977.

Thompson, E.A., and Siiteri, P.K.: Studies on the aromatization of C-19 androgens. *Ann NY Acad Sci* 212:378-388, 1973.

Thrower, S., and Lim, L.: Characterization of rat hypothalamic progestin binding by spheroidal hydroxylapatite chromatography. *Biochem J* 186:295-300, 1980.

Tobias, H., Carr, L., and Vooft, J.: Effects of estradiol on catecholamine synthesizing enzymes, luteinizing hormone (LH) and prolactin in the ovariectomized rat. *Soc Neurosci Abstr* 5:1569, 1979.

Traish, A.M., Muller, R.E., and Wotiz, H.H.: Binding of estrogen receptor to uterine nuclei: Salt-extractable versus salt-resistant receptor estrogen complexes. *J Biol Chem* 252:6823-6830, 1977.

Turner, B.B., and McEwen, B.S.: Hippocampal cytosol binding capacity of corticosterone: no depletion with nuclear loading. *Brain Res* 189:169-182, 1980.

Tyler, J.L., Gordon, J.H., and Gorski, R.A.: Effects of olfactory bulbectomy and estrogen on tyrosine hydroxylase and glutamic acid decarboxylase in the nigrostriatal and mesolimbic dopamine systems of adult female rats. *Pharmacol Biochem Behav* 11:549-552, 1979.

Vacas, M.I., and Cardinali, D.P.: Effect of estradiol on α- and β-adrenoceptor density in medial basal hypothalamus, cerebral cortex and pineal gland of ovariectomized rats. Neurosci Lett 17:73-77, 1980.

Van Doorn, E.J., Burns, B., Wood, D., Bird, C.E., and Clark, A.F.: In vivo metabolism of [³H]dihydrotestosterone and [³H]androstanediol in adult male rats. J Steroid Biochem 6: 1549-1554, 1975.

Vermes, I., Varszegi, M., Toth, E.K., and Telegdy, G.: Action of androgenic steroids on brain neurotransmitters in rats. Neuroendocrinology 28:386-393, 1979.

Vertes, A., Vertes, M., and Kovacs, S.: Hypothalamic effect of oestradiol. Acta Physiol Acad Sci Hung 51:218-219, 1978.

Vijayan, E., and McCann, S.M.: Involvement of γ-aminobutyric acid (GABA) in the control of gonadotropin and prolactin release in conscious female rats. Fed Proc 37:555, 1978.

Vogel, W., Broverman, D.M., and Klaiber, E.L.: EEG responses in regularly menstruating women and in amenorrheic women treated with ovarian hormones. Science 172:388-391, 1971.

Wade, G.N., and Feder, H.H.: [1,2-³H]Progesterone uptake by guinea pig brain and uterus: Differential localization, time-course of uptake and metabolism, and effects of age, sex, estrogen-priming and competing steroids. Brain Res 45: 525-543, 1972a.

Wade, G.N., and Feder, H.H.: Effects of several pregnane and pregnene steroids on estrous behavior in ovariectomized guinea pigs. Physiol Behav 9:773-775, 1972b.

Wade, G.N., and Feder, H.H.: Stimulation of [³H]leucine incorporation into protein by estradiol-17β or progesterone in brain tissues of ovariectomized guinea pigs. Brain Res 73: 545-549, 1974.

Wagner, H.R., Crutcher, K.A., and Davis, J.N.: Chronic estrogen treatment decreases β-adrenergic responses in rat cerebral cortex. Brain Res 171:147-151, 1979.

Walker, M.D., Negreanu, V., Gozes, I., and Kaye, A.M.: Identification of the 'estrogen-induced protein' in brain of untreated immature rats. FEBS Lett 98:187-191, 1979.

Walker, W.A., and Feder, H.H.: Long term effects of estrogen action are crucial for the display of lordosis in female guinea pigs: Antagonism by antiestrogens and correlations with in vitro cytoplasmic binding activity. Endocrinology 104:89-96, 1979.

Wallen, K., Goldfoot, D.A., Joslyn, W.D., and Paris, C.A.: Modification of behavioral estrus in the guinea pig following intracranial cycloheximide. Physiol Behav 8:221-223, 1972.

Wallis, C.J.: Neuroendocrine influences on gamma-aminobutyric acid metabolism in rodent brain tissue. Ph.D. Dissertation, University of Florida, Gainsville, 1976.

Wallis, C.J., and Luttge, W.G.: Maintenance of male sexual behavior by combined treatment with oestrogen and dihydro-testosterone in CD-1 mice. *J Endocrinol* 66:257-262, 1975.
Wallis, C.J., and Luttge, W.G.: Effect of anterior hypothalamic cuts and antibody absorption of luteinizing hormone on glutamic acid decarboxylase activity in discrete regions of rat brain. *Soc Neurosci Abstr* 3:366, 1977.
Wallis, C.J., and Luttge, W.G.: Influence of estrogen and progesterone on glutamic acid decarboxylase activity in discrete regions of rat brain. *J Neurochem* 34:609-613, 1980.
Ward, M.M., Stone, S.C., and Sandman, C.A.: Visual perception in women during the menstrual cycle. *Physiol Behav* 20: 239-243, 1977.
Waremboug, M.: Radioautographic study of the rat brain, uterus and vagina after [^{3}H]R-5020. *Mol Cell Endocrinol* 12:67-79, 1978.
Watson, G.H., and Muldoon, T.G.: Microsomal estrogen receptors in rat uterus and anterior pituitary. *Fed Proc* 36:912, 1977.
Watts, P.D., Quadagno, D.M., and Bast, J.D.: Hypothalamic injection of cycloheximide in the 4-day cyclic rat. Acute suppression of lordotic behavior and ovulation. *Neuroendocrinology* 29:247-254, 1979.
Weichman, B.M., and Notides, A.C.: Analysis of estrogen receptor activation by its [^{3}H]estradiol dissociation kinetics. *Biochemistry* 18:220-225, 1979.
Weichman, B.M., and Notides, A.C.: Estrogen receptor activation and the dissociation kinetics of estradiol, estriol and estrone. *Endocrinology* 106:434-439, 1980.
Weiner, R.I., and Ganong, W.F.: Role of brain monoamines and histamine in regulation of anterior pituitary secretion. *Physiological Rev* 58:905-976, 1978.
Weissman, B.A., and Johnson, D.F.: Possible role of dopamine in diethylstilbestrol-elicited accumulation of cyclic AMP in incubated male rat hypothalamus. *Neuroendocrinology* 21:1-9, 1976.
Weissman, B.A., and Skolnick, P.: Stimulation of adenosine 3',5'-monophosphate formation in incubated rat hypothalamus by estrogenic compounds: Relationship to biologic potency and blockade by anti-estrogens. *Neuroendocrinology* 18: 27-34, 1975.
Weissman, B.A., Daly, J.W., and Sklonick, P.: Deithylstil-bestrol-elicited accumulation of cyclic AMP in incubated rat hypothalamus. *Endocrinology* 97: 1559-1566, 1975.
Weisz, J., and Gibbs, C.: Conversion of testosterone and androstenedione to estrogens *in vitro* by the brain of female rats. *Endocrinology* 94: 616-620, 1974.
Weisz, J., and Gunsalus, P.: Estrogen levels in immature female rats: True or spurious—ovarian or adrenal. *Endocrinology* 93:

1057-1065, 1973.

Whalen, R.E., and Gorzalka, B.B.: The effects of progesterone and its metabolites on the induction of sexual receptivity in rats. *Horm Behav.* 3:221-226, 1972.

Whalen, R.E., and Gorzalka, B.B.: Effects of an estrogen antagonist on behavior and on estrogen retention in neural and peripheral target tissues. *Physiol Behav* 10:35-40, 1973.

Whalen, R.E., and Luttge, W.G.: Differential localization of progesterone uptake in brain: Role of sex, estrogen pretreatment and adrenalectomy. *Brain Res 33*:147-155, 1971a.

Whalen, R.E., and Luttge, W.G.: Testosterone, androstenedione, and dihydrotestosterone: Effects on mating behavior of male rats. *Horm Behav 2*:117-125, 1971b.

Whalen, R.E., and Nakayama, K.: Induction of oestrous behavior: Facilitation by repeated hormone treatments. *J Endocrinol 33*: 525-526, 1965.

Whalen, R.E., and Rezek, D.L.: Localization of androgenic metabolites in the brain of rats administered testosterone or dihydrotestosterone. *Steroids 20*:717-725, 1972.

Whalen, R.E., and Olsen, K.L.: Chromatin binding of estradiol in the hypothalamus and cortex of male and female rats. *Brain Res 152*:121-131, 1978.

Whalen, R.E., Gorzalka, B.B., DeBold, J.F., Quadagno, D.M., Ho, K.-W., and Hough, J.C.: Studies on the effects of intracerebral actinomycin-D implants on estrogen-induced receptivity in rats. *Horm Behav 5*:337-344, 1974.

Whalen, R.E., Yahr, P., and Luttge, W.G.: Hormones, metabolism and sexual behavior. In *Neurobiology of Reproduction*, R. Goy and D. Pfaff, eds. Plenum Press, New York, 1980.

White, J.O., and Lim, L.: The oestrogen receptor in the hypothalamus: Nuclear responses to oestrogen administration. *Biochem Soc Trans 6*:1310-1312, 1978.

White, J.O., Thrower, S., and Lim, L.: Intracellular relationships of the oestrogen receptor in the rat uterus and hypothalamus during the oestrous cycle. *Biochem J 172*:37-47, 1978.

Whitehead, S.A., and Ruf, K.B.: Responses of antidromically identified preoptic neurons in the rat to neurotransmitters and to estrogen. *Brain Res 79*:185-198, 1974.

Wilkinson, M., Herdon, H., Pearce, M., and Wilson, C.: Radioligand binding studies on hypothalamic noradrenergic receptors during the estrous cycle or after steroid injection in ovariectomized rats. *Brain Res 168*:652-655, 1979.

Williams, M.: Protein phosphorylation in the mammalian nervous system. *Trends Biochem Sci 4*:25-28, 1979.

Williams-Ashman, H.G., and Canellakis, Z.N.: Polyamines in mammalian biology and medicine. *Perspect Biol Med 22*:421-453, 1979.

Wilson, E.M., and French, F.S.: Effects of proteases and protease inhibitors on the 4.5S and 8S androgen receptor.
J Biol Chem 254:6310-6319, 1979.
Wirz-Justice, A., Hackmann, E., and Lichtsteiner, M.: The effect of oestradiol dipropionate and progesterone on monoamine
uptake in rat brain. *J Neurochem 22*:187-189, 1974.
Yagi, K., and Sawaki, Y.: Recurrent inhibition and facilita-
tion. Demonstration in the tuberinfundibular system and
effects of strychnine and picrotoxin. *Brain Res 84*:155-159,
1975.
Yaginuma, T., Watanabe, T., Kigawa, T., Nakai, T., Kobayashi,
T., and Kobayashi, T.: Uptake of ^{3}H-leucine in the brain of
the female rats and its change after castration. *Endocrinol
Jpn 16*:591-598, 1969.
Yahr, P.: Data and hypotheses in tales of dihydrotestosterone.
Horm Behav 13:92-96, 1979.
Yamada, Y.: Effects of testosterone on unit activity in hypothalamus and septum. *Brain Res 172*:165-168, 1979.
Yamada, Y., and Veshima, H.: Methods of electrophoretical ap-
plication of estrogen to a single neuron of the brain under
in vivo and *in vitro* conditions. *Endocrinol Jpn 25*:397-401,
1978.
Yamamoto, K.R.: Characterization of the 4S and 5S forms of the
estradiol receptor protein and their interaction with deoxyribonucleic acid. *J Biol Chem 249*:7068-7075, 1974.
Yanase, M.: A possible involvement of adrenaline in the
facilitation of lordosis behavior in the ovariectomized rat.
Endocrinol Jpn 24:507-512, 1977.
Yanase, M., and Gorski, R.A.: Sites of estrogen and progesterone facilitation of lordosis behavior in the spayed rat.
Biol Reprod 15:536-543, 1976a.
Yanase, M., and Gorski, R.A.: The ability of the intracerebral
exposure to progesterone on consecutive days to facilitate
lordosis behavior: An interaction between progesterone and
estrogen. *Biol Reprod 15*: 544-550, 1976b.
Zava, D.T., and McGuire, W.L.: Estrogen receptor: Unoccupied
sites in nuclei of breast tumor cell line. *J Biol Chem 252*:
3703-3708, 1977.
Zigmond, R.E., and McEwen, B.S.: Selective retention of oestradiol by cell nuclei in specific brain regions of the
ovariectomized rat. *J Neurochem 17*:889-899, 1970.
Zolovick, A.J., Pearse, R., Boehlke, K.W., and Eleftheriou,
B.E.: Monoamine oxidase activity in various parts of the rat
brain during the estrous cycle. *Science 154*:649, 1966.

16

Hormones and Brain Development

CYNTHIA KUHN AND SAUL SCHANBERG

The ability of hormones to influence brain development has been known since the 19th century, when the mental retardation associated with thyroid hormone deficiency first was described. Since then many anatomical, electrophysiological, metabolic, and behavioral consequences of exposing the developing central nervous system (CNS) to excessive or insufficient levels of hormones have been described. At present, a large number of hormones, including thyroid hormone, corticosteroids, androgens, estrogens, and growth hormone (GH), have been shown to affect brain development. Recent evidence suggests that even less-studied small peptide hormones, such as vasopressin and adrenocorticotropin (ACTH) have significant effects on certain aspects of CNS development. Whereas some hormones, such as estrogen, affect the development of a limited population of cells that regulates secretion of that particular hormone and related functions, others, such as thyroid hormone, have broad metabolic effects on a wide population of cells. Each of these diverse hormones has a characteristic pattern of effects on development, but they also share an important characteristic. Their effects occur only if the developing brain is exposed to the hormone during a "critical" period of development, the exact time of which varies from hormone to hormone. For example, thyroid hormone markedly affects the development of cerebellar cells if given during the first week of postnatal life in the rat, but it has little effect on these cells earlier or later (Eayrs, 1966; Hamburgh et al., 1964, 1977). Furthermore, although many of these hormones have effects on mature brain cells, these are significantly different from their developmental effects.

Although there is a vast literature describing changes in various aspects of brain development following hormone administration, the mechanisms by which these changes occur and by which tissue sensitivity to the hormone is regulated are in most cases completely unknown. A particular complication in understanding these mechanisms is the difficulty of correlating the changes observed after administration of pharmacological doses of hormone with the role of "physiological" levels of hormones. However, recent advances in our understanding of the maturation of endocrine tissues and the molecular mechanisms of hormone action have provided new insight into many well-known phenomena.

In this chapter, we will try to summarize the information available on the effects of several hormones on brain development. From a necessity to limit the discussion of this vast area of investigation, we will concentrate on those hormones thought to regulate development of the entire CNS. However, the effect of gonadal steroids on hypothalamic differentiation will also be discussed to provide an example of hormonal regulation of specific CNS functions.

THYROID HORMONE

Excess

Behavior

Exposure of developing mammals to excess thyroid hormone is associated with apparent acceleration of many aspects of development. For example, developmental landmarks, such as eye opening, are reached sooner in neonatal rats treated with thyroid hormone during the first few days after birth (Schapiro, 1968). Maturation of behavior is similarly accelerated. Swimming behavior (an index of cerebellar maturation), avoidance learning, motor behavior, and the startle reflex are among the behaviors that appear earlier in thyroxine-treated rats (Schapiro, 1968; Schapiro et al., 1970). However, the apparent acceleration of functional maturation of the CNS is only a temporary condition and is followed by the later development of abnormal behavioral responses. For example, although hyperthyroid neonatal rats learn to avoid electric shock sooner, as they mature, their responses deteriorate to a level below that of untreated controls (Schapiro, 1968). Adult rats that were exposed to excess thyroid hormone as pups are worse at learning certain tasks than untreated controls (Stone and Greenough, 1975; Davenport and Gonzales, 1973). In general, the behavioral sequelae that follow exposure of the devloping CNS to excess thyroxine (T_4) are

those of accelerated maturation resulting in the establishment
of inappropriate behaviors in adults.

Morphology

Anatomical studies of the effects of excess thyroid hormone on
brain maturation have revealed morphological abnormalities that
parallel the reported changes in behavioral development. The
system in which this phenomenon has been studied in the greatest detail is the rat cerebellum. Development of this brain
region occurs mainly postnatally in the rat, and the developmental sequence of this highly organized structure is well
known. Therefore it offers an excellent system for studying
the effects of various hormonal manipulations on development.
Hyperthyroidism in the neonatal period is associated with accelerated morphological development that seems to reflect an
ability of thyroid hormone to accelerate the transition of neurons from proliferation to differentiation (Fig. 16-1). This
accelerated development has different consequences for different cell types. T_4 administration causes accelerated growth
of axons and dendrites of the Purkinje cells and parallel
fibers, cells that are already formed at birth (Rebiere and
Legrand, 1972; Lauder, 1977; Lauder et al., 1974)(Fig. 16-2,
16-3). However, the premature cessation of cell division,
which also occurs, results in a decrease in the number of
basket and granule cells, cells that are not formed until the
end of the first postnatal week (Nicholson and Altman,
1972a,b,c) (Fig. 16-4, 16-5). This pattern of cell development
results in the formation of an abnormal pattern of synaptic
contacts. The longer parallel fibers make more synapses, but
this is offset by the decrease in synaptic contacts resulting
from the smaller number of basket and granule cells. It is
thought that these effects on synaptogenesis result from the
described abnormalities in neurite outgrowth rather than an
effect on synaptogenesis per se. Regardless of the cause,
development of cerebellar structure is determined in part by
the spatial organization of the various cell types, so these
abnormalities in cerebellar maturation resulting from T_4
administration cause permanent changes in cerebellar structure
(Lauder et al., 1974).
 This stimulation of neuronal growth and consequent disruption
of structural development is not restricted to the developing
cerebellum. In the developing hippocampus of rats exposed to
excess thyroid hormone during the first postnatal week, a similar stimulation of axonal growth in the mossy fiber layer is
observed, which is accompanied by a premature end to cell proliferation and an increase in neuronal size (Lauder, 1977;
Lauder and Mugnaini, 1977; Moskovkin and Marshak, 1978). In

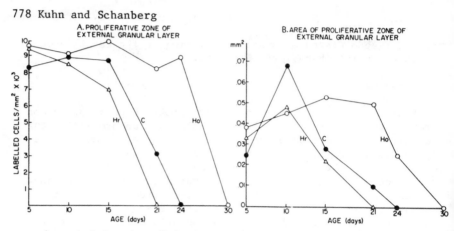

Age at injection (2 h survival) indicated. C, controls; Ho, hypothyroids; Hr, hyperthyroids.

Fig. 16-1. The effects of early hypo- and hyperthyroidism on the development of rat cerebellar cortex. I. Cell proliferation and differentiation. (From Nicholson and Altman, 1972a).

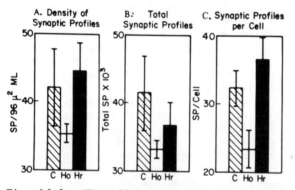

Synaptic content of the molecular layer at 30 days. A: C vs Ho, p<0.0 C vs Hr, p=NS; Ho vs Hr p<0.01. B: C vs Ho, p <0.05; C vs Hr, p=NS; Ho vs Hr, p=NS. C(granu and basket cells): C vs Ho, p<0.01; C vs Hr, p <0.01; Ho vs Hr, p<0.01 C, controls; Ho, hypotl roids; Hr, hyperthyroid

Fig. 16-2. The effects of early hypo- and hyperthyroidism on the development of rat cerebellar cortex. II. Synaptogenesis in the molecular layer. (From Nicholson and Altman, 1972b).

addition, accelerated growth of dendritic spines accompanied by a reduction in cell number has been observed in the cerebral cortex of hyperthyroid rat pups (Schapiro, 1968; Balazs and Motterrell, 1972). Furthermore, the premature termination of cell mitosis caused by excess thyroid hormone results in a large decrease in the number of glial cells in cortex, as in the case for all cell types undergoing postnatal mitosis in rats (Pelton and Bass, 1973).

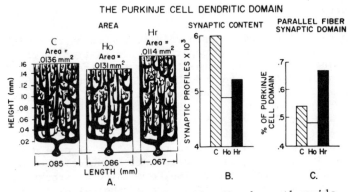

THE PURKINJE CELL DENDRITIC DOMAIN

C, controls; Ho, hypothyroids; Hr, hyperthyroids.

Fig. 16-3. (From Nicholson and Altman, 1972b).

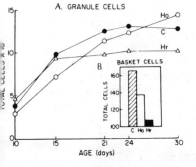

A, developmental increases in granule cells. C vs Ho: (10d) p=NS, (15d) p <0.01, (21d) p<0.05, (30d) p<0.01; C vs Hr: (10,15d) p=NS, (21-30d) p<0.01; Ho vs Hr: (10-30d) p<0.01. B, total basket cells at 30 days. C vs Ho, p=NS; C vs Hr, p<0.01; Ho vs Hr, p=NS. C, controls; Ho, hypothyroids; Hr, hyperthyroids.

Fig. 16-4. (From Nicholson and Altman, 1972a).

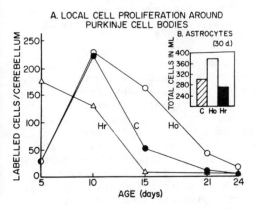

Statistical significance (B): C vs Ho, p<0.01; C vs Hr, p=NS; Ho vs Hr, p<0.01. C, controls; Ho, hypothyroids; Hr, Hyperthyroids.

Fig. 16-5. (From Nicholson and Altman, 1972a).

Biochemistry

The biochemical effects of hyperthyroidism are consistent with the morphological studies just reported. Administration of thyroid hormones to the rat during the first postnatal week results in an increase in protein, DNA and RNA synthesis, but this increase is followed by a decrease in content and synthesis in older pups (Balazs et al., 1968; Clark and Weichsel, 1977; Ardeleanu and Sterescu, 1977; Gourdon et al., 1973; Geel and Timiras, 1967). These findings parallel the reported stimulation of cell growth and premature termination of cell division observed in morphological studies. T_4 stimulation of one biochemical index of brain maturation, ornithine decarboxylase (ODC) activity, correlates not only with morphological findings but also with an index of behavioral maturation. Administration of T_4 during the first few days of life causes a stimulation of both ODC activity and the development of swimming behavior, followed by a suppression of both parameters (Anderson and Schanberg, 1975).

In addition, a similar enhancement and then suppression in ganglioside synthesis is caused by thyroxine administration to neonatal rat pups (Horowitz and Schanberg, 1979b). Neonatal treatment with thyroxine seems to affect myelination and lipid synthesis in the same way that this treatment affects protein and nucleic acid synthesis. Several indices of myelination, including incorporation of ^{32}P into specific lipids, cholesterol content and sulfatide formation, galactosyl transferase activity, and brain content of galacytosyl diacylyglycerols, are enhanced and then suppressed by thyroxine treatment of rat pups during the first postnatal week (Flynn et al., 1977; Ardeleanu and Sterescu, 1978; Schapiro, 1968; Walravens and Chase, 1969; Pelton and Bass, 1973). These effects are consistent with morphological findings of a premature termination of myelination in brains of hyperthyroid rat pups (Balazs et al., 1968; Freundl and Van Wynsberghe, 1978).

Accelerated appearance of several adult-type biochemical functions, including amino acid conversion to glucose, glutamate compartmentation, and $D(-)-\beta$-hydroxybutyric dehydrogenase, as well as premature appearance of several synaptic markers, such as synaptosomal protein, and acetylcholinesterase activity are also caused by hyperthyroidism during the neonatal period in rats (Schapiro, 1968; Rabie and Legrand, 1973; Cocks et al., 1968, 1970; Grave, 1977). However, it must be emphasized that thyroid hormone does not cause an indiscriminate enhancement of the synthesis of all cellular proteins: its effects seem to be quite specific. For example, a number of parameters including enzymes involved in galactosyl diacylglycerol metabolism, cytochrome oxidase, some aspects of phospholipid synthesis, and cyclic-AMP-stimulated protein kinase, are unaffected by excess thyroid hormone, and several lysosomal enzymes are actually decreased (Lau et al., 1977a,b; Schmidt, 1974; Hamburgh and Flexner, 1957).

Insufficiency

Behavior

Thyroid hormone insufficiency during development in mammals results in a retardation of growth and development. The most dramatic effects are on skeletal growth and CNS development (a marked decrease in skeletal size and mental retardation). However, development of almost all tissues is affected, and maturation of other hormone-producing tissues is particularly delayed.

Almost every behavioral index that has been studied in experimental animals is adversely affected by hypothyroidism. In rats made hypothyroid at birth, the normal appearance of reflexes such as the startle response and the placing reflex, is delayed. Although these fairly simple CNS responses gradually do appear, the performance of more complex behavioral tasks by hypothyroid rats never attains normal levels (Eayrs, 1966). Performance in learning paradigms is impaired and behavioral responses to sensory stimuli are blunted. A similar impairment of electrophysiological responses to such stimuli is also observed and the electroencephalogram (EEG) of a hypothyroid animal shows marked abnormalities (Eayrs, 1966; Ford and Cramer, 1977). Like hyperthyroidism, neonatal hypothyroidism results in permanent changes in CNS structure and function, rather than a transitory delay in development. Similar effects are not observed in adult animals that have been rendered hypothyroid (Eayrs, 1966).

Morphology

Abnormal anatomical development of the CNS of hypothyroid animals parallels the altered behavioral development. These changes are characterized by a slowdown in the normal growth of cells and a persistence of cell division past the normal time of termination, resulting in the formation of inappropriate synaptic contacts (Legrand, 1977). The effects of hypothyroidism on the developing rat cerebellum, which have been described in detail, provide a good example of this phenomenon. Although the duration of cell proliferation during the postnatal period is extended (Fig. 16-1), there is an over-all decrease in cell size, synaptic density, and the density of axons and dendrites in the cerebellum of hypothyroid rats (Lewis et al., 1976; Krawiec et al., 1969; Nicholson and Altman, 1972a,b) (Fig. 16-2, 16-3). In addition, although the cell number eventually reaches that of controls, the proportion of different cell types is markedly abnormal. There is a significant decline in the number of basket cells, which is offset by an increase in the number of stellate and granule cells and glia (Nicholson and Altman, 1972a; Balazs, 1977; Lewis et al., 1976). This retardation in cerebellar development in hypothyroid animals is accompanied by delayed functional development, as determined by evoked potential and single cell activity (Crepel, 1974; Clos et al., 1974).

This pattern of changes results from several alterations in the maturation of specific cell types. In a normal cerebellum, Purkinje cells differentiate before birth, and after birth, basket, stellate, and granule cells follow in sequence. The formation of dendrites and the general growth of Purkinje cells is markedly retarded in hypothyroid pups (Legrand, 1977). This effect is believed to mediate the decrease in the number of basket cells, because these cells differentiate only after establishing contact with Purkinje cells. The decreased area of Purkinje cell membrane available for synaptic contact results in a decrease in the number of basket cells formed. The same phenomenon may account for the excess of glia, which multiply to occupy any space on the Purkinje cells not already occupied by synapses. Finally, the decrease in basket cells is probably responsible for the excess number of granule cells, because the latter differentiate last: remaining undifferentiated cells that do not form basket cells can later become granule cells. Interestingly, although the final number of granule cells is increased, at earlier times a decrease occurs that also may result from decreased contact with Purkinje cells (Nicholson and Altman, 1972a,b,c; Lauder, 1977; Lauder et al., 1974). This early retardation in granule cell formation results in abnormal development of mossy fiber-granule cell synaptic contacts. These changes in the proportion of different cell types are shown in Fig. 16-4 and 16-5. These results indicate that the suppressed growth of Purkinje cells and the delayed cell division caused by hypothyroidism disrupt the time course of cell differentiation, and so disrupt the development of cerebellar structure.

Similar disruptions are observed in other brain regions. Prolonged cell division and decreases in cell size and synaptic density occur in both hippocampus and cortex (Moskovkin and Marshak, 1978; Krawiec et al., 1969). This prolongation of division does not result from an increase in generation time, but instead occurs because the number of cells undergoing division is decreased (Lewis et al., 1976). As in the cerebellum, the decreased synaptic density results from decreased growth of axons and dendrites, which markedly decrease the number of synaptic contacts made in these brain regions. These effects contrast with those of hyperthyroidism, which generally enhances neurite outgrowth and results in an increased number of synapses distributed over a larger volume, although in both cases the pattern of synaptogenesis is inappropriate.

Biochemistry

Biochemical studies suggest that impairment of protein synthesis may be the primary biochemical defect that leads to the observed anatomical abnormalities in the brains of hypothyroid rat pups. This hypothesis is consistent with the well-known correlation between the ability of thyroid hormone to stimulate protein synthesis and its role in the development and differentiation of many tissues (Oppenheimer et al., 1979). In young adult rats made hypothyroid at birth, both cell number and brain DNA content are normal, yet both protein content and synthesis are still suppressed (Balazs, 1977; Balazs et al., 1968; Gourdon et al., 1973). RNA content is depressed in most brain regions of hypothyroid rat pups, but this change seems to result from a temporary inhibition of synthesis of rapidly transcribed (messenger) RNA that disappears as they mature (Krawiec et al., 1977; Geel and Timiras, 1967). The general impairment of protein synthesis in brains of hypothyroid rats is reflected in retardation of the normal developmental increases in activity of many enzymes. These include mitochondrial enzymes, such as succinic dehydrogenase and nicotinamide adenine dinucleotide phosphate (NADPH) diaphorase, the transport enzyme, Na^+, K^+-ATPase, synaptosomal proteins, e.g., glutamate decarboxylase and choline acetyltransferase, and cytoplasmic proteins, such as lactate dehydrogenase (Balazs et al., 1968; Ladinsky et al., 1972; Mitskevich and Moskovikin, 1971; Schwark et al., 1972). In addition, decreases in myelin synthesizing enzymes and other indices of myelination, including brain myelin content, cerebrosides, sulfatide, and cholesterol, are observed in brains of rats made hypothyroid at birth (Freundl and Wynsberghe, 1978; Flynn et al., 1977; Walravens and Chase, 1969). These effects are observed in most brain regions, except those like the medulla, which are nearly mature at birth (Mitskevich and Moskovkin, 1971). In general, the effects of hypothyroidism are more widespread than those of hyperthyroidism, which affects protein synthesis more selectively and spares more brain areas.

Mechanism of Effect

The observed changes in the biochemical development of brain associated with hyper- or hypothyroidism suggest that thyroid hormone acts in developing brain as it does in other tissues. Interaction of triiodothyronine (T_3) with a nuclear receptor is thought to trigger some action or actions on gene transcription or translation that stimulates synthesis of specific proteins (Oppenheimer et al., 1979). Thyroid hormone administration seems to cause selective acceleration of cell differentiation, as reflected in the premature differentiation of cerebellar cells, as well as growth of already differentiated cells. The former effect is suggestive of a specific effect on transcription rather than a general stimulation of cell metabolism. The observed specificity in enhancement of protein synthesis and the selective change in synthesis of rapidly labeled (messenger) RNA (see below) are also consistent with this interpretation.

The decreasing sensitivity to T_4 that develops during the first three weeks of life in rats provides further evidence of a specific effect of thyroid hormone on cell differentiation rather than a general stimulation of cellular metabolism. This loss of sensitivity is accompanied by a decline in the number of nuclear T_3 receptors (Macho et al., 1978). It is well known that the CNS of neonatal mammals becomes increasingly resistant to the action of thyroid hormone as the animal matures. In human beings, this is manifested as an inability to reverse the effects of cretinism if therapy is started too late (Brown et al., 1939). In rats, this can be demonstrated both by the inability of thyroxine to stimulate protein synthesis in older pups and by the irreversibility of the effects of neonatal hypothyroidism if treatment is delayed. The parallel loss of receptors and sensitivity to thyroxine is accompanied by the appearance of "nondevelopmental" effects of thyroxine on brain.

Although thyroid hormone does not stimulate energy metabolism or cell differentiation in adult brain, it has characteristic effects distinct from its effects on developing tissues. Thyrotoxicosis in human beings has several specific CNS manifestations, particularly tremor, increased "excitability," and decreased reaction time. These effects appear to be correlated with increased sensitivity of specific sites in the CNS to sensory stimulation, including the reticular formation and spinal cord (Turakulov et al., 1975). These biochemical effects of thyroid hormone on adult brain, like its physiological effects, are far more limited and quite different from those observed in developing animals. For example, neonatal hypo- and hyperthyroidism affect all aspects of catecholamine neuron maturation, from synthetic enzymes to receptor number, whereas the effect of thyroid hormone administration to adults is a fairly specific increase in receptor sensitivity and turnover (Emlen et al., 1972; Rastogi and Singhal, 1974, 1976; Engstrom et al., 1974). Furthermore, the effects of neonatal thyroidectomy are not reversible if therapy is delayed, whereas the effects of manipulation of thyroid status in adults are reversible (Singhal et al., 1975). In general, thyroid hormone does not stimulate increased synthesis of the large number of metabolic and structural proteins in adults that characterizes its effects in neonates.

The maturation of T_3 binding to its receptors shows an interesting parallel to the changing physiological effects of this hormone in developing animals. Although nuclear binding of T_3 decreases as rat pups mature, total uptake of T_3 increases with age (Macho et al., 1978; Ford and Cramer, 1977), suggesting that there are different sites for the actions of this hormone in adult and in developing brain. Synaptosomal uptake of T_3 has been demonstrated (Dratman et al., 1976), a finding that suggests the nerve ending as a potential site for thyroid hormone action in adult brain.

Although some investigators have proposed that the primary effects of thyroid hormone on brain development result from an interaction with the cell nucleus, it has been proposed that the direct action of thyroxine on mitochondria is responsible for these effects. Sokoloff (1977) showed that thyroid hormone can stimulate protein synthesis in a microsomal preparation from adult brain only if mitochondria from neonatal rats are present. Because this effect occurs *in vitro* in the absence of nuclei, he suggested that T_3 affects release or synthesis of a compound in mitochondria that mediates the protein synthesis-stimulating actions of thyroid hormone. However, a role for this compound in brain development has not yet been clearly established.

The effect of thyroid hormones on CNS development has been clearly established, but the active form of hormone is not known. Although much evidence suggests that T_3 is the active form of the hormone in peripheral tissues (Oppenheimer et al., 1979), the role of each form in brain development cannot be determined on the basis of current information. Both hormones have similar effects when administered and both are present in the blood of developing mammels. A number of factors, including permeability of the developing brain and the activity of degradative mechanisms, probably determine the relative contribution of each. However, the greater uptake of T_3 into brain and its greater potency in eliciting effects on protein synthesis (Sokoloff, 1977; Ford and Cramer, 1977) suggest that this form may be the more important.

Role in Normal Development

Although it is not known whether T_3 or T_4 is responsible for thyroid hormone regulation of brain development, the source of circulating hormone is known with more confidence. A variety of findings indicate that the fetal, not maternal, thyroid gland regulates brain development. Firstly, the effects of prenatal hypothyroidism can be discerned in hyman beings, a species with a placenta that is impermeable to T_4 (Nathanielsz, 1976). Secondly, in rats, at least one major T_4-sensitive period occurs postnatally, when the major hormone source is the pup itself. In addition, in most species, sensitivity to exogenous thyroid hormone appears almost simultaneously with the onset of significant fetal thyroid function. For example, in rats, protein and DNA synthesis can first be stimulated on day 18 of gestation, the time when T_4 is first detectable in blood (Zamenhof et al., 1966; Oklund and Timiras, 1977). Although earlier studies suggested that thyroid hormone plays a role earlier in gestation, these were based on manipulations that produced maternal hypothyroidism, and most of the effects observed could be attributed to changes in maternal rather than fetal physiology.

GLUCOCORTICOIDS

Behavior

An absolute requirement for glucocorticoids in specific developmental processes in brain has not been demonstrated because steroid-deficient neonates do not often survive. Therefore most investigations of the role of steroids in development have only observed the result of glucocorticoid administration. Such studies have demonstrated at least two major steroid effects on development. These hormones are best known for their anabolic effects, which in developing animals are manifested as a profound suppression of the growth of all tissues (Field, 1954). However, these effects are accompanied or triggered by steroid-dependent induction of specific enzymes. For example, in the liver and duodenum, a number of enzymes are thought to require the presence of glucocorticoids for their normal development, and premature development of these enzymes occurs following administration of doses too small to inhibit growth (Moog, 1971; Jacquot, 1971).

The effects of glucocorticoids on brain development generally parallel their effects on other tissues. Large doses suppress brain growth, whereas smaller doses are more likely to induce premature development of specific processes (Field, 1954;

Moscona, 1971). The behavioral sequelae of exposure of neonatal mammals to excess glucocorticoids are more subtle than
those of thyroid imbalances. The most often observed behavioral effect of administration of cortisol to neonatal mammals
is delayed development of various behavioral responses, such as
the evoked potential response to sensory stimuli or adult swimming behavior (Schapiro et al., 1970; Schapiro, 1968; Salas and
Schapiro, 1970) (Fig. 16-6). However, this delayed maturation
can appear as an accentuated response rather than a suppressed
one. For example, delayed maturation is thought to account for
the behavioral hyperresponsiveness that is the other characteristic of animals treated with steroids during development
(Howard, 1973; Olton et al., 1974). It has been suggested that
hormone-induced delay in the development of inhibitory processes accounts for this phenomenon (Vernadakis, 1971). As mentioned previously, these effects are both dose and time dependent, that is, they occur only if pups are treated in the first
or second postnatal week (Vernadakis, 1971).

Morphology

Exposure of neonatal mammals to excess glucocorticoid results
in abnormalities in morphological development of the CNS that
parallel the effects of this treatment on behavioral development. In general, an inhibition of cell growth is observed,
characterized by a decrease in cell division and the number of
cells (Howard, 1965; 1968). Although cell size is not affected, the normal growth of dendritic spines is impaired in many
brain areas (Schapiro, 1968; Schapiro et al., 1973; Balazs et
al., 1971; Balazs and Motterrell, 1972). A rebound increase in
cell division is observed when treatment is stopped. However
cells stop dividing at the normal time, so a permanent deficit
in cell number results (Cotterrell et al., 1972). As would be
expected, the brain regions and cell types most severely affected are those dividing actively during the period of treatment. For example, treatment of rats with cortisol during the
first two postnatal weeks affects the cerebellum more than the
cortex, and glial cells in the cortex more than neurons
(Legrand, 1977).
Although suppression of brain growth is usually observed following treatment of developing mammals with steroids, accelerated rather than retarded development of cortical neurons has
been reported following administration of doses too low to suppress growth (Hodge et al., 1976). This suggests the intriguing possibility that physiological levels of steroids might
play a role in normal brain growth that is usually masked by
the pharmacological doses used in most studies.

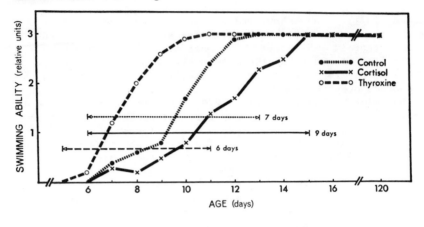

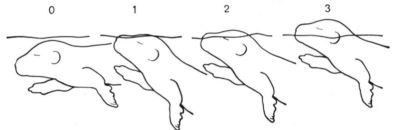

Fig. 16-6. Graphic representation of maturation of swimming
ability in normal and hormone treated rats at
different ages. Drawing illustrates the criteria
used to ascribe arbitrary rating units to swimming behavior. The number of days over which
swimming behavior assumed stable characteristics
is indicated by the horizontal lines. (From
Schapiro et al., 1970).

Biochemistry

Further support for this view is provided by biochemical findings. Doses of glucocorticoids that suppress growth inhibit
DNA synthesis and other indices of cell proliferation, such as
ODC activity, for the duration of treatment (Howard and
Benjamin, 1975; Howard, 1965; Anderson and Schanberg, 1975;
Ardeleanu and Sterescu, 1978). However, the effects of glucocorticoid administration on protein synthesis are somewhat different. A general pattern of suppression at high doses is observed, but induction of specific corticosteroid-dependent enzymes occurs at low doses. Both decreased and unchanged ratios

of RNA:DNA and protein:DNA have been reported (Howard, 1965; Balazs and Richter, 1973) and impaired or accelerated myelination is observed (Schapiro, 1968; Gumbinas et al., 1972; Granich and Timiras, 1971). In general, administration of large doses during the time of active myelination inhibits myelin synthesis as well as synthesis of other membrane constituents, such as gangliosides, whereas administration of small doses at the beginning of active myelination accelerates the process (Granich and Timiras, 1971; Casper et al., 1967; Horowitz and Schanberg, 1979a). Similarly, induction of other specific enzymes, including glutamine synthetase, Na^+,K^+-ATPase, and α-glycerophosphate dehydrogenase, is observed following administration of small doses of glucocorticoids (Moscona, 1971; Balazs and Richter, 1973; Huttenlocher and Amemiya, 1978). Even more persuasive evidence for the role of glucocorticoids in development of specific neural processes is the report that adrenalectomy prevents the normal developmental increase in tryptophan hydroxylase, the rate-limiting enzyme in synthesis of serotonin (Sze et al., 1976). Interestingly, glucocorticoids accelerate maturation of several functions like tryptophan hydroxylase that they also regulate in adult animals.

In summary, the effects of glucocorticoids on developing brain are typical of the effects of these agents on other growing tissues. Administration of large doses suppresses tissue growth by inhibiting DNA synthesis and so decreases cell division. However, small doses do not suppress growth, but instead cause premature maturation of those specific functions that require steroids for normal function.

It should be mentioned that the effects that have been discussed are observed in all brain regions. These contrast markedly with the specific role of glucocorticoids in development of those sites in the hypothalamus and other areas of brain that are involved in the feedback regulation of ACTH secretion. Although little is known about this phenomenon, glucocorticoid sensitivity in specific cells is thought to develop later in ontogeny than the general effects discussed. A marked contrast in the time course of hormone action can be observed between the effects of steroids in developing brain and in adult brain. There is at least a several hour time lag between the time of administration and the first detectable biochemical effects following administration of hormone to neonatal animals. However, some glucocorticoid effects on adult brain, such as acute effects on the firing pattern of specific hypothalamic and hippocampal neurons, occur within minutes (Steiner, 1967; Koranyi and Endroczi, 1969). These effects are thought to be involved in short-term regulation of specific behaviors rather than regulation of cell metabolism.

Mechanism of Effect

The mechanism by which glucocorticoids exert their action on developing brain is not known. However, glucocorticoids are thought to act similarly in all tissues. Therefore extrapolation of information available about glucocorticoid action in liver should provide a starting point for future investigation and so will be mentioned here briefly (see Rees and Gray, chapter 14). Glucocorticoids in the liver bind to a cytoplasmic receptor, which is then translocated to the nucleus. Here the hormone binds to a specific, as yet unidentified, site on the DNA and promotes transcription of specific gene products (Baxter and Ivarie, 1978; Fiegelson et al., 1978). It is thought that glucocorticoids can only promote transcription of already activated genes, rather than activate transcription of previously unexpressed genes. Only a small percentage of cell proteins is affected by administration of glucocorticoids.

The postulated action of glucocorticoids on specific protein synthesis is consistent with experimental findings in brain. Fairly specific effects on DNA synthesis are observed following administration of glucocorticoids, and synthesis of specific proteins, but not total cell protein, is affected. Furthermore, preliminary evidence suggests that the brain contains receptors for glucocorticoid hormones that are similar to those found in liver. Olpe and McEwen (1976) detected glucocorticoid receptors in rat brain on the first postnatal day and suggested that receptors are present even earlier. This mechanism is also consistent with another characteristic of glucocorticoid action on developing brain, that is, there does not seem to be as specific a "critical period" for the action of glucocorticoids as there is for other hormones. For example, whereas administration of thyroid hormone affects cerebellar ODC only during the first few days after birth, cortisol is effective at any time during the first three weeks after birth (Anderson and Schanberg, 1975). This class of hormone seems to affect whatever developmental process is going on at the time of administration, such as cerebellar cell division during the first postnatal week or myelination during the second and third. The mechanism responsible for the anabolic actions of large doses of steroid is not known, although it has been hypothesized that a similar process is involved (Baxter and Ivarie, 1978).

Role in Normal Development

Although pharmacological effects are observed following exogenous steroid administration, the time at which sensitivity to steroid first develops in brain and how this relates to maternal and fetal adrenal function is unclear. Considerable evidence suggests that fetal adrenal function begins at a very early gestational age in all mammalian species. Corticosterone synthesis can be measured as early as the 15th day of gestation (the beginning of the third trimester) in fetal rat adrenals and even earlier in other species, including man (Nathanielsz, 1976; Bloch, 1968). Therefore fetal blood contains glucocorticoids at the time that steroid sensitivity can be detected biochemically in brain. There is also some evidence that CNS control of ACTH and glucocorticoid secretion exists during gestation in many mammals. For example, decapitation of rat fetuses on the 19th day of gestation prevents ether-induced depletion of adrenal ascorbic acid and decreases the size of the fetal adrenal gland (Eguchi and Wells, 1964, 1965; Chatelain et al., 1976; Jost, 1966). However, although fetal adrenal glands definitely synthesize and secrete glucocorticoids, secretion in response to specific physiological signals is not so clearly established. Furthermore, effects of fetal adrenalectomy on brain development have not been described. Therefore glucocorticoid hormones secreted by fetal adrenal tissue have no clearly established role in CNS development.

Although the effect of fetally derived glucocorticoids on CNS development is equivocal, several laboratories have proposed that stress-induced secretion of maternal glucocorticoids affects brain development. This hypothesis derives from the observation that stresses that elevate maternal serum glucocorticoid levels during pregnancy affect behavioral development of the offspring. Specifically, it has been reported that prenatal stress feminizes sexual behavior of male rats and alters offspring "emotionality," learning behavior, acquisition of normal spontaneous locomotor activity, and other indices of behavioral development (Ward, 1972; Thompson, 1957; Ader and Belfer, 1962; Whitney and Herenkohl, 1977; Sobrian, 1977; Archer and Blackman, 1971; Doyle and Yule, 1959; Hockman, 1961; Porter and Wehmer, 1969; Masterpasqua et al., 1976). However, some studies using identical stress paradigms have yielded contradictory results. For example, conditioned avoidance anxiety has been reported to increase, decrease, or not affect open-field activity of offspring (Thompson et al., 1962; Thompson, 1957; Joffe et al., 1972; Smith et al., 1975). Furthermore, although many behavioral effects of prenatal stress have been described, mediation of these effects by altered maternal glucocorticoid secretion has not been established. In fact, prenatal stress effects on maternal glucocorticoid secretion and offspring development have been dissociated by careful control of litter to litter variability in behavioral measures. Chapman and Stern (1979) reported that several prenatal stress paradigms effectively raised maternal steroid levels without af-

fecting the behavioral development of the offspring. The
marked species variability in placental transfer of glucocorticoid hormones casts further doubt on the generalized role of
stress-induced secretion of maternal hormones in control of
fetal CNS development. Although considerable transfer occurs
in rats, transfer in other species, including sheep and human
beings, is minimal (Nathanielsz, 1976).

The ability of stress-induced glucocorticoid secretion from
the fetal adrenal gland to affect CNS maturation is equally
controversial. This hypothesis is based on several assumptions
including: (1) stress elicits glucocorticoid secretion from the
developing adrenal; (2) CNS development is altered by stress;
and (3) increased concentration of glucocorticoids in the brain
mediates the observed alterations in CNS development. The
first two of these assumptions have been verified experimentally, but the last has been challenged by a number of investigators. The evidence supporting or refuting each of these assumptions is summarized.

The hypothesis that stress elicits glucocorticoid secretion
from the developing adrenal has been investigated extensively.
These results show that stress-induced glucocorticoid secretion
is present during at least part of the steroid-sensitive phase
of brain development. However, the presence of stress-induced
secretion during the earlier phases of CNS development has been
more difficult to establish. Stress-induced corticosterone secretion has been reported on the 19th day of gestation in rats
(Eguchi et al., 1964; Eguchi and Wells, 1965; Milkovic and
Milkovic, 1961, 1962, 1963; Jost, 1966; Gray, 1971; Haltmeyer
et al., 1967). However, a large number of studies from other
laboratories have reported that the adrenal of the neonatal rat
is insensitive to stress and that CNS control of ACTH secretion
by the pituitary is minimal until after birth (Hiroshige and
Sato, 1971; Allen and Kendall, 1967; Schapiro et al., 1962).
However, corticosterone secretion following stress is clearly
established by the seventh or eighth postnatal day in rats
(Schapiro, 1962; Levine, 1970; Zarrow, et al., 1966). Furthermore, although steroid effects on brain development during gestation have been described, there have been no reports of
stress-induced corticosterone secretion from fetal adrenals.
Some of these discrepancies result from the varying intensity
of stimulation used to elicit glucocorticoid secretion in different laboratories. Those experimenters who used severe
stresses, such as electric shock or extreme heat, reported
reliable increases in serum glucocorticoids in young rats,
whereas those using milder stimuli did not consistently observe
secretion. In general, the onset of stress responsiveness in
the mammalian hypothalamohypophyseal-adrenal axis cannot be

timed exactly, although some responses during the period of CNS sensitivity to steroids have been demonstrated. Unfortunately, parallel stress-induced alterations in brain glucocorticoid content have not been reported. In fact, Butte et al., (1973) reported that stress elicits measurable increases in serum but not brain corticosterone in neonatal rats. Nevertheless, the possibility that stress changes glucocorticoid content in specific sites has not been eliminated.

Considerable evidence also supports the hypothesis that stress alters CNS development. Again, most evidence has been obtained from studies of postnatal development in rats and similar species. Numerous effects of various "stressful" manipulations, such as handling, on behavioral development have been demonstrated. The literature in this area is too vast to discuss here. However, many effects, including alterations of learning behavior, "emotionality," locomotor activity, and many other behavioral indices, have been described (Denenberg and Zarrow, 1971; Levine, 1969; Weininger, 1953). These effects of "early experience" or "stress" are clearly different from the effects of "prior" experience, as similar stress paradigms do not affect behavior when tested in older animals. "Early experience" or neonatal "stress" paradigms also have been correlated with changes in the anatomical or biochemical development of the brain. For example, Schapiro and Vukovich (1970) reported that subjecting neonatal rats to extreme stress increased the number of synaptic spines on certain cortical neurons. Altered brain weight, DNA, protein, and amino acid content, as well as neuronal morphology and organization, also have been associated with neonatal stress in rats and other mammalian species. Even apparently innocuous environmental manipulations, such as placing animals in an enriched environment, alter behavioral development in many species. The reader should consult the excellent review by Greenough (1977) for a detailed discussion of the subject.

Although glucocorticoid secretion in response to stress appears fairly early in mammalian development, glucocorticoid secretion does not necessarily mediate the observed developmental changes associated with "early experience" or "stress." This evidence will be reviewed briefly here. However, for a more extensive description, the reader should consult the review by Ader (1975). First, most of the evidence indicates that mild stresses, like handling, do not consistently elicit glucocorticoid secretion in experimental animals, although these manipulations do alter behavioral development. Similarly, the parallel decrements in stress-induced glucocorticoid secretion and behavioral that result from stressing neonatal animals have been dissociated under a number of circumstances (Ader, 1968,

1970, 1975). For example, the reduction of adrenocortical reactivity reflects an effect of "prior" experience, as it can be
produced in older animals. In contrast, the behavioral changes
associated with "early stress" occur if animals are stressed
during a specific developmental period (Ader, 1970; Ader and
Grota, 1969). In general, although numerous correlations between steroid secretion and stress-induced alterations in behavioral development exist, a causal relationship between these
two events has not been established. Nevertheless, the possibility that glucocorticoid secretion contributes to stress-
induced alterations in CNS development cannot be discounted entirely, since the ability of early stress to affect behavioral
development in the absence of glucocorticoid secretion has not
been demonstrated.

Although there is no evidence suggesting that glucocorticoid
secretion mediates stress-induced changes in behavioral development, stress-induced glucocorticoid secretion might affect
the appearance of normal CNS control of ACTH secretion. Numerous studies indicate that steroid exposure during a "critical" period affects maturation of diurnal rhythmicity, which
develops much later in ontogeny than basal secretion (Taylor et
al., 1976). Furthermore, daily handling or electric shock during the early postnatal period also affects the appearance of
this rhythm (Ader, 1969). However, administration of corticosterone to female rat pups has been reported to delay the appearance of the diurnal rhythm, whereas stress is reported to
accelerate its appearance (Cost and Mann, 1976; Taylor et al.,
1976). Therefore mediation of neonatal stress effects on control of ACTH secretion by glucocorticoid secretion has not been
definitively established.

In summary, considerable evidence indicates that glucocorticoid secretion begins early in mammalian development and that
exogenous glucocorticoid administration influences CNS development. However, neither basal nor stress-induced glucocorticoid
secretion has been shown to affect specific processes in the
developing CNS. Furthermore, extensive attempts to correlate
effects of stress on glucocorticoid secretion and behavioral
effects have been unsuccessful. Furthermore, the role of
stress-induced secretion of other hormones has been ignored in
these studies, although several recent reports indicate that
the hormonal responses of neonatal mammals include not only
glucocorticoid secretion, but also secretion of catecholamines,
growth hormone, glucagon, and insulin (Kuhn et al., 1978;
McCarty, 1979; Stubbe and Wolfe, 1971; Johnston and Bloom,
1973). These findings suggest that the pattern of hormonal
response might be a more significant factor in stress-induced
alterations of CNS development than secretion of a single hormone.

GROWTH HORMONE

The role of GH in regulating brain development has been a controversial issue. It was reported that administration of GH to
pregnant rats enhanced brain growth, as measured by DNA content, and behavioral development in the offspring (Zamenhof et
al., 1966; Sara et al., 1974; Clendinnen and Eayrs, 1961; Block
and Essman, 1965). However, placental transfer of this large
peptide hormone is probably minimal or nonexistent. On this
basis, the effects of this treatment have been attributed to
changes in maternal metabolism and the hypothesis that brain
development was affected directly by GH has been discounted.
The finding that hypophysectomy of rats during the first week
of life did not affect brain growth supported this conclusion
(Walker et al., 1952; Walker et al., 1950).

However, recent evidence suggests that GH might significantly
affect brain development. First, it was found that the day 6
hypophysectomy mentioned previously was incomplete and that
enough thyroid hormone remained to maintain GH secretion. Second, more careful clinical reports contradicted earlier reports
that the intellectural development of GH-deficient children was
normal. The most convincing of such evidence is the finding
that the IQ's of children who secrete biologically inactive GH
are significantly below those of children from a properly selected control group (Laron et al., 1971).

Finally, by using more sensitive measures of brain development, several investigators have found direct effects of GH on
brain. Administration of GH directly into the brain increases
the activity of ODC, the rate-limiting enzyme in the synthesis
of polyamines and a sensitive index of cell proliferation and
differentiation (Roger et al., 1974; Russell, 1973). In addition, administration to neonatal rats of agents that stimulate
or inhibit GH secretion increases or decreases brain ODC activity respectively (Kuhn et al., 1978). Direct stimulation of
glial proliferation by the GH-dependent peptide somatomedin has
been observed (Fryklund et al., 1974; Westermark and Wasteson,
1975). Finally, Pelton et al. (1977) found that inducing a
specific GH deficiency by injecting neonatal rats with antibody
to GH resulted in a 70 to 80 percent decrease in myelin lipids,
a 65 percent decrease in DNA content, and an abnormal accumulation of undifferentiated glial cells in the subependymal zone.
Although this experimental approach does not distinguish between direct effects of GH on brain and effects secondary to
peripheral metabolic effects, it does indicate that the presence of GH is necessary for normal brain development. It has
also been suggested that some of the effects of hypothyroidism
result from the consequent impaired GH synthesis (Geel and
Timiras, 1967). GH administration can partially reverse the

decreases in cell size and dendritic growth that result from hypothyroidism (Legrand, 1977).

In summary, there is evidence that the presence of GH modulates the biochemical and anatomical development of mammalian brain, but the extent of its effects and its sites of action are unknown. Although the effects of GH antibody administration suggest that the presence of GH is obligatory for normal brain development, it is not clear whether its action is mediated by effects directly on brain, by peripheral metabolic effects, or both.

The small amount of evidence available suggests that the time during which developing brain becomes sensitive to GH follows shortly after maximal GH secretion is established. For example, in rats, serum GH is highest during the last few days of gestation and declines steadily to a minimum at about the time of puberty (Blazquez et al., 1974; Walker et al., 1977; Strosser and Mialhe, 1975). Tissue responsiveness to GH appears right after birth (Hurley et al., 1980) and continues to increase during the first few weeks of life (Roger et al., 1974). The ability of GH to stimulate ODC activity disappears when the animals reach adulthood, at which time GH secretion is extremely low during most of the day, with the exception of short "surges" of secretion every four to five hours (Tannenbaum and Martin, 1976).

Less is known about the "critical period" of GH sensitivity in human beings, although it has been suggested that GH is effective during the later stages of gestation (Laron et al., 1971). As in rats, sensitivity appears in human beings soon after the establishment of a high rate of GH secretion during the fifth month (Grumbach and Kaplan, 1973), and sensitivity disappears when adult, low levels of secretion are reached. Generally, in the few mammalian species that have been studied, GH secretion begins before tissue sensitivity to GH develops, secretion is high during the period of maximal sensitivity and both sensitivity and secretion decline as the animal reaches adulthood. However, at some point, brain tissue loses its ability to respond to GH, a distinct difference from the sensitivities to T_4 and glucocorticoids, which remain throughout adulthood, albeit in altered forms.

GONADAL STEROIDS

Effects on Maturation of Sex Behavior

It is well-known that development of male or female patterns of gonadotropic hormone (GTH) secretion and sexual behavior is determined during an early stage of development, although the appearance of these characteristics occurs during puberty and childhood. Manipulation of serum levels of gonadal steroids during the early stage of development, when male and female patterns are first differentiating, produces permanent changes in these functions. This phenomenon has been studied most completely in rats. In this species, two functions that show distinct sexual dimorphism have been studied exhaustively, GTH secretion and mating behavior. In females, GTH secretion is cyclic, and a predominant and easily quantified behavioral response during mating is a characteristic arching of the back, termed "lordosis." GTH secretion in males is constant, and the characteristic mating response is mounting behavior, which is followed by intromission and ejaculation during normal mating. Males will rarely demonstrate lordosis even if "primed" with estrogen and progesterone, and females will rarely mount other females unless treated for long periods of time with high doses of testosterone. Treatment of female rat pups with testosterone at birth produces females with constant GTH secretion. These animals do not respond to males with the appropriate "lordosis" response (Gorski, 1968, 1971a; Gorski and Barraclough, 1963) and do not demonstrate the other characteristics of female sex behavior in this species. In addition, mounting behavior can be elicited by administration of testosterone much more easily in neonatally androgenized females than normal females (Mullins and Levine, 1968; Hendricks and Gerall, 1970). Conversely, if male rat pups are castrated at birth, they have cyclic GTH release as adults and behave sexually like females. Lordosis can be triggered in these animals by the appropriate hormone treatment, whereas mounting behavior is suppressed (Gorski, 1979). Furthermore, cyclic ovulation occurs in ovarian tissue transplanted into these animals. Similar effects are observed in most other mammalian species, although some species variability in the extent of the effect is observed. For example, this full syndrome is observed in rats, hamsters, mice, and guinea pigs, whereas "androgenization" of female monkeys or human beings prevents development of normal female sexual behavior but not the development of cyclic GTH secretion (Gorski, 1971a; Goy and Pheonix, 1971; Resko, 1977). Neonatal hormonal manipulation also affects a number of "non-sex" related behaviors in which sexual dimorphism is usually observed. For example, female rats usually drink more bitter-flavored substances than males and ambulate more in an open field. Both of these effects are abolished by treatment of neonatal females with testosterone (Denti and Negroni, 1975; Gorski, 1971b; Pfaff and Zigmond, 1971). Similarly, sex differences in maze learning and aggressive behavior are affected by neonatal hormone treatments. Females treated with testosterone during the first postnatal week make fewer errors in learning a maze than normal females (Joseph et al., 1978). In addition, induction of aggressive behavior by administration of testosterone (a "male" response) occurs as fast in females treated with testosterone pre- or postnatally as it does in males (vom Saal, 1979). There have also been reports of disruption of sex differences in food intake, territorial marking, open-field behavior and learning (Pfaff and Zigmond, 1971; Quadagno and Ho, 1972; Swanson, 1967; Denti and Negroni, 1975; Dawson et al., 1975; Goy and Resko, 1972; Beatty and Beatty, 1970; Reinisch, 1976).

These findings all suggest that the CNS mechanisms that regulate GTH secretion, as well as those involved in regulation of sex behavior and other sexually dimorphic behaviors, require the appropriate hormonal environment during differentiation for the correct expression of the genetic program. The pronounced and extremely specific abnormalities produced by neonatal hormone manipulations suggest that gonadal steroids are among the most important hormones determining CNS development. The critical hormonal determinant of the normal development of male and female GTH secretion and sex behavior seems to be the presence of androgen during the critical period. Whereas gonadectomy of males produces a female type of behavior, gonadectomy of females has no marked effects on the development of normal female sex behavior. Administration of exogenous hormone to female, but not male, rats produces a male pattern of GTH secretion and sex behavior (Feder and Whalen, 1965; Feder et al., 1966; Pfaff and Zigmond, 1971; Gorski, 1971a). It has been suggested that the immature brain is basically cyclic or "female," and that the presence of androgen is necessary to trigger differentiation of normal "male" type development (Fig. 16-7). However, there is considerable species variability in the importance of this sexual dimorphism. In many species, learned sexual behavior and other environmental factors play as strong a role as anatomical organization under some circumstances. For example, in human beings, the sex that a child is raised as (and therefore taught to behave as) is a more important determinant of that person's adult sexual orientation than prenatal exposure to androgenizing hormones (Ehrhardt and Meyer-Bahlburg, 1979).

The abnormal maturation of CNS control of GTH secretion and sexual behavior caused by androgen or estrogen occurs only if certain dose and time requirements are met. The hormone must be administered during the short critical period for differentiation of these mechanisms in the CNS, which is at about the same time during gestation in different species. In rats, this period is during the first five days after birth, whereas in those species that are more developed at birth, such as monkeys and guinea pigs, it occurs prenatally (Gorski, 1971a; Resko, 1977). The timing is slightly different for different func-

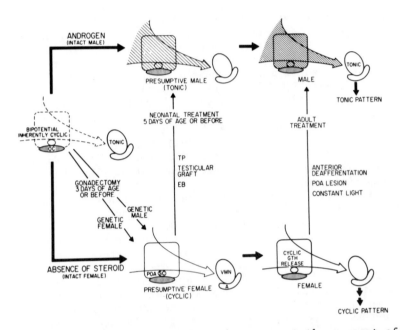

Fig. 16-7.   Summary of the data that support the concept of
the sexual differentiation of neural control of
GTH secretion in the rat. A: arcuate nucleus;
EB: estradiol benzoate; OC: optic chiasm, POA:
preoptic area; SC: suprachiasmatic nucleus; TP:
testosterone propionate; VMN: ventromedial
nucleus. (From Gorski, 1971b).

tions. In rats, absence of cyclic GTH secretion is best produced by administration of testosterone on day 1 or 2. In contrast, disruption of normal female sex behavior is produced
more effectively on days 4 to 6 (Gorski, 1968).

Different doses of androgen produce different degrees of abnormality in GTH secretion. Whereas large doses of testosterone produce the complete behavioral syndrome already described,
smaller doses produce more limited changes. For example, ovulation can still be induced by electrical stimulation of the
hypothalamus in female rats treated with small doses of testosterone, even if the animals do not ovulate spontaneously
(Gorski and Barraclough, 1963). Cyclic GTH secretion seems to
be more sensitive to disruption by testosterone than other parameters in rats, since doses of testosterone that disrupt GTH
secretion do not affect the lordosis response (Gorski, 1971b).
However, the reverse is true in primates; prenatal treatment

with small doses of testosterone "masculinizes" the behavior of monkeys but does not affect GTH secretion (Resko, 1977).

Although the dose and time requirements for "androgenization" of CNS control of GTH secretion and sexual behavior are narrow, the hormonal specificity is quite broad. In fact, either testosterone or estradiol produce these effects when given to female rat pups (Gorski, 1963; Gorski and Barraclough, 1963; Gorski and Wagner, 1965; Doughty et al., 1975). However, derivatives of testosterone, including dihydrotestosterone, are ineffective (Goldfoot and van der Werff ten Bosch, 1975; Whalen and Rezek, 1974; Whalen and Luttge, 1971; Sodersten and Hansen, 1978; Gorski, 1971a; Gorski et al., 1977; Gorski, 1963; Gorski and Wagner, 1965; Gorski and Barraclough, 1963). The changes in GTH secretion and sex behavior that result from manipulation of endocrine status during the critical developmental period represent a developmental effect that is distinct from the effects of such manipulations in adult animals. There are many neuronal and behavioral events that are affected by the presence of androgen and estrogen, including GTH secretion and sex behavior. However, there is a critical distinction between these effects and the effects of endocrine manipulations during the perinatal period. The "sensitizing" effects of hormones in adults are transitory and reflect hormone effects on ongoing neural events. However, the "organizing" effects of gonadal steroids on developing brain are permanent effects on behavior that remain long after the hormone has disappeared from the brain. The presence or absence of the gonads after the critical period does not affect the outcome of these treatments (Gorski, 1971b, 1979). These developmental effects result from changes in the biochemical and anatomical organization of the brain rather than from transitory effects on previously established processes.

Morphologic Effects

These changes in maturation of various aspects of reproductive function caused by manipulation of serum hormone levels in developing animals are thought to result from abnormalities in the development of specific sites in the CNS. The pituitary retains its normal responsivity following neonatal androgen or estrogen treatment. Several findings suggest that disruption of cyclic GTH secretion results from abnormal development in the preoptic area of the hypothalamus, the area thought to be responsible for maintaining cyclic GTH secretion in normal animals. There is no impairment of the pituitary's ability to secrete GTH when the median eminence is stimulated, but stimulation of the preoptic area does not evoke ovulation in androgenized females (Barraclough and Gorski, 1961). Furthermore, implantation of testosterone into this site in neonatal rats causes an ovulatory sterility identical to that caused by peripheral administration of the hormone, whereas implantation at other sites does not (Wagner et al., 1966). However, the effects of neonatal androgen administration are not restricted to the preoptic area and changes in this area alone cannot account for all the effects of this treatment. It has been reported that electrical stimulation near this area can evoke luteinizing hormone (LH) secretion (Kubo et al., 1975; Terasawa et al., 1969) even though ovulation does not result. Changes in mounting behavior resulting from neonatal hormonal manipulation appear to involve changes in other brain areas (McEwen, 1978). Different CNS sites are involved in alterations in different CNS reproductive functions. Similarly, the decreased responsivity of "androgenized" female rats to males is thought to result in part from decreased responsivity to the "priming" effects of progesterone, an effect mediated by the midbrain reticular formation (Clemens, 1974; Clemens et al., 1969). Unfortunately, only a few such places that are affected by neonatal "androgenization" have been identified. Although a large number of sites in the hypothalamus and limbic system are thought to contribute to control of GTH secretion and regulation of reproductive behavior, the effect of neonatal androgenization on these areas has not been described.

It was originally suggested that the effects of perinatal androgen or estrogen administration resulted from changes in sensitivity of certain neurons to hormonal feedback. This hypothesis was proposed because many of the behaviors studied, such as lordosis, are elicited only by "priming" the animal with exogenous steroids, and it was concluded that the animals were insensitive to hormonal priming. However, more recent studies suggest that alterations in the anatomical organization of the hypothalamus are caused by these perinatal endocrine manipulations and it is more likely that the drastic changes in behavior caused by this treatment are the result of this sort of anatomical change rather than from subtle alterations in neuronal sensitivity to hormonal feedback. For example, marked sexual dimorphism in the anatomical organization of the preoptic area in rats has been reported, and perinatal androgen administration produces changes in this anatomical pattern that parallel the effects of this treatment on GTH secretion (Raisman and Field, 1973; Torand-Allerand, 1976). Normally, the density of synapses on dendritic spines and the number of synapses that derive from fibers outside the amygdala are greater in females than in males. Treatment of females with androgen during the "critical period" prevents the development of the greater synaptic density in the preoptic area that is characteristic of females. Although synaptic density of the preoptic area is greater in females, the over-all volume of the medial preoptic area is greater in males. This sexual dimorphism in structure also is affected by hormone manipulation during the critical period for sexual differentiation. Neonatal castration of males decreases the volume of this area and androgenization of females increases it (Gorski, 1979). The findings reported previously indicating that the preoptic nucleus is a primary site for determination of patterns of GTH secretion are consistent with these findings. In addition, they suggest that the structural abnormalities that result from neonatal testosterone administration contribute to the observed behavioral changes. Recently, androgen-induced changes in the anatomical organization of the arcuate nucleus also have been reported. Androgen administration to females increases synaptic density in the arcuate (Arai and Matsumoto, 1978). This finding is interesting because it is thought that the preoptic area triggers cyclic GTH secretion by modulating neural activity in the arcuate nucleus, the site thought to maintain tonic GTH secretion (Gorski, 1971a). It has been suggested that the increased number of synaptic contacts observed in males is caused by ingrowth of fibers from extrahypothalamic brain regions, and that this ingrowth prevents normal synaptogenesis of the projections from the preoptic nucleus (Arai and Matsumoto, 1978). This hypothesis is substantiated by electrophysiological evidence of premature maturation of synaptic contacts in the arcuate nucleus of neonatally "androgenized" females (Curry and Timiras, 1972). The anatomical changes observed in this area suggest that areas other than the preoptic region are involved in the altered GTH secretion caused by neonatal androgen administrations.

Estrogen Receptors

A recent series of studies has provided considerable insight into the biochemical mechanisms responsible for the development of "male" type GTH secretion in either genetic males or females that are exposed to testosterone during development. It is well-known that there are many sites in adult brain that have receptors for gonadal steroid hormones (Plapinger and McEwen, 1978; McEwen, 1978; MacLusky et al., 1979; see Luttge, this volume). Brain tissue from immature mammals also has estrogen receptors (Plapinger and McEwen, 1973; Plapinger et al., 1973; McEwen et al., 1975). In rats, it has been shown that these receptors appear just before the "critical period" for sexual differentiation and increase gradually to adult levels during the first four weeks of life (MacLusky et al., 1979). These receptors are most concentrated in the areas involved in sexual differentiation of GTH secretion and reproductive behavior, the preoptic nucleus, medial basal hypothalamus, and parts of the amygdala (Pfaff and Keiner, 1973; Westley and Salaman, 1977). It is now thought likely that differentiation of certain aspects of sex behavior are mediated by the interaction of estrogen with these receptors.

The estrogen receptors in neonatal rat brain and the intracellular events that follow binding seem to be similar in several characteristics, including sedimentation behavior, affinity constant, hormonal specificity, and subcellular localization to the estrogen receptors in other tissues (McEwen, 1978). Estrogen is thought to bind to a cytosolic receptor that transfers it to the nucleus, where it combines with some as yet unidentified site to elicit changes in gene expression. It appears that in brain, as in other tissues, estrogen interaction with the nucleus results in changes in gene expression that involve synthesis of RNA (Kobayashi and Gorski, 1970). However, aside from the intriguing but unexplained observation that neural activity is also required for the action of estrogen or testosterone (Gorski, 1971a), little else is known about the mechanism by which estrogen acts.

One of the most interesting aspects of the role of these estrogen receptors in CNS maturation is that they are involved in the development of male-type GTH secretion and behavior. Before the presence of these receptors in neonatal brain had been established, it was well known that administration of either estrogen or testosterone produced "masculinizing" effects (see review by Plapinger and McEwen, 1978). It is now thought that testosterone and other androgens probably exert their effects only after conversion to estradiol, which is the hormone that acts on estrogen receptors in neonatal brain. Several findings support this hypothesis. Only those androgens that can be aromatized effect changes in CNS development (Brown-Grant et al., 1971; McDonald and Doughty, 1974). Furthermore, conversion of testosterone to estradiol has been demonstrated in brain tissue of several mammalian species (Naftolin and Brawer, 1978; Weisz and Gibbs, 1974; Lieberburg and McEwen, 1975). There is even some preliminary evidence that directly implicates aromatization with masculinizing effects. Treatment of rats with SKF 525A, a drug that is thought to decrease conversion of testosterone to estradiol, partially blocks the masculinizing effects of this hormone (Clemens, 1974). Finally, it has been shown that treatment of neonatal rats with an estrogen antagonist prevents the development of "male" type GTH secretion and sex behavior (MacLusky et al., 1979; Sodersten, 1978).

Normal females develop both cyclic GTH secretion and female reproductive behavior despite the presence of estrogen because of the protective effect of a steroid-binding protein in blood during the "critical period" which is extremely specific for estradiol (McEwen et al., 1975; Raynaud et al., 1971). This protein allows entrance of testosterone into the CNS but prevents the entrance of estradiol. Therefore testosterone and nonsteroid estrogen derivatives easily enter the CNS and exert their masculinizing effects. In contrast, compounds with a high affinity for the binding protein enter only if present in high concentrations (McEwen, 1978: Doughty et al., 1975).

To summarize the current hypothesis about the mechanisms mediating sexual differentiation of the brain, in normal males, testosterone secreted by the testes enters the CNS where it is converted to estradiol, which is then bound to its receptors and transported to the nucleus. Here it alters gene expression to produce the anatomical and biochemical substrates necessary to maintain tonic GTH secretion and other characteristics of male reproductive function. In contrast, the female brain develops the anatomical and biochemical mechanisms that maintain cyclic GTH secretion because of the absence of hormone during the neonatal period. For this reason, it has been suggested that the undifferentiated hypothalamus is basically female in pattern and that the presence of estradiol formed from testosterone triggers a change or changes in gene expression that cause development of "male" morphology and function.

It should be mentioned that both estradiol and testosterone have effects on adult brain, particularly regulation of neural activity in the hypothalamus and facilitation of reproductive behaviors regulated by these areas. A large amount of evidence suggests that these effects, like the developmental effects of these hormones, are mediated by alterations in gene expression (McEwen, 1978). Furthermore, the estrogen receptors in fetal and adult tissue seem to be similar in many respects (Barley et al., 1974; Fox, 1975; MacLusky et al., 1976). However, there is a vast difference between the reversible temporary effects of these hormones in adult brain and the permanent structural and functional changes that result from interaction of these hormones with neonatal brain tissue. It can be postulated that interaction of estradiol with adult brain merely alters the rate of production of gene products that are already being expressed and so alters neural activity by affecting neurotransmitter receptors, cell metabolism, or other ongoing functions. In contrast, interaction with developing brain can cause differential expression of gene products, leading to formation of new structures and functions.

OTHER HORMONES

The findings reported in the previous sections of this chapter represent the small part of the hormonal control of brain development that is understood. Within the last 10 years, a large number of new hormones have been identified, and the role of these compounds in function of adult or developing brain has yet to be elucidated. However, anecdotal reports suggest that many additional hormones are involved in regulation of brain development. Even maternal hormones may exert effects on brain development through their effects on fetal or maternal physiology. For example, it is thought that the mental development of children of diabetic mothers is subnormal, and the effects of maternal diabetes on CNS development are related to the extent of maternal acetonuria (Chruchill et al., 1969; Haworth et al., 1973). However, the report that insulin increases ODC activity in immature rat brain (Roger et al., 1974) suggests that insulin itself might also play a role in brain development. Defects in brain growth have also been reported in animals with vasopressin deficiency (Boer et al., 1978) and both the pituitary peptide, ACTH, and the mineralocorticoid, aldosterone, have been reported to exert effects on brain protein and nucleic acid synthesis that are distinct from the well-known actions of the glucocorticoids (Jakoubek, 1978; Ardeleanu and Sterescu, 1978).

There are a number of peptide hormones present in the brains of immature mammals, such as MSH, TRH, GnRH, enkephalins, and endorphins that have been shown to exert metabolic and behaviorial effects on adult brain, and recent studies suggest that these compounds may function in developing brain both as neuromodulators and as trophic agents. For example, the endogenous opiate peptides β-endorphin and leu- and met-enkephalin appear early in gestation in many mammalian species, including humans (Begeot et al., 1978; Bayond et al., 1979; Silman et al., 1979), and these substances may influence brain development both directly and indirectly. Direct effects of an enkephalin derivative on DNA synthesis in developing rat brain has been demonstrated (Vertes et al., 1982). In addition, β-endorphin is capable of regulating secretion of growth hormone, prolactin and possibly other growth-regulating hormones in the mammalian fetus (Gluckman et al., 1980). Similarly, a trophic as well as neurotransmitter role for serotonin has been proposed. Serotonin depletion has been shown to delay cell division in developing rat brain, and it has been proposed that serotonin regulates the time of cell formation of cells that eventually receive a serotonergic imput (Lauder and Krebs, 1978).

ACKNOWLEDGMENTS

Dr. Schanberg is supported by Grant No. MH-13688 from the National Institute of Mental Health. Dr. Kuhn is supported by Grant No. DA-02739 from the National Institute of Drug Abuse.

REFERENCES

Ader, R.: Effects of early experience on emotional and physiological reactivity in the rat. *J Comp Physiol Psychol 66:* 267-268, 1968.

Ader, R.: Early experiences accelerate maturation of the 24 hour adrenocortical rhythm. *Science 163:* 1225-1226, 1969.

Ader, R.: The effects of early experience on the adrenocortical response to different magnitudes of stimulation. *Physiol Behav 5:* 837-839, 1970.

Ader, R.: Early experience and hormones: Emotional behavior and adrenocortical function. In *Hormonal Correlates of Behavior,* B.E. Eleftherious and R.L. Sprout, eds. Plenum Press, New York, 1975, pp 6-33.

Ader, R., and Belfer, M.L.: Prenatal maternal anxiety and offspring emotionality in the rat. *Phychol Rep 10:* 711-718, 1962.

Ader, R., and Grota, L.J.: The effects of early experience on adrenocortical reactivity. *Physiol Behav 4:* 303-305, 1969.

Allen, C., and Kendall, J.W.: Maturation of the circadian rhythm of plasma corticosterone in the rat. *Endocrinology 80:* 926-930, 1967.

Anderson, T.R., and Schanberg, S.M.: Effects of thyroxine and cortisol on brain ornithine decarboxylase activity and swimming behavior in developing rat. *Biochem Pharmacol 24:* 495-501, 1975.

Arai, Y., and Matsumoto, A.: Synapse formation of the hypothalamic arcuate nucleus during postnatal development in the female rat and its modification by neonatal estrogen treatment. *Psychoneuroendocrinology 3:* 31-45, 1978.

Archer, J.E., and Blackman, D.E.: Prenatal psychological stress and offspring behavior in rats and mice. *Dev Psychobiol 4:* 193-248, 1971.

Ardeleanu, A., and Sterescu, N.: Hormonal influences on RNA and DNA synthesis in developing rat brain. *Rev Boum Morphol Embryol Physiol Physiologie 3:* 133-140, 1977.

Ardeleanu, A., and Sterescu, N.: RNA and DNA synthesis in developing rat brain; Hormonal influences. *Phychoneuroendocrinology 3:* 93-101, 1978.

Balazs, R.: Effects of thyroid hormone and undernutrition on cell acquistion in the rat brain. In *Thyroid Hormones and Brain Development,* G.D. Grave, ed. Raven Press, New York,

1977, pp 287-302.

Balazs, R., and Motterrell, M.: Effects of hormonal state on cell number and functional maturation of the brain. *Nature* 236:348-350, 1972.

Balazs, R., and Reichter, D.: Effects of hormones on the biochemical maturation of the brain. In *Biochemistry of the Developing Brain*, W. Himwich, ed. Marcel Dekker, New York, 1973, pp 253-299.

Balazs, R., Ko̊vacs, S., Teichgraber, P., Cocks, W.A., and Eayrs, J.T.: Biochemical effects of thyroid deficiency on the developing brain. *J Neurochem* 15:1335-1349, 1968

Balazs, R., Cocks, W.A., Eayrs, J.T. and Ko̊vacs, S.: Effects of thyroid hormones on the developing brain. In *Hormones in Development*, M. Hamburgh and E.J.W. Barrington, eds. Appleton-Century-Crofts, New York, 1971, pp 357-379.

Barley, J., Ginsburg, M., Greeinstei, B.D., MacLusky, N.J., and Thomas, P.J.: A receptor mediating sexual differentiation? *Nature* 252:259-260, 1974.

Barraclough, C.A., and Gorski, R.A.: Evidence that the hypothalamus is responsible for androgen induced sterility in the female rat. *Endocrinology* 68: 68-79, 1961.

Baxter, J.D., and Ivarie, R.D.: Regulation of gene expression by glucocorticoid hormones: Studies of receptors and responses in cultured cells. In *Receptors and Hormone Action*, Vol 2, B. O'Malley and L. Birnbaumer, eds. Academic Press, New York, 1978, pp 252-297.

Bayon, A., Shoemaker, W.J., Bloom, F.E., Mauss, A., and Guillemin, R.: Perinatal development of the endorphin and enkephalin-containing systems in the rat brain. *Brain Res* 179:93-101, 1979.

Beatty, W.W., and Beatty, P.A.: Hormonal determinants of sex differences in avoidance behavior and reactivity to electric shock in the rat. *J Comp Physiol Psychol* 73:446-455, 1970.

Begeot, M., Dubois, M.P., and Dubois, P.M.: Immunologic localization of α- and β-endorphins and β-lipotropin in corticotropic cells of the normal and anencephalic fetal pituitaries. *Cell Tissue Res* 193:413-422. 1978.

Blazquez, E., Simon, F.A., Blazquex, M., and Foa, P.P.: Changes in serum growth hormone levels from fetal to adult age in the rat. *Proc Soc Exp Biol Med* 147:780-783, 1974.

Bloch, E.: Fetal adrenal cortex: Function and steroidogenesis. In *Functions of the Adrenal Cortex*, K.W. McKerns, ed. Appleton-Century-Crofts, New York, 1968.

Block, J.B., and Essman, W.B.: Growth hormone administration during pregnancy: A behavioral difference in offspring rats. *Nature* 205:1136-1137, 1965.

Boer, G.J., Vylings, H.B.M., van Phienen-Verling, C.M.F., and Fisser, G.: Postnatal brain development in rats with

808 Kuhn and Schanberg

hereditary diabetes insipidus (Brattleboro strain). In
Hormones and Brain Development, G. Dorner and M. Kawakami,
eds. Elsevier Holland Press, New York, 1978, pp 253-258.

Brown, A.W., Bronstein, I.P., and Kraenes, R.: Hypothyroidism
and cretinism in childhood VI. influence of thyroid therapy on
mental growth. *Am J Dis Child 57:*517-523, 1939.

Brown-Grant, K., Munck, A., Naftolin, F., and Sherwood, M.R.C.:
The effects of the administration of testosterone and related
steroids to female rats during the neonatal period. *Horm
Behav 2:*173-182, 1971.

Butte, J.C., Kakihana, R., Farnham, M.L., and Noble, E.P.: The
relationship between brain and plasma corticosterone stress
response in developing rats. *Endocrinology 92:*1775-1779,
1973.

Casper, R., Vernadakis, A., and Timiras, P.S.: Influence of estradiol and cortisol on lipids and cerebrosides in the developing brain and spinal cord of the rat. *Brain Res 5:*524-526,
1967.

Chapman, R.H., and Stern, J.M.: Failure of severe maternal
stress or ACTH during pregnancy to affect emotionality of
male rat offspring. Implications of litter effects for prenatal studies. *Dev Psychobiol 12:*255-267, 1979.

Chatelain, A., Dubois, M.P., and Dupuoy, J.P.: Hypothalamus and
cytodifferentiation of the foetal pituitary gland. *Cell Tiss
Res 169:*335-344, 1976.

Churchill, J.A., Berendes, H.W., and Nemore, J.: Neuropsychological deficits in children of diabetic mothers.
*Am J Obstet Gynecol 105:*257-268, 1969.

Clark, B.R., and Weichsel, M.E., Jr.: Correlation of DNA accumulation rate with thymidylate synthetase and thymidine
kinase activities in developing rat cerebellum: Effect of
thyroxine. *J Neurochem 29:*91-100, 1977.

Clemens, L.G.: Neurohormonal control of male sexual behavior.
In *Reproductive Behavior,* W. Montagna and W.A. Sadler, eds.
Plenum Press, New York, 1974, pp 24-53.

Clemens, L.G., Hiroi, M., and Gorski, R.A.: Induction and
facilitation of female mating behavior in rats treated neonatally with low doses of testosterone proprionate.
*Endocrinology 84:*1430-1438, 1969.

Clendinnen, B.G., and Eayrs, J.T.: The anatomical and physiological effects of prenatally administered somatotrophin and
cerebral development. *J Endocrinol 22:*183-193, 1961.

Clos, J., Crepel, F., Legrand, C., Legrand, J., Rabie, A., and
Vigouroux, E.: Thyroid physiology during the postnatal period
in the rat: A study of the development of thyroid function
and of the morphogenetic effects of thyroxine with special
reference to cerebellar maturation. *Gen Comp Endocrinol 23:*
178-192, 1974.

Cocks, J.A., Balazs, R., and Eayrs, J.T.: The effect of thyroid hormones on the biochemical maturation of the rat brain. *Biochem J 111*:18, 1968.

Cocks, J.A., Balazs, R., Johnson, A.L., and Eayrs, J.T.: Effect of thyroid hormone on the biochemical maturation of rat brain: Conversion of glucose carbon into amino acids. *J Neurochem 17*:1275-1285, 1970.

Cost, M.G., and Mann, D.R.: Neonatal corticoid administration: Retardation of adrenal rhythmicity and desynchronization of puberty. *Life Sci 19*:1929-1936, 1976.

Cotterrell, M., Balazs, R., and Johnson, A.L.: Effects of corticosteroids on the biochemical maturation of rat brain: Postnatal cell formation. *J Neurochem 19*:2151-2167, 1972.

Crepel, F.: Excitatory and inhibitory processes acting upon cerebellar Purkinje cells during maturation in the rat: Influence of hypothyroidism. *Exp Brain Res 20*:403-420, 1974.

Curry, J.J., and Timiras, P.S.: Development of evoked potentials in specific brain systems after neonatal administration of estradiol. *Exp Neurol 34*:129-139, 1972.

Davenport, J.W., and Gonzalez, L.M.: Neonatal thyroxine stimulation in rats: Accelerated behavioral maturation and subsequent learning deficit. *J Comp Physiol Psychol 85*: 395-408, 1973.

Dawson, J.L., Cheung, Y.M., and Lau, R.T.S.: Developmental effects of neonatal sex hormones on spatial and activity skills in the white rat. *Biol Psychiatry 3*:213-229, 1975.

Denenberg, V.H., and Zarrow, M.X.: Effects of handling in infancy upon adult behavior and adrenocortical reactivity: Suggestions for a neuroendocrine mechanism. In *Early Childhood: The Development of Self-Regulating Mechanisms,* D.M. Walchiv and D.L. Peters, eds. Academic Press, New York, 1971, pp 39-64.

Denti, A., and Negroni, J.A.: Activity and learning in neonatally hormone treated rats. *Acta Physiol Lat Am 25*: 99-106, 1975.

Doughty, C., Booth, J.E., McDonald, P.G., and Parrott, R.F.: Effects of oestradiol-$\beta17$ benzoate and the synthetic oestrogen R 2858 on sexual differentiation in the neonatal female rat. *J Endocrinol 67*:419-424, 1975.

Doyle, G., and Yule, E.P.: Early experience and emotionality. I. The effects of maternal anxiety on the emotionality of albino rats. *S Afr J Soc Res 10*:57-65, 1959.

Dratman, M.B., Crutchfield, F.L., Axelrod, J., Colburn, R.W., and Nguyen, T.: Localization of triiodothyronine in nerve ending fractions of rat brain. *Proc Natl Acad Sci USA 73*: 941-944, 1976.

Eayrs, J.T.: Thyroid and central nervous development. In *Scientific Basis of Clinical Medicine Annual Reviews,* 1966, pp 317-339.

Eguchi, Y., and Wells, L.J.: Response of the hypothalamo-
hypophyseal axis to adrenal axis to stress: Observations in
fetal and caesarean newborn rats. *Proc Soc Expl Biol Med 120:*
675-678, 1965.

Eguchi, Y., Eguchi, K., and Wells, L.J. Compensatory hypertrophy of right adrenal after left adrenalectomy: Observations in fetal newborn and week old rats. *Proc Soc Expl Biol
Med 116:*89-92, 1964.

Ehrhardt, A., and Meyer-Bahlburg, H.F.L.: Prenatal sex hormones
and the developing brain: Effects on psychosexual differentiation and cognitive function. *Annu Rev Med 30:*417-430,
1979.

Emlen, W., Segal, D.S., and Mandell, A.J.: Thyroid state:
Effects on pre- and post-synaptic central noradrenergic
mechanisms. *Science 175:*79-82, 1972.

Engstrom, G., Svensson, T., and Waldeck, B.: Thyroxine and
brain catecholamines increased transmitter synthesis and increased receptor sensitivity. *Brain Res 77:*471-483, 1974.

Feder, H.H., and Whalen, R.E.: Feminine behavior in neonatally
castrated and estrogen treated male mice. *Science 147:*306-
307, 1965.

Feder, H.H., Phoenix, C.H., and Young, W.C.: Suppression of
feminine behavior by administration of testosterone propionate to neonatal rats. *J Endocrinol 34:*131-132, 1966.

Fiegelson, P., Ramanarayanan-Murphy, L., and Colmann, P.D.:
Studies on the cytoplasmic glucocorticoid receptor and its
nuclear interaction in mediating induction of tryptophan
oxygenase messenger RNA in liver and hepatoma. In *Receptors
and Hormone Action,* B. O'Malley and L. Birnbaumer, eds.
Academic Press, New York, 1978, pp 226-251.

Field, E.J.: Effect of cortisone on the neonatal rat. *Nature
174:*182, 1954.

Flynn, Y.J., Deshmukh, D.S., and Pieringer, R.A.: Effects of
altered thyroid function on galactosyl diacyglycerol metabolism in myelinating rat brain. *J Biol Chem 252:*5864-5870,
1977.

Ford, D.H., and Cramer, E.B.: Developing nervous system in relation to thyroid hormones. In *Thyroid Hormones and Brain
Development,* G.D. Grave, ed. Raven Press, New York, 1977,
pp 1-17.

Fox, T.O.: Oestradiol receptor of neonatal mouse brain.
*Nature 258:*441-444, 1975.

Freundl, K., and Van Wynsberghe, D.M.: The effects of thyroid
hormones on myelination in the developing rat brain.
*Biol Neonate 33:*217-223, 1978.

Fryklund, L., Othne, K., and Sievertisson, H.: Isolation and
characterization of polypeptides from human plasma enhancing

the growth of human normal glial cells in culture. *Biochem Biophys Res Commun* 61:950-956, 1974.

Geel, S.E., and Timiras, P.S.: The influence of neonatal hypothyroidism and of thyroxine on the ribonucleic acid and deoxyribonucleic acid concentrations of rat cerebral cortex. *Brain Res* 4:135-142, 1967.

Gluckman, P.D., Marti-Henneberg, C., Kaplan, S.L., Li, C.H. and Grumback, M.M.: Hormone ontogeny in the ovine fetus. X. The effects of β-endorphin and naloxone on circulating growth hormone, prolactin and chorionic somatomammotropin. *Endocrinology* 107: 76-80, 1980.

Goldfoot, D.A., and van der Werff ten Bosch: Mounting behavior of female guina pigs after prenatal and adult administration of the proprionates of testosterone, dihydrotestosterone and androstenediol. *Horm Behav* 6:139-148, 1975.

Gorski, R.A.: Modification of ovulatory mechanisms by postnatal administration of estrogen to the rat. *Am J Physiol* 205:842-844, 1963.

Gorski, R.A.: Influence of age on the response to perinatal administration of a low dose of androgen. *Endocrinology* 82: 1001-1004, 1968.

Gorski, R.A: Gonadal hormones and the perinatal development of neuroendocrine function. In *Frontiers in Neuroendocrinology*, L. Martini and W.F. Ganong, eds. Oxford University Press, New York, 1971a, pp 237-290.

Gorski, R.A.: Steroid hormones and brain function: Progress, principles, and problems. In *Steroid Hormones and Brain Function*, C.H. Sawyer and R.A. Gorski, ed. University of California Press, Berkeley, 1971b.

Gorski, R.A.: Nature of hormone action in the brain. In *Ontogeny of Receptors and Reproduction Hormone Action*, T.H. Hamilton, J.H. Clark, and W.A. Sadler, eds. Raven Press, New York, 1979, pp 37-392.

Gorski, R.A., and Barraclough, C.A.: Effects of low dosages of androgen on the differentiation of hypothalamic regulatory control of ovulation in the rat. *Endocrinology* 73:210-216, 1963.

Gorski, R.A., and Wagner, J.W.: Gonadal activity and sexual differentiation of the hypothalamus. *Endocrinology* 76:226-239, 1965.

Gorski, R.A., Harlan, R.E., and Chirstensen, L.W.: Perinatal hormonal exposure and the development of neuroendocrine regulatory processes. *J Toxicol Environ Health* 3:97-121, 1977.

Gourdon, J., Clos, J., Coste, C., Dainat, J., and Legrand, J.: Comparative effects of hypothyroidism, hyperthyroidism and undernutrition on the protein and nucleic acid contents of the cerebellum in the young rat. *J Neurochem* 21:861-871,

1973.

Goy, R.W., and Phoenix, C.H.: The effects of testosterone propionate administered before birth on the development of behavior in genetic female rhesus monkeys. In Steroid Hormones
and Brain Function, C.H. Sawyer and R.A. Gorski, eds.
University of California Press, Berkeley, 1971, pp 193-202.

Goy, R.W., and Resko, J.A.: Gonadal hormones and behavior of
normal and pseudohermaphroditic nonhuman female primtes.
Recent Progr Horm Res 28:707-733, 1972.

Granich, M., and Timiras, P.S.: Mechanisms of action of cortisol in maturation of brain lipid patterns in embryonal and
young chicks. In Hormones in Development, M. Hamburgh and
E.J.W. Barrington, eds. Appleton-Century-Crofts, New York,
1971, pp 213-218.

Grave, G.D.: Accelerated appearance of cerebral D(-)-β-
hydroxybutyric dehydrogenase in hyperthyroidism. In Thyroid
Hormones and Brain Development, G.D. Grave, ed. Raven Press,
New York, 1977, pp 303-314.

Gray, P.: Pituitary-adrenocortical response to stress in the
neonatal rat. Endocrinology 89:1126-1128, 1971.

Greenough, W.T.: Enduring brain effects of differential exper-
ience and training. In Neural Mechanism of Learning and
Memory, M.R. Rosenzweig and E.L. Bennett, eds. MIT Press,
Cambridge, Mass., 1977, pp 255-278.

Grumbach, M.M., and Kaplan, S.L.: Ontogenesis of growth hormone, insulin, prolactin and gonadotrophin secretion in the
human fetus. In Foetal and Neonatal Physiology, R.S. Comline,
K.W. Cross, G.D. Dawes, and P.W. Nathaniels, eds. 1973,
pp 462-487.

Gumbinas, M., Oda, M., and Huttenlocher, P.: Effects of corticosteroids on myelination in the developing brain.
Neurology 22:449-450, 1972.

Haltmeyer, G.C., Denenberg, V.H., and Zarrow, M.X.: Modification of the plasma corticosterone response as a function of
infantile stimulation and electric shock parameters.
Physiol Behav 2:61-63, 1967.

Hamburgh, M., and Flexner, L.B.: Effect of hypothyroidism and
hormone therapy on enzyme activities of the developing cerebral cortex of the rat. J Neurochem 1:279-288, 1957.

Hamburgh, M., Lynn, E., and Weiss, E.P.: Analysis of the influence of thyroid hormone on prenatal and postnatal maturation of the rat. Anat Rec 150:147-162, 1964.

Hamburgh, M., Mendoza, L.A., Bennett, I., Krupa, P., Kim, Y.S.,
Kahn, R., Hogreff, K., and Frankfort, H.: Some unresolved
questions of brain-thyroid relationships. In Thyroid Hormones
and Brain Development, G.D. Grave, ed. Raven Press, New York,
1977, pp 49-72.

Haworth, J.C., McRae, K.N., and Dilling, L.A.: Prognosis of in-

fants of diabetic mothers: Effect of epinephrine therapy.
J Pediatr 82:94-97, 1973.

Hendricks, S.E., and Gerall, A.A.: Effect of neonatally administered estrogen on development of male and female rats.
Endocrinology 87:435-439, 1970.

Hiroshige, T., and Sato, T.: Changes in hypothalamic content of
corticotropin-releasing activity following stress during neonatal maturation in the rat. *Neuroendocrinology 7*:257-270,
1971.

Hockman, C.H.: Prenatal maternal stress in the rat: Its effect
on emotional behavior in the offspring. *J Comp Physiol
Psychol 54*:679-684, 1961.

Hodge, G.K., Butcher, L.L., and Geller, E.: Hormonal effects on
the morphological differentiation of layer VI cortical cells
in the rat. *Brain Res 104*:137-141, 1976.

Horowitz, A.J., and Schanberg, S.M.: Hormonal effects on the
development of rat brain gangliosides-I. Cortisol. *Biochem
Pharmacol 28*:881-895, 1979a.

Horowitz, A.J., and Schanberg, S.M.: Hormonal effects on the
development of rat brain gangliosides-II. Thyroxine. *Biochem
Pharmacol 28*:897-903, 1979b.

Howard, E.: Effects of corticosterone and food restriction on
growth and on DNA, RNA and cholesterol contents of the brain
and liver in infant mice. *J Neurochem 12*:181-191, 1965.

Howard, E.: Reductions in size and total DNA of cerebrum and
cerebellum in adult mice after corticosterone treatment in
infancy. *Exp Neurol 22*:191-208, 1968.

Howard, E.: Increased reactivity and impaired adaptability in
operant behavior of adult mice given corticosterone in infancy. *J Comp Physiol Psychol 85*:211-220, 1973.

Howard, E., and Benjamin, J.A.: DNA ganglioside and sulfatide
in brains of rats given corticosterone in infancy, with an
estimate of cell loss during development. *Brain Res 92*:
73-87, 1975.

Hurley, J.T., Kuhn, C., Handwerger, S., and Schanberg, S.:
Differential effects of placental lactogen, growth hormone
and prolactin in the perinatal period. *Life Sci 27*:2269-2275,
1980.

Huttenlocher, P.R., and Amemiya, I.M.: Effects of adrenocortical steroids and of adrenocorticotrophic hormone on
$(NA^+ -K^+)$-ATPase in immature cerebral cortex. *Pediatr Res 12*:
104-107, 1978.

Jacquot, R.: Some hormonally controlled events of liver differentiation in the perinatal period. In *Hormones in Development,* M. Hamburgh and E.J.W. Barrington, eds. Appleton-
Century-Crofts, New York, 1971, pp 587-599.

Jakoubek, B.: The effect of ACTH and/or tranquilizers on the
development of brain macromolecular metabolism. In *Hormones*

and Brain Development, G. Dorner and M. Kawakami, eds. Elsevier Holland Press, New York, 1978, pp 259-264.

Joffe, J.M., Milkovic, K., and Levine, S.: Effects of changes in maternal pituitary-adrenal function on behavior of rat offspring. *Physiol Behav 8:*425-430, 1972.

Johnston, D.I., and Bloom, S.R.: Plasma glucagon levels in the term human infant and effect of hypoxia. *Arch Dis Child 48:* 451-454, 1973.

Joseph, R., Hess, S., and Birecree, E.: Effects of hormone manipulations and exploration on sex differences in maze learning. *Behav Biol 24:*364-377, 1978.

Jost, A.: Anterior pituitary function in foetal life. In *The Pituitary Gland,* Vol 2, G.W. Harris and B.T. Donovan, eds. 1966, pp 299-326.

Jost, A.: Hormonal control of fetal development and metabolism. *Adv Metab Disord 4:*123-184, 1970.

Kobayashi, F., and Gorski, R.A.: Effects of antibiotics on androgenization of the neonatal female rat. *Endocrinology 86:* 285-289, 1970.

Koranyi, L., and Endroczi, E.: Influence of pituitary-adrenocortical hormones on thalamocortical and brain stem limbic circuits. *Progr Brain Res 32:*120-130, 1969.

Krawiec, L., Garcia, A., Gomez, C.A., and Pasquini, J.M.: Hormonal regulation of brain development. III: Effects of triiodothyronine and growth hormone on the biochemical changes in the cerebral cortex and cerebellum of the rat. *Brain Res 6:*621-634, 1969.

Krawiec, L., Montalbano, C.A., Duvilanski, B.H., deGuglilmone, A.E.R., and Gomez, C.J.: Influence of neonatal hypothyroidism on brain RNA synthesis. In *Thyroid Hormones and Brain Development,* G.D. Grave, ed. Raven Press, New York, 1977, pp 315-326.

Kubo, K., Mennin, S.O., and Gorski, R.A.: Similarity of plasma LH release in androgenized and normal rats following electrochemical stimulation of the basal forebrain. *Endocrinology 96:*492-500, 1975.

Kuhn, C.M., Butler, S.R., and Schanberg, S.M.: Selective depression of serum growth hormone during maternal deprivation in rat pups. *Science 201:*1034-1036, 1978.

Ladinsky, H., Consolo, S., Peri, G., and Garattini, S.: Acetylcholine, choline and choline acetyltransferase activity in the developing brain of normal and hypothyroid rats. *J Neurochem 19:*1947-1952, 1972.

Laron, Z., Pertzelan, A., and Frankel, J.: Growth and development in the syndromes of familial isolated absence of HGH or pituitary dwarfism with high serum concentration of an immunoreactive but biologically inactive HGH. In *Hormones in Development,* M. Hamburgh and E.J.W. Barrington, eds.

Appleton-Century-Crofts, New York, 1971, pp 573-586.

Lau, H.C., Horowitz, C., Jumawan, J., and Koldovsky, O.: Effect of cortisone and thyroxine on acid glycosidases in rat forebrain and cerebellum during early postnatal development. *J Neurochem 31*:261-267, 1977a.

Lau, H.C., Horowitz, C., Humawan, J., and Koldovsky, O: Effect of cortisone or triiodothyronine administration to pregnant rats on lysosomal hydrolases in fetal forebrain and cerebellum. *Experientia 34*:566-567, 1977b.

Lauder, J.M.: Effects of thyroid state on development of rat cerebellar cortex. In *Thyroid Hormones and Brain Development,* G.D. Grave, ed. Raven Press, New York, 1977, pp 215-234.

Lauder, J.M., and Krebs, H.: Serotonin as a differentiation signal in early neurogenesis. *Dev. Neurosci. 1*:15-30, 1978.

Lauder, J.M., and Mugnaini, E.: Early hyperthyroidism alters the distribution of mossy fibres in the rat hippocampus. *Nature 268*:335-337, 1977.

Lauder, J.M., Altman, J., and Krebs, H.: Some mechanisms of cerebellar foliation: Effects of early hypo- and hyperthyroidism. *Brain Res 76*:33-40, 1974.

Legrand, J.: Morphologic and biochemical effects of hormones on the developing nervous system in mammals. In *Brain--Fetal and Infant,* S.R. Berenberg, ed. The Hague Nijhoff Medical Division, 1977, pp 137-164.

Levine, S.: An endocrine theory of infantile stimulation. In *Stimulation in Early Infancy,* A. Ambrose, ed. Academic Press, New York, 1969, pp 45-55.

Levine, S.: The pituitary adrenal system and the developing brain. In *Progr Brain Res 32*:79-88, 1970.

Lewis, P.D., Patel, A.J., Johnson, A.L., and Balazs, R.: Effect of thyroid deficiency on cell acquisition in the postnatal rat brain: A quantitative histological study. *Brain Res 104*: 49-62, 1976.

Lieberburg, I., and McEwen, B.: Estradiol-β17: A metabolite of testosterone recovered in cell nuclei from limbic areas of neonatal rat brains. *Brain Res 85*:165-170, 1975.

McCarty, R.: Annual Meeting of the International Society for the Development of Psychobiology (Abstr), p 4.

McDonald, P.G., and Doughty, C.: Effect of neonatal administration of different androgens in the female rat: Correlation between aromatization and the induction of sterilization. *J Endocrinol 61*:95-103, 1974.

McEwen, B.S.: Gonadal steroid receptors in neuroendocrine tissue. In *Receptors and Hormone Action,* Vol VII, B. O'Malley and L. Birnbaumer, eds. Academic Press, New York, 1978, pp 353-400.

McEwen, B.S., Plapinger, L., Chaptal, C., Gerlach, J., and Wallach, G.: Role of fetoneonatal estrogen binding proteins in the associations of estrogen with neonatal brain cell nuclear receptors. *Brain Res 96:*400–406, 1975.

Macho, L., Knopp, J., Brtko, J., and Strbak, V.: Effect of thyroxine on brain protein synthesis and binding of thyroxine to receptors in brain during ontogenesis. In *Hormones and Brain Development,* G. Dorner and M. Kawakami, eds. Elsevier Holland Press, New York, 1978, pp 229–234.

MacLusky, N.J., Chaptal, C., Lieberburg, I., and McEwen, B.S.: Properties and subcellular inter-relationships of presumptive estrogen receptor macromolecules in the brains of neonatal and prepubertal female rats. *Brain Res 114:*158–165, 1976.

MacLusky, N.J., Lieberburg, I., and McEwen, B.S.: Development of steroid receptor systems in the rodent brain. In *Ontogeny of Receptors and Reproductive Hormone Action,* T.H. Hamilton and W.A. Sadler, eds. Raven Press, New York, 1979, pp 393–402.

Masterpasqua, F., Chapman, R.H., and Lore, R.K.: The effects of prenatal psychological stress on the sexual behavior and reactivity of male rats. *Dev Psychobiol 9:*403–411, 1976.

Milkovic, K., and Milkovic, S.: Reactiveness of fetal pituitary to stressful stimuli—does maternal ACTH cross the placenta? *Proc Soc Expl Biol Med 107:*47–49, 1961.

Milkovic, K., and Milkovic, S.: Studies of the pituitary-adrenocortical system in the fetal rat. *Endocrinology 71:*799–802, 1962.

Milkovic, K., and Milkovic, S.: Functioning of the pituitary-adrenocortical axis in rats at and after birth. *Endocrinology 73:*535–539, 1963.

Mitskevich, M.S., and Moskovkin, G.M.: Some effects of thyroid hormones on the development of the central nervous system in early ontogenesis. In *Hormones in Development,* M. Hamburgh and E.J.W. Barrington, eds. Appleton-Century-Crofts, New York, 1971, pp 437–452.

Moog, F.: Corticoids and the enzymic maturation of the intestinal surface. In *Hormones in Development,* M. Hamburgh and E.J.W. Barrington, eds. Appleton-Century-Crofts, New York, 1971, pp 143–160.

Moscona, A.A.: Control mechanisms in hormonal induction of glutamine synthesis in the embryonic neural retina. In *Hormones in Development,* M. Hamburgh and E.J.W. Barrington, eds. Appleton-Century-Crofts, New York, 1971, pp 169–189.

Moskovkin, G., and Marshak, T.: Thyroid hormones in CNS development: Effect of hypo- and hyperthyroidism on early postnatal maturation of the rat hippocampus. In *Hormones and Brain Development,* G. Dorner and M. Kawakami, eds. Elsevier Holland Press, New York, 1978, pp 235–240.

Mullins, R.F., and Levine, S.: Hormonal determinants during infancy of adult sexual behavior in female rats. *Physiol Behave*

3:333-338, 1968.

Naftolin, F., and Brawer, J.R.: The effect of estrogens on hypothalamic structure and function. *Am J Obstet Gynecol 132:* 758-765, 1978.

Nathanielsz, P.W.: *Fetal Endocrinology: An Experimental Approach.* North Holland Publishing Co., Amsterdam, 1976.

Nicholson, J.L., and Altman, J.: Synaptogenesis in the rat cerebellum: Effects of early hypo- and hyperthyroidism. *Science 176*:530-532, 1972a.

Nicholson, J.L., and Altman, J.: The effects of early hypo- and hyperthyroidism on the development of rat cerebellar cortex. I. Cell proliferation and differentiation. *Brain Res 44:* 13-23, 1972b.

Nicholson, J.L., and Altman, J.: The effects of early hypo- and hyperthyroidism on the development of the rat cerebellar cortex. II. Synaptogenesis in the molecular layer. *Brain Res 44:* 25-36, 1972c.

Oklund, S., and Timiras, P.S.: Influences of thyroid levels on brain ontogenesis *in vivo* and *in vitro*. In *Thyroid Hormones and Brain Development,* G.D. Grave, ed. Raven Press, New York, 1977, pp 33-48.

Olpe, H.R., and McEwen, B.S.: Glucocorticoid binding to receptor-like proteins in rat brain and pituitary: Ontogenetic and experimentally-induced changes. *Brain Res 105*:121-128, 1976.

Olton, D.S., Johnson, C.T., and Howard, E.: Impairment of conditioned active avoidance in adult rats given corticosterone in infancy. *Dev Psychobiol 8*:55-61, 1974.

Oppenheimer, J.H., Dillman, W.H., Schwartz, H.L., and Towle, H.C.: Nuclear receptors and thyroid hormone action: A progress report. *Fed Proc 38*:2154-2161, 1979.

Pelton, E.W., and Bass, N.H.: Adverse effects of excess thyroid hormone on the maturation of rat cerebrum. *Arch Neurol 29:* 145-150, 1973.

Pelton, E.W., Grindeland, R.E., Young, E., and Bass, N.H.: Effects of immunologically induced growth hormone deficiency on myelinogenesis in developing rat cerebrum. *Neurology 27:* 282-288, 1977.

Pfaff, D.W., and Keiner, M.: Atlas of estradiol-concentrating cells in the central nervous system of the female rat. *J Comp Neurol 151*:121-158.

Pfaff, D.W., and Zigmond, R.E.: Neonatal androgen effects on sexual and nonsexual behavior of adult rats tested under various hormone regimes. *Neuroendocrinology 7*:129-145, 1971.

Plapinger, L., and McEwen, B.S.: Ontogeny of estradiol-binding sites in rat brain I. Appearance of presumptive adult receptors in cytosol and nuclei. *Endocrinology 93*:1119-1128, 1973.

Plapinger, L., and McEwen, B.S.: Gonadal steroid-brain interactions in sexual differentiation. In *Biological Determinants of Sexual Behavior,* J. Hutchinson, ed. Wiley, New York, 1978, pp 153-218.

Plapinger, L., McEwen, B.S., and Clemens, L.E.: Ontogeny of estradiol-binding sites in rat brain II. Characteristics of neonatal binding macromolecules. *Endocrinology* 93:1129-1139, 1973.

Porter, R.H., and Wehmer, F.: Maternal and infantile influences upon exploratory behavior and emotional reactivity in the albino rat. *Dev Psychobiol* 2:19-25, 1969.

Quadagno, D.M., and Ho, G.K.W.: Influence of gonadal hormones on social, sexual, emergence and openfield behavior in the rat rattus norvegicus. *Anim Behav* 20:732-740, 1972.

Rabie, A., and Legrand, J.: Effects of thyroid hormone and undernourishment on the amount of synaptosomal fraction in the cerebellum of the young rat. *Brain Res* 61:267-278, 1973.

Raisman, G., and Field, P.M.: Sexual dimorphism in the neuropil of the preoptic area of the rat and its dependence on neonatal androgen. *Brain Res* 54:1-29, 1973.

Rastogi, R.B., and Singhal, R.L.: Thyroid hormone control of 5-hydroxytryptamine metabolism in developing rat brain. *J Pharmacol Exp Ther* 191:72-81, 1974.

Rastogi, R.B., and Singhal, R.L.: Influence of neonatal and adult hyperthyroidism on behavior and biosynthetic capacity for norepinephrine dopamine and 5 HT in rat brain. *J Pharmacol Exp Ther* 198:609-618, 1976.

Raynaud, J.P., Mercier-Bodard, C., and Baulieu, E.E.: Rat estradiol binding plasma protein (EBP). *Steroids* 18:767-788, 1971.

Rebiere, P.A., and Legrand, E.J.: Effets compares de la sousalimentation, de l' hypothyroidisme et de l'hyperthyroidisme sur la maturation histologique de la zone moleculaire du cortex cerebelleux chez le jeune rat. *Arch Anat Microsc Morphol Exp* 61:105-126. 1972.

Reinisch, J.M.: Effects of prenatal hormone exposure on physical and psychological development in humans and animals: With a note on the state of the field. In *Hormones, Behavior and Psychopathology,* E.J. Sachar, ed. Raven Press, New York, 1976, pp 69-94.

Resko, J.A.: Fetal hormones and development of the central nervous system in primates. *Adv Sex Horm Res* 3:139-168, 1977.

Roger, L.J., Schanberg, S.M., and Fellows, R.E.: Growth and lactogenic hormone stimulation of ornithine decarboxylase in neonatal rat brain. *Endocrinology* 95:904-910, 1974.

Russell, D.H.: Polyamines in growth, normal and neoplastic. In *Polyamines in Normal and Neoplastic Growth,* Raven Press, New York, 1973, pp 1-14.

von Saal, F.S.: Prenatal exposure to androgen influences morphology and aggressive behavior of male and female mice.

Horm Behav 12: 1-11, 1979.

Salas, M., and Schapiro, S.: Hormonal influences on the maturation of the rat brain's responsiveness to sensory stimuli. *Physiol Behav* 5: 7-11, 1970.

Sara, V.R., Lazarus, L., and Stuart, M.C.: Fetal brain growth: Selective action by growth hormone. *Science* 186: 446-447, 1974.

Sara, V.R., King, T.L., Stuart, M.C., and Lazarus, L.: Hormonal regulation of fetal brain cell proliferation: Presence in serum of a trophin responsive to pituitary growth hormone stimulation. *Endocrinology* 99: 1512-1518, 1976.

Schapiro, S.: Some physiological, biochemical, and behavioral consequences of neonatal hormone administration: Cortisol and thyroxine. *Gen Comp Endocrinol* 10: 214-228, 1968.

Schapiro, S., and Vukovich, K.R.: Early experience effects upon cortical dendrites: A proposed model for development. *Science* 167: 292-294, 1970.

Schapiro, S., Geuer, E., and Eiduson, S.: Neonatal adrenocortical response to stress and vasopressin. *Proc Soc Expl Biol Med* 109: 937-941, 1962.

Schapiro, S., Salas, M., and Vukovich, K.: Hormonal effects on ontogeny of swimming ability in the rat: Assessment of central nervous system development. *Science* 168: 147-150, 1970.

Schapiro, S., Vukovich, K., and Globus, A.: Effects of neonatal thyroxine and hydrocortisone administration on the development of dendritic spines in the visual cortex of rats. *Exp Neurol* 40: 286-296, 1973.

Schmidt, M.J.: Effects of neonatal hyperthyroidism on activity of cyclic AMP-dependent microsomal protein kinase. *J Neurochem* 22: 469-471, 1974.

Schwark, W.S., Singhal, L., and Ling, G.M.: Metabolic control mechanisms in mammalian systems. *J Neurochem* 19: 1171-1182, 1972.

Silman, R.E., Holland, D., Chard, T., Lowry, P.J., Hope, J., Rees, L.H., and Nathanielsz, T.A.: Adrenocorticotrophin-related peptides in adult and foetal sheep pituitary glands. *J Endocrinol* 81: 19-33, 1979.

Singhal, R.L., Rastogi, R.B., and Hrdina, P.D.: Brain biogenic amines and altered thyroid function. *Life Sci* 17: 1617-1626, 1975.

Smith, D.J., Joffe, J.M., and Heseltine, G.F.D.: Modification of prenatal stress effects in rats by adrenalectomy, dexamethasone and chlorpromazine. *Physiol Behav* 15: 461-469, 1975.

Sobrian, S.K.: Aversive prenatal stimulation: Effects on behavioral biochemical and somatic ontogeny in the rat. *Dev Psychobiol* 10: 41-51, 1977.

Sodersten, P.: Effects of antioestrogen treatment of neonatal male rats on lordosis behavior in the adult. *J Endocrinol* 76:

241-249, 1978.

Sodersten, P., and Hansen, S.: Effects of castration and testosterone, dihydrotestosterone or oestradiol replacement treatment in neonatal rats on mounting behavior in the adult. J Endocrinol 76:251-260, 1978.

Sokoloff, L.: Biochemical mechanisms of the action of thyroid hormones. In Thyroid Hormones and Brain Development, G.D. Grave, ed. Raven Press, New York, 1977, pp 73-92.

Steiner, F.A.: Effects of ACTH and corticosteroids on single neurons in the hypothalamus. Prog Brain Res 32:102-197, 1967.

Stone, J.M., and Greennough, W.T.: Excess neonatal thyroxine: Effects on learning in infant and adolescent rats. Dev Psychobiol 8:479-488, 1975.

Strosser, M.T., and Mialhe, P.: Growth hormone secretion in the rat as a function of age. Horm Metab Res 7:275-278, 1975.

Stubbe, P., and Wolf, H.: The effect of stress on growth hormone, glucose and glycerol levels in newborn infancts. Horm Metab Res 3:175-179, 1971.

Swanson, H.H.: Alterations of sex-typical behavior of hamsters in open field and emergence tests by neonatal administration of androgen or estrogen. Anim Behav 15:209-216, 1967.

Sze, P.Y., Neckers, L., and Towle, A.C.: Glucocorticoids as a regulatory factor for brain tryptophan hydroxylase. J Neurochem 26:169-173, 1976.

Tannenbaum, G., and Martin, J.B.: Evidence for an endogenous ultradian rhythm governing growth hormone secretion in the rat. Endocrinology 98:562-570, 1976.

Taylor, A.N., Lorenz, R.J., Turner, B.B., Ronnekleiu, O.K., Casady, R.L., and Branch, B.J.: Factors influencing pituitary-adrenal rhythmicity: Its ontogeny and circadian variations in stress responsiveness. Psychoneuroendocrinology 1: 291-301, 1976.

Terasawa, E., Kawakami, M., and Sawyer, C.H.: Induction of ovulation by electrochemical stimulation in androgenized and spontaneously constant estrous rats. Proc Soc Expl Biol Med 132:497-501, 1969.

Thompson, J.R., Watson, J., and Charlesoorth, W.R.: The effects of prenatal maternal stress on offspring behavior in rats. Psychol Monogr 76:no. 577 (1962).

Torand-Allerand, C.D.: Sex steroids and the development of the newborn mouse hypothalamus and preoptic area in vitro: Implications for sexual differentiation. Brain Res 106:407-412, 1976.

Turakulov, Y.K., Gagelgans, A.I., Salakhova, N.S., and Mirakhmedov, A.K.: Thyroid Hormones: Biosynthesis, Physiological Effects and Mechanisms of Action. Consultants Bureau, New York, 1975.

Vernadakis, A.: Influence of cortisol on brain and spinal cord excitability in developing rats. In *Steroid Hormones and Brain Function,* C.H. Sawyer and R.A. Gorski, eds. University of California Press, Berkeley, 1971, pp 35–42.

Vertes, Z., Melegh, G., Vertes, M., and Kovacs, S.: Effect of naloxone and D-met[5]-pro[5]-enkephalinamide treatment on the DNA synthesis in the developing rat brain. *Life Sci 31:*119–126, 1982.

Wagner, J.W., Erwin, W., and Critchlow, V.: Androgen sterilization produced by intracerebral implants of testosterone in neonatal female rats. *Endocrinology 79:*1135–1142, 1966.

Walker, D.G., Simpson, M.E., Asling, C.W., and Evans, H.M.: Growth and differentiation in the rat following hypophysectomy at 6 days of age. *Anat Rec 106:*539–554, 1950.

Walker, D.G., Asling, C.W., Simpson, M.E., Li, C.H., and Evans, H.M.: Structural alterations in rats hypophysectomized at six days of age and their correction with growth hormone. *Anat Rec 114:*19–36, 1952.

Walker, P., Dussault, J.H., Alvarado-Urbina, G., and Dupont, A.: Development of the hypothalamo-pituitary axis in the neonatal rat: Hypothalamic somatostatin and pituitary and serum growth hormone concentrations. *Endocrinology 101:*782–787, 1977.

Walravens, P., and Chase, H.P.: Influence of thyroid on formation of myelin lipids. *J Neurochem 16:*1477–1484, 1969.

Ward, I.L.: Prenatal stress feminizes and demasculinizes the behavior of males. *Science 175:*82–84, 1972.

Weininger, O.: Mortality of albino rats under stress as a function of early handling. *Can J Psychol 7:*111–114, 1953.

Weisz, J., and Gibbs, C.: Metabolites of testosterone in the brain of the newborn female rat after an injection of tritiated testosterone. *Neuroendocrinology 14:*72–86, 1974.

Westermark, B., and Wasteson, A.: The response of cultural human normal glial cells to growth factors. *Adv Metabol Disoril 8:*85–100, 1975.

Westley, B.R., and Salaman, D.F.: Nuclear binding of the estrogen receptor of neonatal rat brain after injection of estrogens and androgens: Localization and sex differences. *Brain Res 119:*375–388, 1977.

Whalen, R.E., and Luttge, W.G.: Perinatal administration of dihydrotestosterone to female rats and the development of reproductive function. *Endocrinology 89:*1320–1321, 1971.

Whalen, R.E., and Rezek, D.L.: Inhibition of lordosis in female rats by subcutaneous implants of testosterone, androstanedione or dihydrotestosterone in infancy. *Horm Behav 5:*125–128, 1974.

Whitney, J.B., and Hesrenkohl, L.R.: Effects of anterior hypothalamic lesions on the sexual behavior of prenatally-stressed male rats. *Physiol Behav 19:*167–169, 1977.

Zamenhof, S., Mosley, J., and Schuller, E.: Stimulation of the proliferation of cortical neurons by prenatal treatment with growth hormone. *Science 152:*1396-1397, 1966.

Zarrow, M.X., Haltmeyer, G.C., Denenberg, V.H., and Thatcher, J.: Response of the infantile rat to stress. *Endocrinology 79:*631-634, 1966.

17

Role of Monoaminergic Neurons in the Age-Related Alterations in Anterior Pituitary Hormone Secretion

JAMES W. SIMPKINS AND KERRY S. ESTES

INTRODUCTION

The intimate anatomical relationship between the hypothalamus and the anterior pituitary gland (AP) provided the first suggestion that AP secretory activity might be regulated by the central nervous system (CNS). The discovery of an intricate portal vascular system connecting the medial basal hypothalamus (MBH) and the AP and the observation that hypophysial portal blood transports neuronal substances capable of altering the synthesis and release of AP hormones provided convincing evidence of the importance of this close anatomical relationship. It is now well established that the CNS regulates AP function by the production of releasing hormones and release-inhibiting hormones, which are secreted into the vicinity of the capillary plexus of the portal vascular system. Further, volumes of evidence have demonstrated that surgical or pharmacological separation of the CNS or its components from the AP severely compromises the ability of the AP to modify trophic hormone secretion in response to alterations in both internal and external environments.

The central role played by the AP trophic hormones in the maintenance of homeostatic processes in the body has focused attention of students of aging on the possibility that alterations in the AP or its control mechanisms are responsible for the age-related decline in homeostasis and perhaps for the aging of the organism itself (Dilman, 1971; Everitt, 1980). However, there is insufficient evidence to state with certainty that the age-associated decline in homeostatic mechanisms results from altered AP hormone secretion. Some preliminary evidence has been presented that indicates that AP target glands,

for example, ovaries, testes, thyroid, and adrenals, may participate in age-related changes in some parameters of CNS function (see Randall and Finch, chapter 18). Presently, we will
analyze the alterations in AP function which accompany increasing age and consider evidence that indicates that monoaminergic
neurons may participate in these age-related alterations.

Although numerous substances endogenous to the brain meet
sufficient criteria to warrant classification as putative
neurotransmitters (Cooper et al., 1978), only a few will be
considered with respect to their biosynthesis and metabolic
disposition during the aging process. Primary consideration is
given to the noradrenergic, dopaminergic, and serotonergic
systems that reside in or have terminal input to the hypothalamus and preoptic area of the ventral diencephalon. These
three systems have been studied most intensely in aging animals
and man for the following reasons: (1) the neuroanatomical relationships of cell bodies and synaptic terminals are well described for these three systems, (2) the neural and biochemical
factors regulating synaptic transmission in these systems is at
least partially characterized, and (3) perhaps most important,
these three systems are more easily manipulated and thus their
role in the "normal" regulation of AP function in young animals
is better understood than is the involvement of the other putative neurotransmitters in this regard.

Monaminergic input regulating hormone secretion from the AP
arises from two intrahypothalamic dogaminergic systems and projections from mesercephalic noradrenergic and serotonergic cell
groups. In the MBH, dopaminergic cell bodies of the tuberoinfundibular dopamine (TIDA) system are located in the arcuate and ventral periventricular neclei, project short distances to the external layers of the median eminence and terminate adjacent to the
capillary plexus of the hypophysial portal system. The rostral
portion of this system projects caudally into the neurohypophysis to provide the DA innervation of the posterior pituitary
gland (see Moore and Bloom, 1978 for review). A second in-
certohypothalamic DA (IHDA) system has cell bodies in the zona
incerta and adjacent periventricular nuclei and projects rostrally to the anterior hypothalamic nuclei, preoptic area, and
stria terminalis (Bjorklund et al., 1975). The observation
that total hypothalamic deafferentation causes profound depletion of hypothalamic norepinephrine (NE) and serotonin (5-HT)
but does not significantly alter intrahypothalamic DA concentration indicates that the cell bodies of these two DA systems
are confined to the hypothalamus (Weiner et al., 1972). Recently, an extrahypothalamic DA input to the median eminence
has also been suggested (Kizer et al., 1976; Palkovits et al.,
1977).

The noradrenergic input to the hypothalamus arises from the

locus coeruleus (LC) and subceruleal areas. The LC projection
to the hypothalamus travels as part of the dorsal noradrenergic
bundle and terminates largely in the peri- and paraventricular
nuclei of the hypothalamus. The noradrenergic innervation of
the remaining hypothalamic nuclei arises from subceruleal areas
and travels rostrally via the ventral noradrenergic bundle (see
Moore and Bloom, 1979 for review).

Serotonergic projections to the hypothalamus arise from the
narrow band of midbrain raphe nuclei and course via the medial
forebrain bundle to form dense terminal beds in the suprachiasmatic nuclei and other hypothalamic and preoptic areas (Ungerstedt, 1971). A group of cell bodies that concentrates 5-HT
has been identified in the dorsomedial nucleus of the hypothalamus (Fuxe and Ungerstedt, 1968), suggesting the existence
of intrahypothalamic serotonergic neurons.

The vast majority of studies of the neuroendocrinology of
aging have used short-lived laboratory rodents or human patients as experimental subjects. Although rats and mice have a
maximum life-span of greater than three years, 2-year-old
animals are considered aged and are often compared to "young"
sexually mature animals (usually 2 to 4 months of age). Thus,
most aging studies have evaluated the extreme ends of the aging
process. Recently, interest has been focused on the period
between maturity and senescence in an effort to evaluate the
time of onset of alterations in neuroendocrine function. In
these studies, animals ranging in age from 10 to 24 months are
used as experimental subjects. Studies of the aging process in
human subjects have been limited by the lack of available subjects in any one age group. Therefore definitions of young and
old are widely disparate in the literature. The literature
cited is confined to articles published prior to July, 1980,
the time this review was completed.

AGE-RELATED ALTERATIONS IN CENTRAL CATECHOLAMINERGIC
AND SEROTONERGIC SYSTEMS

Nigrostriatal Dopamine System

Although the nigrostriatal DA system (Tables 17-1, 17-2) is not
directly involved in neuroendocrine regulatory mechanisms, the
age-related alterations in this pathway are presently considered because it has been the most intensively studied monoaminergic system in old animals. This DA system, then, serves as
the standard by which to compare the effects of advancing age
on other monoamine systems.

The nigrostriatal DA pathway is particularly labile during

Table 17-1. Age-Related Alterations In Monoaminergic Enzymes Activity In Brain

Age-related change	Species	Ages compared	Reference
Tyrosine Hydroxylase			
Decreased in neostriatum	Wistar rat	4-15 vs 25-28 mo	McGeer et al., 1971
Decreased in caudate nucleus, olfactory tubercle; increased in hypothalamus; no change in locus cerulus	Fischer 344 rat	4 vs 24-26 mo	Reis et al., 1977
No change in locus cerulus, hypothalamus substantia nigra or caudate nucleus	CB6F1 mouse	4 vs 24-26 mo	Reis et al., 1977
Decreased in substantia nigra, caudate nucleus, putamen	Man	15 vs 65 yr[a]	Ordy et al., 1975; Cote and Kremzner, 1975
No change in hypothalamus	Man	15 vs 65 yr[a]	Cote and Kremzner, 1975
Aromatic Amino Acid Dearboxylase			
No change in caudate nucleus	Fischer 344 rat	4 vs 24-26 mo	Reis et al., 1977
No change in locus cerulus, hypothalamus, substantia nigra, caudate nucleus	CB6F1 mouse	4 vs 24-26 mo	Reis et al., 1977
Decreased in hypothalamus, substantia nigra, caudate nucleus, putamen	Man	15 vs 65 yr[a]	Ordy et al., 1975; Cote and Kremzner, 1975

Dopamine-β-hydroxylase

Decreased in hypothalamus, no change in locus cerulus, cortex	Fischer 344 rat	4 vs 24-26 mo	Reis et al., 1977
No change in locus cerulus, hypothalamus	CB6F1 mouse	4 vs 24-26 mo	Reis et al., 1977

Tryptophan Hydroxylase

No change in raphe nuclei	Fischer 344 rat	4 vs 24-26 mo	Reis et al., 1977
Decreased in B, B cell groups, septum, and hippocampus	SD rat[b]	1 vs 24 mo	Meek et al., 1977

Monoamine Oxidase

Increased in hindbrain	Man	25-55 vs >65 yr[a]	Robinson et al., 1972
Increase widespread in brain	Man	21 vs 84 yr[a]	Robinson, 1975
No change in whole brain	Rat	2 vs >35 wk[a]	Prange et al., 1967
Change dependent upon substrate used	Rat	2 vs 24 mo	Shin, 1975

[a] Ages indicate range tested.

[b] SD: Sprague-Dawley.

Table 17-2. Age-Related Alterations in Dopamine Concentration and Turnover in Brain

Age-related change	Species	Ages compared	Reference
Concentration			
No change in hypothalamus	LE[a] and Wistar rat	3 vs 33 mo[b]	Simpkins et al., 1977 Ponzio et al., 1978 Carlsson and Winblad, 1976
No change in preoptic area	LE rat	7-8 vs 16-18 mo	Huang et al., 1977
No change in whole brain minus striatum and cerebellum	Wistar rat	7 vs 24-29 mo	McGeer et al., 1971
Decreased in hypothalamus	LE rat		Miller et al., 1976
Decreased in medial basal hypothalamus	Wistar rat	3-4 vs 21 mo	Simpkins et al., 1977
No change in whole brain, hypothalamus, decreased in striatum	C57BL/6J mouse	12 vs 28 mo	Finch, 1973
Decreased in median eminence	LE rat	4-8 vs 20-24 mo	Wilkes et al., 1979
Decreased in median eminence, arcuate nucleus, area retrochiasmatica	Wistar rat	3 vs 24 mo	Estes and Simpkins, 1980

Decreased in median eminence	C57BL/6J mouse	9-12 vs 25-28 mo	Finch, 1979
Increased in preopticus medialus, preopticus suprachiasmaticus, and anterior hypothalamic nuclei	Wistar rat	3 vs 24 mo	Estes and Simpkins, 1980
Turnover			
Decreased in striatum and hypothalamus	Wistar rat	3 vs 33 mo	Ponzio et al., 1978
Decreased in striatum and hypothalamus	C57BL/6J mouse	12 vs 28 mo	Finch, 1973
Decreased in medial basal hypothalamus	Wistar rat	3-4 vs 21 mo	Simpkins et al., 1977
No change in preoptic area, absence of normal castration-induced increase	LE rat	7-8 vs 16-18 mo	Huang et al., 1977

[a] LE: Long-Evans.

[b] Ages indicate range tested.

the aging process in both man and other mammalian species.
Tyrosine hydroxylase, the rate-limiting enzyme in catecholamine
biosynthesis, exhibits decreased activity with age in the substantia nigra, putamen, and caudate nucleus in nonparkinsonian
human patients (Ordy et al., 1975; Cote and Kremzer, 1975) and
rats (McGeer et al., 1971; Reis et al., 1977), whereas no alteration was observed in the CB6Fl mouse (Reis et al., 1977).
Aromatic amino acid decarboxylase activity is decreased in the
nigrostriatal pathway of human subjects (Ordy et al., 1975;
Cote and Kremzner, 1975) but not in rats or mice (Reis et al.,
1977). The steady-state concentration of striatal DA increases
with increasing age in mice (Finch, 1973), rats (McGeer at al.,
1971), and in nonparkinsonian human patients (Riederer and
Wuketich, 1976). The consistently and widely observed decrease
in striatal tyrosine hydroxylase activity and in striatal DA
concentration with increasing age would suggest that a decrease
in metabolism in the nigrostriatal DA pathway is a common feature of the normal aging process in mammals. Indeed, using
[^{3}H]-tyrosine incorporation into DA as an index of striatal DA
turnover, Finch (1973) and Ponzio et al. (1978) have observed
50 and 25 percent decreases, respectively, in striatal DA turnover with age in laboratory rodents. Since the transport of
tyrosine into brain tissue *in vivo* (Finch, 1973) as well as
into slices of striatum (Finch et al., 1975) remains constant
with increasing age, a decreased availability of precursors
cannot account for the age-related decline in striatal DA
turnover. However, defects in neuronal uptake mechanisms for
DA may contribute to the age-related decline in striatal DA
concentration. Jonec and Finch (1975) have observed decreased
DA uptake in synaptosome preparations obtained from striatal
tissue of old mice.

   A comparison of the behavioral and biochemical consequences
of stimulation of postsynaptic striatal DA receptors in old
animals yields apparently conflicting results. An age-related
decrease in DA-stimulated adenylate cyclase in the striatum of
rat has been reported from three laboratories (Walker and Boas-
Walker, 1973; Puri and Volicer, 1976; Govoni et al., 1977).
However, Joseph et al., (1978) noted that rotational behavior
induced by unilateral lesioning of the substantia nigra was attenuated in old as compared to young rats following amphetamine
administration but was normal in old rats following apomorphine
treatment. Similarly, impaired swimming behavior observed in
24- to 27-month-old Fischer rats is reversed by apomorphine
treatment (Marshall and Berrios, 1979). In view of recent
reports that with increasing age striatal DA receptor number
decreased by about 30 percent in several species (Makman et
al., 1978; Roth, 1979; Severson and Finch, 1980), these data
would suggest a hypersensitivity of the behavioral response to

DA-receptor stimulation. Thus, the maintenance of relatively normal movement behaviors throughout most of postmaturational life in rodents may result from compensatory mechanisms within this extrapyramidal system.

Intrahypothalamic Dopaminergic Systems

Although less well studied than the nigrostriatal pathway, intrahypothalamic DA systems and monoaminergic projections to the hypothalamus and preoptic area show age-related alteration in their function (Tables 17-1, 17-2).

DA concentrations in tissue fragments encompassing the whole hypothalamus and preoptic area appears to be relatively stable with age in several strains of rats and mice (Finch 1973; Miller et al., 1976; Simpkins et al., 1977; Ponzio et al., 1978; Huang et al., 1977). Interestingly, examination of DA concentrations within specific regions of the hypothalamus reveals age-related variations, with DA concentrations decreased in the MBH (Simpkins et al., 1977), median eminence (Estes and Simpkins, 1980; Finch, 1979; Wilkes et al., 1979), arcuate, and area retrochiasmatica (Estes and Simpkins, 1980), but increased in the anterior hypothalamus, and in preoptic nuclei (Estes and Simpkins, 1980). These differential alterations in DA concentrations in the hypothalamus with age probably explain why studies utilizing whole hypothalamic fragments failed to reveal age-related alterations in DA concentration. Interestingly, the nuclei that contain nerve terminal beds of the IHDA system show increased DA concentrations with age, whereas the nuclei that contain nerve terminals of the TIDA system show decreased DA concentrations.

DA turnover as estimated by both steady-state and nonsteady-state methods is decreased in whole hypothalami of male mice and rats (Finch, 1973; Ponzio et al., 1978), MBH of male rats (Simpkins et al., 1977) but not the anterior hypothalami of female Long-Evans rats (Huang et al., 1977). In this latter study, ovariectomy, which increased anterior hypothalamic DA turnover in young mature female rats, failed to accelerate DA turnover in their old counterparts. These results indicate that an age-related decrease in the activity of hypothalamic DA neurons occurs, although the precise cause of the deficiency and whether these alterations are homogeneously distributed among hypothalamic nuclei is not yet known. A systematic analysis of DA turnover in specific regions of the hypothalamus is needed and could feasibly utilize available microdissection and microassay techniques. Until such analyses are conducted, we cannot state with certainty whether the conflicting observations on the age-related alterations in the activity of enzymes

involved in DA biosynthesis (Table 17-1) (Cote and Kremzner, 1975; Ordy et al., 1975; Reis et al., 1977) reflect species differences, as suggested by Finch (1977), or simply differences in the extent of tissue fragments used for the analyses.
   To date, no studies have been conducted that examined age-related alterations in DA receptor number or DA receptor mediated biochemical events in the hypothalamus or AP.

Hypothalamic Noradrenergic Systems

NE concentration has been reported to be decreased in the hindbrain and hypothalamus of human subjects and rhesus monkeys (Tables 17-1, 17-3) (Robinson et al., 1972; Samorajski, 1975), and in the rat hypothalamus (Miller et al., 1976; Simpkins et al., 1977; Huang et al., 1977), MBH (Simpkins et al., 1977), arcuate nucleus, and medial preoptic area (Estes and Simpkins, 1980). In marked contrast to these alterations in NE concentrations seen in men, monkeys, and rats with age, NE concentrations appear very stable in the mouse whole brain (Samorajski et al., 1971; Finch, 1973), hypothalamus, cerebellum, brainstem (Finch, 1973), and median eminence (Finch, 1979; Wilkes et al., 1979) throughout postmaturational life. The mouse appears to be unique among mammals yet tested in this regard. However, in both the mouse and rat, a substantial age-related decline in hypothalamic NE turnover has been observed (Finch, 1973; Simpkins et al., 1977; Ponzio et al., 1978; Huang et al., 1977). Further, the normally observed acceleration in hypothalamic NE turnover induced by ovariectomy is absent in the old female rat (Huang et al., 1977). In the rat, but not the mouse, dopamine-β-hydroxylase activity is decreased with age (Reis et al., 1977).
   Although the age-related alterations in hypothalamic noradrenergic receptors have not yet been specifically studied, there is evidence for a wide-spread decrease in noradrenergic receptor number and NE-receptor stimulated adenylate cyclase activity in other areas of the brain of old animals. Noradrenergic receptor number is decreased in the cerebellum, striatum, and pineal gland in the old rat (Greenberg and Weiss, 1978). An age-related attenuation of the NE-receptor stimulated increase in adenylate cyclase activity is observed in the cerebral cortex, caudate nucleus. cerebellum, and hippocampus of rats (Walker and Boas-Walker, 1973; Berg and Zimmerman, 1975). A determination of the changes in the number of hypothalamic noradrenergic receptors and the responsiveness of NE-receptor mediated biochemical alterations with age is needed. This need for receptor sensitivity studies is particularly apparent, since basal rates of secretion of AP hormones show only modest

alterations with age, whereas rather dramatic alterations in NE
and DA turnover in the hypothalamus are observed. An age-related compensatory hypersensitivity or increase in number of
receptors may in part account for these alteration.

Hypothalamic Serotonergic Systems

Least studied of the three monoaminergic systems in senescent
animals is the serotonergic component innervating the hypothalamus (Tables 17-1, 17-4). In man and the rhesus monkey, hypothalamic 5-HT concentrations are decreased with age (Bertler, 1961;
Samorajski and Rolstein, 1973) but hindbrain concentrations in
human patients were reported to be unchanged (Robinson et al.,
1972). In an extensive study of the senescent rat, Meek et al.
(1977) observed decreases in 5-HT concentrations in midbrain
raphe nuclei and in the hippocampus, although no alteration in
rat hypothalamic 5-HT concentrations were observed in another
study (Simpkins et al., 1977). In the mouse, whole brain 5-HT
concentrations are reported to be stable (Finch, 1973) or
slightly decreased (Samorajski et al., 1971). Clearly, esti-
mation of 5-HT concentration alone is not sufficient to assess
the effects of age on serotonergic neurons.
    Attempts at estimating serotonergic neuronal activity have
been equally unsuccessful in clearly elucidating the effects of
age on this neuronal system. Tryptophan hydroxylase activity
has been reported to decrease in the B_7 and B_9 cell groups, the
septum and hippocampus of the Sprague-Dawley rat (Meek et al.,
1977), but not changed in the raphe nucleus of Fischer 344 rats
(Reis et al., 1977). In both the rat whole brain and human
cerebrospinal fluid (CSF) 5-hydorxyindoleacetic acid (5-
HIAA) concentrations are elevated (Bowers and Gerbode, 1968;
Gottfries et al., 1971; Simpkins et al., 1977). Although in
young mature animals, elevated 5-HIAA concentrations or elevated 5-HIAA/5-HT is often taken as evidence for enhanced
serotonergic activity (Hery et al., 1972), this method does not
apply in the old animal because it assumes that monoamine
oxidase (MAO) activity and the clearance of 5-HIAA from the
brain are constant with increasing age. Neither of these two
assumptions are correct. First, in both the human and rat
brain MAO activity appears to increase with age (Robinson et
al., 1972; Robinson, 1975; Prange et al., 1967; Shin, 1975).
Further, it has been observed that following blockade of MAO
with pargyline, 5-HIAA levels decrease significantly in young
but not old male rats (Simpkins et al., 1977). This latter
observation suggests that the accumulation of 5-HIAA in the
brain and CSF during aging may result from an age-related
decrease in clearance of this 5-HT metabolite rather than an

Table 17-3. Age-Related Alterations in Norepinephrine Concentration and Turnover in Brain

Age-related change	Species	Ages compared	Reference
Concentration			
No change in hypothalamus, cerebellum and brainstem	C57BL/6J mouse	12 vs 28 mo	Finch, 1973
No change in hypothalamus	Wistar rat	3 vs 33 mo	Ponzio et al., 1978
No change in whole brain	Mouse C57BL/10; C57BL/6J	9-12 vs 28 mo	Samorajski et al., 1971; Finch, 1973
Decreased in hypothalamus, brainstem	Rhesus monkey	3-5 vs 12-18 yr	Samorajski, 1975
Decreased in hypothalamus	Man	43-60 vs 73-87 yr[a]	Bertler, 1961
Decreased in hindbrain	Man	25 vs >65 yr[a]	Robinson et al., 1972
Decreased in hypothalamus	LE rat[b]		Miller et al., 1976
Decreased in whole hypothalamus and medial basal hypothalamus	Wistar rat	3-4 vs 21 mo	Simpkins et al., 1977
Decreased in preoptic area	LE rat	7-8 vs 16-18 mo	Huang et al., 1977

Decreased in medial preoptic area, arcuate nucleus	Wistar rat	4 vs 24 mo	Estes and Simpkins, 1980
No change in median eminence	LE and Wistar rat	4-6 vs 24-30 mo	Wilkes et al., 1979 Estes and Simpkins, 1980
No change in median eminence	C57BL/6J mouse	9-12 vs 25-28 mo	Finch, 1979
Turnover			
Decreased in hypothalamus	C57BL/6J mouse	12 vs 28 mo	Finch, 1973
Decreased in hypothalamus	Wistar rat	3-4 vs 21 mo	Simpkins et al., 1977
Decreased in hypothalamus and brainstem	Wistar rat	3 vs 33 mo	Ponzio et al., 1978
Decreased in preoptic area, absence of castration-induced increase	LE rat	7-8 vs 16-18 mo	Huang et al., 1977

[a] Ages listed are the extremes of those tested.

[b] LE: Long-Evans.

Table 17-4. Age-Related Alterations in Serotonin (5-HT) and 5-hydroxyindolacetic acid (5-HIAA) Concentrations in Brain

Age-related change	Species	Ages compared	Reference
5-HT concentration			
Decreased in whole brain	C57BL/10 mouse	3 vs 21 mo	Samorajski et al., 1971
No change in whole brain	C57BL/6J mouse	12 vs 28 mo	Finch, 1973
Decreased in hypothalamus	Rhesus monkey	3-5 vs 12-18 yr	Samorajski and Rolsten, 1973
Decreased in hypothalamus	Man	43-60 vs 73-87 yr	Bertler, 1961
No change in hindbrain	Man	25 vs > 65 yr[a]	Robinson et al., 1972
No change in hypothalamus	Wistar rat	3-4 vs 21 mo	Simpkins et al., 1977
Decreased in B7, B8, B9 cell groups, hippocampus	SD rat[b]	1 vs 24 mo	Meek et al., 1977
5-HIAA concentration			
Increased in brain minus hypothalamus	Wistar rat	3-4 vs 21 mo	Simpkins et al., 1977
No change in hindbrain	Man	25 vs > 65 yr[a]	Robinson et al., 1972
Increased in cerebrospinal fluid	Man	35-55 vs > 55 yr[a]	Bowers and Gerbode, 1968
Increased in cerebrospinal fluid	Man	18-54 vs 71-81 yr[a]	Gottfries et al., 1971

[a] Ages listed are the extremes of those tested.

acceleration in activity of serotonergic neurons. Thus, there is little convincing evidence for a widespread alteration in serotonergic activity with increasing age.

Age-Related Alterations in AP Hormone Secretion

Studies of the "normal" role of monoaminergic systems in AP hormone secretion comprise an overwhelming volume of literature. The present discussion is limited to systems for which monoamine involvement in AP hormone secretion are clearly defined and cases in which large and significant change in monoamine activity are observed with advanced age. The interested reader is referred to two recent and extensive reviews of monoaminergic regulation of AP hormones (Frohman and Berelowitz, chapter 4; Weiner and Ganong, 1978) for a more complete discussion of neural regulation of AP function.

Luteinizing Hormone and Follicle-Stimulating Hormone

A dicotomy exists between human subjects and laboratory rodents in the age-related alteration in basal luteinizing hormone (LH) and follicle-stimulating hormone (FSH) secretion (Tables 17-5, 17-6). In both men and women, basal serum LH and FSH concentrations begin to increase dramatically in the fifth and sixth decade of life and reach levels similar to those found in young mature castrated patients by the eight and ninth decade of life (Harman, 1978). In both sexes this elevation in basal rates of gonadotropin secretion appears to result from a decline in circulating gonadal steriods, which accompanies aging, (Schiff and Wilson, 1978; Harman, 1978). Appropriate steroid treatment results in a prompt decrease in circulating LH and FSH in old subjects (Odell and Swerdloff, 1968). In man then, the neural components mediating the negative feedback of gonadal steroids on LH and FSH secretion appears to remain intact with increasing age.

In marked contrast, basal secretions of LH and FSH in old laboratory rodents show much less dramatic alterations with age. In male rats and mice basal serum LH and FSH levels show modest decreases or do not change with advanced age (Shaar et al., 1975; Simpkins et al., 1977; Bronson and Desjardins, 1977; Estes and Simpkins, 1980). Similarly, in old female rats, LH and FSH are normal or slightly elevated (Huang et al., 1976; Lu et al., 1979; Wilkes et al., 1979).

In contrast to the stability of basal secretory rates of gonadotropins in old rodents, challenge of the gonadotropin secretory mechanism reveals substantial age-related alteration.

Table 17-5. Age-Related Alterations In Luteinizing Hormone Secretion

Age-related change	Species	Ages compared	Reference
Concentrations			
No change	Female LE rat[a]	4-6 vs 23-30 mo	Shaar et al., 1975
Increase	Female LE rat	4-5 vs 22-24 mo	Huang et al., 1976
Decrease	Male LE rat	4-6 vs 23-30 mo	Shaar et al., 1975
Decrease	Male CBF, mouse	2 to 30 mo[a]	Bronson and Desjardins, 1977
No change	Male C57BL/6J mouse	12 vs 28 mo	Finch et al., 1977
No change	Female LE rat	8-9 vs 27 mo	Lu et al., 1979
Decrease	Male Wistar rat	3-4 vs 21 mo	Simpkins et al., 1977
Increase	Human male	25 vs 89 yr[b]	Harman et al., 1979
Decrease	Human male	20 vs 89 yr[b]	Huag et al., 1974
Increased	Human female	50-61 yr	Odell and Swerdloff, 1968
Increase	Human female	23-31 vs 43-49 yr	Tsai and Yen, 1971

Response

Response		Age	Reference
Decreased response to orchiectomy	LE rat	4-6 vs 23-30 mo	Shaar et al., 1975
Decreased response to ovariectomy	LE rat	4 vs 30 mo	Shaar et al., 1975 Huang et al., 1976
Decreased response to hemiovariectomy	SD rat[a]	4 vs 14 mo	Howland and Priess, 1975
Decrease pulses	SD rat	3 vs 18 mo[b]	Estes et al., 1980
Decreased response to estrogen and progesterone	LE rat	3-4 vs 11-13 mo	Lu et al., 1977
Decrease estrogen response	Wistar rat	5-6 vs 18-19 mo	Peluso et al., 1977
Decrease, delay in surge	(RxU)F rat	3-5 vs 8-10 mo	Van der Schoot, 1976
Decreased response to orchiectomy	C57BL/6J mouse	12 vs 28 mo	Finch et al., 1977

[a] LE: Long-Evans; SD: Sprague-Dawley.

[b] Age indicated range tested.

Table 17-6. Age-Related Alterations in Follicle-Stimulating Hormone Secretion

Age-related change	Species	Ages compared	Reference
Concentration			
No change	Female LE rat[a]	4-5 vs 22-30 mo	Huang et al., 1976
No change	Female SD rat	4 vs 14 mo	Howland and Preiss, 1975
Increase	Female CFY rat	2-4 vs 10-14 mo	Gosden and Bancraft, 1976
Increase	Female LE rat	4-5 vs 20-24 mo	Huang et al., 1976
Decrease	Male Wistar rat	3-4 vs 21 mo	Simpkins et al., 1977
No change	Human males	20 vs 89 yr[b]	Huang et al., 1974
No change	Male CBF mice	2 vs 30 mo[b]	Bronson and DesJardins, 1977
No change	Male C57 BL/6J mouse	12 vs 28 mo	Finch et al., 1977
Increase	Human males	25 vs 89 yr[b]	Harman et al., 1979
Response			
No change following hemiovariectomy	SD rat[a]	4 vs 14 mo	Howland and Preiss, 1975

Decreased response to ovariectomy	SD rat	3 vs 22 mo[b]	Howland, 1976
No change following ovariectomy	CFY rat	2-4 vs 10-14 mo	Gosden and Bancroft, 1976
No change following orchiectomy	C57 BL/6J mouse	12 vs 28 mo	Finch et al., 1977
Decreased response to ovariectomy	LE rat	4-5 vs 22-30 mo	Huang et al., 1976

[a]LE: Long-Evans; SD: Sprague-Dawley.

[b]Ages indicated range tested.

In old male and female rats, castration results in a slower onset and lower magnitude of LH and FSH secretion than is observed in young animals (Shaar et al., 1975; Howland and
Preiss, 1975; Huang et al., 1976; Howland, 1976; Peluso et al.,
1977). A similar observation for LH, but not FSH, has been
made in the male mouse (Finch et al., 1977). Estes et al.
(1980) have studied the dynamics of the postcastration rise in
serum LH in young and old rats. They have observed that the
inability of old female rats to increase serum LH following
ovariectomy is due to an age-related decrease in the amplitude
of serum LH pulses. These secretory bursts of LH are dependent
upon the pulsatile release of luteninizing hormone-releasing
hormone (LH-RH) into the portal vessels (Gallo, 1980) and thus
the absence of LH pulses in the old animal suggest a hypothalamic deficiency.

Some reports indicate that responsiveness of the AP to a
single injection of LH-RH is decreased in old rats (Bruni et
al., 1977; Miller and Riegle, 1978). This decreased response
to LH-RH can be overcome by multiple injections of the releasing hormone or by treatment with lose dose of estradiol
(Miller and Riegle, 1978; Peluso et al., 1977). Further, no
age-related alteration in the LH-RH-induced release of LH and
FSH was observed in the C57BL/6J mouse (Finch et al., 1977).
These data indicate that the steroid environment and the frequency of exposure to LH-RH are more important than age in
determining AP responsiveness. Further, since the amount of LH
released during the proestrous surge far exceeds that required
to induce ovulation, a small decline in AP responsiveness to
LH-RH would not likely result in the abnormal reproductive
states observed in senescent animals.

In response to gonadal steroid treatments that induce surges
of LH and FSH secretion in young animals, old female rats (Howland, 1976; Peluso et al., 1977; Lu et al., 1977) and postmenopausal female patients (Tsai and Yen, 1971) show markedly
reduced responses. This age-related decline in the stimulatory
feedback of gonadal steroid on gonadotropin secretion probably
accounts for the initial irregularity and subsequent absence of
estrous cycles in the aging rat (Huang and Meites, 1975), and
the age-related alteration in the profile of the proestrous LH
surge (Van der Schoot, 1976). In rodents, the stimulatory
feedback of gonadal steriods on gonadotropin secretion is
mediated by the preoptic area of the diencephalon (Hillarp,
1949). Electrolytic lesions in this area of the brain in young
female rats produce reproductive patterns similar to that observed in old rats (Clemens and Bennett, 1977), whereas electrical stimulation of the medial preoptic area (Wuttke and
Meites, 1973; Clemens et al., 1969) induces the secretion of
sufficient amounts of LH to stimulate ovulation in old rats.

Similarly, ovulation and a reinitiation of estrous cycles has been demonstrated in old rats following medial preoptic implants of L-dopa (Cooper et al., 1979). Interestingly an age-related decline in brain estradiol receptor number has been demonstrated in the female rat (Kanungo et al., 1975). This decline in estradiol receptor number may contribute to the decreased ability of the CNS to mediate the stimulatory effects of estrogen on gonadotropin secretion.

Several lines of evidence indicate that alterations in central catecholaminergic neurons are important in the impaired gonadotropin responses seen in old rats. In young rats, castration stimulates hypothalamic NE turnover, whereas appropriate steroid replacement blocks this response (Coppola, 1969; Simpkins et al., 1980). In the aged female rat this castration-induced acceleration in NE turnover is absent and the gonadotropin secretion is attenuated (Huang et al., 1976, 1977). Similarly, a two-fold increase in NE turnover precedes the proestrous surge of LH and FSH (Advis et al., 1978) and is observed following stimulatory regimens of gonadal steriods (Munaro, 1977; Simpkins et al., 1979) in the rat. Both the proestrous and the steriod-induced surges of gonadotropin are absent or reduced in old rats, suggesting a deficiency in the noradrenergic mediation of these responses.

The most convincing evidence to date for a catecholaminergic involvement in the age-related decline in LH and FSH secretory capacity comes from studies utilizing catecholaminergic drugs to reinitiate estrous cycles in old rats. Treatment of old rats with the catecholamine precursors, tyrosine (Linnoila and Cooper, 1976) and L-dopa (Quadri et al., 1973; Huang and Meites, 1975; Huang et al., 1976; Linnoila and Cooper, 1976), as well as with the MAO inhibitor iproniazid (Quadri et al., 1973), reinitiate estrous cycles in old rats. Interestingly, ACTH, progesterone, and ether stress are also effective in reinitiating estrous cycles in constant estrous rats (Everett, 1940; Huang et al., 1976). Both ACTH and stress increase adrenal progesterone secretion, and progesterone acts in the hypothalamus to increase NE turnover (Munaro, 1977; Simpkins et al., 1979) and serum LH (Caligaris et al., 1968). Thus, these latter three stimuli may reinstate estrous cycles in old rats through a central noradrenergic mechanism.

A dopaminergic involvement in reproduction senescence is suggested by the observation that the DA receptor blocker pimozide, but not the adrenergic receptor blockers phenoxybenzamine or L-propranolol, blocks the effect of L-dopa on estrous cycles in old rats (Linnoila and Cooper, 1976). Also, the DA receptor agonist lergotrile mesylate has been reported to reinstate estrous cycles in the rat (Clemens and Bennett, 1977).

A serotonergic involvement in the impairment of LH secretion, which accompanies increasing postmaturational age has been proposed (Walker and Timiras, 1980). This conclusion is based largely upon the observation that extending the normal photoperiod, which stimulates the serotonergic system, results in dampened and prolonged LH surges in young rats on the day of proestrus. This pattern of LH secretion appears to mimic the LH profiles reported for aging rats (Van der Shoot, 1976). However, in the absence of a clearly demonstrated change in serotonergic activity in the hypothalamus of old animals, this hypothesis has not yet received general support.

Prolactin

Prolactin (Table 17-7) unlike the other five AP hormones, is maintained under the tonic inhibitory influence of the hypothalamus (Frohman and Berelowitz, chapter 4). Considerable evidence has accumulated that indicates that the TIDA system exerts this inhibitory influence on prolactin secretion (Weiner and Ganong, 1978). However, it is not yet clear whether DA itself or a DA-stimulated prolactin-inhibiting factor (PIF) acts on the AP to inhibit prolactin secretion. Gonadal steroids (Fuxe et al., 1969; Simpkins et al., 1979), enkephalins (Gudelsky and Porter, 1979), and prolactin itself (Gudelsky et al., 1976) influence the activity of TIDA neurons and may thereby influence the release of pituitary prolactin as well as other hormones. The serotonergic system has also been implicated in regulation of prolactin secretion. Lesioning the raphe nuclei decreases, whereas stimulating the raphe nuclei increases, both hypothalamic 5-HT turnover and serum prolactin levels (Advis et al., 1978). Also, inhibiting serotonergic neurotransmission blocks the surges in serum prolactin induced by suckling (Kordon et al., 1973) and estrogen treatment (Caligaris and Taleisnik, 1974). The increase in serotonergic activity observed during the dark phase of the rodent light cycle is accompanied by an increase in serum prolactin concentration (Meites and Clemens, 1972). There is little convincing evidence that noradrenergic neurons participate in the regulation of prolactin secretion (Weiner and Ganong, 1978).

In both male and female rats, increases in serum prolactin concentrations are a common characteristic of increasing age (Huang et al., 1976; Riegle and Meites, 1976; Simpkins et al., 1977; Clemens et al., 1979; Estes and Simpkins, 1980). In man, serum prolactin levels remain normal through the seventh decade of life and appear to increase thereafter (Singer et al., 1979). This age-related increase in serum prolactin in senescent rats correlates well with the decreased median eminence DA

Table 17-7. Age-Related Alterations in Prolactin Secretion

Age-related change	Species	Ages compared	Reference
Concentrations			
Increase	Female LE rat[a]	4-6 vs 23-30 mo	Shaar et al., 1975
Increase	Female LE rat	4-5 vs 22-30 mo	Huang et al., 1977
Increase	Male LE rat	4-6 vs 22-30 mo	Riegle and Meites, 1976
Increase	Male Wistar rat	3-4 vs 21 mo	Simpkins et al., 1977
No change	Male LE rat	4-6 vs 23-30 mo	Shaar et al., 1975
Increase	Human	84 yr[b]	Singer et al., 1979
Response			
Decreased response to L-dopa	Male LE rat	4-6 vs 22-30 mo	Riegle and Meites, 1976
Decreased response to ovariectomy	Female LE rat	4-6 vs 23-30 mo	Shaar et al., 1975
Normal response to orchiectomy	Male LE rat	4-6 vs 23-30 mo	Shaar et al., 1975
Normal response to estradiol	Female LE rat	4-5 vs 22-30 mo	Huang et al., 1976
Delay decline after estradiol withdrawal	Female LE rat	4-5 vs 22-30 mo	Huang et al., 1976

[a] LE: Long-Evans.

[b] Mean age of senescent group; young controls not included.

concentration and turnover (Simpkins et al., 1977; Finch, 1979; Wilkes et al., 1979; Estes and Simpkins, 1980). Treatment of old rats with L-dopa (Riegle and Meites, 1976) or the DA agonist lergotrile mesylate (Clemens et al., 1979) results in a dramatic decrease in circulating prolactin. These data support a DA involvement in the age-related increase in serum pro-lactin.

Although daily long-term (two-month) treatment with lergo-trile mesylate dramatically decreased serum prolactin concen-trations in old rats, the drug was not able to reduce prolactin concentrations to values observed in young rats (Clemens et al., 1979). Similarly, serum prolactin concentrations decrease in old female rats after withdrawal of estrogen stimulation, but do not return to low values observed in young animals (Huang et al., 1976). These data may indicate that DA recep-tors or DA-receptor mediated biochemical events are less sen-sitive in old rats. However, neither DA-receptor number nor receptor sensitivity has been specifically studied in hypotha-lamic or pituitary tissues of old animals. The high incidence of prolactin-secreting pituitary adenomas observed in old rats (Huang et al., 1976), as well as the elevated occurrence of prolactin-containing microadenomas in pituitaries from aged patients (Kovacs et al., 1980), suggests prolactin cells are transformed and therefore less responsive to DA.

The observed hypersecretion of prolactin in old rats raises the possibility that prolactin may be involved in the etiology of reproductive decline in rodents. In young male and female rats, elevated serum prolactin levels suppress the postcastra-tion rise in gonadotropins (Grandison et al., 1977; Gudelsky et al., 1976; Hodson et al., 1980). In human patients, hyper-prolactinemia is often associated with amenorrhea; further, decreasing serum prolactin, through treatment with ergobromo-criptine, re-established menstral cycles (Franz et al., 1972). Thus, it has been proposed that hyperprolactinemia associated with increasing age in the rat may contribute to impaired gonadotropin secretion (Aschheim, 1976). Clemens et al. (1979) tested this possibility by decreasing serum prolactin in old female rats through daily administration of lergotrile mesylate but failed to observe any effect of this treatment on LH secre-tion. Consequently the age-related hypersection of prolactin appears not to be the primary factor involved in the suppres-sion of gonadotropin secretion in old rats.

Growth Hormone

There is considerable evidence that central noradrenergic neu-rons, through an α-receptor mechanism, exert a stimulatory in-

fluence on growth hormone (GH) secretion in young mature
animals of several species (Martin, 1976; Weiner and Ganong
1978; Frohman and Berelowitz, chapter 4). The role of dopaminergic and serotonergic pathways in the regulation of GH
secretion is currently controversial (Table 17-8).

In human subjects pituitary content and basal serum levels of
GH remain normal with increasing age. (Gershberg, 1957; Russfield, 1960; Utiger, 1964; Dudl et al., 1973; Blichert-Toft,
1975). However, old men and women have a decreased capacity to
secrete GH in response to stress, surgical trauma, exercise,
and arginine treatment (Dudl et al., 1973; Blichert-Toft, 1975;
Bazzarre et al., 1976). Further, the increase in GH secretion
observed during sleep in young adults is absent in old subjects
(Finkelstein et al., 1972; Carlson et al., 1972; Blichert-Tolf,
1975). Collectively, these data indicate that the pituitary
contains sufficient GH to maintain basal GH secretion throughout life; but in response to CNS-mediated stimuli, a marked
age-related decline in the GH secretory response is observed.

In human subjects, L-dopa through its conversion to NE, but
not DA increases GH secretion (Frohman and Berelowitz, chapter
4). This L-dopa-induced increase in GH is absent in old subjects (Bazzarre et al., 1975), suggesting a deficiency in the
noradrenergic input to the hypothalalmus with age in human
subjects. In young rats, GH is released in secretory pulses.
These pulses of serum GH can be blocked by phenoxybenzamine and
reinitiated by clonidine, suggesting an α-adrenergic input
into this pulsatile secretion of GH (Martin, 1976). In old
male rats these pulses are severly dampaned or absent (Sonntag
et al., 1980). The interesting similarity between the noradrenergic stimulation of GH and LH pulses in young animals and
the age-related decline in the pulsatile secretion of both AP
hormones suggests that a deficiency in noradrenergic neurons in
old rats may mediate both of these age-related alterations.

Thyroid-Stimulating Hormone

The regulation of thyroid stimulating hormone (TSH), although
less studied than the previously discussed AP hormones, appears
to be under stimulatory influence of NE and may be inhibited by
DA, whereas the role of 5-HT is unresolved (Weiner and Ganong,
1978).

Basal secretory rates of TSH, response to thyrotropin-releasing hormone (TRH), and pituitary content of TSH are normal
or only slightly affected by age in rat, mouse, and man (Table
17-9). However, like the other aforementioned AP hormones,
some CNS-mediated alterations in TSH secretion have been observed in old animals. Klug and Adelman (1979) reported that

Table 17-8. Age-Related Alterations In Growth Hormone Secretion

Age-related change	Species	Ages compared	Reference
Concentrations			
No change	Man	22 to 81 yr[a]	Dudl et al., 1973
No change	Man	18-30 vs 70-94 yr	Blichert-Toft, 1975
No change	Man	12 vs 72 yr[a]	Gershberg, 1957
No change	Man	9 vs 86 yr[a]	Russfield, 1960
Decreased pulsatile secretion	SD rat[b]	5-6 vs 18-20 mo	Sonntag et al., 1980
Responses			
No change in glucose response	Man	22-81 yr[a]	Dudl et al., 1973
Decreased response to stress	Man	18-30 vs 70-94 yr	Blichert-Toft, 1975
Decreased response to surgical trauma	Man	18-30 vs 70-94	Blichert-Toft, 1975
Decreased Response to exercise	Man	23-29 vs 61-69 yr[a]	Bazzarre et al., 1976

Absence of sleep-induced increase	Man	8-42 vs 47-62	Finkelstein et al., 1972
Absence of sleep-induced increase	Man	20-44 vs 53-73 yr	Carlson et al., 1972
Absence of sleep-induced increase	Man	18-30 vs 70-94 yr	Blichert-Toft, 1975
Absence of L-dopa-induced increase	Man	23-29 vs 61-69 yr	Bazzarre et al., 1976

[a] Ages indicate range tested.

[b] SD: Sprague-Dawley.

Table 17-9. Age-Related Alterations in Thyroid-Stimulating Hormone Secretion

Age-related change	Species	Ages compared	Reference
Concentrations			
No change	Man	25-35 vs 60-93 yr	Ohara et al., 1974
No change	Male Winstar rat	3-4 vs 21 mo	Simpkins et al., 1977
No change	Male SD rat[a]	2 vs 24 mo	Klug and Adelman, 1979
No change	Female LE rat[a]	3-4 vs 16-17 mo	Chen and Walfish, 1978
Increase	Male LE rat	3-5 vs 22-24 mo	Chen and Walfish, 1979
Absence of circadian rhythm	SD rat	2 vs 24 mo	Klug and Adelman, 1979
No change	Male C57BL/6J mouse	12 vs 28 mo	Finch et al., 1977
Responses			
Absence of stress-induced decrease	SD rat	3-4 vs 18-20 mo	Simpkins et al., 1978
No change in response to cold	LE rats	6-8 vs 20-24 mo	Huang et al., 1980
No change in response to thyroidectomy	LE rats	6-8 vs 20-24 mo	Huang et al., 1980

the circadian rhythm in serum TSH observed in young male rats
is absent in 24-month-old rats. During the "lights on" phase
of the daily light-dark cycle, serum TSH levels are equal in
young and old rats, although the normal dark phase decline in
TSH is absent in old animals. Relevant to the Klug and Adelman
observation is the relative absence of CNS-mediated inhibition
of TSH secretion in the rat. Simpkins et al. (1978) noted that
several types of stress produced rapid, profound decreases in
serum TSH in young, but not old rats. Old rats are apparently
unable to activate mechanisms that inhibit TSH secretion. Although the neuronal system that mediates this stress-induced
inhibition of TSH secretion is not clearly defined, it is well
known that stress activates central noradrenergic and serotonergic neurons (Stone, 1975; Mueller et al., 1976).

A recent study by Huang et al. (1980) demonstrated that TSH
secretion in old male rats in response to low ambient temperature or thyroidectomy is equal to that observed in young rats.
These data are of particular interest, since it has been reported in several studies that the cold-induced secretion of
TSH is mediated by central noradrenergic neurons (Krulich et
al., 1977; Scapagnini et al., 1977). Thus, in contrast to age-
related deficiencies in NE-mediated secretion of LH and GH, NE-
mediated TSH secretion appears normal in old rats. This differential effect of increasing age on LH, GH, and TSH secretory
response may indicate that components of the central noradrenergic system regulating the secretion of these AP hormones age
at different rates. An alternative explanation is that with
increasing age a compensatory TSH secretory mechanism becomes
active.

Adrenocorticotropic Hormone

The effects of NE on ACTH secretion are thought to be inhibitory in the rat and dog, but are uncertain in other species
(Table 17-10). Similarly, the roles of DA and 5-HT in ACTH
regulation are controversial (Weiner and Ganong, 1978).

Alterations in ACTH secretion have been more difficult to
assess then other AP hormones because of inherent difficulties
in immunoassay procedures for ACTH. Most studies that evaluate
ACTH secretion use as an end point alteration in adrenal corticoid secretion. This presents a major obstacle in interpreting
data concerned with ACTH regulation in aging animals, since an
age-related decline in ACTH-induced increase in corticoid secretion has been noted in the rhesus monkey, cow, and female
rat (Bowman and Wolf, 1969; Shapiro and Leathem, 1970), although human adrenals appear to respond well to ACTH throughout
life (Blichert-Toft, 1975).

Table 17-10. Age-Related Alterations in Adrenocorticotropic Hormone Secretion

Age-related change	Species	Ages compared	Reference
Concentrations[a]			
No change in basal levels	C57BL/6J mouse	8-12 vs 28-32 mo	Latham and Finch, 1976
No change in basal levels of corticosterone	Rat	4 vs 25 mo	Hess and Riegle, 1970
No change in basal levels of cortisol	Man	18-30 vs 70-94 yr	Blichert-Toft, 1975
No change in ACTH and cortisol circadian rhythm	Man	18-30 vs 70-94 yr	Blichert-Toft, 1975
Circadian rhythm of corticosterone present but nadir level elevated	SD rat[b]	2 vs 24 mo	Klug and Adelman, 1979
Responses			
No change in response to insulin	Man	13-50 vs 67-91 yr[c]	Friedman et al., 1969
No change in stress-induced increase	Rat	4 vs 25 mo	Hess and Riegle, 1970

[a] As indicated in the text, serum corticoids are frequently used as an index of ACTH secretion.

[b] SD: Sprague-Dawley.

[c] Ages indicated range tested.

In old human subjects, adrenal and ACTH secretion appear
normal in response to insulin, arginine, and dexamethasone
(Friedman et al., 1969; Jensen and Blichert-Toft, 1970). The
circadian rhythm in serum ACTH and corticoids remains normal
with age in human subjects (Blichert-Toft, 1975).
In contrast, circadian corticoid fluctuations appear altered
in the old male rat. Nadir corticoid concentrations are elevated in aged compared to young rats (Klug and Adelman, 1979).
We have recently observed that adrenal progesterone, which is
secreted in response to ACTH, shows a similar altered circadian
rhythm in old male rats, and nadir concentrations are similarly
elevated (Simplins et al., unpublished observation). In view
of the age-related decrease in adrenal responsiveness to ACTH
and the observation that the half-life of serum corticosterone
does not change with age (Riegle and Hess, 1972), these data
suggest that the old rat may have consistently high serum ACTH
concentrations. Since the central noradrenergic system appears
to exert an inhibitory influence on ACTH release, the observed
deficiency in hypothalamic NE turnover in the rat may contribute to elevated circulating corticosterone. An alternative
possibility is that an age-related decline in the sensitivity
of the hypothalamus to the inhibitory effects of circulating
corticoid occurs. This possibility is supported by the studies
of Riegle and Hess (1972) who demonstrated a moderate age-related decline in the ability of dexamethasone to block the
stress-induced rise in serum corticosterone. Studies utilizing
immunoassay techniques for ACTH may better elucidate the role
of central catecholaminergic and indolaminergic systems in the
release of ACTH in old animals.

CONCLUSIONS

The hypothalamus maintains the capacity to regulate basal secretory rates of AP hormones throughout life in most mammalian
species. However, challenging old animals with CNS-mediated
stimuli reveal substantial alterations in the ability of the
hypothalamus to modify AP hormone secretion. The decrease in
hypothalamic noradrenergic activity that accompanies increasing
age may contribute to the age-related alterations in the secretory patterns of LH, FSH, GH, TSH, and ACTH. Further, there is
strong evidence that the decrease in activity of the TIDA system is responsible for the age-related elevation in prolactin
secretion. The role of the TIDA or IHDA system in the regulation of the secretion of other AP hormones is not well defined in either young or old animals.
No clearly defined alterations in hypothalamic 5-HT metabolism have been observed with advancing age in rodents, al-

though increased MAO activity has been widely observed in man. In view of conflicting reports on the role of 5-HT in the mechanisms regulating AP hormone secretion, the involvement of 5-HT neurons in the aging process is uncertain. A further evaluation of the role of the CNS in the age-related alterations in AP function awaits a clear definition of the normal involvement of 5-HT, as well as other putative neurotransmitters, in the regulation of hormone secretion.

Although much attention has been given to the evaluation of alterations in LH, FSH, and prolactin secretion with increasing age, relatively little effort has been devoted to the assessment of age-related changes in GH, TSH, and ACTH secretion. Studies devoted to these latter three AP hormones have been primarily clinical evaluations and for obvious humanitarian reasons these studies were limited largely to evaluation of plasma hormone concentration. Studies using appropriate animal models to evaluate systemically the role of monoaminergic neurons in the age-related alteration in GH, TSH, and ACTH secretion are needed.

Finally, evalution of monoaminergic metabolism in old animals has been confined largely to determinations of basal concentrations and turnover rates in relatively large brain tissue fragments. Determining the responses of central monoaminergic neurons to CNS-mediated stimuli may reveal age-related deficiencies in these neurons which are not apparent under basal conditions. Further, given the large number of homeostatic processes that are influenced by specific hypothalamic areas, evaluating discrete regions may reveal differential, age-associated alterations in monoamine metabolism. Such an approach would not only improve our understanding of the cause of the altered secretion of AP hormones that accompanies age, but also may contribute to our understanding of the age-related alterations in other homeostatic processes.

REFERENCES

Advis, J.P. Simpkins, J.W., Chen, H.T., and Meites, J.: Relation of biogenic amines to onset of puberty in the remale rat. *Endocrinology* 103:11-16, 1978.

Ascheim, P.: Aging in the hypothalamic-hypophysial ovarian axis in the rat. In *Hypothalamus, Pituitary and Aging*, A.V. Everitt and J.A. Burgess, eds. Charles C Thomas Publishing Company, Springfield, Ill., 1976, pp 376-418.

Bazzarre, T.L., Johanson, A.J., Huseman, C.A., Varma, M.M., and Blizzard, R.M.: Human growth hormone changes with age. In *Growth Hormone and Related Peptides*, A. Pecile and E.E. Muller, eds. Excerpta Medica, Amsterdam, 1976, pp 261-270.

Berg, A.P., and Zimmerman, I.D.: Effects of electrical stimulation and norepinephrine on cyclic AMP levels in the cerebral cortex of the aging rat. *Mech Aging Dev* *4*:377-383, 1975.

Bertler, A.: Occurrence and localization of catecholamines in the human brain. *Acta Physiol Scand* *51*:97-107, 1961.

Bjorklund, A., Lindvall, O., and Nobin, A.: Evidence of an incerto-hypothalamic dopamine neurone system in the rat. *Brain Res* *89*:29-42, 1975.

Blichert-Toft, M.: Secretion of corticotrophin and somatotrophin by the senescent adenohypophysis in man. *Acta Endocrinol (Copenh)* *78*:1-157, 1975.

Bowers, M.B., and Gerbode, R.A.: Relationship of monoamine metabolites in human cerebrospinal fluid to age. *Nature* *219*: 1256-1257, 1968.

Bowman, R.E., and Wolf, R.C.: Plasma 17-hydroxycorticosteroid response to ACTH in M. mullata: dose, age, weight, and sex. *Proc Soc Exp Biol Med* *130*:61-64, 1969.

Bronson, F.H., and Desjardins, C.: Reproductive failure in aged CBF. male mice: Interrelationships between pituitary gonadotropic hormones, testicular function and mating success. *Endocrinology* *101*:939-945, 1977.

Bruni, J.F., Huang, H.H., Marshall, S., and Meites, J.: Effects of single and multiple injections of synthetic GnRH on serum LH, FSH, and testosterone in young and old male rats. *Biol Reprod* *17*:309-312, 1977.

Caligaris, L., Astrada, J.J., and Taleisnik, S.: Stimulatory and inhibiting effects of progesterone on the release of luteinizing hormone. *Acta Endocrinol (Copenh)* *59*:177-185, 1968,

Caligaris, L., and Taleisnik, S.: Involvement of neurons containing 5-hydroxytryptamine in the mechanism of prolactin release induced by oestrogen. *J Endocrinol* *62*:25-33, 1974.

Carlson, H.E., Gillin, J.C., Gorden, P., and Synder, F.: Absence of sleep related growth hormone peaks in aged normal subjects and acromegaly. *J Clin Endocrinol Metab* *34*:1102-1105, 1972.

Carlsson, A., and Winblad, B.: The influence of age and time interval between death and autopsy on dopamine and 3-methosytyramine levels in human basal ganglia. *J Neural Transm* *38*: 271-276, 1976.

Chen, H.T., and Walfish, P.G.: Effects of age and ovarian function on the pituitary-thyroid system in female rats. *J Endocrinol* *78*:225-232, 1978.

Chen, H.T., and Walfish, P.G.: Effects of age and testicular function on the pituitary-thyroid system in male rats. *J Endocrinol* *82*:53-59, 1979.

Clemens, J.A., and Bennett, D.R.: Do aging changes in the preoptic area contribute to loss of cyclic endocrine function?

J Gerontol 32:19-24, 1977.

Clemens, J.A., Amenomori, Y., Jenkins, T., and Meites, J.: Effects of hypothalamic stimulation, hormones and drugs on ovarian function in old female rats. *Proc Soc Exp Biol Med 132*:561-563, 1969.

Clemens, J.A., Fuller, R.W., and Owen, N.V.: Some neuoendocrine aspects of aging. In *Advances in Experimental Medicine and Biology: Parkinson's Disease-II*, C.E. Finch, D.E. Potter, and A.D. Kenny, eds. Plenum Press, New York, 1979, pp 77-100.

Cooper, J.R., Bloom, F.E., and Roth, R.H.: *The Biochemical Basis of Neuropharmacology, 3rd edition*, Oxford University Press, New York, 1978.

Cooper, R.L., Brandt, S.J., Linnoila, M., and Walker, R.F.: Induced ovulation in aged female rats by L-DOPA implants into the medial preoptic area. *Neuroendocrinology 28*:234-240, 1979.

Coppola, J.A.: Turnover of hypothalamic catecholamines during various states of gonadotropin secretion. *Neuroendocrinology 5*:75-80, 1969.

Cote, L.J., and Kremzner, L.T.: Changes in neurotransmitter systems with increasing age in human brain. *Trans Am Soc Neurochem 5*:83, 1975.

Dilman, V.M.: Age-associated elevation of hypothalamic threshold to feedback control, and its role in development, aging and disease. *Lancet 1*:1211-1219, 1971.

Dudl, R.J., Ensinck, J.W., Palmer, H.E., and Williams, R.H.: Effect of age on growth hormone secretion in man. *J Clin Endocrinol Metab 37*:11-16, 1973.

Estes, K.S., and Simpkins, J.W.: Age-related alterations in catecholamine concentrations in discrete preoptic area and hypothalamic regions in the male rat. *Brain Res 192*:556-560, 1980.

Estes, K.S., Simpkins, J.W., and Chen, C.L.: Alteration in pulsatile release of LH in aging female rats. *Proc Soc Exp Biol Med 163*:384-387, 1980.

Everitt, A.V.: The Neuroendocrine system in aging. *Gerontology 26*:108-119, 1980.

Everett, J.W.: The restoration of ovulatory cycles and corpus luteum formation in persistent-estrous rats by progesterone. *Endocrinology 27*:681-686, 1940.

Finch, C.E.: Catecholamine metabolism in the brains of aging male mice. *Brain Res 52*:271-276, 1973.

Finch, C.E.: Neuroendocrine and autonomic aspects of aging. In *Handbook of the Biology of Aging*, C.E. Finch and L. Hayflick, eds. Van Nostrand Reinhold Company, New York, 1977, pp 262-280.

Finch, C.E., Jonec, V., Hody, G., Walker, J.P., Morton-Smith, W., Alper, A., and Dougher, G.J.: Aging and the passage of L-tyrosine, L-DOPA, and insulin into mouse brain slices in vitro. *J Gerontol* 30:33-40, 1975.

Finch, C.E., Jonec, V., Wisner, J.R., Sinha, Y.N., De Vellis, J.S., and Swerdloff, R.S.: Hormone production by the pituitary and testes of male C57BL/6J mice during aging. *Endocrinology* 101:1310-1317, 1977.

Finch, C.E.: Age-related changes in brain catecholamines: A synopsis of findings in C57BL/6J mice and other rodent models. In *Advances in Experimental Medicine and Biology, vol. 113: Parkinson's Diseases-II*, C.E. Finch, D.E. Potter, and A.D. Kenny, eds. Plenum Press, New York, 1979, pp 15-40.

Finkelstein, J.W., Roffwarg, H.P., Boyer, R.M., Kream, J., and Hellman, I.: Age-related change in the twenty-four-hour spontaneous secretion of growth hormone. *J Clin Endocr Metab* 35:665-670, 1972.

Frantz, A.G., Kleinberg, D.L., and Noel, G.L.: Studies on prolactin in man. *Recent Prog Horm Res* 28:527-590, 1972.

Friedman, M., Green, M.F., and Sharland, D.E.: Assessment of hypothalamic-pituitary-adrenal function in the geriatric age group. *J Gerontol* 24:292-297, 1969.

Fuxe, K., and Ungerstedt, U.: Histochemical studies on the distribution of catecholamines and 5-hydroxytryptamine after intraventricular injections. *Histochemistry* 13:16-28, 1968.

Fuxe, K., Hokfelt, T., and Nelsson, O.: Castration, sex hormones and tuberoinfundibular dopamine neurons. *Neuroendocrinology* 5:107-120, 1969.

Gallo, R.: Neuroendocrine regulation of pulsatile luteinizing hormone release in the rat. *Neuroendocrinology* 30:122-131, 1980.

Gershberg, H.: Growth hormone content and metabolic actions of human pituitary gland. *Endocrinology* 61:160-165, 1957.

Gosden, R.G., and Bancroft, L.: Pituitary function in reproductively senescent female rat. *Exp Gerontol* 11:157-160, 1976.

Gottfries, C.G., Gottfries, I., Johansson, R., Olsson, R., Persson, T., Roos, B.E., and Jostrom, R.: Acid monoamine metabolites in human cerebrospinal fluid and their relations to age and sex. *Neuropharmacology* 10:665-672, 1971.

Govoni, S., Loddo, P., Spano, P.F., and Trabuchi, M.: Dopamine receptor sensitivity in brain and retina of rats during aging. *Brain Res* 138:565-570, 1977.

Grandison, L., Hodson, C., Chen, H.T., Advis, J., Simpkins, J., and Meites, J.: Inhibition by prolactin of post-castration rise in LH. *Neuroendocrinology* 23:312-322, 1977.

Greenberg, L.H., and Weiss, B.: β-Adrenergic receptors in aged rat brain: Reduced number and capacity of pineal gland to develop supersensitivity. *Science* 201:61-63, 1978.

Gudelsky, G.A., and Porter, J.C.: Morphine- and opioid peptide-induced inhibition of the release of dopamine from tuber-infundibular neurons. *Life Sci 25*:1697-1702, 1979.

Gudelsky, G.A., Simpkins, J.W., Mueller, G.P., and Meites, J.: Selective effect of prolactin on dopamine turnover in the hypothalamus and on serum LH and FSH. *Neuroendocrinology 22*: 206-215, 1976.

Harman, S.M.: Clinical aspects of aging of the male reproductive system. In *Aging, Vol: The Aging Reproductive System*, E.L. Schneider, ed. Raven Press, New York, 1978, pp 29-58.

Harman, S.M., Martin, C.E., and Tsitaures, P.D.: Gonadotropins, sex steroids, and sexual activity in healthy aging men. In *Endocrine Aspects of Aging*, S.G. Koreman, ed. NIH Publication, Bethesda, (Abstr) 1979.

Haug, E., Aakvaag, A.L., Suadt, T., and Torjesen, P.A.: The gonadotropins response to synthetic gonadotropin releasing hormone in males in relation to age, dose and basil serum levels of testosterone oestradiol 17B and gonadotrophin. *Acta Endocrinol (Copenh) 77*:625-635, 1974.

Hery, F., Rouer, E., and Glowinsky, J.: Daily variations of serotonin metabolism in the rat brain. *Brain Res 43*:445-465, 1972.

Hess, G.O., and Riegle, G.D.: Adrenocortical responsiveness to stress and ACTH in aging rats. *J Gerontol 25*:354-358, 1970.

Hillarp, N.A.: Studies on the localization of hypothalamic centres controlling the gonadotrophic function of the hypophysis. *Acta Endocrinol (Copenh) 2*:11-23, 1949.

Hodson, C.A., Simpkins, J.W., Pass, K.A., Aylsworth, C.F., Steger, R.W., and Meites, J.: Effects of a prolactin-secreting pituitary tumor on hypothalamic, gonadotropic and testicular function in male rats. *Neuroendocrinology 30*:7-10, 1980.

Howland, B.E.: Reduced gonadotropin release in response to progesterone or gonadotropin releasing hormone (GnRH) in old female rats. *Life Sci 19*:219-224, 1976.

Howland, B.E., and Preiss, C.: Effects of aging on basal levels of serum gonadotropins, ovarian compensatory hypertrophy and hypersecretion of gonadotropins after ovariectomy in female rats. *Fertil Steril 26*:271-276, 1975.

Huang, H.H., and Meites, J.: Reproductive capacity of aging female rats. *Neuroendocrinology 17*:289-295, 1975.

Huang, H.H., Marshall, S., and Meites, J.: Capacity of old vs young female rats to secrete LH, FSH and prolactin. *Biol Reprod 14*:538-543, 1976.

Huang, H.H., Simpkins, J.W., and Meites, J.: Hypothalamic norepinephrine (NE) and dopamine (DA) turnover and relation to LH, FSH and prolactin release in old female rat. *Endocrinology 100 (Suppl)*:331, 1977.

Huang, H.H., Steger, R.W., and Meites, J.: Capacity of old vs

young male rats to release thyrotropin (TSH), thyroxine (T)
and triiodothyronine (T_3) in response to different stimuli.
Exp Aging Res 6:3-12, 1980.

Jensen, H.K., and Blichert-Toft, M.: Pituitary-adreneral function in old age evaluated by the intravenous metyrapone test.
Acta Endocrinol (Copenh) 64:431-438, 1970.

Jonec, V., and Finch, C.E.: Aging and dopamine uptake by
subcellular fractions in the C57BL/6J male mouse brain.
Brain Res 91:197-215, 1975.

Joseph, J.A., Berger, R.E., Engel, B.T., and Roth, G.S.: Age-
related changes in the nigrostriatum: A behavioral and biochemical analysis. *J Gerontol 33*:643-649, 1978.

Kanungo, M.S., Patnaik, S.K., and Koul, O.: Decrease in 17 -
oestradiol receptor in brain of aging rats. *Nature 253*:366-
367, 1975.

Kizer, J.S., Palkovits, M., and Brownstein, M.J.: The projections of the A8, A9 and A10 dopaminergic cell bodies: Evidence for a nigro-hypothalamic-median eminence dopaminergic
pathway. *Brain Res 108*:363-370, 1976.

Klug, T.L., and Adelman, R.C.: Altered hypothalamic-pituitary
regulation of thyrotropin in male rats during aging.
Endrocrinology 104:1136-1142, 1979.

Kordon, C., Blake, C.A., Terkel, J., and Sawyer, C.H.: Participation of serotonin-containing neurons in the suckling-induced rise in plasma prolactin levels in lactating rats.
Neuroendocrinology 13:213-223, 1973.

Kovacs, K., Ryan, N., Horvath, E., Singer, W., and Ezrin, C.:
Pituitary adenomas in old age. *J Geronotol 35*:16-22, 1980.

Krulich, L., Giachetti, A., Marchlewskakoj, A., Hefro, E., and
Jameson, H.E.: On the role of the central noradrenergic and
dopaminergic systems in the regulation of TSH secretion in
the rat. *Endocrinology 100*:496-505, 1977.

Latham, K.R., and Finch, C.E.: Hepatic glucocorticoid binders
in mature and senescent C57BL/6J male mice. *Endocrinology 98*:
1480-1489, 1976.

Linnoila, M., and Cooper, R.L.: Reinstatement of vaginal cycles
in aged female rats. *J Pharmacol Exp Ther 199*:477-482, 1976.

Lu, K.H., Huang, H.H., Chen H.T., Kurcz, M., Mioduszewski, R.,
and Meites, J.: Positive feedback by estrogen and progesterone on LH release in old and young rats. *Proc Soc Exp Biol
Med 154*:82-85, 1977.

Lu, K.H., Hopper, B.R., Vargo, T.M., and Yen, S.S.C.: Chronological changes in sex steroid, gonadotropin and prolactin
secretion in aging female rats displaying different reproductive states. *Biol Reprod 21*:193-203, 1979.

McGeer, E.G., Fibiger, H.C., McGeer, P.L., and Wickson, V.:
Aging and brain enzymes. *Exp Gerontol 6*:391-396, 1971.

Makman, M.J., Ahn, H.S., Thal, L., Dvorkin, B., Horowitz, S.G.,

Sharpless, N., and Rosenfeld, M.: Decreased brain biogenic amine-stimulated adenylate cyclase and spiroperidol-binding sites with aging. *Fed Proc 37*:548, 1978.

Marshall, J.F., and Berrios, N.: Movement disorders of aged rats: Reversal by dopamine receptor stimulation. *Science 206*:477-479, 1979.

Martin, J.B.: Brain regulation of growth hormone secretion. In *Frontiers in Neuroendocrinology*, Vol 4, L. Martini and W. Ganong, eds. Raven Press, New York, 1976, pp 129-168.

Meek, J.L., Bertilsson, L., Cheney, D.L., Zsilla, G., and Costa, E.: Aging induced changes in acetylcholine and serotonin content of discrete brain nuclei. *J Gerontol 32*: 129-131, 1977.

Meites, J., and Clemens, J.A.: Hypothalamic control of prolactin secretion. *Vitam Horm 30*:165-221, 1972.

Miller, A.E., and Riegle, G.D.: Serum LH levels following multiple LHRH injections in aging rats. *Proc Soc Exp Biol Med 157*:497-499, 1978.

Miller, A.E., Shaar, C.J., and Riegle, G.D.: Aging effects on hypothalamic dopamine and norepinephrine content in the male rat. *Exp Aging Res 2*:475-480, 1976.

Moore, R.Y., and Bloom, F.E.: Central catecholamine neuron systems: Anatomy and physiology of the dopamine systems. *Annu Rev Neurosci 1*:129-169, 1978.

Moore, R.Y., and Bloom, F.E.: Central catecholamine neuron systems: Anatony and physiology of the norepinephrine and epinephrine systems. *Annu Rev Neurosci 2*:113-168, 1979.

Mueller, G.P., Twohy, C.P., Chen, H.T., Advis, J.P., and Meites, J.: Effect of L-tryptophan and restraint stress on hypothalamic and brain serotonin turnover, and pituitary TSH and prolactin release in rats. *Life Sci 18*:715-724, 1976.

Munaro, N.I.: The effect of ovarian steroids on hypothalamic norepinephrine activity. *Acta Endocrinol (Copenh) 86*:235-242, 1977.

Odell, W.D., and Swerdloff, R.S.: Progesterone-induced luteinizing and follicle stimulating hormone in postmenopausal women: A simulated ovulatory peak. *Proc Natl Acad Sci USA 61*: 529-536, 1968.

Ohara, H., Kobayashi, T., Shiraishi, M., and Wada, T.: Thyroid function of the aged as viewed from the pituitary-thyroid system. *Endocrinol Jpn 21*:377-386, 1974.

Ordy, J.M., Kaach, B., and Brizzee, K.R.: Life-span neurochemical changes in the human and non-human primate brain. In *Aging, Vol I: Clinical, Morphologic and Neurochemical Aspects in the Aging Central Nervous System*, H. Brody, D. Harman and J.M. Ordy, eds. Raven Press, New York, 1975, pp 133-168.

Palkovits, M., Fekete, M., Makara, G.B., and Herman, J.P.: Total and partial hypothalamic deafferentations for topagraphical identification of catecholaminergic innervations of certain preoptic and hypothalamic nuclei. *Brain Res 127*:127-136, 1977.

Peluso, J.J., Steger, R.W., and Hafez, E.S.E.: Regulation of LH secretion in aged female rats. *Biol Reprod 16*:212-215, 1977.

Ponzio, F., Brunell, N., and Algeri, S.: Catecholamine synthesis in the brain of aging rats. *J Neurochem 30*:1617-1620, 1978.

Prange, A.J., Jr., White, J.E., and Lipton, M.A.: Influence of age on monoamine oxidase and catechol-o-methyltransferase in rat tissue. *Life Sci 6*:581-586, 1967.

Puri, S.K., and Volicer, L.: Effect of aging on cyclic AMP levels and adenylate cyclase and phosphodiesterase activities in the rat corpus striatum. *Mech Aging Dev 6*:53-58, 1976.

Quadri, S.K., Kledzik, G.S., and Meites, J.: Reinitiation of estrous cycles in old constant estrous rats by central acting drugs. *Neuroendocrinology II*:248-255, 1973.

Reis, D.J., Ross, R.A., and Joh, T.H.: Changes in the activity and amounts of enzymes synthesizing catecholamines and acetylcholine in brain, adrenal medulla, and sympathetic ganglia of aged rat and mouse. *Brain Res 136*:465-474, 1977.

Riederer P., and Wuketich, S.: Time course of nigrostriatal degeneration in Parkinson's disease. *J Neural Transm 38*:277-301, 1976.

Riegle, G.D., and Hess, G.D.: Chronic and acute dexamethasone suppression of stress activation of the adrenal cortex in young and aged rats. *Neuroendocrinology 9*:175-187, 1972.

Riegle, G.D., and Meites, J.: Effects of aging on LH and prolactin after LHRH, L-DOPA, methyl dopa and stress in the male rat. *Proc Soc Exp Biol Med 151*:507-511, 1976.

Robinson, D.S.: Changes in monoamine oxidase and monoamines with human development and aging. *Fed Proc 34*:103-107, 1975.

Robinson, D.S., Nies, A., Davis, J.M., Bunney, W.E., Davies, J.M., Colburn, R.W., Bourne, H.R., Shaw, D.M., and Copper, A.J.: Aging, monoamines and monoamine oxidase. *Lancet 1*:290-291, 1972.

Roth, G.S.: Hormone action during aging: Alterations and mechanisms. *Mech Aging Dev 9*:497-514, 1979.

Russfield, A.B.: Combined bioassay and histological study of 73 human hypophyses. *Cancer 13*:790-803, 1960.

Samorajski, T.: Age-related changes in brain biogenic amines. In *Aging, Vol I: Clinical, Morphologic and Neurochemical Aspects in the Aging Central Nervous System*, H. Brody, D. Harman, and J.M. Ordy, eds. Raven Press, New York, 1975, pp 199-214.

Samorajski, T., and Rolsten, C.: Age and regional differences in the chemical composition of brains of mice, monkeys and humans. In *Progress in Brain Research, Vol 40: Neurobiolog-*

ical Aspects of Maturation and Aging, D.H. Ford, ed. Elsevier Press, New York, 1973, pp 253-265.

Samorajski, T., Rolsten, C., and Ordy, J.M.: Changes in behavior, brain and neuroendocrine chemistry with age and stress in C57BL/10 male mice. *J Gerontol 26*:168-175, 1971.

Scapagnini, U., Annunziato, L., Clementi, G., Di Renzo, G.F., Schettini, G., Fiore, L., and Preziosi, P.: Chronic depletion of brain catecholamines and thyrotropin secretion in the rat. *Endocrinology 101*:1064-1070, 1977.

Schiff, I., and Wilson, E.: Clinical aspects of aging of the female reproductive system. In *Aging, Vol 4: The Aging Reproductive System*, E.L. Schneider, ed. Raven Press, New York, 1978, pp 9-28.

Severson, J.A., and Finch, C.E.: Age changes in human basal ganglion dopamine receptors. *Fed Proc 39*:508, 1980.

Shaar, C.J., Euker, J.S., Riegle, G.D., and Meites, J.: Effects of castration and gonadal steroids on serum luteinizing hormone and prolactin in old and young rats. *J Endocrinol 66*:45-51, 1975.

Shapiro, B.H., and Leathem, J.H.: Aging and adrenal delta[5]-3β-hydroxy-steroid dehydrogenase in female rats. *Proc Soc Exp Biol Med 136*:19-20, 1970.

Shin, J.C.: Multiple forms of monoamine oxidase and aging. In *Aging, Vol I: Clinical, Morphologic and Neurochemical Aspects in the Aging Central Nervous System*, H. Brody, D. Harman, and J.M. Ordy, eds. Raven Press, New York, 1975, pp 191-198.

Simpkins, J.W., Mueller, G.P., Huang, H.H., and Meites, J.: Evidence for depressed catecholamine and enhanced serotonin metabolism in aging male rats: Possible relation to gonadotropin secretion. *Endocrinology 100*:1672-1678, 1977.

Simpkins, J.W., Hodson, C.A., and Meites, J.: Differential effects of stress on release of thyroid-stimulating hormone in young and old male rats. *Proc Soc Exp Biol Med 157*:144-147, 1978.

Simpkins, J.W., Huang, H.H., Advis, J.P., and Meites, J.: Changes in hypothalamic NE and DA turnover resulting from steriod-induced LH and prolactin surges in ovariectomized rats. *Biol Reprod 20*:625-632, 1979.

Simpkins, J.W., Kalra, P.S., and Kalra, S.P.: Effects of testosterone on catecholamine turnover and LHRH contents in the basal hypothalamus and preoptic area. *Neuroendocrinology 30*:94-100, 1980.

Singer, W., Kovacs, K., Llewellyn, A., Horvath, E., and Gryfe, G.: Pituitary dysfunction in old age. In *Endocrine Aspects of Aging*, S.G. Koreman, ed. NIH Publication, Bethesda. (Abstr), 1979.

Sonntag, W.E., Steger, R.W. Forman, L.J., and Meites, J.: Decreased pulsatile release of growth hormone in old male rats. *Endocrinology 107*:1875–1879, 1980.

Stone, E.A.: Stress and catecholamines. In *Catecholamines and Behavior, vol 2*, A.J. Friedhoff, ed. Plenum Press, New York, 1975, pp 31–72.

Tsai, C.C., and Yen, S.S.C.: Acute effects of intravenous infusion of 17β-estradiol on gonadotropin release in pre- and post-menapausal women. *J Clin Endocrol Metab 32*:766–771, 1971.

Ungerstedt, U.: Sterotaxic mapping of monoamine pathways in the rat brain. *Acta Physiol Scand 367*:1–48, 1971.

Utiger, R.D.: Extraction and radioimmunoassay of growth hormone in human serum. *J Clin Endocrol Metab 24*:60–67, 1964.

Van der Schoot, P.: Changing pro-oestrous surges of luteinizing hormone in aging 5-day cycling rats. *J Endocrinol 69*:287–288, 1976.

Walker, J.P., and Boas-Walker, J.: Properties of adenyl cyclase from senescent rat brain. *Brain Res 54*:391–396, 1973.

Walker, R.F., Cooper, R.L., and timiras, P.S.: Constant estrus: Role of rostral hypothalamic monoamines in development of reproductive dysfunction in aging rats. *Endocrinology 107*:249–255.

Weiner, R.I., and Ganong, W.F.: Role of brain monoamines and histamine in regulation of anterior pituitary secretion. *Physiol Rev 58*:905–976, 1978.

Weiner, R.I., Shryne, J.E., Gorski, R.A., and Sawyer, C.H.: Changes in catecholamine content of rat hypothalamus following deafferentation. *Endocrinology 90*:867–873, 1972.

Wilkes, M.M., Lu, K.H., Fulton, S.L., and Yen, S.S.C.: Hypothalamic-pituitary, ovarian interactions during reproductive senescence in the rat. In *Advances in Experimental Medicine and Biology, Vol 113, Parkinson's Disease - II*, C.E. Finch, D.E. Potter, and A.D. Kenny, eds. Plenum Press, New York, 1979, pp 127–148.

Wuttke, W., and Meites, J.: Effects of electrochemical stimulation of medial preoptic area on prolactin and luteinizing hormone release in old female rats. *Pflugers Arch 341*:1–6, 1973.

18

Neuroendocrine Mechanisms in Rodent Reproductive Aging

PATRICK K. RANDALL AND CALEB FINCH

INTRODUCTION

One of the best characterized postmaturational age changes in
mammals is the decline and eventual cessation of reproductive
activity in the female. Two aspects of this phenomenon in
short-lived laboratory animals have made it a particularly
suitable and popular model for the investigation of neuroendocrine processes during aging. Firstly, in most species,
including the human, reproductive senescence occurs during the
middle third of the life-span, an age when gross pathological
lesions are rare. The general vigor of mammals at this age
minimizes some experimental confounds presented by age-related
pathological alterations that occur in later phases of life
(see Nelson et al., 1975). Secondly, there is strong evidence
that alterations in neuroendocrine control mechanisms in the
central nervous system (CNS) are major determinants of at least
one phase of this process, viz, the cessation of the typical
cyclic pattern of gonadotropins which drives ovarian cycles.

The general interest in the area of age-related changes in
peripheral endocrine organs and the CNS has stimulated several
reviews (Aschheim, 1976; Finch, 1978; Meites et al., 1978;
Miller and Reigle, 1978; Talbert, 1977; Cooper and Walker,
1979). In this chapter, we will critically evaluate the major
hypotheses regarding the causes of estrous cycle disruption in
the aging rodent and will discuss possible approaches to en-
courage progression from the present descriptive to an explanatory phase of research.

DESCRIPTIVE DATA

Estrous Cycle Changes

By 6 to 9 months of age, the frequency of abnormal estrous cycles begins to increase in colonies of female rats and mice of most strains. Such irregularities may include extended diestrus or estrus vaginal smears, leukocytic or cornified, respectively, or of various apparently noncyclic sequences interposed between ovulatory cycles of relatively normal length. Abnormalities of gestational physiology may also occur at this age, such as decreases of litter size associated with fetal resorptions (Fabricant and Schneider, 1978) and increased still births associated with delayed parturition (Holinka et al., 1978, 1979). The number of ova shed at ovulation does not generally decline in most rodent strains before one year (Harman and Talbert, 1970).

Following this period of irregular estrous cycles, animals exhibit reproductive states characterized by one of the following: (1) apparently normal cycles throughout most of life; (2) irregular cycles of varying lengths; (3) extended or continuous periods of vaginal estrus ("constant" estrus (CE) or metestrus); (4) repeated pseudopregnancy (RPP); and (5) continuous anestrus with atrophic ovaries. Since anestrous rats with atrophic ovaries have a high incidence of large prolactin (PRL) secreting pituitary tumors (Huang et al., 1976; Miller and Reigle, 1978; Lu et al., 1979), we will consider it less than the other relatively nonpathological states.

The documentation of the several discrete reproductive syndromes in the aging rodent has been of considerable importance and may necessitate a reinterpretation of studies using animals whose reproductive history was not known. By far the most thoroughly investigated rats are those showing CE and RPP vaginal smear profiles. Few investigators have made extensive use of aging animals that maintain normal estrous cycles (Aschheim, 1976), although the importance of this group is increasingly recognized (Wilkes et al., 1978; Miller and Reigle, 1980).

Few longitudinal data on the transition of individual animals into or between those ovarian states are available, although we are currently studying such transitions (Nelson et al., 1982). Data of this type could be extremely valuable in elucidating the etiology of these senile states, there being a crucial deficiency of detailed documentation on individual patterns. The importance of these prospective studies is underlined because in both man and rodents the rate of decline in reproductive function is extremely variable.

By and large, the RPP state appears to occur at later ages than does CE (Aschheim, 1976; Lu et al., 1979). Although some investigators view sequential transition from CE to RPP as the norm (Huang and Meites, 1975; Lu et al., 1979), a study by Aschheim (1976) indicates that, although a definite population of such animals exists, the transition from CE to RPP is relatively rare. Most rats made a single transition either to CE or to RPP and remained in that state until death or until they became anestrous very late in life. The transition from normal cycling to RPP appears to occur at a later age and can give the impression of the CE to RPP sequence in cross-sectional data. Such patterns, however, may be subject to strain and colony variation.

Neuroendocrine Status of Aging Female Rats

Histological examination of the ovaries and direct measurement of hypophyseal and ovarian hormones have sometimes revealed striking differences between the CE and RPP groups of reproductively senile rats.

The CE rat has small polyfollicular ovaries with follicles in different states of development and often has cystic follicles (Everett, 1940; Huang and Meites, 1975). The absence of corpora lutea accounts for the generally small ovary size and confirms the basically anovulatory nature of this syndrome. The RPP rat, on the other hand, has very large ovaries with several cohorts of corpora lutea (Huang and Meites, 1975; Aschheim, 1976). Interestingly, rats made pseudopregnant by the kidney capsule implantation of pituitary tissue have a single cohort of corpora lutea (Quilligan and Rothschild, 1960), suggesting a possible deficit in luteolysis in the aging animals of this group.

Basal levels of ovarian and pituitary hormones, with the exception of PRL, are similar in senescent and young cycling animals (Table 18-1). The maintenance of significant basal hormone secretion with increasing age may explain why the vaginal cytology of rodents, but not human beings, usually does not show major atrophy after the loss of cycles (Finch and Flurkey, 1977).

Serum PRL concentrations are elevated in both the CE and RPP, to values not usually found in the regularly cycling rats. The elevated PRL levels in the CE rat result to some extent from the continuous estrogen stimulation of the pituitary and from emergent lactotrophic pituitary adenomas. It is not surprising that plasma PRL in CE rats is substantially higher than in the RPP animals, since the continuous pattern of estrogen with low levels of progesterone secretion may contribute to the observed

Table 18-1. Hormone Profiles of Rats during Reproductive Aging[a]

	Progesterone, ng/ml	Estradiol, pg/ml	LH, ng/ml	FSH, ng/ml	PRL, ng/ml
Proestrous, 4 mo	10-50 (pk)	60-80 (pk)	50-800 (pk)	40-175 (pk)	
Estrous, 4 mo	10	30	40	125-150	175
Diestrous, 4 mo	15	30-40	20	25-45	110
Constant Estrous, 20-24 mo	5-10	40	50	120-215	400
Pseudopregnant, 20-24 mo	25-35	40	20	85-100	200
Anestrous	5	30	43[b]	75	800

[a] Computed from Huang, et al. (1976, 1978).

[b] Very recent data on C57BL/6J mice show that eventually plasma LH rises to postovariectomy levels after 20-24 months (Gee, Flurkey, Finch, 1983). Since the increases do not occur if pituitary tumors are present in old mice, the absence of LH elevations in old, anestrus rats may be consequent to their high incidence of pituitary tumors.

hyperprolactinemia (Huang et al., 1976). The old anestrous rat has very high PRL levels that are unresponsive to ovariectomy and are the consequence of the often huge pituitary tumors characteristic of this group (Huang et al., 1976; Lu et al., 1979).

Alterations in the regulation of the hypothalamic-pituitary-ovarian axis are more striking. In both the CE and RPP rat, the elevation of follicle-stimulating hormone (FSH) and luteinizing hormone (LH) following ovariectomy is less rapid and ultimately smaller than in the young cycling rat (Huang et al., 1976). Also, the supression of LH and FSH in response to subsequent estradoil (E_2) replacement may be smaller, possibly indicating alterations in both the initiation of and release from negative feedback mechanisms (Huang et al., 1976). Likewise, the induction of the LH surge (positive feedback) to regimens of estrogen or testosterone and progesterone is diminished or absent (Gosden and Bancroft, 1976; Lu et al., 1977; Peluso et al., 1977; Howland, 1976; Finch, et al., 1980). These latter observations may indicate a mechanism for the onset of the CE state, since the absence of the preovulatory LH surge appears to be the most definitive characteristic of CE animals.

POSSIBLE ETIOLOGIES OF SENILE ESTROUS DEVIATIONS

The possible causes of the senile deviations of the estrous cycle to be considered here are: (1) alterations in CNS sensitivity to ovarian steroids, (2) increased PRL and (3) altered neurotransmitter metabolism, including monoamine balance. These hypotheses are not mutually exclusive and the actual cause of reproductive senescence might include all three. Inasmuch as all of the alterations are sensitive to the hormonal status of the animals, it has been difficult to establish a cause and effect relationship among the many age-related variables in neuroendocrine mechanisms.

Evidence that Ovarian Exhaustion Is Not the Primary Cause

Many current hypotheses are based on the assumption that the locus of the immediate cause of cycle disruption is in the CNS. Although aging changes occur at many loci throughout the reproductive apparatus, such as pituitary, ovary and its vasculature, and uterus, the appearance of the senile deviation in the estrous cycle is probably a result of alterations in the CNS coordination of cyclic gonadotropin output.

Early hypotheses of reproductive aging focused predominantly on the progressive and irreversible depletion of primary oocytes, which begins in mammals at birth and continues throughout the life-span (Talbert, 1977). Recent evidence indicates that complete exhaustion of oocytes has not occurred in the rat and most mouse strains when cycles have ceased, 12 to 15 months. (Exceptions to this are discussed in Finch, 1978). Furthermore, the decreased number of oocytes does not appear to be a crucial factor in the decline in the number of litters born nor the decrease in litter size. Rats with 1000 to 2000 oocytes remaining following irradiation can still pro-

duce pups, whereas the senile rat or mouse with 2000 or more
oocytes remaining is noncyclic and sterile (Ingram, 1958, 1959;
Shelton, 1959). Nonetheless, normal numbers of ova may be shed
at ovulation in female rodents through 12 months of age (Harman
and Talbert, 1970) and in old rats with induced estrous cycles
(Aschheim, 1976). Very recent data show that the ovary of
C57BL/6J mice is approaching exhaustion of its follicular
pools, although in some mice acyclicity arises with considerable
reserves (Gosden et al., 1983). In the human, the loss of menstru
cycles also appears to occur without the complete exhaustion of ov
ian oocytes, about 10,000 remain in the age range of 39 to 45 year
oocytes, about 10,000 remain in the age range of 39 to 45 years
(Block, 1952). At least some remaining primordial follicles
can mature, ovulate, and luteinize after the last regular
menstrual period (Costoff and Mahesh, 1975; Novak, 1970),
which, precluding generalization, raises the possibility that
age changes in extraovarian loci (see later) also causes
reproductive senescence in the human female.

The Role of the Pituitary

Pituitary pathology may underlie some age-related impairments
observed in the response of old rats to releasing hormones,
since PRL can inhibit pituitary responses to luteinizing
hormone-releasing hormone (LH-RH) (Muralidhar et al., 1977). A
clear-cut correlation exists for impaired gonadotropin production in rats with gross PRL-secreting tumors (Huang et al.,
1976): the consequences of lactotropic microadenomas on LH and
FSH output are not known in aging rodents. The maintenance of
basal plasma concentrations of LH, FSH, PRL, growth hormone
(GH), and thryoid-stimulating hormone (TSH), as well as pituitary responses to LH-RH and thyrotropin-releasing hormone (TRH)
in male C57BL/6J mice between the 3 and 28 months (Finch et
al., 1977) may be related to their apparent absence of pituitary tumors (see before). Thus, age-related impairments in the
pituitary responsiveness to releasing hormones do not in-
evitably occur during the life-span of the rodent.

Reinitiation of Ovarian Cycles

There are several ways to initiate ovulation or the resumption
of estrous cycles in the CE rats. Also, the RPP rats normally
ovulate during the interspersed estrous periods, even though
some of them previously passed through a long anovulatory
period in CE. Progesterone, electrical stimulation, adeno-
corticotropic hormone (ACTH), ether stress, social isolation,

alterations in the light period, and a variety of catecholaminergic stimulants produce variable resumption of estrous cycles or ovulation in one or both of the senile deviations. Two score years ago, J.W. Everett (1940) discovered that 95 percent of CE rats aged 8 to 12 months regained regular ovulatory cycles when treated with progesterone. Others have documented this action of progesterone in aging CE rats (Clemens et al., 1969; Wuttke and Meites, 1973; Quadri et al., 1973). One interpretation is that the polyfollicular CE rat has insufficient progesterone to trigger a preovulatory gonadotropin surge. (Although plasma E_2 is not notably deficient in CE rats, being similar to levels in young rats at diestrus, progesterone is less than 50 percent of the lowest values observed at any stage of the cycle in young rats (Huang et al., 1978) and is even lower in CE C57BL/6J mice (Nelson et al., 1981; Flurkey et al., 1982).

Reactivation of cycles in CE rats can also result from systemic administration of L-DOPA and other adrenergic agonists (Clemens et al., 1969; Huang et al., 1976; Quadri et al., 1973), as well as by implantation of L-DOPA into the medial preoptic region (Cooper et al., 1979). One interpretation is that adrenergic agonists act by compensating for age-related deficiencies of hypothalamic cathecholamines (Quadri et al., 1973; Finch, 1973, 1976).

Finally, physical stress, such as ether (Huang et al., 1976), exposure to cold (Ascheim and Latouche, 1975), or even handling (unpublished observation from several laboratories), reactivates cycles in CE, but not RPP, aging rodents. Additionally, the stress-related hormones epinephrine and ACTH (Huang et al., 1976) also reactivate CE rats. Possibly such treatments act by increasing plasma progesterone via the adrenal cortex, since plasma progesterone increases after ACTH (Brown et al., 1976) and since epinephrine releases ACTH (Recant et al., 1950). There is as yet no published data on whether these treatments affect plasma steroids in the aging CE rats. Elevated progesterone or other steroids, by compensating for central catecholaminergic deficiencies in CE rats, could trigger the preovulatory gonadotropin surge. Sex steroids, including progesterone, influence catecholamine synthesis and turnover (Anton-Tay and Wurtman, 1968; Bapna et al., 1971; Beattie and Soyka, 1973; Lofstrom and Backstrom, 1978). It is pertinent that progesterone, as well as deoxycorticosterone, can induce early preovulatory surges of LH in young rats, whether injected systemically or implanted into the median eminence, but not the pituitary (Kobayashi et al., 1970).

The ability of diverse agents to reactivate cycles in noncycling rodents underscores the persistence of some pituitary and hypothalamic functions in aging rats and suggests that pituitary tumors are not the sole cause for the failure of young ovaries to cycle in old hosts. These reactivations of cycling also demonstrate that considerable pituitary functions are retained.

Altered CNS Responses to Ovarian Steroids

The studies mentioned provide strong evidence that aging female rats fail to cycle, whereas their ovaries, pituitaries, and basic ovulatory mechanisms are capable of functioning with appropriate stimuli. The focus, then, of much of the research and the hypotheses discussed here, is on brain regulatory phenomena, not on peripheral organs, nor on the basic process of gonadotropin release at the pituitary level.

Probably the most straightforward hypothesis on the etiology of CE and RPP is that of altered CNS sensitivity and responsivity to steroidal signals. An alteration in sensitivity in either positive or negative feedback could easily account for the cessation of cycles; there is considerable evidence that such alterations do occur with aging, both in the CNS and possibly in the pituitary.

Briefly, old female CE and RPP rats show less gonadotropin increase following ovariectomy and less responsiveness to estrogen replacement following ovariectomy (see before). A more important observation is that of reduced hypothalamic-pituitary sensitivity to estrogen for positive feedback, that is, in the loss of experimentally induced LH surge by E_2 and other steroids (see before). These studies suggest that the hypothalamic–pituitary system cannot respond to rising E_2 levels during follicular maturation after the preceding ovulation; hence, the preovulatory LH surge on the critical day of proestrus fails to occur. Hoffman (1973) has suggested a model of female reproductive aging, in which positive feedback becomes reduced concurrently with an unaltered or more sensitive negative feedback system, that effectively predicts the occurrence of CE. Normally, E_2 levels are sufficiently high during a critical period on proestrus to initiate an LH surge in the afternoon. With the putative age-related increase in threshold for E_2 induction of the LH surge, proestrus E_2 levels are no longer sufficient to trigger the LH surge and the animal remains in estrus. Hoffman suggests that during aging, the E_2 threshold for the LH surge becomes elevated to a point above that of negative feedback. That is, negative feedback of E_2 on gonadotropins would result in a decreased gonadotropin release before the E_2 reached the threshold for stimulation of the LH surge. The animal should then remain in continuous, but not necessarily severe, E_2 stimulation. This is precisely the type of hormonal profile the CE animals show.

If this were the case, or if a decrease in the capabililty of the ovary to synthesize or release E_2 were responsible for failure to reach the threshold for positive feedback, injections of E_2 ought to reinitiate cycling in the CE rat, much as progesterone does. So far as we are aware, such studies have not been published.

Definitive studies aimed at the evaluation of age-related alterations in the sensitivity of the stimulatory action of gonadal steroids on gonadotropin secretion are needed. Such studies must utilize dose-response assessments of sensitivity to establish age-related alterations in the minimal concentrations of steroids necessary to activate the LH secretory mechanism. In this regard, the metabolic clearance of steroids as well as their distribution among body compartments must be determined.

The present evidence then, is that some diminution in sensitivity of E_2 may occur in older animals. Although not conclusive, it seems fair to say that the LH output in response to standard priming doses of steroids is reduced. Evidence from behavioral experiments, however, indicates that age-related decreases in sensitivity to ovarian steroids may not occur in all CNS systems responding to these steroids. The varying steroid levels during the estrous cycle result in major changes in a number of behavioral parameters. This is particularly evident in proestrus, during which food intake decreases to its lowest point during the cycle and locomotor activity levels are the highest. Also during the evening of proestrus and the early morning of estrus, the female becomes behaviorally receptive to sexually active males. Both the effect of E_2 on food intake and sexual behavior appear to be enhanced, rather than diminished, in aging animals (Cooper, 1977; Cooper and Walker, 1979). Thus, alterations in the response of CNS to ovarian steroids is either different in different neural systems or the final expression of the change is altered by some other, possibly neurochemical, variable.

The Possible Role of Prolactin

Both the RPP and the CE rat show some degree of PRL elevation. The apparent covariance between high PRL and disruption of the estrous cycle and the fact that some agents, such as L-DOPA and lergotrile mesylate, which reinstate estrous cycling, also lower PRL levels of the old noncycling rats suggests that the primary dysfunction initiating disruption of regular cycles could be a decrease in the inhibitory control of PRL. Also consistent with this hypothesis is the observation that dopamine (DA) uptake (Jonec and Finch, 1975), turnover, and levels are lower in the median eminence of old mice than in young mice (Finch, 1978; Osterburg et al., 1981); hypothalamic DA metabolism is known to have profound, inverse effects on PRL secretion (Judd et al., 1979). Additionally, PRL is known to have inhibitory effects on gonadotropin secretion both in the rat and the human, at either a hypothalamic or ovarian level (Clemens et al., 1978).

One of the most difficult aspects of evaluating this hypothesis is the interdependence of PRL and ovarian steroids. It is quite possible, particularly in the CE rat, that elevated PRL is the result rather than the cause of cyclic disruption. Estrogens are known to stimulate PRL secretion, probably at the pituitary-median eminence level (Bishop et al., 1972). Rats brought into CE by a variety of techniques other than aging, characteristically have elevated plasma PRL (Ratner and Peake, 1974; Bishop et al., 1972; Negro-Vilar et al., 1968). Ovariectomy in these cases, including aging, reduces PRL levels dramatically.

At present, it appears more likely that PRL is responsible for the RPP than the CE state, even through the levels of this hormone may not be as high in the RPP as in the CE (Huang et al., 1978; Lu et al., 1979). The likely persistence of elevated PRL in RPP rather than a rapid decline, as normally follows ovulation, would tend to maintain the usually transient luteal function of the estrous cycle. The induction of hyperprolactinemia in the normal adult by implanation of a secondary pituitary into the renal capsule also induces an RPP syndrome (Quilligan and Rothchild, 1960). The smaller PRL elevation of the RPP rat relative to CE suggests that the higher progesterone values of the pseudopregnancy period may inhibit the PRL secretion (Chen and Meites, 1970). During the interspersed estrous period between pseudopregnancies, PRL values would remain high until luteal function was re-established. Unfortunately, no data are available on PRL levels during this period in the RPP rat.

The other evidence that PRL may be instrumental in reproductive senescence is the reinitiation of cycles in both CE and RPP rats by agents that among other things, lower PRL. Centrally administered L-DOPA (Cooper et al., 1979) or systemic administration of lergotrile mesylate (Clemens et al., 1978), a more potent and specific inhibitor of PRL, reinitiated cycling in RPP rats, which are apparently much more resistant to initiation of cycling than the CE rat. However, recently Clemens et al. (1978) directly tested the possible involvement of elevated serum PRL in the altered response of old female rats to ovariectomy. They were unable to detect any improvement in the postcastration increases in LH in rats whose serum PRL had been depressed for two months.

In summary, although elevated PRL is indeed a striking characteristic of both major groups of noncycling aging rats (CE or RPP), sufficient data are not yet available to assess whether elevated PRL concentrations are a cause or an effect of reproductive senescence.

The Involvement of Monoamines

There is little doubt that hypothalamic monoamines are important in the control or modulation of gonadotropin release in the female, although the precise mechanisms are controversial (Weiner and Ganong, 1978; see Simpkins and Estes, this volume). In conjunction with this literature are the numerous reports on lowered catecholamine levels, turnover, or synthesis rates in aging laboratory rodents and human beings (reviewed in Finch, 1978; Osterberg, et al., 1981). The age-related deficits in hypothalamic catecholamines would seem a likely mechanism by which to explain cycle irregularity and cessation during aging. Probably the most suggestive evidence for the involvement of catecholamines in female reproductive senescence is the reactivation of cycling in CE and RPP rats by pharmacological treatment with direct or indirectly acting catecholamine agonists. These agents include L-DOPA, iproniazid, epinephrine, and lergotrile mesylate. The action of L-DOPA, the most widely investigated drug, is probably central, as indicated by the potentiation of the effect by the administration of a peripheral decarboxylase inhibitor that would block its conversion to catecholamines in the periphery and for that reason enhance the entry of the drug into the CNS (Linnoila and Cooper, 1976). In addition, reactivation of cycles in CE and RPP rats by placement of L-DOPA via cannula into the medial preoptic area suggests not only a central action but a hypothalamically localized one, in that placements into the medial septum were without effect (Cooper et al., 1979). Lergotrile mesylate, a specific DA agonist, is also effective in producing cycling in RPP rats (Clemens and Bennett, 1977), whereas systemic L-DOPA does not (Huang and Meites, 1975). Also long-term dietary supplements of L-tyrosine (3 percent) beginning prior to cycle cessation remarkably delay the cessation on the order of six months (Cooper and Walker, 1979). This study clearly shows the large ovarian potential remaining when cycles normally cease. Also, it is of much interest that mammary tumors were rare in the still cycling older rats with tyrosine supplements. Possibly the tyrosine supplemented diet aids in maintaining regular estrous cycles via an enhancement of catecholamine metabolism (Wurtman and Fernstrom, 1976).

Several investigators have recently suggested that in addition to decreases in catecholamine activity (primarily DA), age-related alterations in hypothalamic serotonin (5-HT) content or metabolism may be involved in the disruption of the estrous cycle in the rat (Meites et al., 1978; Cooper and Walker, 1979). 5-HT is though to have a facilitory effect on PRL release and may have an inhibitory effect on gonadotropin release (see Frohman and Berelowitz, chapter 4). Thus, increases in serotonergic activity during aging would be consistent with the neuroendocrine profile of aging rats. Hypothalamic levels of 5-HT are either unchanged or increased in the hypothalamus of old male rats, whereas turnover is accelerated (Simpkins, et al., 1977).

It was proposed that the important neurochemical factor in the cessation of the estrous cycle during aging is the balance between catecholamines and 5-HT, rather than the absolute status of either: the first, by providing additional precursor for catecholamine synthesis; and the second, by the competition of tyrosine with tryptophan for neutral amino acid uptake into the brain (Cooper and Walker, 1979). Thus, the tyrosine-enriched diet may promote a catecholaminergic dominance to a greater extent than treatment with catecholaminergic agonists alone.

A difficulty in interpretation of the catecholamine hypothesis of reproductive sensecence is the indirect nature of the present evidence. It is clear that whatever deficit is responsible for the declining cyclicity in the female can be overridden pharmacologically by diverse agents, many of which enhance catecholaminergic functions. The possibility that L-DOPA acts peripherally to reactivate cycles, for example, in the rich sympathetic innervation of the ovary (Bahr et al., 1974), seems to be ruled out by the L-DOPA cannulation study described previously (Cooper et al., 1979). Most would agree, however, that the sequence of neuronal events leading up to ovulation is extraordinarily complex, one involving the interaction of many different neuronal systems in which communication is subject to a variety of hormonal influences. The fact that a dysfunction in this system can be overridden by stimulation of one highly involved component does not prove that that component was responsible by itself for the original dysfunction. It is equally likely that a defect in a system preceding the catecholamine component was simply bypassed, or that the reinitiation of cycling is a result of a relatively nonspecific increase in excitability of a large pool of neurons.

The historical development of the literature has added to this ambiguity. Age-related changes in neurotransmitter function were considered important for many reasons quite independent of reproductive senescence. In an attempt to evaluate the influence of age on these systems, uncontaminated by changing hormonal environments (the catecholamine systems are highly responsive to such alterations), 24- to 30-month-old male C57BL/6J mice (the age of average longevity) were chosen because in some colonies this strain retains a remarkable stability of basal plasma pituitary hormones and testosterone (Finch et al., 1977). However, physiological similarities between the 28-month-old male and the 10- to 15-month-old female showing reproductive senescence is at best vague, and the vast majority of studies on aging and catecholamines have little if any bearing on the problem of reproductive senescence in the female, which occurs at much earlier ages. Recent studies of younger age groups of male C57BL/6J mice (4, 10, and 17 months) did not detect changes in catecholamine levels or turnover in major hypothalamic subdivisions (Osterburg et al., 1981). However, DA receptors (Makman et al., 1978; Severson and Finch, 1980) and DA-sensitive adenylate cyclase in striatum decrease during aging before midlife (Schmidt and Thornberry, 1978).

When we try to examine the same questions in the reproductively senile animal, we are faced with the same problem that those studies avoided, namely, how to distinguish a causative factor from the result of the altered hormonal milieu. One study (Wilkes et al., 1978) suggests that in the aging female, alterations in hypothalamic DA levels may follow elevated PRL. In that study, DA levels were significantly decreased in 24-month-old CE and 30-month-old RPP rats. However, DA levels in the 12-month-old CE rat were identical to those of young cyclers. Importantly, DA levels returned to normal following ovariectomy of older animals, indicating that its decline might be due to some ovarian factor or factors, such as the influence of E_2 on PRL levels. In another study, the fall in DA was confirmed in aging RPP rats (Clemens et al, 1978).

Another complication in the interpretation of catecholamine involvement in reproductive decline is the converse of those interactions with PRL. Catecholamine precursors or DA stimulants decrease PRL and may affect vaginal status through this mechanism.

Finally, one of the most striking findings in catecholamine research has been the extraordinary capacity of catecholaminergic neurons to compensate for partial destruction or pharmacological blockade in young rodents (reviewed in Stricker and Zigmond, 1976). The deficits in even very old rodents in either NE or DA levels and turnover are on the order of 20 to 35 percent, rarely greater (Finch, 1978). It would appear unlikely that any functional deficit in catecholamine neurons responsible for cyclicity disruption at 10 to 15 months would

be a simple presynaptic deficit. It is always possible, however, that there may be drastic alterations in very small brain
areas. The increasing sensitivity of techniques for measuring
catecholamine turnover in very small brain regions when considered with emerging knowledge of the critical anatomical
areas for neuroendocrine control, suggests a very promising
area of research.

Each of these hypotheses suffers not from a lack of plausibility, but from the difficulty of devising specific tests.
Although the interactions of neuronal systems complicate their
evaluation, the most fundamental problem remaining is the
separation of the causes from the consequences of the cycle
deviation. To be established as a cause of the cessation or
irregularity of the estrous cycle, a deficit should be
demonstrable at or prior to cessation. Furthermore, at any
particular age, in those animals still cycling, the deficit
should be absent or present in a diminished form. By way of
analogy, the role of striatal DA in the pathophysiology of
Parkinson's disease was confirmed only by more elaborate
arguments, including the verification that parkinsonian
patients had characteristic reductions in striatal DA, whereas
aged-matched controls, not so afflicted, showed moderate
changes. Even more convincing was the unilateral striatal DA
deficit associated with unilateral parkinsonism (Barolin et
al., 1964).

There are two major experimental paradigms circumventing most
of these problems. The first is, as Aschheim (1976) suggests,
the use of the still cycling old animal as the appropriate
control group for studies examining estrous cycle cessation.
This animal is simply the aged-matched control not showing the
disorder. Secondly, and far more important, is the identification with respect to time of cycle cessation and senile
group, of the pattern of the transition from regular cycling to
one of the senile deviations. Any proposed mechanism should be
consistent not only with the final stable vaginal configuration, but also with the sequence of events leading to it. As
imperfect as it is, the only practical, available nonobtrusive
method of following the cycling performance of an animal over a
long period of time is the vaginal smear.

Since various experimental procedures result in estrous cycle
deviations generally similar to those in aging, there is an
immediate possibility of second-order models, that is, studying
a model for reproductive aging in the rat rather than the aging
rat itself. There are, however, a few stable neuroendocrine
states that result from disruption of the estrous cycle. Moreover it is clear that the rat might ultimately reach the same
hormonal state, for example, RPP or CE, by very disparate
routes. Invoking a model system in many cases simply adds

another complication: the congruence of the model system with the real one. The most promising models, then, are those that, in addition to providing the similar noncycling animal, suggest a direct relationship with the failure of cycling in the aging animal. Two such models will be discussed. The first, the delayed anovulatory syndrome (DAS) is discussed in considerable detail, since sufficient data are already available to understand how the model may represent or simulate mechanisms operating in the aging rat itself.

We have stressed that chronological age is not a clear predictor of the states of reproductive dysfunctions, i.e., rodents aged 18 to 24 months may be in constant estrus or irregularly pseudopregnant, whereas others may be regularly cycling. This surprisingly large variability is also clearly shown in the C57BL/6J mouse (Nelson et al., 1982), despite 40 years of inbreeding. As will be described, the pathways to reproductive aging may depend on various aspects of steroid exposure, including exposure during neonatal life.

STEROID-DEPENDENT ASPECTS OF REPRODUCTIVE AGING

First consider another type of ovarian transplant experiment, a potentially remarkable phenomenon found by Aschheim (1976): if hosts were ovariectomized when young and allowed to age beyond 20 months, then young ovarian transplants cycled normally in the long-term castrate hosts. This finding was repeated in a pilot study of C57BL/6J mice ovariectomized at 3 months and given young ovarian transplants at 26 months. In a majority of the mice, multiple four to six day estrous cycles occurred, and in some cases cycles persisted until just before spontaneous death (Nelson, et al., 1980). This result implies that the ovarian-related functions induce or facilitate expression of aging changes in key hypothalamic-pituitary loci.

The converse phenomenon may also be induced. Recent studies by J.R. Brawer and colleagues suggest that E_2 treatment of adult rats induces changes in ovarian and hypothalamic-pituitary functions that have important similarities to some aspects of reproductive senescences. In two lines of Wistar rats a single SC injection of 2 mg E_2 valerate into "young" (130 to 170 gm, probably seven to nine weeks) adults leads to the loss of "normal" cycles and the emergence of "apparently random" variations in vaginal cytology during the six weeks after injection (Brawer et al., 1978). Most rats showed "persistent vaginal estrus" within the six weeks; the ovaries were polyfollicular without corpora lutea. From morphological criteria, these ovaries appear similar to those from aging rodents in CE.

Morphological changes in the hypothalamus (neuropil degeneration, glial hyperactivity in the arcuate nucleus and median
eminence) progressed during subsequent months and are suggested
to be an "intensification of normal age-related process."
Analysis of plasma hormones indicated that E_2 was elevated to
120 pg/ml during the first week after injection, and then
rapidly declined to a normal range (30 pg/ml) by eight weeks,
when most rats were in CE. PRL was elevated only in the first
week. The absence of persistently elevated PRL is a key point
in evaluating these studies, because PRL may inhibit pituitary
responses to LH-RH (Muralidhar et al., 1977) and may inhibit
ovarian progesterone production (McNatty et al., 1976). PRL-
containing pituitary tumors are readily induced by multiple
injections of E_2 after four months or so (Brawer and
Sonnenschein, 1975) but do not appear to be involved here in
the initial changes induced by E_2. Most recently E_2-treated CE
rats were found to have reduced release of LH in response to
preoptic stimulation in comparison with proestrus controls
(Brawer et al., 1979); however, the output of LH in response to
LH-RH was close to normal. These results were interpreted to
indicate that E_2 treatment induced CE via hypothalamic lesions,
rather than pituitary lesions. These observations suggest that
the E_2 injections have led to accelerated reproductive
senescence in young rats.
    The manipulations of reproductive senescence achieved by
long-term ovariectomy or by E_2 injections in rodents (described
previously) can be viewed in a broader context. It is well
established that neonatal injections of E_2 or testosterone and
other (aromatizable) steroids into female rodents irreversibly
"androgenize or masculinize" their responses to sex steroids as
adults, with respect to the absence of estrous cycles, steroid-
induced LH surges, and lordosis and other sexually differentiated behaviors (Gorski et al., 1979). The ovaries of neonatally androgenized adult rats are generally polyfollicular and
anovulatory (Gorski, 1968): at least superficially, they bear
striking resemblance to the polyfollicular ovaries of aging
constant (met)estrus rats. Sex differences in the size of the
preoptic nucleus (Gorski et al., 1978) and the preoptic
synaptic density (Raisman and Fields, 1973) are also manipulated by neonatal steroids in parallel with the altered
physiology, behavior, and brain morphology of adult rodents.
Although sexual dimorphisms in the preoptic region have been
emphasized (Raisman and Fields, 1973; Greenough et al., 1977;
Gorski et al., 1978), other neuroanatomical loci are probably
altered by neonatal steroids (neonatal androgenization alters
the morphology of putative catecholamine-containing granules in
the arcuate nucleus) (Ratner and Adamo, 1971). However, it is
generally believed that the preoptic region, rather than the

arcuate nucleus and basal hypothalamus, is the most critical in determining sex differences in the regulation of gonadotropins (Halasz and Gorski, 1967; Gorski et al., 1978). The sensitivity of rodents to the masculinizing effects of steroids decreases markedly after day 10 (Gorski, 1968; Lobl and Gorski, 1974), thereby defining a "neonatal critical period." These phenomena are widely interpreted as brain sexual "dimorphisms," which are epigenetically determined by steroids in the neonate.

If smaller testosterone doses are given neonatally, then the onset of impaired reproductive function is delayed. "Lightly androgenized" rats may show some (number unknown) normal estrous cycles and may retain ovarian cyclicity and the steroid-induced LH surge for several months before its eventual loss (DAS). DAS was previously suggested as a model for accelerated reproductive aging (Swanson and van der Werff ten Bosch, 1964; Gorski, 1977). Further studies suggest intriguing parallels of DAS with the ovarian transplant studies in aging long-term ovariectomized mice and rats: if lightly androgenized rats are ovariectomized prepubertally (at 30 days), the delayed anovulatory syndrome can be postponed for $\geq$ 3 months, as assayed by the functions of ovarian transplants given at 3 months (Kikuyama and Kawashima, 1966; Arai, 1971) or by the retention of the steroid-induced LH surge at ages when the intact lightly androgenized controls had lost their responses (Harlan and Gorski, 1978). DAS was also induced in rats that were lightly androgenized with testosterone as neonates, then ovariectomized at 30 days and given daily injections of E_2 or testosterone for 60 days: Such treatments caused a loss of the steroid-induced LH surge as seen in the typical DAS. However, the spayed, neonatally androgenized controls still retained steroid-induced LH surges (Harlan and Gorski, 1978). In contrast to these reports, Hendricks et al. (1977) found no influence of ovariectomy or steroid treatment on DAS. Thus steroids may have an impact on neuroendocrine function which extends beyond the traditional neonatal critical period into the postpubertal period. These studies also suggest that in the peripubertal period (30 to 90 days), steroids can further modify neonatal androgenization. These results suggest similarities to the effects of E_2 injections in young adult rats in the studies of Brawer et al.

In some major respects, DAS with its hypothalamic insensitivity in lightly androgenized rats resembles the spontaneous noncycling state of aging control (met)estrus (polyfollicular, anovulatory) rodents (Table 18-2): (1) loss of steroid-induced LH surges; (2) reactivation of ovulatory cycles in polyfollicular ovaries, by some of the same drugs; (3) failure of young ovarian transplants to cycle in these constant metestrous hosts; (4) the dependence of the phenomena on the presence of

882 Randall and Finch

Table 18-2. Comparison of the Delayed Anovulatory Syndrome (DAS) and
Normal Reproductive Aging

	DAS	Normal aging[a]
Ovarian state Polyfollicular, anovulatory	yes[b]	yes[c]
Vaginal smear Cornified cells and leukocytes	yes[b]	yes[c]
Cycles reactivated Progesterone,	40%[d]	60-95%[e]
Iproniazide,	70%[d]	85%[e]
L-DOPA	20-30%[d]	55-80%[e]
Steroid-induced LH surge	no[i]	no[f]
Cyclic functioning of young (control) ovarian grafts to noncycling hosts	no[g]	no[h]
Dependency of the development of "hypothalamic insensitivity" on continued presence of ovary	yes[i]	yes[h]
Retention of sex steroid induced behavior	yes[j]	yes[k]

[a]Noncycling, constant metestrous.

[b]Gorski, 1978.

[c]Huang et al., 1976, 1978.

[d]Lehman et al., 1978.

[e]Reviewed in Finch, 1978.

[f]Gosden and Bancroft, 1976;
Lu et al., 1977; Peluso et al., 1977.

[g]Arai, 1971.

[h]Aschheim, 1965.

[i]Harlan and Gorski, 1978.

[j]Hendricks et al., 1977; Harlan and Gorski, 1978.

[k]Peng et al. 1977; Cooper, 1977.

ovaries. However, in contrast to the preceding, the sex-
steroid-induced female behaviors are not lost in lightly
androgenized (Hendricks et al., 1977; Harlan and Gorski, 1978)
and intact aging females (Peng et al., 1977; Cooper, 1977).
Retention of sex behavior in aged and lightly androgenized
rodents emphasizes the selective impact of steroids on neural
function. Because the adult long-term ovariectomized rodents
in ovarian transplant studies show parallel phenomena to those
in DAS studies. E_2, and possibly other steroids, may have a
cumulative affect on some aspects of brain-pituitary function
which continues beyond the traditional neonatal critical
period. Possibly, exposure to certain steroids in the adult
continues or extends the effect of neonatal steroid exposure.
If so, some aspects of female reproductive aging in rodents may
be equivalent to "androgenization."

Ultimately, it may be possible to relate individual differences in pre- and neonatal steroid exposure to adult cycling
potential, for example, some rats stop cycling as early as 10
months, whereas others continue to 18 months (Aschheim, 1976).
In utero, the presence of male fetuses on both sides of a female fetus appears to partially "masculinize" some adult behavioral traits in comparison to a female fetus flanked in utero
by other females (vom Saal and Bronson, 1978). The major question of whether postpubertal steroid exposure acts on the same
loci in the neonate cannot be approached until the neonatal
androgenization phenomenon is better understood.

CONSTANT ESTRUS INDUCED BY CONTINUOUS ILLUMINATION

The second potential model for reproductive senescence that appears to interact with age is the CE state produced by exposure
to continuous illumination. The following observations suggest
the possibility that desynchrony of hormonal events or sensitivity to light might be at least partially responsible for the
disappearance of cycling.

Firstly, exposure of female rats to constant light induces a
CE that is similar to that of the aging CE. There is a persistently cornified vaginal smear, elevated PRL levels (Negro-
Vilar et al., 1968), and an absence of ovulation (Critchlow,
1963). (It should be noted that a number of other treatments
produce similar, differentiable CE states.) Electrical stimulation of the preoptic area or copulation (Singh and Greenwald,
1967) induce ovulation, as they do in the senile CE (Clemens et
al., 1969; Wuttke and Meites, 1973). Secondly, the latency for
CE in the constantly illuminated environment decreases markedly
with age (Daane and Parlow, 1971). This effect is evident even

in animals as young as 5 months of age. Thirdly, increasing the dark portion of the light-dark cycle or placing the CE animal in constant darkness reactiviates cycling (Everett, 1943; Aschheim, 1965).

Fourthly, in aging animals that are still cycling the preovulatory surge of LH occurs later in the day than in young cycling animals (Van der Schoot, 1976). Both light-induced CE and old CE animals that have been induced to cycle by pharmacological techniques show LH surges that are less precisely entrained to the circadian signal, that is, more random in time (Kledzik and Meites, 1974; Huang et al., 1976). The young rat during the first few days of exposure to constant light exhibits a lack of synchrony between various events in the cycle, such as uterine ballooning, ovulation, cornification (Lawton* and Schwartz, 1967). Aging males also show alterations in the day-night distribution of both sexual behavior (Larsson, 1958) and food intake (Nisbett et al., 1975).

Fifthly, Naranjo and Greene (1977) have observed severe primary optic tract degeneration in aging males of the Fisher 344 strain. Although retinal degeneration can be dissociated by the use of pigmented rats (Brown-Grant, 1974) or continuous red light (Lambert, 1975), this degenerative change often accompanies the light-induced persistent estrus.

Although there is no direct evidence for the participation of light or photoperiod in the senile CE, a number of relatively simple experiments could readily test its plausibility. What is the effect of blinding, inferior accessory optic tract lesions, or cervical ganglionectomy on the occurrence of CE in aging rats? All of these treatments are known to block the effects of light in the young cycling rat (Wurtman et al., 1964, 1967). The effectiveness of the surgical removal of the superior cervical ganglion suggests the possibility of pineal involvement in the constant light CE. The pineal receives its major, if not total, innervation, from the sympathetic system via this ganglion (Wurtman et al., 1967). Melatonin injections produce short-lasting pseudodiestrus phases in constant illumination CE rats (Wurtman et al., 1964). Pinealectomy however, does not result in a shorter latency for CE in constant lighting conditions, nor does it drastically alter the normal cycle (Hoffman and Cullin, 1975; Hoffman, 1973).

SUMMARY

A detailed analysis of female reproductive aging is described to illustrate the hierarchical interaction of physiological systems of controls during aging. Three major hypotheses regarding the immediate cause of estrous cycle cessation in the aging rodent have been discussed. Although all have a good deal of indirect confirmatory evidence, there is little direct evidence that elevated PRL, alteration in CNS sensitivity to ovarian steroids, or diminished catecholamine function are associated with the onset of estrous cycle irregularities. It is suggested that the evaluation of the heterogeneity of the still cycling population or the use of physiological models of similar hormonal conditions might provide strategies for dissecting the highly interactive properties of systems involved in this phenomenon.

ACKNOWLEDGEMENTS

This work supported by grants from the National Institute on Aging to CEF (AG00117, AG00446) and to PKR (AG00855) and Project grant (AG00760).

AUTHORS ADDENDUM

Aside from minor revisions the major review for this chapter was completed in late 1979. We regret that a substantial amount of recent progress could not be included.

REFERENCES

Anton-Tay, F., and Wurtman, R.J.: Norepinephrine: Turnover in rat brains after gonadectomy. *Science* 159:1245, 1968.

Arai, Y.: Possible process of the secondary sterilization: Delayed anovulation syndrome. *Experientia* 27:463-464, 1971.

Aschheim, P.: La reactivation de l'ovaire des-rattes seniles en oestrus permante au moyen d'hormones ganodotropes ou do la mis a l'obscurite. *CR Acad Sci (Paris)*:5627-5630, 1965.

Aschheim, P.: Aging in the hypothalamic-hypophysial ovarian axis in the rat. In *Hypothalamus, Pituitary and Aging*, A.V. Everitt and F.A. Burgett, eds. Charles C Thomas Publishing Company, Springfield, Ill., 1976, pp 376-418.

Aschheim, P., and Latouche, J.: Les effect du sejour au froid sur le cycle ovarian de la ratte et son age biologigue, In *Les endocrines et le millieu. Problemes actuel d'endocrinologie et de nutrition*, Serie No. 19, H.P. Klotz, ed. Expansion cientifique Francaise, Paris, 1975, pp. 95-110.

Bahr, J., Kao, L., and Nalbandov, A.V.: Role of catecholamines and nerves in ovulation. *Biol Reprod* 10:273-290, 1974.

Bapna, J., Neff, N.H., and Costa, E.: A method for studying norepinephrine and serotonin metabolism in small regions of rat brain: Effect of ovariectomy on amine-metabolism in anterior and posterior hypothalamus. *Endocrinology 89:* 1345-1356, 1971.

Barolin, G.S., Bernheimer, H., and Hornykiewicz, O.: Seiten-
verschiedenes ver halten des Dopamins (3-hydroxytyramin) im
Gehirn eines Falles von Hemiparkinsonismus. *Schweiz Arch
Neurol Psychiat 94:*241-248, 1964.

Beattie, C.W., and Soyka, L.F.: Influence of progestational
steroid on hypothalamic tyrosine hydroxylase *in vitro*.
*Endocrinology 93:*1453-1455, 1973.

Bishop, W., Kalra, P.S., Fawcett, C.P., Krulich, L., and
McCann, S.M.: Effects of hypothalamic lesions on release of
gonadotropins and prolactin in response to estrogen and progesterone treatment in female rats. *Endocrinology 91:*1404-
1410, 1972.

Block, E.: Quantitative morphological investigations of the
follicular system in women. Variations at different ages.
*Acta Anat 14:*197-223, 1952.

Brawer, J.R., and Sonnenschein, C.: Cytopathological effects of
estradiol on the arcuate nucleus of the female rat. A possible mechanism for pituitary tumorigenesis. *Am J Anat 144:*
57-87, 1975.

Brawer, J.R., Naftolin, F., Martin, J., and Sonneschein, C.:
Effects of a single injection of estradiol valerate on the
hypothalamic arcuate nucleus and on reproductive function in
the female rat. *Endocrinology 103:*501-512, 1978.

Brawer, J.R., Ruf, K.B., and Naftolin, F.: Effects of estradiol-
induced lesions of the arcuate nucleus on gonadotropin release in response to preoptic stimulation in the rat.
*Neuroendocrinology 30:*144-149, 1979.

Brown, C.P., Courtney, A., and Marotta, S.F.: Progesterone secretion by adrenal gland of hamsters and comparison of ACTH
influence in rats and hamsters. *Steroids 28:*275-283, 1976.

Brown-Grant, K.: The role of the retina in the failure of
ovulation in female rats exposed to constant light.
*Neuroendocrinology 16:*243-254, 1974.

Chen, C.L., and Meites, J.: Effects of estrogen and progesterone on serum and pituitary prolactin levels in ovariectomized
rats. *Endocrinology 86:*503, 1970.

Clemens, J.A., and Bennett, D.R.: Do aging changes in the preoptic area contribute to loss of cyclic endocrine function.
*J Gerontol 32:*19-24, 1977.

Clemens, J.A., Amenomori, Y., Jenkins, T., and Meites, J.: Effects of hypothalamic stimulation, hormones, and drugs on
ovarian function in old female rats. *Proc Soc Exp Biol Med
132:*561-563, 1969.

Clemens, J.A., Fuller, R.W., and Owen, N.V.: Some neuroendo-
crine aspects of aging. In *Parkinson's Disease II., Aging and
Neuroendocrine Relationships, Adv Exptl Med Biol,* Vol 113,
C.E. Finch, D.E. Potter, and A.D. Kenney, eds. Plenum Press,
New York, 1978, pp 77-100.

Cooper, R.L.: Sexual receptivity in aged female rats. Behavioral evidence for increased sensitivity to estrogen. *Horm Behav* 9:321-333, 1977.

Cooper, R.L., and Walker, R.F.: Potential therapeutic consequences of age-dependent changes in brain physiology, In *CNS Aging and its Neuropharmacology: Experimental and Clinical Aspects. Interdisciplinary Topics of Gerontology*, Vol 15, Bk #08212, W. Meierruge, ed. S. Karger AG, Basel, 1979, pp. 54-76.

Cooper, R.L., Brandt, S.J., Linnoila, K., and Walker, R.F.: Induced ovulation in aged female rats by L-DOPA implants into the medial preoptic area. *Neuroendocrinology* 28:234-240, 1979.

Costoff, A., and Mahesh, V.B.: Primordial follicles with normal oocytes in the ovaries of postmenopausal women. *J Am Geriatr Soc* 23:193-196, 1975.

Critchlow, B.V.: The role of light in the neuroendocrine system. In *Advances in Neuroendocrinology*, A.V. Nalbandov, ed. University of Illinois Press, Urbana, Ill., 1963.

Daane, T.A., and Parlow, A.F.: Serum FSH and LH in constant light-induced persistent estrus: Short-term and long-term studies. *Endocrinology* 88:964-968, 1971.

Everett, J.W.: The restoration of ovulatory cycles and corpus luteum formation in persistent-estrous rats by progesterone. *Endocrinology* 27: 681-686, 1940.

Everett, J.W.: Further studies on the relationship of progesterone to ovulation and lutinization in the persistent estrous rat. *Endocrinology* 32:285-292, 1943.

Fabricant, J.D., and Schneider, E.L.: Studies of the genetic and immunologic components of the maternal age effect. *Dev Biol* 66:337-343, 1978.

Finch, C.E.: Catecholamine metabolism in the brains of aging, male mice. *Brain Res* 52:261-276, 1973.

Finch, C.E.: The regulation of physiological changes during mammalian aging. *Q Rev Biol* 51:49-83, 1976.

Finch, C.E.: Age-related changes in brain catecholamines: A synopsis of finding in C57BL/6J mice and other rodent models. In *Parkinson's Disease II., Aging and Neuroendocrine Relationships, Adv Exp Med Biol*, Vol 113. C.E. Finch, E.E. Potter, and A.D. Kenney, eds. Plenum Press, New York, 1978, pp 15-39.

Finch, C.E., Felicio, L.S., Flurkey, K., Gee, D.M., Mobbs, C., Nelson, J.F., and Osterburg, H.H.: Studies in ovarian-hypothalamic-pituitary interactions during reproductive aging in C57BL/6J mice. *Peptides*, Vol. 1., Suppl. 1:163-176, 1980.

Finch, C.E., and Flurkey, K.: The molecular biology of estrogen replacement therapy. *Contemp Obstet Gynecol* 9:97-107, 1977.

Finch, C.E., Jonec, V., Wisner, J.R., Jr., Sinha, Y.N., de Vellis, J.S., and Swerdloff, R.S.: Hormone production by the pituitary and testes of male C57BL/6J mice during aging. *Endocrinology 101*:1310-1318, 1977.

Flurkey, K., Gee, D.M., Sinha, Y.N., Wisner, J.R. Jr., and Finch, C.E.: Age effects on luteinizing hormone, progesterone, and prolactin in proestrous and acyclic C57BL/6J mice. *Biol Reprod 26*:835-846, 1982.

Gee, D.M., Flurkey K., and Finch, C.E.: Aging and the regulation of luteinizing hormone in C57BL/6J mice-impaired elevations after ovariectomy and spontaneous elevations at advanced ages. *Biol Repro 28*:598-607, 1983.

Gorski, R.A.: Influence of age on the response to perinatal administration of a low dose of androgen. *Endocrinology 82*: 1001-1004, 1968.

Gorski, R.A., Gordon, J.H., Shryne, J.E., and Southam, A.M.: Evidence for morphological sex differences within the medial preoptic area of the rat brain. *Brain Res 148*: 333-346, 1978.

Gorski, R.A., Harlan, R.E., and Christiansen, L.W.: Perinatal hormonal exposure and the development of neuroendocrine processes. *J Toxicol Environ Health 3*:97-121, 1977.

Gosden, R.G., and Bancroft, L.: Pituitary function of reproductively senescent female rats. *Exp Gerontol 11*:157-160, 1976.

Gosden, R.G., Laing, S.C., Felicio, L.S., Nelson, J.F., and Finch, C.E.: Imminent oocyte exhaustion and reduced follicular recruitment mark the transition to acyclicity in aging C57BL/6J mice. *Biol Repro 28*:255-260, 1983.

Greenough, W.T., Carter, C.S., Steerman, C., and DeVoogd, T.J.: Sex-differences in dendritic patterns in hamster preoptic area. *Brain Res 126*:63-72, 1977.

Halasz, B., and Gorski, R.A.: Gonadotrophic hormone secretion in female rats after partial or total interruption of neural afferents to the medial basal hypothalamus. *Endocrinology 80*: 608-622, 1967.

Harlan, R.E., and Gorski, R.A.: Effects of postpubertal ovarian steroids on reproductive function and sexual differentiation of lightly androgenized rats. *Endocrinolgy 102*:1716-1724, 1978.

Harman, M.S., and Talbert, G.B.: The effect of maternal age on ovulation, corpora lutea of pregnancy, and implantation failure in mice. *J Reprod Fertil 23*:33-39, 1970.

Hendricks, S.E., McArthur, D.A., and Pickett, S.: The delayed anovulation syndrome: Influence of hormones and correlation with behavior. *J Endocrinol 75*:15-22, 1977.

Hoffman, J.C.: The influence of photoperiods on reproductive functions in female mammals. In *Handbook of Physiology*, Section 7, *Endocrinology*, Vol II. *Female Reproductive System*, Part 1, R.O. Greep, ed. American Physiological Society, Washington, D.C., 1973.

Hoffman, J.C., and Cullin, A.: Effects of pinealectomy and constant light on the estrus cycles of female rats. *Neuroendocrinology* 17:167-174, 1975.

Holinka, C.F., Tseng, Y.-C., and Finch, C.E.: Prolonged gestation, elevated preparturitional progesterone, and reproductive aging in C57BL/6J mice. *Biol Reprod* 19:807-816, 1978.

Holinka, C.F., Tseng, Y.-C., and Finch, C.E.: Reproductive aging in C57BL/6J mice: Plasma progesterone, viable embryos, and resorption frequency throughout pregnancy. *Biol Reprod* 20:1201-1211, 1979.

Howland, B.E.: Reduced gonadotropin release in response to progesterone or gonadotropin releasing hormone (GnRH) in old female rats. *Life Sci* 19:219-224, 1976.

Huang, H.H., and Meites, J.: Reproductive capacity of aging female rats. *Neuroendocrinology* 17:289-295, 1975.

Huang, H.H., Marshall, S., and Meites, J.: Capacity of old versus young female rats to secrete LH, FSH, and prolactin. *Biol Reprod* 14:538-543, 1976.

Huang, H.H., Steger, R.W., Bruni, J.F., and Meites, J.: Patterns of sex steroid and gonadotropin secretion in aging female rats. *Endocrinology* 103:1855-1859, 1978.

Ingram, D.L.: Fertility and oocyte numbers after x-irradiation of the ovary. *J Endocrinol* 17:81-90, 1958.

Ingram, D.L.: The vaginal smear of senile laboratory rats. *J Endocrinol* 19:182-188, 1959.

Jonec, V., and Finch, C.E.: Aging and dopamine uptake by subcellular fractions in the C57BL/6J male mouse brain. *Brain Res* 91:197-215, 1975.

Judd, S.J., Rigg, L.A., and Yen, S.S.C.: The effects of ovariectomy and estrogen treatment on the dopamine inhibition of gonadotropin and prolactin release. *J Clin Endoc Metab* 49:182-184, 1979.

Kikuyama, S., and Kawashima, S.: Formation of corpora lutea in ovarian grafts in ovariectomized adult rats subjected to early postnatal treatment with androgen. *Sci Pap Coll Gen Educ Jpn* 16:69, 1966.

Kledzik, G.S., and Meites, J.: Reinitiation of estrous cycles in light induced constant estrous female rats by drugs. *Proc Soc Exp Biol Med* 146:980-992, 1974.

Kobayashi, F., Hara, K., and Miyake, T.: Facilitation of luteinizing hormone release by progesterone in proestrous rats. *Endocrinol Jpn* 17:149-155, 1970.

Lambert, H.H.: Continuous red light induces persistent estrous without retinal degeneration in the albino rat. *Endocrinology* 97:208-210, 1975.

Larsson, K.: Age differences in the diurnal periodicity of male sexual behavior. *Gerontologia* 2:64, 1958.

Lawton, I.E., and Schwartz, N.B.: Pituitary-ovarian function in rats exposed to constant light: A chronological study. *Endocrinology* 81:497-508, 1967.

Lehman, J.R., McArthur, D.A., and Hendricks, S.E.: Pharmacological induction of ovulation in old and neonatally androgenized rats. *Exp Geront 13:*107-114, 1978.

Linnoila, M., and Cooper, R.L.: Reinstatement of vaginal cycles in aged female rats. *J Pharm Exp Therap 199:*477-482, 1976.

Lobl, R.T., and Gorski, R.A.: Neonatal intrahypothalamic androgen administration: the influence of dose and age on androgenization of female rats. *Endocrinology 94:*1325-1330, 1974.

Lofstrom, A., and Backstrom, T.: Relationship between plasma estradiol and brain catecholamine content in the diestrus female rat. *Psychoneuroendocrinology 3:*103-107, 1978.

Lu, K.H., Huang, H.H., Chen, H.T., Kurcz, M., Mioduszewski, R., and Meites, J.: Positive feedback by estrogen and progesterone on LH release in old and young rats. *Proc Soc Exp Biol Med 154:*82-85, 1977.

Lu, K.H., Hopper, B.R., Vargo, T.M., and Yen, S.S.C.: Chronological changes in sex steroid, gonadotropin and prolactin secretion in aging female rats displaying different reproductive states. *Biol Reprod 21:*193-203, 1979.

McNatty, K.P., Neal, P., and Baker, T.G.: Effect of prolactin on the production of progesterone by mouse ovaries *in vitro*. *J Reprod Fertil 47:*155-156, 1976.

Makman, M.H., Ahn, H.S., Thal, L.J., Sharpless, N.S., Dvorkin, S., Horowitz, S.G., and Rosenfeld, M.: Aging and monoamine receptors in brain. *Fed Proc 37:*548, 1978.

Meites, J., Huang, H.H., and Simpkins, J.W.: Recent studies on neuroendocrine control of reproductive senescence in rats. In *Reproduction and Aging,* Vol 4, E.L. Schneider, ed. Raven Press, New York, 1978, pp 213-236.

Miller, A.E., and Riegle, G.D.: Serum LH levels following multiple LH-RH injections in aging rats. *Proc Soc Exp Biol Med 57:*497-499, 1978.

Miller, A.E., and Riegle, G.D.: Temporal changes in serum progesterone in aging female rats. *Endocrinology 106:*1579-1583, 1980.

Muralidhar, K., Manackjee, R., and Moudgal, N.R.: Inhibition of *in vivo* pituitary release of luteinizing hormone in lactating rats by exogenous prolactin. *Endocrinology 100:*1137-1142, 1977.

Naranjo, N., and Greene, E.: Use of reduced silver staining to show loss of connections in aged rat brain. *Brain Res Bull 2:* 71-74, 1977.

Negro-Vilar, A., Dickerman, E., and Meites, J.: Effects of continuous light on hypothalamic FSH-releasing factor and pituitary FSH levels in rats. *Proc Soc Exp Biol Med 127:* 751-755, 1968.

Nelson, J.F., Felicio, L.S., and Finch, C.E.: Ovarian hormones and the etiology of reproductive aging in mice. In *Aging--Its chemistry,* A.A. Dietz, ed. American Association of Clinical Chemists, Washington, D.C., 1980, pp 64-81.

Nelson, J.F., Latham, K.R., and Finch, C.E.: Plasma testosterone levels in C57BL/6J male mice: Effects of age and disease. *Acta Endocrinol (Copenh) 80*:744-752, 1975.

Nelson, J.F., Felicio, L.S., Osterburg, H.H., and Finch, C.E.: Altered profiles of estradiol and progesterone associated with prolonged estrous cycles and persistent vaginal cornification in aging C57BL/6J mice. *Biol Reprod 21*:193-203, 1981.

Nelson, J.F., Felicio, L.S., Randall, P.K., Simms, C., and Finch, C.E.: A longitudinal study of estrous cyclicity in aging C57BL/6J mice: I, Cycle frequency, length, and vaginal cytology. *Biol Reprod 27*:327-339, 1982.

Nisbett, R.E., Braver, A., Jusela, G., and Kezur, D.: Age and sex differences mediated by the ventromedial hypothalamus. *J Comp Physiol Psych 88*:735-746, 1975.

Novak, E.R.: Ovulation after fifty. *Obstet Gynecol 36*:903-910, 1970.

Osterburg, H.H., Donahue, H.G., Severson, J.A., and Finch, C.E.: Catecholamine levels and turnover during aging in brain regions of male C57BL/6J mice. *Brain Res 224*:337-352, 1981.

Peluso, J.J., Steger, R.W., and Hafez, E.S.E.: Regulation of LH secretion in aged female rats. *Biol Reprod 16*:212-215, 1977.

Peng, M.T., Chunong, C.F., and Peng, Y.M.: Lordosis response of senile female rats. *Neuroendocrinology 24*:317-324, 1977.

Quadri, S.K., Kledzik, G.S., and Meites, J.: Reinitiation of estrous cycles in old constant-estrous rats by central acting drugs. *Neuroendocrinology 11*:248-255, 1973.

Quilligan, E.J., and Rothchild, I.: The corpus-luteum-pituitary relationship: The leuteotrophic activity of homo transplanted pituitaries in intact rats. *Endocrinology 67*:48-53, 1960.

Raisman, G., and Field, P.: Sexual dimorphism in the neuropil of the preoptic area of the rat and its dependence on neonatal androgen. *Brain Res 54*:1-29, 1973.

Ratner, A., and Adamo, N.J.: Arcuate nucleus region in androgen-sterilized female rats: Ultrastructural observations. *Neuroendocrinology 88*:26-35, 1971.

Ratner, A., and Peake, G.T.: Maintenance of hyperprolactinemia by gonadal steroids in androgen-sterilized and spontaneously constant estrus rats. *Proc Soc Exp Biol Med 146*:680-683, 1974.

Recant, L., Hume, D.M., Forsham, P.H., and Thorn, G.W.: Studies on the effect of epinephrine on the pituitary-adrenal system. *J Clin Endocrinol Metab 10*:187-229, 1950.

Schmidt, M.J., and Thornberry, J.F.: Cyclic-AMP and cyclic-GMP accumulation *in vitro* in brain regions of young, old, and aged rats. *Brain Res 139*:169-177, 1978.

Severson, J.A., and Finch, C.E.: Reduced dopaminergic binding during aging in the rodent striatum. *Brain Res 192*:147-162, 1980.

Shelton, M.: A comparison of the population of oocytes in nulliparous and multiparous senile laboratory rats. *J Endocrinol 18*:451-455, 1959.

892 Randall and Finch

Simpkins, J.W., Mueller, G.P., Huang, H.H., and Meites, J.: Evidence for depressed catecholamine and enhanced serotonin metabolism in aging male rats: Possible relation to gonadotropin secretion. *Endocrinology* 100:1672-1678, 1977.

Singh, K.B., and Greenwald, G.S.: Effects of continuous light on the reproductive cycle of the female rat: Induction of ovulation and pituitary gonadotropins during persistent estrus. *J Endocrinol* 38:389-394, 1967.

Stricker, E.M., and Zigmond, M.J.: Recovery of function following damage to central catecholamine-containing neurons: A neurochemical model for the lateral hypothalamic syndrome. In *Progress in Physiological Psychology and Psychobiology,* J.M. Sprague and A.E. Epstein, eds. Academic Press, New York, 1976, pp 121-188.

Swanson, H.E., and van der Werff ten Bosch, J.J.: The early-androgen syndrome: Its development and response to hemi-spaying. *Acta Endocrinol* 45:1-12, 1964.

Talbert, G.: Aging of the female reproductive system. In *Handbook of the Biology of Aging,* C.E. Finch, and L. Hayflick, eds. Van Nostrand, New York, 1977, pp 318-356.

Van der Schoot, P.: Changing pro-estrous surges of luteinizing hormone in aging 5-day cyclic rats. *J Endocrinol* 69:287-288, 1976.

vom Saal, F.S., and Bronson, B.H.: In utero proximity of female mouse fetuses to males: Effect on reproductive performance during later life. *Biol Reprod* 19:842-853, 1978.

Weiner, R.I., and Ganong, W.F.: Role of brain mono-amines and histamine in regulation of anterior-pituitary secretion. *Physiol Rev* 58:905-976, 1978.

Wilkes, M.M., Lu, K.H., Fulton, S.L., and Yen, S.S.C.: Hypo-thalamic-pituitary-ovarian interactions during reproductive senescence in the rat. In *Parkinson's Disease II. Aging and Neuroendocrine Relationships, Adv Exp Med Biol,* Vol 113, C.E. Finch, D.E. Potter, and A.D. Kenney, eds. Plenum Press, New York, 1978, pp 127-148.

Wurtman, R.J., and Fernstrom, J.D.: Control of brain neuro-transmitter synthesis by precursor availability and nutritional state. *Biochem Pharmacol* 25:1691-1696, 1976.

Wurtman, R.J., Axelrod, J., Chu, E.W., and Fischer, J.E.: Mediation of some effects of illumination of the rat estrus cycle by the sympathetic nervous system. *Endocrinology* 75: 266-272, 1964.

Wurtman, R.J., Axelrod, J., Chu, E.W., Heller, A., and Moore, R.Y.: Medial forebrain bundle lesions: Blockade of effects of light on rat gonads and pineal. *Endocrinology* 81:509-514, 1967.

Wuttke, W., and Meites, J.: Effects of electrochemical stimulation of medial preoptic area on prolactin and luteinizing hormone release in old female rats. *Pflugers Arch* 341: 1-6, 1973.

19

Animal Models
in Psychoneuroendocrinology

RICHARD B. MAILMAN, CLINT KILTS, AND TOM McCOWN

INTRODUCTION

"Behavioral neuroendocrinology" is the study of the role of
hormones in mediating or modulating the behavior of the intact
organism. The intriguing nature of this field is reflected in
the fact that an animal's behavior may be dramatically altered
by a miniscule change in the concentration of a hormone. Added
significance arises from the finding that there is usually defined localization, rather than diffuse distribution, of hormones in the central nervous system (CNS). This suggests that
the function of those brain regions containing a relatively
high concentration of a hormone may be modulated or mediated by
the hormone in question. Indeed, electrophysiological studies
have demonstrated the ability of many hormones to alter the
firing rate of different neuronal populations. Twenty years
ago, many scientists would have considered it science fiction
to be told that animals would eat, drink, copulate, or exhibit
other specific behaviors simply because small amounts of hormone were administered. The historical account of this meta-
morphosis is reviewed, in part, in chapter 1 of this volume.
Specific studies that have been pursued with individual hormones are concisely discussed in other chapters.

However, two basic questions give direction for experimentation in this field and also provide a framework for the critical assessment of the data that result. First, what is the
relevance of the profound changes that underlie behavioral
neuroendocrinology with regard to the neurobiology of the studied organism? Secondly, what significance, if any, do these
observations have to human disease?

Scientists develop hypotheses that their scientific endeavors evaluate. However, neophytes to this field may not be aware of the types or significance of the assumptions and limitations inherent in the initiation, conduction, or evaluation of these experiments. The formalization of this latter process may be called "model development." There have been many books and review articles evaluating aspects of the neurobiological relevance of models (see Suggested Reading). In this chapter a different tack will be taken, and an attempt will be made to detail some of the motivations and limitations that are frequently involved in developing "models" in general and why the application of this strategy to psychoneuroendocrinology is unique. These concepts may be somewhat redundant for experienced investigators, but they will, hopefully, provide a critical framework that will help the neophyte in behavioral neuroendocrinology to understand and to integrate the information presented in this volume.

The term "model" is frequently used in biomedical research and often results in misleading interpretations, depending on the correctness of the framework of application. The dictionary defines model primarily as "a copy in miniature, but to scale" (Wyld and Partridge, 1957). It is important that a secondary definition, "a standard or pattern of excellence," is not substituted. Despite these connotations, a more frequent interpretation of the term "model" is as an analogue of a biologically relevant situation. Rhetorically, it may be asked why anyone would pride themselves on establishing a small copy or imitation of the real thing? The answer is, of course, the ethical limitations and the nature of the technical problems introduced by the complexity of the brain. The purpose of this chapter will be to discuss some criteria that are fundamental to the appropriate use of models in psychoneuroendocrinology. In addition, many of the motives that underlie the establishment of models will be discussed, since these are ultimately relevant to interpretation of data.

Motives for Establishing Models

There are, prima facie, two distinct, although interrelated, motives for the use of models. The first is an attempt to relate hormone-behavior interactions to some aspect of human physiology or pathology. The goal may be an attempt to understand mechanisms that are involved in a disease process, or the development of effective therapeutic or prophylactic measures. The second rationale would attempt to use a model not as "a miniature, to scale, of the original," but rather as "a mini-

ature in terms of inherent complexity." Thus, this latter
direction would result in using a model to provide a means of
studying basic processes that, by virtue of their complexity,
are inherently resistant to analysis by available techniques.
This approach, in psychoneuroendocrinology, would attempt to
relate the model to "real" hormone-induced perturbations of
behaviors or of other neurohormonal systems. These descrip-
tions should encompass most motives for the development of
models in general. Certain criteria are necessary to permit
assessment of the validity of a given model.

EVALUATION OF ANIMAL MODELS

Animal models have been constructed using a number of species,
primarily rodents, such as rats and mice. In evaluating hormone-behavior interactions, it is important to remember that
the differences between species may be as important as their
similarities. It is often assumed that homologous neuronal
systems will be of similar functional importance between species. However, this may not be valid for several reasons.
Since the normal expression of a behavior will be controlled or
modulated by several neuronal systems, these neuronal interrelationships, if different between species, may modify how
perturbation of only one pathway is expressed. In addition,
the stimulation of certain receptors thought to be causally
related to a specific behavior possibly may cause apparently
different responses among several species.
This contention suggests some general guidelines for the
critical evaluation and dynamic application of models in psychoneuroendocrinology. First, the phylogeny of the animal must
be considered, and key to such a consideration is the adaptive
function that the particular model system serves. If an animal's response to stress is to "freeze," then one might expect
different patterns of hormonal or neurochemical effects from
those of an animal whose response is "escape." A more specific
example involves the comparison of the rat's olfactory senses
with that of the human's. In the rat, the sense of smell oc-
cupies the primary sensory input, as vision does for man.
Thus, the homologous neuronal systems of olfaction or vision in
the two species may have entirely different sets of compensatory neuronal mechanisms. This difference must be accounted
for when using cues to a specific sensory modality in behavioral paradigms, especially if global conclusions are to be
drawn. It is evident that an understanding of both the biological and behavioral repertoire of the species being studied
is an essential component of proper modeling. Frequently this

fact is neglected when the motivation for a particular model is
to study a disease process. Most of the diseases of interest
will not have defined etiologies nor "symptoms" that can be detected in lower animals. This means that the model of the disease must be created or artificially induced--any direct analogies to human disease states must be very conservative. For
example, administration of cholecystokinin at various doses can
alter food consumption, yet this observation does not necessarily imply that cholecystokinin is involved in mechanisms of
obesity (see chapter 11). When all these considerations are
thoughtfully applied to an experimental design, seemingly
simple observations can provide a meaningful insight into the
actions of drugs and hormones. Conversely, overzealous anthro-
pomorphic conclusions usually only offer confusion and misleading interpretations about a given area of research.

When interspecies differences are carefully evaluated, they
can often provide important sources of information with respect
to the action of some drug or hormone. Such is the case with
regard to the actions of amphetamine. It has been reported
that high doses of amphetamine produce stereotyped or repetitive behaviors in many species; however the actual behaviors
are qualitatively quite different between species (Randrup and
Munkvad, 1963). Rats exhibit repetitive gnawing, licking,
sniffing, and head movements, whereas man exhibits complex
repetitive behaviors, such as constantly dismantling and as-
sembling a clock. Furthermore, man's repetitive behaviors can
change from day to day. These observations can be used to
study the action of amphetamine across a number of species and
possibly provide some insight into both the site and mechanism
of amphetamine's action, even though the stereotyped behaviors
exhibited by man are far more complex than those of a rat. Additionally, differences in response could provide information
on the neuronal action of the drug. Extension of this type of
approach could include the comparison of centrally acting hormones or drugs between species that either do or do not respond
to the agent. This approach could provide a means to dissect
the anatomical sites and biochemical substrates of drug or
hormone action.

These facts are representative of how various experimental
approaches may be confounded by numerous technical and theoretical factors; however, they also emphasize the potential utility that careful use of animal models may have in psychoneuroendocrinology. The adherence to proper control of such factors
should help both experimenter and observer to be more critical
about the strengths and weaknesses of any model proposed. This
book provides an excellent compendium of some important models
in psychoneuroendocrinology; thus, review of these specific
models is inappropriate. Similarly, the application and evalu-

ation of animal models used in various subdisciplines of psychiatry and pharmacology has been the subject of numerous articles and books, many of which are listed at the end of this chapter as Suggested Reading. However, this chapter will outline some criteria relevant to establishing and evaluating an animal model. These criteria fall into two categories (see Table 19-1). The first are technical aspects related to whether behavioral or biochemical changes do in fact result directly from the hormone of interest. The second are theoretical aspects about the physiological relevance of observed changes and their applicability to more generalized systems.

Chemical-Biochemical Specificity

A major question involved in all models in psychoneuroendocrinology is the chemical identity of the hormones that purportedly affect the behavior being studied. The simplest aspect of this problem, albeit one of marked complexity, is the correct identification of the chemical species responsible for the behavioral effects. The isolation, purification, and characterization of endogenous molecules has been the forte of biochemists. Recently it has also become one of the more dominant themes in neurobiology. One of the most exciting developments in recent years has been the isolation, characterization, and structural definition of several opioid peptides, including the enkephalins, endorphins, and their precursors (see chapters 7 and 8). Two distinct routes often result in the isolation of endogenous humors. The first is the isolation and purification of molecules from specific organs or bodily fluids. Subsequent to this, a functional role for these compounds is sought. This type of protocol was used in developing hypotheses about roles for neurotensin (Carraway and Leeman, 1973; Nemeroff, 1980). The second involves attempts to isolate unknown endogenous molecules for which a physiological role has been postulated. As an example, hypotheses and elucidation about an endogenous "morphine" receptor (Pert and Snyder, 1973; Simon et al., 1973; Terenius, 1973) ultimately led to the isolation and characterization of the opioid peptides (Hughes et al., 1975; Li and Chung, 1976). It should be noted that, although high-affinity specific drug receptors may be identified (as for the benzodiazepines, see Squires and Braestrup, 1977), isolating an endogenous ligand may be more difficult, occasionally leading to erroneous conclusions (Marangos et al., 1979; Fox, 1980).

The isolation and identification of endogenous chemicals then permits more detailed pharmacological and neurobiological studies. Following the chemical synthesis of the pure compound,

Table 19-1. General Criteria to Consider in Evaluating Models

General classification	Some specific considerations

Motivation	As model of human disease
	  Mimics etiology of disease
	  Drug effects on symptoms
	  (i.e., "screen")

	As basic neurobiological tool
Chemical-biological specificity	Isolation and identification of hormone
	Synthesis of analogues
	Biological conversion to other active species
Pharmacological factors	Characteristics of receptors for hormones
	Secondary effects dependent upon route of administration
	Pharmacokinetics of hormone
	Mechanism of action, e.g., direct or indirect
	Interactions with drugs or other hormones
	Time course
Biological factors	Method of quantifying behavior
	Sites of action of hormone
	Associated neural systems
	Biological rhythms
	Validity of behavioral measures

structural analogues having increased or decreased activity, or being more resistant to biological degradation, are sought. In addition, the availability of purified hormone permits the production of antisera that may then be used in a variety of immunological methods (Elde et al., 1979; Brownstein, 1977). It is important to realize that any model in psychoneuroendocrinology is ultimately dependent upon the successful attainment of these goals. Periodic determinations of the purity of hormones to be used as drugs will minimize experimental artifacts.

Finally, consideration of metabolic processes are important—— in pharmacological studies, the effects of drug metabolism are usually evaluated reasonably early. Since the routes, sites, and mechanisms of metabolism for many different agents may vary between various species, the importance of the rate of metabolism and the pharmacology of the metabolites is important in considering the action of a given agent. For example, morphine, and presumably the endorphins and the enkephalins, cause hyperactivity in mice and cats, whereas rats exhibit a period of hypoactivity followed by hyperactivity (Hollinger, 1969; Buxbaum et al., 1973). This apparent difference in drug action might be due to metabolic differences between the species. Only after these data were available could one consider the possibility of different centrally mediated responses between the species.

Receptor Identification

A potent and efficacious pharmacological response to an administered hormone is usually construed to indicate the presence of indigenous receptors, sensitive to that molecule, and mediating their pharmacological effect. Electrophysiological methods might, in ideal circumstances, provide the most direct way of characterization, but these techniques have major technical limitations (Werman, 1972). In recent years, attempts to characterize receptors have been aided by the widespread application of radioligand binding methods. The methods are most often used with membranous (insoluble) receptors, but they are also useful for cytosolic (or solubilized) receptors. These techniques assume a high affinity of the ligand, such as the hormone or its analogue, for the receptor, an acceptable criterion because of the extremely small quantities of hormone that result in behavioral perturbations. There are various factors necessary for such characterizations. These include: selection or synthesis of radioligand, source and preparation of tissue containing the purported receptor, development of suitable incubation conditions, and separation of bound from unbound ligand. The utility of the resulting method often rests on careful pharmacological studies, such as displacement of radio-ligand with compounds of varied biological activity. Success-ful application of these methods may also permit the radio-ligand selected to be used in distribution studies, both bio-chemical and autoradiographic, thus permitting the elucidation of potential loci that mediate the observed behaviors attrib-uted to the ligand. Details of these methods are available in several outstanding compendia or reviews (Pepeu et al., 1980; Cuatrecasas, 1974; Yamamura et al., 1978). However, it is im-perative that these techniques be used conservatively, since they are subject to many biochemical effects that may markedly change interpretation of data. One of the earliest examples of this was the effect of sodium on opiate receptor binding (Pert et al., 1973). Despite this, insufficient concern about buffer or ion composition is sometimes used in studies with radio-ligand binding.

Pharmacological Factors Influencing Hormone Action

The tacit assumption in psychoneuroendocrinology is that a hor-mone, an endogenous molecule, has an important role in one or more behaviors. Invariably, pursuit of this hypothesis in-volves detailed pharmacological studies. As alluded to ear-lier, the usual pharmacokinetic factors, such as absorption, distribution, metabolism, and elimination, that complicate all pharmacological experiments are compounded by the fact that the drug used in psychoneuroendocrinological studies is an endogen-ous molecule. Therefore the existence of specific inactivating mechanisms at the loci of action should be expected. Although these considerations have been discussed elsewhere (Liebeskind et al., 1979), a brief review may be useful.

One important variable is the chosen route of administration. This is significant because many hormones have multiple direct or indirect effects on both central or peripheral tissues, which may complicate the interpretation of the observed behav-ioral effects. Furthermore, the inability of some hormones to penetrate into the CNS from peripheral sites of administration may limit the usefulness of this route. For example, the pres-ence of opioid receptors in the gastrointestinal tract, an ob-servation that has provided a useful bioassay method, must be considered if the opioid peptides are administered periph-erally. For these reasons, and because of the limited avail-ability of these often expensive molecules, administration directly into aspects of the cerebroventricular system or into specific loci in the brain is often necessary, especially when a centrally mediated effect is being studied.

The endogenous nature of hormones requires that even greater concern be given to the bioinactivation of these molecules. For example, the presence of inactivating mechanisms at loci where a hormone might normally act could possibly result in the spurious (low) estimates of potency. This makes time course, dose-response and, when possible, use of structural analogues, of paramount importance. Further, there has been some difficulty in differentiating the specificity of deactivating enzymes, a special problem with peptide hormones. These and other related technical factors will be of concern for all models that are proposed. In themselves, they form a basis only for validating the specific system being investigated. Other factors are also relevant to the biology of hormone action. Although this chapter will emphasize *in vivo* models using mammals, systems using lower animals, cell culture, of even *in vitro* models may provide information important to psychoneuroendocrinological studies.

Site of Hormone Action

One primary decision affecting a model will be the anatomical target selected for examination or perturbation. Hormone action is almost always thought to be associated with considerable anatomical specificity. Immunohistofluorescence and radioligand binding techniques have usually provided converging evidence showing that hormones and hormone receptors in brain are localized in specific regions rather than being diffusely distributed. An important factor to be integrated into a model will be this question of anatomical specificity--in some cases distinct behaviors may be affected by perturbation of a receptor to a specific hormone located in different brain regions. The application of hormone to specific brain sites will also be useful in differentiating specific behavioral effects from changes caused by secondary effects.

These considerations suggest experimentally useful approaches, as well as certain problems. A major pharmacodynamic consideration is the ability of the hormone or drug to traverse biological barriers between purported loci and the site of administration (Kastin et al., 1978). Additionally, lesioning studies may be an essential part of model testing. These may include destruction of perikarya, disruption of afferent or efferent pathways, or inactivation of receptors. These may encompass electrolytic or surgical destruction, or, where applicable, the use of somewhat more specific chemical tools, including neurocytotoxins or immunological methods (Breese, 1975). Because of the lack of specificity inherent with any of these methods, it is desirable to use several in combination to obtain converging lines of evidence.

Biological Interactions with Psychoendocrine Effects

There are many important factors that can influence the behavioral effects caused by a hormone. It should be noted that the developing CNS, and its interactions with peripheral endocrine tissue, presents a unique series of problems and opportunity. Since this area has already been dealt with in this volume (see chapter 16), such interactions will be mentioned here only as a cautionary note in reference to models. Although the administration of hormones during development results in striking behavioral effects, it has also been frequently noted that hormonal interactions with peripheral endocrine tissues may alter the observed behaviors. For example, the estrogen-priming effects that permit the expression of maternal or sexual behavior in ovariectomized rodents otherwise unresponsive to hormonal administration are well known (see chapter 15).

Another major source of biological variability relates to the interaction of biological rhythms with behavioral effects of hormones. Many of these effects may be gleaned from discussions of pineal physiology (Motta et al., 1971; Reiter, 1973). Although it is often possible to control for the effects of diurnal or circadian rhythms on behavioral changes, it is conceivable there may be seasonal effects that may complicate experiments, especially in comparing animals not reared at similar times.

SELECTION AND EVALUATION OF BEHAVIOR FOR INVESTIGATION

Several different routes commonly lead to the selection of a specific behavior for use in psychoneuroendocrinological studies. During the past decade, immunohistofluorescent and radioreceptor distribution studies have localized many hormones, previously thought to be present and act only in peripheral tissues, to specific brain regions and thus suggested roles for these molecules in the CNS. Since various behaviors are believed to be modulated in specific loci, the effects of hormones administered into a given region can then be evaluated in terms of their ability to alter the associated behavior. In this fashion, evidence, albeit indirect, can be obtained concerning the functional interaction of the intracerebrally administered hormone with those putative neuronal systems believed to modulate or mediate the selected behavior. In many cases, the observed changes are so marked that the directions for further study are obvious. However, a limitation of this approach is that the apparatus selected for observations of behavior, or the expected alterations, may obfuscate other relevant changes. Finally, it should be noted that hormone-induced behavioral changes, which form the basis for subsequent models, may result from serendipitous findings during the conduction of seemingly unrelated experiments.

The number of behavioral paradigms that are used as indices of centrally mediated responses are numerous. Three basic points relevant to behavioral paradigms may allow more critical evaluation of the validity of any behaviorial model, and closer scrutiny of the significance of various pharmacological or physiological manipulations to it. These points are: what is physically required of the animal in order to perform the behavior; how do general learning principles influence both the behavior and hormone effects; and, finally, what biological function does the behavior actually reflect? By selecting appropriate behavioral paradigms and evaluating the results in light of these questions, one enhances the power of the behavioral measure and avoids the common interpretational errors that frequently occur in this area of research.

An initial characterization of either a drug or hormone will usually begin by qualitatively assessing the gross behavioral consequences of administration of the drug or hormone by a selected route. Various doses of the agent are given, and some method (automated or observational) is used to rate the behaviors, thus providing a means of quantitation. For example, many times this gross evaluation of an agent entails its effects on the locomotion of a rat or mouse. Detailed observations by trained personnel, ignorant of the treatment condition, can provide better characterization than will automated measuring devices. This is especially true with regard to complex patterns of behavior, although intra-observer variability may create problems in terms of replication of the results. Mechanical measuring devices will usually decrease variability, but the restricted flexibility inherent in automated measuring devices precludes further elaboration of the nature of the locomotor activity. Usually, a combination of both methods across a wide dose range of the drug or hormone in question provides the best delineation of the drug effects.

A seemingly simple behavior, such as locomotion, provides a good example for the application of the principles that have been previously outlined. The locomotor activity exhibited by rats is generally thought to be an index of exploratory behavior. This obviously requires a certain amount of integrated physical function, and clearly this behavior will decrease as the animals become habituated to the novel qualities of apparatus, that is, a new environment. A number of drugs will alter

locomotor activity, so a common experimental tack is to assess
the interactive effects of a given hormone on drug-induced increases in locomotor activity. If a hormone does attenuate
drug-induced locomotor effects, three possible explanations for
the decrease can be posited. The hormone may nonspecifically
debilitate the animals. In this case, no drug will cause increased locomotion because the animal is incapable of responding. Alternatively, the hormone may enhance the effects of the
drug on behaviors that are incompatible with the expression of
locomotor activity. For example, high doses of amphetamine
cause stereotyped behaviors that are registered in an automated
locomotor activity box as low levels of movement. Thus, the
potentiation of an intermediate dose of amphetamine would result in stereotyped behaviors and lower activity levels.
Clearly, to conclude that the hormone attenuated the actions of
amphetamine would be totally erroneous. Finally, the hormone
may indeed block the action of the drug used to induce increased locomotion. These possible drug-hormone interactions
demonstrate that locomotor activity is not a unitary behavior,
but a complex set of behaviors; therefore results must be interpreted with caution. This conclusion is not confined to
this behavior.

A second significant point of concern deals with the use of
locomotor activity measurements, or any behavioral measure, as
an index of chronic drug effects. It is important to always
assess the learning environment of the test animal. When an
animal is chronically exposed to a test situation under the
influence of a drug, conditioning of the drug-induced behaviors
can occur. In other words, the test apparatus becomes associated with a set of behaviors and the subject will exhibit these
behaviors when placed in the test environment. For example,
rats chronically treated with amphetamine prior to test sessions for locomotor activity, exhibit increased levels of
locomotor activity (compared to saline-injected controls) when
subsequently tested with saline (Tilson and Rech, 1973). One
means of avoiding such drug-dependent conditioning is to administer the drug chronically following the test sessions. Then,
the animal is not chronically tested under the influence of the
drug, and the association of drug effects with the test apparatus will not be made.

Investigators may assess the effect of a hormone on a particular ongoing behavior and draw conclusions about the effect of
the hormone upon the generalized system. Two exemplary paradigms are brain self-stimulation and active-avoidance behaviors. With the former paradigm, hormones, such as endorphins
(Dorsa et al., 1979), have been injected into the brain and the
effects upon the animal's self-stimulation behavior evaluated.
When one observes a hormone-associated decrease in the self-

stimulation behavior, the conclusion that the hormone has attenuated the reinforcing properties of the self-stimulation behavior is not implicit. If the animal is performing at or near a maximal rate of responding prior to hormone administration, only rate decreasing but not rate increasing effects will be detected. Likewise, if the animal is responding at a very low rate, the most likely direction of change is upward. These points necessitate evaluating the effects of different doses of the drug or hormone at several different current intensities to control for rate-dependent effects and nonspecific actions of the administered agents.

A genetic determinant of individual response variables has been demonstrated in active avoidance behavior. Whenever shock is used to control response rates, one must first determine if the hormone or drug in question alters an animal's sensitivity to the shock. If so, the contingencies of the paradigm will have changed, because the stimulus parameters have been altered. Another interesting point to consider has been reported by Barrett et al. (1977), who showed that a strain difference in Y maze avoidance performance between rats was explained solely on the basis of their inherent response to shock. One strain of rats remained immobile, whereas the other strain ran. Such a condition meant that the active strain of rats came in contact with the contingencies for avoiding the shock much sooner than the inactive strain. These strain differences in response to shock could clearly confound the results of any study if one were not aware of the effects such changes could produce.

Many purported animal models of human disorders, especially those related to the CNS, are often "validated" by demonstrating the "appropriate" response to drugs used to treat the human illness. These are based upon clinical observations, as well as by demonstrating the lack of response to compounds that are clinically ineffective. For example, approach-avoidance or conflict paradigms have proved effective as screens for anxiolytic drugs, such as diazepam. However, these paradigms by design select for animals that respond to the benzodiazepines. Thus, it is important to emphasize that this serves as a drug screen, rather than as a model of human anxiety. Likewise, the "behavioral despair" swimming test for antidepressants (Porsalt et al., 1977) has been proposed to provide an excellent screen for antidepressant compounds of several chemical classes when these are administered acutely. However, without some a priori hypothesis of the mechanism of action of these drugs, care must be taken in correlating this paradigm with human depression. For example, effects of tricyclic antidepressants on the swimming test are immediate, whereas clinical responses in human beings usually require several weeks of treatment. Therefore it is not surprising that the Porsalt test has recently been shown also to be responsive to drugs with no antidepressant activity (Schechter and Chance, 1979). As theories are advanced, such as involvement of dopamine systems in schizophrenia (Hökfelt et al., 1974; Joseph et al., 1979), models can be created to test the basic hypothesis. It is important to emphasize that these types of models depend upon a correlation with clinical observations for validation. Their application in psychoneuroendocrinology has been difficult because of the small amount of clinical data regarding the therapeutic use of exogenous hormones. This fact makes psychoneuroendocrinology all the more unique and challenging relative to other behavioral disciplines.

Tools to Assist in Psychoneuroendocrinology

There are several excellent series that deal with the use of pharmacological, physiological, and other neurobiological tools to ascertain the role of various neurotransmitters and receptor systems in specific behaviors. In addition, many of the chapters of this book contain useful descriptive comments concerning the application of these methods to specific psychoendocrinological models. For these reasons, it is necessary only to note that the development of a useful model in psychoneuroendocrinology will ultimately require the use of multiple methods to analyze underlying biological mechanisms. Examples of this have been cited in examples in this chapter and in others in this book.

APPLICATION OF ANIMAL MODELS

The discussion with which this chapter was introduced has left major questions unanswered. Why is there a need to conceptualize a model, rather than just establishing testable hypotheses to be evaluated? In actuality, a model is, no more and no less, a framework that dictates the direction that experimentation should pursue, since it will generate several hypotheses to be verified experimentally. It is for this reason that even a model subsequently shown to be of limited utility may result in significant gains in knowledge about the subject being modeled. However, proper adherence to the criteria listed earlier, coupled with well-designed and well-interpreted experimentation, will increase the probability that major theoretical advances will occur. Within this conceptual framework, models are often selected for several distinct motivations.

Models Mimicking Human Disease

Probably the major portion of the literature on models has focused on various human illnesses. Recalling the alternate definitions cited earlier, these models could have as a rationale a "miniature of the original." Unfortunately, although many appealing models have been developed, those dealing with the CNS have been useful only in a limited sense. For example, there have been no generally accepted models for human depression, schizophrenia, or childhood hyperkinesis, although models have been obtained for neurological disorders. Symptoms in animals resembling those seen in human beings with parkinsonism (using 6-hydroxydopamine), Huntington's chorea (kianic acid), or demyelinating disorders (immunologically induced allergic encephalomyelitis) have been used as models of the disorders (see Hanin and Usdin, 1977). However, although they may provide a useful guide to the new therapies, their utility for understanding the underlying etiologies may be more limited.

The relative newness of the field of psychoneuroendocrinology has contributed to a relative paucity of proposed models in this area, although this is rapidly changing. Recently, many models have been proposed involving the opioid peptides. Multiple, and often strikingly different responses have been demonstrated after administration of these agents, and the symptoms in some cases have been strikingly similar to those seen in human beings with particular mental illnesses (for example, Bloom et al., 1976). Similarly, the previously inexplicable actions of therapies, such as acupuncture, have been aligned to effects of these manipulations on "opiate" systems (Pomeranz, 1978). However, the development of psychoneuroendocrinological models of disease states has been primarily limited to thyroid or adrenal dysfunction and the resulting behavioral effects. The chapters by Kuhn and Schanberg (chapter 16) and Loosen and Prange (chapter 13) deal with the former issue and the chapter by Rees and Gray (chapter 14), with the latter issue. Finally, it should be noted that the most widespread use of this type of model is as a drug-screening device. Obviously, if an animal model is believed to mimic human illness, and if it fulfills the criteria outlined earlier, it may be an efficient way to screen for drugs having therapeutic potential. Several paradigms discussed earlier have been used for this purpose. Certain of these may be relevant to psychoneuroendocrinology, since there have been reports that hormonal alterations resulted in dramatic effects in models otherwise considered purely "pharmacological" (chapter 12). For example, the model most relevant to this discussion may be animal models of anxiety. A variety of paradigms have been used to develop "anxiety states" in animals, and it is of interest that a major criterion for acceptability of these models is the relative potency and efficacy of benzodiazepines, drugs considered to be the clinically prototypical anxiolytics. As cited earlier, the recent finding that there are specific benzodiazepine receptors in brain has spurred the search for endogenous ligands for this receptor. Isolation and identification of this molecule will bring these models fully into the realm of psychoneuroendocrinology. It will be apparent from the following section that other models of clinical relevance may soon result from other studies.

Models as "Miniatures": Basic Studies in Psychoneuroendocrinology

By far the greatest use of animal models, and possibly the one of greatest utility, has been the establishment of models in which hormone-induced behavioral changes are felt to result from alterations in psychobiology directly relevant to the biology of the experimental animal. There are several advantages in this conceptual approach. First, unlike using a model to mimic human illness, the necessity of accounting for interspecies differences is eliminated, leaving only the singular, although formidable, problem of making relevant correlations within the one species. Secondly, although this is a problem of immense magnitude, the necessity of defining a behavioral change in terms of the milieu of a lower animal minimizes many of the biases that occur when analogies are made to what may be uniquely human traits. In other words, attempts to establish a phenomenological equivalency between an induced animal behavior and human pathology are unwarranted. Such attempts may in fact place excessive constraint on the model (Carlton, 1978). This latter point is especially important when dealing with aspects of the test animal's repertoire that may not have intrinsic meaning relevant to human behavior. For example, marked alterations of water or food consumption or induction of precopulatory behaviors may be somewhat simpler to deal with than interpretation of "learned helplessness" or "anxiety." It should be noted that even the "simpler" behaviors are often shown to have components that are induced by inexplicable events, for example, tail pinch induced eating (Antelman et al., 1975).

In summary, this latter approach to models is also subject to many limitations that confound the interrelation of psychoneuroendrocrinological effects to a particular behavior. However, this approach does permit the integration of the psychoneuroendocrine effects with data obtained from other fundamental psychopharmacological and psychobiological studies. Thus, the demonstrated involvement of a certain neurotransmitter pathway

in eliciting a behavior may be correlated with the hypothesis
that a specific hormone, with a known receptor, elicits the
same behavior. This interaction may ultimately provide new
testable hypotheses interrelating these results, for example,
the hormone receptor may be located on the presynaptic cell of
the involved neurotransmitter pathway. This may become the
major use of animal models in psychoneuroendocrinology. It
permits the formulation of testable hypotheses of neurological
organization and provides working, simplified concepts that facilitate communication between scientists and laymen. In this
regard, the difficulty in establishing an ideal model that
meets all the criteria outlined earlier will not negate the
utility that such models may have. It is hoped that this discussion will provide a useful foundation for individuals unfamiliar with this field for evaluating and appreciating many
aspects of ongoing research.

REFERENCES

Antelman, S.M., Szechtman, H., Chin, P., and Fisher, A.E.: Tail
pinch-induced eating, gnawing and licking behavior in rats:
Dependence on the nigrostriatal dopamine system. *Brain Res*
99:319-337, 1975.

Barrett, R.J., Leith, N.J., and Ray, O.J.: A behavioral and
pharmacological analysis of variables mediating active-
avoidance behavior in rats. *J Comp Physiol Psychol 82*:
489-500, 1977.

Bloom, F.E., Segal, D., Ling, N., and Guillemin, R.: Endorphins: Profound behavioral effects in rats suggest new
etiological factors in mental illness. *Science 194*:
630-632, 1976.

Breese, G.R.: Chemical and immunochemical lesions by specific
neurocytotoxic substance and antisera. In *Handbook of
Psychopharmacology,* vol 1, L. Iversen, S.D. Iversen, eds.
Plenum Press, New York, 1975, pp 137-189.

Brownstein, M.J.: Biological active peptides in the mammalian
central nervous system. In *Peptides in Neurobiology,* H.
Gainer, ed. Plenum Press, New York, 1977, pp 145-170.

Buxbaum, D.M., Yarbrough, G.G., and Carter, M.E.: Biogenic
amines and narcotic effects. I. Modification of morphine-
induced analgesia and motor activity alter alteration of
cerebral amine levels. *J Pharmacol Exp Ther 185*:317-327,
1973.

Carlton, P.L.: Theories and models in psychopharmacology. In
Psychopharmacology: A Generation of Progress, M.A. Lipton, A.
DiMascio, and K. Killam, eds. Raven Press, New York, 1978,
pp 553-561.

Carraway, R.E., and Leeman, S.E.: The isolation of a new hypotensive peptide neurotensin, from bovine hypothalami. J Biol Chem 248:6854-6861.

Cuatrecasas, P.: Membrane receptors. Annu Rev Biochem 43: 169-214, 1974.

Dorsa, D.M., Van Ree, J.M., and DeWied, D.: Effects of [Des-Tyr[1]]-γ-endorphin and α-endorphin on substantia nigra self-stimulation. Pharmacol Biochem Behav 10:899-905, 1979.

Elde, R., Hökfelt, T., Ho, R., Seybold, V., Coulter, H.D., Micevych, P., and de Lauerolle, N.: Immunohistochemical studies of central and peripheral peptidergic Prog Brain Res 51:221-237, 1979.

Fox, J.L.: Search for antianxiety agent sparks debate. Chem Eng News, Feb. 4, 1980, pp 24-25.

Hanin, I., and Usdin, E., eds.: Animal Models in Psychiatry and Neurology. Pergamon Press, New York, 1977.

Hökfelt, T., Ljungdahl, A., Fuxe, K., and Johansoon, O.: Dopamine nerve terminals in the rat limbic cortex: Aspects of the dopamine hypothesis of schizophrenia. Science 184:177-179, 1974.

Hollinger, M.: Effects of reserpine, -methyl-p-tyrosine, p-chlorophenylalanine and pargyline on levorphanol-induced running activity in mice. Arch Int Pharmacodyn Thér 179: 419-424, 1969.

Hughes, J., Smith, T.W., Kosterlitz, H.W., Fothergill, L.A., Morgan, B.A., and Morris, H.R.: Identification of two related pentapeptides from the brain with potent opiate agonist activity. Nature 258:577-581, 1975.

Joseph, M.H., Frith, C.D., and Waddington, J.L.: Dopamine mechanisms and cognitive deficit in schizophrenia. Psychopharmacology (Berlin) 63:273-280, 1979.

Kastin, A.J., Sandman, C.A., Plotnikoff, N.R., Coy, D.H., Olson, R.D., Schally, A.V., and Miller, L.H.: Central nervous system effects of MSH and hypothalamic peptides. In Clinical Psychoneuroendocrinology in Reproduction, L. Carenza, P. Pancheri, and L. Zichella, eds. Academic Press, London, 1978, pp 47-56.

Li, C.H., and Chung, D.: Isolation and structure of an intriakontapeptide with opiate activity from camel pituitary glands. Proc Natl Acad Sci USA 73:1145-1148, 1976.

Liebeskind, J.C., Dismukes, R.K., Barker, J.L., Berger, P.A., Creese, I., Dunn, A.J., Segal, D.S., Stein, L., and Vale, W.W.: Peptides and behavior: A critical analysis of research strategies. Neurosci Res Program Bull 16:489-635, 1979.

Marangos, P.J., Paul, S.M., and Goodwin, F.K.: Putative endogenous ligands for the benzodiazepine receptors. Life Sci 25:1093-1102, 1979.

Motta, M., Schiaffini, O., Piva, F., and Martini, L.: Pineal principles and the control of adrenocorticotropin secretion. In *The Pineal Gland*, G.E.W. Wolstenholine, and J. Knight, eds. Churchill Livingstone, London, 1971, pp 279-291.

Nemeroff, C.B.: Neurotensin: Perchance an endogenous neuroleptic? *Biol Psychiatry 15*:283-302, 1980.

Pepeu, G., Kuhar, M., and Enna, S.J., eds.: *Receptors for Neurotransmitters and Peptide Hormones*. Raven Press, New York, 1980.

Pert, C.B., and Snyder, S.H.: Opiate receptor: Demonstration in nervous tissue. *Science 179*:1011-1014, 1973.

Pert, C.B., Pasternak, G., and Snyder, S.H.: Opiate agonists and antagonists discriminated by receptor binding in brain. *Science 182*:1359-1361, 1973.

Pomeranz, B.: Do endorphins mediate acupuncture analgesia? *Adv Biochem Psychopharmacol 18*:351-360, 1978.

Porsolt, R.D., LePichon, M., and Jalfre, M.: Depression: A new animal model sensitive to antidepressant treatments. *Nature 266*:730, 1977.

Randrup, A., and Munkvad, I.: Stereotyped activities produced by amphetamine in several animal species and man. *Psychopharmacologia (Berlin) 11*:300-310, 1963.

Reiter, R.J.: Comparative physiology: Pineal Gland. *Annu Rev Physiol 35*:305-328, 1973.

Schechter, M.D., and Chance, W.T.: Non-specificity of "behavioral despair" as an animal model of depression. *Eur J Pharmacol 60*:139-142, 1979.

Simon, E.J., Hiller, J.M., and Edelman, I.: Stereospecific binding of the potent narcotic analgesic H-etorphine to rat brain homogenate. *Proc Natl Acad Sci USA 70*:1947-1949, 1973.

Squires, R.F., and Braestrup, C.: Benzodiazepine receptors in rat brain. *Nature 266*:732-734, 1977.

Terenius, L.: Characteristics of the "receptor" for narcotic analgesics in synaptic plasma membrane fraction from rat brain. *Acta Pharmacol Toxicol (Copenh) 33*:377-384, 1973.

Tilson, H.A., and Rech, R.H.: Conditioned drug effects and absence of tolerance to d-amphetamine induced motor activity. *Pharmacol Biochem Behav 1*:149-153, 1973.

Werman, R.: CNS cellular level: Membranes. *Annu Rev Physiol 34*:337-374, 1972.

Wyld, H.C., and Partridge, E.H., eds.: *The Little and Ives Webster Dictionary*, complete and unabridged. J.J. Little and Ives Co., New York, 1957.

Yamamura, H.I., Enna, S.J., and Kuhar, M.J.: *Neurotransmitter Receptor Binding*. Raven Press, New York, 1978.

ADDITIONAL SUGGESTED READING

Barchas, J.D., Akil, H., Elliott, G.R., Holman, R.B., and Watson, S.J.: Behavioral neurochemistry: Neuroregulators and behavioral states. *Science 200:*964-973, 1978.

Beech, F.A.: Animal models of human sexuality. *Ciba Found Symp 62:*113-142, 1978.

Breese, G.R., Mueller, R.A., Mailman, R.B., Frye, G.D., and Vogel, R.A.: An alternative to animal models of central nervous system disorders: Study of drug mechanisms and disease symptoms in animals. *Prog Neuro Psychopharmacol 2:*313-325, 1978.

Ingle, D., and Shein, H., eds.: *Model Systems in Biological Psychiatry.* MIT Press, Cambridge, 1975.

Krieger, D.T., and Liotta, A.S.: Pituitary hormones in brain: Where, how and why? *Science 205:*366-372, 1979.

Mrosovsky, N., and Sherry, D.F.: Animal anorexias. *Science 207:*837-842, 1980.

Seligman, M.E.P., and Moser, J.D., eds.: *Psychopathology: Experimental Models.* W.H. Freeman, San Francisco, 1977.

Serban, G., and Kling, A., eds.: *Animal Models in Human Psychobiology.* Plenum Press, New York, 1976.

Tedischi, G.H.: Criteria for the selection of pharmacological test procedures useful in the evaluation of neuroleptic drugs. In *The Present Status of Psychotropic Drugs: Pharmacological and Clinical Aspects.* A. Cerletti, and F.J. Bove, eds. Excerpta Medica Foundation, Amsterdam, 1969, pp 145-153.

de Wied, D.: Peptides and behavior. *Life Sci 20:*195-204.

20

Behavioral and Neuroendocrine Interactions with Immunogenesis

Nicholas R. Hall

INTRODUCTION

Correlations between the behavioral state of an individual and
susceptibility to disease have been observed since long before
the advent of modern medicine. As early as the second century
A.D., Galen noted that cancer occurred more frequently in mel-
ancholy women than in sanquine women (Solomon et al., 1969a).
More recently, evidence has accumulated suggesting that this
association is due to an altered state of immunity resulting
from physiological changes in the brain. Psychotic behavior
has been associated with inhibition of both cellular and humoral immunity. Furthermore, neurochemical, morphological, and
electrical changes have been detected in the brain during the
immune response. There is also evidence that manipulations by
direct suggestion under hypnosis and by association using
classical-conditioning can alter the course of immunological
events.

Involvement of the neuroendocrine system is suggested by evidence of bidirectional interaction between this and the immune
system (Besedovsky and Sorkin, 1977). It has been demonstrated
that disturbances involving one system can profoundly alter the
other. For example, animals without a thymus or shielded from
antigenic stimulation have been found to have altered states of
thyroid, gonadal and adrenal functioning. In the example of
animals without a thymus, certain of these changes can be reversed by early thymus implantation (Besedovsky and Sorkin,
1974;1977). An altered hormonal state has also been found to
interfere with immunity. Rejection of transplants, a T-cell
function, is impaired in hypopituitary dwarf mice. However,
exogenous somatotropic hormone (STH) and thyroxine can overcome

this deficit (Fabris et al., 1971). This chapter will review the evidence suggesting that these changes are part of a complex interrelationship that enables the central nervous system (CNS) to modulate the many components of immunity.

A major difficulty in interpreting the results of research concerned with CNS immune interactions is the complexity of the two systems. Figure 20-1 illustrates the differentiation of the cell types involved in the elimination of foreign antigen. All of these cells are derived from a hematopoietic stem cell found in bone marrow. In some instances interactions between these cell types are extremely important in bringing about ultimate elimination of the foreign antigen. For example, measuring a deficiency in antibody production to a T dependent antigen could reflect a deficiency in processing of the antigen by phagocytic cells, a dysfunction involving regulatory T cells or a deficiency in the B lymphocyte. The difficulties are compounded even further when a single manipulation of the CNS could interfere with a multitude of physiologicol processes having indirect influence upon immunity.

PSYCHOIMMUNOLOGY

Psychiatric Disorders

Distinct personality traits have been associated with susceptibility to rheumatoid arthritis (Moos, 1964), recovery from infectious mononucleosis (Greenfield et al., 1966), recovery from influenza (Imboden et al., 1961), and susceptibility to cancer (Greer, 1979). In addition, diseases with autoimmune components have been correlated with distinct personality traits. Ulcerative colitis due to anticolon antibodies has been associated with obsessive-compulsive character traits (Engle, 1953). Disseminated lupus erythematosus has been found to occur following stressful situations and to be associated with an excessive desire for independence and activity (McClary et al., 1955). Allergies and skin disorders have also been associated with psychological states and hyperactivity (Engles and Wittkower, 1975; Marceca, 1978).

Some investigators have reported data suggesting that these observations are due to changes in various parameters of immunity. Abnormal lymphocytes have been reported in patients diagnosed as schizophrenic (Fessel et al., 1965) and lymphocytes with decreased responsiveness to phytohemagglutinin have been reported in autistic children (Stubbs et al., 1977). Differing immunoglobulin (Ig) levels have been found between psychiatric patients and healthy control subjects. Elevated

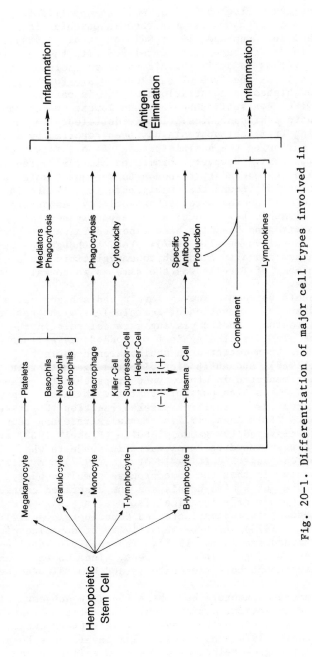

Fig. 20-1. Differentiation of major cell types involved in immunological responses.

gamma globulins and macroglobulins as well as specific increases in IgA and IgM have been reported in psychiatric
patients (Fessel and Grunbaum, 1961; Solomon et al., 1969).
Increased levels of IgM have been correlated with the severity
of psychotic symptoms in schizophrenic patients (Solomon et
al., 1966) as well as the type of schizophrenia. Pulkkinen
(1977) reported highest concentrations of IgM in patients
diagnosed as withdrawn schizophrenics, and lowest concentrations in paranoid schizophrenics. Immune deficiency in patients suffering from schizophrenia has been reported by some
investigators following the administration of pertussis vaccine
(Vaughn et al., 1949). However, Solomon et al. (1970) reported
no alteration in the secondary response to tetanus toxoid, and
Friedman et al. (1967) found that schizophrenic patients injected with cholera vaccine were better antibody responders
than were normal subjects. A decreased responsiveness of
lymphocytes to mitogens has also been reported in patients with
schizophrenia (Kovaleva et al., 1977). But, although the response to both concanavalin A and phytohemagglutinin was decreased, the number of T cells was the same as in normal
control subjects.

The relationship between serum Ig levels and schizophrenia is
difficult to discern. Elevation of certain Ig's might reflect
the presence of autoantibodies playing a causal role in the
disease process; IgG directed against oligodendrocytes in the
septal region has been detected in schizophrenic patients
(Heath et al., 1968), and antibodies directed against brain
tissue have been measured in other types of psychiatric disease
(Fessel, 1963). It is also possible that the changes in Ig
concentration arose as a result of altered reaction to stress
in these patients (Pulkkinen, 1977). Mental stress has been
associated with elevated 19s gamma globulin (Fessel, 1962) and
IgA levels have been found to be elevated in subjects who
habitually suppress anger (Pettingale et al., 1977). A third
potential mechanism is that the same process simultaneously
causes the symptoms of schizophrenia and the described changes
in immunity. A number of etiological factors in schizophrenia
have been suggested, including increased dopaminergic activity
(Johnstone et al., 1978) and hypersensitivity of dopaminergic
receptors (Chouinard and Jones, 1978). The possible involvement of dopamine is of particular interest, since dopamine has
been found to influence both the cellular and humoral branches
of immunity.

Changes in immune parameters have been found in subjects with
other types of neurological dysfunction. Epileptic patients
have been found to have elevated levels of Ig's (Barajaktar-
ovic-Nikoloc et al., 1974) and elevated IgM levels have been
reported in first degree relatives of certain epileptic pa-

tients (Haldorsen and Aarli, 1977). Other investigators have
correlated Ig levels with intellectual performance and mental
retardation. Cohen and Eisdorfer (1977) compared serum Ig
levels in elderly men and women who were in good health. Significant negative correlations were found between performance
on the Wechster Adult Intelligence Scale vocabulary and digit
symbol subtest and serum IgG and IgA in men. The relationship
was not significant in women. The effects of mental retardation are in dispute. Donner (1954) and Siegel (1948) presented evidence that patients with Down's syndrome are more
susceptible to infectious disease, but attempts to correlate
this observation with diminished immunity have been inconclusive. Deficient T cell responses as well as antibody production in Down's syndrome patients have been reported (Gordon et
al., 1971). Other authors, however, have found no difference
between Down's syndrome patients and control subjects in their
ability to produce antibodies to a variety of antigens (Hawkes
et al., 1978). This discrepancy might be due in part to the
institutional environment in which many of these patients live;
Sutnick et al. (1969) reported significant elevations in IgG in
institutionalized versus noninstitutionalized patients.

Emotional Distress

The mechanism by which stressful stimuli alter susceptibility
to disease is thought by some to involve the pituitary-adrenal
axis. Hypertrophy of the adrenal glands has been observed following exposure to shuttle-box and confinement stress (Marsh
and Rasmussen, 1960), and predator exposure has been found to
increase plasma corticosteroid levels (Hamilton, 1974). This
increase has been correlated with decreased resistance to para-
sitic infection. Other studies have correlated increased
levels of corticosteroids with increased susceptibility to
other types of disease and to inhibition of immunogenesis
(Rogers et al., 1979). However, the effects of corticosteroids
upon immune mechanisms are not simply inhibitory. Cortisol is
necessary for antibody formation under certain conditions
(Ambrose, 1964).

Not all investigators agree that stress effects upon immunity
are mediated by corticosteroids. Bartrop et al. (1977) found
that the mitogen responsiveness of T cells was significantly
depressed in bereaved spouses. This depression could not be
attributed to changes in corticosteroids, growth hormone, prolactin, or thyroid hormone, since serum concentrations of these
hormones did not differ from control values. Other investigators have reported an abnormality in the hypothalamopituitary
regulation of luteinizing hormone (LH) and thyroid-stimulating
hormone (TSH) in depressed patients; the LH response to thyro-

tropin releasing hormone/luteinizing hormone-releasing hormone
(TRH/LH-RH) infusion was significantly enhanced in secondary
depression patients (Ettigi et al., 1979). In the same study,
a delayed TSH response was noted using the same paradigm in
primary unipolar depressed subjects.

That stressful stimulation can alter immunity is suggested by
a number of studies. Avoidance learning has been found to decrease susceptibility to anaphylaxis (Rasmussen et al., 1959),
whereas other forms of stress depress antibody formation.
Group housing and overcrowding of rats have been found to inhibit both the primary and secondary antibody responses to
flagellin (Solomon, 1969b). High-density grouped mice had
lower antibody titers than low-density grouped mice (Edwards
and Dean, 1977), with dominant mice having higher titers than
other members of the group (Vessey, 1964). Spleen cells taken
from stressed donors were less responsive than cells from controls when cultured and tested *in vitro* (Gisler et al., 1971).
Monjan and Collector (1977) found an initial decrease in lymphocyte responsivity following stress, but later measured an
increase. Glenn and Becker (1969) found significantly decreased antibody titers to bovine serum albumin (BSA) in individually housed mice compared with group-housed mice. Another
group reported that noxious stimulation has a protective effect
in mice infected with malaria-inducing parasite (Friedman et
al., 1973).

Antibodies are produced by plasma cells derived from B lymphocytes. Nonetheless, the effect of stress upon antibody
titers could be mediated by regulatory T lymphocytes (Fig.
20-1). Delayed hypersensitivity reactions, which are T cell
mediated, can be modified by stress (Pitkin, 1965). A suppressive effect of stress upon T cells is also suggested by the
observation that avoidance training can increase the survival
time of skin transplants in mice (Wistar and Hildemann, 1960).
Solomon et al. (1974) concluded that most stress effects are
due to effects upon T lymphocytes. This conclusion was based
upon the observation that stress effects are dose dependent and
that antibody stimulation using low concentrations of antigen
can be suppressed. With lower antigen concentrations, antibody
production is thought to be more T cell dependent. These authors also found that IgG is more readily depressed by stress
than is IgM. Cellular immunity is implicated, since IgG is
considered to be more T cell dependent than IgM production
(Taylor and Wortis, 1968).

Hypnosis and Classical Conditioning

Several parameters of immunity can be altered by direct suggestion under hypnosis and by classical conditioning. Instant

inhibition of immediate hypersensitivity reactions can be induced by hypnotic suggestion (Black, 1963; Ikemi and Nakagawa,
1963; Kaneko and Takaishi, 1963). Delayed hypersensitivity
reactions are also susceptible to direct suggestion. Black et
al. (1963) reported that the swelling and erythema characteristic of the Mantoux reaction could be inhibited by hypnosis.
The mechanism was thought to involve a reduction in the exu-
dation of fluid since connective tissue in the inhibited response condition was found to be more compact with less fluid
between the fibrillar components of collagen. It was considered unlikely that the dynamics of the cellular response had
been affected, since biopsy revealed no alterations in cellular
infiltration.

Both inhibition and enhancement of antibody production have
been induced by classical conditioning. Lukyanenko (1961)
demonstrated that conditioning could prevent the production of
antibodies to a variety of antigens in rabbits. Similar findings have been reported in rats using sheep red blood cells
(SRBC) (Ader, 1974; Ader and Cohen, 1975; Rogers et al., 1976;
Wayner et al., 1978). Enhancement of antibody production has
also been demonstrated. Lukyanenko et al., (1963) reported an
increase in antibody titers to Pasteurella bovis, P. ovis, and
P. suis, in oxen. They concluded that a conditioned reflex
underlies the regulation of the "non-specific function of the
immunogenetic apparatus."

Wayner et al. (1978) confirmed earlier reports demonstrating
conditioned immunosuppression of the T dependent antigen, SRBC,
but they were not able to modify the response to Brucella abortus, a T independent antigen. However, conditioned enhancement of immunity has been found to include not only cellular
components but also humoral reactions and phagocyte activity.
These changes have been demonstrated using rabbits, guinea
pigs, rats, mice, dogs, monkeys, and man. A review of these
studies was reported by Metalnikov (1934-cited by Spector,
1980).

Behavioral Changes Induced by Immunity

The studies reviewed thus far suggest that altered behavioral
states in psychiatric disorders and neurological dysfunction
can be associated with changes in immunity. At least one report suggests that the opposite relationship can also exist.
Kamrin (1967) described an altered state of behavior that he
attributed to the induction of immune tolerance. Tolerance was
induced in chicks by means of egg parabiosis and alpha globulin
injection. It was found that 100 percent of immune tolerant
fowl consistently followed a stimulus object. Control animals

of similar strain and age exhibited erratic behavior. With-
drawal-escape activity was also demonstrated in a higher percentage of immune tolerant chicks compared with controls.
Electrophoretic analysis of the serum revealed a significant
decrease in circulating gamma globulin in the tolerant birds.
It was concluded that the behavioral and gamma globulin changes
reflected a physiological event in the CNS that came about as a
consequence of immune tolerance.

Insufficient evidence precludes the formulation of a precise
mechanism by which these changes might be mediated. However,
the possibility that adrenocorticotropic hormone (ACTH) is involved has to be considered. Pierpaoli and Maestroni (1977;
1978a; 1978b) implicated this hormone in the induction of tolerance, whereas Martin and van Wimersma-Greidanus (1978) showed
that $ACTH_{1-10}$ facilitates the imprinting approach response.

ANTIBODY FORMATION

Effects of Brain Lesions and Stimulation

Manipulations of the CNS have been correlated with marked
changes in antibody levels. Brain regions implicated by lesion
studies to influence antibody production include the anterior
and posterior hypothalamus as well as the midbrain. However,
there is disagreement among different investigators as to
whether these effects can be associated with specific brain
regions.

Decreased antibody titers have been correlated with lesions
in the anterior hypothalamus of rats (Tyrey and Nalbandov,
1972; Spector, 1980) and guinea pigs (Stein et al., 1976).
Control lesions in the medial or posterior hypothalamus did not
produce a concomitant decrease in antibody titer.

Other investigators have found that posterior hypothalamic
lesions can be correlated with a decrease in circulating antibody titers in rats (Paunovic et al., 1976), guinea pigs
(Mikhailov et al., 1970), and rabbits (Korneva et al., 1978).

Eremina and Devoina (1973) made a lesion in the midbrain
raphe nucleus of rabbits and then challenged the animals with
BSA. In contrast to the effect of hypothalamic lesions, this
paradigm resulted in antibody titers that were twice control
values. Goldstein (1978) also reported an effect of midbrain
lesions in the rat. He found that hemolyzing and hemagglutin-
ating antibodies to SRBC were slightly lower than in intact
animals. Plaque-forming cells (PFC) were decreased in animals
with either midbrain, thalamic, anterior hypothalamic, or
posterior hypothalamic lesions. This reduction was attributed
to a disruption of metabolic processes, since significant

weight loss was found in the animals with lesions.

Complete transection of the thoracic spinal cord has been found to decrease antibody titers in rats immunized with SRBC (Stranton et al., 1942). However, inhibition occurred only when the environmental temperature was not regulated and body temperature fell 5°C below normal. Within 36 hours after being maintained at normal body temperatures, a sharp rise in antibody titer was detected (Kopeloff and Stanton, 1942). At normal temperatures, transected animals were no different from controls in their ability to respond to antigen.

Morphological changes in lymphatic tissues also have been found following brain lesions (Isakovic and Jankovic, 1973). Involution of the thymus was observed after a lesion of the hypothalamus, reticular formation, thalamus, superior colliculus, caudate nucleus, or amygdaloid complex. Other changes were found only after a lesion was made in the hypothalamus. These included a significant depletion of cortical thymocytes as well as morphological changes in the spleen and lymph nodes. The spleen and lymph nodes of rats with lesions in sites other than the hypothalamus were essentially normal. In contrast, Solomon et al. (1974) reported a slight increase in thymus and spleen weight in rats with circumscribed lesions in the posterior hypothalamus. These finding do not necessarily contradict the observations reported by Isakovic and Jankovic (1973), since these authors did not correlate their findings with specific nuclei in the hypothalamus.

Not all investigators agree that CNS lesions can influence antibody production. Lesions in the anterior, medial, and posterior hypothalamus of rabbits were found to have no significant effect on antibody titers to egg albumin (Ado and Goldstein, 1974). Thrasher et al. (1971) reported no effect of hypothalamic lesions in weanling rats, although close inspection of their data reveals a lower end titer after two weeks for rats with posterior hypothalamic lesions. Solomon et al. (1974) found hypothalamic lesions to be ineffective in altering the antibody response to flagellin. Negative findings also were reported following the injection of rat erythrocytes parasitized with Plasmodium berghei into mice (Spector, 1980). Bilateral anterior hypothalamic lesions or lesions of the nucleus locus coeruleus did not alter the primary response to SRBC in mice (Hall et al., 1979).

The effects of electrical stimulation of the hypothalamus on antibody production appear to depend upon the species and where in the hypothalamus the electrode is placed. Decreased antibody titers have been reported following stimulation of the posterior hypothalamus in rabbits (Tsypin and Maltsev, 1967; Korneva and Kai, 1967 – cited by Spector, 1980). In rats, stimulation of the lateral hypothalamus has been shown to

increase gamma globulin levels (Fessel and Forsyth, 1963).
Control rats that received stimulation in other hypothalamic
areas or in the cerebellum showed no change.

CELLULAR IMMUNITY

Effects of Brain Lesions

Manipulations involving the CNS have also been found to influence T lymphocytes. Jankovic and Isakovic (1973) evaluated
delayed skin reactions to BSA in rats with bilateral lesions in
either the hypothalamus, thalamus, superior colliculus, caudate
and putamen, reticular formation, or amygdaloid complex. Delayed reactions were observed 24 hours after intradermal injection of BSA. Lesions in the reticular formation and hypothalamus resulted in a significant reduction in the mean diameter of the lesions as well as a reduction in the number of
responders. Stein et al. (1976) and Macris et al. (1970) reported the same effect with circumscribed lesions in the
anterior hypothalamus of guinea pigs. A decrease in the number
of severe reactions to picryl chloride and small tuberculin
(purified protein derivative) reactions were found. Bilateral
lesions in the medial or posterior basal hypothalamus had no
effect.
  Brain lesions resulting from disease in human beings can also
depress cellular immunity. Patients with Parkinson's disease
have a reduction in the number of T cells, depressed mitogenic
responses, and hyporesponsiveness to skin tests (Hoffman et
al., 1978).

RETICULOENDOTHELIAL SYSTEM
AND STEM CELLS

Phagocytic cells play an important role both prior to and after
the activation of lymphocytes. Antigen processing by these
cells can render the foreign configuration more immunogenic and
thus susceptible to destruction by antibodies. The phagocytic
activity of these cells also is enhanced when antibody coats
antigen, leading to more rapid elimination of the antigen from
the host. The CNS can influence the host's defense system via
these important cells. Lesions in the anterior hypothalamus of
cats have been found to produce a marked decrease in the ability of reticuloendothelial cells to phagocytose carbon particles. Middle and posterior hypothalamic lesions were found
to produce an even greater decrease (Thakur and Manchanda,
1969). Sectioning of the spinal cord at either C3 or L2 de-

creased phagocytic activity following the administration of dysentery, typhoid, or Proteus antigen (Loverdo, 1958).

In addition to altering the activity of differentiated cells, brain lesions also have been found to influence the granulocyte progenitor cell residing in the bone marrow. Lesions in the nucleus locus coeruleus significantly reduced the number of granulocyte-macrophage colony-forming units (GM-CFU) cultured from mouse bone marrow (Hall et al., 1978). Unilateral lesions resulted in a 30 percent reduction of GM-CFU, whereas bilateral lesions resulted in a 60 percent reduction. Animals with locus coeruleus lesions also had a significantly reduced white blood cell count, although a differential count did not reveal a specific decrease in granulocytes. Because of the decrease in total WBC the possibility was considered that the pluripotent stem cell that is capable of giving rise to hematopoietic cells was affected by the lesions. To test this possibility, a subsequent experiment was conducted to evaluate the effect of brain lesions upon the formation of colony-forming units on the spleen (CFU-S) (Hall et al., 1978). These colonies are considered to be of pluripotent origin and in addition to giving rise to leukocytes, also give rise to erythrocytes. A significant reduction in the number of these macroscopic colonies was found in animals with locus coeruleus lesions, but not in animals with anterior or posterior hypothalamic lesions or control animals with lesions in the cerebellum.

The mechanism by which the CNS is able to influence cells residing in bone marrow is not known; however, both an endocrine and autonomic link have to be considered. If an autonomic influence is exerted over the bone marrow, it is consistent with conclusions reached by those studying the effects of brain lesions on erythrocyte production. Electrical stimulation of the posterior hypothalamus in rabbits has been found to cause an increase in both reticulocyte count and blood volume. Stimulation of the anterior hypothalamus abolished the effects of posterior hypothalamic stimulation (Seip et al., 1961). Increased erythropoietic activity also has been reported in rats following electrical stimulation of the posterior hypothalamus (Medado et al., 1967). Large doses of corticosteroids or ACTH had no effect upon the brain-stimulated rise in erythropoiesis; however, 150 mg/kg of atropine blocked the stimulatory effects. It was concluded by the authors that erythropoiesis is under autonomic control.

TUMOR REJECTION

Defense against neoplastic growth is an important function of the immune system (Currie, 1974). Balitskii et al. (1976) examined the growth of Brown-Pearce tumors in rabbits following electrolytic stimulation or destruction of the posterior hypothalamus. Stimulation was found to accelerate resorption of the tumor, whereas lesions suppressed rejection. Lesioned rabbits also had decreased corticosteroid levels. Similar results were reported in a subsequent experiment using dimethylbenzanthracene-induced tumors in Wistar rats (Vinnytskii and Shmalko, 1978). Electrical stimulation of the posterior hypothalamus increased antitumor activity. It was concluded that the antitumor effects of hypothalamic stimulation were mediated by the sympathoadrenal system, since phasic changes in catecholamine metabolism were detected. An earlier study had suggested a relationship between the functional state of the sympathoadrenal system and the immunological status of the animal (Vinnitskii and Shmalko, 1977). The possibility that the brain also might be capable of influencing leukocyte dynamics was suggested by the observation that tumor-bearing mice with bilateral lesions in the locus coeruleus had a 24 percent reduction in spleen weight when compared with unlesioned control mice (Hall and Smith, unpublished observations). The mechanism could have involved either altered proliferation of lymphoid and blastlike cells in the spleen, or the migration of cells to this organ. Both proliferation and migration account for the increased cellularity of spleen tissue in tumor-bearing hosts (Lala and Lind, 1975). There is suggestive evidence that each process could be modified by the CNS. Cells in the bone marrow that give rise to the spleen cells are in close proximity to nerve endings and can be stimulated to undergo proliferation by either β-adrenergic stimulation or cholinergic stimulation (Byron, 1975). Furthermore, alterations in the catecholaminergic activity of the spleen have been detected following the administration of 6-hydroxydopamine (6-OHDA), suggesting that this organ is influenced by the sympathetic nervous system (Kasahara et al., 1977).

In addition to changes in tumor growth associated with CNS manipulation, changes have also been detected within the brain of tumor-bearing hosts. Cotzias and Tang (1977) reported an association between breast cancer in mice and low dopamine-stimulated adenylate cyclase activity in homogenates of caudate. Mice with a low propensity for breast cancer were found to have significantly elevated enzymatic activity when compared with strains having a high propensity for the disease. The possibility that this finding was associated with an altered state of immunity has to be considered, since dopamine has been implicated in modulating both cellular and humoral immunity.

NEUROTRANSMITTERS

Effects Upon Immunity

Manipulation of the serotonergic system by using a variety of procedures has been found to alter the immune responsiveness of animals. Triweekly injections of lysergic acid diethylamide (LSD) given to guinea pigs immunized with brain homogenates has been found to decrease the incidence of paralysis and the mortality rate (O'Brien et al., 1962). In addition, the injection of 5-hydroxytryptophan (5-HTP) has been shown to inhibit circulating antibody levels in rabbits (Devoino et al., 1970) and in mice (Idova and Devoino, 1972). However, a lesion of the raphe nucleus stimulated antibody production and increased the intensity of the primary immune response (Eremina and Devoina, 1973), a finding consistent with an inhibitory role for serotonin upon immunity.

If serotonin is part of a physiological mechanism by which the brain is able to influence the immune system, then changes in brain serotonin levels would be expected during the course of the immune response. Such changes have been observed. Vekshina and Magaeva (1974) measured serotonin levels in the hypothalamus and hippocampus of rabbits after immunization with typhoid Vi antigen. A significant decrease in serotonin concentrations was found in the hypothalamus with a "tendency toward a decrease" reported for the hippocampus. These changes occurred within the first day after antigen injection, during the latent phase of antibody production.

Dopamine also has been implicated in immunity. Cotzias et al. (1977) found that feeding L-dopa to mice led to a more "youthful appearance" and to an increase in life-span of up to 50 percent. This result was attributed in part to improved immunity. Furthermore, some patients with Parkinson's disease have been found to have decreased skin responses and total T lymphocytes, as well as decreased mitogen responsiveness. However, serum levels of IgA and IgG have been found to be elevated in these patients (Hoffman et al., 1978). Furthermore, changes in brain dopaminergic activity have been correlated with immunity. Dopamine-stimulated adenylate cyclase activity in caudate homogenates increased significantly following the injection of bacillus Calmette-Guérin antigen (BCG). Similar increases were reported following the injection of Corynebacterium parvum.

Additional evidence implicating dopamine, norepinephrine, and serotonin in mediating immunological activity has been reported

by investigators using reserpine. Treatment of rats with this
drug has been found to prolong the survival of skin grafts, decrease the intensity of delayed hypersensitivity reactions to
BSA and tuberculin, cause involution of the thymus, and impair
antibody production (Draskoci and Jankovic, 1964; Devoino et
al., 1970).

Interactions with Neuroendocrine-Immune Axis

Neurotransmitters could exert their influence over immunity indirectly by modifying the synthesis or release of hormone-releasing factors in the hypothalamus or by acting directly upon
cells involved in immunity. A central role for at least serotonin and dopamine was suggested by the detection of fluctuations of these neurotransmitters concomitant with changes involving immunological parameters (Vekshina and Magaeva, 1974;
Cotzias and Tang, 1977). Both of these neurotransmitters have
been found to act upon hypothalamic hypophysiotropic systems
implicated in regulating immunity. That they might be acting
in this capacity to modulate immunity has been suggested by a
number of investigations.

Devoino et al. (1970) reported that the inhibitory effects of
5-HTP and reserpine upon the primary immune response were not
observed in rabbits that had been hypophysectomized or lesioned
in the hypophysial stalk region. Pierpaoli and Maestroni
(1977, 1978a, 1978b) evaluated interactions between pharmacological manipulations of neurotransmitters and specific hormones. 5-HTP, phentolamine, and haloperidol in combination
were found to completely inhibit both the primary and secondary
immune response to SRBC. A decrease was found in both direct
and indirect PFC's suggesting a suppression of both IgM and
IgG. However, this suppression was specific, since the animals
that had received this regimen of drugs were able to respond
normally to a different antigen (B. abortus). The inhibitory
effect was prevented when LH, follicle-stimulating hormone
(FSH), and ACTH were administered either prior to or simultaneously with the drug regimen. When dopamine was added, a
prolongation of skin allograft retention was noted, suggesting
a suppressive effect upon T cells as well as upon antibody
production. As was true of antibody production, the suppression of T cells was specific to the allograft transplanted when
drug administration commenced.

If the hormones discussed play an important role in mediating
immune functioning, hypophysectomy should result in profound
effects upon immunity. Depression of both T and B cell functioning could be expected due to the loss of gonadotropins,

growth hormone, and/or TSH, since all of these hormones appear to exert a facilitatory influence over immunity (Ahlquist, 1976). However, it could be reasoned that enhanced immunity would follow hypophysectomy, since concentrations of inhibitory steroid hormones would be reduced. Experiments with hypophysectomized animals have revealed little or no effect of his procedure on various parameters of immunity. Working with rats, Molomut (1939) found that complete ablation of the hypophysis had no effect on antibody titers to SRBC, ovalbumin, or horse serum. Kalden et al. (1970) hypophysectomized 21-day-old rats and subsequently challenged the animals with the T dependent antigen SRBC. No significant differences in either the number of PFC's or in serum antibody titers were observed between the hypophysectomized animals and sham-operated controls. There was a decrease in body size and the size of lymphoid tissues, however. The authors also reported a depletion of cells in the perfollicular zone of the spleen, but no histological changes were noted in the thymus. Thrasher et al. (1971) reported no effect of hypophysectomy on antibody production to SRBC. Similarly, Nagareda (1954) reported no change in hemolysin titers to SRBC following hypophysectomy alone, but combining hypophysectomy with 500 r of x-ray did result in a suppression of antibody titers. A similar experiment was conducted by Duquesnoy et al. (1969). In addition to depressed hemagglutinin titers to SRBC, these authors reported a depression of skin allograft rejection and a decrease in the total leukocyte count. The data demonstrating an effect of gonadotropins and ACTH upon immunity are not necessarily contradicted by the observation that hypophysectomy by itself has no effect. Instead, the data suggest an involvement of ACTH synthesized within the brain rather than within the adenohypophysis (Krieger and Liotta, 1979). The synthesis of gonadotropins within the brain has not been studied; however, such a source is suggested by the findings mentioned.

In addition to the complexities resulting from the elimination of a variety of hormones, some with opposing or unknown actions upon the immune system, the hypophysectomy experiments were also being conducted in animals with severly altered metabolism. Depressed cellular metabolism in general could have accounted for the decreased immunoresponsiveness observed by some of the authors. This possibility was considered by both Duquesnoy et al. (1969) and Kalden et al. (1970). When just the adrenal glands and gonads were removed, enhanced immunity was reported by Streng and Nathan (1973). Increased spleen weights, spleen cell counts, and 19S PFC's during both the primary and secondary immune response occurred following combined adrenalectomy-gonadectomy.

PINEAL GLAND

The pineal gland contains a number of polypeptides and indoles that influence hormone systems, shown by some to modify immunity. These include adrenal, thyroid, and gonadal functioning. Jankovic et al. (1969; 1970) reported a partial and transient impairment of delayed hypersensitivity reactions and a slight delay in homograft rejection following pinealectomy. The same procedure had little effect upon hemolytic plaque formation and antibody responsiveness to SRBC (Rella and Lapin, 1976). Spleen cell responsiveness to phytohemagglutinin was slightly accelerated.

CHANGES IN THE CNS

The evidence that a feedback system provides for bidirectional information flow between the immune system and the brain is inconclusive. Changes within the brain have been detected during the course of the immune response. Changes in serotonin levels and dopamine activity during antibody formation have been discussed. The firing rate of neurons in the hypothalamus also has been found to increase during antibody formation.
   Korneva and Klimenko (1976-cited by Spector, 1980) recorded single-unit activity in several hypothalamic sites. Changes in the firing patterns of more than 1400 units were detected in the posterior, ventromedial, and supramamillary nuclei. Maximal changes were observed during the first six days following the administration of antigen. Besedovsky et al. (1977) also measured a significant increase in the firing rate of neurons in the ventromedial hypothalamus one, two, four, and five days after the injection of TNP-hemocyanin. Similar results were reported following a subsequent study using SRBC (Besedovsky and Sorkin, 1977). In this experiment, no increase was observed on day 1, but a threefold increase in firing rate was observed on day 5 at the time of maximum PFC activity.
   Changes in the brain also have been reported during the rejection of skin allografts in mice. A statistically significant increase in the nuclear volume in neurosecretory cells in the supraoptic nucleus was reported during allograft rejection but not in control mice with autografts (Srebro et al., 1974).
   Although the neurotransmitter, physiological and morphological changes detected in the CNS might be manifestations of a feedback system, all of these changes could represent the cause of immunological events rather than the consequence. They might also reflect changes occurring within the CNS in response to hormonal adaptations with no direct bearing upon immunity.

SUMMARY

Manipulations of the CNS have been found to alter the course of several parameters of the immune system (Table 20-1). Lesions in circumscribed areas of the midbrain and hypothalamus have been found to inhibit *in vitro* proliferation of granulocytes as well as the phagocytic activity of macrophages. The possibility that these findings are the consequence of disrupted autonomic function is suggested by morphological as well as pharmacological evidence. Noradrenergic nerve endings are present in bone marrow and are found in close proximity to cellular elements. Futhermore, the hematopoietic stem cell is responsive to both catecholaminergic and cholinergic stimulation. Certain hormones controlled by the brain have been found to influence phagocytic cells, but no investigations of neuroendocrine mechanisms regulating this important aspect of immunity have been reported.

Lymphocytes appear to be under autonomic and neuroendocrine influence. Peripheral symphathectomy using 6-OHDA has been found to inhibit antibody production, and administration of LH, FSH, and ACTH reverses the inhibitory effects of 5-HTP, phentolamine, and haloperidol upon antibody production. In addition to the gonadotropins and ACTH, it is possible that an additional hormone system exists that has as its primary function the regulation of immunity. The report of a small peptide in the adenohypophysis that markedly stimulates the uptake of [^{3}H]thymidine into thymocytes supports such a mechanism (Saxena and Talwar, 1977).

In summary, experiments have shown that: (1) manipulation of the CNS can alter several parameters of immunogenesis; (2) morphological, pharamacological and physiological changes can be detected within the brain during the course of immune reactions; and (3) neuroendocrine and autonomic output channels appear to be involved. Other studies have found an association between abnormal states of behavior and susceptibility to disease. That the laboratory and clinical findings are functionally related would be an overly presumptuous conclusion in view of conflicting findings by some investigators and the absence of appropriate control procedures by some others. However, because of the diverse nature of the evidence implicating neuroendocrine and autonomic influences over the immune system, to exclude such a possibility would be equally presumptuous.

Part of the difficulty in interpreting the results of neuroimmunological reseach can be attributed to the complexity of the two systems. Interactions between the brain and a multitude of physiological processes require that elaborate control procedures be performed to isolate individual variables for

Table 20-1. Summary of Reported Effects of CNS Manipulation upon
Immunological Processes

Manipulation and site	Antibody formation	Delayed hypersensitivity	RES and phagocytosis
Electrolytic lesion			
Anterior Hypothalamus	↓ or 0		
Posterior Hypothalamus	↓ or 0	0	
Midbrain	↑ or ↓		ND
Electrolytic stimulation			
Posterior Hypothalamus	↓	ND	ND
Lateral Hypothalamus	↑	ND	ND
Pharmacological agents			
6-OHDA	↓ or ↑	ND	ND
5-HT	↓	ND	ND
DA	↑	ND	ND
Reserpine	↓		ND

  ↓ Decrease in process being measured.
  ↑ Increase in process being measured.
0 No change in process being measured.
ND Insufficient or no data available.

measurement or manipulation. Similarly, the cascade of interrelated events comprising immunogenesis makes it difficult to
isolate a single parameter without involving others. As the
experimental protocols available to those studying interactions
between these two systems become more precise, the nature and
biological importance of CNS modulation of immunity will undoubtedly become known.

ACKNOWLEDGMENTS

The author wishes to thank Dr. Novera H. Spector for permission
to cite his unpublished findings and for providing a preprint
of his chapter, "The Central State of the Hypothalamus in
Health and Disease: Old and New Concepts." Appreciation is
also extended to Dr. Robert D. Schimpff for his insights and
for encouragement to pursue this area of study.

NOTE

The data cited in this chapter were compiled in 1979 and submitted in final form during the winter of 1980. During the intervening years, considerable progress has been made in this rapidly evolving field.

REFERENCES

Ader, R.: Behaviorally conditioned immunosuppression. *Psychosom Med 36*:183-184, 1974.

Ader, R., and Cohen, N.: Behaviorally conditioned immunosuppression. *Psychosom Med 37*:333-340, 1975.

Ado, A.D., and Goldstein, M.M.: Are the posterior hypothalamic areas the center of regulation of antibody production? *Sechenov Physiol USSR 60*:548-555, 1974.

Ahlquist, J.: Endocrine influences on lymphatic organs, immune response, inflammation and immunity. *Acta Endocrinol (Copenh) 83 (Suppl 206)*:1-136, 1976.

Ambrose, C.T.: The requirement for hydrocortisone in antibody-forming tissue cultivated in serum free medium. *J Exp Med 119*:1027-1049, 1964.

Balitskii, K.P., Vinnitskii, V.B., Takaishvili, K.A., and Umanskii, Y.A.: The role of the hypothalamus in the formation of antitumor resistance. *Vopr Onkol 22*:69-75, 1976.

Barjaktarovic-Nikolic, K., Nikolic, V.P., Nikolic, B., and Vitic, J.: Serum immunoglobulins and beta lipoproteins in epilepsy. *Anali Zovoda za Metalno Adraulje 6*:73-82, 1974.

Bartrop, R.W., Luckhurts, E., Lazarus, L., Kiloh, L.G., and Penny, R.: Depressed lymphocyte function after bereavement. *Lancet 1*:834-836, 1977.

Besedovsky, H.O., and Sorkin, E.: Thymus involvement in female sexual maturation. *Nature 249*:356-358, 1974.

Besedovsky, H., and Sorkin, E.: Network of immune-neuroendocrine interactions. *Clin Exp Immunol 27*:1-12, 1977.

Besedovsky, H., Sorkin, E., Felix, D., and Haas, H.: Hypothalamic changes during the immune response. *Eur J Immunol 7*:323-325, 1977.

Black, S.: Inhibition of immediate-type hypersensitivity response by direct suggestion under hypnosis. *Br Med J 1*:925-929, 1963.

Black, S., Humphrey, J.H., and Niven, J.S.: Inhibition of the mantoux reaction by direct suggestion under hypnosis. *Br Med J 6*:1649-1652, 1963.

Byron, J.W.: Manipulation of the cell cycle of the hemopoietic stem cell. *Exp Hematol 3*:44-53, 1975.

Chouinard, G., and Jones, B.D.: Schizophrenia as dopamine-deficiency disease. *Lancet 2*:99-100, 1978.

Cohen, D., and Eisdorfer, D.: Behavioral-immunologic relationships in older men and women. *Exp Aging Res 3*:225-229, 1977.

Cotzias, G.C., and Tang, L.C.: An adenylate cyclase of brain reflects propensity for breast cancer in mice. *Science 197*: 1094-1096, 1977.

Cotzias, G.C., Miller, S.T., Tang, L.C., Papavasilious, P.S., and Wang, Y.Y.: Levadopa, fertility and longevity. *Science 196*:549-551, 1977.

Currie, G.A.: *Cancer and the Immune Response*. Edward Arnold, London, 1974.

Devoino, L.V., Eremina, O.F., and Yu Ilyutchenok, R.: The role of the hypothalamo-pituitary system in the mechanism of action of reserpine and 5-hydroxytryptophan on antibody production. *Neuropharmacology 9*:67-72, 1970.

Donner, M.: A study of the immunology and biology of mongolism. *Ann Med Exp Biol Fenn 32 (Suppl 9)*:1-80, 1954.

Draskoci, M. and Jankovic, B.D.: Involution of thymus and suppression of immune response in rats treated with reserpine. *Nature 202*:408-409, 1964.

Duquesnoy, R.J., Mariani, T., and Good R.A.: Effect of hypophysectomy on immunological recovery after sublethal irradiation of adult rats. *Proc Soc Exp Biol NY 131*:1176-1183, 1969.

Edwards, E.A., and Dean, L.M.: Effects of crowding of mice on humoral antibody formation and protection to lethal antigenic challenge. *Psychosom Med 39*:19-24, 1977.

Engel, G.: Studies of ulcerative colitis III. Nature of the psychologic process. *Am J Med 19*:231-256, 1953.

Engels, W.D., and Wittkower, E.D.: Psychological allergic and skin disorders. In *Comprehensive Textbook of Psychiatry*, A.M. Freedman, H.J., Kaplan, and B.J. Sadock, Eds. Williams & Wilkins, Baltimore, 1975, pp. 1685-1694.

Eremina, O.F., and Devoina, L.V.: Production of humoral antibodies in rabbits with destruction of the nucleus of the midbrain raphe. *Bull Exp Biol Med 75*:149-151, 1973.

Ettigi, P.G., Brown, G.M., and Seggie, J.A.: TSH and LH responses in subtypes of depression. *Psychosom Med 41*:203-208, 1979.

Fabris, N., Pierpaoli, W., and Sorkin, E.: Hormones and the immunological capacity. IV. Restorative effects of developmental hormones or of lymphocytes on the immunodeficiency syndrome of the dwarf mouse. *Clin Exp Immunol 9*:227-240, 1971.

Fessel, W.J.: Mental stress, blood proteins, hypothalamus: Experimental results showing effect of mental stress upon 4S and 19S protein; speculation that functional behavior disturbances may be expressions of general metabolic disorders. *Arch Gen Psychiatry 7*:427-435, 1962.

Fessel, W.J.: "Antibrain" factors in psychiatric patients sera. I. Further studies with hemagglutination technique. *Arch Gen*

Psychiatry 8:614-621, 1963.

Fessel, M.J., and Forsyth, R.P.: Hypothalamic role in control of gamma globulin levels. (Abstr) *Arthritis Rheum 6*:771-772, 1963.

Fessel, W.J., and Grunbaum, B.W.: Electrophoretic and analytical ultracentrifuge studies in sera of psychotic patients: Elevation of gamma globulins and macroglobulins and splitting of alpha 2 globulins. *Ann Intern Med 54*:1134-1145, 1961.

Fessel, W.J., Hirata-Hibi, M., and Shapiro, I.M.: Genetic and stress factors affecting the abnormal lynphocyte in schizophrenis. *J Psychiat Res 3*:275-283, 1965.

Friedman, S.B., Ader, R., and Grota, L.J.: Protective effect of noxious stimulation in mice infected with rodent malaria. *Psychosom Med 35*:535-537, 1973.

Friedman, S.B., Cohen, J., and Iker, H.: Antibody response to cholera vaccine: Differences between depressed, schizophrenic and normal subjects. *Arch Gen Psychiat 16*:312-315, 1967.

Gisler, R.H., Bussad, A.E., Maxie, J.C., and Hess, R.: Hormonal regulation of the immune response I. Induction of an immune response *in vitro* with lymphoid cells from mice exposed to acute systemic stress. *Cell Immunol 2*:634-645, 1971.

Glenn, W.G., and Becker, R.E.: Individual versus group housing in mice: Immunological responses to time phased injections. *Physiol Zool 42*:411-416, 1969.

Goldstein, M.M.: Antibody-forming cells of the rat spleen after injury to the midbrain. *Bull Exp Biol Med 85*:183-187, 1978.

Gordon, M.C., Sinha, S.K., and Carlson, S.C.: Antibody responses to influenza vaccine in patients with Down's syndrome. *Am J Ment Defic 75*:391-399, 1971.

Greenfield, N.S., Roessler, R., and Crosley, A.P.: Ego strength and length of recovery from infectious mononucleosis. *J Nerv Ment Dis 128*:125-128, 1966.

Greer, S.: Psychological enquiry: A contribution to cancer research. *Psychol Med 9*:81-89, 1979.

Haldorsen, T., and Aarli, J.A.: Immunoglobulin concentrations in first degree relatives of epileptic patients with drug induced IgA deficiency. *Acta Neurol Scand 56*:608-612, 1977.

Hall, N.R., Lewis, J.K., Schimpff, R.D., Smith, R.T., Trescot, A.M., Gray, H.E., Wenzel, S.E., Abraham, W.C., and Zornetzer, S.F.: Effects of diencephalic and brainstem lesions on haemopoietic stem cells. *Soc Neurosci Abstr 4*:20, 1978.

Hall, N.R., Lewis, J.K., Smith, R.T., and Zornetzer, S.F.: Effects of locus coeruleus and anterior hypothalamic brain lesions on antibody formation in mice. *Soc Neurosci Abstr 5*: 511, 1979.

Hamilton, D.R.: Immunosuppressive effects of predator induced stress in mice with acquired immunity to *Hymenolepis nana*. *J Psychosom Res 18*:143-153, 1974.

Hawkes, R.A., Boughton, C.R., and Schroeter, D.R.: The antibody response of institutionalized Down's syndrome patients to seven microbial antigens. *Clin Exp Immunol 31*:298-304, 1978.

Heath, R.G., Fitzjarrel, A., and Krupp, I.M.: Schizophrenic gamma G immunoglobulin at specific brain sites. *Proceedings of the 124th Annual Meeting American Psychiatric Association.* (Abst) 1968.

Hoffman, P.M., Robbins, D.S., Nolte, M.T., Gibbs, Jr., C.J., and Gajdusek, D.C.: Immunity and immunogenetics in Guamanians with amyotrophic lateral sclerosis and Parkinsonism-dementia. (Abstr) *J Supramol Struct 8 (Suppl 2)*: (1978).

Idova, G.V., and Devoino, L.V.: Dynamics of formation of IgM- and IgG-anti-bodies in mice after administration of serotonin and its precursor 5-Hydroxytryptophan. *Bull Exp Biol Med 73*: 294-296, 1972.

Ikemi, Y., and Nakagawa, S.: Psychosomatic study of so called allergic disorders. *Jpn J Med Progr 50*:451-474, 1963.

Imboden, J.B., Canter, A., and Cluff, L.E.: Convalescence from influenza: A study of the psychological and clinical determinants. *Arch Intern Med 108*:115-121, 1961.

Isakovic, K., and Jankovic, B.D.: Neuro-endocrine correlates of immune response. II. Changes in the lymphatic organs of brain-lesioned rats. *Int Arch Allergy Appl Immunol 45*:373-384, 1973.

Jankovic, B.D., and Isakovic, K.: Neuro-endocrine correlates of immune response. I. Effects of brain lesions on antibody production, arthus reactivity and delayed hypersensitivity in the rat. *Int Arch Allergy 45*:360-372, 1973.

Jankovic, B.D., Isakovic, K., and Petrovic, S.: Immune capacity of pinealectomized rats. *Proc Yugosl Immunol Soc 1*:31, 1969.

Jankovic, B.D., Isakovic, K., and Petrovic, S.: Effect of pinealectomy of immune reactions in rat. *Immunology 18*:1-6, 1970.

Johnstone, E.C., Crow, T.J., Frith, C.D., Carney, M.W.P., and Price, J.S.: Mechanism of the antipsychotic effect in the treatment of acute schizophrenia. *Lancet 1*:848-851, 1978.

Kalden, J.R., Evans, M.M., and Irvine, W.J.: The effect of hypophysectomy on the immune response. *Immunology 18*:671-679, 1970.

Kamrin, B.B.: The effect of the immune tolerant state on early behavior of domestic fowl. *Anim Behav 15*:217-222, 1967.

Kaneko, Z., and Takaishi, N.: Psychometric studies on chronic urticaria. *Folia Psychiatr Neurol Jpn 17*:16-24, 1963.

Kasahara, K., Tanaka, S., Ito, T., and Hamashima, Y.: Suppression of the primary immune response by chemical sympathectomy. *Res Commun Chem Pathol Pharmacol 16*:687-694, 1977.

Kopeloff, L.M., and Stanton, A.H.: The effect of body temperature upon hemolysin-production in the rat. *J Immunology 44*: 247-250, 1942.

Korneva, E.A., and Kai, L.M.: Effect of the stimulation of different structures of the mesencephalon on the course of immunological reactions. *Fiziol Zh SSSR 53(1)*:42-47, 1967.

Korneva, E.A., Klimenko, V.M., and Shhinek, A.K.: *Neurohumoral Regulation of Immune Homeostasis*. Nauda, Leningrad, 1978.

Kovaleva, E.S., Bonartsev, P.D., and Prilipko, L.L.: Lymphocyte reaction in schizophrenia patients to the phytomitogens, concanavalin and phytohemagglutinin. *Bull Exp Biol Med 84*: 1136-1139, 1977.

Krieger, D.T., and Liotta, A.S.: Pituitary hormones in brain: Where, how and why? *Science 205*:366-372, 1979.

Lala, P.K., and Lind, C.: Cell kinetics in the spleen of tumor-bearing hosts. *Anat Rec 181*:403, 1975.

Loverdo, T.V.: Neuroreflex mechanisms in the regulation of phagocytosis. In *Control of Immunogenesis by the Nervous System*, A.N. Gordienko, ed. 1958, pp 125-135.

Lukyanenko, V.I.: The functional structure of the immunogenetic process and its nervous regulation. *Folia Biol 7*:379-389, 1961.

Lukyanenko, V.I., Flerov, B.A., Megreladze, O.Y., and Kuznetsov, S.M.: Microbiology and immunity: Conditioned reflex regulation of the serum globulin level. *Bull Exp Biol Med 55*:664-666, 1963.

McClary, A.R., Meyer, E., and Weitzman, D.J.: Observations on the role of the mechanism of depression in some patients with disseminated lupus erythematosus. *Psychosom Med 17*:311-321, 1955.

Macris, N.T., Schiavi, R.C., Camerino, M.S., and Stein, M.: Effect of hypothalamic lesions on immune processes in the guinea pig. *Am J Physiol 219*:1205-1209, 1970.

Marceca, A.: The Problem of hyperactivity. *Academic Therapy 13*: 277-284, 1978.

Marsh, J.T., and Rasmussen, A.F.: Response of adrenals, thymus, spleen and leukocytes to shuttlebox and confinement stress. *Proc Soc Exp Biol Med 104*:180-183, 1960.

Martin, J.T. and van Wimersma-Greidanus, Tj.B.: Imprinting behavior: Influence of vasopressin and ACTH analogues. *Psychoneuroendocrinology 3*:261-269, 1978.

Medado, P., Izak, G., and Feldman, S.: The effect of electrical stimulation of the central nervous system on erythropoiesis in the rat. II. Localization of a specific brain structure capable of enhancing red cell production. *J Lab Clin Med 69*: 776-786, 1967.

Mikhailov, V.V., Astafeva, N.G., and Soloveva, V.Y.: Role of the posterior hypothalamic nuclei in development of experimental allergic encephalomyelitis and postdiphtheric polyneuritis. *Bull Exp Biol Med 69(2)*:124-127, 1970.

Molomut, N.: The effect of hypophysectomy on immunity and hypersensitivity in rats with a brief description of the

936 Hall

operative technique. *J Immunol* 37:113-131, 1939.

Monjan, A.A., and Collector, M.I.: Stress-induced modulation of the immune response. *Science* 196:307-308, 1977.

Moos, R.H.: Personality factors associated with rheumatoid arthritis: A review. *J Chronic Dis* 17:41-55, 1964.

Nagareda, S.C.: antibody formation and the effect of X-radiation on circulating antibody levels in the hypophysectomized rat. *J Immunol* 73:88-94, 1954.

O'Brien, D.J., Hughes, F.W., and Newberne, J.: Influence of lysergic acid diethylamide on experimental allergic encephalomyelitis. *Proc Soc Exp Biol* 111:490-493, 1962.

Paunovic, V.R., Petrovic, S., and Jankovic, B.D.: Influence of early postnatal hypothalamic lesions on immune responses of adult rats. *Periodicum Biologorum* 78:50, 1976.

Pettingale, K.W., Greer, S., and Tee, D.E.: Serum IgA and emotional expression in breast cancer patients. *J Psychosom Res* 21:395-399, 1977.

Pierpaoli, W., and Maestroni, G.J.M.: Pharmacological control of the immune response by blockade of the early hormonal changes following antigen injection. *Cell Immunol* 31:355-363, 1977.

Pierpaoli, W., and Maestroni, G.J.: Pharmacological control of the hormonally modulated immune response. II. Blockade of antibody production by a combination of drugs acting on neuro-endocrine functions. Its prevention by gonadotropins and corticotrophin. *Immunology* 34:419-430, 1978a.

Pierpaoli, W., and Maestroni, G.J.: Pharmacologic control of the hormonally modulated immune response. III. Prolongation of allogenic skin graft rejection and prevention of runt disease by a combination of drugs acting on neuroendocrine function. *J Immunol* 120:1600-1603, 1978b.

Pitkin, D.H.: Effect of physiological stress on the delayed hypersensitivity reaction. *Proc Soc Exp Biol Med* 120:350-351, 1965.

Pulkkinen, E.: Immunoglobulins, psychopathology and prognosis in schizophrenia. *Acta Psychiatr Scand* 56:173-182, 1977.

Rasmussen, A.F., Spencer, E.S., and Marsh, J.T.: Decrease in susceptibility of mice to passive anaphylaxis following avoidance-learning stress. *Proc Soc Exp Biol Med* 100:878-879, 1959.

Rella, W., and Lapin, V.: Immunocompetence of pinealectomized and stimultaneously pinealectomized and thymectomized rats. *Oncology* 33:3-6, 1976.

Rogers, M.P., Reich, P., Strom, T.B., and Carpenter, C.B.: Behaviorally conditioned immunosuppression: Replication of a recent study. *Phychosom Med* 38:447-451, 1976.

Rogers, M.P., Dubey, D., and Reich, P.: The influence of the psyche and the brain on immunity and disease susceptibility:

A critical review. *Psychosom Med 41*:147-164, 1979.

Saxena, R.K., and Talwar, G.P.: An anterior pituitary factor stimulates thymidine incorporation in isolated thymocytes. *Nature 268*:57-58, 1977.

Seip, M., Halvorsen, S., Andersen, P., and Kaada, B.R.: Effects of hypothalamic stimulation on erythropoiesis in rabbits. *Scand J Clin Lab Invest 13*:553-563, 1961.

Siegel, M.: susceptibility of mongoloids to infection. I. Incidence of pneumonia, influenza A and shigella dysenteriae. *Am J Hyg 48*:53-62, 1948.

Solomon, G.F.: Emotions, stress, the central nervous system and immunity. *Annu NY Acad Sci 164*:335-343, 1969a.

Solomon, G.F.: Stress and antibody response in rats. *Int Arch Allergy Appl Immunol 35*:97-104, 1969b.

Solomon, G.F., Moos, R.H., Fessel, W.J., and Morgan, E.E.: Globulins and behavior in schizophrenia. *Int J Neuropsychiat 2*:20-26, 1966.

Solomon, G.F., Allansmith, M., McClellan, B., and Amkraut, A.: Immunoglobulins in psychiatric patients. *Arch Gen Psychiatry 20*:272-277, 1969.

Solomon, G.F., Rubbe, S.D., and Batchelder, E.: Secondary immune response to tetanus toxoid in psychiatric patients. *J Psychiatr Res 7*:201-207, 1970.

Solomon, G.F., Amkraut, A.A., and Kasper, P.: Immunity, emotions and stress. *Psychother Psychosom 23*:209-217, 1974.

Spector, N.H.: A central state of the hypothalamus in health and disease: Old and new concepts. In *Handbook of the Hypothalamus*, P. Morgane and J. Panksepp, eds. Dekker, New York, 1980.

Srebro, Z., Spisak-Plonka, I., and Szirmai, E.: Neurosecretion in mice during skin allograft rejection. *Agressologie 15*:125-130, 1974.

Stanton, A., Meunning, L., Kopeloff, L.N., and Kopeloff, N.: Spinal cord section and hemolysin-production in the rat. *J Immunol 44*:237-246, 1942.

Stein, M., Schiavi, R.C., and Camerino, M.: Influence of brain and behavior on the immune system. *Science 191*:435-440, 1976.

Streng, C.B., and Nathan, P.: The immune response in steroid deficient mice. *Immunology 24*:559-565, 1973.

Stubbs, E.G., Crawford, M.L., Burger, D.R., and Vandenbark, A.A.: Depressed lymphocyte responsiveness in autistic children. *J Autism Child Schizophr 7*:49-55, 1977.

Sutnick, A.I., London, W.T., and Blumbert, B.S.: Effects of host and environment on immunoglobulins in Down's syndrom. *Arch Intern Med 124*:772-775, 1969.

Taylor, R.B., and Wortis, H.H.: Thymus dependence of antibody response: Variability with dose of antigen and class of antibody. *Nature 220*:927-928, 1968.

Thakur, P.K., and Manchanda, S.K.: Hypothalamic influence on the activity of reticuloendothelial system of cat. *Indian J Physiol Pharmacol (Abstr) 13:*10, 1969.

Thrasher, S.G., Bernardis, L.L., and Cohen, S.: The immune response in hypothalamic-lesioned and hypophysectomized rats. *Int Arch Allergy Appl Immunol 41:*813-820, 1971.

Tsypin, A.B., and Maltsev, V.N.: The effect of hypothalamic stimulation on the serum content of normal antibodies. *Patol Fiziol Eksp Ter 11:*83-84, 1967.

Tyrey, L., and Nalbandov, V.: Influence of anterior hypothalamic lesions on circulating antibody titers in the rat. *Am J Physiol 222:*179-185, 1972.

Vaughn, W.J., Sullivan, J.C., and Elmadjian, F.: Immunity and schizophrenia: Survey of ability of schizophrenic patients to develop active immunity following injection of pertussis vaccine. *Psychosom Med 11:*327-333, 1949.

Vekshina, N.L., and Magaeva, S.V.: Changes in the serotonin concentration in the limbic structures of the brain during immunization. *Bull Exp Biol Med 77:*625-627, 1974.

Vessey, S.H.: Effects of grouping on circulating antibodies in mice. *Proc Soc Exp Biol Med 115:*252-255, 1964.

Vinnytskii, V.B., and Shmalko, Y.P.: Influence of different immunologic processes on urinary excretion of noradrenaline and adrenaline in rats. *Fiziol Zh 23:*789-793, 1977.

Vinnytskii, V.B., and Shmalko, Y.P.: Effect of electrostimulation of the area of posterior hypothalamic nuclei on the excretion of catecholamines with urine and the development of DMBA (dimethylbenzanthracene) induced tumors in rats. *Fiziol Zh 24:*401-406, 1978.

Wayner, E.A., Flanner, G.R., and Singer, G.: The effects of taste aversion conditioning on the primary antibody response to sheep red blood cells and Brucella abortus in the albino rat. *Physiol Behav 21:*995-1000, 1978.

Wistar, R., and Hildemann, W.H.: Effect of stress on skin transplantation immunity in mice. *Science 131:*159-160, 1960.